NATIONAL ACADEMIES *Sciences Engineering Medicine*

NATIONAL ACADEMIES PRESS
Washington, DC

Nutrient Requirements of Poultry

Tenth Revised Edition
Animal Nutrition Series

Committee on Nutrient Requirements of Poultry

Board on Agriculture and Natural Resources

Division on Earth and Life Studies

Consensus Study Report

NATIONAL ACADEMIES PRESS 500 Fifth Street, NW Washington, DC 20001

This study was supported by AB Vista, Addisseo, Ajinomoto Animal Nutrition North America, Inc.; Archer Daniels Midland; Danisco Inc.; DSM; Eli Lilly and Company; Evonik Industries AG; Huvepharma; Hy-Line International; IFEEDER (Institute for Feed Education and Research); Illinois Corn Marketing Association; Micronutrients Inc.; Minnesota Corn Growers Association; National Academies' Nutrient Requirements Series; Novus International; PennAg Industries Association; United Soybean Board; U.S. Department of Health and Human Services; U.S. Food and Drug Administration. Any opinions, findings, conclusions, or recommendations expressed in this publication do not necessarily reflect the views of any organization or agency that provided support for the project.

International Standard Book Number-13 (hardcover): 978-0-309-49052-8
International Standard Book Number-13 (paperback): 978-0-309-49049-8
Digital Object Identifier: https://doi.org/10.17226/25395
Library of Congress Control Number: 2025951172

This publication is available from the National Academies Press, 500 Fifth Street, NW, Keck 360, Washington, DC 20001; (800) 624-6242; https://nap.nationalacademies.org.

The manufacturer's authorized representative in the European Union for product safety is Authorised Rep Compliance Ltd., Ground Floor, 71 Lower Baggot Street, Dublin D02 P593 Ireland; www.arccompliance.com.

Printed in the United States of America.

Suggested citation: National Academies of Sciences, Engineering, and Medicine. 2026. *Nutrient Requirements of Poultry: Tenth Revised Edition*. Washington, DC: National Academies Press. https://doi.org/10.17226/25395.

Consensus Study Reports published by the National Academies of Sciences, Engineering, and Medicine document the evidence-based consensus on the study's statement of task by an authoring committee of experts. Reports typically include findings, conclusions, and recommendations based on information gathered by the committee and the committee's deliberations. Each report has been subjected to a rigorous and independent peer-review process and it represents the position of the National Academies on the statement of task.

Proceedings published by the National Academies of Sciences, Engineering, and Medicine chronicle the presentations and discussions at a workshop, symposium, or other event convened by the National Academies. The statements and opinions contained in proceedings are those of the participants and are not endorsed by other participants, the planning committee, or the National Academies.

Rapid Expert Consultations published by the National Academies of Sciences, Engineering, and Medicine are authored by subject-matter experts on narrowly focused topics that can be supported by a body of evidence. The discussions contained in rapid expert consultations are considered those of the authors and do not contain policy recommendations. Rapid expert consultations are reviewed by the institution before release.

For information about other products and activities of the National Academies, please visit www.nationalacademies.org/about/whatwedo.

COMMITTEE ON NUTRIENT REQUIREMENTS OF POULTRY

KIRK C. KLASING, *Chair*, University of California, Davis
RYAN N. DILGER, University of Illinois Urbana-Champaign
WILLIAM DOZIER, Auburn University, Auburn
ROBERT G. ELKIN, The Pennsylvania State University, University Park
PETER FERKET,[1] North Carolina State University, Raleigh
DOUG KORVER, University of Alberta, Edmonton, Alberta
JASON LEE,[2] Texas A&M University, College Station
GONZALO G. MATEOS, Universidad Politécnica de Madrid, Madrid
MICHAEL PERSIA, Virginia Tech, Blacksburg
SHEILA PURDUM, University of Nebraska–Lincoln
ROSEMARY L. WALZEM, Texas A&M University, College Station

Study Staff

ROBIN A. SCHOEN, Study Director
ALBARAA H. SARSOUR, Study Director
SAMANTHA SISANACHANDENG, Senior Program Assistant

Consultant

GERALD HUNTINGTON, North Carolina State University, Raleigh

[1] Served on the committee until August 2023.
[2] Served on the committee until May 2018.

BOARD ON AGRICULTURE AND NATURAL RESOURCES

Reviewers

This Consensus Study Report was reviewed in draft form by individuals chosen for their diverse perspectives and technical expertise. The purpose of this independent review is to provide candid and critical comments that will assist the National Academies of Sciences, Engineering, and Medicine in making each published report as sound as possible and to ensure that it meets the institutional standards for quality, objectivity, evidence, and responsiveness to the study charge. The review comments and draft manuscript remain confidential to protect the integrity of the deliberative process.

We thank the following individuals for their review of this report:

OLAYIWOLA ADEOLA, Purdue University
TODD APPLEGATE, University of Georgia
MINGAN CHOCT, University of New Zealand
STEVE LEESON, University of Guelph (retired)
MICHAEL LILBURN, The Ohio State University (retired)
CARL PARSONS, University of Illinois Urbana-Champaign
VELMURUGU RAVINDRAN, Massey University
MAMDUH SIFRI, ADM (retired)
KELLEY WAMSLEY, Mississippi State University
MARTIN ZUIDHOF, University of Alberta

Although the reviewers listed above provided many constructive comments and suggestions, they were not asked to endorse the conclusions or recommendations of this report nor did they see the final draft before its release. The review of this report was overseen by **DONALD BEITZ,** Iowa State University. He was responsible for making certain that an independent examination of this report was carried out in accordance with the standards of the National Academies and that all review comments were carefully considered. Responsibility for the final content rests entirely with the authoring committee and the National Academies.

Acknowledgments

The committee would like to acknowledge the robust participation of Dr. Nilva Kazui Sakomrua and her students and post-doctoral scholars for their highly instructive help in modeling the amino acid requirements of broilers and layers. We are also grateful to Dr. Sally L. Noll for her guidance in our attempts at modeling turkey amino acid requirements. Prior to the establishment of the committee, Drs. Rosalina Angel and Todd Applegate took actions to encourage the poultry industry and ingredient suppliers to supplement funding from the National Academies of Sciences, Engineering, and Medicine and from federal agencies and foundations. The committee is most appreciative of their efforts. Lastly, the committee would like to thank Robin Schoen, Dr. Gerald Huntington, and Dr. Albaraa Sarsour for their guidance during the process of writing this report, and Samantha Sisanchandeng, Paige Jacobs, Sarah Kwon, and Jenna Briscoe for administrative support.

Contents

Figures and Tables

FIGURES

TABLES

Preface

The National Research Council (NRC) Nutrient Requirements reports have periodically been published since the 1940s and are a series of nutrient requirement recommendations that cover most agriculturally important animal species. The first report on the *Nutrient Requirements of Poultry* was published in 1944, and the most recent was published in 1994 (NRC, 1994). This new 2026 edition builds on the 1994 report and those that preceded it. Reorganization at the National Academies of Sciences, Engineering, and Medicine has resulted in new reports referenced as National Academies instead of NRC reports.

Summary

During the past 80 years, the *Nutrient Requirements of Poultry* reports have sequentially documented our understanding of most commonly used practices in poultry nutrition and have continually become longer and better referenced, and they include more background information, but their frequency of publication has declined (Applegate and Angel, 2014; Bailey, 2020). The scientific rigor and transparency of decision processes for specific recommendations have also increased with time. The committee agreed to follow the previous standard of relying exclusively on peer-reviewed sources and the committee's own original analysis for their recommendations, but sometimes readers are referred to other sources for background information.

It has been 31 years since the publication of the previous update (NRC, 1994), and the committee had a sizable assignment. The research, computer modeling, and writing of this report took a little over 7 years, during which the committee had five in-person meetings and several dozen video conferences. The previous edition of this report (NRC, 1994) provides documentation for peer-reviewed studies that were used for their recommendations, and that report can be consulted for access to the older literature.

The target audience of these reports continues to be professional nutritionists in industry and academia, veterinarians, government regulators, the judicial system, and instructors and students in poultry science. This new report has 21 chapters and updates nutrient recommendations for the same poultry species covered in the previous report. There is also increased attention to broiler breeders, ducks, geese, and Japanese quail. This new report provides considerably more background information specific to poultry, including some recent global production statistics, a nutritional perspective on their domestication as well as strain characteristics, gastrointestinal physiology, nutritional quality of products, feed processing, environmental considerations of poultry diet formulation, and the effects of stress on nutrient requirements. Additionally, computer models based on the application of the relationship between rates of tissue deposition or egg production and tissue and egg nutrient composition were developed to predict amino acid requirements of broilers and layers. Models for turkeys were considered, but ultimately the committee decided that there was insufficient knowledge and expertise to develop that approach. The committee also allied with the U.S. Department of Agriculture–supported National Animal Nutrition Program-9 (NANP-9) for managing databases involving modeling efforts and nutrient composition of feedstuffs that are provided in this report with the goal of enabling greater access, transparency, and frequency of updates.

The committee also focused on increasing the transparency of how publications were selected to develop or update nutrient recommendations and digestibilities and bioavailabilities of nutrients. The committee established general criteria for publication suitability. The committee's general criteria are summarized subsequently and were adhered to when an adequate number of high-quality research publications were available. However, literature gaps were frequently encountered by the committee, and individual chapters provide additional qualifications and guidance for inclusion or exclusion of research reports used to develop recommendations as needed. Furthermore, the committee agreed with the approach used in the previous reports that more speculative guidance should be provided where necessary to give safe levels of dietary nutrient concentrations for situations where there were few publications that met the stringent list of inclusion criteria. Recommendations based on speculative guidance are clearly identified in the text and by the fonts used in the tables.

TRANSPARENCY OF SELECTION CRITERIA AND CONFIDENCE LEVEL OF RECOMMENDATIONS

To the extent possible, the committee adhered to strict criteria for deciding if a research study should be included as supportive data to develop a recommendation on a nutrient requirement. The criteria were based on details of experimental design and data analysis including the following: (1) A common basal diet was employed across all treatments; (2) a sufficient number and range of nutrient additions to the common basal diet were included to establish a dose–response relationship below the requirement followed by a plateau indicating that the requirement had been met; (3) dietary concentrations of the nutrients of interest were chemically analyzed using appropriate techniques, including the amount determined in the unsupplemented basal diet; (4) the experimental design and statistical analyses were appropriate for estimating nutrient requirements, and sufficient methodological information was provided to allow for replication of the experiment; (5) management practices and environmental conditions were employed that did not impair animal performance; and (6) the response criteria of these studies were based on the dietary amount that resulted in optimal performance and avoidance of signs of deficiency, whichever was lowest. Differences in the availability of research reports to meet the committee's criteria for inclusion led to differences in confidence for individual nutrient recommendations.

There is a high degree of confidence for those nutrient recommendations based on three or more research reports that met all the committee's stated criteria. Furthermore, the committee asserts that such recommendations are accurate for the genetic, management, and dietary conditions under which the experiments were done. The dietary concentrations recommended with a high degree of confidence are indicated by normal font within tables and were set by averaging the estimates from all studies that met the inclusion criteria. Conditions that produced nutrient recommendations with a lower degree of confidence are described subsequently. Recommendations with a lower degree of confidence are explained in the text, and this lower level of confidence is signified in bold and italicized fonts within the tables.

For cases where there were insufficient studies available that met the foregoing inclusion criteria, but there were one or more studies that met most of the stated criteria, the Committee provides guidance at a moderate level of confidence. These exceptions were made so that a rough estimate of a nutrient requirement could be provided. There is not a high degree of confidence that these recommendations provide an optimal concentration in a least-cost ration formulation, but they are likely to support adequate bird performance. Other situations from which guidance values were derived include the following: (1) For some nutrients, there were studies that did meet the foregoing inclusion criteria, but supplementation of the basal diet produced no beneficial response and the nutrient concentration within the basal diet was acceptably analyzed. In that setting, the committee concluded that the basal diet had an adequate amount of the nutrient under study. In these cases, nutrient recommendations are based on the amount of the nutrient in the basal diet. (2) Many of the published experiments designed to titrate the requirement for various nutrients analyzed the results using analysis of variance followed by mean comparisons and did not use analysis by broken line regressions or other valid regression approaches to quantitatively estimate a requirement. In those cases, the committee used the highest level that was statistically superior to lower levels. (3) For species such as geese, ducks, and pheasants, additional accommodations were made as described in those chapters due to the paucity of well-designed studies.

NUTRIENT REQUIREMENTS

The National Academies' recommendations differ from most other sources of nutrient requirements in that they are meant to be minimal levels for optimal production and avoidance of signs of deficiencies. The recommendations from many other organizations or companies typically include a margin of safety to account for problems related to feed mixing errors, ingredient variation, environmental stresses, genetic variation, nutrient degradation during storage, and undocumented field or research experience. The impacts of many of these factors are reviewed in Chapters 8 to 11, and nutritionists are accordingly advised to add a margin of safety to their requirement matrix. Moreover, the reader is reminded that the derivation of the committee's recommended nutrient requirements was constrained by only considering data reported in peer-reviewed publications.

In the case of broilers, requirement recommendations are based on weight instead of age to allow more accurate use across breeds that have different growth rates and across the many different management systems that result in different growth patterns. This was not done for the other species, mostly because research publications often reported the beginning and ending age and the rate of gain but failed to report either the beginning or final weights. A significant conclusion of the committee was that the research community must always report beginning weights in addition to rates of gain so that requirements can be made on a weight basis instead of an age basis, thus permitting more accurate applications with changes in genetics and differences in production practices. Adoption of this simple obligation would support needed efforts in developing nutrient requirement models.

REQUIREMENT TABLES

The nutrient requirement tables are designed so that the level of confidence the committee has in specific recommendations can be quickly discerned. This is done by the use of different fonts (normal font versus bolded and italicized font) and shading. Values with a higher degree of confidence are provided using normal font, and the text indicates the reason for this confidence. Guidance recommendations are signified by bold and italicized font. Where no relevant new information exists since the previous report (NRC, 1994), the recommendations were carried over from that report and are signified by shading.

DOCUMENTATION OF THE LITERATURE TABLES

For broiler chickens, broiler breeders, laying hens, turkeys, ducks, geese, and minor species, there is a table at the end of each respective chapter that documents all of the studies that were considered in the compilation of the recommended nutrient requirements, including the ages, weights, sexes, and breeds of the birds used in the studies. Additional information on the composition and content of metabolizable energy and protein of the basal diets, range of nutrient amounts supplemented, statistical approaches, response criteria examined, and confounding factors are also reported. These tables provide much more detail on specific studies than the text and are meant to allow the reader to adjust their requirement recommendations to their specific applications.

REFERENCES

Applegate, T.J., and R. Angel. 2014. Nutrient requirements of poultry publication: History and need for an update. *Journal of Applied Poultry Research*. 23:567–575.

Bailey, C.A. 2020. Chapter 21—Precision poultry nutrition and feed formulation. In *Animal Agriculture*. F.W. Bazer, G.C. Lamb, and G. Wu, editors. Academic Press. 367–378.

NRC (National Research Council). 1994. *Nutrient Requirements of Poultry*, 9th rev. ed. Washington, DC.: National Academy Press.

1

Energy

INTRODUCTION

Feed is the most expensive input in poultry production with energy representing the largest component of that cost. Energy is not a nutrient but becomes a property of the diet when lipids, proteins, or carbohydrates are oxidized in the body. Accurate estimation of the energy content of ingredients and diets is fundamental to meeting nutrient requirements and optimizing feed cost and poultry production.

The energy value of ingredients and diets can be expressed using several systems (i.e., digestible energy, metabolizable energy, and net energy). Metabolizable energy (ME) is the system most used in practice to define the energy available from the feeds and to express the energy requirements for poultry. The methodology has changed little in the last 50 years (Vohra, 1972; NRC, 1977). Net energy (NE) describes the energy for maintenance plus the energy retained in bodyweight (BW) gain, feathers, and eggs. Net energy values from prediction equations, based on analytical data obtained by wet chemistry or near-infrared spectroscopy (NIRS), are gaining interest as a tool to estimate the energy content of ingredients and diets (FEDNA, 2018; Barzegar et al., 2019; CVB, 2019b; van der Klis and Jansman, 2019; Wu et al., 2019; Barzegar et al., 2020; Noblet et al., 2022). At the present time, however, its use in practical poultry feeding is very limited (FEDNA, 2018; CVB, 2019b; van der Klis and Jansman, 2019) due to inadequacies in available data.

This chapter will focus on research and energy concepts disseminated since the last revision of the document (NRC, 1994). Critical research published prior to 1994 will be discussed as needed. Factors that affect the ability of a given feed to satisfy the energy needs of the bird, particularly those findings that help to explain differences in energy values observed in practice, will be addressed.

ENERGY TERMINOLOGY

The energy content of feedstuffs and diets can be expressed as calories (cal) or joules (J). One calorie is the amount of heat energy required to raise the temperature of 1 g of water from 16.5°C to 17.5°C. Because the specific heat of water changes with temperature, energy is defined more precisely as J, with a conversion factor of 4.184 J = 1 cal. The J has been selected by Le Systéme International d'Unites (International System of Units) and the U.S. National Bureau of Standards (NBS, 1986), now known as the National Institute of Standards and Technology (NIST), as the preferred unit for expressing all forms of energy. Terms used to describe the energy content of feeds and the energy requirements in poultry species are defined and discussed in detail in the document "Nutritional Energetics of Domestic Animals and Glossary of Energy Terms" (NRC, 1981). Energy components defined hereafter will be expressed in kilocalories (kcal). For further details of terms and definitions, the reader can review these reports (NRC, 1994; Wu et al., 2020).

ENERGY SYSTEMS

Energy systems express the bioavailability of the energy of ingredients and diets in relation to the energy requirement of the animal. A diagram illustrating the classical partitioning of the gross energy (GE) of a theoretical feed in laying hens is shown in Figure 1-1.

Figure 1-1 expresses the energy lost at different stages of digestion and metabolism, following the consumption of 1,000 g of feed (NRC, 1994). The partitioning of energy, as depicted in this figure, divides energy intake into three general categories: products formed (tissues and eggs), heat produced, and waste products that reflect unretained energy. Based on this diagram, ingredients and diets can be evaluated in at least three systems: digestible energy (DE), ME, and NE. In this example, 1 kg of feed yields 4,000 kcal when completely oxidized in a bomb calorimeter. Thus, 4,000 kcal is the maximal amount of energy that could be obtained from this diet. However, birds are not able to retain all this energy, and losses occur through feces (800 kcal), urine (300 kcal),

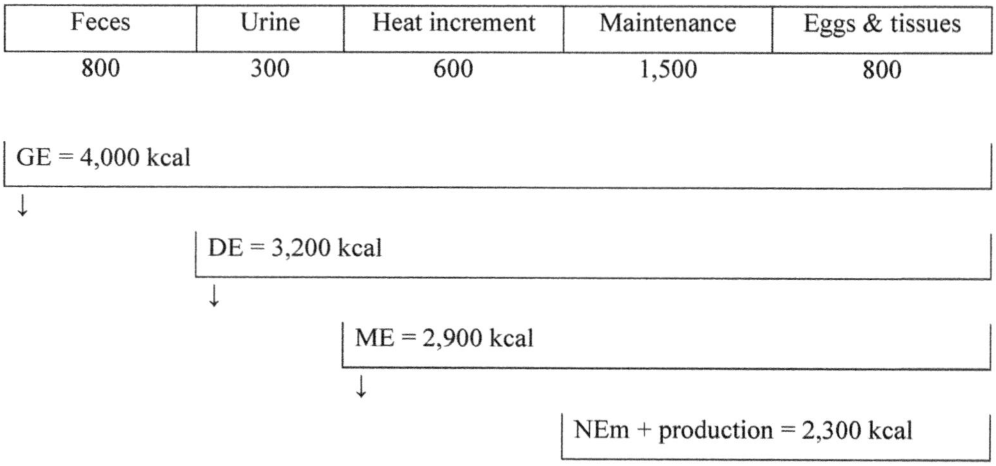

FIGURE 1-1 Example partition of energy in a diet providing 4,000 kcal GE/kg when fed to a laying hen.

and heat production (600 kcal). Consequently, only 2,300 kcal (NE) will be available to the bird to be used for maintenance (1,500 kcal) and egg production and tissue growth (800 kcal). As a result, the real efficiency of the energy contained in an ingredient to form poultry products is 20 percent (800 kcal in eggs and new body tissues from the original 4,000 kcal of GE). The efficiency of the process is affected by numerous factors including those described in Figure 1-1, bird genetics and life stage, environmental conditions, and the characteristics of the diet (feed processing, ingredient composition, nutrient content, and supplemental additives). Each factor will be described in detail later in the chapter.

Gross Energy

Gross energy is the amount of energy released as heat when, under standard conditions, a substance is completely oxidized to carbon dioxide and water. The GE content of a feedstuff is determined directly by burning a sample in a bomb calorimeter. Alternatively, the GE can be estimated by adding the GE of each of the specific macronutrients present in its composition. In this respect, the heat of combustion (kcal/kg) of the components of an ingredient or a diet is estimated as 3,700 for glucose and other simple sugars, 4,200 for starch and cellulose, 5,600 for proteins, and 9,400 for pure fats (Atwater and Bryant, 1900). The GE of starch and sugars depends upon the degree of polymerization of the carbon chain and might vary from 4,135 kcal/kg for starch to less than 3,600 kcal/kg for simple sugars (van Milgen et al., 2018). Similarly, the GE of the crude protein (CP) fraction of an ingredient or diet depends on the amino acid profile. Some amino acids, such as glycine and serine, have GE values below 4,100 kcal/kg whereas the branched and aromatic amino acids have GE values above 7,000 kcal/kg.

The GE of feed ingredients and complete feeds can be predicted accurately from their chemical composition (Ewan, 1989; Sauvant et al., 2004). Ewan (1989) estimated the GE of diets using the following equation:

$$\text{GE (kcal/kg DM)} = 4{,}143 + (1.5 \times \text{CP}) + (5.6 \times \text{EE}) - (4.4 \times \text{ash})$$

where all nutrients are presented in g/kg DM and EE is ether extract.

Sauvant et al. (2004), based on more than 2,000 results of GE measurements on feed ingredients for poultry, proposed the following equation:

$$\text{GE (kcal/kg DM)} = 4{,}143 + (1.47 \times \text{CP}) + (5.24 \times \text{EE}) + (9.3 \times \text{CF}) - (4.46 \times \text{ash}) + \Delta\Delta$$

where all nutrients are presented in g/kg DM, CF is crude fiber, and Δ is a correction factor that depends on the type of ingredient tested, with values that range from –231 for soy hulls to +308 for corn gluten meal.

The GE value of a given feed gives only an indication of the maximum amount of energy that can be evolved but is of little value for discriminating among feed ingredients or diets based on their real energy availability to the bird. In any case, the determination of the GE content of the test material and of the different compounds voided during the process of digestion is needed to elucidate the amount of energy that is lost in each step of the digestive process.

Digestible Energy

Digestion is the process by which the ingested feed is broken down to smaller components that can be absorbed in the gastrointestinal tract (GIT) of the bird. Digestible energy results from subtracting the GE voided in feces from the dietary GE (Figure 1-1). Birds excrete feces and urine together via the cloaca, and consequently it is difficult to separate each component of the excreta to measure digestibility without surgical treatment to separate the two waste streams (Paulson, 1969). As a result, DE values are seldom determined or used in practical poultry feed formulation. During the process of digestion, the microbial population of the GIT modifies to a certain extent the end products present in the excreta. In addition, part of the digesta of the distal part of the GIT corresponds to mucosa desquamation, endogenous enzymes, and other tissue losses. Most studies on the DE content of ingredients in poultry do not partition the energy of the excreta into the fraction originating from the feed and that originating from the bird (endogenous losses). Consequently, published values on DE are expressed generally as apparent DE.

Metabolizable Energy

The ME content of ingredients and diets corresponds to the difference between the DE and the energy losses in urine and gases resulting from enteric fermentation processes. In poultry, most of the N losses are due to the excretion of uric acid through the urine, and consequently the ME:DE ratio of ingredients and diets decreases as the CP content increases. In contrast, the amounts of gaseous products produced by the bird during feed digestion are negligible and usually are not determined.

The ME can be described as apparent (AME) and true (TME) and as corrected or uncorrected for N balance, each of them with its own advantages and disadvantages.

Apparent Metabolizable Energy

AME is the GE contained in the feed consumed minus the GE of the feces, urine, and gases originated by the digestion processes. AME does not account for the endogenous losses from digestive secretions, mucosa desquamations, intestinal bacteria, and other constituents originated from the metabolism of nutrients and the maintenance of the GIT. Therefore, AME underestimates the energy value of the feed when feed intake (FI) is low, because under these conditions a relatively greater proportion of the excreta is of endogenous origin and not from the feed.

For comparative purposes, the AME value of poultry diets is also frequently corrected (AMEn) for a zero N balance (Bourdillon et al., 1990a,b; Farrell et al., 1991) to eliminate the differences among ingredients that differ in CP content. Consequently, AMEn decreases the variability observed among AME estimates (Lopez and Leeson, 2007, 2008). The correction for N retention was practiced originally for AME data obtained in non-growing adult roosters and applied then to growing chicken as a comparison (Farrell, 1999). The most widely used procedure to estimate N correction assumes that the amino groups liberated during the oxidation of tissue protein will release ammonia that ends up in uric acid, which has a GE of 8.22 kcal/g N (Hill and Anderson, 1958). However, this assumption is criticized, because only 60 percent to 90 percent of the N in chicken urine is in the form of uric acid (Coulson and Hughes, 1930; Krogdahl and Dalsgard, 1981). Nitrogen retention is estimated also by measuring the increase in BW of the birds during a balance experiment and assuming that in young broilers, the protein in the gain is 20 percent of BW gain (Bourdillon et al., 1990a,b; Carré et al., 2014).

Numerous authors (Sibbald and Slinger, 1963; Choct, 2004; Lopez and Leeson, 2008; van der Klis and Jansman, 2019; Barzegar et al., 2020; Wu et al., 2020; Abdollahi et al., 2021; Noblet et al., 2022) have questioned the merits and validity of correcting the AME content of a feed for N retention. The excretion of N in the urine depends on the difference between digestible N and retained N, which in turn depends on the quantity of amino nitrogen in the feed and the capacity of the bird to retain or export energy (meat and eggs) as protein. It is difficult to accept a penalization of an ingredient that maximizes N retention, since protein deposition in tissues and eggs is among the main objectives of poultry production. The correction for N assumes that all absorbed amino acids are deaminated, and thus a 100 percent turnover of all protein retained over the time of the bioassay is accepted. In practice, the ME of a given amino acid ranges between two values: (1) lowest value when the amino acid is fully deaminated and all the N is derived for uric acid synthesis and (2) highest value when the amino acid is deposited in full as body tissue or egg, with no losses in uric acid. Consequently, the efficiency of the energy content of the protein fraction of the diet in relative terms (AMEn/GE) is greater when used for deposition of body tissue or egg protein than when used to yield fat depots or for maintenance purposes. In modern poultry strains (broilers, turkeys, and laying hens) over 40 percent of the ingested protein might be used for muscle accretion or egg protein production rather than deaminated and stored as fat. Therefore, N correction, as currently practiced, might penalize the real contribution of protein sources to the energy content of the diet (Farrell, 1999; Wu, 2018). Moreover, the penalty might be more pronounced in modern birds fed diets formulated on an ileal digestible amino acid basis, with moderate levels of CP, and using the ideal protein concept. In this respect, Abdollahi et al. (2021) reported that the correction for zero N retention reduced the AME of 28 samples of soybean meal from different origins by 9.9 percent to 17.8 percent with the greater differences observed for higher quality soybean meals. Lopez and Leeson (2008) concluded that the correction for N imposes a penalty in the energy content of corn of 3 percent to 5 percent while that of soybean meal varied from 7 percent to 12 percent. Similarly, the correction for N reduces the metabolizable energy content of high protein diets more than it does that of low protein diets (Dale and Fuller, 1984).

When the birds are at similar rates of N retention, a situation that applies to all modern broilers, turkeys, and laying hen strains, AME may be a more appropriate measure of energy than AMEn in many production situations (Noblet, 2015; Noblet et al., 2022). However, availability of peer reviewed information on the AME content of ingredients and diets is relatively scarce, which makes it difficult to implement AME values as a criterion for energy evaluation of poultry diets in practice. Consequently, despite the recognized merits of using AME as an energy unit, the energy content of ingredients and diets will be presented hereafter as AMEn values, albeit these also suffer from methodological errors as do AME determinations. Appendix C provides estimates of AME and AMEn values for 15 frequently used ingredients drawn from the literature that demonstrates the variability of reported values. Chapter 20 provides links to the database maintained by the National Animal Nutrition Program (NANP, https://animalnutrition.org) as an evergreen source of compositional information on a wide variety of ingredients.

True Metabolizable Energy

True metabolizable energy is the GE of the feed consumed minus the GE of the excreta of feed origin. The concept of TME has been developed in poultry considering that a fraction of the excreta energy losses is of endogenous origin at both intestinal (i.e., bile, enzymes, mucosal cells) and urinary levels. Thus, TME values include a correction for endogenous energy and can be calculated from the following equation:

$$TME = (GE_{intake} - GE_{excreta} + GE_{EEL})/FI$$

where GE_{EEL} is the GE of the endogenous energy loss voided by the bird during the digestive balance experiment, determined directly using fasted birds (Sibbald, 1976). It should be recognized that AME and TME values are not interchangeable. Only when FI approaches the maximum intake capacity of the birds do both values tend to be similar. A correction for N retention may be applied to TME values to originate TMEn values, with the same advantages and disadvantages indicated for AME values (Sibbald, 1982; Choct, 2004).

The TME system allows the evaluation of the energy content of ingredients for which only small amounts of samples are available. Moreover, the cost of determining TME is low, and the results are obtained more quickly than AME values. The TME procedure, however, has been severely criticized (Pesti and Edwards, 1983; Wu et al., 2019) because it imposes long periods of feed deprivation (of up to 48 h), which affects the physiological status of the bird. Originally, it was assumed that under standard conditions, the excretion of fecal metabolic and urinary endogenous energy, and thus TME values, were constant and independent of the level of FI. However, the assumption appears incorrect (Farrell et al., 1991). In fact, it has been shown that fasting overestimates the endogenous losses for TME determination (Sibbald, 1975; Härtel, 1986). Practical problems of the use of the TME technique in poultry, as indicated by Mateos et al. (2019a), are summarized here:

- The test ingredient is often force-fed alone, without other diet ingredients. Thereby, the system prevents all potential synergistic or antagonistic effects among ingredients on energy utilization (Mateos and Sell, 1981).
- The methodology involves the utilization of roosters kept in individual cages and force-fed different amounts of feed after fasting. The fasting period is abnormally long, especially in the birds used to estimate endogenous losses. Crucially, as mentioned previously, there are physiological reasons to expect errors with the use of this approach in the estimation of these losses. In addition, because of animal welfare concerns, these practices are not allowed in numerous countries, including those of the European Union.
- The agreement between AME and TME values is acceptable for traditional ingredients and when the amount of feed ingested is close to the voluntary FI of the bird. However, the agreement is not consistent in many cases, particularly for fibrous ingredients fed at low levels.
- The scientific information on the TME content of ingredients and diets published in the last 20 years in the different poultry species is scarce, which limits its utilization in practical formulation. Most of the information available belongs to the data set published by the NRC (1994).
- No recommendations have been published on the TME estimates for any poultry species. Consequently, the method cannot be used under most practical situations.

Methods for Estimating the Metabolizable Energy Content of Feeds

The AMEn content of ingredients and diets can be estimated from in vivo trials, prediction equations derived from such trials, and table values published by different institutions.

In Vivo Bioassays

In vivo trials conducted at research institutions and feed companies provide valid estimates of the energy content of raw materials, especially of nontraditional ingredients for which available data are scarce (Bourdillon et al., 1990a). The AMEn values can be determined by total collection, by methods that measure directly total FI and excreta output, or by the marker technique, in which the ratio of FI to excreta output is determined indirectly by using an indigestible, inert material. Scientifically, the "total collection" method is more accurate and reproducible than the methods involving a marker (van Milgen et al., 2018; Abdollahi et al., 2021).

However, because of the work involved, and despite the potential problems encountered, including higher variability of the final data, most research conducted in the recent past used indigestible markers, such as celite, TiO_2, or Cr_2O_3, in the determination of the AMEn content of ingredients and diets.

Three methods, namely direct, substitution (by difference), and regression, have been recommended and used to determine AME of ingredients for poultry, each of them with its own merits and drawbacks (Noblet et al., 2022). The methodology to be used varies depending on the material to test: complete diets or individual ingredients. When the objective is to determine the AMEn of a diet, a single experimental feed is used (direct feeding). For ingredients, however, the studies must include at least two levels of the test material in substitution of either the same amount of the test diet or of an ingredient of known ME value used as a control. The number of diets used depends on the nature and characteristics of the ingredient tested. In the simplest method, the test ingredient is substituted for glucose/dextrose (Anderson et al., 1958) or for the same amount of the whole control diet (Sibbald and Slinger, 1963). The basal diet used by Anderson et al. (1958) contained 50 percent glucose and assumed a standard AMEn value of 3,650 kcal/kg for this sugar. Samples of glucose from different suppliers, processed and evaluated under different conditions, might show different AMEn values (Mateos and Sell, 1981; CVB, 2021; FEDNA, 2021). Consequently, it is best to determine the energy of the glucose, prior to its use, under a common set of experimental conditions (Mateos and Sell, 1980a,b).

Regardless of the basal diet used, the accuracy of the AMEn values obtained depends on the level of inclusion of the test ingredient. In the extrapolation to calculate AMEn values, any error is multiplied by a factor of 100 and divided by the percentage of substitution. Consequently, it is convenient to include in the experimental diets the highest proportion possible of the test ingredient. Some ingredients, however, might have reduced flowability (i.e., lipid sources) or might not be palatable because of dustiness (i.e., manioc) or excess of fiber (i.e., palm kernel meal). Also, the inclusion of some ingredients might result in very high protein diets, leading to excessive protein catabolism, or may contain variable amounts of anti-nutritional factors (ANFs) (i.e., condensed tannins in sorghum grains, glucosinolates in *Brassica* seeds, and trypsin inhibitors in certain legume seeds). Also, soybean meal and other high-protein ingredients might create problems in AMEn determination when included at high levels, because of the excess of protein present in the experimental diets. Under these circumstances, it is not possible to use high levels of the material to test with, which will result in an increase in the uncertainty of the values obtained.

Potter et al. (1960) proposed a linear regression procedure for the calculation of AMEn values for ingredients substituted at several levels. The energy of the ingredient is derived from a regression equation relating the AMEn of the test diet and the proportion of the test ingredient in such diets. Then, the AMEn of the test ingredient is derived by extrapolation to 100 percent inclusion. The criticism of this approach, often used for lipid sources, is that the extrapolation of the data, beyond the range of values tested in the experiment, introduces a new systematic error that is difficult to control.

Fats and oils are the most difficult ingredients to evaluate in vivo (Wiseman et al., 1991; Ravindran et al., 2016; Aardsma et al., 2017; Mandalawi et al., 2017; Mateos et al., 2019a). The main factors affecting the amount of energy available to the bird are age, chemical quality of the lipid source (including GE, peroxide value, and moisture, impurity, unsaponifiable, and non-eluted fat contents) and the structure of the molecule (specifically the proportion of free fatty acids, the degree of unsaturation, the average length of the carbon chain, and the position of the fatty acids within the triacylglycerols).

Initially, the response to added fat was measured using simple linear models, often at a single level, and the energy content of the fat was estimated by difference between the energy of the control diet and that of the fat-supplemented diet. However, in the determination of the energy content of a lipid source, the amount of fat included in the experimental diet is quite limited (usually less than 6 percent to 9 percent), and therefore any small error in the technique is magnified, resulting in wide confidence intervals for the estimated AME value. In addition, the variability among laboratories is large, despite using the same protocol. The methodology creates considerably uncertainties and often results in values for the test fat that are greater than its GE value. The use of the multiple slope–ratio analysis of AME regressed on dietary fat level better estimates the real contribution of lipid sources to the energy of the diet (Wiseman et al., 1991; Aardsma et al., 2017; von Schaumburg et al., 2019). However, the multilevel approach is onerous and requires extra work, which limits its use under most commercial and research circumstances.

An additional problem in the determination of the energy content of lipid sources is that their AME is often greater when determined by difference between the AME of the control and the supplemental fat diets than when calculated from the GE and the digestibility of the supplemental fat, suggesting that fat increases the utilization of other dietary components (Mateos and Sell, 1980a; Aardsma and Parsons, 2017). In fact, as indicated previously, the determined energy value of fat sources using this approach is often higher than the corresponding GE values (Blanch et al., 1996; Pesti et al., 2002; Cao and Adeola, 2016; Aardsma et al., 2017), a finding that does not have any biological basis. Consequently, these extremely high values are most likely caused by uncontrolled methodological circumstances such as the beneficial effects of supplemental fat on the utilization of the nonfiber components of the diet. In this respect, supplemental fat reduces rate of feed passage, facilitating the contact between nutrients and digestive enzymes, thereby increasing the utilization of other components of the diet. However, because of the methodology used for the calculation of the AME of the lipid source, the increase is often attributed to the supplemental fat (Mateos et al., 1982; Mandalawi et al., 2017). Consequently, the accuracy of the available data on the real energy contribution of supplemental fat to the diet is not very high.

Prediction Equations

The AMEn of ingredients and diets can also be estimated from their proximate composition, using appropriate prediction equations (WPSA, 1989; Rostagno et al., 2017; CVB, 2019a) than from table values. Regression equations based on data obtained from NIRS technology, wet chemistry, and in vitro and in vivo tests, best accommodate the inherent variation among samples obtained at different periods and under different circumstances (Losada et al., 2010). The methodology predicts the contribution of each of the chemical components of the feedstuff based on their previously determined GE and digestibility coefficients (Carpenter and Clegg, 1956; Pesti et al., 1986; Dale et al., 1990; Dale and Jackson, 1994; Rochell et al., 2011).

Digestibility trials have been the basis for deriving AMEn values of feedstuffs for adult roosters for many years. The information obtained from these trials was used to develop equations for calculating AMEn of the ingredients tested. As early as 1956, Carpenter and Clegg (1956) proposed to estimate the AME of ingredients for laying hens using an equation based on the proximal chemical composition:

$$\text{AME (kcal/kg DM)} = 59 + 3.8 \times [\text{CP} + \text{sugars} + (2.25 \times \text{EE}) + (1.1 \times \text{starch})]$$

where the nutrients are presented in g/kg DM.

More recently, Rochell et al. (2011) proposed the following equation ($P \leq 0.01$; $R^2 = 0.87$) for broilers to predict the AMEn content of corn co-products (Distillers Dried Grains with Solubles (DDGS), corn germ, corn gluten feed, and corn bran, among others) from their chemical composition:

$$\text{AMEn (kcal/kg DM)} = (0.81 \times \text{GE}) - (3.02 \times \text{NDF}) - (1.23 \times \text{CP})$$

where the nutrients are presented in g/kg DM and GE is expressed as kcal/kg DM.

Meloche et al. (2013) proposed to predict the AMEn content of corn DDGS varying in EE content for broilers using the following equation ($P < 0.01$; $R^2 = 0.86$):

$$\text{AMEn (kcal/kg DM)} = -12.28 + (2.60 \times \text{GE}) + (8.98 \times \text{CP}) + (12.58 \times \text{starch}) - (4.07 \times \text{total dietary fiber})$$

where the nutrients are expressed in g/kg DM and GE is expressed as kcal/kg DM.

The subcommittee of the European Federation of the World's Poultry Science Association (WPSA, 1989) combined information from multiple European laboratories to develop a set of equations, based on multiple regression analysis, to estimate the energy value of ingredients from digestibility data. The resulting equations, based on experiments with adult roosters, provided AMEn values for a series of traditional ingredients, such as cereals, cereal by-products, and protein meals. The AMEn values for common ingredients recommended by the WPSA (1989) used the Weende analyses (proximate analysis) as the explanatory variables (CP, EE, and NFE) in most cases (Janssen et al., 1979). The CP was calculated as $N \times 6.25$, the EE was determined by petroleum or diethyl ether extraction without prior acid hydrolysis, and the NFE was estimated as the difference between the DM content and the sum of CP, EE, CF, and ash. The WPSA (1989) proposed the use of the following coefficients (kcal/g of digestible nutrient) for adult cocks: 9.28 for fat, 4.31 for CP, and 4.14 for NFE, to generate the following general equation:

$$\text{AMEn (kcal/kg DM)} = (4.31 \times \text{digestible CP}) + (9.28 \times \text{digestible EE}) + (4.14 \times \text{digestible NFE})$$

where each of the digestible components is based on total tract digestibility studies and is expressed in g/kg DM. In these equations, the energy contribution of the fiber fraction was assumed to be negligible, acting as a simple diluent.

The specific regression equation proposed by the WPSA (1986) for corn was:

$$\text{AME (kcal/kg DM)} = (3.62 \times \text{CP}) + (8.54 \times \text{EE}) + (3.73 \times \text{NFE})$$

where all components are presented in g/kg DM. For the calculations, it was estimated that the total tract digestibility coefficient of the CP, EE, and nitrogen free extract (NFE) fractions are 0.84, 0.92, and 0.90, respectively.

For solvent-extracted soybean meal, the equation recommended was:

$$\text{AME (kcal/kg DM)} = (3.75 \times \text{CP}) + (4.64 \times \text{EE}) + (1.49 \times \text{NFE})$$

where all components are presented as g/kg DM. The apparent fecal digestibility coefficients estimated for CP, EE, and NFE were 0.87, 0.50, and 0.36, respectively.

The CVB (CVB, 2016; 2021) provided information to estimate the AMEn content of feedstuffs for broilers based on fecal digestibility data, using previous HCl hydrolysis for determining EE digestibility and including starch + sugars rather than NFE in the equation. The new general equation, derived from multiple regression analyses in studies conducted with 15 common ingredients, was:

$$\text{AMEn (kcal/kg DM)} = (4.31 \times \text{Dig. CP}) + (9.28 \times \text{Dig. EE}_h) + [4.14 \times \text{Dig. (starch}_a + \text{sugars)}] + (3.52 \times \text{lactic acid})$$

where "Dig." means fecal digestibility of the components, "EE_h" is EE content after HCl hydrolysis, and "$starch_a$" is the starch content obtained by the amylase procedure. All data are expressed in g/kg DM.

The specific prediction equation proposed by the CVB (2019a) for soybean meal and soybean expellers for poultry, derived from data from adult cocks, was:

$$\text{AMEn (kcal/kg DM)} = 1{,}838 - (1.84 \times \text{ash}) + (1.54 \times \text{CP}) + (7.03 \times \text{EE}) - (3.85 \times \text{CF})$$

where ash, CP, EE, and CF are expressed in g/kg DM.

For pure fats and oils the CVB (2021) proposed the following equation:

$$\text{AMEn (kcal/kg DM)} = \text{GE (kcal/kg DM)} \times \text{DCEE (g/100 g)}$$

where the digestibility coefficient (DC) of the EE can be estimated as:

$$\text{DCEE (\%)} = 96.1 - 0.3746 \times (\text{C16:0} + \text{C18:0})$$

with (C16:0 + C18:0) expressed in percent units of the total fatty acid fraction.

Under some circumstances, accurate prediction equations might be obtained by quantifying the presence of key ANF in the ingredient, such as condensed tannins in sorghum (Sibbald, 1977; Elkin et al., 1996; Mandal et al., 2006; Selle et al., 2018; CVB, 2019a) and trypsin inhibitors and heat damaged protein in soybean meals (Palliyeguru et al., 2011). For sorghum grains with a condensed tannin content in catechin-equivalents between 0.4 and 1.0 percent, the equation proposed by the WPSA (1989) is:

$$\text{AMEn (kcal/kg DM)} = 3{,}855 - (39.46 \times \text{g tannins/kg DM})$$

A summary with detailed information of the prediction equations used to estimate the ME value of feed ingredients from their proximate composition and digestible values is presented as an appendix in the previous report (NRC, 1994).

In practice, one of the main problems encountered when using prediction equations that rely upon chemical composition data to determine the AMEn content of an ingredient is the lack of information on the procedures used to analyze the different components of the material tested. For example:

- Starch content is greater when determined by polarimetry, using the Ewers method (ISO, 1997), than when determined using the enzyme amylase method (AOAC, 2000).
- Ether extract values are higher when analyzed after HCl hydrolysis (AOCS, 2004) than when analyzed directly (AOAC, 2000).
- Neutral detergent fiber (NDF) values are lower when the enzyme amylase is used in the procedure (Van Soest et al., 1991).
- Crude protein content of most ingredients is higher when analyzed using the Dumas method than when analyzed using the Kjeldahl procedure (AOAC, 2000).
- Calculating crude protein content as N × 6.25 for all ingredients by assuming a fixed N content for all protein sources. True N conversion coefficient varies widely depending on the protein's amino acid profile. In general, the coefficient has lower values for proteins of vegetable origin than for proteins of animal origin.
- The ANF content of the ingredients and its effects on AMEn values are not taken into consideration.

The use of NFE to estimate the energy content of ingredients in many prediction equations and diets is of concern. Nitrogen free extract is not a chemical component of the diet, and thus it is not analyzed but calculated by difference between 1,000 and the sum (g/kg) of moisture, CP, ash, EE, and CF. Consequently, the existing variability in laboratory analyses and potential mistakes associated with CP, ash, EE, and CF determinations will affect the estimated energy value of the ingredient. Moreover, the NFE concept gives the same energy value to starch, sugars, and organic acids as it does to hemicellulose, soluble fiber, and lignin, resulting in poor estimation of the energy content of many ingredients (e.g., soybean products, acid whey, and several co-products from the corn industry). In this respect, the WPSA (1986) proposed to utilize the term "(starch + sugar)/NFE" in the prediction equations to obtain a more reliable estimate of the digestibility of the NFE fraction. In addition, the method used for the determination of the digestibility of the protein fraction (fecal, apparent ileal, or standardized ileal) has a greater effect on ingredients and diets with high protein contents.

The main assumption in the determination of the AMEn value of poultry diets based on prediction equations is the additivity of the energy content of the ingredients. However, dietary energy depends on potential interactions between the feed and the bird. Consequently, the assumption might not be correct (Mateos et al., 2019a,b; Wu et al., 2019). For example, type and amount of fiber (Hetland et al., 2005; González-Alvarado et al., 2007; Mateos et al., 2012; Jiménez-Moreno et al., 2019); fat supplementation (Mateos et al., 1982; Aardsma et al., 2017); presence of ANF, contaminants, and toxins in the diet (Palliyeguru et al., 2011; García-Rebollar et al., 2016); and the inclusion of enzymes, emulsifiers, organic acids, and other additives (Wenk, 2000; Lázaro et al., 2003, 2004) modify the energy contribution of each particular ingredient to the diet. Some of the problems reported when estimating the AMEn of ingredients and diets under practical situations, as indicated by Mateos et al. (2019a), are described next.

- The sample in evaluation does not belong to the same population as the samples used to create the prediction equation.
- The influence of feed processing (feed form and particle size) and the presence of ANF on energy utilization are not taken into consideration.
- The sum of the major components (moisture, ash, EE, CP, starch, sugars, NDF, soluble fiber, and organic acids) obtained by chemical analyses does not add to 100 percent.
- The analytical methods used in the determination of the starch, CP, EE, and NDF of the samples are not specified, resulting in discrepancies among labs.
- Nitrogen-free extract is used as a key variable to estimate the energy value of the feed.
- The prediction equation used was obtained with a limited number of samples, resulting in high variability and low predictability value of the estimation.

Data from Table Values

The AMEn content of the ingredients used in feed formulation can be obtained from ingredient composition tables published by different institutions and university laboratories (WPSA, 1989; EVONIK, 2016, 2021; Rostagno et al., 2017; INRAE-CIRAD, 2020; CVB, 2021; FEDNA, 2021). In fact, table values are the basis of feed formulation under most practical situations. However, the high variability among research institutions on the energy value of many common ingredients limits their confidence and reliability for practical application in poultry production (Mateos et al., 2019a; Ibáñez et al., 2020). This conclusion will be amply shown in Appendix C Table C3 (a–d) that summarizes literature values for AME and AMEn for common feed ingredients; variability and imprecision continued to be found despite the committee's rigorous inclusion criteria for publication use.

Factors Affecting the Energy Value of Ingredients and Diets

The contribution of a given ingredient to the energy content of the diet depends on numerous factors, including bird species and age, environmental and management conditions, and the physico-chemical characteristics of the diet (heat processing, feed form and particle size, ingredient composition, nutrient content, and feed additives supplementation). As a result, the AMEn value assigned to individual ingredients varies considerably among research laboratories (Pesti et al., 1986; Choct, 2004; Mateos et al., 2019a,b; Wu et al., 2020; Khalil et al., 2021, 2022). The lack of a common protocol makes it difficult to give a single value to the energy content of a given ingredient (Wu, 2018; Veluri and Olukosi, 2020; Wu et al., 2020; Khalil et al., 2021, 2022; Olukosi, 2021). Consequently, two priorities in research, aimed at developing more accurate predictions of the energy content of ingredients and feeds, are to standardize the in vivo procedures used in trials and the implementation on-line of simple methods based on NIRS technology.

Bird Effects

Energy values of ingredients and diets from research conducted in broilers or adult roosters are applied frequently to other types of birds, such as rearing pullets, laying hens, turkeys, ducks (see Chapter 17), quail, and game birds (see Chapter 19). However, the AMEn content of feed ingredients and formulated diets varies widely among species (i.e., broilers vs. turkeys vs. ducks), bird age, and sex (i.e., chicks vs. pullets vs. laying hens), genetic background, and health status of the birds. Consequently, AMEn data from different laboratories might not be similar (Wu, 2018). In general, adult birds extract more energy from raw materials than young birds, with greater differences observed for ingredients that are difficult to digest, such as those with a high content in fiber, protected starch, and saturated fat (Bourdillon et al., 1990a; Black et al., 2005; Gracia et al., 2009; Jiménez-Moreno et al., 2011, 2016; Mandalawi et al., 2017). The AMEn values presented in tables for laying hens and minor species are often based on research with adult roosters when values for the species of interest are not available. However, hens might utilize the energy of the lipid fraction of the diet more efficiently than adult roosters. In this respect, the CVB (2019a) proposed the following equation for laying hens:

$$\text{AMEn (kcal/kg)} = (4.31 \times \text{Dig. CP}) + (10.67 \times \text{Dig. EE}) + (4.14 \times \text{Dig. NFE})$$

where it is assumed that laying hens have a 15 percent higher capacity to digest the lipid fraction of the feed than adult roosters, whereas no differences are given for starch, CP, or other constituents of the feed. In all cases, studies comparing the energy content of ingredients commonly used in poultry diets in birds differing by species, breed, sex, and age are sporadic and unavailable for most local ingredients (i.e., cereal by-products, native legumes, and commercial lipid mixtures).

Diet Effects

Processing, physical characteristics, and ingredient composition of the diet affect the proportion of the GE available to the bird. Feed form (mash vs. pellets), particle size (fine vs. coarse grinding), and heat processing are the most relevant factors related to the physical-chemical characteristics of the feed that affect the energy content of ingredients and diets (Gracia et al., 2003; Abdollahi et al., 2011; Jiménez-Moreno et al., 2019). In general, pelleting increases FI and reduces feed wastage while creating often negative effects on nutrient digestibility and energy content, especially prevalent in diets based on wheat (Abdollahi et al., 2013; Serrano et al., 2013; Abdollahi et al., 2014, 2018; Jiménez-Moreno et al., 2019). Fine grinding and heat processing might be of further benefit when the starch or the lipid fraction of the ingredients are protected or encapsulated and therefore are difficult to digest (i.e., starch in peas and other legumes and oil in toasted soybeans) (González-Alvarado et al., 2007; Frikha et al., 2009).

Fat and fiber content of the diet (Mateos and Sell, 1980a; Jiménez-Moreno et al., 2009a,b; Cao and Adeola, 2016), presence of ANF (García-Rebollar et al., 2016; Woyengo et al., 2017), and the inclusion of additives such as organic acids, enzymes, emulsifying agents, prebiotics, and probiotics affect the AMEn of the diet (Bedford, 2000; Wenk, 2000; Gracia et al., 2009). ANFs, such as trypsin inhibitors in soybean meal and peas; glucosinolates in rapeseed, camelina, and mustard meals; condensed tannins in sorghum grains, peas, and fava beans; oligosaccharides in legumes; xylans and β-glucans in small grains; and phytate in all seeds, reduce the AMEn content of the ingredients under most practical conditions. The presence of ANFs not only reduces the palatability and the energy content of the ingredient per se but also affects the integrity of the digestive mucosa, increasing endogenous losses and reducing the utilization of the energy contained in other constituents of the diet. Wu et al. (2020) reviewed the accuracy of available methodologies in depth and detailed their inherent limitations to accurate measurements the AMEn of ingredients and diets.

Net Energy

Net energy is calculated as the difference between the ME of the feed and the heat increment (HI) associated with its utilization (i.e., energy cost of ingestion, digestion, and metabolism and a normal level of physical activity). In practice, the NE content of a feedstuff corresponds to the energy used for maintenance plus the energy retained (ER) in eggs and body tissue gains. The ER can be estimated directly, according to the comparative slaughter technique or calculated as the difference between ME intake and heat production. In broilers, up to 50 percent to 60 percent of the ME may go toward heat production (Ao et al., 2009; van Milgen et al., 2018; Wu et al., 2019). Therefore, ER, represented by the difference between ME and heat production, might reach up to 40 percent to 50 percent of the ME.

In theory, the NE system describes more precisely than the AMEn system the amount of energy available for metabolic functions; therefore, NE might predict more accurately the energy value of feeds for the animal (Noblet et al., 1994; Noblet, 2015; Barzegar et al., 2020). Based on the information available for pigs, numerous authors (De Groote, 1974a,b; Pirgozliev and Rose, 1999; Choct, 2004; Carré et al., 2013; Swick et al., 2013; Carré et al., 2014; Barzegar et al., 2019; Wu et al., 2019; Noblet et al., 2022) have advocated for the use of the NE system to evaluate the energy content of ingredients and diets in poultry. However, the benefits of a NE system compared with an ME system are less clear in poultry than in pigs (Noblet et al., 2007, 2010; van der Klis et al., 2010; Noblet, 2015; van der Klis and Jansman, 2019), which limits its use in practical feeding.

Methods of Estimating Net Energy in Feeds

In Vivo Bioassays

The two procedures most frequently used to measure the NE content of ingredients and diets under laboratory conditions are the comparative slaughter technique and indirect (respiration) calorimetry. The slaughter technique measures the ER by the bird by comparing the composition and energy content of the carcasses of birds slaughtered at different BW after receiving a known amount of ME (Carré et al., 2014). Heat production is estimated as the difference between the ME intake and the ER in eggs or body tissues. Alternatively, heat production can be predicted by indirect calorimetry based on oxygen consumption and carbon dioxide production of the bird, using adequate equations (Wu et al., 2019). Usually, heat production is higher when measured by the comparative slaughter technique than when measured by indirect calorimetry, because birds decrease their activity when they are confined in a respiration chamber. Consequently, maintenance needs and heat production are reduced. At present, indirect calorimetry is the technique used most commonly to measure heat production and ER.

Prediction Equations

Because of cost and time needed, NE measurements are not typically determined directly for individual ingredients or diets. Adoption of the NE approach, derived from indirect calorimetry studies, led to the development of NE prediction equations based on the relationship between the chemical composition of the diet and the ER in the carcass or in the eggs (Pirgozliev and Rose, 1999; Wu et al., 2019). The most common methodology consists of determining in vivo the AMEn content of many diets meeting the nutrient specifications of the birds but at the same time differing widely in nutrient composition (i.e., starch + sugars, CP, fat, and NDF contents). Measurements are made after the birds are placed in respiration chambers where the ER and the heat produced can be measured. Alternatively, a comparative slaughter technique can be used. Linear regression analyses are performed to generate prediction equations on the efficacy of ME for NE based on the chemical composition of the experimental diets. In all cases, the accuracy of the NE values depends on the accuracy of the AME values determined for the ingredients, because these last values are used as predictors of the NE values.

Several concerns have been raised in relation to the application of NE prediction equations for practical diets and feed ingredients for poultry. The major criticisms are the tedious nature of determining the HI of ingredients and diets and the substantial errors in the estimation of fasting heat production upon which NE values rely. In addition, most NE prediction equations available were developed using complete diets, and caution is essential when applying these equations to individual ingredients (Noblet et al., 2022). Finally, the NE value of ingredients and diets includes the energy for maintenance and for production. Consequently, NE efficiency varies with the final energy distribution (maintenance vs. tissue growth vs. egg production) and will be higher when used for fat deposition than when used for lean tissue growth. Moreover, HI tends to be lower when the AMEn is supplied below maintenance energy need.

The AMEn system is currently accepted and used in practice to describe the energy value of feedstuffs for poultry, and therefore AMEn will be the system used in this document. However, the implementation of a NE system, after methodological fine-tuning of total HP determinations, might result in more accurate estimates of the energy value of ingredients for poultry diets. Diets formulated on NE values have less protein and more fat and crystalline amino acids than diets based on AMEn, increasing the sustainability of poultry production systems. Consequently, it is advisable to produce robust NE data of feedstuffs from research in all poultry species (Barzegar et al., 2019; van der Klis and Jansman, 2019; Wu et al., 2019; Barzegar et al., 2020; Wu et al., 2020; Noblet et al., 2022).

Setting Dietary Energy Levels

A basic principle of poultry nutrition is that, in general, birds eat to satisfy their energy requirements (NRC, 1994; Plavnik et al., 1997; Brickett et al., 2007). Consequently, in formulating poultry diets, energy level is usually selected as the starting point of the process. Once the energy content of the diet is defined, the concentration for all other nutrients is set as ratios to energy. Consequently, the energy content selected for a particular diet will influence the amount of essential nutrients consumed by the birds.

The mechanisms by which poultry appetite responds to the energy concentration of the diet are not well understood (NRC, 1987; Nielsen, 2004) and therefore the assumption that chickens eat to satisfy their energy requirements must be taken with caution. Dozier et al. (2007) reported that broilers primarily consume feed to meet amino acid needs from 1 to 5 weeks of age, with a shift to appetite driven by both energy and amino acid density at 6 to 9 weeks of age. In fact, chickens tend to over-consume or under-consume energy as the ingredient composition and the energy concentration of the diet increase or decrease, respectively (Bouvarel et al., 2010; Pérez-Bonilla et al., 2012). Moreover, factors such as husbandry (i.e., water availability, ambient temperature, and lighting program), feed texture (i.e., feed form and particle size), storage capacity of the GIT (i.e., young vs. adult birds), feed palatability (i.e., fat supplementation and absence of ANF), and uncontrolled deficiency in any specific nutrient will affect the eating behavior and FI of the birds.

Husbandry

Ambient temperature and humidity affect FI in birds, and under heat stress situations BW gain and egg mass production will decrease. In laying hens maintained in cages, each 1°C above 30°C reduces FI by between 2.5 and 4.0 g per day (Sykes, 1979), a decrease that depends largely on the range of temperature studied, the humidity, and the feather coverage of the birds (Peguri and Coon, 1991, 1993). In any case, the negative effects of high temperatures on chicken performance are more evident when the other nutrients of the diet are not increased proportionally to compensate for the expected reduction in FI.

Chickens tend to eat when light is available and compensate with extra FI immediately before the start and after the end of the dark period (Fondevila et al., 2020). Consequently, a reduction in day length of 6 h or more may reduce energy intake, especially when the birds are not trained (i.e., unexpected lack of availability of feed) or are fed low-energy diets in mash form (Brickett et al., 2007). However, birds can adapt their FI patterns to partially compensate for reduced day length or can learn to eat in the dark (Griffin et al., 2005).

Feed Form and Mean Particle Size

In general, pelleting increases FI, growth rate, and feed efficiency in poultry (McKinney and Teeter, 2004; Guzmán et al., 2015; Saldaña et al., 2015) with benefits that are greater in young birds fed low-energy, high-quality pellets (Brickett et al., 2007; Abdollahi et al., 2018). Fine particles tend to leave the gizzard faster than coarse particles, and consequently FI will increase when finely ground diets are presented in pellet form (Jiménez-Moreno et al., 2019).

Energy Concentration of the Diet

Nutrient density of the diet affects the economics of poultry production because of its impact on cost per ton of feed, growth potential, feed efficiency, and health of the birds. An appropriate energy level is one that most likely will result in the lowest feed cost per unit of the final product. Consequently, the most appropriate energy level will depend on the relative cost of the ingredients locally available at the time the recommendations are made. In general, low relative prices for corn versus wheat, high-protein soybean meal versus sunflower meal, and fat sources versus starch sources will favor the use of high-energy diets levels.

The energy concentration of diets for poultry indicated in the headings of this chapter's tables should not be regarded as a recommendation to meet an energy requirement, indeed birds do not have an energy needs per kg of diet per se. The committee chose these values as bases of reference because they represent the energy concentration commonly used under practical conditions in corn-soybean meal–based diets for the different poultry species and in experiments designed to determine nutrient requirements. The response of most birds (i.e., broilers, turkeys, and laying hens) to an increase in energy concentration of the diet is smaller than expected when the control diet is high in energy, especially when the feed is presented as pellets (Morris, 1968; Latshaw et al., 1990; dePersio et al., 2015; Guzmán et al., 2015; Abdollahi et al., 2018).

The AMEn concentration of diets for broilers typically varies from 2,875 to 3,200 kcal/kg, with lower values more appropriate in pelleted diets fed to slow-growth-rate chickens, or when the cost difference between high- and low-energy sources is high. For turkeys, the range of values varies between 2,800 and 3,200 kcal/kg, probably because of the high protein content and high pellet quality required by this species (Sell and Owings, 1981). In pullets, the energy content of the diets varies widely with age, with most values close to 2,950 kcal/kg from hatch to 5 weeks of age and to 2,750 kcal/kg or less from 10 to 16 weeks of age, a period in which the inclusion of high levels of insoluble fiber might be of benefit for the development of the GIT (see Chapter 2). For laying hens, the range of AMEn values varies between 2,550 and 2,850 kcal/kg. The higher values are more frequently observed in small-frame birds fed mash diets under hot weather conditions and during the early phase of the egg production cycle. The lower values are more frequent in hens in the last stage of the laying cycle, when egg size does not determine egg selling price or when shell quality is a main problem. Very low-energy diets are bulky and can result in undesirable reductions in feed intake. Pérez-Bonilla et al. (2012) reported higher egg production when the AMEn concentration of the diet increased from 2,650 to 2,850 kcal/kg. However, a further increase to 2,950 kcal/kg did not affect egg mass production, and the only effect observed was an increase in BW of the hens.

Future Research Needs

Common flaws were identified in methodologies and processes that underlie variability and inaccuracies in energy values determined for many ingredients. Criteria used by the committee to include or exclude publications used in Appendix C sought to avoid the most egregious methodological and assay process errors, but additional flaws were found for many individual ingredients, some resulting in the exclusion of values. Despite imposition of rigorous definition of qualities for usable publications, significant variation and overt inaccuracies in energy values for ingredients reported in Appendix C are apparent. Among the most frequent flaws in properly designed bioassays was inadequate characterization of the test ingredient's chemical composition. Another common compromising factor was failure to adequately describe the basal diet in which the ingredient is being tested. In studies testing fat sources, other shortcomings include inconsideration of digestive physiology and the effect of fat to slow transit time and allow improvement in the digestibility of other dietary components and the subsequent inappropriate assignment of apparent increase in available energy content to fat being tested—in some cases to values above those obtained by bomb calorimetry! These sources of variability and inaccuracy could be minimized or eliminated by adoption of consensus analytical methodologies, study design, and reporting requirements. Methods that are validated through round-robin testing should be encouraged. Given the importance of dietary energy value determinations for the poultry industry, establishment of valid consensus methodology and reporting requirements should be a priority.

REFERENCES

Aardsma, M.P., R.D. Mitchell, and C.M. Parsons. 2017. Relative metabolizable energy values for fats and oils in young broilers and adult roosters. *Poultry Science*. 96(7):2320–2329.

Aardsma, M.P., and C.M. Parsons. 2017. A slope-ratio precision-fed rooster assay for determination of relative metabolizable energy values for fats and oils. *Poultry Science*. 96(1):108–117.

Abdollahi, M.R., V. Ravindran, and B. Svihus. 2013. Influence of grain type and feed form on performance, apparent metabolisable energy and ileal digestibility of nitrogen, starch, fat, calcium and phosphorus in broiler starters. *Animal Feed Science and Technology*. 186(3):193–203.

Abdollahi, M.R., V. Ravindran, and B. Svihus. 2014. Influence of feed form on growth performance, ileal nutrient digestibility, and energy utilisation in broiler starters fed a sorghum-based diet. *Livestock Science*. 165:80–86.

Abdollahi, M.R., V. Ravindran, T.J. Wester, G. Ravindran, and D.V. Thomas. 2011. Influence of feed form and conditioning temperature on performance, apparent metabolisable energy and ileal digestibility of starch and nitrogen in broiler starters fed wheat-based diet. *Animal Feed Science and Technology*. 168(1):88–99.

Abdollahi, M.R., M. Wiltafsky-Martin, and V. Ravindran. 2021. Application of apparent metabolizable energy versus nitrogen-corrected apparent metabolizable energy in poultry feed formulations: A continuing conundrum. *Animals*. 11(8):2174.

Abdollahi, M.R., F. Zaefarian, V. Ravindran, and P.H. Selle. 2018. The interactive influence of dietary nutrient density and feed form on the performance of broiler chickens. *Animal Feed Science and Technology*. 239:33–43.

Anderson, D.L., F.W. Hill, and R. Renner. 1958. Studies of the metabolizable and productive energy of glucose for the growing chick. *The Journal of Nutrition*. 65(4):561–574.

Ao, T., A.H. Cantor, A.J. Pescatore, M.J. Ford, J.L. Pierce, and K.A. Dawson. 2009. Effect of enzyme supplementation and acidification of diets on nutrient digestibility and growth performance of broiler chicks. *Poultry Science*. 88(1):111–117.

AOAC International. 2000. *Official Methods for Analysis of AOAC International*, 17th ed. Gaithersburg, MD: AOAC International.

AOCS (American Oil Chemists' Society). 2004. *Official Methods and Recommended Practices of the AOCS*, 5th ed. Champaign, IL: Journal of the American Oil Chemists' Society.

Atwater, W.O., and A.P. Bryant. 1900. *The Availability and Fuel of Food Materials*. Storrs, CT: Connecticut Agricultural Experimental Station.

Barzegar, S., S.B. Wu, J. Noblet, M. Choct, and R.A. Swick. 2019. Energy efficiency and net energy prediction of feed in laying hens. *Poultry Science*. 98(11):5746–5758.

Barzegar, S., S.B. Wu, M. Choct, and R.A. Swick. 2020. Factors affecting energy metabolism and evaluating net energy of poultry feed. *Poultry Science*. 99(1):487–498.

Bedford, M. 2000. Removal of antibiotic growth promoters from poultry diets: Implications and strategies to minimise subsequent problems. *World's Poultry Science Journal*. 56(4):347–365.

Black, J.L., R.J. Hughes, S.G. Nielsen, A.M. Tredrea, R. MacAlpine, and R.J. van Barneveld. 2005. The energy value of cereal grains, particularly wheat and sorghum, for poultry. *Proceedings of the 17th Australian Poultry Science Symposium*. 21–29.

Blanch, A., A.C. Barroeta, M.D. Baucells, X. Serrano, and F. Puchal. 1996. Utilization of different fats and oils by adult chickens as a source of energy, lipid and fatty acids. *Animal Feed Science and Technology*. 61(1):335–342.

Bourdillon, A., B. Carré, L. Conan, J. Duperray, G. Huyghebaert, B. Leclercq, M. Lessire, J. McNab, and J. Wiseman. 1990a. European reference method for the in vivo determination of metabolisable energy with adult cockerels: Reproducibility, effect of food intake and comparison with individual laboratory methods. *British Poultry Science*. 31(3):557–565.

Bourdillon, A., B. Carré, L. Conan, M. Francesch, M. Fuentes, G. Huyghebaert, W.M. Janssen, B. Leclercq, M. Lessire, J. McNab, et al. 1990b. European reference method of in vivo determination of metabolisable energy in poultry: Reproducibility, effect of age, comparison with predicted values. *British Poultry Science*. 31(3):567–576.

Bouvarel, I., Y. Nys, M. Panheleux, and P. Lescoat. 2010. Comment l'alimentation des poules ingluence la qualité des oeufs? *INRA Production Animals*. 23:167–182.

Brickett, K.E., J.P. Dahiya, H.L. Classen, and S. Gomis. 2007. Influence of dietary nutrient density, feed form, and lighting on growth and meat yield of broiler chickens. *Poultry Science*. 86(10):2172–2181.

Cao, M.H., and O. Adeola. 2016. Energy value of poultry byproduct meal and animal-vegetable oil blend for broiler chickens by the regression method. *Poultry Science*. 95(2):268–275.

Carpenter, K.J., and K.M. Clegg. 1956. The metabolizable energy of poultry feeding stuffs in relation to their chemical composition. *Journal of the Science of Food and Agriculture*. 7(1):45–51.

Carré, B., M. Lessire, and H. Juin. 2013. Prediction of metabolisable energy value of broiler diets and water excretion from dietary chemical analyses. *Animal*. 7(8):1246–1258.

Carré, B., M. Lessire, and H. Juin. 2014. Prediction of the net energy value of broiler diets. *Animal*. 8(9):1395–1401.

Choct, M. 2004. *The Net Energy Value for Commonly Used Plant Ingredients for Poultry in Australia*. Kingston, Australia: Rural Industries Research and Development Corporation.

Coulson, E.J., and J.S. Hughes. 1930. Collection and analysis of chicken urine. *Poultry Science*. 10(1):53–58.

CVB. 2016. Chemical composition and nutritional values of feedstuffs. In *CVB Feed Table*. Wageningen, The Netherlands: UR Livestock Research.

CVB. 2019a. Chemical composition and nutritional values of feedstuffs. In *CVB Feed Table*. Wageningen, The Netherlands: UR Livestock Research.

CVB. 2019b. *CVB Table Booklet: Feeding of Poultry in CVB, Series no 61*. Federatie Nederlandse Diervoederketen. Wageningen, The Netherlands: UR Livestock Research.

CVB. 2021. *CVB Feed Table*. Chemical composition and nutritional values of feedstuffs in. Wageningen, The Netherlands: Livestock Research (WLR) and Research Institute of Agriculture, Fisheries and Food (ILVO).

Dale, N., and H.L. Fuller. 1984. Correlation of protein content of feedstuffs with the magnitude of nitrogen correction in true metabolizable energy determinations. *Poultry Science*. 63(5):1008–1012.

Dale, N., and D. Jackson. 1994. True metabolizable energy of corn fractions. *Journal of Applied Poultry Research*. 3(2):179–183.

Dale, N.M., G.M. Pesti, and S.R. Rogers. 1990. True metabolizable energy of dried bakery product. *Poultry Science*. 69(1):72–75.

De Groote, G. 1974a. A comparison of a new net energy system with the metabolisable energy system in broiler diet formulation, performance and profitability. *British Poultry Science*. 15(1):75–95.

De Groote, G., ed. 1974b. Utilisation of metabolisable energy. B.M. Freeman, editor. Edinburgh, UK: British Poultry Science.

dePersio, S., P.L. Utterback, C.W. Utterback, S.J. Rochell, N. O'Sullivan, K. Bregendahl, J. Arango, C.M. Parsons, and K.W. Koelkebeck. 2015. Effects of feeding diets varying in energy and nutrient density to Hy-Line W-36 laying hens on production performance and economics. *Poultry Science*. 94(2):195–206.

Dozier III, W., A. Corzo, M. Kidd, and S. Branton. 2007. Dietary apparent metabolizable energy and amino acid density effects on growth and carcass traits of heavy broilers. *Journal of Applied Poultry Research*. 16(2):192–205.

Elkin, R.G., M.B. Freed, B.R. Hamaker, Y. Zhang, and C.M. Parsons. 1996. Condensed tannins are only partially responsible for variations in nutrient digestibilities of sorghum grain cultivars. *Journal of Agricultural and Food Chemistry*. 44(3):848–853.

EVONIK. 2016. *AMINODat 5.0. Amino Acid Analysis*. M.S. Redshaw, J. Fickler, J. Fontaine, W. Heimbeck, V. Hess, and I. Reinmann, editors. Hanau-Wolfgang, Germany: Evonik Degussa GmbH.

EVONIK. 2021. *AMINODat 6.0. Feed raw material database*. Hanau-Wolfgang, Germany: Evonik Operations GmbH Nutrition & Care.

Ewan, R.C. 1989. Predicting the energy utilization of diets and feed ingredients by pigs. *Proceedings of the 11th Symposium of Energy Metabolism of Farm Animals*. 43:215–218.

Farrell, D.J. 1999. In vivo and in vitro techniques for the assessment of the energy content of feed grains for poultry: A review. *Australian Journal of Agricultural Research*. 50:881–888.

Farrell, D.J., E. Thomson, J.J. Du Preez, and J.P. Hayes. 1991. The estimation of endogenous excreta and the measurement of metabolisable energy in poultry feedstuffs using four feeding systems, four assay methods and four diets. *British Poultry Science*. 32(3):483–499.

FEDNA (Fundación Española para el Desarrollo de la Nutrición Animal). 2018. Necesidades nutricionales para avicultura. Normas FEDNA. In *FEDNA (Fundación Española para el Desarrollo de la Nutrición Animal)*. G. Santomá and G.G. Mateos, editors. Madrid, Spain: FEDNA.

FEDNA 2021. FEDNA tables on the composition and nutritional value of raw materials for the production of compound animal feeds, 4th ed. C. de Blas, P.G. Rebollar, M. Gorrachategui, and G.G. Mateos, editors. Madrid, Spain: FEDNA.

Fondevila, G., J.L. Archs, L. Cámara, A.F. de Juan, and G.G. Mateos. 2020. The length of the feed restriction period affects eating behavior, growth performance, and the development of the proximal part of the gastrointestinal tract of young broilers. *Poultry Science*. 99(2):1010–1018.

Frikha, M., H.M. Safaa, M.P. Serrano, X. Arbe, and G.G. Mateos. 2009. Influence of the main cereal and feed form of the diet on performance and digestive tract traits of brown-egg laying pullets. *Poultry Science*. 88(5):994–1002.

García-Rebollar, P., L. Cámara, R.P. Lázaro, C. Dapoza, R. Pérez-Maldonado, and G.G. Mateos. 2016. Influence of the origin of the beans on the chemical composition and nutritive value of commercial soybean meals. *Animal Feed Science and Technology*. 221:245–261.

González-Alvarado, J.M., E. Jiménez-Moreno, R. Lázaro, and G.G. Mateos. 2007. Effect of type of cereal, heat processing of the cereal, and inclusion of fiber in the diet on productive performance and digestive traits of broilers. *Poultry Science*. 86(8):1705–1715.

Gracia, M.I., M.A. Latorre, M. Garcia, R. Lazaro, and G.G. Mateos. 2003. Heat processing of barley and enzyme supplementation of diets for broilers. *Poultry Science*. 82(8):1281–1291.

Gracia, M.I., R. Lázaro, M.A. Latorre, P. Medel, M.J. Araníbar, E. Jiménez-Moreno, and G.G. Mateos. 2009. Influence of enzyme supplementation of diets and cooking–flaking of maize on digestive traits and growth performance of broilers from 1 to 21 days of age. *Animal Feed Science and Technology*. 150(3):303–315.

Griffin, A.M., R.A. Renema, F.E. Robinson, and M.J. Zuidhof. 2005. The influence of rearing light period and the use of broiler or broiler breeder diets on forty-two-day body weight, fleshing, and flock uniformity in broiler stocks. *Journal of Applied Poultry Research*. 14(2):204–216.

Guzmán, P., B. Saldaña, H.A. Mandalawi, A. Pérez-Bonilla, R. Lázaro, and G.G. Mateos. 2015. Productive performance of brown-egg laying pullets from hatching to 5 weeks of age as affected by fiber inclusion, feed form, and energy concentration of the diet. *Poultry Science*. 94(2):249–261.

Härtel, H. 1986. Influence of food input and procedure of determination on metabolisable energy and digestibility of a diet measured with young and adult birds. *British Poultry Science*. 27(1):11–39.

Hetland, H., B. Svihus, and M. Choct. 2005. Role of insoluble fiber on gizzard activity in layers. *Journal of Applied Poultry Research*. 14(1):38–46.

Hill, F.W., and D.L. Anderson. 1958. Comparison of metabolizable energy and productive energy determinations with growing chicks. *The Journal of Nutrition*. 64(4):587–603.

Ibáñez, M.A., C. de Blas, L. Cámara, and G.G. Mateos. 2020. Chemical composition, protein quality and nutritive value of commercial soybean meals produced from beans from different countries: A meta-analytical study. *Animal Feed Science and Technology*. 267:114531.

INRAE-CIRAD. 2020. INRAE-CIRAD-AFZ feed tables. Composition and nutritive values of feeds for cattle, sheep, goats, pigs, poultry, rabbits, horses and salmonids. Paris, France: INRAE.

ISO (International Organization for Standardization). 1997. *Native Starch—Determination of Starch Content—Ewers Polarimetric Method*. Geneva, Switzerland: International Organization for Standardization.

Janssen, W.M.M.A., K. Terpstra, F.F.E. Beeking, and A.J.N. Bisalsky. 1979. *Feeding Values for Poultry*, 2nd ed. Beekbergen, The Netherlands: Spelderholt Center for Poultry Research and Information Center.

Jiménez-Moreno, E., S. Chamorro, M. Frikha, H.M. Safaa, R. Lázaro, and G.G. Mateos. 2011. Effects of increasing levels of pea hulls in the diet on productive performance, development of the gastrointestinal tract, and nutrient retention of broilers from one to eighteen days of age. *Animal Feed Science and Technology*. 168(1):100–112.

Jiménez-Moreno, E., A. de Coca-Sinova, J.M. González-Alvarado, and G.G. Mateos. 2016. Inclusion of insoluble fiber sources in mash or pellet diets for young broilers. 1. Effects on growth performance and water intake. *Poultry Science*. 95(1):41–52.

Jiménez-Moreno, E., J.M. González-Alvarado, A. de Coca-Sinova, R.P. Lázaro, L. Cámara, and G.G. Mateos. 2019. Insoluble fiber sources in mash or pellets diets for young broilers. 2. Effects on gastrointestinal tract development and nutrient digestibility. *Poultry Science*. 98(6):2531–2547.

Jiménez-Moreno, E., J.M. González-Alvarado, A. González-Serrano, R. Lázaro, and G.G. Mateos. 2009a. Effect of dietary fiber and fat on performance and digestive traits of broilers from one to twenty-one days of age. *Poultry Science*. 88(12):2562–2574.

Jiménez-Moreno, E., J.M. González-Alvarado, R. Lázaro, and G.G. Mateos. 2009b. Effects of type of cereal, heat processing of the cereal, and fiber inclusion in the diet on gizzard pH and nutrient utilization in broilers at different ages. *Poultry Science*. 88(9):1925–1933.

Khalil, M.M., M.R. Abdollahi, F. Zaefarian, P.V. Chrystal, and V. Ravindran. 2021. Apparent metabolizable energy of cereal grains for broiler chickens is influenced by age. *Poultry Science*. 100(9):101288.

Khalil, M.M., M.R. Abdollahi, F. Zaefarian, P.V. Chrystal, and V. Ravindran. 2022. Influence of broiler age on the apparent metabolizable energy of cereal grains determined using the substitution method. *Animals*. 12(2):183.

Krogdahl, A., and B. Dalsgard. 1981. Estimation of nitrogen digestibility in poultry: Content and distribution of major urinary nitrogen compounds in excreta. *Poultry Science*. 60(11):2480–2485.

Latshaw, J.D., G.B. Havenstein, and V.D. Toelle. 1990. Energy level in the laying diet and its effects on the performance of three commercial leghorn strains. *Poultry Science*. 69(11):1998–2007.

Lázaro, R., M. García, M.J. Araníbar, and G.G. Mateos. 2003. Effect of enzyme addition to wheat-, barley- and rye-based diets on nutrient digestibility and performance of laying hens. *British Poultry Science*. 44(2):256–265.

Lázaro, R., M.A. Latorre, P. Medel, M. Gracia, and G.G. Mateos. 2004. Feeding regimen and enzyme supplementation to rye-based diets for broilers. *Poultry Science*. 83(2):152–160.

Lopez, G., and S. Leeson. 2007. Relevance of nitrogen correction for assessment of metabolizable energy with broilers to forty-nine days of age. *Poultry Science*. 86(8):1696–1704.

Lopez, G., and S. Leeson. 2008. Assessment of the nitrogen correction factor in evaluating metabolizable energy of corn and soybean meal in diets for broilers. *Poultry Science*. 87(2):298–306.

Losada, B., P. García-Rebollar, C. Álvarez, P. Cachaldora, M.A. Ibáñez, J. Méndez, and J.C. De Blas. 2010. The prediction of apparent metabolisable energy content of oil seeds and oil seed by-products for poultry from its chemical components, in vitro analysis or near-infrared reflectance spectroscopy. *Animal Feed Science and Technology*. 160(1):62–72.

Mandal, A.B., P.K. Tyagi, A.V. Elangovan, P.K. Tyagi, S. Kaur, and A.K. Johri. 2006. Comparative apparent metabolisable energy values of high, medium and low tannin varieties of sorghum in cockerel, guinea fowl and quail. *British Poultry Science*. 47(3):336–341.

Mandalawi, H.A., J.J. Mallo, D. Menoyo, R. Lázaro, and G.G. Mateos. 2017. Metabolizable energy content of traditional and re-esterified lipid sources: Effects of inclusion in the diet on nutrient retention and growth performance of broilers from 7 to 21 days of age. *Animal Feed Science and Technology*. 224:124–135.

Mateos, G.G., L. Cámara, G. Fondevila, and R.P. Lázaro. 2019a. Critical review of the procedures used for estimation of the energy content of diets and ingredients in poultry. *Journal of Applied Poultry Research*. 28(3):506–525.

Mateos, G.G., G. Fondevila, and L. Cámara. 2019b. The importance of the fibre fraction of the feed in non-ruminant diets. *Agricultural and Food Sciences*. 3:61–83.

Mateos, G.G., E. Jiménez-Moreno, M.P. Serrano, and R.P. Lázaro. 2012. Poultry response to high levels of dietary fiber sources varying in physical and chemical characteristics. *Journal of Applied Poultry Research*. 21(1):156–174.

Mateos, G.G., and J.L. Sell. 1980a. Influence of carbohydrate and supplemental fat source on the metabolizable energy of the diet. *Poultry Science*. 59(9):2129–2135.

Mateos, G.G., and J.L. Sell. 1980b. True and apparent metabolizable energy value of fat for laying hens: Influence of level of use. *Poultry Science*. 59(2):369–373.

Mateos, G.G., and J.L. Sell. 1981. Influence of fat and carbohydrate source on rate of food passage of semipurified diets for laying hens. *Poultry Science*. 60(9):2114–2119.

Mateos, G.G., J.L. Sell, and J.A. Eastwood. 1982. Rate of food passage (transit time) as influenced by level of supplemental fat. *Poultry Science*. 61(1):94–100.

McKinney, L.J., and R.G. Teeter. 2004. Predicting effective caloric value of nonnutritive factors: I. Pellet quality and II. Prediction of consequential formulation dead zones. *Poultry Science*. 83(7):1165–1174.

Meloche, K.J., B.J. Kerr, G.C. Shurson, and W.A. Dozier. 2013. Apparent metabolizable energy and prediction equations for reduced-oil corn distillers dried grains with solubles in broiler chicks from 10 to 18 days of age. *Poultry Science*. 92(12):3176–3183.

Morris, T.R. 1968. The effect of dietary energy level on the voluntary calorie intake of laying birds. *British Poultry Science*. 9(3):285–295.

NBS (National Bureau of Standards). 1986. *The International System of Units (SI)*. C.C.H. Page and P. Vigoreaux, editors. Washington, DC: U.S. Government Printing Office.

Nielsen, B.L. 2004. Behavioural aspects of feeding constraints: Do broilers follow their gut feelings? *Applied Animal Behaviour Science*. 86(3):251–260.

Noblet, J. 2015. Comparative interest and limits of metabolizable energy and net energy for evaluating poultry and pig feeds. *Proceedings of the 20th European Symposium on Poultry Nutrition*. 46–54.

Noblet, J., S. Dubois, J. van Milgen, M.B. Warpechowski, and B. Carré. 2007. Heat production in broilers is not affected by dietary crude protein. In *Energy and Protein Metabolism and Nutrition, I.* Ortigues-Marty, N. Miraux, and W. Brand-Williams, eds. Institut national de la recherche agronomique, European Association for Animal Production. Wageningen, The Netherlands: Wageningen Academic Publishers.

Noblet, J., H. Fortune, X.S. Shi, and S. Dubois. 1994. Prediction of net energy value of feeds for growing pigs. *Journal of Animal Science*. 72(2):344–354.

Noblet, J., J.V. Milgen, and S. Dubois. 2010. Utilisation of metabolisable energy of feeds in pigs and poultry: interest of net energy systems? *Proceedings of the 21st Australian Poultry Science Symposium*. 26–35.

Noblet, J., S.-B. Wu, and M. Choct. 2022. Methodologies for energy evaluation of pig and poultry feeds: A review. *Animal Nutrition*. 8:185–203.

NRC (National Research Council). 1977. *Nutrient Requirement of Poultry*, 7th rev. ed. Washington, DC: National Academy Press.

NRC. 1981. *Nutritional Energetics of Domestic Animals and Glossary of Energy Terms*, 2nd rev. ed. Washington, DC: National Academy Press.

NRC. 1987. *Predicting Feed Intake of Food-Producing Animals*. Washington, DC: National Academy Press.

NRC. 1994. *Nutrient Requirement of Poultry*, 9th rev. ed. Washington, DC: National Academy Press.

Olukosi, O.A. 2021. Investigation of the effects of substitution levels, assay methods and length of adaptation to experimental diets on determined metabolisable energy value of maize, barley and soya bean meal. *British Poultry Science*. 62(2):278–284.

Palliyeguru, M.W., S.P. Rose, and A.M. Mackenzie. 2011. Effect of trypsin inhibitor activity in soya bean on growth performance, protein digestibility and incidence of sub-clinical necrotic enteritis in broiler chicken flocks. *British Poultry Science*. 52(3):359–367.

Paulson, G.D. 1969. An improved method for separate collection of urine, feces, and expiratory gases from the mature chicken. *Poultry Science*. 48(4):1331–1336.

Peguri, A., and C. Coon. 1991. Effect of temperature and dietary energy on layer performance. *Poultry Science*. 70(1):126–138.

Peguri, A., and C. Coon. 1993. Effect of feather coverage and temperature on layer performance. *Poultry Science*. 72(7):1318–1329.

Pérez-Bonilla, A., S. Novoa, J. García, M. Mohiti-Asli, M. Frikha, and G.G. Mateos. 2012. Effects of energy concentration of the diet on productive performance and egg quality of brown egg-laying hens differing in initial body weight. *Poultry Science*. 91(12):3156–3166.

Pesti, G.M., R.I. Bakalli, M. Qiao, and K.G. Sterling. 2002. A comparison of eight grades of fat as broiler feed ingredients. *Poultry Science*. 81(3):382–390.

Pesti, G.M., and H.M. Edwards. 1983. Metabolizable energy nomenclature for poultry feedstuffs. *Poultry Science*. 62(7):1275–1280.

Pesti, G.M., L.O. Faust, H.L. Fuller, N.M. Dale, and F.H. Benoff. 1986. Nutritive value of poultry by-product meal. 1. Metabolizable energy values as influenced by method of determination and level of substitution. *Poultry Science*. 65(12):2258–2267.

Pirgozliev, V., and S.P. Rose. 1999. Net energy systems for poultry feeds: A quantitative review. *World's Poultry Science Journal*. 55(1):23–36.

Plavnik, I., E. Wax, D. Sklan, I. Bartov, and S. Hurwitz. 1997. The response of broiler chickens and turkey poults to dietary energy supplied either by fat or carbohydrates. *Poultry Science*. 76(7):1000–1005.

Potter, L.M., L.D. Matterson, A.W. Arnold, W.J. Pudelkiewicz, and E.P. Singsen. 1960. Studies in evaluating energy content of feeds for the chick: I. The evaluation of the metabolizable energy and productive energy of alpha cellulose. *Poultry Science*. 39(5):1166–1178.

Ravindran, V., P. Tancharoenrat, F. Zaefarian, and G. Ravindran. 2016. Fats in poultry nutrition: Digestive physiology and factors influencing their utilisation. *Animal Feed Science and Technology*. 213:1–21.

Rochell, S.J., B.J. Kerr, and W.A. Dozier. 2011. Energy determination of corn co-products fed to broiler chicks from 15 to 24 days of age, and use of composition analysis to predict nitrogen-corrected apparent metabolizable energy. *Poultry Science*. 90(9):1999–2007.

Rostagno, H.S., L.F.T. Albino, M.I. Hannas, J.L. Donzele, N.K. Sakomura, F.G. Perazzo, A. Saraiva, M.L. Teixeira, P.B. Rodrigues, R.F. Oliveira, S.L. Barreto, and C. de Oliveira Brito. 2017. *Brazilian Tables for Poultry and Pork. Food Composition and Nutritional Requirements*, 4th ed. Viçosa, Brazil: Federal University of Viçosa.

Saldaña, B., P. Guzmán, L. Cámara, J. García, and G.G. Mateos. 2015. Feed form and energy concentration of the diet affect growth performance and digestive tract traits of brown-egg laying pullets from hatching to 17 weeks of age. *Poultry Science*. 94(8):1879–1893.

Sauvant, D., J.M. Perez, and G. Tran. 2004. *Tables of Composition and Nutritional Value of Feed Materials: Pig, Poultry, Sheep, Goats, Rabbits, Horses, Fish.* Wageningen, The Netherlands: Wageningen Academic Press.

Sell, J.L., and W.J. Owings. 1981. Supplemental fat and metabolizable energy-to-nutrient ratios for growing turkeys. *Poultry Science*. 60(10):2293–2305.

Selle, P.H., A.F. Moss, H.H. Truong, A. Khoddami, D.J. Cadogan, I.D. Godwin, and S.Y. Liu. 2018. Outlook: Sorghum as a feed grain for Australian chicken-meat production. *Animal Nutrition*. 4(1):17–30.

Serrano, M.P., M. Frikha, J. Corchero, and G.G. Mateos. 2013. Influence of feed form and source of soybean meal on growth performance, nutrient retention, and digestive organ size of broilers. 2. Battery study. *Poultry Science*. 92(3):693–708.

Sibbald, I.R. 1975. The effect of level of feed intake on metabolizable energy values measured with adult roosters. *Poultry Science*. 54(6):1990–1997.

Sibbald, I.R. 1976. A bioassay for true metabolizable energy in feedingstuffs. *Poultry Science*. 55(1):303–308.

Sibbald, I.R. 1977. The true metabolizable energy values of some feedingstuffs. *Poultry Science*. 56(1):380–382.

Sibbald, I.R. 1982. Measurement of bioavailable energy in poultry feedingstuffs: EASU review. *Canadian Journal of Animal Science*. 62(4):983–1048.

Sibbald, I.R., and S.J. Slinger. 1963. A biological assay for metabolizable energy in poultry feed ingredients together with findings which demonstrate some of the problems associated with the evaluation of fats. *Poultry Science*. 42(2):313–325.

Swick, R.A., S. Wu, J. Zuo, N. Rodgers, M.R. Barekatain, and M. Choct. 2013. Implications and development of a net energy system for broilers. *Animal Production Science*. 53(11):1231–1237.

Sykes, A.H. 1979. Environmental temperature and energy balance in the laying hens. In *Food Intake Regulation in Poultry*. Edinburgh, UK: British Poultry Science.

van der Klis, J.D., and A.J.M. Jansman. 2019. Net energy in poultry: Its merits and limits. *Journal of Applied Poultry Research*. 28(3):499–505.

van der Klis, J.D., C. Kwakernaak, A. Jansman, and M.C. Blok. 2010. Energy in poultry diets: Adjusted AME or net energy. *Proceedings of the 21st Proceedings of the Australian Poultry Science Symposium*. 44–49.

van Milgen, J.J., E. Labussière, J. Noblet, and C.F.M. de Lange. 2018. Bioavailability of the dietary energy component. In *Feed Evaluation Science*. Wageningen, The Netherlands: Wageningen Academic Publishers.

Van Soest, P.J., J.B. Robertson, and B.A. Lewis. 1991. Methods for dietary fiber, neutral detergent fiber, and nonstarch polysaccharides in relation to animal nutrition. *Journal of Dairy Science*. 74(10):3583–3597.

Veluri, S., and O.A. Olukosi. 2020. Metabolizable energy of soybean meal and canola meal as influenced by the reference diet used and assay method. *Animals*. 10(11):2132.

Vohra, P. 1972. Evaluation of metabolizable energy for poultry. *World's Poultry Science Journal*. 28(2):204–214.

von Schaumburg, P., P. Utterback, and C.M. Parsons. 2019. Further evaluation of a slope-ratio precision-fed rooster assay and a limit-fed broiler chicken growth assay for relative metabolizable energy and relative bioavailable energy values of fats and oils. *Poultry Science*. 98(4):1723–1731.

Wenk, C. 2000. Recent advances in animal feed additives such as metabolic modifiers, antimicrobial agents, probiotics, enzymes and highly available minerals—review. *Asian-Australasian Journal of Animal Sciences*. 13:86–95.

Wiseman, J., F. Salvador, and J. Craigon. 1991. Prediction of the apparent metabolizable energy content of fats fed to broiler chickens. *Poultry Science*. 70(7):1527–1533.

Woyengo, T.A., E. Beltranena, and R.T. Zijlstra. 2017. Effect of anti-nutritional factors of oilseed co-products on feed intake of pigs and poultry. *Animal Feed Science and Technology*. 233:76–86.

WPSA (World's Poultry Science Association). 1986. European table of energy values for poultry feedstuffs: Sub-committee of the Working Group No. 2 (Nutrition) of the European Federation of Branches of the World's Poultry Science Association. Beekbergen, The Netherlands: European Federation of Branches of the World's Poultry Science Association.

WPSA. 1989. *European Tables of Energy Values for Poultry Feedstuffs*, 3rd ed. Beekbergen, The Netherlands: WPSA Beekbergen.

Wu, S.-B. 2018. *Review of AME Bioassay for Ingredient Matrix Values of Poultry*. Publication No. 19007. Wagga Wagga, New South Wales, Australia: AgriFutures Australia.

Wu, S.B., M. Choct, and G. Pesti. 2020. Historical flaws in bioassays used to generate metabolizable energy values for poultry feed formulation: A critical review. *Poultry Science*. 99(1):385–406.

Wu, S.B., R.A. Swick, J. Noblet, N. Rodgers, D. Cadogan, and M. Choct. 2019. Net energy prediction and energy efficiency of feed for broiler chickens. *Poultry Science*. 98(3):1222–1234.

2

Carbohydrates and Fiber

INTRODUCTION

Carbohydrates (CHOs) comprise a highly diverse array of compounds ranging from simple monosaccharides to highly complex and heterogeneous polymers. Although CHOs are the dominant source of metabolizable energy and fiber in most poultry diets, they are not considered to be nutritionally essential for poultry when fed practical diets. Glucose is the most important CHO in intermediary metabolism. Diets lacking both CHO and gluconeogenic precursors result in depressed plasma glucose concentrations and slow growth rates (Renner and Elcombe, 1967). Chickens can synthesize glucose from dietary fructose, galactose, sorbitol, xylose, glycerol, and a variety of amino acids (Dickson and Langslow, 1978; Brady et al., 1979; Niwa et al., 1986; Klasing, 1998). Thus, CHOs are considered conditionally essential because signs of deficiency are expressed only if gluconeogenic precursors are insufficient to produce enough glucose. Quantitatively, the requirement for glucose or its precursors in modern breeds of poultry is not known, but no natural or commercial diets for poultry have been identified that are deficient in both digestible CHO and gluconeogenic precursors.

Dietary fiber in poultry diets consists of carbohydrates in the form of non-starch polysaccharides as the major component and lignin as a minor part. Lignin is a polyphenolic compound, not a carbohydrate. Although dietary fiber is not a required nutrient, its physicochemical properties contribute to the maturation of the gastrointestinal tract (GIT) and can profoundly influence the digestibility of other nutrients and energy, both positively and negatively. Therefore, fiber levels in the diet are often maintained in a range that optimizes its positive attributes and minimizes its negative properties unless economic considerations dictate otherwise.

Nutritionally, dietary CHOs diverge into three general categories: absorbable CHOs, digestible CHOs, and fiber. Absorbable CHOs do not require enzymatic digestion prior to absorption. Digestible CHOs can be hydrolyzed by enzymes that poultry can produce themselves (i.e., endogenous enzymes) and then absorbed in the monosaccharide form. By definition, dietary fiber (DF) cannot be digested by the bird. Usually, fiber has been considered as an anti-nutritional factor, especially in young chicks, because of its negative effects on palatability, which reduces feed intake when fed at high levels. However, dietary fiber is highly variable in both concentration and physicochemical properties in feedstuffs fed to poultry. Some components of fiber are readily fermented by the microbiota, whereas other components are refractory to fermentation.

CHEMICAL CLASSIFICATION OF CARBOHYDRATES

Simple Sugars

Simple sugars include monosaccharides and sugar alcohols. They are the only CHOs that can be absorbed by intestinal epithelial cells without further hydrolysis. Monosaccharides include glucose, fructose, galactose, mannose, xylose, arabinose, rhamnose, fucose, ribose, and sugar alcohols that include sorbitol, mannitol, and xylitol. Most feed ingredients fed to poultry have very low levels of free simple sugars such as glucose and fructose.

Disaccharides

Nutritionally important disaccharides in poultry diets include sucrose, maltose, trehalose, and lactose. Sucrose and maltose can be hydrolyzed by enzymes on the apical membrane of enterocytes in the brush border of the intestinal epithelium to provide absorbable monosaccharides (Kedinger et al., 1981). Sucrase and maltase activity is sustained throughout the growth of chicks (Sell et al., 1989; Chotinsky et al., 2001) but this may not be the case for trehalase. In most poultry diets, soybean meal is a predominant source of sucrose with ~5–7 percent by weight. Other sources include coproducts of sugar beets, sugar cane, or fruits. Sometimes sucrose is added to poultry feeds as an energy source or as a flavoring agent. However, poultry have very poor ability to taste sweet flavors because they lack the T1R2 gene that codes for the sugar taste receptor (Roura et al., 2013). Maltose is found in germinating seeds and in brewery by-products, but it is also an important product of starch digestion. Trehalose is found in insects and fungi, which are sometimes used in diets for the organic market. Chickens do not express lactase activity on their brush border membrane and are therefore unable to hydrolyze lactose to glucose and galactose (Kedinger et al., 1981), but this disaccharide is readily fermented by the microbiota to lactic acid and volatile fatty acids (Hume et al., 1992), and low levels of inclusion may have beneficial effects on GIT health. However, high levels of lactose may result in fluid secretion into the intestinal lumen causing osmotic diarrhea, especially in young chicks.

Oligosaccharides

Oligosaccharides are low molecular weight polymers of monosaccharides. Digestible oligosaccharides in poultry feeds include maltodextrins from the partial hydrolysis of starch. These are found in corn syrup and other sweeteners that are present in ingredients from food wastage such as bakery by-products.

Most of the oligosaccharides present in poultry feedstuffs can't be digested by endogenous enzymes but are readily fermented by the microbiota (Coon et al., 1990; Carré et al., 1995). These resistant oligosaccharides usually have glycoside bonds and include raffinose, stachyose, verbascose, fructooligosaccharides, galactooligosaccharides, lactulose, and mannanoligosaccharides. Legumes, such as soybeans, lupines, faba beans, and peas, have high amounts but variable levels of α-(1–6)-galactooligosaccharides, raffinose, verbascose, and stachyose.

Carré et al. (1990) estimated that the apparent digestibility of oligosaccharides in cockerels fed a diet based on soybean meal and corn starch was 82 percent, indicating a high rate of fermentation. The fermentation of oligosaccharides results in lower AMEn values relative to that of absorbed monosaccharides. Genetically selected soybean strains with low concentrations of raffinose, sucrose, and stachyose have higher AMEn (Parsons et al., 2000; Perryman and Dozier, 2012; Chen et al., 2013), although there are many differences in composition among the strains that could contribute to the higher energy content reported. Both naturally occurring and synthetic fermentable oligosaccharides are sometimes added to poultry feeds for their prebiotic effects on the microbiota or for their influence on the immune system (Choct et al., 2010; Iji and Tivey, 1998).

Polysaccharides

Polysaccharides are the main CHOs in most feedstuffs fed to commercial poultry. The dominant digestible polysaccharide in most poultry diets is starch. Non-starch polysaccharides (NSPs) are refractory to digestion by endogenous enzymes, are highly heterogeneous in structure, and contribute to dietary fiber.

Starch

Starch is a polymer of linked glucose monomers that is stored by plants for later use as an energy source. It is found at high levels in underground storage tissues, such as tubers (e.g., manioc), and in seeds, especially cereals (e.g., rice, corn, sorghum, wheat, and barley) and some legumes (e.g., peas and fava beans), to support germination. It is intermediate to low in plant by-products, such as corn gluten meal, rice bran, wheat middlings, and distiller's dried grains with solubles. Currently, starch is the dominant energy source in most commercial poultry diets due to the use of high grain diets (Svihus, 2014). Starch accumulates in intracellular granules in two primary structural forms, amylose and amylopectin. Amylose is a mostly linear helical polymer of glucose with predominantly α-(1–4) glycoside bonds, whereas amylopectin consists of an α-(1–4) polymer with frequent branches using α-(1–6) glycoside bonds and has a much higher molecular weight. Depending on the feedstuff, the ratio between amylose and amylopectin varies, but the majority is usually 75–80 percent amylopectin in feedstuffs such as corn. However, selective breeding has resulted in some grains where the starch is almost entirely amylopectin, which is referred to as waxy starch.

Feed ingredients used in poultry production vary in the characteristics of the matrix of starch granules, which affect the rate and extent of digestion (Bergh et al., 1999). In general, rice starch is more digestible than the starch of other cereals. Also, cereals starch is more digestible than legume starch and both starches are more digestible than the starch of raw potatoes (Weurding et al., 2001; Herwig et al., 2020). The susceptibility of starch to digestion by endogenous enzymes such as amylase and amylopectinase in poultry depends on a variety of physical and chemical attributes including granule size, shape and size of pores in the granule surface, ratio between amylose and amylopectin, proportions of crystalline and non-crystalline forms, chain length of amylose, linearization of amylopectin, and associations with lipids and other macromolecules that affect the density, orientation, and crystalline structure of the starch helices in storage granules (Englyst et al., 1992; Parada and Aguilera, 2011; Zaefarian et al., 2015; Ma and Boye, 2018). In cereal grains, most of the starch is densely packed in an orthogonal pattern (type A configuration) and is efficiently hydrolyzed by α-amylase in the small intestine. Starch in tubers, such as potatoes, is less densely packed and in a hexagonal pattern (type B configuration) and is mostly resistant to digestion unless the starch is hydrated during processing. Legumes such as peas and fava beans contain both patterns (type C configuration), and the digestibility of their starch is lower than that of cereals unless the seeds are properly processed (Carré et al., 1991).

Feed processing (e.g., grinding, milling, cooking, extrusion, pelleting) can increase the digestibility of starch in some but not all feedstuffs (Svihus, 2014; Itani and Svihus, 2019; Moran, 2019; Itani et al., 2021). Increased digestibility due to processing may be due to reduced surface-to-volume ratio increasing the accessibility of enzymes to the starch granules and also the gelatinization of starch. Extrusion is often more effective than pelleting at gelatinizing wheat and increasing its digestibility (Abd El-Khalek and Janssens, 2010). Addition of some fiber sources also increases starch digestibility. The inclusion of oat hulls, but not sugar beet pulp, at levels of up to 7.5 percent of the diet increases the coefficient of apparent ileal digestibility of starch in a low-fiber diet based on rice and fermented soybean meal (Jiménez-Moreno et al., 2013). See Chapter 9 (Feed Processing and Quality) for additional information on the impact of processing on nutrient digestibility.

Starch can be classified as rapidly digested starch, slowly digested starch, or resistant starch depending on its in vitro digestion rate using α-amylase during a predetermined period of time (Englyst et al., 1992; Lee et al., 2011) that is adjusted as appropriate to reflect the physiology of different species or ages of animals. Slowly digestible starches have nutritional properties that differ from rapidly digested starch. In humans and pigs, slowly digested starches result in more gradual glucose absorption that may more closely match the rates and capacity of their glucose clearance and metabolism. When fed to broilers, slowly digested starch from peas results in more starch reaching the ileum, an increase in the concentration of short chain fatty acids (SCFAs) in both the ileum and cecum, and increased rate of passage through the GIT (Herwig et al., 2020). Slowly digested starch can have a positive effect on performance when substituted for rapidly digested starch in broilers (Herwig et al., 2019b; Weurding et al., 2003) and potentially in laying hens (Herwig et al., 2019a), although the exact mechanism requires further study.

Most of the resistant starch and a small portion of slowly digested starch are fermented in the distal intestine and contribute to the energy requirement via the absorption of SCFA. However, the apparent metabolizable energy corrected for zero N balance (AMEn) obtained from fermenting starch is less than that from absorbed glucose (Wiseman, 2006), so starch disappearance from the terminal ileum should be corrected for the proportion that is fermented to obtain accurate AMEn values for high starch feedstuffs.

Non-Starch Polysaccharides

Plant cell walls typically consist of 70–90 percent NSP and provide structural support. Also, some feedstuffs have nutritionally important amounts intracellularly (e.g., seeds from legumes). Structural NSPs are those that are integrated components of multi-molecular complexes and include celluloses, hemicelluloses, pectins, gums, and mucilages. Chemically, the primary monosaccharides found in these polysaccharides are hexose sugars (glucose, galactose, and mannose), pentose sugars (arabinose and xylose), uronic acids (glucuronic and galacturonic), and 6-deoxyhexoses (fucose and rhamnose). Cellulose is the only major constituent of NSP that has a well-defined chemical structure, consisting of a linear chain of β-(1$\rightarrow$4) linked D-glucose units, which makes it relatively easy to identify, purify, and study. The chemical structures of the other NSPs are highly heterogeneous, which confers variable physical properties, such as solubility in water and other solvents, viscosity, and water binding properties (Bach Knudsen, 2001). For example, the hemicelluloses have traditionally been defined as alkaline extractable components of the cell wall and include xyloglucans, xylans, mannans, and galactomannans. Pectins are also a diverse group of compounds based on historical extraction techniques. Nutritionally, categories based on extraction by solvents are not very informative but are often used due to the difficulty and expense of more accurate techniques. Progress in analytical methods would facilitate new advances of the role of NSP in poultry nutrition (Choct, 2015; De Vries, 2015).

Non-starch polysaccharides are a significant part of plant ingredients used in poultry feeds. The soluble fraction of the NSPs are fermented by GIT microbiota and provide some energy to the bird, but they also exhibit anti-nutritional activities leading to changes in gut health. In contrast, the insoluble fiber proportion acts very differently and at low levels may impart beneficial effects on the development of the GIT (Mateos et al., 2012; Choct, 2015).

The proportion of NSP that disappears from the GIT is often referred to as the coefficient of apparent total tract digestibility and has been measured for many feedstuffs. In general, apparent digestibility is correlated with the proportion of water soluble NSPs because they are more easily fermented than insoluble NSPs (Carré et al., 1990; De Vries, 2015). However, fermentation contributes a very small portion of the ME requirement of poultry (Jørgensen et al., 1996). Additionally, soluble NSPs bind large quantities of water in the digesta and contribute to its viscosity and many of the anti-nutritional effects of fiber (Bedford and Classen, 1992). Conversely, water insoluble NSP are poorly fermented, but they enhance the development of the proventriculus and gizzard of young chicks (Sacranie et al., 2012). Corn has a low amount of soluble NSP with an apparent digestibility of around 10 percent, whereas rye and barley have higher levels of soluble NSP and higher apparent digestibility. This relationship likely exists because the physical size of NSPs when they travel through the intestines is an important determinant of entrance into the ceca due to the filtering mechanism that excludes particles greater than 0.2 mm. Soluble and insoluble fibers that are not embedded in the plant cell wall are most likely to enter the ceca and be fermented by the cecal microbiota (De Vries, 2015; Jozefiak et al., 2004).

DIETARY FIBER

Dietary fiber is defined in many ways depending on physicochemical and nutritional properties. In poultry nutrition, dietary fiber usually means CHO from plants that are refractory to endogenously produced enzymes plus lignin, gums, and waxes. Note that this definition does not include non-digestible components of animal origin (e.g., collagen), which may be quantified as dietary fiber depending on the analytical method applied. Lignin is a cross-linked polyphenol and not a polysaccharide. However, it is often grouped with the CHO components of dietary fiber due to its covalent bonding with polysaccharides within the cell wall and somewhat comparable nutritional properties as insoluble NSP. Crude fiber (CF) is an old term that refers to the remnants of plant material remaining after extraction with mild acid and alkali solutions during a proximate analysis procedure. Crude fiber greatly underestimates the amount of hemicellulose and almost completely misses pectic polysaccharides in feedstuffs and consequently underestimates the amount of dietary fiber. For example, the CF content in soybean meal is ~4 percent to 6 percent, whereas its content of NSP and oligosaccharides is about 25 percent (Choct et al., 2010). Measures of crude fiber also fail to subdivide functionally disparate types of fiber, so they are of limited use in modern poultry nutrition yet are commonly reported.

Plant cell walls are highly enriched in fiber (Bach Knudsen, 2014). The structure of plant cell walls is highly ordered and consists of fibrillar polysaccharides (mainly cellulose), matrix polysaccharides (mainly hemicellulose and pectin), and encrusting substances (mainly lignin). The ratio among these components changes markedly with maturation of the plant. The water soluble portion of dietary fiber can provide some nutritional value to the bird (Carré et al., 1995; Jørgensen et al., 1996) when it is digested by microbes that are predominantly located in the distal ileum and the ceca. The microbiota predominantly uses anaerobic fermentation resulting in the production of SCFAs that can be absorbed by the bird and used for energy.

In poultry, highly fermentable fibers include soluble pectins, gums, and oligosaccharides. Moderately fermentable fibers include resistant starches and some forms of hemicellulose. Non-fermentable or poorly fermentable fibers include lignin, cellulose, and some forms of hemicelluloses. Feed ingredients rich in fiber are similarly classified: highly fermentable fiber sources (e.g., sugar beet pulp), moderately fermentable fiber sources (e.g., soybean hulls), and non-fermentable or slowly fermentable fiber sources (e.g., cereal straws and palm kernel meal). However, specific fermentation characteristics, such as efficiency and speed, that allow evaluation of various fiber sources for poultry have yet to be standardized. In general, the fermentability of fiber in an ingredient depends on the proportions of its constituent oligosaccharides and NSP, their crosslinking, and their association with lignin. Fiber sources with chemical and physical properties that permit high solubility and water holding capacity are typically the most fermentable because these properties allow microbes to penetrate the fiber matrix at a faster rate and initiate a faster rate of fiber degradation (Bach Knudsen and Hansen, 1991).

Although insoluble NSPs are not appreciably fermented, they have important physical, physiological, and nutritional effects in the bird that must be considered. Physical effects include their impacts on the caloric density (ME/kg) of the diet, digesta volume, fecal volume, and the osmolality, viscosity, buffering capacity, water binding capacity, and swelling properties of the digesta (Slama et al., 2019). Physiological effects include influences on the rate of feed intake, motility of the GIT, efficiency of nutrient absorption, makeup of the microbiota, and influences on the intestinal epithelium (Montagne et al., 2003; Choct et al., 2010; Mateos et al., 2012). Together, these properties influence weight gain, efficiency of feed utilization, intestinal morphology and health, immune function, reproductive performance, and optimal formulation of the diet. Dietary fiber can have either a positive or a negative impact on animal growth, egg production, and health status depending on its composition and level of dietary inclusion. Correlations between chemical, physical, and physiological characteristics of dietary fiber and their impact on nutritionally important outcomes is a critical area of current research. For example, the role of insoluble fiber in improving gizzard function and health, development of the GIT, and nutrient digestibility has resulted in feedstuffs high in insoluble NSP and lignin being supplemented to poultry diets (Sacranie et al., 2012; Svihus, 2011).

As described in the next section, the specific effects of dietary fiber on poultry can be either positive or negative depending on the type and level of fiber and also on the age of the bird. In general, feed ingredients that are high in soluble fiber and low in insoluble NSP and lignin have a preponderance of negative attributes that limit their use in poultry diets. Conversely, those with low levels of soluble fiber and higher amounts of insoluble fiber can have positive effects when included at low levels. For example, sugar beet pulp is rich in soluble fiber and low in lignin and is used at high levels in research to examine negative effects of soluble fiber on growth, feed intake, and nutrient absorption (González-Alvarado et al., 2010; Jiménez-Moreno et al., 2011, 2013). Conversely, oat hulls are rich in insoluble lignified fiber, have fewer anti-nutritional effects, and, at low levels, may have positive effects on the development of the digestive tract and nutrient digestion (González-Alvarado et al., 2010; Jiménez Moreno et al., 2011; Scholey et al., 2020).

Positive Effects of Fiber

Historically, dietary fiber was considered a diluent of dietary energy that caused decreased nutrient digestibility, especially in younger birds that have low fermentation capacity

(NRC, 1994). However, poultry diets lacking the structural properties provided by insoluble fiber result in decreased size of the gizzard (Svihus, 2011) and dilation of the walls of the proventriculus with hypertrophy of its sub-mucosal glands (O'Dell et al., 1959; Riddell, 1976). For practical aspects on structural properties of feeds and their practical implications, see Chapter 9, Feed Processing.

In broiler chicks, the beneficial effects of the addition of low levels of fiber containing mostly insoluble NSP (e.g., oat hulls) to a grain-based diet include increases in growth rate, digestibility of some nutrients, especially starch and fat, and the retention of water, which increases litter quality (González-Alvarado et al., 2010; Hetland et al., 2004; Mateos et al., 2012, 2020). The mechanism for these effects include increased retention time of digesta in the crop and gizzard, which increases gizzard weight, decreased gizzard pH, increased growth of intestinal villi, intensified reflux of digesta between organs, and increased enzyme production by the pancreas (Jørgensen et al., 1996; Starck, 1999; Jiménez-Moreno et al., 2010; Sacranie et al., 2012; Yokhana et al., 2015). Increased gizzard function increases the secretion of HCl, bile-acids, and digestive enzymes, and regulates the flow of digesta to the small intestine, which may increase nutrient digestion overall (Jiménez-Moreno et al., 2019; Svihus, 2011). In addition, the inclusion of insoluble fiber in the diet may change the growth and composition of the microbiota, with effects that suggest increased health status of the bird (Jha and Mishra, 2021).

In general, the beneficial effects of dietary fiber are most pronounced in younger broilers (<21 d) and occur when fiber-rich sources of NSP and lignin, such as oat hulls, are added to the diet at around 2–3 percent (González-Alvarado et al., 2010; Svihus, 2011; Mateos et al., 2012; Jiménez-Moreno et al., 2019; Mateos et al., 2020). Often the physical characteristics of fiber influence the size and nutritional properties of the particles in milled feed ingredients. The interactive effects of fiber amount and type with particle size in mash or pellet forms of feed is an active area of research (Kheravii et al., 2018a,b; Jiménez-Moreno et al., 2019) and is further described in Chapter 9 (Feed Processing).

The value of adding fiber to diets of pullets and laying hens has received less attention than for broilers, likely because their energy requirements are low relative to broilers and commercial diets typically have more fiber. In young pullets, the addition of dietary fiber to grain-based diets increases nutrient digestibility and growth performance, and the amount required depends on the fiber source used and the health status of the bird. During the first part of the rearing period, young pullets showed increased performance when 2 percent sunflower hulls were added to a control diet based on corn, wheat, and soybean meal (Guzmán et al., 2015b). However, after the first 5 weeks of age, the most relevant effects of insoluble fiber in pullets are related to development of the GIT with limited effect on bird performance (Guzmán et al., 2015a). The inclusion of moderate amounts of insoluble fiber (e.g., 2% to 4% oat hulls or straw) in the diet at 10 weeks of age increased the digestive capacity of the bird at 15 weeks of age (García et al., 2019). It is likely that increased GIT development facilitates the adaptation of the digestive tract of pullets to accommodate higher feed intake required as they transition to egg production. Research in growing pullets (Guzmán et al., 2015a) and adult quail (Starck, 1999) indicate that the gizzard responds to an increase in NDF with a rapid increase in size. In both studies, the phenotypic responses observed were directional and scaled to demand with an increase in gizzard size as the proportion of NDF in the diet was raised, and it was a reversible response, with a reduction in gizzard size as the NDF of the diet was reduced. In short, the effects of dietary fiber on the size and development of the GIT are not long lasting and revert quickly once the bird changes to a diet with a lower digestible fiber (DF) content (Mateos et al., 2012).

The effects of high-fiber diets on hen performance are not clear, with some data showing benefits and some showing no effect at all (Roberts et al., 2007; Sozcu and Ipek, 2020). When the fiber content of the diet is increased, some extra fat is typically added to compensate for the lower energy content of DF that increases the palatability of the feed, resulting in an increase in feed intake and egg size. In contrast, under most circumstances, the use or moderate levels of insoluble fiber sources (3% to 6%) in adult birds does not affect egg production and has limited effects on GIT development.

Cannibalism is an important issue for the egg industry because its incidence affects animal welfare and causes economic losses to the egg producer. The problem has increased in the last years because of the increased pressure to ban beak-trimming practices in poultry. The problem is multifactorial, but there is evidence that a minimum level of insoluble fiber might help in the prevention of this behavioral disorder (Bearse et al., 1940; Wahlström et al., 1998). It is likely that the increased rate of passage of digesta due to insoluble fiber results the birds spending more time eating with less time available for pecking (Mens et al., 2020). In fact, Hartini et al. (2002) observed that cannibalism was reduced in a dose-dependent manner as the fiber content of the diet increased.

Most plant-based feedstuffs differ in their proportions of the various components of fiber, including oligosaccharides, NSP, and lignin. Therefore, when they are used in a diet to provide similar fiber levels, they often will have different effects on digesta pH, viscosity, rate of passage, and rates and locations of SCFA production (Montagne et al., 2003; Raninen et al., 2011). Consequently, two feedstuffs that have similar amounts of the main fiber types may cause different effects in the intestine, which translate into different effects on feed intake and growth. This is because each of the fiber categories that are measured are comprised of a very large number of complex molecules, each with differing nutritional and physiochemical properties. Complicating the picture, even the same feedstuff harvested at different stages of maturity or in different locations may have different amounts and composition of dietary fiber. For these reasons, making specific recommendations on the type and amount of fiber that provide optimal positive (and negative) effects on poultry nutrition is extremely difficult.

Fermentation of fiber provides energy to poultry, but the amount is small relative to pigs and many other omnivorous animals, likely due to their relatively small area in the digestive tract for fermentation and their relatively short retention time of digesta (Klasing, 2005). The susceptibility of a fiber source to fermentation includes its water solubility, water holding capacity, and presence in small versus large particles. Fiber that is water soluble or that is present in small particles is able to pass through the filter in the neck of the ceca and be retained for a much longer time in the ceca than insoluble fiber or fiber in larger particles, which are rapidly excreted. In general, highly lignified fiber sources such as cereal straw or cereal hulls are poorly fermented, and fiber sources high in oligosaccharides and soluble NSP (e.g., sugar beet pulp) are more easily and rapidly fermented. In broiler chicks, Jørgensen et al. (1996) estimated that the amount of metabolizable energy (ME) derived from the fermentation of NSP fraction in oat bran, wheat bran, and pea fiber was only 3–4 percent of their daily ME intake and 6 percent of their daily energy production. Jamroz et al. (2002) observed that the total tract disappearance of NSP was similar in growing (3–5 weeks of age) dual breed chickens, mulard ducks, and Landaise geese fed a barley–wheat–soy-based diet. Presumably the energy obtained from their fermentation would also be similar. They calculated that the ME of NSP would contribute 3.5 percent of the daily ME needs of these three species. Taken together, these studies indicate the fermentation of fiber likely makes a minimal contribution to the energy requirement of a growing chick.

The microbial mass that grows on dietary fiber is rich in essential amino acids, vitamins, and trace minerals, and they can make an important nutritional contribution, especially in adult gamebirds. Gamebirds consume nutritionally important amounts of feces that originate from their ceca, a process called cecotrophy, but this process is less important in commercial chicken production (Klasing, 2005).

Negative Effects of Fiber

Historically, DF has been considered as an energy diluent and as an anti-nutritional factor in poultry diets, with negative effects on feed intake and nutrient digestibility (Sklan et al., 2003; Rougière and Carré, 2010). Consequently, commercial diets for most domestic poultry species were usually formulated with a limited fiber content. Feedstuffs that have high levels of fiber typically cannot be used in poultry diets at high inclusion levels without diminishing performance (Ravindran, 2013; Scholey et al., 2020). This includes coproducts from the milling of human foods such as wheat middlings, rice bran, palm kernel meal, and cereal hulls. When used at high levels, negative effects on intestinal physiology, growth, and efficiency of feed utilization are most evident with feedstuffs high in soluble oligosaccharides and soluble NSP relative to those high in insoluble NSP. The exact amount of dietary fiber tolerated is highly dependent on the composition of the fiber, breed, and age of the bird. The mechanism underlying the negative effects of fiber can be separated into two general categories: dilution of energy and nutrients and anti-nutritional effects.

Diet Dilution, Bulk, and Feed Intake

Diets with high levels of fiber usually have lower ME levels unless a compensatory amount of lipid is added, which is a strategy that has nutritional, physiological, practical, and economic limits. Some dietary fibers, such as pectins, gums, and β-glucans, have high water holding capacity, which increases the bulk of the digesta and feces. Increased bulk markedly decreases feed intake (Nascimento et al., 2020). Much of the fiber in the diet that contributes to bulk are soluble fibers that also increase viscosity of the digesta (Bedford and Classen, 1992; Classen, 2017). Some insoluble fibers, such as cellulose, can however absorb a considerable amount of water that contributes to bulk.

The reduction in the energy content of the diet when moderate amounts of fiber are added at the expense of the whole diet generally results in increased feed intake by the bird in an attempt meet its energy needs (Jiménez-Moreno et al., 2013; Mateos et al., 2020; Nascimento et al., 2020; Taylor et al., 2020). However, compensation by increasing intake has limits due to the capacity of the digestive system to process and digest feed. For this reason, productivity decreases if the decrease in dietary ME concentration exceeds the bird's threshold to compensate by increasing consumption. This threshold reflects the maximum anatomical and physiological capacity for feed consumption of the bird. Typically, poultry bred for fast growth or high rates of egg production are more sensitive to high dietary fiber than are strains that have lower rates (Gous, 2007; Mbajiorgu et al., 2011; Classen, 2017; Tallentire et al., 2018). In general, the tolerance for dietary fiber increases with age. Sanchez et al. (2019) reported that the inclusion of 11 percent rice bran (9.5% NDF) in a broiler diet from 24 to 35 days of age did not affect broiler performance. During the rearing phase, brown egg-laying hens tolerated diets containing 4.0 percent wheat straw with up to 14 percent NDF from 0 to 5 weeks and up to 17 percent NDF from 10 to 17 weeks of age (Guzmán et al., 2015a). In layers, the tolerance for fiber is even higher if the energy content of the diet is maintained and tolerance increases as egg production declines (Roberts et al., 2007).

Anti-Nutritional Effects

Fiber, and in particular soluble fiber, has chemical and physical properties that result in decreased absorption or utilization of most of the other nutrients. Part of this is due to increased nutrient binding, which prevents absorption of dietary nutrients and increases endogenous nutrient losses (Adedokun and Adeola, 2020). Some NSP molecules have hydrophobic surface properties that attract fats, bile acids, and cholesterol, causing lipid malabsorption and a lower rate and efficiency of growth (Smits and Annison, 1996). Some types of fiber, such as pectins, have regions of negative charges that form complex dietary cations (e.g., Mn, Fe, Zn, Cu), which usually decreases their digestibility (Smits and Annison, 1996; Greger, 1999). As indicated previously, most of these negative effects are greater for the water soluble fraction than for the insoluble fraction of fiber.

Negative attributes of fiber also include increased viscosity, water holding capacity, and bulk of the digesta (Choct, 2015; Nascimento et al., 2020). Viscosity occurs due to the dissociation of NSP in the gastrointestinal tract to form high molecular weight aggregates that have adhesive properties and result in decreased growth rate and efficiency of feed utilization and increased water consumption, sticky droppings, and wet litter (Bedford and Classen, 1992; Choct, 2015; Maisonnier et al., 2001; Montagne et al., 2003). For example, the inclusion of 3 percent soluble high-methylated pectins reduced feed efficiency up to 28 percent (Langhout and Schutte, 1996). Increased viscosity of the digesta is highly correlated with decreased absorption of fatty acids, amino acids, and, to a lesser extent, glucose (Choct et al., 1996; Maisonnier et al., 2001; Carré et al., 2002). The decrease in digestibility is partially due to reduced mixing of dietary components with endogenous digestive enzymes, especially in young birds. It is also due to gel-forming components, such as gums and pectins, interacting with the glycocalyx of the brush border and increasing the thickness of the unstirred water layer. This slows diffusion of digesta and decreases its interaction with the intestinal microvilli and access to their digestive enzymes and nutrient transporters (Montagne et al., 2003). Increased viscosity also reduces the intestinal pool size of bile salts, which decreases micelle formation and lipid absorption (Maisonnier et al., 2003). Finally, higher viscosity-inducing fiber fractions decrease digesta passage rate, which can result in increased microbial growth in proximal areas of the intestines, resulting in increased populations of pathogens in highly absorptive areas of the intestine (Choct et al., 1996; Pan and Yu, 2014; Tellez et al., 2014).

Pectins, pentosans, and β-glucans are the primary viscous components of small grains such as wheat, rye, barley, oats, and triticale. For example, when the pentosans extracted from wheat were fed to broilers, performance decreased and digesta viscosity increased markedly, but feeding the same amount of depolymerized pentosans had no effect (Choct and Annison, 1992). Purified wheat pentosans also decreased the apparent digestibility of amino acids (Angkanaporn et al., 1994). Exogenous enzymes of bacterial or fungal origin that hydrolyze soluble fiber are commonly supplemented to commercial diets containing these ingredients in order to decrease the viscosity of problematic feedstuffs (Adeola and Cowieson, 2011; Aftab and Bedford, 2018).

Xylose, arabinose, glucuronic acid, and galacturonic acid are found in NSPs (e.g., pectin, hemicellulose) of many plant-based ingredients and can be released by microbial digestion or exogenous enzymes. Broiler chickens fed diets with 5 percent D-xylose or L-arabinose had decreased growth and efficiency of feed utilization along with a concomitant increase in urinary output and watery droppings (Regassa et al., 2017; Schutte et al., 1992). At higher dietary concentrations, D-galacturonic acid and D-glucuronic acid are considerably more growth depressing than either D-xylose or L-arabinose (Longstaff et al., 1988).

CONCLUSIONS

Dietary CHOs constitute an important category of nutrients across all poultry breeds and compel consideration of the variety of plant-based ingredients used and the concentration and forms of CHO contained in those feedstuffs (see Table 2-1). Starch remains the predominant source of digestible energy in poultry diets, and dietary fiber has historically been considered as predominantly a negative attribute in feedstuffs (NRC, 1994). However, there is growing evidence that fractions of insoluble dietary fiber enhance the development of the intestinal tract. Evidence suggests that growing poultry should have sufficient fiber, especially the insoluble fraction, to provide structural properties to the diet to facilitate the development of the proventriculus and gizzard. However, the potential negative effects of soluble fiber on nutrient digestibility and performance compel the inclusion of maximum limits in the diet, especially in young chicks.

In summary, poultry species respond differently to the inclusion of dietary fiber, depending primarily on the level and type of fiber, the composition of the control diet, and the spe-

TABLE 2-1 Nutritional and Chemical Categories of Carbohydrates for Poultry

Chemical Attribute	Nutritional Attributes and Example Carbohydrates		
	Absorbable[a]	Digestible	Indigestible[b]
Monosaccharide	Glucose, fructose, galactose		
Disaccharide		Sucrose, maltose, trehalose	Lactose
Oligosaccharide		Maltodextrins	Raffinose, stachyose, verbascose, fructooligosaccharides,
Polysaccharide		Most starch	Cellulose, hemicellulose, pectins, β-glucans, resistant starch

[a] Absorbable without the aid of digestive enzymes.
[b] Food components that cannot be digested with the bird's own enzymes.

cies, age, and health status of the birds. Under commercial conditions, diets for all types of poultry should be formulated with a maximum and, likely, a minimum level of fiber.

REFERENCES

Abd El-Khalek, E., and G. Janssens. 2010. Effect of extrusion processing on starch gelatinisation and performance in poultry. *World's Poultry Science Journal*. 66:53–64.

Adedokun, S.A., and O. Adeola. 2020. Regression-derived ileal endogenous amino acid losses in broiler chickens and cannulated pigs fed corn fiber, wheat bran, and pectin. *Animals*. 10:2145.

Adeola, O., and A. Cowieson. 2011. Opportunities and challenges in using exogenous enzymes to improve nonruminant animal production. Board invited review. *Journal of Animal Science*. 89:3189–3218.

Aftab, U., and M. Bedford. 2018. The use of NSP enzymes in poultry nutrition: Myths and realities. *World's Poultry Science Journal*. 74:277–286.

Angkanaporn, K., M. Choct, W.L. Bryden, E.F. Annison, and G. Annison. 1994. Effects of wheat pentosans on endogenous amino acid losses in chickens. *Journal of the Science of Food and Agriculture*. 66:399–404.

Bach Knudsen, K., and I. Hansen. 1991. Gastrointestinal implications in pigs of wheat and oat fractions: 1. Digestibility and bulking properties of polysaccharides and other major constituents. *British Journal of Nutrition*. 65:217–232.

Bach Knudsen, K. 2001. The nutritional significance of "dietary fibre" analysis. *Animal Feed Science and Technology*. 90:3–20.

Bach Knudsen, K.E. 2014. Fiber and nonstarch polysaccharide content and variation in common crops used in broiler diets. *Poultry Science*. 93:2380–2393.

Bearse, G.E., V. Miller, and C. McClary. 1940. The cannibalism preventing properties of the fiber fraction of oat hulls. *Poultry Science*. 19:210–215.

Bedford, M.R., and H.L. Classen. 1992. Reduction of intestinal viscosity through manipulation of dietary rye and pentosanase concentration is effected through changes in the carbohydrate composition of the intestinal aqueous phase and results in improved growth rate and food conversion efficiency of broiler chicks. *The Journal of Nutrition*. 122:560–569.

Bergh, M., A. Razdan, and P. Åman. 1999. Nutritional influence of broiler chicken diets based on covered normal, waxy and high amylose barleys with or without enzyme supplementation. *Animal Feed Science and Technology*. 78:215–226.

Brady, L.J., D.R. Romsos, and G.A. Leveille. 1979. Gluconeogenesis in isolated chicken (Gallus domesticus) liver cells: Comparative biochemistry and physiology. *Comparative Biochemistry B*. 63:193–198.

Carré, B., E. Beaufils, and J.P. Melcion. 1991. Evaluation of protein and starch digestibility and energy value of pelleted or unpelleted pea seeds from winter or spring cultivars in adult and young chickens. *Journal of Agricultural and Food Chemistry*. 39:468–472.

Carré, B., L. Derouet, and B. Leclercq. 1990. The digestibility of cell-wall polysaccharides from wheat (bran or whole grain), soybean meal, and white lupin meal in cockerels, muscovy ducks, and rats. *Poultry Science*. 69:623–633.

Carré, B., J. Gomez, and A. Chagneau. 1995. Contribution of oligosaccharide and polysaccharide digestion, and excreta losses of lactic acid and short chain fatty acids, to dietary metabolisable energy values in broiler chickens and adult cockerels. *Journal of British Poultry Science*. 36:611–630.

Carré, B., A. Idi, S. Maisonnier, J.-P. Melcion, F.-X. Oury, J. Gomez, and P. Pluchard. 2002. Relationships between digestibilities of food components and characteristics of wheats (Triticum aestivum) introduced as the only cereal source in a broiler chicken diet. *British Poultry Science*. 43:404–415.

Chen, X., C.M. Parsons, and N. Bajjalieh. 2013. Nutritional evaluation of new reduced oligosaccharide soybean meal in poultry. *Poultry Science*. 92:1830–1836.

Choct, M. 2015. Feed non-starch polysaccharides for monogastric animals: Classification and function. *Journal of Animal Production Science*. 55:1360–1366.

Choct, M., and G. Annison. 1992. Anti-nutritive effect of wheat pentosans in broiler chickens: Roles of viscosity and gut microflora. *British Poultry Science*. 33:821–834.

Choct, M., Y. Dersjant-Li, J. McLeish, and M. Peisker. 2010. Soy oligosaccharides and soluble non-starch polysaccharides: A review of digestion, nutritive and anti-nutritive effects in pigs and poultry. *Asian-Australasian Journal of Animal Sciences*. 23:1386–1398.

Choct, M., R.J. Hughes, J. Wang, M. Bedford, A. Morgan, and G. Annison. 1996. Increased small intestinal fermentation is partly responsible for the anti-nutritive activity of non-starch polysaccharides in chickens. *British Poultry Science*. 37:609–621.

Chotinsky, D., E. Toncheva, and Y. Profirov. 2001. Development of disaccharidase activity in the small intestine of broiler chickens. *British Poultry Science*. 42:389–393.

Classen, H.L. 2017. Diet energy and feed intake in chickens. *Journal of Animal Feed Science Technology*. 233:13–21.

Coon, C.N., K. Leske, O. Akavanichan, and T. Cheng. 1990. Effect of oligosaccharide-free soybean meal on true metabolizable energy and fiber digestion in adult roosters. *Poultry Science*. 69:787–793.

De Vries, S. 2015. Fiber in poultry nutrition: Bonus or burden? In *20th European Symposium on Poultry Nutrition*. Prague, Czech Republic.

Dickson, A.J., and D.R. Langslow. 1978. Hepatic gluconeogenesis in chickens. *Journal of Molecular and Cellular Biochemistry*. 22:167–182.

Englyst, H.N., S. Kingman, and J. Cummings. 1992. Classification and measurement of nutritionally important starch fractions. *European Journal of Clinical Nutrition*. 46:S33– S50.

García, J., G. Fondevila, L. Cámara, R. Scappaticcio, L. Aguirre, and G. Mateos. 2019. Influence of egg weight and inclusion of oat hulls in the diet on digestive tract traits and growth performance of brown pullets reared under stress conditions. *Poultry Science*. 98:5767–5777.

González-Alvarado, J., E. Jiménez-Moreno, D. González-Sánchez, R. Lázaro, and G. Mateos. 2010. Effect of inclusion of oat hulls and sugar beet pulp in the diet on productive performance and digestive traits of broilers from 1 to 42 days of age. *Animal Feed Science and Technology*. 162:37–46.

Gous, R. 2007. Predicting nutrient responses in poultry: Future challenges. *Animal*. 1:57–65.

Greger, J. 1999. Nondigestible carbohydrates and mineral bioavailability. *The Journal of Nutrition*. 129:1434S–1435S.

Guzmán, P., B. Saldaña, M. Kimiaeitalab, J. García, and G. Mateos. 2015a. Inclusion of fiber in diets for brown-egg laying pullets: Effects on growth performance and digestive tract traits from hatching to 17 weeks of age. *Poultry Science*. 94:2722–2733.

Guzmán, P., B. Saldaña, H. Mandalawi, A. Pérez-Bonilla, R. Lázaro, and G. Mateos. 2015b. Productive performance of brown-egg laying pullets from hatching to 5 weeks of age as affected by fiber inclusion, feed form, and energy concentration of the diet. *Poultry Science*. 94:249–261.

Hartini, S., M. Choct, G. Hinch, A. Kocher, and J.V. Nolan. 2002. Effects of light intensity during rearing and beak trimming and dietary fiber sources on mortality, egg production, and performance of ISA brown laying hens. *Journal of Applied Poultry Research*. 11:104–110.

Herwig, E., D. Abbott, K.V. Schwean-Lardner, and H.L. Classen. 2019a. Assessing the effect of rate and extent of starch digestion on laying hen performance. *Poultry Science*. 98:2940–2947.

Herwig, E., D. Abbott, K.V. Schwean-Lardner, and H.L. Classen. 2019b. Effect of rate and extent of starch digestion on broiler chicken performance. *Poultry Science*. 98:3676–3684.

Herwig, E., K. Schwean-Lardner, A. Van Kessel, R.K. Savary, and H.L. Classen. 2020. Assessing the effect of starch digestion characteristics on ileal brake activation in broiler chickens. *PloS One*. 15:e0228647.

Hetland, H., M. Choct, and B. Svihus. 2004. Role of insoluble non-starch polysaccharides in poultry nutrition. *World's Poultry Science*. 60:415–422.

Hume, M.E., L.F. Kubena, R.C. Beier, A. Hinton Jr., D.E. Corrier, and J.R. Deloach. 1992. Fermentation of [14 C] lactose in broiler chicks by cecal anaerobes. *Poultry Science*. 71:1464–1470.

Iji, P., and D. Tivey. 1998. Natural and synthetic oligosaccharides in broiler chicken diets. *World's Poultry Science Journal*. 54:129–143.

Itani, K., J. Hansen, B. Kierończyk, A. Benzertiha, P. Kurk, R. Ånestad, F. Sundby, L. Mydland, M. Øverland, and B. Svihus. 2021. Interactions between starch source and gelatinisation degree on performance and small intestinal digestion in broiler chickens. *British Poultry Science*. 62:424–434.

Itani, K., and B. Svihus. 2019. Feed processing and structural components affect starch digestion dynamics in broiler chickens. *British Poultry Science*. 60:246–255.

Jamroz, D., K. Jakobsen, K.E.B. Knudsen, A. Wiliczkiewicz, and J. Orda. 2002. Digestibility and energy value of non-starch polysaccharides in young chickens, ducks and geese, fed diets containing high amounts of barley. *Comparative Biochemistry and Physiology Part A: Molecular & Integrative Physiology*. 131:657–668.

Jha, R., and P. Mishra. 2021. Dietary fiber in poultry nutrition and their effects on nutrient utilization, performance, gut health, and on the environment: A review. *Journal of Animal Science and Biotechnology*. 12:1–16.

Jiménez-Moreno, E., M. Frikha, A. de Coca-Sinova, J. García, and G. Mateos. 2013. Oat hulls and sugar beet pulp in diets for broilers 1. Effects on growth performance and nutrient digestibility. *Animal Feed Science and Technology*. 182:33–43.

Jiménez-Moreno, E., J. González-Alvarado, A. de Coca-Sinova, R. Lázaro, L. Cámara, and G. Mateos. 2019. Insoluble fiber sources in mash or pellets diets for young broilers. 2. Effects on gastrointestinal tract development and nutrient digestibility. *Poultry Science*. 98:2531–2547.

Jiménez-Moreno, E., J. González-Alvarado, D. González-Sánchez, R. Lázaro, and G. Mateos. 2010. Effects of type and particle size of dietary fiber on growth performance and digestive traits of broilers from 1 to 21 days of age. *Poultry Science*. 89:2197–2212.

Jiménez Moreno, E., C. Romero, J. Berrocoso, M. Frikha, and G. Gonzalez Mateos. 2011. Effects of the inclusion of oat hulls or sugar beet pulp in the diet on gizzard characteristics, apparent ileal digestibility of nutrients, and microbial count in the ceca in 36 day old broilers reared on floor. *Animal Feed Science and Technology*. 162:37–46.

Jørgensen, H., X.-Q. Zhao, K.E.B. Knudsen, and B.O. Eggum. 1996. The influence of dietary fibre source and level on the development of the gastrointestinal tract, digestibility and energy metabolism in broiler chickens. *British Journal of Nutrition*. 75:379–395.

Jozefiak, D., A. Rutkowski, and S.A. Martin. 2004. Carbohydrate fermentation in the avian ceca: A review. *Animal Feed Science and Technology*. 113:1–15.

Kedinger, M., P. Simon, J. Grenier, and K. Haffen. 1981. Role of epithelial-mesenchymal interactions in the ontogenesis of intestinal brush-border enzymes. *Developmental Biology*. 86:339–347.

Kheravii, S.K., N.K. Morgan, R.A. Swick, M. Choct, and S.B. Wu. 2018a. Roles of dietary fibre and ingredient particle size in broiler nutrition. *World's Poultry Science Journal*. 74:301–316.

Kheravii, S.K., R.A. Swick, M. Choct, and S.-B.J.B.g. Wu. 2018b. Upregulation of genes encoding digestive enzymes and nutrient transporters in the digestive system of broiler chickens by dietary supplementation of fiber and inclusion of coarse particle size corn. *BMC Genomics*. 19:208.

Klasing, K. 2005. Poultry nutrition: A comparative approach. *Journal of Applied Poultry Research*. 14:426–436.

Klasing, K.C. 1998. *Comparative Avian Nutrition*. Wallingford, Oxfordshire, United Kingdom: CAB International.

Langhout, D., and J. Schutte. 1996. Nutritional implications of pectins in chicks in relation to esterification and origin of pectins. *Poultry Science*. 75:1236–1242.

Lee, T.-T., Y.-F. Huang, C.-C. Chiang, T.-K. Chung, P.W.-S. Chiou, and B. Yu. 2011. Starch characteristics and their influences on in vitro and pig prececal starch digestion. *Journal of Agricultural and Food Chemistry*. 59:7353–7359.

Longstaff, M., A. Knox, and J. McNab. 1988. Digestibility of pentose sugars and uronic acids and their effect on chick weight gain and caecal size. *British Poultry Science*. 29:379–393.

Ma, Z., and J.I. Boye. 2018. Research advances on structural characterization of resistant starch and its structure-physiological function relationship: A review. *Critical Reviews in Food Science and Nutrition*. 58:1059–1083.

Maisonnier, S., J. Gomez, A. Bree, C. Berri, E. Baeza, and B. Carre. 2003. Effects of microflora status, dietary bile salts and guar gum on lipid digestibility, intestinal bile salts, and histomorphology in broiler chickens. *Poultry Science*. 82:805–814.

Maisonnier, S., J. Gomez, and B. Carré. 2001. Nutrient digestibility and intestinal viscosities in broiler chickens fed on wheat diets, as compared to maize diets with added guar gum. *British Poultry Science*. 42:102–110.

Mateos, G., E. Jiménez-Moreno, M. Serrano, and R. Lázaro. 2012. Poultry response to high levels of dietary fiber sources varying in physical and chemical characteristics. *Journal of Applied Poultry Research*. 21:156–174.

Mateos, G.G., G. Fodevila, and L. Camara. 2020. The importance of the fibre fraction of the feed in non-ruminant diets. In *The value of fibre: Engaging the second brain for animal nutrition*. M.R.B. G. González-Ortiz, K.E. Bach Knudsen, C.M. Courtin, and H.L. Classen, editors. Wageningen, The Netherlands: Wageningen Academic Publishers. 63–85.

Mbajiorgu, C.A., J. Ng'Ambi, and D. Norris. 2011. Voluntary feed intake and nutrient composition in chickens. *Asian Journal of Animal and Veterinary Advances*. 6:20–28.

Mens, A., M. Van Krimpen, and R. Kwakkel. 2020. Nutritional approaches to reduce or prevent feather pecking in laying hens: Any potential to intervene during rearing? *World's Poultry Science Journal*. 76:591–610.

Montagne, L., J. Pluske, and D. Hampson. 2003. A review of interactions between dietary fibre and the intestinal mucosa, and their consequences on digestive health in young non-ruminant animals. *Animal Feed Science and Technology*. 108:95–117.

Moran Jr., E. 2019. Starch: Granule, amylose-amylopectin, feed preparation, and recovery by the fowl's gastrointestinal tract. *Journal of Applied Poultry Research*. 28:566–586.

Nascimento, M., R.M. Gous, M. Reis, J.B.K. Fernandes, and N. Sakomura. 2020. Prediction of maximum scaled feed intake in broiler chickens based on physical properties of bulky feeds. *British Poultry Science*. 61:676–683.

Niwa, H., T. Yamano, T. Sugano, and R.A. Harris. 1986. Hormonal effects and the control of gluconeogenesis from sorbitol, xylitol and glycerol in perfused chicken liver. *Journal of Comparative Biochemistry & Physiology B, Comparative Biochemistry*. 85:739–745.

NRC (National Research Council). 1994. *Nutrient Requirements of Poultry*, 9th rev. ed. Washington, DC: National Academy Press.

O'Dell, B., P. Newberne, and J. Savage. 1959. An abnormality of the proventriculus caused by feed texture. *Poultry Science*. 38:296–301.

Pan, D., and Z. Yu. 2014. Intestinal microbiome of poultry and its interaction with host and diet. *Gut Microbes*. 5:108–119.

Parada, J., and J. Aguilera. 2011. Starch matrices and the glycemic response. *Food Science and Technology International*. 17:187–204.

Parsons, C., Y. Zhang, and M. Araba. 2000. Nutritional evaluation of soybean meals varying in oligosaccharide content. *Poultry Science*. 79:1127–1131.

Perryman, K., and W. Dozier III. 2012. Apparent metabolizable energy and apparent ileal amino acid digestibility of low and ultra-low oligosaccharide soybean meals fed to broiler chickens. *Poultry Science*. 91:2556–2563.

Raninen, K., J. Lappi, H. Mykkänen, and K. Poutanen. 2011. Dietary fiber type reflects physiological functionality: Comparison of grain fiber, inulin, and polydextrose. *Nutrition Reviews*. 69:9–21.

Ravindran, V. 2013. Poultry feed availability and nutrition in developing countries. *Poultry Development Review*. 2:60–63.

Regassa, A., E. Kiarie, J. Sands, M. Walsh, W. Kim, and C. Nyachoti. 2017. Nutritional and metabolic implications of replacing cornstarch with

D-xylose in broiler chickens fed corn and soybean meal-based diet. *Poultry Science*. 96:388–396.

Renner, R., and A.M. Elcombe. 1967. Metabolic effects of feeding "carbohydrate-free" diets to chicks. *The Journal of Nutrition*. 93:31–36.

Riddell, C. 1976. The influence of fiber in the diet on dilation (hypertrophy) of the proventriculus in chickens. *Avian Diseases*. 20:442–445.

Roberts, S.A., H. Xin, B.J. Kerr, J.R. Russell, and K. Bregendahl. 2007. Effects of dietary fiber and reduced crude protein on nitrogen balance and egg production in laying hens. *Poultry Science*. 86:1716–1725.

Rougière, N., and B. Carré. 2010. Comparison of gastrointestinal transit times between chickens from D+ and D– genetic lines selected for divergent digestion efficiency. *Animal*. 4:1861–1872.

Roura, E., M.W. Baldwin, and K. Klasing. 2013. The avian taste system: Potential implications in poultry nutrition. *Animal Feed Science and Technology*. 180:1–9.

Sacranie, A., B. Svihus, V. Denstadli, B. Moen, P. Iji, and M. Choct. 2012. The effect of insoluble fiber and intermittent feeding on gizzard development, gut motility, and performance of broiler chickens. *Poultry Science*. 91:693–700.

Sanchez, J., A. Thanabalan, T. Khanal, R. Patterson, B.A. Slominski, and E. Kiarie. 2019. Growth performance, gastrointestinal weight, microbial metabolites and apparent retention of components in broiler chickens fed up to 11% rice bran in a corn-soybean meal diet without or with a multi-enzyme supplement. *Animal Nutrition*. 5:41–48.

Scholey, D., A. Marshall, and A. Cowan. 2020. Evaluation of oats with varying hull inclusion in broiler diets up to 35 days. *Poultry Science*. 99:2566–2572.

Schutte, J., J. de Jong, E. Van Weerden, and M. Van Baak. 1992. Nutritional value of D-xylose and L-arabinose for broiler chicks. *British Poultry Science*. 33:89–100.

Sell, J., L, O. Koldovsky, and B.L. Reid. 1989. Intestinal disaccharidases of young turkeys: Temporal development and influence of diet composition. *Poultry Science*. 68:265–277.

Sklan, D., A. Smirnov, and I. Plavnik. 2003. The effect of dietary fibre on the small intestines and apparent digestion in the turkey. *British Poultry Science*. 44:735–740.

Slama, J., K. Schedle, G.K. Wurzer, and M. Gierus. 2019. Physicochemical properties to support fibre characterization in monogastric animal nutrition. *Journal of the Science of Food and Agriculture*. 99:3895–3902.

Smits, C.H.M., and G. Annison. 1996. Non-starch plant polysaccharides in broiler nutrition—towards a physiologically valid approach to their determination. *World's Poultry Science Journal*. 52:203–221.

Sozcu, A., and A. Ipek. 2020. The effects of lignocellulose supplementation on laying performance, egg quality parameters, aerobic bacterial load of eggshell, serum biochemical parameters, and jejunal histomorphological traits of laying hens. *Poultry Science*. 99:3179–3187.

Starck, J.M. 1999. Phenotypic flexibility of the avian gizzard: rapid, reversible and repeated changes of organ size in response to changes in dietary fibre content. *Journal of Experimental Biology*. 202:3171–3179.

Svihus, B. 2011. The gizzard: function, influence of diet structure and effects on nutrient availability. *World's Poultry Science Journal*. 67:207–224.

Svihus, B. 2014. Starch digestion capacity of poultry. *Poultry Science*. 93:2394–2399.

Tallentire, C., I. Leinonen, and I. Kyriazakis. 2018. Artificial selection for improved energy efficiency is reaching its limits in broiler chickens. *Scientific Reports*. 8:1–10.

Taylor, J., P. Sakkas, and I. Kyriazakis. 2020. What are the limits to feed intake of broilers on bulky feeds? *Poultry Science*. 100:100825.

Tellez, G., J.D. Latorre, V.A. Kuttappan, M.H. Kogut, A. Wolfenden, X. Hernandez-Velasco, B.M. Hargis, W.G. Bottje, L.R. Bielke, and O.B. Faulkner. 2014. Utilization of rye as energy source affects bacterial translocation, intestinal viscosity, microbiota composition, and bone mineralization in broiler chickens. *Frontiers in Genetics*. 5:339.

Wahlström, A., R. Tauson, and K. Elwinger. 1998. Effects on plumage condition, health and mortality of dietary oats/wheat ratios to three hybrids of laying hens in different housing systems. *Acta Agriculturae Scandinavica A—Animal Sciences*. 48:250–259.

Weurding, R., H. Enting, and M. Verstegen. 2003. The relation between starch digestion rate and amino acid level for broiler chickens. *Poultry Science*. 82:279–284.

Weurding, R.E., A. Veldman, W.A. Veen, P.J. van der Aar, and M.W. Verstegen. 2001. Starch digestion rate in the small intestine of broiler chickens differs among feedstuffs. *The Journal of Nutrition*. 131:2329–2335.

Wiseman, J. 2006. Variations in starch digestibility in non-ruminants. *Animal Feed Science and Technology*. 130:66–77.

Yokhana, J., G. Parkinson, and T. Frankel. 2015. Effect of insoluble fiber supplementation applied at different ages on digestive organ weight and digestive enzymes of layer-strain poultry. *Poultry Science*. 95:550–559.

Zaefarian, F., M. Abdollahi, and V. Ravindran. 2015. Starch digestion in broiler chickens fed cereal diets. *Animal Feed Science and Technology*. 209:16–29.

3

Protein and Amino Acids

INTRODUCTION

The goal of this chapter is to review approaches for evaluating the amino acid requirements of poultry. In addition, classification, sources, bioavailability of D-amino acids, analysis of feedstuffs and physiological fluids, empirical estimates, factors influencing variations in requirements, deficiencies, imbalance, antagonism, toxicity, and special relationships of amino acids will be briefly described. In addition, amino acid digestibility coefficients for 14 feed ingredients are summarized. Methodology needed to properly design amino acid requirement and digestible amino ratio experiments and factors influencing amino acid requirements will be discussed.

PROTEINS

Amino acids obtained from dietary protein are used by poultry to fulfill a diversity of functions. For example, amino acids, as components of proteins, are primary constituents of structural and protective tissues, such as skin, feathers, bone matrix, and ligaments, as well as of the soft tissues, including organs and muscles. Also, amino acids and small peptides resulting from digestive and absorptive processes may serve a variety of metabolic functions and as precursors of many important non-protein body constituents. Adequate intake of amino acids is necessary to regularly replace those lost from skin, feather, and intestinal tissues (i.e., maintenance) and permit the accretion of protein in tissues for growth and eggs during reproduction. If dietary protein (i.e., as a source of amino acids) is inadequate, a reduction or cessation of growth or productivity and a withdrawal of protein from some body tissues to maintain the functions of more vital tissues will result. Proteins (amino acids) can also be oxidized to generate energy, and Tillman (2019) recently compiled tables of gross energy, nitrogen-corrected apparent metabolizable energy, and net energy values for 20 amino acids.

CLASSIFICATION OF AMINO ACIDS

There are 22 amino acids in body proteins, and all are physiologically essential. Nutritionally, these amino acids can be divided into two categories: those that poultry cannot synthesize at all or rapidly enough to meet metabolic requirements (essential or indispensable) and those that can be synthesized from other nutrients (nonessential or dispensable). The essential amino acids must be derived from the diet, whereas nonessential amino acids not supplied by the diet must be synthesized in vivo. Nonessential amino acids can be considered semi-essential under certain circumstances when the rate of amino acid synthesis in vivo is unable to meet the demand for protein synthesis. Dietary need for both essential and nonessential amino acids is an appropriate way to ensure that all amino acids needed physiologically are provided.

SOURCES OF AMINO ACIDS

Carbonaceous and proteinaceous concentrates are intact sources that provide amino acids to the diet. Examples of carbonaceous concentrates include cereal grains, dried bakery products, cassava, buckwheat, and potato products. These feedstuffs typically contain less than 20 percent crude protein and 18 percent fiber. Cereal grains typically represent approximately 60 percent to 73 percent of the total ingredient composition for poultry diets, with the exception of geese (Chapter 18). These feedstuffs are relatively low in amino acid content and display a poor amino acid balance. Due to their high inclusion rate, cereal grains provide approximately 12 percent to 25 percent of the requirement for a given amino acid in poultry diets. Examples of proteinaceous concentrates include oilseed meals, grain co-products, and meals of animal origin. These products contain more than 20 percent crude protein and are a good source of essential amino acids for poultry. Soybean meal, a co-product from soy oil production, is an excellent source of amino acids from an intact protein for poultry. Based on its amino acid profile, soybean meal complements that of grains because of its relatively high concentration of lysine. It can contribute approximately 67 percent of the essential amino acid requirements in poultry diets with the exception of methionine plus cyst(e)ine (40%). Grain co-products and animal meals typically provide a lower amount of the essential amino acid requirements for poultry, being in the range of 5 percent to 15 percent depending on their inclusion rate and amino acid composition. In poultry diets, cereal grains and oilseed meals have a complementary relationship to provide significant amounts of energy and amino acids to the bird. For example, corn is high in starch (an energy source) and low in lysine and tryptophan. In contrast, soybean meal is a good source of lysine, threonine, valine, isoleucine, arginine, and tryptophan but is somewhat low in energy. Amino acid digestibility coefficients were determined from 14 intact feed ingredients used in poultry production (Table 3-1). Criteria selection consisted of summarizing amino acid digestibility coefficients from peer-review literature: (1) Only ingredient data published after 1992 were used, (2) published data were accompanied with adequate methodological detail to ensure data integrity, (3) strong preference was given for ingredients that are commonly fed to poultry species in the U.S. poultry industry, (4) amino acid digestibility data were surveyed for macro ingredients, and (5) typical inclusion of 5 percent or higher. Twenty-two journals were used for the initial selection of amino acid digestibility coefficients of ingredients.

Amino acids used in protein synthesis are L-isomers, but some amino acids in the D- or DL-configurations are also utilized by converting to their L-form through deamination and transamination processes (Baker, 1994; Lewis and Baker, 1995). Notably, D-lysine cannot be utilized for protein synthesis in poultry. Feed-grade amino acids used in the poultry industry include L-lysine HCl, L-threonine, L-valine, L-isoleucine, L-arginine, L-tryptophan, DL-methionine, and DL-2-hydroxy-4-(methylthio) butanoic acid (free acid or calcium salt). Although the molar equivalency of dietary methionine sources for poultry has been actively investigated since the early 1950s (Bird, 1952), a plethora of studies have yielded variable results, and the topic remains controversial today. Feed-grade sources of methionine and lysine are marketed in both solid and liquid forms to the feed industry. The price spread between the primary intact amino acid source (e.g., soybean meal) and the primary grain source (e.g., corn) along with oil/fat price largely dictates the inclusion rate of feed-grade amino acids in poultry diets. Over the last 15 years, formulating diets on a digestible amino acid basis has encouraged the use of feed-grade amino acids. This technology has allowed nutritionists to take advantage of alternative feedstuffs to formulate diets more economically without compromising growth performance, egg production, meat yield, or environmental sustainability.

Organic poultry diets pose a unique challenge to the poultry nutritionist due to the limitations of synthetic methionine inclusion (0.10% of the diet for broiler and layers and 0.15% of the diet for turkeys in the United States) and a complete ban of synthetic methionine use in European organic diets. In most cases, nutritionists supply dietary protein in amounts greater than requirements to meet methionine and total sulfur-containing amino acids (TSAA; i.e., methionine plus cyst(e)ine) requirements of poultry. Over-supply of nitrogen leads to foot pad lesions, excessive nitrogen excretion, and increased ammonia emissions from poultry housing (Burley et al., 2016). Ingredients particularly higher in methionine that can be sourced as organic include animal proteins, potentially including dried egg products (Burley and Patterson, 2017), fish meal (Fanatico et al., 2018), and emerging insect proteins (Fanatico et al., 2018). Common organic plant sources relatively high in methionine include expelled sesame seed, sainfoin seed, expelled sunflower seed, rapeseed, soybean meal (van Krimpen et al., 2016), and brazil nut meal (Burley and Patterson, 2017). Expelling and other plant processing techniques may also assist in concentrating methionine and other amino acids in plant-based organic methionine sources. Digestibility of methionine is more variable in organic sourced plant proteins than inorganic sources (van Krimpen et al., 2016).

BIOAVAILABILITY OF D-AMINO ACIDS

The bioavailability of supplemented amino acids is based on their role in metabolic functions, including their direct use for protein synthesis, and their efficiency of conversion to an L-enantiomer that can be incorporated into proteins or used in metabolic pathways that provision intermediates for antioxidants and many other biologically active compounds. Except for glycine, all amino acids have two enantiomers, D and L, which refers to the position of the amino group in the Fisher projection. As such, L-amino acids have amino groups on the left side, which equates to polarized light being rotated in a levorotary fashion, and D-amino acids have amino groups on the right side (i.e., dextrorotary rotation of polarized light). Vertebrate tRNA binds L-amino acids with very high affinity but has a very low affinity for D-amino acids, and protein synthesis cannot be accomplished with D-amino acids unless they can first be converted to L-amino acids.

Conversion of a D-amino acid to an L-amino acid requires oxidative deamination to the α-keto analog and subsequent enzymatic transamination of an amino group from a donor such as glutamate. Amino acids differ greatly in the extent to which they can undergo this process, which most likely relates to their relative specificity as substrates of hepatic and renal forms of D-amino acid oxidase (DAAO). For example, with very few exceptions, D-phenylalanine and D-proline have the greatest DAAO substrate specificities across vertebrate species ranging from ducks and chickens to rats, mice, pigs, monkeys, and humans, with methionine ranking between second and seventh (Elkin et al., 1988b). Consequently, the bioavailability of D-amino acids in poultry ranges from nearly 100 percent to 0 percent (Lewis and Baker, 1995). This is important because D-amino acids are

TABLE 3-1 Apparent and Standardized Amino Acid Digestibility (%) of Feed Ingredients Used in Poultry Diets

Ingredient	Corn grain, dry						Corn distillers dried grains with solubles					
AAFCO #	48.4						27.6					
	AID			SID			AID			SID		
Item	x̄	n	SD	x̄	n	SD	x̄	n	SD	x̄	n	SD
Crude protein (N × 6.25)	79.0	1	–	91.3	1	–	–	–	–	70.1	1	–
Essential												
Arginine	85.8	5	1.30	90.3	16	5.04	77.9	3	1.81	77.1	12	8.19
Histidine	84.5	4	2.38	88.0	15	6.44	70.6	3	1.93	71.4	12	8.50
Isoleucine	82.4	5	2.07	87.2	16	8.64	71.1	3	1.92	73.6	12	7.96
Leucine	90.0	5	1.22	92.2	16	4.12	81.3	3	1.91	82.3	12	6.25
Lysine	75.8	5	4.44	82.6	16	9.35	52.2	3	2.62	60.8	11	6.99
Methionine	88.6	5	1.82	90.7	16	6.36	76.4	3	3.67	78.3	12	9.87
Phenylalanine	86.2	5	1.30	91.0	16	5.39	77.8	3	1.82	78.7	12	6.97
Threonine	65.4	5	3.65	79.6	16	6.83	58.0	3	2.75	63.8	12	8.70
Tryptophan	–	–	–	86.0	5	5.20	73.6	3	2.96	76.4	8	7.20
Valine	81.0	5	2.35	86.9	16	6.44	71.1	3	1.91	72.8	12	8.76
Nonessential												
Alanine	88.2	5	1.48	90.3	14	4.97	78.3	3	1.97	77.3	10	7.25
Aspartic acid	77.2	5	2.77	83.6	14	8.63	61.2	3	2.55	65.1	10	10.36
Cysteine	71.0	1	–	83.1	11	5.88	64.4	3	3.46	70.7	12	9.00
Glutamic acid	90.2	5	0.84	91.5	14	4.96	77.3	3	2.34	76.8	10	8.56
Glycine	75.0	4	2.45	83.1	11	8.36	64.1	3	2.06	68.5	10	10.73
Proline	–	–	–	85.9	7	6.50	75.1	3	2.48	74.0	10	9.05
Serine	77.4	5	2.61	86.0	14	6.85	68.1	3	2.45	73.2	11	8.58
Tyrosine	77.0	5	2.35	86.3	14	3.07	76.7	3	2.80	79.1	10	6.72

Ingredient	Corn gluten meal						Feather meal					
AAFCO #	48.14						9.15					
	AID			SID			AID			SID		
Item	x̄	n	SD	x̄	n	SD	x̄	n	SD	x̄	n	SD
Crude protein (N × 6.25)	–	–	–	–	–	–	59.0	1	–	–	–	–
Essential												
Arginine	94.0	2	1.41	73.3	6	17.53	67.8	3	2.36	66.2	6	5.74
Histidine	–	–	–	91.9	6	3.69	60.9	3	7.07	58.9	6	6.74
Isoleucine	92.0	2	0.00	86.5	3	4.50	72.4	3	2.12	72.6	6	2.22
Leucine	95.5	2	2.12	90.5	6	3.93	67.7	3	3.21	66.8	6	4.20
Lysine	85.5	2	0.71	93.7	6	4.36	55.5	3	1.29	55.3	6	3.54
Methionine	94.0	2	1.41	84.6	6	3.34	61.3	3	6.46	59.1	6	14.80
Phenylalanine	95.0	2	1.41	93.0	6	3.39	72.7	3	0.58	71.4	6	5.53
Threonine	89.0	2	0.00	92.9	6	4.13	54.1	3	3.59	52.6	6	6.34
Tryptophan	83.5	2	0.71	87.1	6	4.70	–	–	–	45.0	1	–
Valine	93.0	2	1.41	79.5	3	11.70	68.0	3	3.00	67.4	6	3.61
Nonessential				89.5	6	5.19						
Alanine	94.5	2	0.71				64.7	3	3.11	63.2	5	4.74
Aspartic acid	89.5	2	0.71	92.5	4	4.68	35.8	3	4.33	32.2	5	8.48
Cysteine	83.5	2	4.95	89.2	5	5.51	42.7	2	10.89	42.0	5	7.57
Glutamic acid	94.5	2	2.12	84.9	6	6.28	58.1	3	3.77	55.8	5	6.57
Glycine	–	–	–	90.7	5	7.60	66.4	2	3.32	67.3	2	3.54
Proline	95.5	2	0.71	–	–	–	66.5	2	9.26	61.0	1	–
Serine	92.5	2	0.71	90.8	4	5.95	63.4	3	2.16	62.5	5	6.47
Tyrosine	–	–	–	91.2	5	3.67	67.0	2	1.41	65.4	4	3.78

TABLE 3-1 Continued

Ingredient	Field peas						Fish meal, mechanical extracted					
AAFCO #	none						51.14					
	AID			SID			AID			SID		
Item	x̄	n	SD	x̄	n	SD	x̄	n	SD	x̄	n	SD
Crude protein (N × 6.25)	78.8	7	4.80	88.3	1	–	79.0	1	–	–	–	–
Essential												
Arginine	89.1	19	2.75	91.2	12	3.21	82.0	2	2.83	83.7	8	7.86
Histidine	83.9	16	5.57	84.6	8	4.38	76.0	2	0.00	80.0	8	6.58
Isoleucine	80.2	19	4.54	83.8	12	4.84	83.5	2	3.54	81.7	8	9.52
Leucine	81.5	20	4.47	83.9	12	4.91	84.0	2	1.41	82.8	8	8.45
Lysine	87.3	20	2.84	89.5	12	2.59	85.0	2	1.41	83.5	8	7.66
Methionine	79.9	19	5.27	85.3	12	5.41	85.0	2	2.83	82.9	8	8.37
Phenylalanine	81.3	19	5.04	86.0	12	5.78	82.0	2	1.41	80.9	8	9.67
Threonine	76.5	20	4.60	81.9	12	4.73	75.0	2	1.41	78.3	8	8.59
Tryptophan	76.5	8	3.97	76.3	2	1.84	–	–	–	84.0	2	8.41
Valine	79.4	19	4.77	83.2	12	5.40	82.5	2	0.71	80.4	8	8.67
Nonessential												
Alanine	80.5	20	4.49	83.8	10	5.79	79.0	2	2.83	79.8	7	8.67
Aspartic acid	82.5	19	5.12	83.7	10	4.65	72.5	2	3.54	72.6	7	8.41
Cysteine	66.7	20	4.53	74.3	12	6.08	57.0	1	–	64.6	7	8.92
Glutamic acid	88.1	19	2.96	88.3	10	7.51	81.0	2	1.41	80.0	7	9.33
Glycine	77.0	11	5.81	84.2	11	5.87	71.0	1	–	75.2	2	4.41
Proline	77.6	14	7.21	82.8	5	10.60	–	–	–	81.5	2	5.09
Serine	79.0	19	4.48	83.3	11	4.64	72.0	2	2.83	74.6	7	9.16
Tyrosine	77.1	6	6.34	83.9	6	4.26	79.5	2	3.54	78.6	7	9.84

Ingredient	Meat and bone meal						Poultry by-product meal					
AAFCO #	9.41						9.1					
	AID			SID			AID			SID		
Item	x̄	n	SD	x̄	n	SD	x̄	n	SD	x̄	n	SD
Crude protein (N × 6.25)	76.0	8	5.53	–	–	–	–	–	–	–	–	–
Essential												
Arginine	73.2	11	7.63	77.3	30	5.54	81.2	1	–	80.8	2	2.47
Histidine	68.0	11	10.33	72.3	30	8.49	75.5	1	–	75.6	2	2.19
Isoleucine	66.7	11	11.74	71.9	30	9.32	75.2	1	–	76.4	2	1.91
Leucine	69.8	11	9.45	74.2	30	8.03	77.8	1	–	78.4	2	1.91
Lysine	70.2	11	9.83	73.0	30	8.85	76.7	1	–	74.5	2	4.95
Methionine	70.5	11	11.39	73.5	30	9.59	80.3	1	–	79.4	2	3.32
Phenylalanine	72.3	11	9.32	75.5	30	7.94	78.2	1	–	81.4	2	6.22
Threonine	59.5	11	11.53	68.7	30	9.64	70.2	1	–	73.2	2	3.04
Tryptophan	79.6	9	5.90	60.1	2	7.21	–	–	–	78.0	1	–
Valine	68.8	11	11.06	72.3	30	8.47	75.7	1	–	76.1	2	2.97
Nonessential												
Alanine	71.4	11	8.38	75.3	29	6.36	78.9	1	–	80.2	1	–
Aspartic acid	55.0	11	11.81	59.7	29	10.37	59.0	1	–	61.3	1	–
Cysteine	25.8	8	8.56	45.5	29	16.13	50.6	1	–	56.8	2	–
Glutamic acid	70.5	11	10.80	72.6	29	6.65	75.8	1	–	77.4	1	–
Glycine	67.5	11	8.86	71.1	26	6.14	75.7	1	–	76.7	1	–
Proline	62.5	9	8.59	68.2	29	7.24	71.3	1	–	73.1	1	–
Serine	57.7	11	10.12	66.8	29	8.30	68.4	1	–	72.1	1	–
Tyrosine	66.8	11	11.55	71.1	22	9.89	–	–	–	–	–	–

continued

TABLE 3-1 Continued

Ingredient	Ground grain sorghum, dry						Soybean meal, solvent extracted, 48% CP					
AAFCO #	42.1						84.61					
	AID			SID			AID			SID		
Item	x̄	n	SD	x̄	n	SD	x̄	n	SD	x̄	n	SD
Crude protein (N × 6.25)	79.0	1	–	–	–	–	78.2	3	7.83	89.0	2	1.70
Essential												
Arginine	80.6	5	0.89	87.8	7	2.24	87.8	7	3.03	93.5	4	1.77
Histidine	75.0	5	2.45	79.8	7	4.33	86.0	4	3.13	90.9	4	1.69
Isoleucine	81.2	5	1.64	87.5	7	2.19	82.9	7	5.34	90.1	4	4.25
Leucine	84.6	5	1.67	87.6	7	1.42	82.7	7	5.14	89.1	4	1.56
Lysine	73.0	5	4.18	85.4	7	5.55	86.1	7	4.14	90.9	4	3.39
Methionine	84.0	5	2.24	88.8	7	2.96	87.4	7	8.56	91.9	4	4.66
Phenylalanine	83.2	5	1.79	88.8	7	1.78	83.9	7	4.49	88.3	4	3.76
Threonine	68.4	5	2.51	82.3	7	4.46	77.5	7	4.92	90.4	4	5.14
Tryptophan	–	–	–	84.1	2	4.10	72.0	1	–	86.5	3	3.27
Valine	79.8	5	2.28	85.8	7	2.89	81.7	7	5.74	89.0	4	2.89
Nonessential												
Alanine	84.8	5	1.30	87.9	6	1.32	80.8	7	5.57	89.0	3	2.36
Aspartic acid	79.0	5	1.58	85.8	6	2.43	81.3	7	4.13	89.0	3	2.54
Cysteine	80.0	1	–	81.6	3	9.72	66.0	3	13.06	81.6	3	4.82
Glutamic acid	84.4	5	2.51	87.5	6	2.18	86.5	7	3.09	91.7	3	0.39
Glycine	73.0	4	2.71	84.6	5	6.44	79.5	4	6.41	87.8	3	2.39
Proline	–	–	–	79.1	1	–	83.7	4	4.41	89.9	3	2.06
Serine	75.8	5	2.59	86.1	6	4.31	80.5	7	4.49	90.2	3	4.23
Tyrosine	75.0	5	1.58	82.8	6	1.90	86.1	5	1.34	91.7	1	–

Ingredient	Soybean meal, solvent extracted, 44% CP						Sunflower meal, solvent extracted					
AAFCO #	84.61						71.221					
	AID			SID			AID			SID		
Item	x̄	n	SD	x̄	n	SD	x̄	n	SD	x̄	n	SD
Crude protein (N × 6.25)	78.2	3	7.83	89.0	2	1.70	76.1	2	4.10	–	–	–
Essential												
Arginine	87.8	7	3.03	93.5	4	1.77	90.3	3	3.61	92.0	4	2.99
Histidine	86.0	4	3.13	90.9	4	1.69	78.1	2	6.93	82.6	3	6.77
Isoleucine	82.9	7	5.34	90.1	4	4.25	83.7	3	5.03	86.9	4	4.49
Leucine	82.7	7	5.14	89.1	4	1.56	82.5	3	4.77	85.7	4	4.23
Lysine	86.1	7	4.14	90.9	4	3.39	85.2	3	9.65	87.8	4	7.86
Methionine	87.4	7	8.56	91.9	4	4.66	90.0	2	7.07	91.7	3	4.93
Phenylalanine	83.9	7	4.49	88.3	4	3.76	82.1	3	8.00	86.0	4	7.11
Threonine	77.5	7	4.92	90.4	4	5.14	72.5	3	3.72	78.2	4	3.84
Tryptophan	72.0	1	–	86.5	3	3.27	–	–	–	87.0	1	–
Valine	81.7	7	5.74	89.0	4	2.89	82.1	3	4.85	85.2	4	4.42
Nonessential												
Alanine	80.8	7	5.57	89.0	3	2.36	82.6	3	4.13	85.3	3	4.11
Aspartic acid	81.3	7	4.13	89.0	3	2.54	79.1	3	5.80	81.3	3	5.73
Cysteine	66.0	3	13.06	81.6	3	4.82	92.0	1	–	87.7	2	10.88
Glutamic acid	86.5	7	3.09	91.7	3	0.39	88.9	3	3.78	90.1	3	3.76
Glycine	79.5	4	6.41	87.8	3	2.39	68.3	2	8.06	71.1	2	7.92
Proline	83.7	4	4.41	89.9	3	2.06	–	–	–	–	–	–
Serine	80.5	7	4.49	90.2	3	4.23	72.6	3	4.22	76.5	3	3.66
Tyrosine	86.1	5	1.34	91.7	1	–	82.9	3	4.80	85.1	3	6.41

TABLE 3-1 Continued

Ingredient	Wheat grain						Wheat middlings					
AAFCO #	none						96.5					
	AID			SID			AID			SID		
Item	$\bar{x}$	n	SD	$\bar{x}$	n	SD	$\bar{x}$	n	SD	$\bar{x}$	n	SD
Crude protein (N × 6.25)	82.0	1	–	–	–	–	73.0	1	–	66.0	1	–
Essential												
Arginine	77.6	5	4.39	83.0	8	4.59	77.0	1	–	75.2	3	9.06
Histidine	77.8	5	5.89	84.3	8	5.60	74.0	1	–	75.3	3	7.81
Isoleucine	81.2	5	3.03	84.6	8	11.12	75.0	1	–	77.0	3	9.33
Leucine	83.8	5	2.59	89.2	8	4.23	77.0	1	–	77.2	3	8.10
Lysine	69.2	5	8.81	81.2	5	3.03	74.0	1	–	75.3	3	8.90
Methionine	81.6	5	6.62	84.4	8	9.86	79.0	1	–	79.1	3	9.19
Phenylalanine	84.4	5	2.70	90.3	8	4.32	77.0	1	–	78.7	3	5.48
Threonine	65.0	5	5.83	79.4	8	10.58	64.0	1	–	68.7	3	11.55
Tryptophan	–	–	–	86.0	1	–	–	–	–	79.0	1	–
Valine	78.0	5	3.81	84.8	8	6.94	73.0	1	–	73.5	3	9.49
Nonessential												
Alanine	76.2	5	4.66	79.8	7	11.12	74.0	1	–	69.0	2	13.28
Aspartic acid	69.8	5	5.97	73.5	7	13.34	72.0	1	–	69.5	2	10.47
Cysteine	87.0	1	–	87.3	4	7.19	71.0	1	–	74.3	3	5.88
Glutamic acid	93.4	5	2.61	94.8	7	3.24	87.0	1	–	85.0	2	6.08
Glycine	74.0	4	5.35	84.7	4	4.93	69.0	1	–	66.4	2	12.01
Proline	–	–	–	–	–	–	–	–	–	77.5	1	–
Serine	76.6	5	5.32	84.3	7	7.18	72.0	1	–	71.6	2	10.66
Tyrosine	68.8	5	9.09	80.8	7	5.46	70.0	1	–	76.6	1	–

present in feedstuffs of microbial origin, and chemically synthesized amino acids have an equal racemic mixture the D- and L-forms. For example, most chemically synthesized commercial sources of supplemental methionine contain approximately 50 percent of the D-enantiomer, and the final product is referred to as D,L-methionine. There are also chemically synthesized sources of lysine that contain approximately 50 percent of the D-enantiomer (i.e., D,L-lysine). In the case of methionine, chickens can convert most of the D-form to the L-form, but in the case of lysine its conversion is negligible, which is in keeping with its relatively poor DAAO substrate specificity (Elkin et al., 1988a). Along with specific amino acid idiosyncrasies, these biological underpinnings have direct implications for determining the bioavailability of supplemental amino acid sources. Details regarding the experimental design and statistical methods required to appropriately interpret bioavailability results quantitatively are explained in Chapter 8 (Digestibility and Bioavailability).

Research on conversion of D-amino acids to bioavailable L-amino acids was mostly done prior to 1995 and was summarized by Lewis and Baker (1995) and more recently by Tomonaga and Furuse (2019). With the exception of methionine, most of these historical studies do not meet the criteria for quantitatively assessing bioavailability of an amino acid as described in Chapter 8. However, the committee provides guidance levels for the bioavailability of D- relative to L-amino acid enantiomers for chickens based on growth rates and efficiency of feed conversion as follows: arginine ≈ 0 percent, cysteine ≈ 0 percent, histidine < 20 percent, isoleucine ≈ 0 percent, leucine > 90 percent, lysine ≈ 0 percent, phenylalanine > 50 percent, threonine ≈ 0 percent, tryptophan < 50 percent, and valine < 50 percent (Lewis and Baker, 1995). These are provisional bioavailability estimates, and more research is needed to provide quantitative recommendations for specific D-amino acids.

The first limiting amino acid in corn–soybean meal-based poultry diets is methionine, and the bioavailability of D- and L-enantiomers of methionine has received considerably more research attention than other amino acids. Several experiments compared D-methionine directly to L-methionine in chickens (Marrett et al., 1964; Tipton et al., 1966; Katz and Baker, 1975; Tipton et al., 1966); however, these studies utilized slow-growing stains of chickens and purified diets and therefore are difficult to interpret and apply to commercial applications. They also had insufficient supplementation levels to analyze by slope ratio or regression analysis as described in Chapter 8 (Digestibility and Bioavailability). In general, these studies indicated that D-methionine was used less efficiently compared with L-methionine in slow-growing chickens. In support of this notion, Elkin et al. (1988a) showed that D-methionine was also used less efficiently in White Pekin ducks, which was recently confirmed by Zhang et al. (2019) and Wu et al. (2021).

Most research on the supplementation of methionine has compared the bioavailability of D,L-methionine to

L-methionine because D,L-methionine is commercially available and L-methionine is the form of methionine supplied by intact protein in most feedstuffs. Comparisons between D,L-methionine and L-methionine have revealed variable results in poultry. Seven studies indicate that D-methionine is less bioavailable (Marrett et al., 1964; Smith, 1966; Noll et al., 1984; Shen et al., 2015; Esteve-Garcia and Khan, 2018; Park et al., 2018; Wang et al., 2019). Five of these studies (Noll et al., 1984; Shen et al., 2015; Esteve-Garcia and Khan, 2018; Park et al., 2018; Wang et al., 2019) used a sufficient number of inclusion levels and appropriate statistical analysis of the data as recommended in Chapter 8 (Digestibility and Bioavailability) for a quantifiable bioavailability study. Additionally, five studies indicate no statistical difference between D,L-methionine and L-methionine (Guttridge and Lewis, 1964; Kuzmicky et al., 1977; Parsons and Potter, 1981; Dilger and Baker, 2007; Jankowski et al., 2017), of which only one (Parsons and Potter, 1981) used sufficient numbers of levels and statistical analysis recommended in Chapter 8. Paradoxically, three studies found that D,L-methionine had a higher bioavailability than L-methionine (Guttridge and Lewis, 1964; Tipton et al., 1966; Elkin and Hester, 1983). In general, the foregoing studies indicate that the bioavailabilities of D-methionine and D,L-methionine are lower than that of L-methionine, but there is strong evidence that the cyst(e)ine content of the diet plays a pivotal role in conversion and use of D-methionine (Dilger and Baker, 2007).

Based on the three studies done in broiler chicks that met our criteria described in Chapter 8, the average bioavailability of D,L-methionine relative to L-methionine was 77.4 percent and 66.0 percent for body weight gain and feed utilization, respectively (Shen et al., 2015; Esteve-Garcia and Khan, 2018; Wang et al., 2019). Based on the three studies done in turkey poults that met our criteria described in Chapter 8, the average bioavailability of D,L-methionine relative to L-methionine was 79.6 percent for body weight gain (Parsons and Potter, 1981; Noll et al., 1984; Park et al., 2018).

In many of the foregoing studies, the difference in bioavailability decreased as the level of methionine in the diet approached the requirement, indicating that the rate of conversion was a limiting factor when large amounts of D-methionine replaced L-methionine. For this reason, in practical diets the bioavailability of D,L-methionine may be higher than the foregoing estimates using diets with high levels of supplemental D-methionine. Similar observations have been reported for the conversion of the hydroxyl analogue of methionine (2-hydroxy-4-methylthiobutyric acid) to L-methionine (Baker and Boebel, 1980).

AMINO ACID ANALYSIS

Studies of amino acid nutrition and metabolism require reliable methods for the analysis of amino acids in feedstuffs, complete mixed feeds, excreta, animal tissues, and blood serum or plasma. Automated amino acid analysis began over 60 years ago with the publication of the companion reports of Moore et al. (1958) and Spackman et al. (1958), which respectively described the chromatography of amino acids on sulfonated polystyrene resins and an automatic recording apparatus for use in the chromatographic analysis of amino acids (Ettre and Gehrke, 2006). Soon thereafter, commercial instrumentation became available, and the analytical technique and ion exchange resins were further improved, permitting the use of a single column instead of one long column for acidic/neutral amino acids and one short column for basic amino acids (Ettre and Gehrke, 2006). With these developments came marked reductions in analytical run times for both complex physiological samples and simpler hydrolysates. Over the years, other methods for the analysis of amino acids emerged (reviewed by Kaspar et al., 2009), including gas chromatography (GC; Kaiser et al. 1974), GC/mass spectrometry (MS) and GC/MS/MS (Kvitvang et al., 2011), high-performance liquid chromatography (HPLC), which employed small particle size resins, small-bore columns, and high-pressure pumps and injection valves (Bidlingmeyer et al., 1984; Qureshi and Qureshi, 1989; Dai et al., 2014), ultra-performance liquid chromatography (UPLC; Boogers et al., 2008; Meussen et al., 2014; Thera et al., 2017), HPLC/MS/MS (Le et al., 2014), and capillary electrophoresis (CE) and CE/MS (Poinsot et al., 2018). Analytical values for amino acid in feed ingredients vary among validated wet chemistry methods. This associated variability among sources of feed ingredients may be due to a myriad of factors such as geographical location, agronomic conditions, sampling, particle size, sample preparation, and laboratory assay variability. Differences in analytical methods can lead to variable results between laboratories, justifying the need for validating strict standards for conducting analytical assays (Llames and Fontaine, 1994). An alternative to wet chemistry methods is near-infrared reflectance spectroscopy (NIRS). When combined with chemometric algorithms for calibration, which are based on wet chemical reference analyses, it enables the accurate and fast prediction of the total contents of many indispensable amino acids in feedstuffs (Fontaine et al., 2001; Fontaine, 2003). In addition, NIRS offers other advantages such as being nondestructive and a more environmentally friendly (no lab chemicals) method than wet chemistry. Because NIRS is based on wet chemistry values, its use does not solve the analytical variability problem.

Feedstuffs

Prior to determining the amino acid content of a feedstuff, amino acids must be released from their protein-bound form by hydrolysis. However, no single hydrolysis method can release amino acids completely and quantitatively from all proteins (Darragh and Moughan, 2005; Otter, 2012). Standard hydrolysis conditions (Fontaine, 2003) involve refluxing the samples in 6 N HCl for 24 h at 110°C, but even that rarely provides maximal amino acid values because of the destruction of some amino acids or the requirement of different hydrolysis times for optimal peptide bond cleavage between certain amino acids (Albin et al., 2000; Darragh and Moughan, 2005). For example, under standard conditions, both asparagine and glutamine are deaminated, tryptophan is destroyed, and methionine and cysteine are converted by partial oxidation into several derivatives (detailed in Fontaine, 2003). Therefore, for a complete and accurate amino acid analysis of a feedstuff, three separate procedures must be employed: (1) a standard hydrolysis (Fontaine, 2003); (2) alkaline hydrolysis for tryptophan (Fontaine et al., 2003); and (3) oxidation of the sulfur-containing amino acids prior to hydrolysis, typically with performic acid, which converts methionine to methionine dioxide and cysteine to cysteic acid but totally or partially destroys phenylalanine, tyrosine, and histidine (Elkin and Griffith, 1985a; Thera et al., 2018). Details for each of the three procedures are also described in AOAC Official Methods 982 30E(a), 988.15, and 994.12, respectively (AOAC International, 2000). Sequential hydrolysis curves (Albin et al., 2000; Darragh and Moughan, 2005) can be used to correct for variability in the ease of peptide bond cleavage and differences in the stability of certain amino acids in acid; however, they are typically not conducted for feedstuff or tissue analyses because they depend heavily on the structure of individual proteins. Therefore, the standard 24-h hydrolysis time at 110°C represents a compromise to optimize hydrolysis and stability of amino acids (Darragh and Moughan, 2005; Peace and Gilani, 2005).

Following hydrolysis, amino acids typically are separated via cation-exchange HPLC or HPLC using post-column derivatization with ninhydrin and absorbance detection (i.e., the reference method), while alternative approaches employ pre-column derivatization of amino acids with o-phthalaldehyde (OPA), 9-fluorenylmethylchloroformate (FMOC), dimethylamino-azo-benzenesulfonyl chloride (DABSYL-Cl), dimethylaminonaphthalensulfonyl chloride (DANSYL-Cl), 7-fluoro-4-nitrobenzo-2-oxa-1,3-diazo le (NBD-F), phenylisothiocyanate (PITC), or 6-aminoquino lyl-N-hydroxysuccinimidyl carbamate (AQC), separated by reversed-phase HPLC or UPLC, and then detected either by fluorescence or ultraviolet (UV) absorbance (Bidlingmeyer et al., 1984; Elkin, 1984a; Elkin and Griffith, 1985b; Elkin and Wasynczuk, 1987; Boogers et al., 2008; Kaspar, 2009; Otter, 2012; Dai et al., 2014). Methods for the determination of free, non-protein-bound amino acids (e.g., feed-grade lysine, methionine, and threonine) that are typically supplemented to poultry diets are reviewed by Fontaine (2003) and described in AOAC Official Method 999.13 (AOAC International, 2000).

Physiological Fluids

As compared with feedstuff hydrolysates, the measurement of free amino acids in physiological fluids provides a more difficult challenge for the analyst because of the greater number of analytes and the longer analytical times required. In addition, analysis of free amino acids in biological samples requires removal of intact proteins prior to analysis, typically by precipitation with acid or organic solvent, followed by ultrafiltration (Qureshi and Qureshi, 1989; Kaspar et al., 2009). This denaturation and removal of intact proteins is critical not only to preserve the analytical equipment (e.g., protect the HPLC column) but also because free amino acids in biological samples are at far lower concentrations than the amino acids present in common body proteins (e.g., albumins and globulins in blood). Although amino acid analyzers employing long columns packed with cation exchange resin, post-column ninhydrin derivatization, and absorbance detection were the original "gold standard" instrumentation for conducting physiological amino acid analyses during the 1960s to 1980s; they were eventually replaced by HPLC-based systems employing pre-column or post-column derivatization techniques and UV or fluorescence detection as described previously. Examples of some of the early reports of avian blood serum or plasma amino acid analyses by HPLC employed pre-column derivatization with OPA to measure free amino acids in chicken, turkey, and duck serum (Elkin, 1984b), while pre-column derivatization with PITC also was used to measure circulating amino acids in two methionine utilization studies with ducks (Elkin et al., 1988a,b). More recently, methods utilizing HPLC/MS and HPLC/MS/MS have been reported for the measurement of amino acids in physiological samples (Le et al., 2014; Yoshida et al., 2015). Chiral separations for the analysis of D-amino acids in biological samples were recently reviewed by Szökő et al. (2016).

EMPIRICAL ESTIMATES OF AMINO ACID REQUIREMENTS

Amino acid requirements are estimated by either factorial modeling or empirical methods. The factorial method estimates requirements by summing the requirements for maintenance along with growth and egg production (Hurwitz and Bornstein, 1973; Hurwitz et al., 1983). Requirements estimated by the factorial method are based on variables such as feed intake, energy intake, body weight, growth curve, sex, and efficiency of utilization for a particular amino acid. Empirical estimates are based on animal experiments in which graded concentrations of a dietary amino acid are provided to poultry based on a given set of response criteria (body weight, feed conversion, egg production, etc.). These experiments determine the optimum response to the amino acid of interest based on linear and nonlinear models.

Amino acid requirements for poultry are typically expressed either as a percentage of the diet or as an absolute quantity consumed per day. Expressing requirements as a percentage of the diet is used when dietary variations in metabolizable energy are minimal. This approach works well for corn–soybean meal-based diets, because metabolizable energy values do not vary to a great extent among those feedstuffs. Even though laying hens are also primarily fed corn–soybean meal diets (in North America), requirements are more commonly expressed on a daily intake basis because hens exhibit more consistent feed intake patterns than broilers or turkeys. In addition, expressing amino acid requirements on a mg per day basis accounts for feed intake differences and ensures that the bird is achieving the daily dietary requirements of nutrients for support of productive purposes (e.g., growth). If feed intake for a given phase of growth or production is known, requirements can be expressed on a daily intake basis, thereby also accounting for differences in environment and management practices to better meet the amino acid need of the bird.

Empirical requirements are specific to the environment, sex, management practices, and diets used during experimentation. Proper experimental design is paramount when conducting amino acid requirement studies to ensure that an accurate requirement is obtained. Careful consideration should be placed on the following criteria to determine validity of empirical requirement estimates: range and number of experimental treatments (i.e., number of graded concentrations of a nutrient), adequacy of other nutrients in the test diet, amino acid analysis of each experimental diet in order to confirm the calculated values, duration of experimental growth phase, and appropriate level of biological replication. Dietary amino acid concentrations should be titrated across a range extending from deficiency to summit (i.e., beyond the expected requirement). It is important to ensure that the basal diet is adequate in amino acids other than the test amino acid so the response of the test amino acid is not influenced by amino acid limitation in the basal diet.

Data can be analyzed by either linear or nonlinear models, and experiments should be designed differently depending on the desired analysis. As a general rule, treatments should range from 70 percent to 130 percent of their anticipated amino acid requirement in order to capture the complete response interval. Within this range, at least seven treatment concentrations should be used, as centered on the purported requirement, to generate an adequate slope and estimate an accurate requirement when using nonlinear models (Dozier et al., 2008, 2010). The first four dietary increments are used to create an adequate slope leading up to the purported requirement, with the requirement ideally being estimated between the fourth and fifth amino acid concentration. Then, the remaining points above the requirement are used to establish a plateau, which is especially critical for accurate requirement estimates generated using a quadratic response. Moreover, it is important to have at least a 0.05 percent difference between consecutive dietary concentrations in order to be able to confirm the concentrations by analysis. Given this guidance, it should be noted that additional considerations may apply for specific amino acids. For example, lysine is known to elicit an acute response with growth and breast meat yield and may warrant additional dietary treatments to ensure generation of an accurate requirement estimate based on a quadratic analysis (Dozier et al., 2010).

Requirement studies have also been designed to estimate an amino acid requirement through linear broken-line analysis. These designs typical utilize only five to six points to estimate a requirement with a 0.10 percent difference between dietary treatments (Baker et al., 2002). Caution should be exercised, however, as linear broken-line models tend to underestimate the requirement when evaluating data that provide a good fit with quadratic broken-line analysis (Dozier et al., 2008). In addition, statistical analysis can directly influence the requirement estimate (Dozier et al., 2008), so it is imperative that an appropriate analysis be used to best describe the pivotal response. Lower and upper 95 percent confidence limits, coefficient of determination, and standard error of the requirement estimate are all important parameters that should be used to determine the appropriate analysis when reporting requirement estimates. For reference, Robbins et al. (2006) reviewed methodological details and considerations necessary for the estimation of nutrient requirements using broken-line analysis. Moreover, adequate replication is needed to conduct a dose-titration experiment in order to accurately estimate requirements. Bird age and response criterion are factors that can influence the number of replicates when conducting amino acid requirement experiments (Baker et al., 2002; Dozier et al., 2008, 2010).

Lean tissue deposition requires specific proportions of essential amino acids (Wang and Fuller, 1989), and these proportions can be used to estimate the requirement for all of the essential amino acids based on the requirement for a single amino acid. This is the foundation of the ideal amino acid ratio concept, which expresses amino acid requirements relative to lysine (Chung and Baker, 1992). Use of the ideal amino acid concept provides essential amino acids in similar proportions as needed for optimum growth rate, thereby reducing amino acid excesses and the risk of nitrogen excretion in the manure. Theoretically, genetic strain, body weight, and sex should not affect the proportions of amino acids used in the ideal amino acid concept (Chung and Baker, 1992). Once the ideal amino acid pattern has been established for a given category of poultry, this formulation strategy can be applied across various ages and weight classes with fast growing birds. This method of formulation has become popular due to the commercial availability of feed-grade amino acids. Lysine is typically considered the reference amino acid and other amino acids can then be expressed proportional to lysine for a given category of poultry (Baker et al., 2002). Lysine is considered the reference amino acid due to several reasons, including its primary function being for lean tissue accretion, its requirement is known for various weight and age classes of poultry, and extensive knowledge of lysine composition in feed ingredients. When formulating using the ideal amino acid concept, the goal is for the dietary amino acid supply to match the exact proportions required by the bird to the extent economically possible.

Amino acid intake is central in meeting the physiological requirements of poultry. It is time-consuming and expensive to evaluate amino acid requirements under a variety of factors that influence amino acid requirements such as strain of the bird, sex, rate of growth, egg production, temperature/humidity, response criteria, and dietary amino acid profile. The application of digestible amino acid ratios provides an ease in formulation to enable nutritionists to meet amino acid requirements as percentage of the diet. It is important to ensure that the appropriate digestible lysine value is used to ensure that amino acids are provided in adequate amounts since other amino acids are formulated as ratios to lysine. In addition, understanding feed intake for a given weight category of bird for a given location that accounts for factors that impact physiological requirements of the bird is central for amino acid requirements expressed as intake (mg/d). In addition to feed intake, applying accurate digestibility coefficients for ingredients used in diet formulation is key to meet amino acid needs of the bird.

VARIATIONS IN REQUIREMENTS

Amino acid requirements vary considerably according to the productive state of the bird (i.e., the rate of growth or egg production). For example, turkey poults and broiler chicks have high amino acid requirements to meet the needs for rapid lean tissue accretion. The mature rooster has lower daily amino acid requirements than does the laying hen, even though its body size is greater and its feed consumption is similar. Body size, growth rate, and egg production of poultry are determined by their genetics. Amino acid requirements, therefore, also differ among types, breeds, and strains of poultry, as can be seen by comparing the values shown in the requirement tables provided in this report for the different types of poultry. Genetic differences in amino acid requirements may reflect differences in efficiency of nutrient digestion, absorption, metabolism, lean tissue deposition, and egg production rates (NRC, 1975). Although dietary requirements for amino acids are typically expressed as percentages of the diet, the quantitative needs of poultry must be provided by a balanced source to obtain maximum productivity. Thus, factors that affect feed consumption also will affect quantitative intakes of amino acids and, consequently, will influence the dietary concentration of these nutrients needed to provide adequate nutrition. Factors affecting feed consumption are discussed in the section on "Setting Dietary Levels" and have been reviewed in the NRC (1987) publication *Predicting Feed Intake of Food-Producing Animals*. As related to this point, adjustments in dietary amino acid concentrations may also be necessary to compensate for differences in energy concentrations between diets. This is especially true for White Leghorn chickens (Morris, 1968; Byerly et al., 1980) and turkey hens (Kratzer et al., 1976).

Ambient temperature also affects feed intake of poultry (Hurwitz et al., 1980). Amino acid recommendations listed herein generally pertain to poultry maintained at moderate temperatures as birds advance in age (i.e., decreasing from 33°C to 19°C). Ambient temperatures outside of this range cause an inverse response in feed consumption; that is, the lower the temperature, the greater the feed intake and vice versa (NRC, 1981). Consequently, percentage requirements of amino acids may be increased in warmer environments and decreased in cooler environments in accordance with expected differences in feed intake. These adjustments may aid in ensuring required daily intakes of amino acids. Some precautions, however, should be used in increasing the dietary amino acid concentrations for poultry subjected to high ambient temperature. Heat stress conditions may decrease plasma thyroxine and triiodothyroxine concentrations, which could decrease protein synthesis and negate the need to increase the dietary amino acid percentage requirements (Geraert et al., 1996; Tao et al., 2006). Waldroup et al. (1976) reported that performance of broiler chicks was improved by minimizing excess dietary amino acids. Readers should refer to Chapter 10 for more details on the relationships of nutrition and environmental stressors in poultry production.

Information available from research documenting the influence of dietary energy concentration and ambient temperature on feed intake has been integrated with data describing amino acid needs for maintenance, body growth (such as for muscle and feathers), or egg production to derive mathematical models to predict the dietary amino acid requirements of poultry (Fisher et al., 1973; Hurwitz and Bornstein, 1973; Hurwitz et al., 1978; Emmans, 1981; Slagter and Waldroup, 1984). Prediction models may be useful in feed formulation, and they also provide valuable insight into areas of amino acid nutrition where more definitive information is needed on requirements. Generally, as the dietary protein level increases, essential amino acid requirements (expressed as a percentage of the diet) also increase, although when expressed as a percentage of dietary protein, essential amino acid requirements are little affected (Almquist, 1952; Boomgaardt and Baker, 1971, 1973; Morris et al., 1987; Robbins, 1987; Mendonca and Jensen, 1989). This phenomenon demonstrates the importance of maintaining a balance among the proportions of essential and nonessential amino acids in poultry diets. Optimal balance is important for efficient utilization of dietary amino acids. Achieving maximum growth and production, however, may not always ensure maximum economic returns, particularly when prices of protein or amino acid sources are high (Roland et al., 1998). If decreased performance can be tolerated, dietary concentrations of amino acids may, accordingly, be reduced to maximize economic returns.

AMINO ACID DEFICIENCIES

The need for the indispensable (essential) amino acids, and deficiencies of several indispensable amino acids, creates general symptoms, with productive activities being most profoundly impacted. For example, the energy used by growing birds is heavily committed to assembling the contractile elements in muscle cells (hypertrophy) but not to increasing cell number (hyperplasia); thus, protein inadequacies readily affect muscle (anterior latissimus dorsi) size but not fiber number (Timson et al., 1983). Similarly, the effect of protein inadequacies on protein synthesis in the liver and oviduct is greatest with the laying hen (Muramatsu et al., 1987).

In contrast to the specific signs that may occur as a result of vitamin or mineral deficiencies, the effects of indispensable amino acid deficiencies are generally nonspecific and include reduced growth rate or loss of body mass, decreased feed consumption, decreased meat yield, increased body fat deposition, decreased circulating thyroid hormone levels, altered thyroid morphology, and decreased egg mass and production rate (Elkin et al., 1980, 1981; Scott et al., 1983; Corzo et al., 2007; Klasing and Korver, 2020). In addition, deficiencies or excesses of individual indispensable amino acids distort plasma amino acid concentrations and are rapidly sensed in the brain, resulting in altered performance objectives (Tobin and Boorman, 1979; Alam et al., 2012; Kurpad, 2018).

Additional symptoms in chicks may appear that characterize deficiency of specific amino acids. For example, attenuated (Konashi et al., 2000) or enhanced (Humphrey et al., 2006) immune responses were observed with multiple amino acid deficiencies or lysine deficiency, respectively. Lysine deficiency also resulted in increased muscle (pectoralis major) protein turnover (Tesseraud et al., 1996), reduced circulating hemoglobin and hematocrit values, and changes in femur chemical and histological composition in chickens (Braham et al., 1961). Inadequate lysine is also known to cause depigmentation of the wing feathers in Bronze turkey poults (Fritz et al., 1946; Vohra and Kratzer, 1959; Parker et al., 2011) and certain colored chicks (Klain et al., 1957). A variety of abnormalities in feather development of growing chicks occur with deficiencies of other amino acids, including arginine, valine, leucine, isoleucine, tryptophan, phenylalanine, and tyrosine (Newberne et al., 1960; Robel, 1977; Penz and Kratzer, 1984; Farran and Thomas, 1992). Peng et al. (2018) reported that methionine deficiency decreased hepatic lipid exportation and induced liver lipid accumulation in chicks, while Chavez and Kratzer (1974) observed a foot pad dermatitis in poults when methionine was deficient, but cysteine had to be adequate for the dermatitis to occur. Because threonine plays an important role in the maintenance of intestinal barrier defense and in mucin synthesis, deficient dietary levels of this amino acid can alter small intestinal morphology and intestinal mucin dynamics in chicks and ducklings (Horn et al., 2009; Wils-Plotz et al., 2013; Wils-Plotz and Dilger, 2013; Chen et al., 2017). Grau (1945) reported a tongue deformity in White Leghorn chicks fed a purified diet (crystalline amino acids, cellulose, glucose, soybean oil, vitamins, and minerals) deficient in leucine, isoleucine, or phenylalanine, but these observations were not confirmed by Bragg (1953) in Beltsville White turkey poults fed a corn–wheat–oats–soybean meal-based diet.

SPECIFIC AMINO ACID RELATIONSHIPS

Although each amino acid can be metabolized independently of others, there are relationships between certain amino acids (Harper et al., 1970). In some instances, the relationship may be beneficial. For example, one amino acid may be converted to another to fulfill a metabolic need. In other instances, a metabolic antagonism may exist with undesirable consequences. A brief description of amino acid relationships that may be of importance in poultry nutrition is given in the following section.

Methionine plus Cyst(e)ine

Methionine can donate its labile methyl group as part of many biological processes, and the resulting sulfur-containing compound, homocysteine, together with serine, can be used to synthesize cysteine with cystathionine as the intermediate. It should be noted that the molecular weight difference between methionine (149.2 mg/mmol) and cysteine (121.2 mg/mmol) must be taken into consideration when discussing this metabolic interrelationship. Though transsulfuration conversion of methionine sulfur to cysteine sulfur is close to 100 percent efficient, 149.2 mg of methionine (1 mmol) theoretically yields only 121.2 mg of cysteine (1 mmol). Methionine exists in a single chemical form, but the same is not true for cysteine. Whereas only the reduced form, cysteine, can be incorporated into polypeptide chains, the high chemical reactivity of the free sulfhydryl groups allows two moles of cysteine to be condensed to form the oxidized disulfide compound cystine. Formation of such a disulfide bridge within an intact protein produces a stable bond that is integral to protein structure and function. Technically, every mole of cystine (oxidized) provides two moles of cysteine (reduced) upon ingestion by the bird, but it has been clearly established that cysteine and cystine are equally efficacious on a weight or concentration basis when provided as low-level supplements to support physiological outcomes (e.g., growth) (Graber and Baker, 1971; Baker, 2006).

The matter is further complicated when considering analytical specificity because acid hydrolysis partially destroys methionine, cysteine, and cystine (Elkin and Griffith, 1985a). Thus, the sulfur amino acids can only be reliably quantified using a modification of the standard methodology, namely AOAC (2006) International Official Method 982.30 E(b), chapter 45.3.05. The modified procedure includes a performic acid oxidation step prior to the standard acid hydrolysis to convert methionine to methionine dioxide and cysteine and cystine to cysteic acid, thereby preventing the chemical destruction of the sulfur amino acids. In addition, all of the reduced and oxidized forms of cysteine are converted to the reduced form, which makes cysteine and cystine in the originating sample indistinguishable. For this reason, analytical laboratory reports make it clear that the reported values are total cysteine plus cystine, and to keep the terminology simple, this value can be denoted as cyst(e)ine.

Two methionine molecules are needed to ultimately supply the two sulfur atoms incorporated into cystine (du Vigneaud, 1952; Creek, 1968; Baker, 1976). The dietary requirement for methionine can be satisfied only by methionine or a methionine analog/precursor [e.g., 2-hydroxy-(4-methylthio) butanoic acid, 2-keto-4-(methylthio) butyric acid], whereas the dietary requirement for cysteine can be met with a combination of methionine and cyst(e)ine. Thus, it is imperative that dietary methionine and cysteine requirements be considered simultaneously when formulating diets for avian species.

The catabolism of methionine and cyst(e)ine largely leads to conversion of the associated sulfur into sulfate, which is used in multiple metabolic pathways, particularly in connective tissues. Similarly, methyl groups of methionine may be used in transmethylation reactions and the de novo synthesis of sarcosine, betaine, and choline. While not a sulfur-containing amino acid, choline is a major constituent of phospholipids, and its incorporation into membranes is extensive. During rapid growth, when accrual of connective tissue and expansion of membrane surfaces are great, there may be increased sensitivity to methionine at concentrations marginal to the requirement if dietary betaine, choline, and sulfate are not sufficient (Baker et al., 1983; Miles et al., 1983; Blair et al., 1986).

Phenylalanine plus Tyrosine

Tyrosine is the initial product formed during the biological catabolism of phenylalanine. In turn, phenylalanine can be used to meet the bird's need for tyrosine on a mole-for-mole basis (Creek, 1968; Sasse and Baker, 1972). Although this conversion may be reversed to a small extent and tyrosine used to form phenylalanine, its contribution is too small to be of practical significance (Ishibashi, 1972).

Glycine plus Serine

Glycine is a conditionally essential amino acid for poultry (Dean et al., 2006). Although glycine can be synthesized by avian species, the rate is not adequate to support maximal growth (Featherston, 1976). In addition to the role of glycine used in protein synthesis, including collagen and feathers (Robel, 1977; Stilborn et al., 1997), which are highly enriched in glycine, it is also used in the synthesis of purines (Bender, 2012), creatine (Ospina-Rojas et al., 2013), uric acid (Leeson and Summers, 2001), and glutathione (Bender, 2012). Glycine can be produced from catabolizing choline (Soloway and Stetten, 1953), threonine (Malkin and Greenberg, 1964; Davis and Austic, 1994), glyoxylate (Salido et al., 2006), alanine (Salido et al., 2006), and trimethyllysine (Meléndez-Hevia et al., 2009). Serine can be converted to glycine and is effective in meeting the glycine needs on an equimolar basis. This reaction is reversible, and glycine can be used to form serine; 0.71 is the ratio of the molar weight between glycine and serine (Baker et al., 1968; Sugahara and Kandatsu, 1976; Siegert et al., 2015). Because glycine and serine are interconvertible, the requirement is typically expressed as the sum of glycine plus serine.

Imbalance, Antagonism, and Toxicity

The essential amino acids are related to one another by virtue of their need to fulfill production plus maintenance functions. Poultry cannot metabolically adapt to some ratios of dietary amino acids that differ greatly from the ideal amino acid balance, resulting in decreased appetite and changes in amino acid catabolism and transport that increase maintenance losses and decrease rates of growth or egg production. An amino acid imbalance is defined as ratios of dietary amino acids that result in a skewed supply of amino acids in the brain that mediates a decrease in appetite (D'Mello and Lewis, 1970). Amino acid imbalances occur when the first limiting amino acid is near or below the requirement and many other amino acids are in excess of the requirement. The bird's response will vary based on the essential amino acid, the extent of its inadequacy, and existing relationships among the remaining amino acids. As an example, Sugahara et al. (1969) fed chicks a purified amino acid diet corresponding to 100 percent of the requirement for all essential amino acids as the positive control and compared the performance response to when all amino acids were reduced to 60 percent of the requirement as opposed to 60 percent reduction with each one alone. Weight gain was greater with individual decreases of methionine–cysteine, leucine, lysine, and arginine than when a total reduction was imposed, whereas additional weight loss occurred with individual decreases of phenylalanine, tyrosine, tryptophan, isoleucine, valine and threonine. A reduction in dietary histidine gave a similar response to that observed when all amino acids were decreased. Deficiencies of any one of the essential amino acids can be exaggerated by adding purified amino acids and/or combining complete proteins such that the extent of difference between the first and second limiting amino acid increases. The response is generally an additional impairment of body weight gain. Accentuation of the deficiency is greatest with low protein diets, and a decrease in feed intake is the fundamental reason for poor weight gain rather than alteration in efficiency of using the first limiting amino acid (Fisher et al., 1960; Fisher and Shapiro, 1961; Netke et al., 1969).

Amino acid antagonisms may also accentuate a deficiency of the first limiting amino acid, but these differ from imbalances because utilization of the limiting amino acid is reduced due to shared metabolic pathways with other essential amino acids. This can occur when two or more amino acids have side chains that exhibit similar structural and/or chemical characteristics resulting in competition for the same catabolic enzymes or the same membrane transporters. Thus, increasing the dietary concentration of one that is in excess of productive use adversely affects metabolism of the other. In a situation in which one essential amino acid is first limiting, increasing the other's concentration to enlarge the difference antagonizes the use of the first limiting amino acid and induces or exacerbates a deficiency. Antagonisms have been shown to exist for leucine–isoleucine–valine, arginine–lysine, and threonine–tryptophan (D'Mello and Lewis, 1970). Antagonism with leucine and isoleucine can impact growth efficiencies. Certain feedstuff combinations (e.g., corn plus corn gluten meal) can lead to practical diets in which leucine is at particularly high concentrations if isoleucine is marginal in adequacy (Waldroup et al., 2002). Waldroup et al. (2002) reported that a branch chain antagonism with adverse performance responses may not occur with practical diets if leucine, isoleucine, and valine are above minimum requirements. An excess in lysine has been shown to increase the arginine requirement in chicks (Jones, 1964; Chamruspollert et al., 2002). The use of alternative protein sources (e.g., peanut meal) may result in an arginine inadequacy with the use of feed-grade lysine, resulting in a potential arginine–lysine antagonism. Hence, dietary arginine concentration or ratio should be considered when using peanut meal in practical diets. Amino acid concentrations that would be likely to provoke the other antagonisms probably would not occur in practice unless high concentrations of supplemental amino acids were used in low-protein diets.

When poultry are fed high levels of well-balanced amino acids in excess of their requirements, the surplus amino acids are catabolized, and the nitrogen that is released is incorporated primarily into uric acid (Klasing and Korver, 2020). Although generally well tolerated, excess dietary protein may cause hyperuricemia or metabolic acidosis and exacerbate articular gout or visceral gout in growing chickens (Klasing and Korver, 2020).

An amino acid toxicity requires a particularly high concentration of one amino acid relative to all others. Such an occurrence is unlikely under practical circumstances because differences of sufficient magnitude do not exist in most protein feedstuffs. Excesses of individual amino acids, which sometimes result from feed mixing errors or when experimentally formulated, can be especially toxic and severely attenuate food intake and growth rate and increase mortality (Snetsinger and Scott, 1961; Dean and Scott, 1968; Edmonds and Baker, 1987; Han and Baker, 1993; Carew et al., 1998). Methionine, the first limiting amino acid for poultry fed corn–soybean meal-based diets, is the one indispensable amino acid most likely to be supplemented in chicken, turkey, and duck diets. Methionine, cysteine, and cystine are among the most toxic amino acids studied (Edmonds and Baker, 1987; Baker, 2006; Dilger et al., 2007), and methionine has a low safety margin, with toxicity occurring at approximately two to three times the requirement in growing chicks (Klasing and Korver, 2020). Ueda et al. (1981) observed severe depression in feed consumption and growth of chicks given ad libitum access to a diet containing 10 percent protein and 1.5 percent L-methionine. Force-feeding this high-methionine, low-protein diet in amounts equal to the feed intake of controls resulted in death of the chicks. Edmonds and Baker (1987) added excesses of several amino acids to a 23 percent protein corn–soybean meal diet for chicks. Methionine at 4 percent of the diet led to a 92 percent reduction in weight gain, whereas similar excesses of tryptophan, lysine, and threonine were far less toxic. Acute mortality has also been reported in chick assays due to cyst(e)ine toxicity, where it was noted that L-cysteine was more acutely toxic than either L-cystine or N-acetyl-L-cysteine (Dilger et al., 2007). Feeding chicks diets containing supplemental cystine of higher than 2.5 percent of the diet for five days resulted in mortality greater than 50 percent. The basis for the lethality may be related to the free sulfhydryl group of cysteine or catabolic products thereof, but this sulfur amino acid was not as acutely toxic in rats or swine. Of the various commercially available forms of methionine, DL-2-hydroxy-4-methylthiobutanoic acid (HMB) is less toxic than either the L-methionine or DL-methionine (Baker and Boebel, 1980; Xie et al., 2007).

The toxicity of phenylalanine received much early attention because of the interest in developing animal models of inborn errors of aromatic amino acid metabolism, such as phenylketonuria, alkaptonuria, and tyrosyluria in humans (Harper et al., 1970). Numerous studies in chicks have shown that, besides growth depression, impaired feed conversion efficiency, and elevated circulating levels of phenylalanine and tyrosine, phenylalanine toxicity can lead to lower brain serotonin levels, reduced cerebral energy utilization, and several external defects such as poor feather development, swollen hocks, and discolored shanks, face, and eyelids (Grau, 1947; Grau and Kamei, 1950; Snetsinger and Scott, 1961; Tamimie, 1966, 1967a,b, 1968; Tamimie and Pscheidt, 1966a,b,c; Elkin and Rogler, 1983; Lartey and Austic, 2008). Feeding diets containing 5 percent L-phenylalanine to 20-week-old pullets housed with cockerels (3:1 ratio) for 24 weeks resulted in decreased egg production, egg weight, fertility, hatchability, and day-old chick weight (Tamimie and Pscheidt, 1966c). The latter three variables were determined when the pullets were 27 and 29 weeks of age. Results of radiotracer studies in chicks orally administered L-[side chain 3–14C] tryptophan and fed a diet containing 2 percent phenylalanine suggested that brain uptake of tryptophan was inhibited by hyperphenylalaninemia, while dietary supplemental tryptophan partially alleviated the growth-depressive effects of phenylalanine toxicity (Elkin and Rogler, 1983). Snetsinger and Scott (1961) reported that 4 percent supplemental glycine overcame the growth depression caused by feeding chicks 3 percent DL-phenylalanine from 7–21 d of age; however, Elkin and Rogler (1983) failed to confirm their finding when chicks were fed a diet containing 2 percent L-phenylalanine plus 3 percent glycine.

When Anderson and Combs (1952) fed DL-lysine-HCl at 5 percent of the diet to day-old chicks for 6 d, or at 4 percent to chicks for 4 weeks, a growth depression was observed that was more severe in the younger birds. Since that initial report, other researchers reported that high levels of dietary lysine depressed growth (Jones, 1961; Han and Baker, 1993) and that a lysine–arginine antagonism existed (Jones, 1964). Ultimately, it was shown that excess dietary lysine inhibits renal reabsorption of arginine, induces kidney arginase activity, and decreases hepatic glycine transamidinase activity, and that electrolyte balance influenced the dietary lysine–arginine antagonism (Jones et al., 1967; Boorman et al., 1968; Calvert and Austic, 1981; Balnave and Brake, 2002). Supplemental dietary arginine and/or glycine have been shown to overcome the growth depression associated with feeding chickens excess lysine (Snetsinger and Scott, 1961; Dean and Scott, 1968).

The relative order of toxicity of amino acids for mixed-sex growing chickens fed a 23 percent crude protein, corn–soybean meal-based diet separately supplemented with 4 percent of each amino acid and fed from 8 to 16 d of age was methionine > phenylalanine > tryptophan > histidine > lysine > tyrosine > threonine > isoleucine > arginine > valine > leucine (Edmonds and Baker, 1987). In contrast, Carew et al. (1998) investigated the effects of feeding male broiler chicks a corn–soybean meal-based diet with individual excesses [2.84 times that of the NRC (1994) recommended levels] of 10 indispensable amino acids from 10 to 24 d of age and reported that lysine was most toxic, followed by, in order, methionine, threonine, phenylalanine, isoleucine, tryptophan, and valine. Excesses of histidine and leucine had no effect on chick growth. With regard to feed intake, an excess of methionine was most suppressive, followed by (in order) lysine, threonine, arginine, isoleucine, and phenylalanine. Excesses of histidine, leucine, tryptophan, and valine had no such effect. The lack of agreement between these two well-designed studies could have been due to a number of variables, including sex of the chicks, dietary levels of amino acids (constant 4% vs. 2.84 times the NRC requirement), duration of the study, the diet fed prior to the initiation of the studies, and so on. Moreover, in the ordering of amino acid toxicity, neither Edmonds and Baker (1987) nor Carew et al. (1998) considered the possibility that a dispensable amino acid, like cysteine (Dilger et al., 2007; Dilger and Baker, 2008), could be toxic when fed in excess.

AMINO ACID CONVERSION TO VITAMINS

Niacin is the only vitamin that can be synthesized from an amino acid. Tryptophan can be used to alleviate a dietary niacin deficiency, but the rate of conversion is poor (Baker et al., 1973). When methionine is provided at levels exceeding use for protein synthesis, the additional methyl groups may decrease the dietary choline requirement (Pesti et al., 1980).

REFERENCES

Alam, M.R., F. Yoshizawa, and K. Sugahara. 2012. Oral administration of lysine restores food intake and ventromedial hypothalamic dopamine in chicken on a lysine-free diet. *British Poultry Science*. 53:651–657.

Albin, D.M., J.E. Wubben, and V.M. Gabert. 2000. Effect of hydrolysis time on the determination of amino acids in samples of soybean products with ion-exchange chromatography or precolumn derivatization with phenyl isothiocyanate. *Journal of Agricultural and Food Chemistry*. 48:1684–1691.

Almquist, H.J. 1952. Amino acid requirements of chickens and turkeys—A review. *Poultry Science*. 31:966–981.

Anderson, J.O., and G.F. Combs. 1952. Effect of single amino acid excesses on glucose metabolism and chick growth, as influenced by the dietary amino acid balance. *Journal of Nutrition*. 46:161–170.

AOAC. 2006. *Official Methods of Analysis of AOAC International in Association of Official Analytical Chemists*, 18th ed. Gaithersburg, MD: AOAC International.

AOAC International. 2000. *Official Methods of Analysis of AOAC International*, 17th ed. Gaithersburg, MD: AOAC International.

Baker, D.H. 1976. Nutritional and metabolic interrelationships among sulfur compounds in avian nutrition. *Federation of American Societies for Experimental Biology*. 35:1917–1922.

Baker, D.H. 1994. Utilization of precursors for L-amino acids. In *Amino Acids in Farm Animal Nutrition*. J.P.F. D'Mello, editor. Wallingford, Oxfordshire, United Kingdom: International. 37–61.

Baker, D.H. 2006. Comparative species utilization and toxicity of sulfur amino acids. *Journal of Nutrition*. 136:1670S–1675S.

Baker, D.H., N.K. Allen, and A.J. Kleise. 1973. Efficiency of tryptophan as a niacin precursor in the young chick. *Journal of Animal Science*. 36:299–302.

Baker, D.H., A.B. Batal, T.M. Parr, N.R. Augspurger, and C.M. Parsons. 2002. Ideal ratio (relative to lysine) of tryptophan, threonine, isoleucine and valine for chicks during the second and third week of life. *Poultry Science*. 81:485–494.

Baker, D.H., and K.P. Boebel. 1980. Utilization of the D- and L-isomers of methionine and methionine hydroxy analogue as determined by chick bioassay. *Journal of Nutrition*. 110:959–964.

Baker, D.H., K.M. Halpin, G.L. Czarnecki, and C.M. Parsons. 1983. The choline–methionine interrelationship for growth of the chick. *Poultry Science*. 62:133–137.

Baker, D.H., M. Sugahara, and H.M. Scott. 1968. The glycine–serine interrelationship in chick nutrition. *Poultry Science*. 47:1376–1377.

Balnave, D., and J. Brake. 2002. Re-evaluation of the classical dietary arginine:lysine interaction for modern poultry diets: A review. *World's Poultry Science*. 58:275–289.

Bender, D.A. 2012. Glycine, serine and the one-carbon pool. *Amino Acid Metabolism*. Oxford, UK: Wiley-Blackwell. 127–155.

Bidlingmeyer, B.A., S.A. Cohen, and T.L. Tarvin. 1984. Rapid analysis of amino acids using pre-column derivatization. *Journal of Chromatography A*. 336:93–104.

Bird, F.H. 1952. A comparison of methionine and two of its analogues in the nutrition of the chick. *Poultry Science*. 31:1095–1096.

Blair, M.E., L.M. Potter, B.A. Bliss, and J.R. Shelton. 1986. Methionine, choline and sulfate supplementation of practical-type diets for young turkeys. *Poultry Science*. 65:130–137.

Boogers, I., W. Plugge, Y.Q. Stokkermans, and A.L.L. Duchateau. 2008. Ultra-performance liquid chromatographic analysis of amino acids in protein hydrolysates using an automated pre-column derivatisation method. *Journal of Chromatography A*. 1189:406–409.

Boomgaardt, J., and D.H. Baker. 1971. Tryptophan requirement of growing chicks as affected by dietary protein level. *Journal of Animal Science*. 33:595–599.

Boomgaardt, J., and D.H. Baker. 1973. The lysine requirement of growing chicks fed sesame meal–gelatin diets at three protein levels. *Poultry Science*. 52:586–591.

Boorman, K.N., I.R. Falconer, and D. Lewis. 1968. The effect of lysine infusion on the renal reabsorption of arginine in the cockerel. *Proceedings of the Nutrition Society*. 27:61A–62A.

Bragg, D.D. 1953. An attempt to determine the cause of curled or deformed tongues in young Beltsville White turkeys. *Poultry Science*. 31:294–303.

Braham, J.E., C. Teiada, M.A. Guzman, and R. Bressani. 1961. Chemical and histological changes in the femurs of chicks fed lysine-deficient diets. *Journal of Nutrition*. 74:363–375.

Burley, H.K., K.E. Anderson, P.H. Patterson, and P.B. Tillman. 2016. Formulation challenges of organic poultry diets with readily available ingredients and limited synthetic methionine. *Journal of Applied Poultry Research*. 25:443–454.

Burley, H.K., and P.H. Patterson. 2017. Brazil nut and spray-dried egg powders as alternative to synthetic methionine in organic laying hen diets. *Poultry Science*. 96:3994–4005.

Byerly, J.C., J.W. Kessler, R.M. Gous, and O.P. Thomas. 1980. Feed requirements for egg production. *Poultry Science*. 59:2500–2507.

Calvert, C.C., and R.E. Austic. 1981. Lysine–chloride interactions in the growing chick. *Poultry Science*. 60:1468–1472.

Carew, L.B., K.G. Evarts, and F.A. Alster. 1998. Growth, feed intake, and plasma thyroid hormone levels in chicks fed dietary excesses of essential amino acids. *Poultry Science*. 77:295–298.

Chamruspollert, M., G.M. Pesti, and R.I. Bakalli. 2002. Dietary interrelationships among arginine, methionine, and lysine in young broiler chicks. *British Journal of Nutrition*. 88:655–660.

Chavez, E., and F.H. Kratzer. 1974. Effect of diet on foot pad dermatitis in poults. *Poultry Science*. 53:755–760.

Chen, Y.P., Y.F. Cheng, X.H. Li, W.L. Yang, C. Wen, S. Zhuang, and Y.M. Zhou. 2017. Effects of threonine supplementation on the growth performance, immunity, oxidative status, intestinal integrity, and barrier function of broilers at the early age. *Poultry Science*. 96:405–413.

Chung, T.K., and D.H. Baker. 1992. Ideal amino acid pattern for 10-kilogram pigs. *Journal of Animal Science*. 70:3102–3111.

Corzo, A., M.T. Kidd, W.A. Dozier III, and S.L. Vieira. 2007. Marginality and needs of dietary valine for broilers fed certain all-vegetable diets. *Journal of Applied Poultry Research*. 16:546–554.

Creek, R.D. 1968. Nonequivalence in mass in the conversion of phenyl alanine to tyrosine and methionine to cystine. *Poultry Science*. 47:1385–1386.

Dai, Z., Z. Wu, S. Jia, and G. Wu. 2014. Analysis of amino acid composition in proteins of animal tissues and foods as pre-column *o*-phthaldialdehyde derivatives by HPLC with fluorescence detection. *Journal of Chromatography B*. 964:116–127.

Darragh, A.J., and P.J. Moughan. 2005. The effect of hydrolysis time on amino acid analysis. *AOAC International*. 88:888–893.

Davis, A.J., and R.E. Austic. 1994. Dietary threonine imbalance alters threonine dehydrogenase activity in isolated hepatic mitochondria of chicks and rats. *Journal of Nutrition*. 124:1667–1677.

Dean, D.W., T.D. Bidner, and L.L. Southern. 2006. Glycine supplementation to low protein, amino acid-supplemented diets supports optimal performance of broiler chicks. *Poultry Science*. 85:288–296.

Dean, W.F., and H.M. Scott. 1968. Ability of arginine to reverse the growth depression induced by supplementing a crystalline amino acid diet with excess lysine. *Poultry Science*. 47:341–342.

Dilger, R.N., S. Toue, T. Kimura, R. Sakai, and D.H. Baker. 2007. Excess dietary L-cysteine, but not L-cystine, is lethal for chicks but not rats or pigs. *Journal of Nutrition*. 137:331–338.

Dilger, R., and D. Baker. 2007. DL-Methionine is as efficacious as L-methionine, but modest L-cystine excesses are anorexigenic in sulfur amino acid-deficient purified and practical-type diets fed to chicks. *Poultry Science*. 86:2367–2374.

Dilger, R.N., and D.H. Baker. 2008. Excess dietary L-cysteine causes lethal metabolic acidosis in chicks. *Journal of Nutrition*. 138:1628–1633.

D'Mello, J.P.F., and D. Lewis. 1970. Amino acid interactions in chick nutrition. 3. Interdependence in amino acid requirements. *British of Poultry Science*. 11:367–385.

Dozier, W.A. III, A. Corzo, M.T. Kidd, and M. Schilling. 2008. Dietary digestible lysine requirements of male and female broilers from forty-nine to sixty-three days of age. *Poultry Science*. 87:1385–1391.

Dozier, W.A. III, A. Corzo, M.T. Kidd, P.B. Tillman, and S.L. Branton. 2010. Digestible lysine requirements of male broilers from 28 to 42 days of age. *Poultry Science*. 89:2173–2183.

du Vigneaud, V. 1952. *A Trail of Research: Sulfur Chemistry and Metabolism*. Ithaca, NY: Cornell University Press.

Edmonds, M.S., and D.H. Baker. 1987. Comparative effects of individual amino acid excesses when added to a corn-soybean meal diet: Effects on growth and dietary choice in the chick. *Journal of Animal Science*. 65:699–705.

Elkin, R.G. 1984a. Quantitative amino acid analysis of feedstuff hydrolysates by reverse phase liquid chromatography and conventional ion-exchange chromatography. *Association of Official Analytical Chemists*. 67:1024–1026.

Elkin, R.G. 1984b. Measurement of free amino acids in avian blood serum by reverse-phase high-performance liquid chromatography as compared to ion-exchange chromatography. *Journal of Agricultural and Food Chemistry*. 32:53–57.

Elkin, R.G., C.L. Chrisman, J.C. Rogler, and W.R. Featherston. 1981. Influence of phenylalanine and lysine deficiencies on thyroid morphology in the chick. *Poultry Science*. 60:878–881.

Elkin, R.G., W.R. Featherston, and J.C. Rogler. 1980. Effects of dietary phenylalanine and tyrosine on circulating thyroid hormone levels and growth in the chick. *Journal of Nutrition*. 110:130–138.

Elkin, R.G., and J.E. Griffith. 1985a. Amino acid analysis of feedstuff hydrolysates by cation exchange high performance liquid chromatography. *Association of Official Analytical Chemists*. 68:1028–1032.

Elkin, R.G., and J.E. Griffith. 1985b. Hydrolysate preparation for analysis of amino acids in sorghum grains: Effect of oxidative pretreatment. *Association of Official Analytical Chemists*. 68:1117–1121.

Elkin, R.G., and P.Y. Hester. 1983. A comparison of methionine sources for broiler chickens fed corn-soybean meal diets under simulated commercial grow-out conditions. *Poultry Science*. 62:2030–2043.

Elkin, R.G., M.L. Lyons, and J.C. Rogler. 1988a. Comparative utilization of D- and L-methionine by the White Pekin duckling (*Anas platyrhynchos*). *Comparative Biochemistry and Physiology*. 91B:325–329.

Elkin, R.G., M.L. Lyons, and J.C. Rogler. 1988b. Utilization of DL-2-hydroxy-4-(methylthio) butanoic acid as a dietary source of methionine for White Pekin ducks. *Nutrition Reports International*. 37:1189–1198.

Elkin, R.G., and J.C. Rogler. 1983. Partial alleviation of phenylalanine toxicity in the chick by supplemental dietary tryptophan. *Poultry Science*. 62:647–658.

Elkin, R.G., and A.M. Wasynczuk. 1987. Amino acid analysis of feedstuff hydrolysates by precolumn derivatization with phenylisothiocyanate and reversed-phase high-performance liquid chromatography. *Cereal Chemistry*. 64:226–229.

Emmans, G.C. 1981. Alternative feeding systems. In *Proceedings of the Scottish Poultry Conference*. Edinburgh, UK: East of Scotland Agricultural College. 29.

Esteve-Garcia, E., and D.R. Khan. 2018. Relative bioavailability of DL and L-methionine in broilers. *Open Journal of Animal Sciences*. 8:151–162.

Ettre, L.S., and C.W. Gehrke. 2006. The development of the amino acid analyzer. *LCGC North America*. 24:390–400.

Fanatico, A.C., K. Arsi, I. Upadhaya, J. Morales Ramos, D. Donoghue, and A.M. Donoghue. 2018. Sustainable fish and invertebrate meals for methionine and protein feeds in organic poultry production. *Journal of Applied Poultry Research*. 27:437–448.

Farran, M.T., and O.P. Thomas. 1992. Valine deficiency. 1. The effect of feeding a valine-deficient diet during the starter period on performance and feather structure of male broiler chicks. *Poultry Science*. 71:1879–1884.

Featherston, W.R. 1976. Glycine–serine interrelations in the chick. *Federal of American Societies for Experimental Biology*. 35:1910–1913.

Fisher, C., T.R. Morris, and R.C. Jennings. 1973. A model for the description and prediction of the response of laying hens to amino acid intake. *British Poultry Science*. 14:469–484.

Fisher, H., P. Griminger, G.A. Leville, and R. Shapiro. 1960. Quantitating aspects of lysine deficiency and amino acid imbalance. *Journal of Nutrition*. 71:213–220.

Fisher, H., and R. Shapiro. 1961. Amino acid imbalance: Rations low in tryptophan, methionine or lysine and the efficiency of utilization of nitrogen in imbalanced rations. *Journal of Nutrition*. 75:395–401.

Fontaine, J. 2003. Amino acid analysis of feeds. *Amino Acids in Animal Nutrition*, 2nd ed. J.P.F. D'Mello, editor. Wallingford, Oxfordshire, UK: CABI Publishing. 15–40.

Fontaine, J., J. Hörr, and B. Schirmer. 2001. Near-infrared reflectance spectroscopy enables the fast and accurate prediction of the essential amino acid contents in soy, rapeseed meal, sunflower meal, peas, fishmeal, meat meal products, and poultry meal. *Journal of Agricultural and Food Chemistry*. 49:57–66.

Fritz, J.C., J.H. Hooper, J.L. Halpin, and H.P. Moore. 1946. Failure of feather pigmentation in bronze poults due to lysine deficiency. *Journal of Nutrition*. 31:387–396.

Geraert, P.A., J.C.F. Padiha, and S. Guillaumin. 1996. Metabolic and endocrine changes induced by chronic heat exposure in broiler chickens: Biological and endocrinological variables. *British Journal of Nutrition*. 75:205–216.

Graber, G., and D.H. Baker. 1971. Sulfur amino acid nutrition of the growing chick: Quantitative aspects concerning the efficacy of dietary methionine, cysteine and cystine. *Journal of Animal Science*. 33(5):1005–1011.

Grau, C.R. 1945. Deformity of the tongue associated with amino acid deficiencies in the chick. *Proceedings of the Society for Experimental Biology and Medicine*. 59:177–178.

Grau, C.R. 1947. Interrelations of phenylalanine and tyrosine in the chick. *Journal of Biological Chemistry*. 170:661–669.

Grau, C.R., and M. Kamei. 1950. Amino acid imbalance and the growth requirements for lysine and methionine. *Journal of Nutrition*. 41:89–101.

Guttridge, D., and D. Lewis. 1964. Chick bio-assay of methionine and cystine: II. Assay of soyabean meals, groundnut meals, meat meals, methionine isomers and methionine analogue. *British Poultry Science*. 5:193–200.

Han, Y., and D.H. Baker. 1993. Effects of excess methionine or lysine for broilers fed a corn-soybean meal diet. *Poultry Science*. 72:1070–1074.

Harper, A.E., N.J. Benevenga, and R.M. Wohlhueter. 1970. Effects of ingestion of disproportionate amounts of amino acids. *Physiological Reviews*. 50:428–558.

Horn, N.L., S.S. Donkin, T.J. Applegate, and O. Adeola. 2009. Intestinal mucin dynamics: Response of broiler chicks and White Pekin ducklings to dietary threonine. *Poultry Science*. 88:1906–1914.

Humphrey, B.D., C.B. Stephensen, C.C. Calvert, and K.C. Klasing. 2006. Lysine deficiency and feed restriction independently alter cationic amino acid transporter expression in chickens (*Gallus gallus domesticus*). *Comparative Biochemistry and Physiology*. 143A:218–227.

Hurwitz, S., and S. Bornstein. 1973. The protein and amino acid requirements of laying hens: Suggested models for calculation. *Poultry Science*. 52:1124–1134.

Hurwitz, S., Y. Frisch, A. Bar, U. Eisner, I. Bengal, and M. Pines. 1983. The amino acid requirements of growing turkeys. 1. Model construction parameter estimation. *Poultry Science*. 62:2208–2217.

Hurwitz, S., D. Sklan, and I. Bartov. 1978. New formal approaches to the determination of energy and amino acid requirements of chicks. *Poultry Science*. 57:197–205.

Hurwitz, S., M. Weiselberg, U. Eisner, I. Bartov, G. Riesenfield, M. Sharvit, N. Niv, and S. Bornstein. 1980. The energy requirements and performance of growing chickens and turkeys as affected by environmental temperature. *Poultry Science*. 59:2290–2299.

Ishibashi, T. 1972. Protein metabolism in the fowl. IV. Possibility of conversion of tyrosine to phenylalanine in the adult rooster. *Agricultural and Biological Chemistry*. 36:596–603.

Jankowski, J., K. Ognik, M. Kubińska, A. Czech, J. Juśkiewicz, and Z. Zduńczyk. 2017. The effect of DL-, L-isomers and DL-hydroxy analog administered at 2 levels as dietary sources of methionine on the metabolic and antioxidant parameters and growth performance of turkeys. *Poultry Science*. 96:3229–3238.

Jones, J.D. 1961. Lysine toxicity in the chick. *Journal of Nutrition*. 73:107–112.

Jones, J.D. 1964. Lysine-arginine antagonism in the chick. *Journal of Nutrition*. 84:313–321.

Jones, J.D., S.J. Petersburg, and P.C. Burnett. 1967. The mechanism of the lysine-arginine antagonism in the chick: Effect of lysine on digestion, kidney arginase, and liver transamidinase. *Journal of Nutrition*. 93:103–116.

Kaiser, F.E., C.W. Gehrke, R.W. Zumwalt, and K.C. Kuo. 1974. Amino acid analysis: Hydrolysis, ion-exchange cleanup, derivatization, and quantitation by gas-liquid chromatography. *Journal of Chromatography A*. 94:113–133.

Kaspar, H., K. Dettmer, W. Gronwald, and P.J. Oefner. 2009. Advances in amino acid analysis. *Analytical and Bioanalytical Chemistry*. 393:445–452.

Katz, R., and D. Baker. 1975. Efficacy of D-, L- and DL-methionine for growth of chicks fed crystalline amino acid diets. *Poultry Science*. 54:1667–1674.

Klain, G.J., D.C. Hill, J.A. Gray, and H.D. Branion. 1957. Achromatosis in the feathers of chicks fed lysine-deficient diets. *Journal of Nutrition*. 61:317–328.

Klasing, K.C., and D.R. Korver. 2020. Nutritional diseases. In *Diseases of Poultry*, 20th ed. D.E. Swayne, J.R. Glisson, L.P. McDougald, L.K. Nolan, D.L. Suarez, and V.I. Nair, editorss. Hoboken, NJ: John Wiley & Sons. 1257–1285.

Konashi, S., K. Takahashi, and Y. Akiba. 2000. Effects of dietary essential amino acid deficiencies on immunological variables in broiler chickens. *British Journal of Nutrition*. 83:449–456.

Kratzer, F.H., R.A. Ernst, B.J. Marquez, P. Schroeder, C.H. Brown, and S.A. Peoples. 1976. The effect of a low energy diet on the concentration of DDT in adipose tissue of turkeys. *Poultry Science*. 55:365–369.

Kurpad, A.V. 2018. 90th anniversary commentary: Amino acid imbalances: still in balance. *Journal of Nutrition*. 148:1647–1649.

Kuzmicky, D., G. Kohler, H. Walker Jr., and B. Mackey. 1977. Availability of oxidized sulfur amino acids for the growing chick. *Poultry Science*. 56:1560–1565.

Kvitvang, H.F.N., T. Andreassen, T. Adam, S.G. Villas-Bôas, and P. Bruheim. 2011. Highly sensitive GC/MS/MS method for quantitation of amino and nonamino organic acids. *Analytical Chemistry*. 83:2705–2711.

Lartey, F.M., and R.E. Austic. 2008. Phenylalanine requirement, imbalance, and dietary excess in one-week-old chicks: Growth and phenylalanine hydroxylase activity. *Poultry Science*. 87:291–297.

Llames, C.R., and J. Fontaine. 1994. Determination of amino acids in feeds: Collaborative study. *Association of Official Analytical Chemists*. 77:1362–1402.

Le, A., A. Ng, T. Kwan, K. Cusmano-Ozog, and T.M. Cowan. 2014. A rapid, sensitive method for quantitative amino acid analysis of underivatized amino acids by liquid chromatography-tandem mass spectrometry (LC-MS/MS). *Journal of Chromatography B*. 944:166–174.

Leeson, S., and J.D. Summers. 2001. Uric acid synthesis. In *Nutrition of the Chicken*, 4th ed. Guelph, Ontario, Canada: University Books. 120.

Lewis, A.J., and D.H. Baker. 1995. Bioavailability of D-amino acids and DL-hydroxy-methionine. In *Bioavailability of Nutrients for Animals*. Elsevier. Amsterdam, The Netherlands. 67–81.

Malkin, L.I., and D.M. Greenburg. 1964. Purification and properties of threonine or allothreonine aldolase from rat liver. *Journal of Agricultural and Food Chemistry*. 49:1799–1804.

Marrett, L., H. Bird, and M. Sunde. 1964. The effects of different isomers of methionine on growth of chicks fed amino acid diets. *Poultry Science*. 43:1113–1118.

Meléndez-Hevia, E., P. De Paz-Laugo, A. Cornish-Bowden, and M.L. Cárdenas. 2009. A weak link in metabolism: The metabolic capacity for glycine biosynthesis does not satisfy the need for collagen synthesis. *Journal of Biosciences*. 34:853–872.

Mendonca, C., and L.S. Jensen. 1989. Influence of protein concentration on the sulfur-containing amino acid requirement of broiler chickens. *British Poultry Science*. 30:889–898.

Meussen, B.J., A.N.T. van Zeeland, M.E. Bruins, and J.P.M. Sanders. 2014. A fast and accurate UPLC method for analysis of proteinogenic amino acids. *Food Analytical Methods*. 7:1047–1055.

Miles, R.D., P.T. Costa, and R.H. Harms. 1983. The influence of dietary phosphorus level on laying hen performance, egg shell quality and various blood parameters. *Poultry Science*. 62:1033–1037.

Moore, S., D.H. Spackman, and W.H. Stein. 1958. Chromatography of amino acids on sulfonated polystyrene resins. *Analytical Chemistry*. 30:1185–1190.

Morris, T.R. 1968. The effect of dietary energy level on the voluntary caloric intake of laying birds. *British Poultry Science*. 9:285–295.

Morris, T.R., K. Al-Azzawi, R.M. Gous, and G.L. Simpson. 1987. Effects of protein concentration on responses to dietary lysine by chicks. *British Poultry Science*. 28:185–195.

Muramatsu, T., K. Hiramoto, I. Tasaki, and J.-I. Okumura. 1987. Effect of protein starvation on protein turnover in liver, oviduct and whole body of laying hens. *Comparative Biochemistry and Physiology*. 87B:227–232.

Netke, S.P., H.M. Scott, and G.L. Alee. 1969. Effect of excess amino acids on the utilization of first limiting amino acids in chick diets. *Journal of Nutrition*. 99:75–81.

Newberne, P.M., J.E. Savage, and B.L. O'Dell. 1960. Pathology of arginine deficiency in the chick. *Journal of Nutrition*. 72:347–352.

Noll, S.L., P.E. Waibel, R.D. Cook, and J.A. Witmer. 1984. Biopotency of methionine sources for young turkeys. *Poultry Science*. 63:2458–2470.

NRC (National Research Council). 1975. *The Effect of Genetic Variance on Nutritional Requirements of Animals*. Washington, DC: National Academy Press.

NRC. 1981. Poultry. In *Effect of Environment on Nutrient Requirements of Domestic Animals*. Washington, DC: National Academy Press. 109.

NRC. 1987. *Predicting Feed Intake of Food Producing Animals*. Washington, DC: National Academy Press.

NRC, 1994. *Nutrient Requirements of Poultry*, 9th rev. ed. Washington, DC: National Academy Press.

Ospina-Rojas, I.C., A.E. Murakami, I. Moreira, K.P. Picoli, R.J.B. Rodrigueiro, and A.C. Furlan. 2013. Dietary glycine+serine responses of male broilers given low-protein diets with different concentrations of threonine. *British Poultry Science*. 54:486–493.

Otter, D.E. 2012. Standardised methods for amino acid analysis of food. *British Journal of Nutrition*. 108:S230–S237.

Park, I., T. Pasquetti, R. Malheiros, P. Ferket, and S. Kim. 2018. Effects of supplemental L-methionine on growth performance and redox status of turkey poults compared with the use of DL-methionine. *Poultry Science*. 97:102–109.

Parker, C., C. Nixey, and S. Urwin. 2011. Feathering abnormalities in bronze turkeys associated with amino acid deficiencies in organic diets. *Veterinary Record*. 169:470.

Parsons, C., and L. Potter. 1981. Utilization of amino acid analogues in diets of young turkeys. *British Journal of Nutrition*. 46:77–86.

Peace, R.W., and G.S. Gilani. 2005. Chromatographic determination of amino acids in foods. *AOAC International*. 88:877–887.

Peng, J.L., S.P. Bai, J.P. Wang, X.M. Ding, Q.F. Zeng, and K.Y. Zhang. 2018. Methionine deficiency decreases hepatic lipid exportation and induces liver lipid accumulation in broilers. *Poultry Science*. 97:4315–4323.

Penz Jr. A.M., and F.H. Kratzer. 1984. Effect of excess leucine on feather structure and feather composition in the chick. *Nutrition Reports International*. 29:991–995.

Pesti, G.M., A.E. Harper, and M.L. Sunde. 1980. Choline/methionine nutrition of starting broiler chicks: Three models for estimating the choline requirement with economic considerations. *Poultry Science*. 59:1073–1081.

Poinsot, V., V. Ong-Meang, A. Ric, P. Gavard, L. Perquis, and F. Couderc. 2018. Recent advances in amino acid analysis by capillary electromigration methods: June 2015–May 2017. *Electrophoresis*. 39:190–208.

Qureshi, G.A., and A.R. Qureshi. 1989. Determination of free amino acids in biological samples: Problems of quantification. *Journal of Chromatography*. 491:281–289.

Robbins, K.R. 1987. Threonine requirement of the broiler chick as affected by protein level and source. *Poultry Science*. 66:1531–1534.

Robbins, K.R., A.M. Saxton, and L.L. Southern. 2006. Estimation of nutrient requirements using broken-line regression analysis. *Journal of Animal Science*. 84(E. Suppl.):E155–E165.

Robel, E.J. 1977. A feather abnormality in chicks fed diets deficient in certain amino acids. *Poultry Science*. 56:1968–1971.

Roland, D.A. Sr., M.M. Bryant, J.X. Zhang, D.A. Roland Jr., S.K. Rao, and J. Self. 1998. Econometric feeding and management. 1. Maximizing profits in Hy-Line W-36 hens by optimizing total sulfur amino acid intake and environmental temperature. *Journal of Applied Poultry Research*. 7:403–411.

Salido, E.C., X.M. Li, Y. Lu, X. Wang, A. Santana, N. Roy-Chowdhury, A. Torres, L.J. Shapiro, and J. Roy-Chowdhury. 2006. Alanine-glyoxylate aminotransferase-deficient mice: A model for primary hyperoxaluria that responds to adenoviral gene transfer. *Proceedings of the National Academy of Sciences*. 103:18249–18254.

Sasse, C.E., and D.H. Baker. 1972. The phenylalanine and tyrosine requirements and their interrelationship for the young chick. *Poultry Science*. 51:1531–1536.

Scott, M.L., M.C. Nesheim, and R.J. Young. 1983. *Nutrition of the Chicken*, 3rd ed. Ithaca, NY: M.L. Scott & Associates.

Shen, Y.B., P. Ferket, I. Park, R.D. Malheiros, and S.W. Kim. 2015. Effects of feed grade L-methionine on intestinal redox status, intestinal development, and growth performance of young chickens compared with conventional DL-methionine. *Journal of Animal Science*. 93:2977–2986.

Siegert, W., H. Ahmadi, A. Helmbrecht, and M. Rodehutscord. 2015. A quantitative study of the interactive effects of glycine and serine with threonine and choline on growth performance in broilers. *Poultry Science*. 94:1557–1568.

Slagter, P.J., and P.W. Waldroup. 1984. Calculation and evaluation of energy: amino acid ratios for the egg production type hen. *Poultry Science*. 63:1810–1822.

Smith, R.E. 1966. The utilization of L-methionine, DL-methionine and methionine hydroxy analogue by the growing chick. *Poultry Science*. 45:571–577.

Snetsinger, D.C., and H.M. Scott. 1961. Efficacy of glycine and arginine in alleviating the stress induced by dietary excesses of single amino acids. *Poultry Science*. 40:1675–1680.

Soloway, S., and D. Stetten Jr. 1953. The metabolism of choline and its conversion to glycine in the rat. *Journal of Biological Chemistry*. 204:207–214.

Spackman, D.H., W.H. Stein, and S. Moore. 1958. Automatic recording apparatus for use in the chromatography of amino acids. *Analytical Chemistry*. 30:1190–1206.

Stilborn, H.L., E.T. Moran Jr., R.M. Gous, and M.D. Harrison. 1997. Effect of age on feather amino acid content in two broiler strain crosses and sexes. *Journal of Applied Poultry Research*. 6:205–209.

Sugahara, M., and M. Kandatsu. 1976. Glycine-serine interconversion in the rooster. *Agricultural and Biological Chemistry*. 40:833–837.

Sugahara, M., D.H. Baker, and H.M. Scott. 1969. Effect of different patterns of excess amino acids on performance of chicks fed amino acid deficient diets. *Journal of Nutrition*. 97:29–32.

Szökő, É., I. Vincze, and T. Tábi. 2016. Chiral separations for D-amino acid analysis in biological samples. *Journal of Pharmaceutical and Biomedical Analysis*. 130:100–109.

Tamimie, H.S. 1966. Quantitative and periodic restriction of feed and its relationship to the feeding of chicks a 2 percent L-phenylalanine supplemented diet. *Life Science*. 5:1447–1457.

Tamimie, H.S. 1967a. Feeding chicks high levels of L-phenylalanine and L-methionine supplemented diets in the study of experimental aspects of phenylketonuria and homocystinuria. *Poultry Science*. 46:100–107.

Tamimie, H.S. 1967b. Influence of niacin and L-tryptophan on the growth depressive performance of chicks fed high levels of L-phenylalanine and L-methionine. *Life Science*. 6:587–594.

Tamimie, H.S. 1968. Feeding 5 percent L-phenylalanine to chicks of pullets fed same levels of L-phenylalanine. *Comparative Biochemistry and Physiology*. 24:417–425.

Tamimie, H.S., and G.R. Pscheidt. 1966a. Effect of feeding 5 percent L-phenylalanine to young chicks. *Comparative Biochemistry and Physiology*. 17:609–621.

Tamimie, H.S., and G.R. Pscheidt. 1966b. Effect of feeding 5 percent L-phenylalanine to day-old chicks and its relationship to sex and breed. *Comparative Biochemistry and Physiology*. 17:1159–1166.

Tamimie, H.S., and G.R. Pscheidt. 1966c. Influence of feeding 5 percent L-phenylalanine on reproductive performance of pullets. *American Physiological Society*. 211:955–958.

Tao, X., Z.Y. Zhang, H. Dong, H. Zhang, and H. Xin. 2006. Responses of thyroid hormones of market-size broilers to thermoneutral constant and warm cyclic temperature. *Poultry Science*. 85:1520–1528.

Tesseraud, S., R. Peresson, J. Lopes, and A.M. Chagneau. 1996. Dietary lysine deficiency greatly affects muscle and liver protein turnover in growing chickens. *British Journal of Nutrition*. 75:853–865.

Thera, J.C., K.A. Kidd, M.E. Dodge-Lynch, and R.F. Bertolo. 2017. Quantification of sulphur amino acids by ultra-high performance liquid chromatography in aquatic invertebrates. *Analytical Biochemistry*. 539:158–161.

Thera, J.C., K.A. Kidd, M.G. Nosworthy, and R.F. Bertolo. 2018. Evaluation of a performic acid oxidation method for quantifying amino acids in freshwater species. *Limnology and Oceanography: Methods*. 16(12):803–813.

Tillman, P.B. 2019. Determination of nutrient values for commercial amino acids. *Journal of Applied Poultry Research*. 28:526–530.

Timson, B.F., M.S. Chi, and B.K. Bowlin. 1983. The effect of reduced dietary protein on the anterior latissimus dorsi muscle fibers in the single comb White Leghorn pullet. *Poultry Science*. 62:2230–2233.

Tipton, H.C., B.C. Dilworth, and E.J. Day. 1966. A comparison of D-, L-, DL-methionine and methionine hydroxy analogue calcium in chick diets. *Poultry Science*. 45:381–387.

Tobin, G., and K.N. Boorman. 1979. Carotid or jugular amino acid infusions and food intake in the cockerel. *British Journal of Nutrition*. 41:157–162.

Tomonaga, S., and M. Furuse. 2019. Nutritional characteristics and functions of D-amino acids in the chicken. *Journal of Poultry Science*. 0190062.

Ueda, H., S. Yabuta, H.-O. Yokota, and I. Tasaki. 1981. Involvement of feed intake and feed utilization in the growth retardation of chicks given the excessive amounts of leucine, lysine, phenylalanine or methionine. *Nutrition Reports International*. 24:135–144.

van Krimpen, M.M., F. Leenstra, V. Mauerer, and M. Bestman. 2016. How to fulfill EU requirements to feed organic laying hens 100% organic ingredients. *Journal of Applied Poultry Research*. 25:129–138.

Vohra, P., and F.H. Kratzer. 1959. Specificity of lysine for the growth of turkey poults and prevention of feather depigmentation. *Poultry Science*. 38:280–281.

Waldroup, P.W., R. Mitchell, J.R. Payne, and K.R. Hazen. 1976. Performance of chicks fed diets formulated to minimize excess levels of essential amino acids. *Poultry Science*. 55:243–253.

Waldroup, P.W., J.H. Kersey, and C.A. Fritts. 2002. Influence of branched-chain amino acid balance in broiler diets. *International Journal of Poultry Science*. 5:136–144.

Wang, T.C., and M.C. Fuller. 1989. The optimum dietary amino acid pattern for growing pigs. *British Journal of Nutrition*. 62:77–79.

Wang, W., J. Wang, S. Wu, X. Dong, C. Guo, H. Zhang, and G. Qi. 2019. Bioavailability of L-methionine relative to DL-methionine in broiler chickens. *Italian Journal of Animal Science*. 18:1231–1238.

Wils-Plotz, E.L., and R.N. Dilger. 2013. Combined dietary effects of supplemental threonine and purified fiber on growth performance and intestinal health of young chicks. *Poultry Science*. 92:726–734.

Wils-Plotz, E.L., M.C. Jenkins, and R.N. Dilger. 2013. Modulation of the intestinal environmental, innate immune response, and barrier function by dietary threonine and purified fiber during a coccidiosis challenge in floor chicks. *Poultry Science*. 92:735–745.

Wu, Y., J. Tang, J. Cao, B. Zhang, Y. Chen, M. Xie, Z. Zhou, and S. Hou. 2021. Effect of dietary L-methionine supplementation on growth performance, carcass traits, and plasma parameters of starter Pekin ducks at different dietary energy levels. *Animals*. 1:144.

Xie, M., S.S. Hou, W. Huang, and H.P. Fan. 2007. Effect of excess methionine and methionine hydroxy analogue on growth performance and plasma homocysteine of growing Pekin ducks. *Poultry Science*. 86:1995–1999.

Yoshida, H., K. Kondo, H. Yamamoto, N. Kageyama, S. Ozawa, K. Shimbo, T. Muramatsu, A. Imaizumi, T. Mizukoshi, J. Masuda, D. Nakayama, Y. Hayakawa, K. Watanabe, K. Mukaibatake, and H. Miyano. 2015. Validation of an analytical method for human plasma free amino acids by high-performance liquid chromatography ionization mass spectrometry using automated precolumn derivatization. *Journal of Chromatography B*. 998–999:88–96.

Zhang, Y.N., R.S. Xu, L. Min, D. Ruan, H.Y. Kim, Y.G. Hong, W. Chen, S. Wang, W.G. Xia, X. Luo, C.Y. Xie, X.G. Shang, and C.T. Zheng. 2019. Effects of L-methionine on growth performance, carcass quality, feather traits, and small intestinal morphology of Pekin ducks compared with conventional DL-methionine. *Poultry Science*. 98:6866–6872.

4

Lipids

INTRODUCTION

Lipids serve a variety of purposes in poultry diets. For the bird, they provide a concentrated source of energy with a low heat increment, contain essential fatty acids, and facilitate the absorption of fat-soluble vitamins. For the feed manufacturer and poultry producer, they lubricate the feed during pelleting and improve air quality by reducing feed dust during transport and distribution and so improve housing air quality to the benefit of bird health.

Feed-grade lipids may come from many different sources. Feed ingredients such as corn and whole soybeans contain lipids as part of their structure. Used oils (grease) from restaurants, lipids from the rendering of animal carcasses, and the refuse from vegetable oil refining are examples of supplemental lipids used in formulated diets. These sources represent several types and categories. The Association of American Feed Control Officials (AAFCO) publishes definitions and compositional information on feed-grade lipids annually in the AAFCO Official Publication (OP) (https://www.aafco.org/Regulatory/Committees/Ingredient-Definitions). The AAFCO definitions, which are used by U.S. regulatory officials, indicate lipid components and limits of non-fat or non-oil materials in product suitable for use in feed manufacture. Description of moisture (M) and those compounds that are either insoluble in ether (I) or unsaponifiable (U) are usually of no energetic or essential nutritional value, and in composite (MIU) act principally as a diluent. The percentage MIU and percentage digestibility combine to influence the ME_n value. Moisture, insoluble, unsaponifiables, and free fatty acid contents of fats were shown to influence performance responses in broilers (Pesti et al., 2002), with increasing amounts reducing productive performance. All feed fats should be stabilized by an antioxidant to preserve unsaturated fatty acids and routinely monitored for the possible presence of undesirable residues such as insolubles, chlorinated hydrocarbons, unsaponifiables, and peroxides (Blaszczyk et al., 2013). The National Animal Nutrition Program (NANP; https://animalnutrition.org) provides accurate nutrient composition data through a cross-species database of nutrient composition of feedstuffs. More complete information on NANP and the National Research Support Project (NSRP-9) though which the U.S. Department of Agriculture National Institute of Food and Agriculture supports the development and continuous update and maintenance of the database and its relationship to AAFCO definitions is provided in Chapter 20 (Ingredients).

LIPID NUTRITION

In diet formulation, the total fatty acids contributed by all lipid categories, the proportion that are in free form, and the types of fatty acids present provide information related to expected digestibility as well as how the fat may be used subsequently by the bird. Fatty acid chain length, extent of unsaturation, as well as the number and positional distribution of fatty acids esterified on the glycerol backbone of the triacylglycerol molecule all influence the extent of intestinal absorption (Tancharoenirat et al., 2014). When describing specific fatty acids, it is desirable to provide the carbon chain length and the number and position of double bonds in addition to a common name. For example, stearic acid has 18 carbon atoms and no double bonds and is chemically described as 18:0, while the chemical description of linoleic acid (LNA), a fatty acid with 18 carbon atoms and two double bonds, is 18:2 n-6. The n-6 designation indicates that the first double bond is located at the sixth carbon–carbon bond from the methyl carbon of the chain. A second important fatty acid category is the n-3 family in which the first double bond is located at the third carbon-carbon bond from the methyl carbon of the fatty acid. This family includes fatty acids such as α-linolenic acid (ALA, 18:3, n-3) and others with important biological activities such as docosahexaenoic acid (DHA; 22:6 n-3) and eicosapentaenoic acid (EPA; 20:5 n-3). Fatty acids lacking double bonds are termed saturated fatty acids (SFAs), those with a single double bond as monounsaturated fatty acids (MUFAs), while those with two or more double bonds are termed polyunsaturated (PUFA). Polyunsaturated fatty acids with chain lengths greater than 20 carbons and two or more double bonds are termed long-chain PUFAs (LC-PUFA).

The term "fat" refers to a substance composed chiefly of triacylglycerol containing sufficient saturated fatty acids (SFAs) to exist as a solid or semi-solid material at room temperature (23°C). Conversely, the term "oil" refers to a substance composed chiefly of triacylglycerol, comprising predominantly of MUFA and PUFA and which is liquid at room temperature (Gunstone, 2011). The melting points of SFA, which are greater than those of MUFA, which in turn are higher than those of PUFA, determine the physical properties of fats and oils. In triacylglycerol, MUFAs and PUFAs are predominantly esterified to the "sn-2" position of the glycerol backbone while SFAs are esterified to the "sn-1" and "sn-3" position of the glycerol moiety (Figure 4-1). In phospholipids, the "sn-3" position is esterified to a phosphate group with the "sn-1" and "sn-2" positions being esterified to fatty acids in accordance to the predominance observed in triacylglycerol sn-1, sn-2 positions (Gurr et al., 2002). Acylglycerols, for which physical state does not influence nutritional use or metabolism, are generally termed lipids.

Metabolizable Energy Value

Chapter 1 on energy concluded that net energy (NE) may be a more appropriate methodology to use in diet formulation, but that information on NE for feedstuffs is limited although data acquisition is ongoing (Barzegar et al., 2019, 2020; Wu et al., 2019a,b). At present, apparent metabolizable energy (AME) or AME corrected for zero N balance (AMEn) values are the most readily available and commonly used in poultry diet formulation. Flaws in methodology used for AME definitions were summarized in 2019 (Wu et al., 2019a,b) and are detailed in the chapter on energy. The chemical compositions of dietary fats and oils provide a seemingly ready means to calculate AME values (Appendix Table B-1 in NRC, 1994). However, correctness of AME or NE values of fats and oils are the most difficult to evaluate (Mateos et al., 2019).

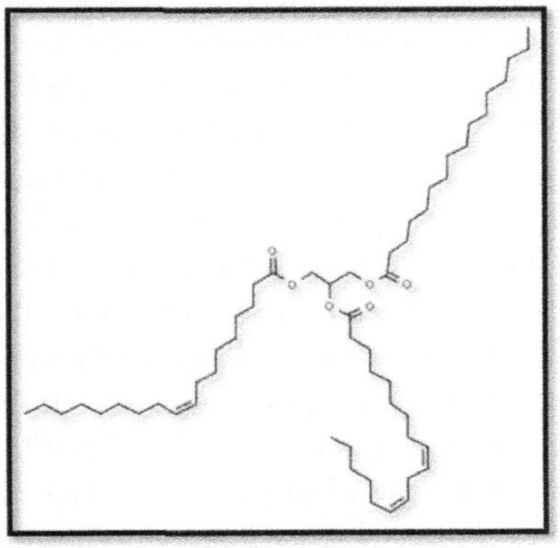

FIGURE 4-1 Structure of common triacylglycerol in poultry. Fatty acids esterified to 1, 2, and 3 position are oleic (18:1, n-9), linoleic (18:2, n-6), and palmitic (16:0), respectively.

Factors influencing the AMEn value of dietary lipid that are not directly associated with lipid quality include bird type and age as well as the method of measurement. Dietary fat and oil utilization increases 2 to 6 weeks post-hatch in chickens, quails, and turkeys (Klasing, 1998; Polawski, 2018; Ravindran et al., 2016; Tancharoenrat et al., 2013; Sell et al., 1986) due to increases in digestive enzyme activity and bile acid formation in the maturing gut. This increase is particularly evident with long-chain SFAs (LC-SFAs) and fats containing substantial proportions of LC-SFAs (Tancharoenrat et al., 2014; Sell et al., 1986). Age was found to influence saturated fat digestibility more in chickens (Shaver broilers) than turkeys (BTU9) (Mossab, 2000). Both increasing dietary fatty acid saturation and amount of unesterified fatty acids present in the diet decrease dietary lipid digestibility (Rodriguez-Sanchez et al., 2019). Emulsifying agents have been included in broiler diets to increase fat digestibility (Siyal et al., 2017). Reports tend to show that emulsifier inclusion increases fat digestibility but does not always translate into increased performance.

It was well recognized in the 1994 NRC publication that the methodology used in obtaining feedstuff energy values influenced the values obtained (NRC, 1994). The information in Appendix Table B-1 in NRC (1994) provided a description of fats that may be used in feeds and their MEn values observed under a variety of circumstances. Data indicate that considerable variation existed, and several factors must be considered in determining feeding value. Some of these factors are included in the equations listed in Appendix Table B-1 of the 1994 report and can be used as a starting point to predict the MEn value of fats. When the effects of method of determination and age of the bird are superimposed on factors associated with the fat, it becomes evident that use of current MEn estimates should take into consideration the methodological limitations of their determinations. An empirical approach to uncertainty surrounding the energy contribution of fat in specific situations will require the producer to monitor the flock's bodyweight or other key production variables and adjust feed composition or delivery rate. This is a cumbersome but pragmatic solution in the absence of an improved understanding of digestive physiology in relation to feed formulations involving dietary lipids.

Blending Fats

Animal tallow contains a high proportion of LC-SFA, and its utilization is often increased when blended with a small amount of oil containing a greater proportion of unsaturated fatty acids. The resulting MEn value of blends is often greater than can be explained from the arithmetic combination. The presence of feed ingredients that contain MUFA and PUFA can also enhance the nutritional properties of animal tallows that primarily contain SFA. Corn is particularly advantageous in this respect because its fatty acids are mostly unsaturated, and it usually constitutes a large portion of feed. Sibbald and Kramer (1980) noted that the true metabolizable energy (TME) for beef tallow was greater when a corn-based carrier was used during measurement than when wheat was used. This difference is due to the lower total fat content of wheat (~1.7% by weight) compared to corn (4.7% by weight), rather than a difference in fatty acid composition and degree of saturation (unsaturated/saturated), which are 5.11 and 3.19 for corn and wheat, respectively (USDA, 2019). Lipase enzymes act at the lipid/water interface of digesta (Gargouri et al., 1983). Increased absorption of the SFA when mixed with MUFA and PUFA arises through the structure of partial glycerides that form during the digestion of MUFA and PUFA. These compounds include mono-acylglycerols that act as amphiphiles and work in combination with biliary phospholipids and lyso-phospholipids to promote SFA absorption. These unsaturated molecules expand the surface area available for lipase activity as well as increasing the carrying capacity of bile salt micelles that solubilize SFA more effectively (Hofmann and Borgstrom, 1962; Krogdahl, 1985). Fatty acid and partial glyceride delivery beyond the unstirred water layer to the apical surface of the enterocyte occurs via bile salt micelles (Krogdahl, 1985).

Within poultry diets, phospholipids intrinsic to diet ingredients also provide amphiphiles. For example, phospholipids such as lecithin in soybeans are sources of phosphoric acid, choline, and fatty acids and, in conjunction with bile acids, help to emulsify dietary fats in the intestine. Dietary soybean lecithin has been shown to be a good source of phosphorus, choline, and energy for broiler chickens but, as reviewed by Vinado et al. (2020), soybean oil replacement by soybean lecithin in starter broiler diets impairs growth performance, reduces feed AME content, and lowers the total tract apparent digestibility PUFA. However, such was not the case for growing or finishing broilers. To confirm and extend the latter findings, Vinado et al. (2020) reported that replacement of soybean oil with soybean lecithin in grower and finisher broiler diets did not modify bird performance metrics; ileal digestibility of total fatty acids, SFA, and MUFA; or jejunal morphology. However, the total replacement of soybean oil with soybean lecithin (at 20 g/kg diet) reduced ileal absorption of PUFA and reduced the PUFA content of abdominal fat pads. Despite the latter changes, the authors concluded that soybean lecithin can either partially or totally replace soybean oil as an energy source in broiler grower diets (up to 10 g/kg diet) or broiler finisher diets (up to 20 g/kg diet).

Extra-Caloric Effect

Employing high levels of added lipid in diets often leads to a greater MEn than can be accounted for from the summation of values obtained for ingredients when assayed in a lower fat dietary context. The reason for this effect may involve an increase in the intestinal retention time of feed that allows for more complete digestion and absorption of nutrients (Summers, 1984). In Single-Comb White Leghorn hens in production, increasing supplemental yellow grease from 0 to 15 percent at the expense of corn increased transit time by ~13 percent, while addition of 20–30 percent yellow grease increased transit time by ~32 percent (Mateos et al., 1982). However, age, strain, diet composition, production status, and fat composition appear to influence observable delays in transit time (Rochell et al., 2012). Studies with mixed-sex broiler chicks in which corn/soy diets containing 0–20 percent supplemental poultry fat showed no effect on intestinal transit time (Golian and Maurice, 1992).

Increased Net Energy of Production

All body tissues have an energy value that corresponds to their heat of combustion. The net energy of production corresponds to this energy gained from either body growth or egg formation. Increased productive efficiency associated with lipid feeding is metabolically based (Parks, 2002). The provision of lipids in feed obviates the cost of fatty acid synthesis from carbohydrate, a process termed de novo lipogenesis (DNL). Re-esterification of preformed dietary fatty acids into acyl-glycerols is more energy efficient than DNL. Isoenergetic amino acid adequate diets having a higher percentage of energy (En %) from fat or oil at the expense of carbohydrate usually result in an increased productive energy (NEp) when the same level of MEn is realized. For example, Sell and Owings (1984) noted that added fat increased the body weight gain of large turkeys, with the greatest advantage occurring between 12 and 20 weeks of age. After 20 weeks, the favorable effect of fat on body weight progressively dissipated as mature bodyweight was attained, but the effect on feed efficiency remained (Moran, 1982). As birds reach mature frame size and maximum lean tissue gain, enhanced energetic efficiency is more likely to encourage adipose deposition unless new productive demands such as egg production occur. Relating energy intake to body energy content rather than simple bodyweight gain would show this more clearly as composition of the gain would be considered.

Laying hens can also respond to added dietary lipid; the provision of preformed fatty acids decreases the need for hepatic fatty acid synthesis via DNL and generally increases yolk and egg weights (March and MacMillan, 1990; Whitehead, 1981). However, in order to gain the advantage of preformed lipid, adequate amounts of dietary LNA (18:2 n-6) must be available. A consistent observation is that with dietary LNA concentrations of at least 1.0 g/100 g diet (1.0% by weight), supplemental lipid up to 6.0 percent by weight will increase egg size. Increased egg weight in these studies ranged from 0.44 g to 1.56 g/egg. It seems that young hens benefit more from supplemental lipid than older hens; stable isotope studies in feed-restricted broiler breeders showed that hens rely more on dietary LNA and DNL than older hens that are more reliant on adipose stores of fatty acids (Salas et al., 2017). The durability of that observation may lapse if continued genetic selection for whole-body leanness continues and alters adult as well as juvenile body composition further. Supplemental lipid can provide a metabolic advantage to hens experiencing high environmental temperatures as the heat increment due to lipid digestion is lower than that of carbohydrate while providing an equivalent amount of energy needed for yolk lipid (and egg) formation (Valencia et al., 1980). The lower heat increment of dietary lipid means that less heat is produced by the bodies of hens, which in turn increases the difference between hen body temperature and environmental temperatures and facilitates thermoregulation by the hens. Diets high in n-3 fatty acids can reduce egg size (Cherian, 2008), although the effect is not consistent (Ao et al., 2015; Park et al., 2015) and variables are incompletely understood.

Fatty Acid Composition

The fatty acid composition of poultry meat and eggs depends upon both DNL and dietary fatty acids. Fatty acid synthesis within fowl occurs primarily, but not exclusively, in the liver (Leveille et al., 1975; Griffin et al., 1992; Hillgartner et al., 1995). Immediately preceding sexual maturity, the rate of synthesis increases dramatically and the rate at which the body's depots accrue fat is great (Resnyk et al., 2017). Fatty acids synthesized by poultry are derived from dietary carbohydrates that are metabolized to acetate prior to synthesis into longer chain (≥16 carbons) SFA and unsaturated fatty acids of the n-9 series (Harwood et al., 2016). The molecular details of fatty acid elongases (lengthening fatty acid chain by two carbons) of chicken, turkey, and duck were reported in 2012–2014 (Gregory et al., 2013; Gregory and James, 2014). Shifts in dietary macronutrient composition can alter the predominance of DNL in determining whole-body fatty acid composition. Poultry diets have macronutrient distributions that promote DNL because they are, relatively speaking, low in fat (2–5% by wt.) and rich in carbohydrate (>60% by wt.). Reduced dietary carbohydrate, increased protein, and increased lipid each reduce DNL (Leveille et al., 1975; Rosebrough and Steele, 1982, 1985; Saadoun and Leclercq, 1987; Griffin et al., 1992; Hermier, 1997), which would promote the incorporation of dietary fatty acids into tissue lipids. Studies from the mid-1970s to late 1980s demonstrated that the extent of influence that each fatty acid source has on body composition increases with the level of intake, duration of feeding, and stage of maturity (Bartov et al., 1974; Salmon and O'Neil, 1973). The hen's adipose depots respond to dietary fat in the same way as do those of growing birds, and the yolk lipid generally exhibits a fatty acid pattern resembling that of the dietary fat (Guenter et al., 1971; Sim et al., 1973).

Fat absorbed from the fowl's intestine is transported to the liver in the form of portomicrons (Bensadoun and Rothfield, 1972), where some modifications may occur. For example, SFA may undergo desaturation, especially stearic acid (18:0), which is converted to oleic acid (18:1 n-9). In addition, although unsaturated fatty acids are generally unchanged, elongation and further desaturation (addition of a double bond) of 18:2 n-6 and 18:3 n-3 may occur in the liver (see subsequent discussion). Depot fat of both broiler chickens (Schuler and Essary, 1971; Edwards et al., 1973). and turkeys (Moran et al., 1973; Salmon and O'Neil, 1973; Moran, 1982) are more influenced by vegetable oils having high proportions of PUFA than by animal fats with a higher proportion of SFA and MUFA, which the bird can also synthesize.

Essential Fatty Acids

An essential nutrient is one that the bird cannot make but must consume in the diet to prevent deficiency and ultimately death (Combs et al., 2016). Linoleic acid (18:2 n-6) is the only essential fatty acid for which a dietary requirement has been demonstrated. However, both LNA and ALA are further metabolized into 20- and 22-carbon "long-chain" bioactive PUFA (LC-PUFA), and so both are recognized as essential precursor fatty acids for LC-PUFA and their bioactive metabolites (Watkins, 1991). The position of the double bonds in these n-6 and n-3 PUFAs, respectively, are unique because they are not formed in the fowl.

Inadequacies of LNA are uncommon in commercially reared poultry, but symptoms that result are due to a loss of cell membrane integrity. Increased needs for water and decreased resistance to disease are characteristic deficiency symptoms observed in poultry (Balnave, 1970). A deficiency of LNA in the male can impair spermatogenesis and reduce fertility (Lillie and Menge, 1968). Insufficient deposition of LNA in the egg will adversely affect embryonic development (Calvert, 1967). As noted in an earlier section, egg production and egg weights are also reduced (Menge et al., 1965). The essential fatty acid requirements of growing and adult birds appear to be amply satisfied by feeding a diet with 1.0 percent by weight of LNA (Whitehead, 1981; Grobas et al., 1999a,b; Keshavarz and Nakajima, 1995; Safaa et al., 2008; Balnave, 1970). Others have found that as little as 0.2 percent of the diet provides for adequate growth in broilers (Zornig et al., 2001), albeit 0.67 percent was needed to limit mortality and reduce liver weight. A 17 g egg yolk from hens fed corn/soy-based diets typically contains 600 mg of LNA (Library, 2018). This amount of LNA would be contained in ~55 g of whole, day-old broiler body (Poureslami et al., 2010), suggesting that there is an immediate post-hatch need for dietary LNA in young, rapidly growing poultry.

The controversy and technical challenges of determining an LNA requirement *per se* create an uncertainty that is pragmatically addressed by considering what has been demonstrated to be an adequate intake. For poultry, adequate amounts of LNA are usually present in diets based on commonly used cereal grains, particularly corn, and protein sources such as soybeans that are good sources of LNA (Safaa et al., 2008). In quail, 0.8 percent LNA in the diet was shown to be adequate for reproduction, and palmitic acid (16:0) was shown to spare LNA (Vilchez, 1990). Thus, continued rigorous efforts to define a LNA requirement for poultry of any kind must consider total diet lipid available in addition to added LNA and preexisting body stores of LNA (Balnave and Weatherup, 1974). Modeling approaches provide methods to determine nutrient requirements in such situations (Clifford and Müller, 2012; Salas et al., 2017).

Cholesterol

Cholesterol is not an acylglycerol, and birds do not metabolize it for energy. Eggs are among the most cholesterol-rich foods in the human diet. The inability to greatly decrease the cholesterol content of chicken eggs through genetic selection programs or dietary (nonpharmacological) approaches led researchers to hypothesize the chickens have a physiological control mechanism that causes a cessation of egg production when yolk cholesterol deposition is inadequate for embryonic survival (the so-called critical concentration hypothesis [CCR]; Hargis, 1988). In contrast, others (Noble and Cocchi, 1990; Griffin, 1992) have suggested that the amount of cholesterol in egg yolk exceeds the needs of the developing chick. Indeed, the developing embryo synthesizes cholesterol in a regulated manner (Wong and Lennarz, 1982). The development of "statin" drugs (3-hydroxy-3-me thylglutaryl-coenzyme A [HMG-CoA] reductase inhibitors), beginning in the late 1980s, allowed for testing the CCR. Statins block the conversion of HMG-CoA to mevalonate, which is considered the rate-limiting reaction of the hepatic cholesterol biosynthetic pathway. Elkin and Yan (1999) fed laying hens a control diet or the control diet supplemented with either 3 or 6 mg/kg of three statin drugs (lovastatin, simvastatin, and atorvastatin) for 5 weeks. The hens were artificially inseminated once per week, and eggs that were not analyzed for cholesterol content were incubated. When averaged across dietary groups, the incidence of unfertilized eggs was greatest in the atorvastatin-fed group (29%) with the other groups ranging from 1.6 percent (controls) to 7.9 percent (simvastatin), with no differences among groups in the hatchability of fertilized eggs. Maximum decreases in egg cholesterol contents from control values were greatest in the atorvastatin-fed group (46%) versus 22 percent and 7 percent for the hens fed simvastatin and lovastatin, respectively. Interestingly, despite an almost constant cholesterol content of ~105 mg per yolk in the eggs from hens fed 6 mg/kg atorvastatin, from 2 to 5 weeks, hatchability ranged from 67 percent (week 2) to 0 percent (week 3) to 20 percent (week 4) to 19 percent (week 5). In addition to being the first study to indicate that a mevalonate-derived product(s) plays an important role in avian fertility, this study challenged the contention that virtually all the cholesterol in chicken egg yolk is essential for embryonic development and survival.

Lipophilic Pigments

Poultry require dietary sources of carotenoids for pigmentation as these lipophilic molecules are not synthesized by animals. Plants, fungi, and microbes can synthesize carotenoids, which are concentrated in blossoms (e.g., marigold petals), seeds (e.g., corn), leaves (e.g., alfalfa), seaweed, and algae. Many animals (e.g., crustacea, fish) concentrate and further metabolize their dietary carotenoids, and meals produced from them can be an important carotenoid source in poultry diets. Carotenoids are subdivided into two categories: carotenes and xanthophylls. Carotenes have unmodified cyclohexenyl rings and have low pigmentation value in poultry, but β-carotene is a precursor for vitamin A (see Chapter 5). Xanthophylls are oxycarotenoids because they have alcohol, keto, or ester groups on the terminal cyclohexenyl rings, and they have higher pigmentation activity than carotenes (Hencken, 1992). The positioning and type of these groups determines the color of xanthophylls, although complexing of xanthophylls with proteins may shift the color (Fox, 1979). The carotenoid content of various feedstuffs is highly variable. For example, the xanthophyll concentration in wheat and sorghum is low (~2 mg/kg), corn has moderate levels (~50 mg/kg), alfalfa meal has moderately high levels (~225 mg/kg), and marigold petals have very high levels (~8,000 mg/kg). Carotenoids are commonly supplemented to poultry diets from marigold flower meal, marigold flower extract, or synthetic sources (Marounek and Pebriansyah, 2018). Corn, alfalfa meal, and marigold petals have high concentrations of lutein and zeaxanthin. Lutein and zeaxanthin impart a lemon-yellow color and a golden yellow color, respectively (Breithaupt, 2007). Canthaxanthin imparts an orange-red color and is enriched in some feedstuffs derived from crustaceans and algae, although in poultry diets it is often provided from chemically synthesized sources (Breithaupt, 2007).

The utilization of dietary carotenoids for pigmentation differs among the species and breeds of poultry and has a strong genetic basis (Klasing, 1998; Surai et al., 2001). There are three primary factors that affect pigmentation (Fox, 1979; Hencken, 1992; Surai et al., 2001; Na et al., 2004): (1) the type and quantity of the carotenoids in the diet; (2) the bird's capacity to digest and absorb specific carotenoids and, in some cases, to metabolize them to the correct chemical form; and (3) the capacity of specific tissues (e.g., epidermal cells, ovarian follicles) to take up carotenoids and use them for pigmentation. Dietary and health factors that affect fat absorption also impact carotenoid utilization (Surai et al., 2001). In addition, stressors such as pro-oxidants, mycotoxins (Surai et al., 2001), and infectious diseases (Chapter 11) can decrease pigmentation of poultry.

In general, poultry are not highly discriminatory in the type or quantity of carotenoids that they absorb and deposit, and the hue and its intensity of their skin, beak, adipose tissue, and egg yolk directly reflects the mixture of xanthophylls in the diet (Marounek and Pebriansyah, 2018). The intensity of pigmentation of the yolk and skin of chickens increases linearly and then plateaus as the dietary xanthophyll concentration increases (Sunde, 1992; Klasing, 1998). Poultry nutritionists pay careful attention to the amounts and concentrations of specific carotenoids provided in the feed so that commercial products are pigmented with the intensity and hue that is appropriate for various consumer preferences (Sunde, 1992). Preferences of consumers vary based on cultural and regional factors. For example, some consumers want a "sun" colored yolk, while others prefer a more yellow color. For this reason, poultry nutritionists have developed empirical equations that permit the formulation of specific xanthophyll combinations and levels to give specific coloration of eggs, legs, or carcasses destined for specific markets (Marusich and Bauernfeind, 1970; Fletcher, 1989).

In addition to their role in providing color, carotenoids can serve an anti-oxidative function (Surai et al., 2001) and modulate immune responses (Meriwether et al., 2010). However, poultry can live, reproduce, and appear healthy when consuming a diet devoid of carotenoids (but with vitamin A), and they are not considered nutritionally essential even though they are commercially important for product quality (Klasing, 1998).

LIPID METABOLISM AND PHYSIOLOGY

Importantly, lipids serve additional physiological functions within the bird other than those noted at the outset of this section. Those functions include the provision of tissue structure in the form of membranes, physical cushioning by adipose tissue, as well as signaling within and between tissues. Lipids considered in this context encompass a wide range of compounds; the International Lipid Classification and Nomenclature Committee (Fahy et al., 2005) defined the following categories containing distinct classes and subclasses of molecules: fatty acids, glycerolipids, glycerophospholipids, sphingolipids, sterol lipids, prenol lipids, saccharolipids, and polyketides. These eight categories cover both eukaryotic and prokaryotic sources, and the classification system, which was designed to be compatible with informatics requirements, has been adopted internationally and widely accepted by the lipidomics community (Fahy et al., 2009).

Within the body, lipoproteins carry out bulk transport of lipids for structural and energetic purposes. These nanoparticles are also responsible for the selective redistribution of specific lipids and lipophilic vitamins, hormones, and signaling molecules among tissues. Lipoproteins are distinctive macromolecular structures assembled in a structurally specific manner that reflects their physiological purpose (Walzem, 1996). While poultry typically consume low-fat diets, lipid transport and tissue lipid composition are of great interest to poultry biologists due to the roles such processes play in defining the nutritional value of poultry products and bird health. Past reviews have summarized the general biochemical features and physiological processes of avian lipoprotein metabolism (Hermier, 1997; Walzem et al., 2012; Chapman, 1980; Gruffat et al., 1996; Speake et al., 1998; Schneider, 2016).

Lipoprotein particle composition is often presented as the percentage of particle weight that protein, unesterified (free) cholesterol (UC), phospholipid (PL), triacylglycerol (TG), and cholesteryl esters (CE) comprise. This type of data has limited value unless accompanied by concentrations per volume plasma or serum. In the absence of information on concentration, information on such gross fractional composition data is not readily comparable to other studies and provides limited insight into why a particular outcome occurred or how to develop a plan to achieve a desired outcome. For studies designed to discover highly detailed nutritional/metabolic relationships in lipid metabolism, newer techniques such as targeted metabolomics are advantageous (Walzem et al., 2007; Kadambini, 2011).

Lipoprotein Particle Types

There are three broad classes of lipoproteins defined by their hydrated density. Hydrated density is also referred to as buoyant density as it reflects the particle's ability to float in increasingly dense salt solutions (Chapman, 1980); this property is established by particle composition and diameter (Walzem, 1996). Regardless of density class assignment, lipoprotein structure is generally spherical, with the outer surface comprised of apolipoproteins and amphipathic lipids such as PL, UC, and a core region of the nonpolar lipids CE and TG. Those particles with larger core components are less dense (more buoyant) than particles with small cores. Particle diameter provides another means to separate particle types (Kieft et al., 1991) albeit separations based on particle size are related to density nomenclature. The three classes are triacylglycerol-rich lipoproteins (TRLs), low-density lipoproteins (LDLs), and high-density lipoproteins (HDLs). Within the TRL class, there are further distinctions based on the tissue site of particle assembly and degree of lipolysis the parent particle has undergone. However, all TRLs in poultry contain $ApoB_{100}$, a minimum of 40 percent by weight of TG, and have a hydrated density of less than 1.019 g/mL (Chapman, 1980). Parent TRL particles are portomicrons, assembled in the intestine (Bensadoun and Rothfield, 1972), and very-low-density lipoprotein (VLDL), assembled primarily in liver. Hens in lay assemble a specific yolk-targeted lipoprotein, VLDLy, to carry lipid nutrients to the yolk (Walzem, 1996; Walzem et al., 1999). Within the hatching egg, the yolk sac membrane assembles VLDL for embryonic use. The kidney also appears to secrete VLDL (Tarugi et al., 1998). Particles arising from the partial lipolysis of portomicrons and VLDL can be termed remnants or intermediate density lipoproteins (IDLs). Particles in the TRL fraction comprise a continuum of progressively smaller, TG depleted particles.

Continued metabolism of TRLs to their metabolic endpoint gives rise to the second broad class of lipoproteins termed LDLs. The LDL particle class is CE-rich compared to either TRL or the third broad class of lipoproteins termed HDLs. High-density lipoproteins are protein-rich compared to LDL or TRL.

Triacylglycerol-Rich Lipoproteins

Assembled in the intestine, portomicrons transport dietary fat into the body through what some have termed the exogenous pathway. Portomicrons enter the portal circulation directly and deliver dietary fatty acids primarily to the liver (Noyan et al., 1964). However, the presence of the coccygeomesenteric vein allows for portomicron diversion through the renal veins and vena cava (Sturkie, 1976) so that portomicrons could also arrive at the liver following passage through the peripheral tissues. Information on whether directional variation in the blood flow within the coccygeomesenteric vein has a practical impact on lipid nutrient delivery is scarce to nonexistent, although VLDL assembly in the kidney might exist to accommodate this possibility (Tarugi et al., 1998).

Very low-density lipoproteins are assembled primarily in the liver and serve to deliver lipids formed in that tissue as a part of what some have referred to as the endogenous pathway. In chickens, the liver is the primary organ for fat synthesis via DNL (Leveille et al., 1975; Hillgartner et al., 1995), as well as reutilize fatty acids obtained as components of portomicron remnants or free fatty acids following release from adipose. These fatty acids give rise to a large variety of acylglycerols such as TG, PL, and CE. Chickens also assemble and secrete VLDL from the yolk sac membrane and kidney (Lazier et al., 1994; Kanai et al., 1996; Tarugi et al., 1998).

During egg production, hen liver assembles and secretes a highly specialized TRL called VLDLy to signify its specific role in targeting lipid nutrients for deposition into yolk (Walzem, 1996; Williams et al., 2002; Walzem et al., 2007). The VLDLy particle is a TRL being 70–85 percent TG by weight (Walzem et al., 1999), and it contains $ApoB_{100}$ and ApoVLDL-II, a small apoprotein that is made in response to increases in estrogen concentration (Wiskocil et al., 1980, 1981). While ApoB100 mediates binding of the VLDLy particle to LR8, the 95-kDa oocyte VLDL/VTG receptor (Schneider, 2016), ApoVLDL-II is thought to reduce VLDLy diameter and allow its passage through the basal lamina of the oocyte (Perry et al., 1984) where the particle can bind to LR8 (Stifani et al., 1990). The ApoVLDL-II protein is known to inhibit lipase (Schneider et al., 1990; Griffin et al., 1982), likely by changing the surface of the particle to exclude water needed for lipolysis (Boyle-Roden and Walzem, 2005). These properties act to preserve TG energy for embryonic use (Walzem et al., 1999). Vitellogenin (gene name VTG) is another apolipoprotein synthesized by the liver in response to estrogen (Wiskocil et al., 1981; Mann et al., 1999); it forms a PL-rich particle that can be isolated from blood as a very high-density lipoprotein (VHDL, 1.195–1.240 g/mL) (Deeley et al., 1975; Hermier et al., 1989; Mann et al., 1999). There are multiple vitellogenin isoforms (Wang et al., 1983). Vitellogenin also binds to LR8 on the surface of the oocyte and is taken up by receptor-mediated endocytosis (Speake et al., 1998; Wiskocil et al., 1981). Thus, together, VLDL and VTG respectively constitute approximately 60 percent and 24 percent of egg yolk dry matter (Burley et al., 1993) and ~95 percent and 5 percent of yolk cholesterol (Griffin, 1992), respectively, with most of the cholesterol in free form (Bitman and Wood, 1980; Kuksis, 1992). Riboflavin-binding protein associates with VTG-VHDL in order to be deposited into yolk (Mac Lachlan et al., 1994). Selective distribution of lipid classes during yolk formation is accomplished using an elaborate system of highly selective receptors (Schneider, 2016) and specialized lipoprotein ligands (Walzem et al., 1999). This system results in the deposition of TRL VLDLy to the yolk while CE-rich LDL is delivered to steroid hormone-producing granulosa, theca, and adrenal cells (Bacon et al., 1978; Hummel et al., 2003; Schneider, 2016). Recent studies found that ovarian tissues assemble generic (non-VLDLy) VLDL during egg production as a protective mechanism to secure follicle development and/or nutritional quality control of yolk for embryo development (Chen et al., 2025).

Low-Density Lipoproteins

LDL particles are the products of TRL metabolism. Importantly, LDL particles are depleted of TG and their lipid composition is extensively remodeled during circulation by intravascular enzymes. The enzymes involved in the remodeling process include the vascular lipases, lipid transfer proteins, and the intravascular cholesterol esterification protein lecithin cholesterol esterase (LCAT) (Walzem et al., 2012). The chicken LCAT prefers phosphatidyl cholines (lecithins) containing linoleate (18:2 n-6) or oleate (18:1 n-9) as fatty acid donors (Grove and Pownall, 1991).

Whether intravascular remodeling of TRL gives rise to LDL as a product is likely of less importance to production and bird health than the role it plays in linking lipoprotein-mediated nutrient delivery to peripheral tissues with the formation and metabolism of HDL. TRL hydrolysis creates an excess of surface PL as core TG diminishes (Bensadoun and Koh, 1977). The remodeling of TRL to IDL and ultimately LDL includes transfer of excess PL to HDL via phospholipid transfer protein (PLTP) (Behr et al., 1981; Guyard-Dangremont et al., 1998).

High-Density Lipoproteins

HDLs arise initially with the transfer of cellular PL and UC to the primary HDL apolipoprotein, Apo A1, by the efflux protein ABCA1 to form nascent HDL. The primary action ascribed to HDL is reverse lipid transport in which cholesterol in peripheral tissues is returned to the liver for reuse or biliary excretion (Oram and Lawn, 2001). This process is selectively invoked during molt (Barron et al., 1999), and defects in this process made White Carneau pigeons susceptible to atherosclerosis (Yancey and St Clair, 1994). However, an increasing number of studies in mammalian literature report that HDL or ApoA1 are also involved in antioxidant, anti-inflammatory, and anti-apoptotic actions and that HDL functionality may be more important to health benefits than HDL cholesterol concentration *per se* (Kontush and Chapman, 2006). These potential aspects of HDL functionality are unstudied in poultry.

Perhaps somewhat surprisingly, HDLs are important to the delivery of lutein and other carotenoids to the retina (Connor et al., 2007; Wang et al., 2007) and skin (Bitgood and Smyth, 1991) of poultry. There is competition among carotenoids for "space" on HDL, and high dietary levels of lutein will depress zeaxanthin and β-carotene content of HDL, while high supplementation rates of zeaxanthin or β-carotene will likewise reduce the concentrations of the other two carotenoids (Jia et al., 2010).

A dietary need for ALA is not proven for the fowl. However, ALA appears to be important in the development of specialized membranes found in the retina and nervous system (Stark et al., 2016). These membranes contain relatively high concentrations of n-3 PUFA that can originate from dietary ALA (Neuringer and Connor, 1986), albeit conversion of ALA to LC-PUFA is limited (Shahidi and Ambigaipalan, 2018). Substrate-level competition among fatty acids of the different families (i.e., n-9, n-6, and n-3) further complicate conversion rates (Gurr et al., 2002; Poureslami et al., 2010; Elkin et al., 2018).

It is increasingly recognized that lipids can serve signaling functions that control or modify immune and inflammatory processes (Chiurchiù et al., 2018). Several stimuli can induce the release of fatty acids from cell membranes through the action of phospholipases. Liberated LNA and ALA fatty acids can be desaturated and elongated to arachidonic acid (ARA, 20:4 n-6) and eicosapentaenoic acid (EPA, 20:5 n-3), respectively, which are the precursors for a class of bioactive lipid derivatives called eicosanoids. Eicosanoids can belong to prostaglandin (PG), prostacyclin (PGI), thromboxane (TX), and leukotriene (LT) families. Formation of eicosanoids is widespread in the body, and these hormone-like compounds affect nearly every physiological system. The eicosanoids are important in embryonic development, reproduction, immunological responses, and bone development in poultry (Watkins, 1991).

The potency of eicosanoids is dependent on the parent fatty acids. Eicosanoids derived from the n-6 LNA (e.g., PGE_2, LTB_4) have more potent inflammatory effects than the corresponding eicosanoids derived from the n-3 ALA (e.g., PGE_3, LTB_5) (Cherian, 2008, 2015). Therefore, feeding a diet enriched in n-3 PUFA increases the proportion of n-3 PUFA in cell membranes, which in turn tends to have an anti-inflammatory effect in the bird (Watkins, 1991). Elevating the n-3 PUFA content of the diet relative to that for n-6 PUFA alters eicosanoid production in immunocompetent cells (Kinsella et al., 1990). To maintain the full spectrum of eicosanoid effects in the body, a balanced intake of n-3 and n-6 PUFA is recommended (Simopoulos, 2002). Inflammation can also be provoked by other classes of bioactive lipids that include ceramides and sphingolipid metabolites (Kolak et al., 2007; Maceyka and Spiegel, 2014) that were demonstrated to accumulate in broiler breeder hens and negatively affect their health and productivity when inadequately feed restricted (Chen et al., 2006, 2017; Pan et al., 2012).

LIPID ANALYSIS

Categorization of Lipids

A complete lipid analysis encompasses the isolation, separation, identification, and determination of the absolute amount of each lipid type (Christie and Han, 2010). Fatty acid and cholesterol analyses are of primary concern due to the importance of these lipids to the nutritional quality of poultry products.

Fatty Acids

Methods for the extraction of fatty acids from a variety of sample matrices and their subsequent derivatization is the subject of several recent reviews (Christie and Han, 2010; Aldai et al., 2012; Pati et al., 2016; Lin et al., 2017; Brenna et al., 2018). However, because the content and chemical form of lipids in feedstuffs (Palmquist and Jenkins, 2003) and animal tissues are heterogeneous, no one extraction method is adequate under all circumstances. Moreover, analytical errors in fatty acid analysis of feedstuffs and digesta samples can result from the misuse of internal standards (Jenkins, 2010). A number of technologies are currently employed for fatty acid determinations, including gas chromatography (GC), GC-mass spectroscopy (MS) high-performance liquid chromatography (HPLC)-MS systems, and ultraHPLC-MS systems (for reviews see Aldai et al., 2012; Pati et al., 2016; Lin et al., 2017; Mossoba et al., 2017; Seppanen-Laakso et al., 2017).

Regardless of the analytical method used, the fatty acid contents of feedstuffs, complete mixed feeds, eggs, tissues, and organs should be reported as a weight unit of fatty acid per weight unit of sample (e.g., mg LNA/kg diet, mg LNA/100 g of fresh tissue, or mg LNA/g egg yolk). In addition, for eggs, yolk weight *per se* should also be reported to allow fatty acid contents to be expressed as a weight unit of fatty acid/whole yolk and to consider whether, in a research setting, the treatment being studied affected yolk weight. This can become important to estimating whether research outcomes can translate into nutritionally significant changes in amount per standardized edible egg as are reported in national food databases such FoodData Central (USDA, 2019). Expressing fatty acid data only as a percentage of total fatty acids is inadequate as the lipid content and/or weight of the egg yolk, tissue, or organ may be influenced by the experimental treatment. Therefore, if fatty acid composition data are reported as a percentage of total fatty acids in a sample, the lipid content of the egg yolk, tissue, organ, or feed sample must also be provided (Elkin, 2009). Finally, total fatty acids, even when accurately measured, provide limited information on a product's nutrient profile as fatty acids are typically part of different acyl-glycerol classes, such as TG versus PL, whose metabolic fates differ following consumption by humans.

Cholesterol

Eggs

Yolk cholesterol can be reliably determined by spectrophotometric or chromatographic methods when properly validated by inclusion of a standard reference material such as SRM1845a from the National Institute of Standards and Technology in each sample run (AOAC, 2016; Rudel and Morris, 1973). Egg cholesterol contents are best reported on both a relative basis (mg cholesterol/g yolk) and on a total basis (mg cholesterol/yolk) in conjunction with yolk weight as experimental treatments may alter egg or yolk weights. Seventy percent of per capita egg consumption in the United States is currently in the form of shell eggs (UERS, 2018); therefore, reporting cholesterol (and fatty acid) contents on a whole egg basis is of practical significance relative to human nutrition (Elkin, 2009).

Plasma and Tissues

Colorimetric, enzymatic, or chromatographic separation and detection methods can be used to analyze lipids in total plasma and tissues when properly standardized and validated.

The use of kits or autoanalyzers for serum/plasma or tissue analysis of lipids need to be validated for the specific analysis, and it may not be assumed that the use of kits for analyses of lipids in serum/plasma or tissues is validated for the measurement of lipids in all tissues (Rider et al., 2016). Additionally, the solvent used for extraction of lipids and the solvent used for the analysis of the kit must be compatible with the reactions in the kit and, secondly, the concentration range in the analyzed solution must be within the range for which the kit is used.

Lipophilic Pigments

Mass spectral methods are prferred for analysis and quantitation of lipophilic pigments (Maoka, 2023).

REFERENCES

Aldai, N., J.K.G. Kramer, C. Cruz-Hernandez, V. Santercole, P. Delmonte, M.M. Mossoba, and M.E.R. Dugan. 2012. Appropriate extraction and methylation techniques for lipids analysis. *Fat and Fatty Acids in Poultry Nutrition and Health.* 249–278.

Ao, T., L.M. Macalintal, M.A. Paul, A.J. Pescatore, A.H. Cantor, M.J. Ford, B. Timmons, and K.A. Dawson. 2015. Effects of supplementing microalgae in laying hen diets on productive performance, fatty-acid profile, and oxidative stability of eggs. *Journal of Applied Poultry Research.* 24(3):394–400.

AOAC International. 2016. *Official Methods of Analysis of AOAC International*, 20th ed. G.W. Latimer Jr., editor. Rockville, MD: AOAC International.

Bacon, W.L., B. Leclercq, and J.C. Blum. 1978. Difference in metabolism of very low density lipoprotein from laying chicken hens in comparison to immature chicken hens. *Poultry Science.* 57(6):1675–1686.

Balnave, D., and S.T.C. Weatherup. 1974. The necessity of supplementing laying hen diets with linoleic acid. *British Poultry Science.* 15(3):325–331.

Balnave, D. 1970. Essential fatty acids in poultry nutrition. *World's Poultry Science.* 26(1):442–460.

Barron, L.G., R.L. Walzem, and R.J. Hansen. 1999. Plasma lipoprotein changes in hens (Gallus domesticus) during an induced molt. *Comparative Biochemistry and Physiology Part B: Biochemistry and Molecular Biology.* 123(1):9–16.

Bartov, I., B. Lipstein, and S. Bornstein. 1974. Differential effects of dietary acidulated soybean oil soapstock, cottonseed oil soapstock and tallow on broiler carcass fat characteristics. *Poultry Science.* 53(1):115–124.

Barzegar, S., S.B. Wu, M. Choct, and R.A. Swick. 2020. Factors affecting energy metabolism and evaluating net energy of poultry feed. *Poultry Science.* 99(1):487–498.

Barzegar, S., S.B. Wu, J. Noblet, M. Choct, and R.A. Swick. 2019. Energy efficiency and net energy prediction of feed in laying hens. *Poultry Science.* 98(11):5746–5758.

Behr, S.R., J.R. Patsch, T. Forte, and A. Bensadoun. 1981. Plasma lipoprotein changes resulting from immunologically blocked lipolysis. *Journal of Lipid Research.* 22(3):443–451.

Bensadoun, A., and T.L. Koh. 1977. Identification of an adipose tissue-like lipoprotein lipase in perfusates of chicken liver. *Journal of Lipid Research.* 18(6):768–773.

Bensadoun, A., and A. Rothfeld. 1972. The form of absorption of lipids in the chicken, Gallus domesticus. *Proceedings of the Society for Experimental Biology and Medicine.* 141(3):814–817.

Bitgood, J.J., and J.R. Smyth Jr. 1991. Albinism in white Leghorn chickens. *Poultry Science.* 70(9):1861–1863.

Bitman, J., and D.L. Wood. 1980. Cholesterol and cholesteryl esters of eggs from various avian species. *Poultry Science.* 59(9):2014–2023.

Błaszczyk, A., A. Augustyniak, and J. Skolimowski. 2013. Ethoxyquin: An antioxidant used in animal feed. *International Journal of Food Science.* 31:222–235.

Boyle-Roden, E., and R.L. Walzem. 2005. Integral apolipoproteins increase surface-located triacylglycerol in intact native ApoB-100-containing lipoproteins. *Journal of Lipid Research.* 46(8):1624–1632.

Breithaupt, D.E. 2007. Modern application of xanthophylls in animal feeding—a review. *Trends in Food Science & Technology.* 18(10):503–506.

Brenna, J.T., M. Plourde, K.D. Stark, P.J. Jones, and Y.H. Lin. 2018. Best practices for the design, laboratory analysis, and reporting of trials involving fatty acids. *The American Journal of Clinical Nutrition.* 108(2):211–227.

Burley, R.W., A.J. Evans, and J.A. Pearson. 1993. Molecular aspects of the synthesis and deposition of hens' egg yolk with special reference to low density lipoprotein. *Poultry Science.* 72(5):850–855.

Calvert, C.C. 1967. Studies on hatchability of fertile eggs from hens receiving a linoleic acid deficient diet. *Poultry Science.* 46(4):967–973.

Chapman, M.J. 1980. Animal lipoproteins: Chemistry, structure, and comparative aspects. *Journal of Lipid Research.* 21(7):789–853.

Chen, C.Y., Y.F. Huang, Y.J. Ko, Y.J. Liu, Y.H. Chen, R.L. Walzem, and S.E. Chen. 2017. Obesity-associated cardiac pathogenesis in broiler breeder hens: Development of metabolic cardiomyopathy. *Poultry Science.* 96(7):2438–2446.

Chen, S.E., J.P. McMurtry, and R.L. Walzem. 2006. Overfeeding-induced ovarian dysfunction in broiler breeder hens is associated with lipotoxicity. *Poultry Science*. 85(1):70–81.

Chen, Y.H., P. Dettipponpong, M.Y. Sin, L.C. Chang, C.Y. Cheng, S.Y. Huang, R.L. Walzem, H.C. Cheng, and S.E. Chen. 2025. Ovarian expression of functional MTTP and ApoB for VLDL assembly and secretion in chickens. *Poultry Science*. 104(5):104993.

Cherian, G. 2015. Nutrition and metabolism in poultry: Role of lipids in early diet. *Journal of Animal Science and Biotechnology*. 6(1):1–9.

Cherian, G. 2008. Egg quality and yolk polyunsaturated fatty acid status in relation to broiler breeder hen age and dietary n-3 oils. *Poultry Science*. 87(6):1131–1137.

Chiurchiù, V., A. Leuti, and M. Maccarrone. 2018. Bioactive lipids and chronic inflammation: Managing the fire within. *Frontiers in Immunology*. 9:38.

Christie, W.W., and X. Han. 2010. *Lipid Analysis—Isolation, Separation, Identification and Lipidomic Analysis*, 4th ed. Bridgwater, UK: The Oily Press.

Clifford, A.J., and H.G. Müller, eds. 2013. *Mathematical Modeling in Experimental Nutrition*. Vol. 445. Springer Science & Business Media.

Combs Jr., G.F., and J.P. McClung. 2016. *The Vitamins: Fundamental Aspects in Nutrition and Health*. New York: Academic Press.

Connor, W.E., P.B. Duell, R. Kean, and Y. Wang. 2007. The prime role of HDL to transport lutein into the retina: Evidence from HDL-deficient WHAM chicks having a mutant ABCA1 transporter. *Investigative Ophthalmology & Visual Science*. 48(9):4226–4231.

Deeley, R.G., D.P. Mullinix, W. Wetekam, H.M. Kronenberg, M. Meyers, J.D. Eldridge, and R.F. Goldberger. 1975. Vitellogenin synthesis in the avian liver: Vitellogenin is the precursor of the egg yolk phosphoproteins. *Journal of Biological Chemistry*. 250(23):9060–9066.

Edwards Jr., H.M., F. Denman, A. Abou-Ashour, and D. Nugara. 1973. Carcass composition studies: 1. Influences of age, sex and type of dietary fat supplementation on total carcass and fatty acid composition. *Poultry Science*. 52(3):934–948.

Elkin, R.G. 2009. Additional perspectives on analytical techniques and standardization: Cholesterol and fatty acid contents of eggs, tissues, and organs. *Poultry Science*. 88(2):249.

Elkin, R.G., and Z. Yan. 1999. Relationship between inhibition of mevalonate biosynthesis and reduced fertility in laying hens. *Journal of Reproduction and Fertility*. 116:269–275.

Elkin, R.G., A.N. Kukorowski, Y. Ying, and K.J. Harvatine. 2018. Dietary high-oleic acid soybean oil dose dependently attenuates egg yolk content of n-3 polyunsaturated fatty acids in laying hens fed supplemental flaxseed oil. *Lipids*. 53(2):235–249.

Fahy, E., S. Subramaniam, H.A. Brown, C.K. Glass, A.H. Merrill, R.C. Murphy, C.R. Raetz, D.W. Russell, Y. Seyama, W. Shaw, and T.A. Shimizu. 2005. A comprehensive classification system for lipids. *Journal of Lipid Research*. 46(5):839–861.

Fahy, E., S. Subramaniam, R.C. Murphy, M. Nishijima, C.R.H. Raetz, T. Shimizu, F. Spener, G.V. Meer, M.J.O. Wakelam, and E.A. Dennis. 2009. Update of the LIPID MAPS comprehensive classification system for lipids. *Journal of Lipid Research*. 50:S9–S14.

Fletcher, D.L. 1989. Factors influencing pigmentation in poultry. *Poultry Biology*. 2:149–169.

Fox, D.L. 1979. Pigment transactions betwen animals and plants. *Biology Review*. 54:237–268.

Gargouri, Y., R. Julien, A.G. Bois, R. Verger, and L. Sarda. 1983. Studies on the detergent inhibition of pancreatic lipase activity. *Journal of Lipid Research*. 24(10):1336–1342.

Golian, A., and D.V. Maurice. 1992. Dietary poultry fat and gastrointestinal transit time of feed and fat utilization in broiler chickens. *Poultry Science*. 71(8):1357–1363.

Gregory, M.K., and M.J. James. 2014. Functional characterization of the duck and turkey fatty acyl elongase enzymes ELOVL5 and ELOVL2. *The Journal of Nutrition*. 144(8):1234–1239.

Gregory, M.K., M.S. Geier, R.A. Gibson, and M.J. James. 2013. Functional characterization of the chicken fatty acid elongases. *The Journal of Nutrition*. 143(1):12–16.

Griffin, H.D. 1992. Manipulation of egg yolk cholesterol: A physiologist's view. *World's Poultry Science*. 48(2):101–112.

Griffin, H., G. Grant, and M. Perry. 1982. Hydrolysis of plasma triacylglycerol-rich lipoproteins from immature and laying hens (Gallus domesticus) by lipoprotein lipase in vitro. *Biochemical Journal*. 206(3):647–654.

Griffin, H.D., K. Guo, D. Windsor, and S.C. Butterwith. 1992. Adipose tissue lipogenesis and fat deposition in leaner broiler chickens. *The Journal of Nutrition*. 122(2):363–368.

Grobas, S., J. Mendez, C. De Blas, and G.G. Mateos. 1999a. Influence of dietary energy, supplemental fat and linoleic acid concentration on performance of laying hens at two ages. *British Poultry Science*. 40(5):681–687.

Grobas, S., J. Mendez, C. De Blas, and G.G. Mateos. 1999b. Laying hen productivity as affected by energy, supplemental fat, and linoleic acid concentration of the diet. *Poultry Science*. 78(11):1542–1551.

Grove, D., and H.J. Pownall. 1991. Comparative specificity of plasma lecithin: Cholesterol acyltransferase from ten animal species. *Lipids*. 26(6):416–420.

Gruffat, D., D. Durand, B. Graulet, and D. Bauchart. 1996. Regulation of VLDL synthesis and secretion in the liver. *Reproduction Nutrition Development*. 36(4):375–389.

Guenter, W., D.B. Bragg, and P.A. Kondra. 1971. Effect of dietary linoleic acid on fatty acid composition of egg yolk, liver and adipose tissue. *Poultry Science*. 50(3):845–850.

Gunstone, F., ed. 2011. *Vegetable Oils in Food Technology: Composition, Properties and Uses*. Hoboken, NJ: John Wiley & Sons.

Gurr, M.I., J.L. Harwood, and K.N. Frayn. 2002. *Lipid Biochemistry*. Vol. 409. Oxford, UK: Blackwell Science.

Guyard-Dangremont, V., C. Desrumaux, P. Gambert, C. Lallemant, and L. Lagrost. 1998. Phospholipid and cholesteryl ester transfer activities in plasma from 14 vertebrate species. Relation to atherogenesis susceptibility. *Comparative Biochemistry and Physiology Part B: Biochemistry and Molecular Biology*. 120(3):517–525.

Hargis, P.S. 1988. Modifying egg yolk cholesterol in the domestic fowl: A review. *World Poultry Science Journal*. 44:17–29.

Harwood, J.L., M.I. Gurr, K.N. Frayn, D.J. Murphy, and R.H. Michell. 2016. *Lipids: Biochemistry, Biotechnology and Health*. Hoboken, NJ: John Wiley & Sons.

Hencken, H. 1992. Chemical and physiological behavior of feed carotenoids and their effects on pigmentation. *Poultry Science*. 71(4):711–717.

Hermier, D. 1997. Lipoprotein metabolism and fattening in poultry. *The Journal of Nutrition*. 127(5):805S–808S.

Hermier, D., P. Forgez, J. Williams, and M.J. Chapman. 1989. Alterations in plasma lipoproteins and apolipoproteins associated with estrogen-induced hyperlipidemia in the laying hen. *European Journal of Biochemistry*. 184(1):109–118.

Hillgartner, F. Bradley, L.M. Salati, and A.G. Goodridge. 1995. Physiological and molecular mechanisms involved in nutritional regulation of fatty acid synthesis. *Physiological Reviews*. 75(1):47–76.

Hofmann, A.F., and B. Borgstrom. 1962. Physico-chemical state of lipids in intestinal content during their digestion and absorption. In *Federation Proceedings*. 21(1):43–50.

Hummel, S., E.G. Lynn, A. Osanger, S. Hirayama, J. Nimpf, and W.J. Schneider. 2003. Molecular characterization of the first avian LDL receptor: Role in sterol metabolism of ovarian follicular cells. *Journal of Lipid Research*. 44(9):1633–1642.

Jenkins, T.C. 2010. Common analytical errors yielding inaccurate results during analysis of fatty acids in feed and digesta samples. *Journal of Dairy Science*. 93(3):1170–1174.

Jia, Y., S.Y. Hwang, J.D. House, M.R. Ogborn, H.A. Weiler, K. O, and H.M. Aukema. 2010. Long-term high intake of whole proteins results in renal damage in pigs. *The Journal of Nutrition*. 140(9):1646–1652.

Kadambini, T. 2011. Lipidomics: A promising area in "omics" research. *Journal of Computer Science and Systems Biology*. 4:93–98.

Kanai, M., T. Soji, E. Sugawara, N. Watari, H. Oguchi, M. Matsubara, and D.C. Herbert. 1996. Participation of endodermal epithelial cells on the synthesis of plasma LDL and HDL in the chick yolk sac. *Microscopy Research and Technique*. 35(4):340–348.

Keshavarz, K., and S. Nakajima. 1995. The effect of dietary manipulations of energy, protein, and fat during the growing and laying periods on early egg weight and egg components. *Poultry Science*. 74(1):50–61.

Kieft, K.A., T.M. Bocan, and B.R. Krause. 1991. Rapid on-line determination of cholesterol distribution among plasma lipoproteins after high-performance gel filtration chromatography. *Journal of Lipid Research*. 32(5):859–866.

Kinsella, J.E., B. Lokesh, S. Broughton, and J. Whelan. 1990. Dietary polyunsaturated fatty acids and eicosanoids: Potential effects on the modulation of inflammatory and immune cells: An overview. *Nutrition (Burbank)*. 6(1):24–44.

Klasing, K.C. 1998. *Comparative Avian Nutrition*. Wallingford, Oxfordshire, United Kingdom: CAB International.

Kolak, M., J. Westerbacka, V.R. Velagapudi, D. Wagsater, L. Yetukuri, J. Makkonen, A. Rissanen, A.M. Hakkinen, M. Lindell, R. Bergholm, and A. Hamsten. 2007. Adipose tissue inflammation and increased ceramide content characterize subjects with high liver fat content independent of obesity. *Diabetes*. 56(8):1960–1968.

Kontush, A., and M.J. Chapman. 2006. Functionally defective high-density lipoprotein: A new therapeutic target at the crossroads of dyslipidemia, inflammation, and atherosclerosis. *Pharmacological Reviews*. 58(3):342–374.

Krogdahl, Å. 1985. Digestion and absorption of lipids in poultry. *The Journal of Nutrition*. 115(5):675–685.

Kuksis, A. 1992. Yolk lipids. *Biochimica et Biophysica Acta (BBA)—Lipids and Lipid Metabolism*. 1124(3):205–222.

Lazier, C.B., M. Wiktorowicz, G.E. DiMattia, D.A. Gordon, R. Binder, and D. L. Williams. 1994. Apolipoprotein (Apo) B and ApoII gene expression are both estrogen-responsive in chick embryo liver but only ApoII is estrogen-responsive in kidney. *Molecular and Cellular Endocrinology*. 106(1–2):187–194.

Leveille, G.A., D.R. Romsos, Y.Y. Yeh, and E.K. O'Hea. 1975. Lipid biosynthesis in the chick: A consideration of site of synthesis, influence of diet and possible regulatory mechanisms. *Poultry Science*. 54(4):1075–1093.

Library, N.A. 2018. *National Nutrient Database for Standard Reference*, v.3.9.4.4.

Lillie, R.J., and H. Menge. 1968. Effect of linoleic acid deficiency on the fertilizing capacity and semen fatty acid profile of the male chicken. *The Journal of Nutrition*. 95(2):311–315.

Lin, Y. 2017. Mass spectrometry in the analysis of fatty acids and derivatives. In *Fatty Acids—Chemistry, Synthesis, and Applications*, M.U. Ahmad, editor. San Diego, CA: Academic Press and AOCS Press. 529–540.

MacLachlan, I., J. Nimpf, and W.J. Schneider. 1994. Avian riboflavin binding protein binds to lipoprotein receptors in association with vitellogenin. *Journal of Biological Chemistry*. 269(39):24127–24132.

Maceyka, M.L., and S. Spiegel. 2014. Sphingolipid metabolites in inflammatory disease. *Nature*. 510(7503):58–67.

Mann, C.J., T.A. Anderson, J. Read, S.A. Chester, G.B. Harrison, S. Köchl, P.J. Ritchie, P. Bradbury, F.S. Hussain, J. Amey, and B. Vanloo. 1999. The structure of vitellogenin provides a molecular model for the assembly and secretion of atherogenic lipoproteins. *Journal of Molecular Biology*. 285(1):391–408.

Maoka, T. 2023. Carotenoids: Distribution, function in nature, and analysis using LC-photodiode array detector (DAD)-MS and MS/MS system. *Mass Spectrometry (Tokyo)*. 12:A0133.

March, B.E., and C. MacMillan. 1990. Linoleic acid as a mediator of egg size. *Poultry Science*. 69(4):634–639.

Marounek, M., and A. Pebriansyah. 2018. Use of carotenoids in feed mixtures for poultry: A review. *Agricultura Tropica et Subtropica*. 51(3):107–111.

Marusich, W.L. and J.C. Bauernfeind. 1970. Oxycarotenoids in poultry pigmentation 1. Yolk studies. *Poultry Science*. 49:1555–1566.

Mateos, G.G., L. Cámara, G. Fondevila, and R.P. Lázaro. 2019. Critical review of the procedures used for estimation of the energy content of diets and ingredients in poultry. *Journal of Applied Poultry Research*. 28(3):506–525.

Mateos, G.G., J.L. Sell, and J.A. Eastwood. 1982. Rate of food passage (transit time) as influenced by level of supplemental fat. *Poultry Science*. 61(1):94–100.

Menge, H., C.C. Calvert, and C.A. Denton. 1965. Further studies of the effect of linoleic acid on reproduction in the hen. *The Journal of Nutrition*. 86(2):115–119.

Meriwether, L., B. Humphrey, D. Peterson, K. Klasing, and E. Koutsos. 2010. Lutein exposure, in ovo or in the diet, reduces parameters of inflammation in the liver and spleen laying-type chicks (Gallus gallus domesticus). *Journal of Animal Physiology and Animal Nutrition*. 94(5):e115–e122.

Moran Jr., E.T. 1982. Production and carcass quality responses of early and late marketed large toms to added dietary fat during the finishing period. *Poultry Science*. 61(5):919–924.

Moran Jr., E.T., E. Larmond, and J. Somers. 1973. Full-fat soybeans for growing and finishing large white turkeys: 2. Effect on tissue fatty acids and organoleptic evaluation. *Poultry Science*. 52(5):1942–1948.

Mossab, A., J.M. Hallouis, and M. Lessire. 2000. Utilization of soybean oil and tallow in young turkeys compared with young chickens. *Poultry Science*. 79(9):1326–1331.

Mossoba, M.M., S.R. Karunathilaka, J.K. Chung, and C.T. Srigley. 2017. Advancement in chromatographic and spectroscopic analyses of dietary fatty acids. In *Fatty Acids*. Champaign, IL: AOCS Press. 505–528.

Na, J.-C., J.-Y. Song, B.-D. Lee, S.-J. Lee, C.-Y. Lee, and G.-H. An. 2004. Effect of polarity on absorption and accumulation of carotenoids by laying hens. *Animal Feed Science and Technology*. 117(3–4):305–315.

Neuringer, M., and W.E. Connor. 1986. n-3 fatty acids in the brain and retina: Evidence for their essentiality. *Nutrition Reviews*. 44(9):285–294.

Noble, R.C., and M. Cocchi. 1990. Lipid metabolism and the neonatal chicken. *Progress in Lipid Research*. 29:107–140.

Noyan, A., W.J. Lossow, N. Brot, and I.L. Chaikoff. 1964. Pathway and form of absorption of palmitic acid in the chicken. *Journal of Lipid Research*. 5(4):538–541.

NRC (National Research Council). 1994. *Nutrient Requirements of Poultry*, 9th rev. ed. Washington, DC: National Academy Press.

Oram, J.F., and R.M. Lawn. 2001. ABCA1: The gatekeeper for eliminating excess tissue cholesterol. *Journal of Lipid Research*. 42(8):1173–1179.

Palmquist, D.L., and T.C. Jenkins. 2003. Challenges with fats and fatty acid methods. *Journal of Animal Science*. 81(12):3250–3254.

Pan, Y.E., Z.C. Liu, C.J. Chang, Y.L. Xie, C.Y. Chen, C.F. Chen, R.L. Walzem, and S.E. Chen. 2012. Ceramide accumulation and up-regulation of proinflammatory interleukin-1β exemplify lipotoxicity to mediate declines of reproductive efficacy of broiler hens. *Domestic Animal Endocrinology*. 42(3):183–194.

Park, J.H., S.D. Upadhaya, and I.H. Kim. 2015. Effect of dietary marine microalgae (Schizochytrium) powder on egg production, blood lipid profiles, egg quality, and fatty acid composition of egg yolk in layers. *Asian-Australasian Journal of Animal Sciences*. 28(3):391–397.

Parks, E.J. 2002. Changes in fat synthesis influenced by dietary macronutrient content *Proceedings of the Nutrition Society*. 61(2):281–286.

Pati, S., B. Nie, R.D. Arnold, and B.S. Cummings. 2016. Extraction, chromatographic and mass spectrometric methods for lipid analysis. *Biomedical Chromatography*. 30(5):695–709.

Perry, M.M., H.D. Griffin, and A.B. Gilbert. 1984. The binding of very low density and low density lipoproteins to the plasma membrane of the hen's oocyte: A morphological study. *Experimental Cell Research*. 151(2):433–446.

Pesti, G.M., R.I. Bakalli, M. Qiao, and K.G. Sterling. 2002. A comparison of eight grades of fat as broiler feed ingredients. *Poultry Science*. 81(3):382–390.

Poławska, E. 2018. Changes in fat and individual fatty acids digestibility during the growth of Japanese quails. *Animal Science Papers and Reports*. 36(4):421–429.

Poureslami, R., K. Raes, G.M. Turchini, G. Huyghebaert, and S. De Smet. 2010. Effect of diet, sex and age on fatty acid metabolism in broiler chickens: n-3 and n-6 PUFA. *British Journal of Nutrition*. 104(2):189–197.

Ravindran, V., P. Tancharoenrat, F. Zaefarian, and G. Ravindran. 2016. Fats in poultry nutrition: Digestive physiology and factors influencing their utilisation. *Animal Feed Science and Technology*. 213:1–21.

Resnyk, C.W., W. Carré, X. Wang, T.E. Porter, J. Simon, L. Bihan-Duval, M.J. Duclos, S.E. Aggrey, and L.A. Cogburn. 2017. Transcriptional analysis of abdominal fat in chickens divergently selected on bodyweight at two ages reveals novel mechanisms controlling adiposity: Validating visceral adipose tissue as a dynamic endocrine and metabolic organ. *BMC Genomics*. 18(1):1–31.

Rider, T., R.C. LeBoeuf, P. Tso, and R.J. Jandacek. 2016. The use of kits in the analysis of tissue lipids requires validation. *Lipids*. 51(4):497–504.

Rochell, S.J., T.J. Applegate, E.J. Kim, and W.A. Dozier III. 2012. Effects of diet type and ingredient composition on rate of passage and apparent ileal amino acid digestibility in broiler chicks. *Poultry Science*. 91(7):1647–1653.

Rodriguez-Sanchez, R., A. Tres, R. Sala, C. Garcés-Narro, F. Guardiola, J. Gasa, and A.C. Barroeta. 2019. Effects of dietary free fatty-acid content and saturation degree on lipid-class composition and fatty-acid digestibility along the gastrointestinal tract in broiler starter chickens. *Poultry Science*. 98(10):4929–4941.

Rosebrough, R.W., and N.C. Steele. 1985. Effect of dietary energy on lipogenesis in turkey hens prior to and after sexual maturity. *Growth*. 49(2):218–228.

Rosebrough, R.W., and N.C. Steele. 1982. The role of carbohydrate and protein level in the regulation of lipogenesis by the turkey poult. *Poultry Science*. 61(11):2212–2217.

Rudel, L.L., and M.D. Morris. 1973. Determination of cholesterol using o-phthalaldehyde. *Journal of Lipid Research*. 14(3):364–366.

Saadoun, A., and B. Leclercq. 1987. In vivo lipogenesis of genetically lean and fat chickens: Effects of nutritional state and dietary fat. *The Journal of Nutrition*. 117(3):428–435.

Safaa, H.M., M.P. Serrano, D.G. Valencia, X. Arbe, E. Jiménez-Moreno, R. Lázaro, and G.G. Mateos. 2008. Effects of the levels of methionine, linoleic acid, and added fat in the diet on productive performance and egg quality of brown laying hens in the late phase of production. *Poultry Science*. 87(8):1595–1602.

Salas, C., R.D. Ekmay, J.A. England, S. Cerrate, and C.N. Coon. 2017. Mechanisms of lipid mobilization towards egg formation in broiler breeder hens using stable isotopes. *Poultry Science*. 96(2):383–387.

Salmon, R.E., and J.B. O'Neil. 1973. The effect of the level and source and of a change of source of dietary fat on the fatty acid composition of the depot fat and the thigh and breast meat of turkeys as related to age. *Poultry Science*. 52(1):302–314.

Schneider, W.J. 2016. Lipid transport to avian oocytes and to the developing embryo. *Journal of Biomedical Research*. 30(3):174.

Schneider, W.J., R. Carroll, D.L. Severson, and J. Nimpf. 1990. Apolipoprotein VLDL-II inhibits lipolysis of triglyceride-rich lipoproteins in the laying hen. *Journal of Lipid Research*. 31(3):507–513.

Schuler, G.A., and E.O. Essary. 1971. Fatty acid composition of lipids from broilers fed saturated and unsaturated fats. *Journal of Food Science*. 36(3):431–434.

Sell, J.L., A. Krogdahl, and N. Hanyu. 1986. Influence of age on utilization of supplemental fats by young turkeys. *Poultry Science*. 65(3):546–554.

Sell, J.L., and W.J. Owings. 1984. Influence of feeding supplemental fat by age sequence on the performance of growing turkeys. *Poultry Science*. 63(6):1184–1189.

Seppänen-Laakso, T., H. Nygren, and H. Rischer. 2017. UPLC-ELSD analysis of algal lipid classes and derivatization of bound and free fatty acids and sterols for GC-MS methods. In *Biofuels from Algae*. New York: Humana. 223–232.

Shahidi, F., and P. Ambigaipalan. 2018. Omega-3 polyunsaturated fatty acids and their health benefits. *Annual Review of Food Science and Technology*. 9:345–381.

Sibbald, I.R., and J.K.G. Kramer. 1980. The effects of fractions of yellow corn on the true metabolizable energy value of beef tallow. *Poultry Science*. 59(7):1505–1509.

Sim, J.S., D.B. Bragg, and G.C. Hodgson. 1973. Effect of dietary animal tallow and vegetable oil on fatty acid composition of egg yolk, adipose tissue and liver of laying hens. *Poultry Science*. 52(1):51–57.

Simopoulos, A.P. 2002. Omega-3 fatty acids in inflammation and autoimmune diseases. *Journal of the American College of Nutrition*. 21(6):495–505.

Siyal, F.A., D. Babazadeh, C. Wang, M.A. Arain, M. Saeed, T. Ayasan, L. Zhang, and T. Wang. 2017. Emulsifiers in the poultry industry. *World's Poultry Science Journal*. 73(3):611–620.

Speake, B.K. 1998. Transport and transformations of yolk lipids during development of the avian embryo. *Progress in Lipid Research*. 37:1–32.

Stark, A.H., R. Reifen, and M.A. Crawford. 2016. Past and present insights on alpha-linolenic acid and the omega-3 fatty acid family. *Critical Reviews in Food Science and Nutrition*. 56(14):2261–2267.

Stifani, S., D.L. Barber, J. Nimpf, and W.J. Schneider. 1990. A single chicken oocyte plasma membrane protein mediates uptake of very low density lipoprotein and vitellogenin. *Proceedings of the National Academy of Sciences*. 87(5):1955–1959.

Sturkie, P. 2014. *Sturkie's Avian Physiology*. London, UK; Waltham, MA: Academic Press; Elsevier.

Summers, J.D. 1984. Extra caloric value of fats in poultry diets. In *Proceedings—Easter School in Agricultural Science, University of Nottingham*.

Sunde, M. 1992. The scientific way to pigment poultry products: Introduction to the symposium. *Poultry Science*. 71(4):709–710.

Surai, P., B. Speake, and N. Sparks. 2001. Carotenoids in avian nutrition and embryonic development. 1. Absorption, availability and levels in plasma and egg yolk. *The Journal of Poultry Science*. 38(1):1–27.

Tancharoenrat, P., V. Ravindran, F. Zaefarian, and G. Ravindran. 2013. Influence of age on the apparent metabolisable energy and total tract apparent fat digestibility of different fat sources for broiler chickens. *Animal Feed Science and Technology*. 186(3–4):186–192.

Tancharoenrat, P., V. Ravindran, F. Zaefarian, and G. Ravindran. 2014. Digestion of fat and fatty acids along the gastrointestinal tract of broiler chickens. *Poultry Science*. 93(2):371–379.

Tarugi, P., G. Ballarini, B. Pinotti, A. Franchini, E. Ottaviani, and S. Calandra. 1998. Secretion of ApoB- and ApoA-I-containing lipoproteins by chick kidney. *Journal of Lipid Research*. 39(4):731–743.

UERS (USDA Economic Research Service). 2018. Eggs: Per capita availability. Food Availability (Per Capita) Data System.

U.S. Department of Agriculture, Agricultural Research Service. 2019. *FoodData Central*. fdc.nal.usda.gov.

Valencia, M.E., P.M. Maiorino, and B.L. Reid. 1980. Energy utilization by laying hens.: II. Energetic efficiency and added tallow at 18.3 and 35 C. *Poultry Science*. 59(9):2071–2076.

Vilchez, C., S.P. Touchburn, E.R. Chavez, and C.W. Chan. 1990. The influence of supplemental corn oil and free fatty acids on the reproductive performance of Japanese quail (Coturnix coturnix japonica). *Poultry Science*. 69(9):1533–1538.

Vinado, A., L. Castillejos, and A.C. Barroeta. 2020. Soybean lecithin as an alternative energy source for grower and finisher broiler chickens: Impact on performance, fatty acid digestibility, gut health, and abdominal fat saturation degree. *Poultry Science*. 99:5653–5662.

Walzem, R.L., Cherian, G., and R. Poureslami, eds. 2012. Fats and fatty acids in poultry nutrition and health. *Context*.

Walzem, R.L., R.A. Baillie, M. Wiest, R. Davis, S.M. Watkins, T.E. Porter, J. Simon, and L.A. Cogburn. 2007. Functional annotation of genomic data with metabolic inference. *Poultry Science*. 86(7):1510–1522.

Walzem, R.L., R.J. Hansen, D.L. Williams, and R.L. Hamilton. 1999. Estrogen induction of VLDLy assembly in egg-laying hens. *The Journal of Nutrition*. 129(2):467S–472S.

Walzem, R.L. 1996. Lipoproteins and the laying hen: Form follows function. *Poultry and Avian Biology Reviews*. 7:31–64.

Wang, Y., S.L. Connor, W. Wang, E.J. Johnson, and W.E. Connor. 2007. The selective retention of lutein, meso-zeaxanthin and zeaxanthin in the retina of chicks fed a xanthophyll-free diet. *Experimental Eye Research*. 84(3):591–598.

Wang, S., D.E. Smith, and D.L. Williams. 1983. Purification of avian vitellogenin III: comparison with vitellogenins I and II. *Biochemistry*. 22(26):6206–6212.

Watkins, B.A. 1991. Importance of essential fatty acids and their derivatives in poultry. *The Journal of Nutrition*. 121(9):1475–1485.

Whitehead, C.C. 1981. The response of egg weight to the inclusion of different amounts of vegetable oil and linoleic acid in the diet of laying hens. *British Poultry Science*. 22(6):525–532.

Williams, T.D., W.L. Reed, and R.L. Walzem, 2002. Egg size variation: mechanisms and hormonal control. In *Avian Endocrinology*. A. Dawson and C.M. Chaturvedi, Editors. New Delhi, India: Narosa. 458.

Wiskocil, R., P. Goldman, and R.G. Deeley. 1981. Cloning and structural characterization of an estrogen-dependent apolipoprotein gene. *Journal of Biological Chemistry*. 256(18):9662–9667.

Wiskocil, R., P. Bensky, W. Dower, R.F. Goldberger, J.I. Gordon, and R.G. Deeley. 1980. Coordinate regulation of two estrogen-dependent genes in avian liver. *Proceedings of the National Academy of Sciences*. 77(8):4474–4478.

Wong, T.K., and W.J. Lennarz. 1982. Biosynthesis of dolichol and cholesterol during embryonic development of the chicken. *Biochimica et Biophysica Acta (BBA)—Lipids and Lipid Metabolism*. 710:32–38.

Wu, S.B., R.A. Swick, J. Noblet, N. Rodgers, D. Cadogan, and M. Choct. 2019a. Net energy prediction and energy efficiency of feed for broiler chickens. *Poultry Science*. 98(3):1222–1234.

Wu, S.B., R.A. Swick, J. Noblet, N. Rodgers, D. Cadogan, and M. Choct. 2019b. Corrigendum to "Net energy prediction and energy efficiency of feed for broiler chickens." *Poultry Science*. 98(3):1222–1234.

Yancey, P.G., and R.W. St Clair. 1994. Mechanism of the defect in cholesteryl ester clearance from macrophages of atherosclerosis-susceptible White Carneau pigeons. *Journal of Lipid Research*. 35(12):2114–2129.

Zornig, W.O., G.M. Pesti, and R.I. Bakalli. 2001. The essential fatty acid requirements of broilers. *Journal of Applied Poultry Research*. 10(1):41–45.

5

Vitamins

INTRODUCTION

Vitamins are a diverse group of organic molecules found naturally in foods and required in small amounts by poultry. Although related functions exist between several of the vitamins, they are grouped together as a nutrient category by virtue of their dissimilarity from other nutrient categories (water, carbohydrates, lipids, proteins, and minerals). Vitamins are considered essential in poultry when one or more specific functions has been demonstrated, when that function is impaired when the diet is lacking, and when normal function is restored upon supplementation (Combs and McClung, 2017a). Since multiple chemically-related compounds may share a function, albeit with different efficacy, a unit conversion table for relevant vitamins is provided (Table 5-1). Although the dietary levels of vitamins required by poultry are very low, moderate deficiencies are often more debilitating than those of protein/amino acids, energy, or fat (Klasing, 1998a). Biochemical and physical measurements for the diagnosis of vitamin deficiencies in chickens and turkeys are summarized in Table 5-2. In breeder hens, marginal deficiencies of vitamins may not decrease egg production but are likely to negatively influence embryonic development and viability (Table 5-3).

Vitamins are divided into two categories: fat-soluble and water-soluble. Much of our understanding of vitamin nutrition and requirements is derived from studies conducted from 1930 to 1970, and with a few notable exceptions (e.g., some of the fat-soluble vitamins), few vitamin requirement studies have been conducted in recent years. As such, much of our recent knowledge has been derived from the mammalian literature, with assumptions that in many regards, avian absorption and metabolism of vitamins reflects that of mammals. In many cases, properly designed and controlled vitamin requirement studies have not been completed since the previous edition of this report (NRC, 1994), and therefore very few changes to nutrient requirements are recommended. The impact of genetic selection of poultry for productivity and efficiency traits on vitamin requirements is largely unknown (Leeson, 2007). Research is also needed to determine vitamin requirements in the context of practical commercial-type diets.

Vitamin requirements for poultry have traditionally been set based on maximizing production criteria (NRC, 1994). In commercial poultry production, vitamins are typically supplemented as a premix, and due to ingredient- and storage-related variability in their concentration and bioavailability, contributions from the other dietary ingredients are not normally considered during feed formulation. In the past, vitamin and trace mineral premixes were often added separately because of the negative impact of trace minerals, in particular divalent metal cations, on vitamin stability (Zhuge and Klopfenstein, 1986; Yang et al., 2021). The presence of choline chloride in the premix reduced the stability of vitamins A, E, and K and water-soluble vitamins (Tavčar-Kalcher and Vengušt, 2007; Yang et al., 2021). However, modern encapsulation technology that protects vitamins from interactions with other feed components (Tolve et al., 2021; Wedergaertner et al., 2020) has led to the widespread use of combined vitamin–trace mineral premixes. Dietary vitamins are generally well-tolerated, and supplemental levels of most vitamins in commercial diets are normally at levels many times in excess of the minimum requirement because of the minimal cost relative to many of the macronutrients, the importance of avoiding deficiencies, and the wide margin between the minimum requirement and toxic levels. The approach of the committee has been to base requirement values on the minimum levels necessary to obtain specific production criteria. As such, these values will typically be lower than those commonly used in industry or recommended by the poultry breeding companies, in which case substantial safety margins are included in the diet.

Substantial increases in production efficiency have resulted in a lower amount of feed consumed per unit of egg or meat output. Therefore, if consistent levels of vitamin supplementation over time are used, the result will be a reduction in vitamin intake per unit of egg or meat output. It was estimated that

TABLE 5-1 Vitamin Unit Conversions

Vitamin	Unit	Conversion	Notes	Reference[a]
Vitamin A	IU	0.3 μg retinol (crystalline vitamin A alcohol)	1 mg retinol = 3,333 IU of vitamin A for poultry	(NRC, 1994)
	IU	0.344 μg vitamin A acetate	1 mg vitamin A acetate = 2,907 IU of vitamin A for poultry	(NRC, 1994)
	IU	0.55 μg vitamin A palmitate	1 mg vitamin A palmitate = 1,818 IU of vitamin A for poultry	(NRC, 1994)
	IU	0.6 μg β-carotene	1 mg β-carotene = 1,667 IU of vitamin A for poultry	(NRC, 1994)
Vitamin D_3	IU	0.025 μg vitamin D_3	1 μg = 40 IU vitamin D activity	(NRC, 1994)
25-hydroxycholecalciferol	IU	0.025 μg 25-hydroxycholecalciferol	1 μg = 40 IU vitamin D activity	
Vitamin D_2	IU		8- to 11-fold less active than vitamin D_3 in chicks	(Chen and Bosmann, 1964)
Vitamin E	IU	1 mg synthetic DL-α-tocopheryl acetate		(NRC, 1994)
	IU	0.735 mg D-α-tocopheryl acetate	1 mg D-α-tocopheryl acetate = 1.361 IU of vitamin E activity for poultry	(NRC, 1994)
	IU	0.671 mg D-α-tocopherol	1 mg D-α-tocopherol = 1.490 IU of vitamin E activity for poultry	(NRC, 1994)
	IU	0.909 mg DL-α-tocopherol	1 mg DL-α-tocopherol = 1.100 IU of vitamin E activity for poultry	(NRC, 1994)
Vitamin K	mg	Vitamin K_3 (Menadione; 2-methyl-1,4-naphthoquinone)	A fat-soluble, synthetic compound considered the reference standard for vitamin K activity	(NRC, 1994)
	mg	1 mg vitamin K_1 or phylloquinone (2-methyl-3-phytyl-1,4-naphthoquinone)	Found in plants	(NRC, 1994)
	mg	1 mg vitamin K_2 or menaquinone (K1 substituted with 2 to 7 isoprene units)	Produced by bacteria	(NRC, 1994)
	mg	2.0 mg menadione sodium bisulfite complex (MSBC)	Water-soluble form	(NRC, 1994)
	mg	2.2 mg menadione dimethylpyrimidol (MPB)	Water-soluble form	(NRC, 1994)
	mg	3.0 mg menadione sodium bisulfite (MSB)	Water-soluble form	(NRC, 1994)
Niacin	mg	45 mg tryptophan	Conversion for chickens	(Baker et al., 1973)
	mg	172 to 181 mg	Conversion for ducks	(Chen et al., 1996)

[a] References are included in the individual vitamin sections.

TABLE 5-2 Biochemical and Physiological Measurements for Diagnosis of Vitamin Deficiencies in Chickens and Turkeys

Nutrients	Biochemical and Physiological Measurements	References[a]
Vitamin A	Hepatic vitamin A is indicative of a deficiency, but blood level is not. Liver xanthine dehydrogenase and kidney arginase both increase even in the first stages of a deficiency. Reduced glycogen phosphorylase in liver, and red and white muscles. Increased thyroid size and reduced T_3 and T_4.	(Rogers, 1969; Nockels and Phillips, 1971; Jensen, 1974; Bruckental and Ascarelli, 1975; Nockels et al., 1984)
Vitamin D	Calcium-binding protein of intestine; 1,25-dihydroxy D_3 versus 24,25-dihydroxy-D_2 in serum (complicated by dietary calcium and phosphorus); plasma alkaline phosphatase; non-proteolipid phospholipid content of rachitic cartilage.	(Bar et al., 1972; Omdahl and DeLuca, 1973; Morrissey et al., 1977; Boyan and Ritter, 1984; Kaetzel and Soares, 1985)
Vitamin E	Superoxide dismutase; glutamic-oxaloacetic transaminase; plasma and tissue vitamin E concentration (all measurements affected by selenium as well).	(Walter and Jensen, 1964; Arnold et al., 1974; Sklan et al., 1981; Sklan and Donoghue, 1982)
Vitamin K	Prothrombin clotting time of plasma.	(Griminger et al., 1970)
Thiamin	Transketolase in erythrocytes and leucocytes; plasma pyruvic acid.	(Lofland et al., 1963; Cheng et al., 1976)
Riboflavin	Liver xanthine dehydrogenase; erythrocyte glutathione reductase.	(Chou, 1971; Lee, 1982)
Niacin	Level and ratio of niacin excretion products N'-methylnicotinamide and N'-methyl-2-pyridone-5-carboxyamide (untested for fowl).	(Darby et al., 1975)
Biotin	Blood pyruvate carboxylase; ratio of C 16:1 to C 18:0 fatty acids in blood.	(Edwards, 1974; Whitehead and Bannister, 1980)
Pantothenic acid	Hepatic coenzyme A.	(Cupo and Donaldson, 1986)
Pyridoxine	Serum glutamic oxaloacetic transaminase; plasma glycine:serine ratio aspartic aminotransferase.	(Daghir and Balloun, 1963; Sifri et al., 1972; Lee et al., 1976)
Folic acid	Dihydrofolic acid reductase in liver; serine hydroxymethyl transferase in liver.	(Rabbani et al., 1973; Zamierowski and Wagner, 1977)
Vitamin B_{12}	B_{12} in blood; excretion of methylmalonic acid.	(Cox and White, 1962; Lau et al., 1965)
Choline	Serum phospholipids.	(Seifter et al., 1972)

[a] References are included in the individual vitamin sections.

TABLE 5-3 Signs of Vitamin Deficiencies in the Embryo

Nutrients	Deficiency Signs	References[a]
Vitamin A	Death at about 48 hr of incubation from failure to develop the circulatory system; abnormalities of kidneys, eyes, and skeleton.	(Asmundson and Kratzer, 1952; Thompson et al., 1965; Heine et al., 1986)
Vitamin D	Death at about day 18 or day 19 of incubation, with malpositions, soft bones, and a defective upper mandible prominent.	(Sunde et al., 1978; Narbaitz and Tsang, 1989)
Vitamin E	Early death at about 84 to 96 hr of incubation, with hemorrhaging and circulatory failure (implicated with selenium).	(Card et al., 1930; Latshaw and Osman, 1974)
Vitamin K	No physical deformities from a simple deficiency, nor can they be provoked by antivitamins, but mortality occurs between day 18 of incubation and hatching, with variable hemorrhaging.	(Griminger, 1964; Hauschka and Reid, 1978b)
Thiamin	High embryonic mortality during emergence but no obvious symptoms other than polyneuritis in those that survive.	(Polin et al., 1962; Charles et al., 1972)
Riboflavin	Mortality peaks at 60 hr, day 14, and day 20 of incubation, with peaks prominent early as deficiency becomes severe. Altered limb and mandible development, dwarfism, and clubbing of down are defects expressed by embryo.	(Romanoff and Bauernfeind, 1942; Landauer, 1967)
Niacin	Embryo readily synthesizes sufficient niacin from tryptophan. Various bone and beak malformations occur when certain antagonists are administered during incubation.	(Snell and Quarles, 1941; Landauer, 1957; Caplan, 1972);
Biotin	High death rate at day 19 to day 21 of incubation, and embryos have parrot beak, chondrodystrophy, several skeletal deformities, and webbing between the toes.	(Cravens et al., 1944; Couch et al., 1947)
Pantothenic acid	Deaths appear around day 14 of incubation, although marginal levels may delay problems until emergence. Variable subcutaneous hemorrhaging and edema; wiry down in poults.	(Kratzer et al., 1955; Beer et al., 1963)
Pyridoxine	Early embryonic mortality based on antivitamin use.	(Landauer, 1967)
Folic acid	Mortality at about day 20 of incubation. The dead generally appear normal, but many have bent tibiotarsus, syndactyly, and mandible malformations. In poults, mortality at day 26 to day 28 of incubation with abnormalities of extremities and circulatory system.	(Sunde et al., 1950; Kratzer et al., 1956)
Vitamin B_{12}	Mortality at about day 20 of incubation, with atrophy of legs, edema, hemorrhaging, fatty organs, and head between thighs malposition.	(Olcese et al., 1950; Ferguson et al., 1955)

[a] References are included in the individual vitamin sections.

between 1985 and 2006, vitamin E intake per egg decreased by 1.1 percent per year for laying hens, and per kg body weight gain by 0.8 percent and 0.6 percent per year for broiler chickens and turkeys, respectively (Leeson, 2007). In the context of the industry practice of including substantial safety margins for most vitamins, these changes are likely not important for commercial poultry production. Additionally, selection for increased productivity may have also selected for increased efficiency of vitamin absorption and utilization.

Therefore, the default vitamin requirement values provided are the values from the previous version of this report (NRC, 1994). For a change to be made to a particular vitamin requirement, the following criteria were necessary: inclusion of a sufficient number of dietary levels of the vitamin above and below the determined requirement (i.e., linear or curvilinear response above and below the requirement, with a plateau or decreasing response above the requirement); analyzed vitamin content(s) of all experimental diets; experiment conducted under typical production conditions (e.g., no added stressors such as high environmental temperatures); performance of control birds comparable to that under commercial conditions; adequate replication of experimental treatments, and appropriate statistical analyses. Based on these criteria, few true vitamin requirement studies have been conducted since 1994. Many of the papers published also investigated the effects of increasing levels of a particular vitamin in the context of changing levels of some other nutrient(s), which confounds the interpretation of the results to establish a vitamin requirement. In some cases, published papers that do not meet the foregoing criteria may either indicate a lack of disagreement with the NRC (1994) values or suggest that a higher or lower requirement value may be indicated. Therefore, without solid experimental data warranting a change, recommended vitamin requirement values remain largely unchanged from the previous version of this report (NRC, 1994) with the exception of several minor species.

FAT-SOLUBLE VITAMINS

The fat-soluble vitamins (A, D, E, and K) are absorbed from the digestive tract along with fat, and factors that influence lipid digestion and absorption will have a corresponding influence on fat-soluble vitamin absorption. Fat-soluble vitamins can be stored in the liver and adipose tissue, and as such, birds with replete body stores of vitamins A, D, and E can withstand long periods of depletion before deficiency symptoms are manifested (Klasing and Korver, 2020). Conversely, this ability to store fat-soluble vitamins can also lead to hypervitaminoses with prolonged durations of extremely high dietary levels.

Vitamin A

The term "vitamin A" is used to refer to the retinoids, a group of compounds possessing the biological activity of retinol (Combs and McClung, 2017b). The three forms of vitamin A are retinol, the alcohol form; retinal is the substitution of the alcohol group with an aldehyde; and substitution with a carboxylic acid group results in retinoic acid. Vitamin A has different isomeric forms because of the double bonds in the side chain. Vitamin A plays important roles in vision, the regulation of gene transcription, epithelial differentiation, immune function, reproduction, bone metabolism, hematopoiesis and skin health (Combs and McClung, 2017b). Relatively large amounts of vitamin A can be stored in the liver as retinyl-fatty acid esters; depletion to the point that deficiency signs develop in adult chickens or turkeys typically takes several months (Klasing and Korver, 2020).

Tissue vitamin A is derived from dietary supplementation of vitamin A or from carotenes with pro-vitamin A activity. The xanthophyll carotenoids contain oxygen and are not efficiently used as precursors for vitamin A, whereas the carotenes are pure hydrocarbons (contain no oxygen) and can serve as vitamin precursors. Carotenes having a β-ionone ring structure can be converted to vitamin A, allowing poultry to convert dietary β-carotene and other carotenes to the active form of vitamin A (Tajima et al., 2001). Although there are approximately 600 naturally occurring carotenoids, the number having provitamin A activity is only approximately 50, and relatively few of those are common in feeds and foods (Combs and McClung, 2017b).

Plant-source feedstuffs usually provide carotenoid pigments that may be converted into vitamin A. The most favorable such pigment in this respect is β-carotene (Flegal et al., 1971), which is converted to retinol in the intestine during absorption (Sklan, 1983). As compared to wheat or sorghum grain, corn has moderately high levels of β-carotene, but corn-based diets for poultry still require supplementation of vitamin A (Klasing and Korver, 2020). Because of the susceptibility of vitamin A sources to oxidative losses, synthetic antioxidants often are included in premixes and complete feeds (Grundboeck et al., 1977).

Natural and synthetic carotenoid sources are available for commercial use in poultry diets, although only some are provitamins A. Because different forms of provitamins and vitamin A have different potencies, vitamin A activity is expressed as international units (IUs). For poultry, 1 IU is defined as the vitamin A activity of 0.3 µg of crystalline retinol, 0.344 µg of vitamin A acetate, 0.55 µg vitamin A palmitate, and 0.6 µg β-carotene (Table 5-1).

During the digestion of a meal, carotenoids associate with the fat phase of the ingesta and are emulsified into lipid droplets and incorporated into mixed micelles. In humans, esterified forms of vitamin A are hydrolyzed prior to absorption primarily in the duodenum, since only the free forms of vitamin A and carotenoids appear to be absorbed (Reboul, 2013). Absorption of carotenoids and vitamin A involves both specific and nonspecific enterocyte transporters (Reboul, 2013).

Vitamin A Deficiency

From a nutritional standpoint, vitamin A is the most challenging of the vitamins because it is deficient in many feedstuffs but is also among the most likely to elicit toxicity when over-supplemented (Klasing, 1998b). A vitamin A deficiency affects multiple physiological systems. Deficiency signs include loss of vision, abnormal bone formation, impaired reproduction, and abnormal growth and differentiation of epithelial tissue leading to keratinization. Substitution of the body's secretory epithelia by keratinized surfaces is the most important change with a vitamin A deficiency. Corneal, conjunctival, esophageal, and tracheal secretory membranes are all altered in chickens (Aydelotte, 1963). Loss of membrane integrity, in turn, alters water retention (Lopez et al., 1973) and impairs the ability to withstand infection (Singh and Donovan, 1973; Sijtsma et al., 1989). Vitamin A deficiency also reduces the immune system's response to challenge and further contributes to disease susceptibility (Davis and Sell, 1989; Sklan et al., 1989). In rats, a severe deficiency of vitamin A results in hyperproliferation and decreased maturation of enterocytes, decreased mucin production (De Luca et al., 1971), a decrease in the number of goblet cells, blunting of villi with decreased expression of brush-border enzymes, and diminished digestive function, which negatively impacts growth rate (Klasing and Korver, 2020). In ducks, insufficient dietary vitamin A resulted in growth depression, xerophthalmia, and reduced liver and plasma retinol (Feng et al., 2019).

The appearance of keratinized secretory surfaces is followed by a typical ataxia. Alterations in bone growth create several areas of compression on the central nervous system that cause a loss in mobility (Thompson et al., 1965). Vitamin A deficiency also adversely affects the pituitary–gonadal axis to create other symptoms that are not readily obvious (Fletcher, 1971). Nockels et al. (1984) reported that hypothyroidism is an early indication of vitamin A deficiency in chicks. Reductions in testes size, circulating testosterone, and fertility have been reported during vitamin A deficiency in cockerels (Paredes and Garcia, 1959; Damjanov et al., 1980).

Muscles in vitamin A–deficient birds have a high level of glycogen, which cannot be readily used because phosphorylase activity is inordinately low (Nockels and Phillips, 1971; Sundeen et al., 1980). Alternatively, glucose is provided by extensive gluconeogenesis from protein (Nir and Ascarelli, 1967; Bruckental et al., 1974), and nitrogen end-products increase such that deposits of uric acid appear in the kidneys and ureters (Bruckental and Ascarelli, 1975; Chandra et al., 1984). Vitamin A deficiency often results in embryonic death during the first week of incubation, while embryos that survive to term may be too weak to hatch or die shortly thereafter (West et al., 1992).

Vitamin A Toxicity

As a fat-soluble vitamin, vitamin A can accumulate in the tissues of the bird and lead to hypervitaminosis. In chicks, retinyl esters rise in the serum with hypervitaminosis A, leading to liver disorders, bone fragility, and increased erythrocyte fragility (Friedman and Sklan, 1997). Both low and high levels of dietary vitamin A reduce various measures of immune response in birds (Friedman and Sklan, 1997).

As noted previously, vitamin A in feedstuffs is subject to oxidation; therefore, concentrated supplements are normally given to ensure that the requirement is met. Vitamin A has the potential to act as a cumulative toxicant in most species studied, although the tremendous storage capacity of the liver affords great protection against toxicity as well as against dietary deficiency (NRC, 1987). Mistakes in the formulation of vitamin premixes have led to occasional toxicosis problems. Skin lesions at the commissure of the beak, nares, and eyes attributable to mucus membrane hyperplastic activity have been shown to occur in chicks within 72 hours after oral dosing with 60,000 IU (Kriz and Holman, 1969). The appearance of rachitic bones together with a hyperplastic parathyroid is due to the known antagonism with vitamin D (Metz et al., 1985; Tang et al., 1985; Veltmann et al., 1987). In broilers, signs of toxicity include slow growth, an unsteady gait, reluctance to walk, reduced bone mineralization, and a higher incidence of tibial dyschondroplasia; in cases of severe vitamin A toxicity, anorexia, conjunctivitis, adhesions of the eyelids, encrustations around the mouth, and thinning of the frontal bones of the skull with thickened osteoid seams have been observed (Tang et al., 1985; Li et al., 2008; Klasing and Korver, 2020). At dietary levels above 45,000 IU/kg diet, vitamin A as retinyl acetate reduced egg production and weight, yolk color, and shell quality in broiler breeder hens, and above 135,000 IU/kg diet, it reduced peripheral blood lymphocyte proliferation (Yuan et al., 2014). Excessive vitamin A has also been shown to antagonize vitamin E (Vahl and van't Klooster, 1987) and increase the likelihood of a deficiency when vitamin E and selenium nutriture is marginal (Combs, 1976). The presumed upper safe level of vitamin A for various poultry species ranges from 15,000 IU/kg diet for growing chickens, turkeys, and geese to 24,000 IU/kg diet for breeding turkeys, to 25,000 IU/kg diet for quail, and to 40,000 IU/kg diet for laying chickens and ducks (NRC, 1987).

Vitamin D

Vitamin D describes a group of steroids having the biological activity of cholecalciferol (Combs and McClung, 2017c). Active vitamin D (1,25 dihydroxy cholecalciferol) is a secosteroid hormone. Vitamin D can be synthesized in the skin from 7-dehydrocholesterol following exposure to ultraviolet (UVB) light (Beadle, 1977; Kühn et al., 2015, 2019). Exposure of laying hens to UVB light allowed hens fed a diet devoid of cholecalciferol to maintain egg production over 4 weeks at a level similar to hens fed 3,000 IU of cholecalciferol/kg diet (Lietzow et al., 2012), although egg 25-hydroxy cholecalciferol status was very low. Because poultry housed in confinement are not exposed directly to sunlight, a dietary requirement for vitamin D exists in these birds. Vitamin D can be supplemented in the diet as vitamin D_3 or its natural or synthetic metabolites. Poultry diets commonly are supplemented with vitamin D_3; however, 25-hydroxy D_3 or 1,25-dihdyroxy D_3 are also sometimes used because they have higher bioavailability and can reduce the incidence of tibial dyschondroplasia and osteomyelitis in broilers and egg quality decline in older hens (Wideman et al., 2015; Han et al., 2017; Swiatkiewicz et al., 2017).

Vitamin D_2 (ergocalciferol) derived from plants can be converted (albeit with less than 100 percent efficiency; Houghton and Vieth, 2006) to 1,25 dihydroxy ergocalciferol by mammals, but it is poorly utilized by birds (Hurwitz et al., 1967; Mattila et al., 2004). This is due largely to a greater clearance from the plasma of ergocalciferol metabolites compared to cholecalciferol metabolites (Hoy et al., 1988), and therefore vitamin D_2 should not be used as a supplement in poultry diets.

Whether derived from the skin or the diet, cholecalciferol must be metabolized through a two-step process to the active form of vitamin D. Most circulating cholecalciferol is converted in the liver to 25-hydroxy cholecalciferol (Soares et al., 1995). This is the main circulating form of vitamin D and thus is used as a biomarker of vitamin D status. The conversion to the active hormone (1,25 dihydroxy cholecalciferol; Norman, 2008) is highly regulated and takes place in the kidney (Soares et al., 1995). Autocrine conversion of 25-hydroxy cholecalciferol to the active hormone takes place in extrarenal tissues including liver, immune tissues, intestine, and muscle (Shanmugasundaram and Selvaraj, 2012).

The active form of vitamin D functions as a secosteroid hormone with genomic and nongenomic functions. The classical function of active vitamin D involves, in concert with parathyroid hormone, the metabolic modulation and regulation of the intestinal absorption, bone accretion and resorption, and renal excretion of calcium and phosphorus.

Physiologically, 1,25 dihydroxy cholecalciferol stimulates the active absorption of calcium (Centeno et al., 2004) and phosphorus (Shao et al., 2019) from the intestine, the resorption of bone (Shao et al., 2019), and the resorption of calcium (Bronner and Stein, 1995) and phosphorus (Khanal et al., 2007) in the kidney. Thus, the kidney plays an important role in regulating plasma calcium and phosphorus levels. Vitamin D also is involved in muscle protein synthesis (Vignale et al., 2015) and immune function (Morris and Selvaraj, 2014; Morris et al., 2014, 2015).

In humans, it has been suggested that vitamin D is incorporated into mixed micelles and passively absorbed from the lumen along the length of the small intestine in a nonsaturable process (Combs and McClung, 2017c). However, others have suggested that there may be specific transport-mediated absorption of vitamin D in humans (Reboul, 2015).

Vitamin D Deficiency

Gross symptoms of a vitamin D deficiency can largely be attributed to a reduction of intestinal binding protein and lack of calcium recovered from feed (McCarthy et al., 1984). A vitamin D deficiency results in similar clinical signs to deficiencies of calcium (Ramp et al., 1989), phosphorus, or both. As the deficiency advances, rickets become evident as severe fragility and bending of long bones due to poor mineralization. Beaks and claws become soft and pliable, feathering is poor, and birds walk with obvious effort and take a few unsteady steps before squatting on their hocks, upon which they rest while swaying slightly from side to side (Klasing and Korver, 2020). For some reason, an abnormal blackening of the feathers also occurs with some pigmented chicks (Glazener and Briggs, 1948). During vitamin D deficiency, growing birds develop hypocalcemia, which, in turn, stunts skeletal development through widened cartilage at epiphyses of long bones and weakened shafts (Noff et al., 1982; Long et al., 1984). Feeding a diet devoid of vitamin D and deficient in Ca and P reduced performance and bone development, and it increased mortality of broiler chicks compared to those fed various vitamin D metabolites (Han et al., 2017). Broiler chicks fed a diet without supplemental vitamin D had reduced growth rate, tibia ash, cutaneous basophil hypersensitivity response, thymus weight and decreased abdominal macrophage phagocytic response compared to chicks fed an adequate diet (Aslam et al., 1998). Once the skeleton has assumed adult size, hypovitaminosis D is obvious only with hens in production.

Hens in production cyclically release estrogen from the ovary to maximize 1,25-dihydroxy D_3 production concurrent with eggshell formation (Castillo et al., 1979). As a result, levels of calcium-binding protein in the uterus (Navickis et al., 1979) and calcium in the medullary bone (Takahashi et al., 1983) are altered to facilitate eggshell formation. In confined laying hens, signs of deficiency begin to occur as soon as 2 weeks after they are deprived of vitamin D. In addition to decreased egg production and egg weight, an increase in the number of thin-shelled and soft-shelled eggs is observed in the initial stages of vitamin D deficiency as bone reserves are progressively depleted (Vohra et al., 1979; Klasing and Korver, 2020). Biochemical indicators include a rapid decrease in the concentrations of 25-hydroxy D_3 in the blood, followed by a decrease in blood calcium concentration. Egg production and eggshell strength may vary in a cyclic manner. Many of the signs of vitamin D deficiency may be similar to calcium deficiency, and analysis of levels of these two nutrients in the diet, or 25-hydroxy D_3 in blood plasma, will confirm the cause (Klasing and Korver, 2020).

Vitamin D nutriture of the hen influences its content in egg yolk and the subsequent need for this vitamin by the chick (Bethke et al., 1936; Griminger, 1966; Stevens and Blair, 1985; Saunders-Blades and Korver, 2015). Vitamin D in the form of 25-hydroxy D_3 obtained from the yolk is metabolized by the embryo to 1,25-dihydroxy D_3 as it is by the adult (Bishop and Norman, 1975). An additional activity for this metabolite is recovery of calcium from the shell at the chorioallantoic membrane to support skeletal mineralization prior to hatching (Narbaitz, 1987). The yolk sac membrane also responds to 1,25-dihydroxy D_3 at the same time, and a portion of the calcium from the shell is transferred into the yolk for later use upon hatching (Clark et al., 1989); however, one or more of the other metabolites must also be present if complete embryonic development and emergence from the shell is to occur (Ameenuddin et al., 1982). Hatchability is reduced markedly by vitamin D deficiency mainly due to increased embryonic mortality late in incubation. Chicks and poults that do not hatch have a high incidence of chondrodystrophy in which the upper or lower mandible is shortened or deformed (Klasing and Korver, 2020).

The very low content of vitamin D in feedstuffs (Morrissey et al., 1977) is generally ignored in feed formulation, and the complete requirement is satisfied by using concentrated premixes. Moreover, supplemental vitamin D in feed may not be fully available to poultry. This vitamin is susceptible to destruction by oxidation, and significant losses may occur unless supplemental antioxidants are used (Fritz et al., 1942). Also, mycotoxins in feeds interfere with the utilization of dietary vitamin D (Bird, 1978; Gedek et al., 1978; Kohler et al., 1978). Losses of vitamin D because of oxidation and poor utilization may result in a deficiency of the vitamin even though initial dietary concentrations of vitamin D substantially exceed known requirements.

Vitamin D Toxicity

Overuse of vitamin D concentrates can lead to toxicity. High levels of circulating 1,25-dihydroxy D_3 occur with toxicosis, along with hypercalcemia and soft tissue mineralization (Morrissey et al., 1977; Ratzkowski et al., 1982). Leg problems may arise with growing birds because of bone calcium loss (Cruickshank and Sim, 1987), but few obvious changes occur with hens other than a general depression in performance (Ameenuddin et al., 1986). Hypervitaminosis D induced by oral gavage of 2.5 mg (i.e., 100,000 IU) vitamin D_3/kg body weight per d in broiler chicks from 7 to 28 d of age resulted in anorexia, weakness, increased water consumption and excretion, dehydration and emaciation, as well as necrosis and metastatic calcification of the kidney (Kumar et al., 2017). Toxic levels of vitamin D may be transferred into the egg to create similar problems for the embryo; however, the hypercalcemia caused by shell resorption and bone mineralization are enhanced (Narbaitz and Fragiskos, 1984).

Young chicks have a high tolerance of excess vitamin D and showed no negative effects when fed 1.25 mg (i.e., 50,000 IU) cholecalciferol/kg diet for 11 to 12 d beginning at 7 or 8 d of age (Baker et al., 1998).

The presumed upper safe level of vitamin D_3 for various poultry species depends upon exposure time: for <60 d, estimates ranged from 40,000 IU/kg diet for chickens, to 90,000 IU/kg of diet for turkeys, and to 120,000 IU/kg diet for Japanese quail (NRC, 1987). For >60 d exposure, much lower upper safe levels were estimated: 2,800 IU/kg diet for chickens to 3,500 IU/kg diet for turkeys, and to 4,700 IU/kg diet for Japanese quail (NRC, 1987). In many countries, the legal limit for dietary supplementation of 5,000 IU vitamin D/kg diet may be too low, since levels as high as 10,000 IU vitamin D/kg diet increased broiler performance and reduced the incidence of tibial dyschondroplasia (Whitehead et al., 2004).

Vitamin E

Vitamin E comprises a group of tocopherols and tocotrienols having the biological activity of α-tocopherol (Combs and McClung, 2017d). All forms of vitamin E have a hydroquinone nucleus and an isoprenoid side chain; the tocopherols have a saturated side chain, whereas the tocotrienols contain three double bonds in the side chain. α-tocopherol is the predominant form found in feedstuffs and has the greatest vitamin E activity. The other tocopherols have far less vitamin E activity in preventing muscular dystrophy in the chicken, and the tocotrienols have essentially none (Machlin, 1991).

Dietary vitamin E is absorbed from the intestine with fat, and its dissemination in the body follows depletion of lipoprotein contents from circulation (Massey, 1984). The preferentially absorbed form of vitamin E is α-tocopherol (Combs and McClung, 2017d), although the other isomers are also absorbed at lower levels. In turn, tissue vitamin E content parallels feed vitamin E levels, and tissues receiving the highest proportions are intestine, liver, adipose, and muscle (Astrup, 1979).

The most important role of vitamin E is as a biological antioxidant (Combs and McClung, 2017d) protecting biological membranes. Vitamin E integrates into cellular membranes (Erin et al., 1984) and protects against oxidative damage that leads to cell dysfunction and death. Cell membranes are also defended against oxidative damage by the selenium-dependent enzyme glutathione peroxidase (Xu and Diplock, 1983). Thus, vitamin E and selenium have interactive effects on maintaining cell viability (Klasing and Korver, 2020). Vitamin E is a chain-breaking antioxidant by virtue of its phenol ring structure, which allows it to quench unpaired electrons from pro-oxidants such as reactive oxygen species (Combs and McClung, 2017d). Antioxidant activity of vitamin E has been demonstrated in broiler chickens (Niu et al., 2018), laying hens (Irandoust and Ahn, 2015), turkeys (Ognik and Wertelecki, 2012), ducks (Xie et al., 2018), and quail (Sahin et al., 2005). Normal reproduction in roosters and laying hens is dependent on vitamin E (Panda and Cherian, 2014; Rengaraj and Hong, 2015). Vitamin E also appears to have an indirect antioxidant effect by increasing the expression of antioxidant enzyme genes (Niu et al., 2018).

A meta-analysis determined that increasing levels of dietary vitamin E increased several aspects of immune function in male broiler chicks (Pompeu et al., 2018). Increasing supplemental dietary vitamin E from 0 to 100 IU/kg diet increased antibody production to Newcastle disease virus and avian influenza virus following vaccination, but also interleukin-1 levels following a bacterial lipopolysaccharide challenge (Zhao et al., 2011). However, increasing levels of dietary vitamin E supplementation eventually decrease indices of immune function in broilers. Antibodies to infectious bronchitis were maximized at 25 IU/kg diet, antibodies to sheep red blood cells were maximized at 50 IU/kg diet, while vitamin E was without effect on response to *Brucella abortus* antigens (Leshchinsky and Klasing, 2001). Mitogenic responses to phytohemagglutinin A and concanavalin A were suppressed by 25, or 25 to 50, IU/kg diet, respectively, with no effect on cutaneous basophil hypersensitivity or acute phase protein production (Leshchinsky and Klasing, 2001).

Vitamin E Deficiency

The amount of vitamin E needed to avoid a deficiency largely depends on the adequacy of dietary selenium and on circumstances presenting oxidative threats to the system. A deficiency of both vitamin E and selenium leads to exudative diathesis, which is a subdermal accumulation of viscous, blue-green-colored exudate from endothelial failures in portions of the vascular system (Scott, 1966). Myopathies of the gizzard, heart, and, to a lesser extent, the skeletal muscles are also apparent. Skeletal muscles, particularly the breast, become more myopathic when the sulfur amino acids are also deficient. Exudative diathesis can be eliminated, and most myopathies can be greatly relieved, when selenium alone is increased (Combs and Scott, 1974).

Moderate vitamin E deficiencies do not markedly reduce growth but do negatively impact a variety of reproductive and immunological processes (Klasing and Korver, 2020). Reproductive effects of deficiency include decreased fertility of males and egg production in hens (Rengaraj and Hong, 2015). With severe vitamin E deficiency, oxidative damage may cause encephalomalacia, exudative diathesis, and nutritional myopathy (muscular dystrophy) in chicks; enlarged hocks and dystrophy of the ventricular musculature in turkeys; and nutritional myopathy in ducks (Klasing and Korver, 2020). Vitamin E deficiency is exacerbated by low levels of dietary selenium and vice versa (Mueller et al., 2009; Huang et al., 2011). Exudative diathesis in chicks and myopathies of the gizzard and heart in young poults may not occur during vitamin E deficiency when dietary selenium is adequate. Meat and eggs from vitamin E–deficient birds have increased susceptibility to oxidative rancidity during processing and storage (Klasing and Korver, 2020).

Vitamin E deficiency symptoms that do not respond to increased dietary selenium are encephalomalacia (Hassan et al., 1985) and the susceptibility of red blood cells to hemolysis (Dobsinska et al., 1982). Degeneration of the Purkinje layer of cells in the cerebellum results in nervous symptoms typified as sudden prostration with toes and legs outstretched, toes flexed, and head outstretched (Ames, 1956). High concentrations of dietary polyunsaturated fatty acids lead to increased contents in cell membranes and, in turn, the additional susceptibility to oxidative stress may enhance the possibilities of encephalomalacia (Budowski and Crawford, 1986). Other stressors such as ozone in the environment (Bartov et al., 1981) or peroxidized fat (Budowski et al., 1979) or medium-chain fatty acids (Ikumo, 1980) contained in the feed also increase the possibility of a vitamin E deficiency.

Adult fowl are less susceptible to a vitamin E deficiency than are actively growing chicks, and the symptoms differ. Testicular degeneration occurs in males deprived of vitamin E for prolonged periods (Klasing, 2008, 2013). Males become infertile because sperm cells become incompetent (Friedrichsen and Willis, 1980). Reduced egg production and hatchability occur when hens are fed diets deficient in both vitamin E and selenium over a prolonged period (Latshaw and Osman, 1974). Embryonic mortality occurs as early as day 4 of incubation, depending on the severity of the deficiency. Symptoms in the vitamin E–deficient embryo include cloudy spots in the eyes, blindness, abnormal vascular system, hemorrhages, and stunting. Turkey embryos may have bilateral cataracts that can cause blindness (Klasing and Korver, 2020). Although supplemental selenium can completely overcome these problems, chicks from these eggs are particularly susceptible to encephalomalacia (Bartov and Bornstein, 1980) and muscular dystrophy (Ewen and Jenkins, 1967).

Because of the similarities of vitamin E and selenium deficiency syndromes, measurement of their concentrations in blood plasma will reveal if either or both are deficient (Huang et al., 2011). Measuring vitamin E levels in feed may not be indicative of levels consumed by birds due to the rapid loss during storage (Klasing and Korver, 2020).

Vitamin E Toxicity

Although vitamin E is generally considered one of the least toxic of the vitamins (NRC, 1987), adding excessive vitamin E to feed can have adverse effects. Chicks fed 1,800 IU/kg diet were not affected, but 2,200 IU/kg diet resulted in growth depression, reduced feed intake, reduced hematocrit, and reticulocytosis. Nockels et al. (1976) reported that feeding 8,000 IU/kg diet reduced body weight gain and gave a waxy appearance to the feathers. Should either vitamin D or vitamin K be marginal when high levels of vitamin E are being fed, rachitic bones and blood clotting failures, respectively, may occur (March et al., 1973; Murphy et al., 1981; Franchini et al., 1988). However, dietary levels in excess of the requirement (100 to 500 IU/kg diet) are advantageous to the oxidative stability of broiler (Lin et al., 1989) and turkey (Sheldon, 1984) meat products. A dietary level of 1,000 IU/kg is presumed to be the safe upper level of vitamin E for the chick (NRC, 1987). Vitamin E can also interfere with vitamin A absorption.

Vitamin K

Vitamin K represents 2-methyl-1,4-naphthoquinone and its derivatives, each of which possess the anti-hemorrhagic function of phylloquinone (Combs and McClung, 2017e). Three main forms of vitamin K exist: phyloquinones (vitamin K_1) synthesized by plants; menaquinones (vitamin K_2) synthesized by bacteria, including intestinal bacteria; and menadione (vitamin K_3), which is a synthetic form that does not occur naturally. The phylloquinones and menaquinones vary in structure based on the number of isoprene units in the side chain.

Adult birds usually have a well-developed intestinal microflora, and vitamin K deficiencies are unusual. Vitamin K_2 is not readily absorbed from the large intestine, but it is digested after coprophagy of cecal excreta (Berdanier and Griminger, 1968). The caging of hens minimizes coprophagy, and minimal amounts of vitamin K reach the egg (Cravens et al., 1941) in the absence of supplementation (Naber, 1993). Griminger and Brubacher (1966) observed that dietary vitamin K_3 is converted and transferred to the yolk as vitamin K_2, but vitamin K_1 is most effectively transferred and is not converted to vitamin K_2.

Vitamin K is a cofactor to synthesize γ-carboxyglutamic residues from glutamic acid in proteins located in the liver and bone. This liver protein is involved in the synthesis of several blood clotting factors, including prothrombin (Factor II) (Suttie, 1987). In addition, four blood clotting factors (plasma clotting factors VII, IX, and X) are produced by the liver in their inactive form and require vitamin K for activation. A diet without vitamin K supplementation increased prothrombin time in young turkey poults relative to diets containing 0.5 to 2.0 mg of phylloquinone (vitamin K_1) per kg diet. Vitamin K is essential for the function of osteocalcin (bone γ-carboxyglutamic protein; Hauschka et al., 1989) and bone matrix protein (BMP) in bone metabolism. Supplementation of vitamin K above the requirement increased cancellous bone volume at 25 weeks of age (Fleming et al., 1998) and at 15 weeks of age and throughout a full laying cycle (Fleming et al., 2003).

Specific carriers for vitamin K have not been identified; rather, it is absorbed from the lumen of the intestine with dietary fat (Combs and McClung, 2017e). Poultry are more susceptible to vitamin K deficiency than other species because of their short digestive tracts and limited hindgut microbial synthesis of menaquinones (Combs and McClung, 2017e). Additionally, feed ingredients that alter intestinal microbial composition may reduce microbial synthesis of vitamin K in the intestine and increase the dietary vitamin K requirement of pheasants and quail (Scott et al., 1961). Low vitamin K quinone reductase activities in poultry lead to inefficient recycling of vitamin K and subsequently a relatively high dietary requirement (Combs and McClung, 2017e).

Vitamin K Deficiency

Although deficient dietary vitamin K reduces bone osteocalcin function, symptoms associated with the skeletal system are not as apparent as blood clotting problems (Scott, 1967; Hauschka and Reid, 1978a). Hemorrhaging may occur subcutaneously, intramuscularly, and internally and may lead to anemia and the appearance of hypoplastic bone marrow. A greatly extended blood clotting time may result in death from exsanguination. Vitamin K adequacy is usually measured in terms of prothrombin clotting time in decalcified plasma (Griminger et al., 1970). Antivitamin K compounds, whether synthetic (Lowenthal and MacFarlane, 1965) or natural (Griminger, 1987), act as anticoagulants. Menadione generally exhibits the greatest vitamin K activity (Dua and Day, 1966), except when anticoagulants are given and the converse occurs (Griminger, 1965). Dietary anticoagulants lead to vitamin K deficiency symptoms commensurate with the extent of toxicity (Veltmann et al., 1981; Bai and Krishnakumari, 1986).

Deficient dietary vitamin K under practical circumstances is most likely to occur during the early post-hatching period; therefore, supplementation of the feed at this time is advantageous (Fritz, 1969). Growing chicks exhibit signs of deficiency as early as 2 weeks of age; hemorrhages appear externally at areas that receive abrasions, such as the feet and wings (Klasing and Korver, 2020). Internally, petechial hemorrhages in the liver and erosion of the kaolin lining of the gizzard may occur. Chicks show an anemia that may result partly from loss of blood but also from development of a hypoplastic bone marrow. Although blood-clotting time is a fairly good measure of vitamin K deficiency, prothrombin time is more accurate (Klasing and Korver, 2020).

Starting feeds seldom contain forage meals high in vitamin K, and a poorly developed intestinal microflora, together with the use of antimicrobials, further reduces access to vitamin K (Bornstein and Samberg, 1954). Nelson and Norris (1961) showed that the inclusion of 0.1 percent sulfaquinoxaline increased the chick's need for supplemental vitamin K by 4- to 7-fold.

Use of vitamin K by embryos parallels that by adults, and deficient levels of vitamin K in breeder diets causes increased embryo mortality late in incubation, with dead embryos appearing hemorrhagic (Klasing and Korver, 2020). Adverse effects on blood clotting also appear after hatching, when hemorrhaging and mortality occur should trauma be encountered (Griminger, 1964).

Vitamin K Toxicity

Information on vitamin K toxicity is virtually nonexistent. Phylloquinone exhibits no adverse effects when administered to animals in massive doses by any route (NRC, 1987). Diets containing menadione nicotinamide bisulfite (MNB) or menadione dimethylpyrimidinol bisulfite (MPB) at 6,000 mg menadione/kg diet decreased body weight gain, feed intake, and gain:feed of New Hampshire X Columbian chicks. However, at the equivalent of 3,000 mg menadione/kg diet, only MNB decreased performance measures (Oduho et al., 1993).

WATER-SOLUBLE VITAMINS

The water-soluble vitamins are not stored in the body for long periods of time, with the exception of cobalamin, and are normally required continually in the diet. In mammals, specific tissue transport carriers have been demonstrated for the absorption of ascorbate, biotin, folate, niacin, pantothenic acid, pyridoxine, riboflavin, and thiamin (Said, 2011). Cobalamin is highly polar and requires the participation of intrinsic factor contained in the gastric juice and a receptor for the intrinsic-factor-cobalamin complex in the distal ileum (Quadros, 2010).

Despite distinct biochemical functions, deficiency symptoms of water-soluble vitamins are often similar (Table 5-4) and may overlap with deficiency signs of other nutrients. Unlike the fat-soluble vitamins A and D, and with the exceptions of niacin and choline, the water-soluble vitamins are generally not toxic when fed in great excess of their nutritional requirements (NRC, 1987; McDowell, 2008).

Thiamin (Vitamin B_1)

Thiamin is the common name for 3-[(4-amino-2-methyl-5-pyrimidinyl)methyl]-5-(2-hydroxyethyl)-4-methylthiazolium (Combs and McClung, 2017f). Thiamin was initially discovered when it was noticed that chickens fed polished rice developed polyneuritis, whereas those fed whole rice did not. Thiamin was ultimately isolated as the active factor (Suzuki and Shimamura, 1911; Jansen and Donath, 1926; Carpenter, 2012).

The function of thiamin is dependent on the presence of an aromatic pyrimidine ring with an amino group on C-4 joined to an aromatic thiazole ring containing a quaternary N, an open C-2, and a phosphorylatable hydroxyethyl group on C-5 (Combs and McClung, 2017f). Thiamin is active as thiamin pyrophosphate, which acts as a coenzyme in the metabolism of amino acids, sugars, and lipids. Thiamin pyrophosphate

TABLE 5-4 Nutrients Associated with Various Signs of Deficiency in Growing Birds

Deficiency Signs	Descriptions	Species	Associated Nutrients
Skin lesions	Crusting and scab formation around eyes and beak	Chick, poult	Biotin, pantothenic acid
	Bottoms of feet rough and calloused with hemorrhagic cracks	Chick, poult	Biotin, pantothenic acid
	Scaliness on feet	Chick	Zinc, niacin
	Lesions around eyes, eyelids stuck together	Chick, poult	Vitamin A
	Mouth, inflammation of oral mucosa (chicken black tongue)	Poult, chick	Niacin
Feather abnormalities	Uneven feather growth, abnormally long primary feathers, feathers not lying smoothly	Chick, poult	Protein, amino acid imbalance
	Frizzled and rough	Chick, poult	Zinc, niacin, pantothenic acid, folic acid, lysine
	Black pigmentation in breeds with red and brown feathers	Chick	Vitamin D
	Depigmentation	Chick, poult	Copper, iron, folic acid
Nervous disorders	Convulsions with head retraction	Chick, pigeon	Thiamin
	Convulsions with hyperexcitability	Chick, poult, duckling	Pyridoxine
	Hyperirritability	Chick, poult, duckling	Magnesium, sodium chloride
	Characteristic fright reaction with tetanic spasms	Chick	Chloride
	Spastic cervical paralysis, neck extended with birds appearing to look down	Poult	Folic acid
	Curled-toe paralysis, gross enlargement of sciatic and brachial nerves with myelin degeneration	Chick	Riboflavin
	Encephalomalacia, tetanic spasms with head retraction, hemorrhagic lesions in cerebellum	Chick	Vitamin E
Blood and vascular system	Anemia	All poultry	
	Macrocytic		Vitamin B_{12}
	Macrocytic, hyperchromic		Folacin
	Microcytic, hypochromic		Iron, copper
	Microcytic		Pyridoxine
	Hemorrhage, intramuscular, subcutaneous, internal from aortic rupture	Chick, poult	Vitamin K, copper
	Exudative diathesis	Chick, poult	Selenium, vitamin E
	Enlarged heart	Chick, poult	Copper
Muscle	Muscular dystrophy, white areas of degeneration in skeletal muscle	Chick, duck, poult	Vitamin E, selenium
	Cardiac myopathy	Poult	Vitamin E, selenium
	Gizzard myopathy	Poult	Vitamin E, selenium
Bone disorders	Soft, easily bent bones and beak (rickets)	All poultry	Vitamin D, calcium, or phosphorus deficiency or imbalance
	Hock enlargement	Poult, chick, gosling, duckling	Niacin, zinc
	Perosis	Chick, poult	Biotin, choline, vitamin B_{12}, manganese, zinc, folacin
	Bowed legs	Duck	Niacin
	Shortening and thickening of leg bones	Chick	Zinc, manganese
	Curled toes	Chick	Riboflavin
Diarrhea		Chick, duck, poult	Niacin, riboflavin, biotin

NOTES: Slow growth and general lack of vigor are generally associated with malnutrition. The signs listed in this table are more specific indications of deficiencies of particular nutrients.

is involved in the citric acid cycle, the pentose phosphate pathway, and the hexose monophosphate shunt.

Thiamin pyrophosphate can act as a phosphate donor for phosphorylation of certain proteins (Combs and McClung, 2017f). This compound is also an essential cofactor for transketolase, pyruvate dehydrogenase complex, α-ketoglutaric dehydrogenase complex, branched chain α-keto acid dehydrogenase complex, and 2-hydroxyacyl CoA lyase (Combs and McClung, 2017f).

At low levels in the intestine, thiamin is actively transported and is passively absorbed at higher concentrations (Combs and McClung, 2017f). An increase in thiamin-requiring enzyme activity can induce a thiamin deficiency, as can dietary factors that interfere with thiamin absorption and metabolism such as aflatoxin (Voigt et al., 1980), tannins (Begovic et al., 1980), or drugs such as amprolium (Anjali et al., 2000). Dietary aflatoxin at 5 μg/kg diet decreased bile, plasma, and liver thiamin by 20 percent to 70 percent in the chick (Voigt et al., 1980).

Thiamin Deficiency

Thiamin is a coenzyme for several enzymes catalyzing decarboxylation and transketolation-type reactions, and a deficiency can lead to less acetyl CoA entering the citric acid cycle, reduced formation of ATP, and a reliance on non-glucose substrates for energy (Remus and Firman, 1990). Although the activity of all these enzymes is depressed in a thiamin deficiency, the accrual of pyruvic acid from decreased brain pyruvic oxidase seems to manifest the most symptoms (Lofland et al., 1963). A decreased ATP/ADP ratio in the hindbrain of thiamin-deficient turkeys led to anorexia (Remus and Firman, 1990), which is typically the initial deficiency symptom observed and, when severe, results in weight loss and leg weakness (Klasing and Korver, 2020). Neurological symptoms of a thiamin deficiency in chicks include lethargy, head tremors, ataxia and opisthotonus, and eventually, death (Gries and Scott, 1972b). Opisthotonus, which involves ataxia and awkward backward flexions of the head and neck ("star gazing"), are typical nervous symptoms (Gries and Scott, 1972b). Neither gross nor histological lesions were observed in the brain, but the duodenal crypts of Lieberkühn were dilated in deficient chicks (Gries and Scott, 1972b). In parrots, thiamin deficiency causes anorexia, decreased motor control, and potentially, death (Macwhirter, 1994). Deficient birds can rapidly detect and discriminate against feeds that do not provide the vitamin (Hughes and Wood-Gush, 1971) and are high in carbohydrate content (Thornton and Shutze, 1960). Chickens suffering from thiamin deficiency respond in a matter of hours to an oral administration of the vitamin.

Most complete feeds satisfy the thiamin requirement because grains and their by-products usually contain adequate amounts. Thiamin is unstable to heat at neutral and alkaline pH (Dwivedi and Arnold, 1973); feed pelleting (Guo and Summers, 1969) or extrusion (Beetner et al., 1974) under these circumstances facilitates loss. Amaranth is very low in thiamin, and the level is reduced further if it is heated to destroy growth-inhibiting properties (Laovoravit et al., 1986). Inclusion of certain fish meals having enzymes capable of destroying thiamin may also decrease dietary content (Ishihara et al., 1974; Bryan et al., 1975). Use of medicants or compounds with thiamin antagonist activity such as amprolium, oxythiamine, pyrithiamine, and chloroethylthiamine can also cause a deficiency (Ott et al., 1965; Shindo and Komai, 1972).

The hen transfers thiamin to the egg in proportion to dietary content (Polin et al., 1963). Although the dietary deficiencies possible under practical terms do not affect breeder flock productivity, high mortality of embryos occurs prior to hatching, and chicks that hatch express a polyneuritis (Polin et al., 1962; Charles et al., 1972).

Thiamin Toxicity

Excess dietary intakes of thiamin, even at levels up to 1,000 times the requirement, are well tolerated and apparently safe for most animal species (NRC, 1987); any unneeded by tissues is excreted in the urine and not stored in any appreciable amounts in the body (Klasing and Korver, 2020). There is no published evidence of thiamin toxicity in poultry.

Riboflavin (Vitamin B_2)

Riboflavin exists in three forms: as riboflavin and as the metabolically active derivatives flavin adenine dinucleotide (FAD) and flavin mononucleotide (FMN). Riboflavin coenzymes include a wide range of enzymes involved in carbohydrate, lipid, and protein metabolism. The flavoproteins FAD and FMN are essential intermediaries in many biological reduction-oxidation reactions (Combs and McClung, 2017g). Erythrocyte glutathione reductase (Lee, 1982) and liver xanthine dehydrogenase (Chou, 1971) have a functional need for riboflavin, and their activities in fowl reflect dietary adequacy. Prior to the development of concentrated riboflavin sources, milk products were incorporated in feed to avoid deficiencies (Culton and Bird, 1940). Riboflavin also plays a role in skin development and strength and in wound healing.

In humans, riboflavin is absorbed from the lumen of the intestine into circulation via riboflavin transporters (Said and Nexo, 2012). Such transporters have not yet been identified in the avian intestine.

Riboflavin Deficiency

General signs of a riboflavin deficiency include decreased feed intake and weight gain (Tang et al., 2013). Riboflavin deficiencies lead to neurological problems, particularly with the sciatic and brachial nerves, where myelin degeneration, Schwann cell proliferation, and axis cylinder fragmentation have been observed (Phillips and Engel, 1938). The legs of deficient chickens appear as splay and hock resting postures, as well as curling of the toes (Wyatt et al., 1973; Ruiz and Harms, 1988b). The characteristic lesion of a riboflavin deficiency in growing chicks is curled-toe paralysis (Gries and Scott, 1972b), and deficiency in ducks leads to leg paralysis with a different presentation (Tang et al., 2013). Turkey poults (Ruiz and Harms, 1989b) and pheasants (Scott et al., 1959) exhibit similar symptoms as the chick, whereas ducks (Fritz et al., 1939) and geese (Serafin, 1981) are more likely to have a bowing of the legs in conjunction with perosis (abnormal bone formation at the hock joint, the slipping of gastrocnemius from its condyles, resulting in a twisting of the leg). Broiler chicks fed a riboflavin-deficient diet developed tomaculous neuropathies of the brachial and sciatic nerves (Cai et al., 2006) or fibroblastic onion bulb–like structures in peripheral nerves (Cai et al., 2007) by 11 days of age. Goff et al. (1953) noted that increased hematocrit, increased mean corpuscular volume, decreased mean hemoglobin concentration, and a marked heterophil leucocytosis appeared in the chick prior to neurological manifestations.

Adult cockerels can endure a riboflavin-deficient feed for a prolonged period before neurological and blood problems similar to those of the growing chick appear (Arscott, 1972). Deficiency symptoms can be reversed upon riboflavin administration to adults, but correction in growing birds becomes increasingly difficult as expression progresses.

Laying hens transfer riboflavin into the yolk and albumen by hormonally induced binding proteins in the liver and oviduct, respectively (Hamazume et al., 1984; White and Merrill, 1988). Saturation of these carriers is dependent on dietary riboflavin content (White et al., 1986), and although deficiency is more likely to adversely affect embryonic development than harm the hen (Tarhay et al., 1975), reduced egg production and increased liver size and fat content are observed (Klasing and Korver, 2020). Severe deficiencies cause death of embryos at 60 hours of incubation because of circulatory system failures (Romanoff and Bauernfeind, 1942). Moderate deficiencies result in deaths at day 14 of incubation, with the appearance of shortened limbs, malformed mandibles, and clubbing of the down. Marginal deficiencies further delay mortality until pipping, and symptoms are largely dwarfism with clubbed down.

Riboflavin Toxicity

Because riboflavin does not appear to be well absorbed from the intestine, it is unlikely to present a hazard to animals when included at high dietary levels (NRC, 1987). Riboflavin toxicity has not been reported in poultry.

Niacin (Vitamin B_3)

Niacin is the collective term for nicotinic acid and nicotinamide, each of which can be incorporated into the coenzymes nicotinamide adenine dinucleotide (NAD) and nicotinamide adenine dinucleotide phosphate (NADP). Both nicotinic acid and nicotinamide have similar activity in fowl (Ruiz and Harms, 1988a). Niacin has wide-ranging involvement in almost all aspects of metabolism via NAD and NADP; each is essential in the metabolism of carbohydrates, fatty acids, and amino acids (Combs and McClung, 2017h). These coenzymes are essential as electron transporters and in most hydrogen transfer reactions in metabolism. Niacin appears to be absorbed via a specific transporter in humans (Nabokina et al., 2005), although this has not yet been described in birds.

Niacin can be produced within the body from the conversion of tryptophan, although the conversion rate is relatively low and variable among species. Therefore, niacin is considered to be essential in the diet of poultry. Niacin requirement studies must therefore control for the conversion of tryptophan to niacin. In Pekin ducklings, mule ducklings, and broiler chicks, 1 mg of niacin can be produced from 181, 172, and 47 mg of tryptophan, respectively (Chen et al., 1996). Because the conversion efficiency is poor, especially in turkeys and ducks, tryptophan is not recommended as a substitute for dietary niacin supplementation (Ruiz and Harms, 1990). This species difference is due to ducks having three to four times greater hepatic picolinic acid carboxylase activities than chickens (Scott and Heuser, 1952; Chen et al., 1996), which diverts 2-amino-3-acroleyl fumaric acid toward the glutaryl-CoA pathway rather than allowing it to condense to quinolinic acid, the immediate precursor of nicotinic acid (McDowell, 2008).

Niacin Deficiency

Availability of niacin in grain and grain by-products is generally low (Manoukas et al., 1968; Yen et al., 1977), and their contribution in determining dietary adequacy is usually ignored in diet formulation. Chicks at hatch have considerable yolk tryptophan reserves; thus, a niacin deficiency will not readily occur unless the feed is low in both this amino acid and the vitamin (Snell and Quarles, 1941). A moderate niacin deficiency reduced performance of chicks, but a severe deficiency caused a dilation of the duodenal crypts of Lieberkühn, similar to that observed with thiamin, riboflavin, and pantothenic acid deficiencies (Gries and Scott, 1972b). Briggs et al. (1943) reported that 2 weeks were required to elicit a deficiency with chicks and that an inflammation of the oral cavity and occasional poor feathering, dermatitis, and perosis were the primary symptoms. Perosis is a malformation of the bones involving an enlargement of the hock, twisted metatarsi, and a resulting slippage of the tendon off the condyle. Niacin deficiency perosis has been reported as a primary deficiency symptom in chicks (Patterson et al., 1956), turkeys (Scott, 1953; Ruiz and Harms, 1988a), ducks (Scott and Heuser, 1952; Heuser and Scott, 1953), pheasants (Scott et al., 1959), and goslings (Serafin, 1981). A niacin-deficient diet reduced quail pectoral muscle protein synthesis, resulting in severe atrophy (Park and Marquardt, 1982).

Niacin Toxicity

Although niacin toxicity in poultry is unlikely, dietary levels of nicotinamide exceeding 5,000 mg/kg diet depressed chick growth from 8 to 16 days (Baker et al., 1976). Levels exceeding 7,500 mg/kg diet have been reported to cause decreased bone thickness and strength (Klasing and Korver, 2020). In contrast, dietary levels as great as 20,000 mg nicotinic acid/kg diet did not affect growth (Baker et al., 1976).

Pantothenic Acid (Vitamin B_5)

Pantothenic acid is found in a wide variety of feedstuffs and is the precursor of coenzyme A (CoA) and acyl carrier protein (ACP) (Combs and McClung, 2017k). It is thereby essential in energy metabolism. CoA and ACP are involved in a large number of pathways involved in the metabolism of fat, carbohydrates, and amino acids and the synthesis of various lipids, neurotransmitters, steroid hormones, porphyrins, and hemoglobin. In metabolic reactions involving the carboxyl group or the methyl group of an acyl CoA, a high energy thioester bond is formed with carboxylic acids from the metabolism of fatty acids, amino acids, and carbohydrates (Combs and McClung, 2017k). ACP functions as part of the multi-enzyme complex fatty acid synthase.

Dietary pantothenic acid is primarily found as part of coenzyme A, which is hydrolyzed to release free pantothenic acid. In mammals, the liberated pantothenic acid is carried into the intestinal cells by the sodium-dependent multivitamin transporter, which also transports biotin (Said and Nexo, 2012). The mechanisms of pantothenic acid transport have not been identified in poultry.

Pantothenic Acid Deficiency

Grains contain low concentrations of pantothenic acid, and complete feeds are usually marginal in satisfying the requirement (Southern and Baker, 1981; Ruiz and Harms, 1989a). Pantothenic acid deficiency not only reduces the productive use of available energy (Beagle and Begin, 1976; Cupo and Donaldson, 1986) but also impairs detoxification mechanisms that depend upon acetylation (Kietzmann, 1981). Deficiency symptoms are associated with the skin and nervous system of growing chicks (Gries and Scott, 1972b). Skin lesions include crusts and scabs, which first appear at the angles of the eyes and beak. Lesions on the feet are seldom and slight. Biotin deficiency symptoms are similar except lesions on the feet are more severe and appear before those on the head. Although an extensive ataxia also occurs, lesions associated with the nervous system are difficult to detect. Turkey poults present the same signs as chicks (Kratzer and Williams, 1948a), but poor feathering is the most prevalent deficiency sign in pheasants and quail (Scott et al., 1964).

Heating cereals used to make hen diets reduced hatchability to 0 percent and induced footpad dermatoses; supplementation of pantothenic acid subsequently increased hatchability to 30 percent. This indicates that heat can destroy pantothenic acid, which is partly responsible for the loss in hatchability (Gillis et al., 1942). Adult cockerels receiving diets deficient in pantothenic acid have reduced semen volume and fertility as well as skin lesions (Goeger and Arscott, 1984). Considerably higher levels of pantothenic acid are needed by chicken and turkey hens to maintain hatchability than for egg production (Kratzer et al., 1955; Balloun and Phillips, 1957). Embryonic mortality occurs from about day 14 of incubation or thereafter, depending on the extent of pantothenic acid deficiency (Beer et al., 1963). Chicks that hatch are of poor quality and have variable degrees of subcutaneous hemorrhaging and edema ("stunted chick disease").

Pantothenic Acid Toxicity

Animals generally tolerate high levels of pantothenic acid (NRC, 1987), and there are no reports of pantothenic acid toxicity in poultry.

Pyridoxine (Vitamin B_6)

Pyrodoxine occurs in three natural forms: pyridoxine, pyridoxal, and pyradoxamine, each of which is converted to the active coenzyme pyridoxal phosphate (Combs and McClung, 2017i). The primary activity of pyridoxine is mediated by pyridoxal phosphate, with a lesser role played by pyridoxamine phosphate. Pyridoxal phosphate is essential for virtually all reactions of amino acid metabolism, including decarboxylations, transamination, racemization, elimination, and replacement reactions, as well as β-group interconversions (Combs and McClung, 2017i). Decarboxylations lead to at least four amines that affect nervous system function. Transaminations of certain glycolysis and Krebs cycle intermediates lead to the formation of most of the nonessential amino acids, whereas the reverse is the basis of gluconeogenesis from amino acids. Pyridoxal phosphate is a necessary cofactor for two enzymes needed for the conversion of tryptophan to niacin, as well as in single-carbon metabolism (Combs and McClung, 2017i). Aspartic transaminase activity in the liver (Lee et al., 1976) and the plasma glycine–serine ratio (Sifri et al., 1972) have been employed to evaluate pyridoxine nutriture.

The various phosphorylated and non-phosphorylated forms of pyridoxine are absorbed as pyridoxamine, pyridoxal, and pyridoxine. Uptake of the various forms of pyridoxine appears to be via simple diffusion in mammals (Said and Nexo, 2012), although there is evidence suggesting that a transporter may be involved (Said et al., 2003). The specifics of pyridoxine absorption in poultry are not well understood.

Pyridoxine Deficiency

The pyridoxine content of complete feeds usually satisfies most requirements (Scheiner and De Ritter, 1968). However, the vitamin availability is dependent on the digestibility of each feedstuff (Heard and Annison, 1986). The dietary requirement level may increase as dietary protein increases (Daghir and Shah, 1973) or due to the presence of linatine when linseed (flaxseed) meal is fed (Kratzer and Williams, 1948b; Klosterman et al., 1967). The inclusion of certain drugs that act as competitive inhibitors may also increase the dietary requirement (Fuller and Dunahoo, 1959).

A dietary pyridoxine deficiency reduces chick feed intake and growth rate (Baker et al., 1999); plumage development is abnormal, and affected chicks show a general weakness characterized by squatting with the wings spread and the head on the ground (Lepkovsky and Kratzer, 1942). A pyridoxine deficiency also increases the incidence of pendulous crops (Miller, 1963). An excess of dietary methionine (12 g/kg diet) increased the pyridoxine requirement for maximal growth from 0.73 mg/kg diet to 1.4 mg/kg diet (Scherer and Baker, 2000). Symptoms exhibited by pyridoxine-deficient chicks differ with the extent of the deficiency (Daghir and Balloun, 1963; Gries and Scott, 1972a). A severe deficiency produces an ataxia in combination with nervousness and intermittent episodes of hyperactivity. Prominent pathological findings include hemorrhages at various locations, particularly primary wing feather follicles and gizzard erosions. Marginal pyridoxine deficiencies are most likely to be expressed as a perosis because of problems with bone growth resulting from collagen fiber defects, including increased solubility of proteoglycans and collagen (Klasing and Korver, 2020). Severe deficiencies of pyridoxine result in microcytic, polychromatic, hypochromic anemia with splenic and thymic atrophy (Asmar et al., 1968), whereas marginal deficiencies result in a microcytic, normochromic polycythemia (Blalock and Thaxton, 1984). Deficient chicks show a decreased immunoglobulin M and immunoglobulin G response to antibody challenge (Blalock and Thaxton, 1984).

Although specific symptoms of pyridoxine deficiency are not obvious in adult chickens, deficient hens lose body weight and exhibit reduced egg production (Attar et al., 1967). Deficient hens also have relatively low serum glutamic-oxaloacetic acid transaminase activities and high serum nonprotein nitrogen levels (Attar et al., 1967). The pyridoxine content of eggs reflects that in the feed, and the level necessary to maintain egg production is one-half of that required for hatchability (Fuller et al., 1961). Characteristics of pyridoxine-deficient embryos have not been reported, but antivitamins injected into eggs cause early deaths (Landauer, 1967).

Pyridoxine Toxicity

Little work has been done examining pyridoxine toxicity in poultry. Studies in pigs and dogs suggest that animals can tolerate dietary pyridoxine at up to 50 mg/kg body weight for prolonged periods, and rats can tolerate up to 1,430 mg pyridoxine/kg diet (NRC, 1987). Falcons were able to survive a single intramuscular injection of 5 mg pyridoxine/kg body weight, or a single oral dose of 25 mg pyridoxine/kg body weight; injected doses at 10 mg or above, and oral doses at 50 mg pyridoxine/kg body weight, were fatal (Samour et al., 2016).

Biotin (Vitamin B_7)

Biotin is the common name of hexahydro-2-oxo-1*H*-thieno[3,4-*d*]imidazole-4-pentanoic acid (Combs and McClung, 2017j). It is a required cofactor for biotin-dependent carboxylases and transcarboxylases, is a regulator of gene expression, and via biotinylation, is a substrate for post-translational modification of proteins (Combs and McClung, 2017j). As a cofactor for carboxylases, biotin is involved in carbohydrate, fat, and protein metabolism and the interconversion of carbohydrates and amino acids. Acetyl coenzyme A carboxylase, which participates in fatty acid synthesis, and pyruvate carboxylase, which enables gluconeogenesis from intermediates in the Krebs cycle, are both affected by biotin nutriture (Whitehead and Bannister, 1980; Watkins and Rogel, 1989). Biotin tends to concentrate in liver, kidney, and bone, the primary sites of activity of enzymes requiring this vitamin (Frigg and Torhorst, 1982).

The specific mechanisms of biotin absorption in birds have not been fully verified. However, in mammals, ingested protein-bound biotin is broken down in the lumen of the gastrointestinal tract by proteases and peptidases to biocytin (biotinyl-L-lysine) and biotin-linked short peptides. These biotin compounds are further cleaved to release free biotin in the intestinal lumen by biotinidase (Wolf et al., 1984). Absorption of free biotin is mediated by the same sodium-dependent transporter as pantothenic acid (sodium-dependent multivitamin transporter) (Said and Nexo, 2012).

Biotin Deficiency

Analysis of typical complete feeds indicates that an adequate amount of total biotin is usually present. However, availability of biotin from milo, oats, and barley is low, and it is almost nonexistent from wheat. Although biotin from corn is almost completely available, it is present at a very low level. Therefore, the use of these cereal grains may result in marginal dietary uptake in comparison with biotin requirements (Frigg, 1976). The major clinical signs of biotin deficiency are associated with severe dermatitis on the foot pad, shank, and toes, together with eye exfoliation and exudative dermatitis (Marusich et al., 1970).

Dermatitis of the footpad and head can result from reduced activity of biotin-dependent enzymes, although more recent studies have not shown an effect of biotin on these signs (Cengiz et al., 2012), or only when fed at high levels. High levels of biotin (2,000 μg/kg) reduced severity of footpad dermatitis severity of female turkeys on dry litter but not on wet litter (Youssef et al., 2012), but levels ranging from 0 to 1,600 μg biotin/kg diet had no effect on gross or histological footpad dermatitis lesions in growing turkeys (Mayne et al., 2007). In chickens housed at a high stocking density, 1,521 μg biotin/kg diet increased growth and reduced severity of footpad dermatitis (Sun et al., 2017). Skin lesions can be related to alterations in the fatty acid composition of associated waxes (Logani et al., 1977). Low dietary fat and the necessity for fatty acid synthesis lead to an abnormal array of fatty acids that predisposes poultry to fatty liver and kidney syndrome (FLKS) (Whitehead and Randall, 1982). Subjecting these birds to a fast such that gluconeogenesis is accelerated precipitates a high death rate from lack of glucose (Whitehead and Siller, 1983).

Biotin deficiency can also result in longitudinally distorted bones and perosis (Bryden, 1991; Quarantelli et al., 2007) and FLKS (Bryden, 1991). Presumably, suboptimal dietary biotin reduces formation of prostaglandins from essential fatty acids, and bone growth fails to respond to stresses during development (Watkins et al., 1989).

Biotin-binding proteins are found in the yolk and albumen of eggs (Bush et al., 1988). The amount of biotin associated with the yolk binding protein changes with biotin content in the feed, and hatchability is reduced when the feed is deficient (White et al., 1987). Embryonic mortality due to biotin deficiency occurs largely during the last 3 day of incubation. Dwarfing, chondrodystrophy, and deformities of the mandibles and skeleton become apparent at that time (Couch et al., 1947). Chicks hatched from breeder hens given marginal dietary biotin have increased risk of deficiency (Whitehead et al., 1985). Inducing a deficiency depends on many factors, particularly those affecting supplementary biotin synthesis by microbes in the ceca and coprophagy. Caging and use of probiotics and medicants in the feed are influential in this respect (Leeson, 1982).

Biotin Toxicity

Poultry can safely tolerate dietary levels of biotin that are 4 to 10 times their nutritional requirements, but because the vitamin is not well retained, the maximal tolerable level may be much greater (NRC, 1987). Biotin toxicity is unlikely, even at high dietary levels of inclusion.

Folate

Folate is the generic term for folic acid and the array of extended glutamic acid conjugates possessing the activity of folic acid. Pure folic acid is the compound pteroyl-γ-monoglutamic acid. The folates are enzyme co-substrates in the metabolism of amino acids and nucleotides, single-carbon-metabolism (methyl group transfer), and S-adenosylmethionine-dependent methylations (Combs and McClung, 2017l). Folate is involved in the metabolism of pyridoxine and vitamin B_{12}; the synthesis of many important metabolites including choline, methionine, purines, and uric acid; as well as in the formation of the purine and pyrimidine nucleotides.

Although folate can be present in the gut as folate monoglutamate, dietary folate is mainly present as conjugated polyglutamate forms, which must be hydrolyzed to the monoglutamate form before absorption because of their size, negative charge, and hydrophilicity (Said and Nexo, 2012). Dietary folacin is absorbed and converted to the reduced form (5-methyl-tetrahydrofolic acid) by the intestine and is distributed throughout the body. The mechanisms of folate absorption in poultry are not known, but in mammals they involve a specific, Na^+-dependent carrier mediated mechanism (Said and Nexo, 2012). Bacterial folate monoglutamate can be absorbed from the digestive tract.

Folate Deficiency

Although most complete feeds provide sufficient folate from their natural ingredients, marginal deficiencies are possible (Cropper and Scott, 1967). Folate is required for normal nucleic acid metabolism and formation of the nucleoproteins required for cell replication (Klasing and Korver, 2020); therefore, the requirement decreases with age because diminished growth rate reduces the need for deoxyribonucleic acid synthesis (Naber et al., 1957; Balek and Morse, 1976). Accentuated formation of uric acid with excessive dietary protein increases the folate requirement (Creek and Vasaitis, 1963), as does choline deficiency (Young et al., 1955) and serine (Rabbani et al., 1973). Use of medicants that antagonize folate formation by cecal microflora and management that prevents coprophagy also increases the dietary requirement (Stokstad and Jukes, 1987).

The most obvious symptom of folate deficiency is perosis with the chick (Daniel et al., 1946) and cervical paralysis with turkey poults (Miller and Balloun, 1967). Macrocytic anemia, abnormal nuclear bodies in erythrocytes, and numerous mitoses and hypersegmented granulocytes occur with marginal deficiencies when no physical symptoms are manifested; a severe deficiency results in megaloblasts in the bone marrow and liver damage (Maxwell et al., 1988). Diets low in folate reduced chick growth rate and feed efficiency by 18 days of age, as well as increased liver weight and blood hemoglobin levels (Pesti et al., 1991).

Hens fed deficient levels of folate have impaired oviduct response to estrogen and ability to form albumen (Anderson and Jackson, 1975; Burns and Jackson, 1979). More folate is needed to sustain hatchability than egg production; thus, the embryo will suffer before the hen (Sunde et al., 1950). High embryonic mortality occurs around day 20 of incubation, and the dead embryos from severely depleted hens exhibit a marked bending of the tibiotarsus, and, to a lesser extent, syndactyly and deformed mandibles. Chicks that successfully emerge are stunted and have poorly developed feathers, with weak and brittle shafts, and are abnormally pigmented (Lillie et al., 1950). Folate, lysine, copper, and iron appear to be required for the prevention of achroma of feathers of breeds such as Rhode Island Red and Black Leghorn chicks (Klasing and Korver, 2020).

Folate Toxicity

Few data are available for poultry, but folate is well tolerated by other animals at extremely high dietary levels (NRC, 1987).

Cobalamin (Vitamin B_{12})

The cobalamins are compounds containing a cobalt-centered corrin nucleus, having the biological activity of cyanocobalamin. Cobalamin is synthesized by prokaryotes and is unique among water-soluble vitamins in that it can be stored in animal tissues for prolonged periods of time (Combs and McClung, 2017m). Cobalamins function in animal metabolism as adenosylcobalamin and methylcobalamin, cofactors for enzymes transferring one-carbon units and catalyzing rearrangements in the carbon skeleton of several metabolic intermediates in nucleic acid, carbohydrate, and fat metabolism (Combs and McClung, 2017m). The functions of cobalamin and folate in single-carbon metabolism are interrelated. Some tissues, particularly the liver, can store relatively large amounts of cobalamin and protect the animal from deficiency for long periods of time. Early work showed that the inclusion of a liver paste in chick diets could substitute as a methyl donor for betaine or choline (Gillis and Norris, 1949a), and that the compound responsible was vitamin B_{12} (Gillis and Norris, 1949b).

In mammals, the proteins haptocorrin in the saliva, intrinsic factor in the stomach and the receptor for intrinsic factor -cobalamin in the distal ileum are necessary for intestinal active transport of vitamin B_{12} from the diet (Quadros, 2010). Intestinal bacteria synthesize cobalamin, which can be subsequently absorbed by the chick, and treatment of chickens with a *Lactobacillus* probiotic increased tissue deposition of cobalamin (Al-Fataftah et al., 2013).

In fowl, vitamin-B_{12}-mediated one-carbon transfers involve methionine, serine, choline, and thymidine (Gillis and Norris, 1949b; Henderson and Henderson, 1966; Langer and Kratzer, 1967), whereas the interconversion of methylmalonyl coenzyme A to succinyl coenzyme A is one of the rearrangement reactions requiring vitamin B_{12} (Ward et al., 1988).

Cobalamin Deficiency

The spleen, bone marrow, liver, kidney, and skin have high concentrations of vitamin B_{12} (Monroe et al., 1952). Although plant feedstuffs are devoid of vitamin B_{12}, its availability from animal products and cecal microflora after coprophagy makes deficiencies unlikely (Milligan et al., 1952); all-vegetable diets fed to cage-reared birds must be supplemented with cobalamin. Deficiencies in chicks have been created by greatly increasing dietary protein content such that carbon rearrangement enzyme activities are accentuated (Rys and Koreleski, 1974; Patel and McGinnis, 1980; Ward et al., 1985). Poor feathering and mortality are the most obvious symptoms of a vitamin B_{12} deficiency, and gizzard erosions may also appear (Mushett and Ott, 1949; Milligan et al., 1952). Vitamin B_{12} deficiency causes myelin degeneration in chicks, while chondrodystrophy may occur in chicks and poults when their diets also lack choline, methionine, or betaine as sources of methyl groups (Klasing and Korver, 2020). Because of the intimate relationship between cobalamin and folate in metabolism, the two vitamins have similar deficiency signs.

Yacowitz et al. (1952) fed a high-protein all-vegetable diet devoid of vitamin B_{12} to hens in cages and reported a reduction in hatchability. Olcese et al. (1950) observed that most embryonic mortality due to vitamin B_{12} deficiency in hens occurs at about day 17 of incubation, with atrophy of the leg musculature and hemorrhaging common. Ferguson et al. (1955) further observed fatty organs, dwarfing, and edema.

Cobalamin Toxicity

Leghorn chicks fed 30 μg of vitamin B_{12}/kg of diet to 4 weeks of age did not show any adverse effects (Schaefer et al., 1949), and dietary cobalamin is unlikely to result in a toxicity (NRC, 1987).

Choline

Choline is involved in cell membrane structure and function as phosphatidylcholine and sphingomyelin, as a precursor in second-messenger signaling, in lipoprotein synthesis, in neurotransmission, in bile formation, and in osmoregulation (Combs and McClung, 2017n). Choline also acts as a methyl donor (including a role as an epigenetic effector), and its use in this respect becomes important when de novo synthesis of one-carbon units cannot meet demand. Choline is required for normal growth and prevents perosis in young birds (Jukes, 1941).

Choline is synthesized by poultry, although particularly in young birds, and in quail, perhaps not at sufficient quantities to meet the metabolic requirements (Jukes et al., 1945; Baker and Sugahara, 1970) because of a limited ability to methylate aminoethanol. Other methyl donors can replace part, but not all, of the requirement for choline. Betaine can replace up to 50 percent of the choline requirement; however, the choline-sparing ability of betaine is minimal when the diet is devoid of choline (Dilger et al., 2007).

Choline is released from dietary phosphatidylcholine by phospholipases in the intestine, largely in the form of lysophosphatidylcholine. Choline can also be absorbed from the gut in the form of sphingomyelin or as free choline (e.g., from supplemental sources). Choline is absorbed via a saturable, carrier-mediated process that is effective at low luminal concentrations and via passive diffusion at high concentrations (Combs and McClung, 2017n).

Choline Deficiency

The need for supplemental choline is the greatest in the early post-hatching bird because all facets of use are likely to be maximal (Seifter et al., 1972; Pesti et al., 1980). As growth diminishes, the necessity for choline supplementation may disappear (Molitoris and Baker, 1976), although supplemental choline increased carcass yield (Jahanian and Ashnagar, 2018), and a growth response to increasing dietary choline was observed to at least 21 days of age in broiler chicks (de Lima et al., 2018) and to 28 days of age in White Pekin ducks (Wen et al., 2016). Perosis is the primary symptom of a choline deficiency in chicks (Fritz et al., 1967) and turkey poults (Evans et al., 1943), whereas Bobwhite quail develop enlarged hocks and bowed legs (Serafin, 1974). Although perosis can also be caused by deficiencies of other nutrients such as manganese, niacin, folate, biotin, or pyridoxine, it is a common sign of choline deficiency (Settle et al., 1969).

Estrogenic hormones greatly accentuate the choline requirement for phospholipid synthesis in the hen's liver to support yolk formation (Vigo and Vance, 1981). Supplemental choline may relieve the hepatic accumulation of fat and improve egg yolk formation (Schexnailder and Griffith, 1973; Tsiagbe et al., 1988). Minimal dietary choline does not affect hatchability with either chickens (Gish et al., 1949) or turkeys (Ferguson et al., 1975), but Japanese quail and their developing embryos readily express general signs of deficiency (Latshaw and Jensen, 1971, 1972), and there was an increase in egg weight with increasing choline supplementation relative to an unsupplemented basal diet (Griep et al., 2020).

Choline Toxicity

The toxicity of choline is generally considered to be low, and choline is considered to be safe when fed to chickens at up to twice the dietary requirement (McDowell, 2008), and levels well above the requirement are tolerated by poultry (NRC, 1987). At 2,200 mg/kg diet, pyridoxine utilization may be impaired (Deeb and Thornton, 1959; Saville et al., 1967). Derilo and Balnave (1980) reported that feeding excess dietary levels of choline depressed growth and feed efficiency in broilers.

Vitamin C (ascorbic acid)

Vitamin C is generally not considered to be essential for poultry, as they can synthesize sufficient quantities of it under normal circumstances (Nagórna-Stasiak et al., 1999a,b; Hooper et al., 2000; Whitehead and Keller, 2003). Aged laying hens laying thin-shelled eggs produced similar levels of ascorbic acid as those laying eggs with thicker, strong-shells (Maurice et al., 2004). Different types of poultry exposed to stressful conditions (e.g., heat, immunological, psychological) may show a reduced effect of the stressor when supplemented with vitamin C (Whitehead and Keller, 2003), and under heat stress, supplemental vitamin C may be beneficial.

Implications of Removal of Antibiotic Growth Promotors on Microbial Vitamin Synthesis and Absorption

Intestinal synthesis and direct absorption of many of the vitamins can contribute to the overall vitamin nutrition of the bird, and coprophagy can make further contributions (Coates et al., 1968). For example, the dietary vitamin K requirement of pheasant chicks was approximately 8-fold greater when 0.1 percent sulfaquinoxaline was fed relative to a diet without antibiotic (Scott et al., 1961). As the global poultry industry moves away from the use of in-feed antibiotic growth promotors, the determined water-soluble vitamin requirements may need to be reevaluated and potentially decreased.

REFERENCES

Al-Fataftah, A.R.A., S.M. Herzallah, K. Alshawabkeh, and S.A. Ibrahim. 2013. Administration of lactic acid bacteria to enhance synthesis of vitamin B_{12} and B_6 and lower cholesterol levels in poultry meat. *Journal of Food Agriculture & Environment*. 11(2):604–609.

Ameenuddin, S., M. Sunde, H. DeLuca, and M. Cook. 1986. Excessive cholecalciferol in a layers diet: Decline in some aspects of reproductive performance and increased bone mineralisation of progeny. *British Poultry Science*. 27(4):671–677.

Ameenuddin, S., M. Sunde, H.F. DeLuca, N. Ikekawa, and Y. Kobayashi. 1982. 24-hydroxylation of 25-hydroxyvitamin D_3: Is it required for embryonic development in chicks? *Science*. 217(4558):451–452.

Ames, S.R. 1956. Role of vitamin E (α-Tocopherol) in poultry nutrition and disease: A review of recent literature. *Poultry Science*. 35(1):145–159.

Anderson, R., and N. Jackson. 1975. The effect of folate deficiency on the in vitro incorporation of serine into the oviduct of oestrogen-treated, immature female domestic fowl. *International Journal of Biochemistry*. 6(9):649–652.

Anjali, S., A.K. Pruthi, and P.C. Verma. 2000. Sequential pathological changes in experimental thiamine deficiency in Japanese quail (*Coturnix coturnix japonica*). *Indian Journal of Veterinary Pathology*. 24:16–18.

Arnold, R., O. Olson, and C. Carlson. 1974. Tissue selenium content and serum tocopherols as influenced by dietary type, selenium and vitamin E. *Poultry Science*. 53(6):2185–2192.

Arscott, G. 1972. Effect of riboflavin on reproductive performance of adult White Leghorn male chickens. *Nutrition Reports International* 5(4):287–291.

Aslam, S.M., J.D. Garlich, and M.A. Qureshi. 1998. Vitamin D deficiency alters the immune responses of broiler chicks. *Poultry Science*. 77(6):842–849.

Asmar, J.A., N.J. Daghir, and H.A. Azar. 1968. Effects of pyridoxine deficiency on the lymphatic organs and certain blood components of the neonatal chicken. *The Journal of Nutrition*. 95(2):153–159.

Asmundson, V., and F. Kratzer. 1952. Observations on vitamin A deficiency in turkey breeding stock. *Poultry Science*. 31(1):71–73.

Astrup, H. 1979. The ubiquitousity of labelled tocopherol in the cockerel body. *International Journal for Vitamin and Nutrition Research*. 49(2):186–188.

Attar, M., N. Daghir, and J. Asmar. 1967. Influence of vitamin B_6 deficiency on certain serum components in mature female chickens. *Poultry Science*. 46(4):838–843.

Aydelotte, M.B. 1963. Vitamin A deficiency in chickens. *British Journal of Nutrition*. 17(1):205–210.

Bai, K.M., and M. Krishnakumari. 1986. Acute oral toxicity of warfarin to poultry, *Gallus domesticus*: A non-target species. *Bulletin of Environmental Contamination and Toxicology*. 37(1):544–549.

Baker, D., and M. Sugahara. 1970. Nutritional investigation of the metabolism of glycine and its precursors by chicks fed a crystalline amino acid diet. *Poultry Science*. 49(3):756–760.

Baker, D., J. Yen, A. Jensen, R. Teeter, E. Michel, and J. Burns. 1976. Niacin activity in niacinamide and coffee. *Nutrition Reports International*. 14:115–120.

Baker, D.H., N.K. Allen, and A.J. Kleiss. 1973. Efficiency of tryptophan as a niacin precursor in young chick. *Journal of Animal Science*. 36(2):299–302.

Baker, D.H., R.R. Biehl, and J.L. Emmert. 1998. Vitamin D_3 requirement of young chicks receiving diets varying in calcium and available phosphorus. *British Poultry Science*. 39(3):413–417.

Baker, D.H., H.M. Edwards, C.S. Strunk, J.L. Emmert, C.M. Peter, I. Mavromichalis, and T.M. Parr. 1999. Single versus multiple deficiencies of methionine, zinc, riboflavin, vitamin B-6 and choline elicit surprising growth responses in young chicks. *Journal of Nutrition*. 129(12):2239–2245.

Balek, J., and L. Morse. 1976. The effect of age and sex on dihydrofolic acid reductase activity in chicken liver. *Nutrition Reports International*. 13:101.

Balloun, S., and R. Phillips. 1957. Interaction effects of vitamin B_{12} and pantothenic acid in breeder hen diets on hatchability, chick growth and livability. *Poultry Science*. 36(5):929–934.

Bar, A., S. Hurwitz, and I. Cohen. 1972. Relationship between duodenal calcium-binding protein, parathyroid activity and various parameters of mineral metabolism in the rachitic and vitamin D-treated chick. *Comparative Biochemistry and Physiology Part A: Physiology*. 43(3):519–526.

Bartov, I., and S. Bornstein. 1980. Susceptibility of chicks to nutritional encephalopathy: Effect of fat and α-tocopherol content of the breeder diet. *Poultry Science*. 59(2):264–267.

Bartov, I., P. Budowski, Y. Dror, U. Sandbank, and J. Bubis. 1981. Effect of ozone exposure on growth, nutritional encephalopathy, and fatty acid composition of cerebellum and lungs in the young chick. *Poultry Science*. 60(3):532–540.

Beadle, P.C. 1977. The epidermal biosynthesis of cholecalciferol (vitamin D_3). *Photochemistry and Photobiology*. 25(6):519–527.

Beagle, W.S., and J.J. Begin. 1976. Effect of pantothenic acid in diet of growing chicks on energy utilization and body composition. *Poultry Science*. 55(3):950–957.

Beer, A., M. Scott, and M. Nesheim. 1963. The effects of graded levels of pantothenic acid on the breeding performance of White Leghorn pullets. *British Poultry Science*. 4(3):243–253.

Beetner, G., T. Tsao, A. Frey, and J. Harper. 1974. Degradation of thiamine and riboflavin during extrusion processing. *Journal of Food Science*. 39(1):207–208.

Begovic, S., J. Klancevic, A. Lubovac, and A. Sacirbegovic. 1980. Effects of tannin and pyrogallol on the course of experimental B-avitaminosis in fowls, with particular reference to nervous disorders. *Veterinaria Yugoslavia*. 29:177–179.

Berdanier, C., and P. Griminger. 1968. In vitro and in vivo absorption of three vitamin K analogs by chick intestine. *Internationale Zeitschrift fur Vitaminforschung*. 38:376–382.

Bethke, R., P. Record, O. Wilder, and C. Kick. 1936. Effect of different sources of vitamin D on the laying bird: II. Storage of vitamin D in the egg and chick and mineral composition of the mature embryo. *Poultry Science*. 15(4):336–344.

Bird, F.H. 1978. The effect of aflatoxin B_1 on the utilization of cholecalciferol by chicks. *Poultry Science*. 57(5):1293–1296.

Bishop, J.E., and A.W. Norman. 1975. Studies on calciferol metabolism: Metabolism of 25-hydroxy-vitamin D_3 by the chicken embryo. *Archives of Biochemistry and Biophysics*. 167(2):769–773.

Blalock, T., and J. Thaxton. 1984. Hematology of chicks experiencing marginal vitamin B_6 deficiency. *Poultry Science*. 63(6):1243–1249.

Bornstein, S., and Y. Samberg. 1954. Field cases of vitamin K deficiency in Israel. *Poultry Science*. 33(4):831–836.

Boyan, B., and N. Ritter. 1984. Proteolipid-lipid relationships in normal and vitamin D-deficient chick cartilage. *Calcified Tissue International*. 36(1):332–337.

Briggs Jr., G., T. Luckey, L. Teply, C. Elvehjem, and E. Hart. 1943. Studies on nicotinic acid deficiency in the chick. *Journal of Biological Chemistry*. 148:517–522.

Bronner, F., and W.D. Stein. 1995. Calcium homeostasis—an old problem revisited. *The Journal of Nutrition*. 125(Suppl 7):1987S–1995S.

Bruckental, I., and I. Ascarelli. 1975. Influence of vitamin A on formation and excretion of end products of nitrogen catabolism in chicks. *International Journal for Vitamin and Nutrition Research*. 45(4):378–387.

Bruckental, I., I. Ascarelli, and A. Bondi. 1974. Effect of vitamin A deficiency on protein catabolism in chicks. *British Journal of Nutrition*. 31(1):1–11.

Bryan, T., D. Blamberg, R. Gerry, P. Harris, and D. O'Meara. 1975. Observations of possible thiaminase activity in scallop viscera fed in broiler diets: Case report. *Poultry Science*. 54(4):1299–1301.

Bryden, W.L. 1991. Tissue depletion of biotin in chickens and the development of deficiency lesions and the fatty liver and kidney syndrome. *Avian Pathology*. 20(2):259–269.

Budowski, P., I. Bartov, Y. Dror, and E. Frankel. 1979. Lipid oxidation products and chick nutritional encephalopathy. *Lipids*. 14(9):768–772.

Budowski, P., and M. Crawford. 1986. Effect of dietary linoleic and α-linolenic acids on the fatty acid composition of brain lipids in the young chick. *Progress in Lipid Research*. 25:615–618.

Burns, R., and N. Jackson. 1979. The effects of folate deficiency and oestradiol administration on the plasma free amino acid concentrations of the immature hen. *British Poultry Science*. 20(2):131–138.

Bush, L., T.J. McGahan, and H. White III. 1988. Purification and characterization of biotin-binding protein II from chicken oocytes. *Biochemical Journal*. 256(3):797–805.

Cai, Z., P.C. Blumbergs, J.W. Finnie, J. Manavis, and P.D. Thompson. 2007. Novel fibroblastic onion bulbs in a demyelinating avian peripheral neuropathy produced by riboflavin deficiency. *Acta Neuropathology*. 114(2):187–194.

Cai, Z., J.W. Finnie, and P.C. Blumbergs. 2006. Avian riboflavin deficiency: An acquired tomaculous neuropathy. *Veterinary Pathology*. 43(5):780–781.

Caplan, A.I. 1972. The site and sequence of action of 6-aminonicotinamide in causing bone malformations of embryonic chick limb and its relationship to normal development. *Developmental Biology*. 28(1):71–83.

Card, L., H. Mitchell, and T. Hamilton. 1930. Further studies on the vitamin E requirements of poultry. II. The performance of pullets raised on a vitamin E free diet. *Proceedings of the 22nd Meeting, Poultry Science Association*. 29.

Carpenter, K.J. 2012. The discovery of thiamin. *Annals of Nutrition and Metabolism*. 61(3):219–223.

Castillo, L., Y. Tanaka, M.J. Wineland, J.O. Jowsey, and H.F. DeLuca. 1979. Production of 1,25-dihydroxyvitamin D_3 and formation of medullary bone in the egg-laying hen. *Endocrinology*. 104(6):1598–1601.

Cengiz, O., J.B. Hess, and S.F. Bilgili. 2012. Dietary biotin supplementation does not alleviate the development of footpad dermatitis in broiler chickens. *Journal of Applied Poultry Research*. 21(4):764–769.

Centeno, V.A., G.E. Diaz de Barboza, A.M. Marchionatti, A.E. Alisio, M.E. Dallorso, R. Nasif, and N.G. Tolosa de Talamoni. 2004. Dietary calcium deficiency increases Ca2+ uptake and Ca2+ extrusion mechanisms in chick enterocytes. *Comparative Biochemistry and Physiology Part A: Molecular & Integrative Physiology*. 139(2):133–141.

Chandra, M., B. Singh, G. Soni, and S. Ahuja. 1984. Renal and biochemical changes produced in broilers by high-protein, high-calcium, urea-containing, and vitamin-A-deficient diets. *Avian Diseases*. 28(1):1–11.

Charles, O., D. Roland, and H. Edwards Jr. 1972. Thiamine deficiency identification and treatment in commercial turkeys and coturnix quail. *Poultry Science*. 51(2):419–423.

Chen, B.J., T.F. Shen, and R.E. Austic. 1996. Efficiency of tryptophan-niacin conversion in chickens and ducks. *Nutrition Research*. 16(1):91–104.

Chen Jr., P.S., and H.B. Bosmann. 1964. Effect of vitamins D_2 and D_3 on serum calcium and phosphorus in rachitic chicks. *The Journal of Nutrition*. 83(2):133–139.

Cheng, C.H., M. Koch, and R.E. Shank. 1976. Leukocyte transketolase activity as an indicator of thiamin nutriture in rats. *The Journal of Nutrition*. 106(11):1678–1685.

Chou, S.T. 1971. Effect of riboflavin deficiency on metabolism of oxypurines in chicks. *Canadian Journal of Physiology and Pharmacology*. 49(12):1059–1062.

Clark, N.B., M.J. Murphy, and S.K. Lee. 1989. Ontogeny of vitamin D action on the morphology and calcium transport properties of the chick embryonic yolk sac. *Journal of Developmental Physiology*. 11(4):243.

Coates, M.E., J. Ford, and G. Harrison. 1968. Intestinal synthesis of vitamins of the B complex in chicks. *British Journal of Nutrition*. 22(3):493–500.

Combs, G.F. Jr., and M.L. Scott. 1974. Dietary requirements for vitamin E and selenium measured at the cellular level in the chick. *The Journal of Nutrition*. 104(10):1292–1296.

Combs, G.F., and J.P. McClung. 2017a. Chapter 1—What is a vitamin? In *The Vitamins*, 5th ed. G.F. Combs and J.P. McClung, editors. Amsterdam, The Netherlands: Academic Press. 3–6.

Combs, G.F., and J.P. McClung. 2017b. Chapter 6—Vitamin A. In *The Vitamins*, 5th ed. G.F. Combs and J.P. McClung, editors. Amsterdam, The Netherlands: Academic Press. 109–159.

Combs, G.F., and J.P. McClung. 2017c. Chapter 7—Vitamin D. In *The Vitamins*, 5th ed. G.F. Combs and J.P. McClung, editors. Amsterdam, The Netherlands: Academic Press. 161–206.

Combs, G.F., and J.P. McClung. 2017d. Chapter 8—Vitamin E. In *The Vitamins*, 5th ed. G.F. Combs and J.P. McClung, editors. Amsterdam, The Netherlands: Academic Press. 207–242.

Combs, G.F., and J.P. McClung. 2017e. Chapter 9—Vitamin K. In *The Vitamins*, 5th ed. G.F. Combs and J.P. McClung, editors. Amsterdam, The Netherlands: Academic Press. 243–265.

Combs, G.F., and J.P. McClung. 2017f. Chapter 11—Thiamin. In *The Vitamins*, 5th ed. G.F. Combs and J.P. McClung, editors. Amsterdam, The Netherlands: Academic Press. 297–314.

Combs, G.F., and J.P. McClung. 2017g. Chapter 12—Riboflavin. In *The Vitamins*, 5th ed. G.F. Combs and J.P. McClung, editors. Amsterdam, The Netherlands: Academic Press. 315–329.

Combs, G.F., and J.P. McClung. 2017h. Chapter 13—Niacin. In *The Vitamins*, 5th ed. G.F. Combs and J.P. McClung, editors. Amsterdam, The Netherlands: Academic Press. 331–350.

Combs, G.F., and J.P. McClung. 2017i. Chapter 14—Vitamin B_6. In *The Vitamins*, 5th ed. G.F. Combs and J.P. McClung, editors. Amsterdam, The Netherlands: Academic Press. 351–370.

Combs, G.F., and J.P. McClung. 2017j. Chapter 15—Biotin. In *The Vitamins*, 5th ed. G.F. Combs and J.P. McClung, editors. Amsterdam, The Netherlands: Academic Press. 371–385.

Combs, G.F., and J.P. McClung. 2017k. Chapter 16—Pantothenic acid. In *The Vitamins*, 5th ed. G.F. Combs and J.P. McClung, editors. Academic Press. 387–398.

Combs, G.F., and J.P. McClung. 2017l. Chapter 17—Folate. In *The Vitamins*, 5th ed. G.F. Combs and J.P. McClung, editors. Academic Press. 399–429.

Combs, G.F., and J.P. McClung. 2017m. Chapter 18—Vitamin B_{12}. In *The Vitamins*, 5th ed. G.F. Combs and J.P. McClung, editors. Amsterdam, The Netherlands: Academic Press. 431–452.

Combs, G.F., and J.P. McClung. 2017n. Chapter 19—Vitamin-like factors. In *The Vitamins*, 5th ed. G.F. Combs and J.P. McClung, editors. Amsterdam, The Netherlands: Academic Press. 453–498.

Combs Jr., G. 1976. Differential effects of high dietary levels of vitamin A on the vitamin E-selenium nutrition of young and adult chickens. *The Journal of Nutrition*. 106(7):967–975.

Couch, J.R., W.W. Cravens, C.A. Elvehjem, and J.G. Halpin. 1947. Biotin deficiency in the newly hatched chick. *Poultry Science*. 26(5):536. (Abstr.).

Cox, E., and A. White. 1962. Methyl-malonic acid excretion: An index of vitamin-B_{12} deficiency. *Lancet*. 2:853–856.

Cravens, W., W. McGibbon, and E. Sebesta. 1944. Effect of biotin deficiency on embryonic development in the domestic fowl. *Anatomical Record*. 90:55–64.

Cravens, W., S. Randle, C. Elvehjem, and J. Halpin. 1941. Vitamin K studies: I. Effect of the vitamin K content of the hen's ration on the clotting ability of chick blood. *Poultry Science*. 20(4):313–316.

Creek, R., and V. Vasaitis. 1963. The effect of excess dietary protein on the need for folic acid by the chick. *Poultry Science*. 42(5):1136–1141.

Cropper, W., and M. Scott. 1967. Studies on folic acid nutrition in chicks and poults. *British Poultry Science*. 8(1):65–73.

Cruickshank, J., and J. Sim. 1987. Effects of excess vitamin D_3 and cage density on the incidence of leg abnormalities in broiler chickens. *Avian Diseases*. 332–338.

Culton, T.G., and H. Bird. 1940. The effect of some riboflavin supplements on chick growth and curled-toe paralysis. *Poultry Science*. 19(6):424–430.

Cupo, M.A., and W.E. Donaldson. 1986. Effect of pantothenic-acid deficiency on lipogenesis in the chick. *Nutrition Reports International*. 33(1):147–155.

Daghir, N., and S. Balloun. 1963. Evaluation of the effect of breed on vitamin B_6 requirements of chicks. *The Journal of Nutrition*. 79(3):279–288.

Daghir, N., and M. Shah. 1973. Effect of dietary protein level on vitamin B_6 requirement of chicks. *Poultry Science*. 52(4):1247–1252.

Damjanov, I., S. Neilsen, and H. Eaton. 1980. Testicular changes of acute vitamin A deficiency of cockerels. *American Journal of Veterinary Research*. 41(4):586–590.

Daniel, L.J., F.A. Farmer, and L. Norris. 1946. Folic acid and perosis. *Journal of Biological Chemistry*. 163:349–350.

Darby, W.J., K.W. McNutt, and E.N. Todhunter. 1975. Niacin. *Nutrition Reviews*. 33(10):289–297.

Davis, C.Y., and J.L. Sell. 1989. Immunoglobulin concentrations in serum and tissues of vitamin A-deficient broiler chicks after Newcastle disease virus vaccination. *Poultry Science*. 68(1):136–144.

de Lima, M.B., E.P. da Silva, R. Pereira, G.G. Romano, L.W. de Freitas, C.T.S. Dias, and J.F.M. Menten. 2018. Estimate of choline nutritional requirements for chicks from 1 to 21 days of age. *Journal of Animal Physiology and Animal Nutrition*. 102(3):780–788.

De Luca, L., M. Schumacher, and D.P. Nelson. 1971. Localization of the retinol-dependent fucose-glycopeptide in the goblet cell of the rat small intestine. *Journal of Biological Chemistry*. 246(18):5762–5765.

Deeb, S.S., and P.A. Thornton. 1959. The choline requirement of the chick. *Poultry Science*. 38(5):1198–1198.

Derilo, Y.L., and D. Balnave. 1980. The choline and sulphur amino acid requirements of broiler chickens fed on semi-purified diets. *British Poultry Science*. 21(6):479–487.

Dilger, R.N., T.A. Garrow, and D.H. Baker. 2007. Betaine can partially spare choline in chicks but only when added to diets containing a minimal level of choline. *The Journal of Nutrition*. 137(10):2224–2228.

Dobsinska, E., L. Polasek, V. Zidek, M. Karmazin, and M. Sandova. 1982. The influence of diet enriched by vitamin-E on some hematological and biochemical indicators in chickens. *Veterinarni Medicina*. 27(9):557–565.

Dua, P., and E.J. Day. 1966. Vitamin K activity of menadione dimethylpyrimidmol bisulfite in chicks. *Poultry Science*. 45(1):94–96.

Dwivedi, B.K., and R.G. Arnold. 1973. Chemistry of thiamine degradation on food products and model systems. Review. *Journal of Agricultural and Food Chemistry*. 21(1):54–60.

Edwards Jr., H.M. 1974. Uses and limitations of biochemical methods in diagnosing nutritional deficiencies. *Proceedings of the Georgia Nutrition Conference*. 1.

Erin, A.N., M.M. Spirin, L.V. Tabidze, and V.E. Kagan. 1984. Formation of α-tocopherol complexes with fatty acids. A hypothetical mechanism of stabilization of biomembranes by vitamin E. *Biochimica et Biophysica Acta -Biomembranes*. 774(1):96–102.

Evans, R.J., M. Rhian, and C.I. Draper. 1943. Perosis in turkey poults and the choline content of their diets. *Poultry Science*. 22(1):88–93.

Ewen, L.M., and K. Jenkins. 1967. Antidystrophic effect of selenium and other agents on chicks from vitamin E-depleted hens. *The Journal of Nutrition*. 93(4):470–474.

Feng, Y.L., M. Xie, J. Tang, W. Huang, Q. Zhang, and S.S. Hou. 2019. Effects of vitamin A on growth performance and tissue retinol of starter White Pekin ducks. *Poultry Science*. 98(5):2189–2192.

Ferguson, T., R. Atkinson, J. Bradley, and D. Miller. 1975. Reproductive performance of caged Beltsville Small White turkeys as affected by choline, bird density and forced molting. *Poultry Science*. 54(5):1679–1684.

Ferguson, T.M., R.H. Rigdon, and J.R. Couch. 1955. A pathologic study of vitamin B_{12}-deficient chick embryos. *Archives of Pathology*. 60(4):393–400.

Flegal, C.J., P.J. Schaible, and H. Hall. 1971. The relative biopotency of fermentation beta-carotene, crystalline beta-carotene and vitamin A for poultry. *Poultry Science*. 50(2):349–358.

Fleming, R., H. McCormack, and C. Whitehead. 1998. Bone structure and strength at different ages in laying hens and effects of dietary particulate limestone, vitamin K and ascorbic acid. *British Poultry Science*. 39(3):434–440.

Fleming, R.H., H.A. McCormack, L. McTeir, and C.C. Whitehead. 2003. Effects of dietary particulate limestone, vitamin K_3 and fluoride and photostimulation on skeletal morphology and osteoporosis in laying hens. *British Poultry Science*. 44(5):683–689.

Fletcher, R.A. 1971. Effects of vitamin A deficiency on the pituitary-gonad axis of the California quail, *Lophortyx californicus*. *Journal of Experimental Zoology*. 176(1):25–34.

Franchini, A., A. Meluzzi, S. Bertuzzi, and G. Giordani. 1988. High doses of vitamin E in the broilers diets. *Archiv fuer Gefluegelkunde* (Germany, FR).

Friedman, A., and D. Sklan. 1997. Effects of retinoids on immune responses in birds. *World's Poultry Science Journal*. 53(2):185–195.

Friedrichsen, J., and D. Willis. 1980. Improvement in fertility of White Leghorn males by vitamin E following a prolonged deficiency. *Nutrition Reports International*. 22(1):41–47.

Frigg, M. 1976. Bio-availability of biotin in cereals. *Poultry Science*. 55(6):2310–2318.

Frigg, M., and J. Torhorst. 1982. Autoradiographic localisation of 3H-biotin in chick tissues. *International Journal for Vitamin and Nutrition Research*. 52(4):417–422.

Fritz, J.C. 1969. Vitamin K deficiency in chicks fed practical diets. *Poultry Science*. 48(2):736–737.

Fritz, J.C., W. Archer, and D. Barker. 1939. Riboflavin requirements of ducklings. *Poultry Science*. 18(6):449–454.

Fritz, J.C., W.F. Archer, and D.K. Barker. 1942. Observations on the stability of vitamin D. *Poultry Science*. 21(4):361–369.

Fritz, J.C., T. Roberts, and J.W. Boehne. 1967. The chick's response to choline and its application to an assay for choline in feedstuffs. *Poultry Science*. 46(6):1447–1454.

Fuller, H.L., and W.S. Dunahoo. 1959. The effect of various drug additives on the vitamin B_6 requirement of chicks. *Poultry Science*. 38(5):1150–1154.

Fuller, H.L., R. Field, R. Roncalli-Amici, W. Dunahoo, and H. Edwards Jr. 1961. The vitamin B_6 requirement of breeder hens. *Poultry Science*. 40(1):249–253.

Gedek, B., B. Hüttner, D. Kahlau, H. Köhler, and E. Vielitz. 1978. Rickets in broilers due to the contamination of feed with *Fusarium moniliforme* Sheldon. 1. Field observations, reproduction of the disease picture and therapeutic studies. *Zentralblatt fur Veterinarmedizin, Reihe B*. 25(1):29.

Gillis, M., G. Heuser, and L. Norris. 1942. The need for pantothenic acid and an unidentified factor in reproduction in the domestic fowl. *The Journal of Nutrition*. 23(2):153–163.

Gillis, M.B., and L.C. Norris. 1949a. Effect of the animal protein factor on the requirement for methylating compounds. *Journal of Biological Chemistry*. 179(1):487–488.

Gillis, M.B., and L.C. Norris. 1949b. Vitamin B_{12} and the requirement of the chick for methylating compounds. *Poultry Science*. 28(5):749–750.

Gish, C., F. Kummerow, and L. Payne. 1949. Choline and ethanolamine requirements in the laying ration. *Poultry Science*. 28(2):305–306.

Glazener, E., and G. Briggs. 1948. Further studies in abnormal blackening of feathers of vitamin D deficient chicks. *Poultry Science*. 27(4):462–465.

Goeger, M., and G. Arscott. 1984. Effect of pantothenic acid on reproductive performance of adult White Leghorn cockerels. *Nutrition Reports International*. 30(5):1193–1196.

Goff, S., W.C. Russell, and M.W. Taylor. 1953. Hematology of the chick in vitamin deficiencies. 1. Riboflavin. *Poultry Science*. 32(1):54–60.

Griep, D.N., A.A. Pereira, G.M. Oliveira, V.V.S. de Almeida, E.F. da Silva, W.A. da Silva, R.S. de Alcantara, W.A. da Silva, G.S. Cardoso, and C.G. de Oliveira. 2020. Choline and digestible methionine plus cystine supplementation for quail in the laying phase. *Semina-Ciencias Agrarias*. 41(1):305–312.

Gries, C.L., and M. Scott. 1972a. The pathology of pyridoxine deficiency in chicks. *The Journal of Nutrition*. 102(10):1259–1267.

Gries, C.L., and M. Scott. 1972b. The pathology of thiamin, riboflavin, pantothenic acid and niacin deficiencies in the chick. *The Journal of Nutrition*. 102(10):1269–1285.

Griminger, P. 1964. Effect of vitamin K nutrition of the dam on hatchability and prothrombin levels in the offspring. *Poultry Science*. 43(5):1289–1290.

Griminger, P. 1965. Vitamin K activity in chickens: Phylloquinone and menadione in the presence of stress agents. *The Journal of Nutrition*. 87(3):337–343.

Griminger, P. 1966. Influence of maternal vitamin D intake on growth and bone ash of offspring. *Poultry Science*. 45(4):849–851.

Griminger, P. 1987. Vitamin K antagonists: The first 50 years. *The Journal of Nutrition*. 117(7):1325–1329.

Griminger, P., and G. Brubacher. 1966. The transfer of vitamin K_1 and menadione from the hen to the egg. *Poultry Science*. 45(3):512–519.

Griminger, P., Y. Shum, and P. Budowski. 1970. Effect of dietary vitamin K on avian brain thromboplastin activity. *Poultry Science*. 49(6):1681–1686.

Grundboeck, J., K. Honory, and D. Honory. 1977. Influence of vitamin E on the intestinal resorption of carotenoids and vitamin A in adult hens. *Bulletin of the Veterinary Institute in Pulawy*. 21(1–2):55–60.

Guo, L., and J. Summers. 1969. Stability and effect on energy metabolism of thiamine in poultry rations. *Poultry Science*. 48(4):1471–1478.

Hamazume, Y., T. Mega, and T. Ikenaka. 1984. Characterization of hen egg white- and yolk-riboflavin binding proteins and amino acid sequence of egg white-riboflavin binding protein. *The Journal of Biochemistry*. 95(6):1633–1644.

Han, J.C., G.H. Chen, J.L. Zhang, J.G. Wang, H.X. Qu, Y.F. Yan, X.J. Yang, and Y.H. Cheng. 2017. Relative biological value of 1α-hydroxycholecalciferol to 25-hydroxycholecalciferol in broiler chicken diets. *Poultry Science*. 96(7):2330–2335.

Hassan, S., L. Jönsson, and J. Hakkarainen. 1985. Morphological studies on nutritional encephalomalacia in chicks, with special reference to mineralized deposits in the cerebellum. *Zentralblatt für Veterinärmedizin Reihe A*. 32(1–10):662–675.

Hauschka, P.V., J.B. Lian, D. Cole, and C.M. Gundberg. 1989. Osteocalcin and matrix Gla protein: Vitamin K-dependent proteins in bone. *Physiological Reviews*. 69(3):990–1047.

Hauschka, P., and M.L. Reid. 1978a. Vitamin D dependence of a calcium-binding protein containing gamma-carboxyglutamic acid in chicken bone. *Journal of Biological Chemistry*. 253(24):9063–9068.

Hauschka, P.V., and M.L. Reid. 1978b. Timed appearance of a calcium-binding protein containing γ-carboxyglutamic acid in developing chick bone. *Developmental Biology*. 65(2):426–434.

Heard, G.S., and E. Annison. 1986. Gastrointestinal absorption of vitamin B-6 in the chicken (*Gallus domesticus*). *The Journal of Nutrition*. 116(1):107–120.

Heine, U., A. Roberts, E. Munoz, N. Roche, and M. Sporn. 1986. Effects of retinoid deficiency on the development of the heart and vascular system of the quail embryo. *Virchows Archiv B*. 50(1):135.

Henderson, R.F., and T.R. Henderson. 1966. Vitamin B_{12} and synthesis of thymine and choline in chick. *Journal of Nutrition*. 88(1):151–156.

Heuser, G., and M. Scott. 1953. Studies in duck nutrition: 5. Bowed legs in ducks, a nutritional disorder. *Poultry Science*. 32(1):137–143.

Hooper, C.L., D.V. Maurice, S.F. Lightsey, and J.E. Toler. 2000. Factors affecting ascorbic acid biosynthesis in chickens. I. Adaptation of an assay and the effect of age, sex and food deprivation. *Journal of Animal Physiology and Animal Nutrition*. 84(1–2):48–56.

Houghton, L.A., and R. Vieth. 2006. The case against ergocalciferol (vitamin D_2) as a vitamin supplement. *The American Journal of Clinical Nutrition*. 84(4):694–697.

Hoy, D.A., C.F. Ramberg Jr., and R.L. Horst. 1988. Evidence that discrimination against ergocalciferol by the chick is the result of enhanced metabolic clearance rates for its mono- and dihydroxylated metabolites. *The Journal of Nutrition*. 118(5):633–638.

Huang, J.Q., D.L. Li, H. Zhao, L.H. Sun, X.J. Xia, K.N. Wang, X. Luo, and X.G. Lei. 2011. The selenium deficiency disease exudative diathesis in chicks is associated with downregulation of seven common selenoprotein genes in liver and muscle. *Journal of Nutrition*. 141(9):1605–1610.

Hughes, B.O., and D.G.M. Wood-Gush. 1971. Investigations into specific appetites for sodium and thiamine in domestic fowls. *Physiology & Behavior*. 6(4):331–339.

Hurwitz, S., H. Harrison, E. Bull, and H. Harrison. 1967. Comparison of the actions of vitamins D_2 and D_3 in the chick with their retention in serum, liver and intestinal mucosa. *The Journal of Nutrition*. 91(2):208–212.

Ikumo, H. 1980. Comparison in inducing effect on vitamin E deficiency symptoms in chicks between dilauryl succinate and unsaturated fatty acids. *The Journal of Nutrition*. 110(10):2045–2050.

Irandoust, H., and D.U. Ahn. 2015. Influence of soy oil source and dietary supplementation of vitamins E and C on the oxidation status of serum and egg yolk, and the lipid profile of egg yolk. *Poultry Science*. 94(11):2763–2771.

Ishihara, T., H. Kinari, and M. Yasuda. 1974. Studies on thiaminase I in marine fish. III. Vitamin B_1 deficiency disease of chicken caused by inclusion of anchovy in diet. *Bulletin of the Japanese Society of Scientific Fisheries*. 40:309–316.

Jahanian, R., and M. Ashnagar. 2018. Effects of dietary supplementation of choline and carnitine on growth performance, meat oxidative stability and carcass composition of broiler chickens fed diets with different metabolisable energy levels. *British Poultry Science*. 59(4):470–476.

Jansen, B., and W. Donath. 1926. The isolation of anti-beriberi vitamin. *Geneeskundig Tijdschrift voor Nederlandsche-Indie*. 66(4).

Jensen, L. 1974. *Fat Soluble Vitamin Problems in Biochemical Diagnosis*. Athens, GA: University of Georgia. 14.

Jukes, T., J. Oleson, and A. Dornbush. 1945. Observations on monomethylaminoethanol and dimethylaminoethanol in the diet of chicks. *The Journal of Nutrition*. 30(3):219–223.

Jukes, T.H. 1941. Effects of choline, gelatin and creatine on perosis in chicks. *Proceedings of the Society for Experimental Biology and Medicine*. 46:155–157.

Kaetzel, D.M., and J.H. Soares. 1985. Effect of dietary calcium stress on plasma vitamin D_3 metabolites in the egg-laying Japanese quail. *Poultry Science*. 64(6):1121–1127.

Khanal, R.C., N.M. Smith, and I. Nemere. 2007. Phosphate uptake in chick kidney cells: Effects of 1,25(OH)2D3 and 24,25(OH)2D3. *Steroids*. 72(2):158–164.

Kietzmann, M. 1981. Influence of pantothenic acid on the metabolism and excretion of sulfonamides in chicken. *Archiv fuer Gefluegelkunde* (Germany, FR). 45(6):233–239.

Klasing, K.C. 1998a. *Comparative Avian Nutrition*. Wallingford, Oxfordshire, UK: CAB International.

Klasing, K.C. 1998b. Nutritional modulation of resistance to infectious diseases. *Poultry Science*. 77(8):1119–1125.

Klasing, K.C. 2008. Nutritional diseases. In *Diseases of Poultry*, 12th ed. Y.M. Saif, editor. Ames, IA: Blackwell Publishing. 1121–1148.

Klasing, K.C. 2013. Nutritional diseases. In *Diseases of Poultry*. D.E. Swayne, M. Boulianne, C.M. Logue, L.R. McDougald, V. Nair, D.L. Suarez, S. de Wit, T. Grimes, D. Johnson, M. Kromm, T.Y. Prajitno, I. Rubinoff, and G. Zavala, editors. Hoboken, NJ: John Wiley & Sons. 1203–1232.

Klasing, K.C., and D.R. Korver. 2020. Nutritional diseases. In *Diseases of Poultry*, 14th ed. D.E. Swayne, M. Boulianne, C.M. Logue, L.R. McDougald, V. Nair, D.L. Suarez, S. de Wit, T. Grimes, D. Johnson, M. Kromm, T.Y. Prajitno, I. Rubinoff, and G. Zavala, editors. Hoboken, NJ: John Wiley & Sons. 1257–1285.

Klosterman, H., G. Lamoureux, and J. Parsons. 1967. Isolation, characterization, and synthesis of linatine. A vitamin B_6 antagonist from flaxseed (*Linum usitatissimum*). *Biochemistry*. 6(1):170–177.

Kohler, H., B. Huttner, E. Vielitz, D. Kahlau, and B. Gedek. 1978. Rickets in broilers by food contamination with *Fusarium-moniliforme* Sheldon. 2. Histological and mycotoxicological studies. *Zentralblatt Fur Veterinarmedizin Reihe B*. 25(2):89–109.

Kratzer, F., P. Davis, and U.K. Abbott. 1956. The folic acid requirements of turkey breeder hens. *Poultry Science*. 35(3):711–716.

Kratzer, F., P. Davis, B. Marshall, and D. Williams. 1955. The pantothenic acid requirement of turkey hens. *Poultry Science*. 34(1):68–72.

Kratzer, F., and D. Williams. 1948a. The pantothenic acid requirement of poults for early growth. *Poultry Science*. 27(4):518–523.

Kratzer, F., and D. Williams. 1948b. The relation of pyridoxine to the growth of chicks fed rations containing linseed oil meal. *The Journal of Nutrition*. 36(2):297–305.

Kriz, H., and J. Holman. 1969. Histology of onset of skin changes in hypervitaminosis A in chickens. *Internationale Zeitschrift fur Vitaminforschung*. 39:3–12.

Kühn, J., A. Schutkowski, F. Hirche, A.C. Baur, N. Mielenz, and G.I. Stangl. 2015. Non-linear increase of vitamin D content in eggs from chicks treated with increasing exposure times of ultraviolet light. *Journal of Steroid Biochemistry and Molecular Biology*. 148:7–13.

Kühn, J., C. Wassermann, S. Ebschke, A. Schutkowski, K. Thamm, M. Wensch-Dorendorf, E. von Borell, and G.I. Stangl. 2019. Feasibility of artificial light regimes to increase the vitamin D content in indoor-laid eggs. *Poultry Science*. 98(10):5177–5187.

Kumar, R., R.S. Brar, and H.S. Banga. 2017. Hypervitaminosis D_3 in broiler chicks: Histopathological, immunomodulatory and immunohistochemical approach. *Iranian Journal of Veterinary Research*. 18(3):170–176.

Landauer, W. 1957. Niacin antagonists and chick development. *Journal of Experimental Zoology*. 136:509–530.

Landauer, W. 1967. The hatchability of chicken eggs as influenced by environment and heredity. Monograph 1. (Revised). Storrs, CT: Connecticut Agricultural Experiment Station.

Langer Jr., B., and F. Kratzer. 1967. The vitamin B_{12}-formaldehyde complex as a one-carbon unit precursor in the biosynthesis of methionine in turkey poult liver homogenates. *Poultry Science*. 46(3):749–754.

Laovoravit, N., F. Kratzer, and R. Becker. 1986. The nutritional value of amaranth for feeding chickens. *Poultry Science*. 65(7):1365–1370.

Latshaw, J., and L. Jensen. 1971. Choline level and its effect on egg weight in the Japanese quail. *Poultry Science*. 50(3):790–794.

Latshaw, J., and L. Jensen. 1972. Choline deficiency and synthesis of choline from precursors in mature Japanese quail. *The Journal of Nutrition*. 102(6):749–755.

Latshaw, J., and M. Osman. 1974. A selenium and vitamin E responsive condition in the laying hen. *Poultry Science*. 53(5):1704–1708.

Lau, K.-S., C. Gottlieb, L.R. Wasserman, and V. Herbert. 1965. Measurement of serum vitamin B_{12} level using radioisotope dilution and coated charcoal. *Blood*. 26(2):202–214.

Lee, D., A. Robblee, and R. Blair. 1976. The use of growth and liver aspartate aminotransferase to assess the effect of source of non-essential nitrogen on pyridoxine depletion, repletion and requirements of chicks. *British Poultry Science*. 17(1):43–51.

Lee, D.J.W. 1982. Growth, erythrogyte glutathione redugtase and liver flavin as indicators of riboflavin status in turkey poults. *British Poultry Science*. 23(3):263–272.

Leeson, S. 1982. Factors affecting biotin specification of poultry diets. *Proceedings of the Eighteenth Annual Nutrition Conference for Feed Manufacturers*. 97.

Leeson, S. 2007. Vitamin requirements: Is there basis for re-evaluating dietary specifications? *World's Poultry Science*. 63(2):255–266.

Lepkovsky, S., and F.H. Kratzer. 1942. Pyridoxine deficiency in chicks. *Journal of Nutrition*. 24(6):515–521.

Leshchinsky, T.V., and K.C. Klasing. 2001. Relationship between the level of dietary vitamin E and the immune response of broiler chickens. *Poultry Science*. 80(11):1590–1599.

Li, J., D. Bi, S. Pan, Y. Zhang, and D. Zhou. 2008. Effects of high dietary vitamin A supplementation on tibial dyschondroplasia, skin pigmentation and growth performance in avian broilers. *Research in Veterinary Science*. 84(3):409–412.

Lietzow, J., H. Kluge, C. Brandsch, N. Seeburg, F. Hirche, M. Glomb, and G.I. Stangl. 2012. Effect of short-term UVB exposure on vitamin D concentration of eggs and vitamin D status of laying hens. *Journal of Agricultural and Food Chemistry*. 60(3):799–804.

Lillie, R.J., G. Combs, and G. Briggs. 1950. Folic acid in poultry nutrition: II. Effect of maternal diet and chick diet upon mortality, growth and feathering of progeny. *Poultry Science*. 29(1):122–129.

Lin, C., J. Gray, A. Asghar, D. Buckley, A. Booren, and C. Flegal. 1989. Effects of dietary oils and α-tocopherol supplementation on lipid composition and stability of broiler meat. *Journal of Food Science*. 54(6):1457–1460.

Lofland Jr., H.B., H.O. Goodman, T.B. Clarkson, and R.W. Prichard. 1963. Enzyme studies in thiamine-deficient pigeons. *The Journal of Nutrition*. 79(2):188–194.

Logani, M.K., D.B. Nhari, P.D. Forbes, and R.E. Davies. 1977. Diester waxes from skin lipids of the feet of biotin depleted and biotin supplemented turkey poults. *Lipids*. 12(7):626–628.

Long, P.H., S. Lee, G. Rowland, and W. Britton. 1984. Experimental rickets in broilers: gross, microscopic, and radiographic lesions. III. Vitamin D deficiency. *Avian Diseases*. 933–943.

Lopez, G., R. Phillips, and C. Nockels. 1973. Body water kinetics in vitamin A-deficient chickens. *Proceedings of the Society for Experimental Biology and Medicine*. 144(1):54–55.

Lowenthal, J., and J. MacFarlane. 1965. Vitamin K-like and antivitamin K activity of substituted para-benzoquinones. *Journal of Pharmacology and Experimental Therapeutics*. 147(1):130–138.

Machlin, L.J. 1991. *Handbook of Vitamins*, 2nd ed. New York: Marcel Dekker.

Macwhirter, P. 1994. Malnutrition. In *Avian Medicine: Principles and Application*. B.W. Ritchie and L.R. Harrison, editors. Lake Worth, FL: Wingers Publishing. 842–861.

Manoukas, A., R. Ringrose, and A. Teeri. 1968. The availability of niacin in corn, soybean meal and wheat middlings for the hen. *Poultry Science*. 47(6):1836–1842.

March, B., E. Wong, L. Seier, J. Sim, and J. Biely. 1973. Hypervitaminosis E in the chick. *The Journal of Nutrition*. 103(3):371–377.

Marusich, W., E. Ogrinz, M. Brand, and M. Mitrovic. 1970. Induction, prevention and therapy of biotin deficiency in turkey poults on semi-purified and commercial-type rations. *Poultry Science*. 49(2):412–421.

Massey, J.B. 1984. Kinetics of transfer of α-tocopherol between model and native plasma lipoproteins. *Biochimica et Biophysica Acta (BBA)—Lipids and Lipid Metabolism*. 793(3):387–392.

Mattila, P., J. Valaja, L. Rossow, E. Venalainen, and T. Tupasela. 2004. Effect of vitamin D_2- and D_3-enriched diets on egg vitamin D content, production, and bird condition during an entire production period. *Poultry Science*. 83(3):433–440.

Maurice, D.V., S.F. Lightsey, and J.E. Toler. 2004. Ascorbic acid biosynthesis in hens producing strong and weak eggshells. *British Poultry Science*. 45(3):404–408.

Maxwell, M.H., C.C. Whitehead, and J. Armstrong. 1988. Haematological and tissue abnormalities in chicks caused by acute and subclinical folate deficiency. *British Journal of Nutrition*. 59(1):73–80.

Mayne, R.K., R.W. Else, and P.M. Hocking. 2007. High dietary concentrations of biotin did not prevent foot pad dermatitis in growing turkeys and external scores were poor indicators of histopathological lesions. *British Poultry Science*. 48(3):291–298.

McCarthy, J.T., S.S. Barham, and R. Kumar. 1984. 1, 25-Dihydroxyvitamin D_3 rapidly alters the morphology of the duodenal mucosa of rachitic chicks: Evidence for novel effects of 1, 25-dihydroxyvitamin D_3. *Journal of Steroid Biochemistry*. 21(3):253–258.

McDowell, L.R. 2008. *Vitamins in Animal and Human Nutrition*, 2nd ed. New York: Wiley-Blackwell.

Metz, A.L., M.M. Walser, and W.G. Olson. 1985. The interaction of dietary vitamin A and vitamin D related to skeletal development in the turkey poult. *The Journal of Nutrition*. 115(7):929–935.

Miller, B.F. 1963. Pendulous crop in pyridoxine deficient chicks. *Poultry Science*. 42(3):795–796.

Miller, D., and S. Balloun. 1967. Folacin requirements of turkey breeder hens. *Poultry Science*. 46(6):1502–1508.

Milligan, J.L., G.H. Arscott, and G.F. Combs. 1952. Vitamin B_{12} requirement of New Hampshires: 2. Influence of maternal ration on requirement of progeny. *Poultry Science*. 31(5):830–837.

Molitoris, B.A., and D.H. Baker. 1976. The choline requirement of broiler chicks during the seventh week of life. *Poultry Science*. 55(1):220–224.

Monroe, R.A., H. Patrick, C.L. Comar, and O.E. Goff. 1952. Metabolism of vitamin B_{12}: 1. The comparative excretion and distribution in the chick of CO^{60} and vitamin B_{12} labeled with CO^{60}. *Poultry Science*. 31(1):79–84.

Morris, A., and R.K. Selvaraj. 2014. In vitro 25-hydroxycholecalciferol treatment of lipopolysaccharide-stimulated chicken macrophages increases nitric oxide production and mRNA of interleukin-1beta and 10. *Veterinary Immunology Annopathology*. 161(3–4):265–270.

Morris, A., R. Shanmugasundaram, M.S. Lilburn, and R.K. Selvaraj. 2014. 25-Hydroxycholecalciferol supplementation improves growth performance and decreases inflammation during an experimental lipopolysaccharide injection. *Poultry Science*. 93(8):1951–1956.

Morris, A., R. Shanmugasundaram, J. McDonald, and R.K. Selvaraj. 2015. Effect of in vitro and in vivo 25-hydroxyvitamin D treatment on macrophages, T cells, and layer chickens during a coccidia challenge. *Journal of Animal Science*. 93(6):2894–2903.

Morrissey, R.L., R.M. Cohn, R. Empson Jr., H.L. Greene, O.D. Taunton, and Z.Z. Ziporin. 1977. Relative toxicity and metabolic effects of cholecalciferol and 25-hydroxycholecalciferol in chicks. *The Journal of Nutrition*. 107(6):1027–1034.

Mueller, A.S., J. Fischer, E. Most, and J. Pallauf. 2009. Investigation into selenium requirement of growing turkeys offered a diet supplemented with two levels of vitamin E. *Journal of Animal Physiology and Animal Nutrition*. 93(3):313–324.

Murphy, T., K. Wright, and W. Pudelkiewicz. 1981. An apparent rachitogenic effect of excessive vitamin E intakes in the chick. *Poultry Science*. 60(8):1873–1878.

Mushett, C.W., and W.H. Ott. 1949. Influence of crystalline vitamin B_{12} on gizzard erosions in chicks. *Poultry Science*. 28(6):850–854.

Naber, E.C. 1993. Modifying vitamin composition of eggs: A review. *Journal of Applied Poultry Research*. 2(4):385–393.

Naber, E.C., H. Bird, and C. Morgan. 1957. Nucleic acids as growth factors for the folacin deficient chick. *Poultry Science*. 36:480–485.

Nabokina, S.M., M.L. Kashyap, and H.M. Said. 2005. Mechanism and regulation of human intestinal niacin uptake. *American Journal of Physiology—Cell Physiology*. 289(1):C97–C103.

Nagórna-Stasiak, B., M. Kowalczyk, and J. Lechowski. 1999a. Synthesis of vitamin C in geese in embrional and postnatal life. *Medycyna Weterynaryjna*. 55:619–622.

Nagórna-Stasiak, B., J. Lechowski, and M. Kowalczyk. 1999b. Synthesis of vitamin C in turkeys. *Medycyna Weterynaryjna*. 55(3):195–198.

Narbaitz, R. 1987. Role of vitamin D in the development of the chick embryo. *The Journal of Experimental Zoology*. 1:15–23.

Narbaitz, R., and B. Fragiskos. 1984. Hypervitaminosis D in the chick embryo: Comparative study on the activity of various vitamin D 3 metabolites. *Calcified Tissue International*. 36(1):392–400.

Narbaitz, R., and C. Tsang. 1989. Vitamin D deficiency in the chick embryo: Effects on prehatching motility and on the growth and differentiation of

bones, muscles, and parathyroid glands. *Calcified Tissue International.* 44(5):348–355.
Navickis, R.J., B.S. Katzenellenbogen, and A.V. Nalbandov. 1979. Effects of the sex steroid hormones and vitamin D_3 on calcium-binding proteins in the chick shell gland. *Biology of Reproduction.* 21(5):1153–1162.
Nelson, T., and L. Norris. 1961. Studies on the vitamin K requirement of the chick. 2. Effect of sulfaquinoxaline on the quantitative requirements of the chick for vitamin K_1, menadione and menadione sodium bisulfite. *Journal of Nutrition.* 73:135–142.
Nir, I., and I. Ascarelli. 1967. The effect of thyroid status and restriction of dietary protein on survival of vitamin A deprived chicks. *British Poultry Science.* 8(3):169–175.
Niu, Z.Y., Y.N. Min, and F.Z. Liu. 2018. Dietary vitamin E improves meat quality and antioxidant capacity in broilers by upregulating the expression of antioxidant enzyme genes. *Journal of Applied Animal Research.* 46(1):397–401.
Nockels, C., D. Menge, and E. Kienholz. 1976. Effect of excessive dietary vitamin E on the chick. *Poultry Science.* 55(2):649–652.
Nockels, C., and R. Phillips. 1971. Influence of vitamin A deficiency on tissue glycogen metabolism in growing chickens. *Poultry Science.* 50(1):174–181.
Nockels, C.F., D.L. Ewing, H. Phetteplace, K.A. Ritacco, and K.N. Mero. 1984. Hypothyroidism: an early sign of vitamin A deficiency in chickens. *The Journal of Nutrition.* 114(9):1733–1736.
Noff, D., A. Simkin, and S. Edelstein. 1982. Effect of cholecalciferol derivatives on the mechanical properties of chick bones. *Calcified Tissue International.* 34(1):501–505.
Norman, A.W. 2008. From vitamin D to hormone D: Fundamentals of the vitamin D endocrine system essential for good health. *The American Journal of Clinical Nutrition.* 88(2):491S–499S.
NRC (National Research Council). 1987. *Vitamin Tolerance of Animals.* Washington, DC: National Academy Press.
NRC. 1994. *Nutrient Requirements of Poultry*, 9th rev. ed. Washington, DC: National Academy Press.
Oduho, G.W., T.K. Chung, and D.H. Baker. 1993. Menadione nicotinamide bisulfite is a bioactive source of vitamin K and niacin activity for chicks. *The Journal of Nutrition.* 123(4):737–743.
Ognik, K., and T. Wertelecki. 2012. Effect of different vitamin E sources and levels on selected oxidative status indices in blood and tissues as well as on rearing performance of slaughter turkey hens. *The Journal of Applied Poultry Research.* 21(2):259–271.
Olcese, O., J.R. Couch, J.H. Quisenberry, and P.B. Pearson. 1950. Congenital anomalies in the chick due to vitamin B_{12} deficiency. *The Journal of Nutrition.* 41(3):423–431.
Omdahl, J., and H. DeLuca. 1973. Regulation of vitamin D metabolism and function. *Physiological Reviews.* 53(2):327–372.
Ott, W.H., A.M. Dickinson, and A. Van Iderstine. 1965. Amprolium: 8. Comparison with oxythiamine and pyrithiamine as antagonists of thiamine in the chick. *Poultry Science.* 44(4):920–925.
Panda, A.K., and G. Cherian. 2014. Role of vitamin E in counteracting oxidative stress in poultry. *Poultry Science.* 51(2):109–117.
Paredes, J.R., and T.P. Garcia. 1959. Vitamin A as a factor affecting fertility in cockerels. *Poultry Science.* 38(1):3–7.
Park, I.-K., and R.R. Marquardt. 1982. Effects of niacin deficiency on pyridine nucleotide levels and enzyme activities in various organs of young growing quail. *The Journal of Nutrition.* 112(5):863–873.
Patel, M., and J. McGinnis. 1980. The effect of vitamin B_{12} on the tolerance of chicks for high levels of dietary fat and carbohydrate. *Poultry Science.* 59(10):2279–2286.
Patterson, E., J. Hunt, P. Vohra, L. Blaylock, and J. McGinnis. 1956. The niacin and tryptophan requirements of chicks. *Poultry Science.* 35(3):499–504.
Pesti, G., A. Harper, and M. Sunde. 1980. Choline/methionine nutrition of starting broiler chicks: Three models for estimating the choline requirement with economic considerations. *Poultry Science.* 59(5):1073–1081.
Pesti, G.M., G.N. Rowland, and K.S. Ryu. 1991. Folate-deficiency in chicks fed diets containing practical ingredients. *Poultry Science.* 70(3):600–604.
Phillips, P., and R. Engel. 1938. Neuromalacia associated with low riboflavin diets, a preliminary report. *Poultry Science.* 17(6):463–465.
Polin, D., W.H. Ott, E.R. Wynosky, and C.C. Porter. 1963. Estimation of thiamine requirement for optimum hatchability from the relationship between dietary and yolk levels of the vitamin. *Poultry Science.* 42(4):925–928.
Polin, D., E.R. Wynosky, and C.C. Porter. 1962. Amprolium 10. Influence of egg yolk thiamine concentration on chick embryo mortality. *Proceedings of the Society for Experimental Biology and Medicine.* 110(4):844–846.
Pompeu, M.A., L.F.L. Cavalcanti, and F.L.B. Toral. 2018. Effect of vitamin E supplementation on growth performance, meat quality, and immune response of male broiler chickens: A meta-analysis. *Livestock Science.* 208:5–13.
Quadros, E.V. 2010. Advances in the understanding of cobalamin assimilation and metabolism. *British Journal of Haematology.* 148(2):195–204.
Quarantelli, A., A. Cacchioli, S. Romanelli, F. Righi, I. Alpigiani, and C. Gabbi. 2007. Effects of different levels of dietary biotin on the performance and bone structure of broilers. *Italian Journal of Animal Science.* 6(1):5–17.
Rabbani, A., S. Akrabawi, and N. Daghir. 1973. Effect of dietary folic acid deficiency on serine-glycine interconversion and the activity of liver serine hydroxymethyl transferase in the chick. *Poultry Science.* 52(5):1962–1967.
Ramp, W.K., R.L. Baker, and L.S. Trinkle. 1989. Interrelationships of vitamin D, bone metabolism and blood calcium concentration in the chick. *Bone and Mineral.* 5(2):117–128.
Ratzkowski, C., N. Fine, and S. Edelstein. 1982. Metabolism of cholecalciferol in vitamin D intoxicated chicks. *Israel Journal of Medical Sciences.* 18(6):695–700.
Reboul, E. 2013. Absorption of vitamin A and carotenoids by the enterocyte: focus on transport proteins. *Nutrients.* 5(9):3563–3581.
Reboul, E. 2015. Intestinal absorption of vitamin D: From the meal to the enterocyte. *Food & Function.* 6(2):356–362.
Remus, J.C., and J.D. Firman. 1990. Effect of thiamin deficiency on energy metabolites in the turkey. *Journal of Nutritional Biochemistry.* 1(12):636–639.
Rengaraj, D., and Y. Hong. 2015. Effects of dietary vitamin E on fertility functions in poultry species. *International Journal of Molecular Sciences.* 16(5):9910–9921.
Rogers Jr., W. 1969. Reexamination of enzyme activities thought to show evidence of a coenzyme role for vitamin A. *The American Journal of Clinical Nutrition.* 22(8):1003–1013.
Romanoff, A.L., and J.C. Bauernfeind. 1942. Influence of riboflavin-deficiency in eggs on embryonic development. *The Anatomical Record.* 82:11–21.
Ruiz, N., and R. Harms. 1988a. Niacin requirement of turkey poults fed a corn-soybean meal diet from 1 to 21 days of age. *Poultry Science.* 67(5):760–765.
Ruiz, N., and R. Harms. 1988b. Riboflavin requirement of broiler chicks fed a corn–soybean diet. *Poultry Science.* 67(5):794–799.
Ruiz, N., and R. Harms. 1989a. Pantothenic acid studies with turkey poults. *Nutrition Reports International.* 40(4):639–643.
Ruiz, N., and R. Harms. 1989b. Riboflavin requirement of turkey poults fed a corn–soybean meal diet from 1 to 21 days of age. *Poultry Science.* 68(5):715–718.
Ruiz, N., and R. Harms. 1990. Research note: The lack of response of broiler chickens to supplemental niacin when fed a corn-soybean meal diet from 3 to 7 weeks of age. *Poultry Science.* 69(12):2231–2234.
Rys, R., and J. Koreleski. 1974. The effect of dietary propionic acid on the requirement of chicks for vitamin B_{12}. *British Journal of Nutrition.* 31(2):143–146.
Sahin, N., K. Sahin, M. Onderci, M. Karatepe, M. Smith, and O. Kucuk. 2005. Effects of dietary lycopene and vitamin E on egg production, antioxidant status and cholesterol levels in Japanese quail. *Asian-Australasian Journal of Animal Sciences.* 19(2):224–230.
Said, H.M. 2011. Intestinal absorption of water-soluble vitamins in health and disease. *Biochemical Journal.* 437(3):357–372.

Said, H.M., and E. Nexo. 2012. Mechanisms and regulation of intestinal absorption of water-soluble vitamins: Cellular and molecular aspects. In *Physiology of the Gastrointestinal Tract*. L.R. Johnson, editor. Amsterdam, The Netherlands: Elsevier. 1711–1756.

Said, H.M., A. Ortiz, and T.Y. Ma. 2003. A carrier-mediated mechanism for pyridoxine uptake by human intestinal epithelial Caco-2 cells: Regulation by a PKA-mediated pathway. *American Journal of Physiology—Cell Physiology*. 285(5):C1219–C1225.

Samour, J., J. Perlman, J. Kinne, V. Baskar, U. Wernery, and G. Dorrestein. 2016. Vitamin B_6 (pyridoxine hydrochloride) toxicosis in falcons. *Journal of Zoo and Wildlife Medicine*. 47(2):601–608.

Saunders-Blades, J.L., and D.R. Korver. 2015. Effect of hen age and maternal vitamin D source on performance, hatchability, bone mineral density, and progeny in vitro early innate immune function. *Poultry Science*. 94(6):1233–1246.

Saville, D.G., A. Solvyns, and C. Humphries. 1967. Choline-induced pyridoxine deficiency in broiler chickens. *Australian Veterinary Journal*. 43:346–348.

Schaefer, A.E., W.D. Salmon, and D.R. Strength. 1949. Interrelationship of vitamin B_{12} and choline; effect on growth of the chick. *Proceedings of the Society for Experimental Biology and Medicine*. 71(2):202–204.

Scheiner, J., and E. De Ritter. 1968. Vitamin B_6 content of feeds. *Journal of Agricultural and Food Chemistry*. 16(5):746–751.

Scherer, C.S., and D.H. Baker. 2000. Excess dietary methionine markedly increases the vitamin B-6 requirement of young chicks. *Journal of Nutrition*. 130(12):3055–3058.

Schexnailder, R., and M. Griffith. 1973. Liver fat and egg production of laying hens as influenced by choline and other nutrients. *Poultry Science*. 52(3):1188–1194.

Scott, M. 1953. Prevention of the enlarged hock disorder in turkeys with niacin and vitamin E. *Poultry Science*. 32(4):670–677.

Scott, M. 1966. Studies on the interrelationship of selenium, vitamin E and sulfur amino acids in a nutritional myopathy of the chick. *Annals of the New York Academy of Sciences*. 138(1):82.

Scott, M., and G. Heuser. 1952. Studies in duck nutrition: 4. Bowed legs in ducks caused by niacin deficiency. *Poultry Science*. 31(4):752–754.

Scott, M., E. Holm, and R. Reynolds. 1964. Studies on the pantothenic acid and unidentified factor requirements of young ringnecked pheasants and Bobwhite quail. *Poultry Science*. 43(6):1534–1539.

Scott, M., E.R. Holm, and R. Reynolds. 1961. Studies on the vitamin K requirements of young pheasants and quail. *Poultry Science*. 40(6):1593–1597.

Scott, M.L. 1967. Vitamin K in animal nutrition. In *Vitamins & Hormones*. Amsterdam, The Netherlands: Elsevier. 633–647.

Scott, M.L., E.R. Holm, and R.E. Reynolds. 1959. Studies on the niacin, riboflavin, choline, manganese and zinc requirements of young ringnecked pheasants for growth, feathering and prevention of leg disorders. *Poultry Science*. 38(6):1344–1350.

Seifter, E., G. Rettura, and S.M. Levenson. 1972. Effect of choline deficiency on serum phospholipids in chicks. *Poultry Science*. 51(2):695–697.

Serafin, J. 1974. Studies on the riboflavin, niacin, pantothenic acid and choline requirements of young Bobwhite quail. *Poultry Science*. 53(4):1522–1532.

Serafin, J. 1981. Studies on the riboflavin, pantothenic acid, nicotinic acid, and choline requirements of young embden geese. *Poultry Science*. 60(8):1910–1915.

Settle, E.A., F.R. Mraz, C.R. Douglas, and J.K. Bletner. 1969. Effect of diet and managanese level on growth perosis and ^{54}Mn uptake in chicks. *Journal of Nutrition*. 97(1):141.

Shanmugasundaram, R., and R.K. Selvaraj. 2012. Vitamin D-1α-hydroxylase and vitamin D-24-hydroxylase mRNA studies in chickens. *Poultry Science*. 91(8):1819–1824.

Shao, Y., Q. Wen, S. Zhang, L. Lu, L. Zhang, X. Liao, and X. Luo. 2019. Dietary supplemental vitamin D_3 enhances phosphorus absorption and utilisation by regulating gene expression of related phosphate transporters in the small intestine of broilers. *British Journal of Nutrition*. 121(1):9–21.

Sheldon, B. 1984. Effect of dietary tocopherol on the oxidative stability of turkey meat. *Poultry Science*. 63(4):673–681.

Shindo, H., and T. Komai. 1972. Metabolic fate and mechanism of action of chloroethylthiamine. *The Journal of Vitaminology*. 18(1):41–47.

Sifri, M., N. Daghir, and J. Asmar. 1972. Serum aminotransferase activities and plasma free amino acid concentrations as criteria for pyridoxine nutritional status of chicks. *British Journal of Nutrition*. 28:173.

Sijtsma, S.R., C.E. West, J.H. Rombout, and A.J. Van Der Zijpp. 1989. The interaction between vitamin A status and Newcastle disease virus infection in chickens. *The Journal of Nutrition*. 119(6):932–939.

Singh, S., and G. Donovan. 1973. A relationship between coccidiosis and dietary vitamin A levels in chickens. *Poultry Science*. 52(4):1295–1301.

Sklan, D. 1983. Carotene-cleavage activity in chick intestinal mucosa cytosol: association with a high-molecular-weight lipid-protein aggregate fraction and partial characterization of the activity. *British Journal of Nutrition*. 50(2):417–425.

Sklan, D., and S. Donoghue. 1982. Vitamin E response to high dietary vitamin A in the chick. *The Journal of Nutrition*. 112(4):759–765.

Sklan, D., H.D. Rabinowitch, and S. Donoghue. 1981. Superoxide dismutase: Effect of vitamin A and E. *Nutrition Reports International*. 24:551–555.

Sklan, D., T. Yosefov, and A. Friedman. 1989. The effects of vitamin A, beta-carotene and canthaxanthin on vitamin A metabolism and immune responses in the chick. *International Journal of Vitamin and Nutrition Research*. 59(3):245–250.

Snell, E.E., and E. Quarles. 1941. The effect of incubation on the vitamin content of eggs. *The Journal of Nutrition*. 22(5):483–489.

Soares, J.H. Jr., J.M. Kerr, and R.W. Gray. 1995. 25-hydroxycholecalciferol in poultry nutrition. *Poultry Science*. 74(12):1919–1934.

Southern, L., and D. Baker. 1981. Bioavailable pantothenic acid in cereal grains and soybean meal. *Journal of Animal Science*. 53(2):403–408.

Stevens, V., and R. Blair. 1985. Effects of supplementary vitamin D_3 on egg production of two strains of Japanese quail and growth of their progeny. *Poultry Science*. 64(3):510–519.

Stokstad, E., and T. Jukes. 1987. Sulfonamides and folic acid antagonists: A historical review. *The Journal of Nutrition*. 117(7):1335–1341.

Sun, Z.W., Q.H. Fan, X.X. Wang, Y.M. Guo, H.J. Wang, and X. Dong. 2017. High dietary biotin levels affect the footpad and hock health of broiler chickens reared at different stocking densities and litter conditions. *Journal of Animal Physiology and Animal Nutrition*. 101(3):521–530.

Sunde, M., W. Cravens, C. Elvehjem, and J. Halpin. 1950. The effect of folic acid on embryonic development of the domestic fowl. *Poultry Science*. 29(5):696–702.

Sunde, M., C. Turk, and H. DeLuca. 1978. The essentiality of vitamin D metabolites for embryonic chick development. *Science*. 200(4345):1067–1069.

Sundeen, G., J. Richards, and D. Bragg. 1980. The effect of vitamin A deficiency on some postmortem parameters of avian muscle. *Poultry Science*. 59(10):2225–2236.

Suttie, J.W. 1987. Recent advances in hepatic vitamin K metabolism and function. *Hepatology*. 7(2):367–376.

Suzuki, U., and T. Shimamura. 1911. Active constituent of rice grits preventing bird polyneuritis. *Tokyo Kagaku Kaishi*. 32:4–7, 144–146, 335–358.

Swiatkiewicz, S., A. Arczewska-Wlosek, D. Bederska-Lojewska, and D. Jozefiak. 2017. Efficacy of dietary vitamin D and its metabolites in poultry—review and implications of the recent studies. *World's Poultry Science*. 73(1):57–68.

Tajima, S., T. Goda, and S. Takase. 2001. Co-ordinated induction of beta-carotene cleavage enzyme and retinal reductase in the duodenum of the developing chicks. *Comparative Biochemistry and Physiology B—Biochemistry & Molecular Biology*. 128(3):425–434.

Takahashi, N., T. Shinki, E. Abe, N. Horiuchi, A. Yamaguchi, S. Yoshiki, and T. Suda. 1983. The role of vitamin D in the medullary bone formation in egg-laying Japanese quail and in immature male chicks treated with sex hormones. *Calcified Tissue International*. 35(1):465–471.

Tang, J., M. Xie, J. Yang, Z. Wen, Y. Zhu, W. Huang, and S. Hou. 2013. Riboflavin requirements of white Pekin ducks from hatch to 21 d of age. *British Poultry Science*. 54(3):407–411.

Tang, K.-N., G.N. Rowland, and J. Veltmann Jr. 1985. Vitamin A toxicity: Comparative changes in bone of the broiler and leghorn chicks. *Avian Diseases*. 29(2):416–429.

Tarhay, S., E. Buss, and C. Clagett. 1975. Avian riboflavinuria: 10. Quantitative changes of riboflavin-binding protein in individual egg tissues during incubation. *Poultry Science*. 54(2):562–571.

Tavčar-Kalcher, G., and A. Vengušt. 2007. Stability of vitamins in premixes. *Animal Feed Science and Technology*. 132(1–2):148–154.

Thompson, J., J.M. Howell, G. Pitt, and C.I. Houghton. 1965. Biological activity of retinoic acid ester in the domestic fowl: production of vitamin A deficiency in the early chick embryo. *Nature*. 205(4975):1006–1007.

Thornton, P., and J. Shutze. 1960. The influence of dietary energy level, energy source and breed on the thiamine requirement of chicks. *Poultry Science*. 39(1):192–199.

Tolve, R., F. Tchuenbou-Magaia, M. Di Cairano, M.C. Caruso, T. Scarpa, and F. Galgano. 2021. Encapsulation of bioactive compounds for the formulation of functional animal feeds: The biofortification of derivate foods. *Animal Feed Science and Technology*. 279:115036.

Tsiagbe, V.K., M.E. Cook, A.E. Harper, and M.L. Sunde. 1988. Alterations in phospholipid composition of egg yolks from laying hens fed choline and methionine-supplemented diets. *Poultry Science*. 67(12):1717–1724.

Vahl, H., and A.T. van't Klooster. 1987. Effects of excessive vitamin A levels in broiler rations. *Journal of Animal Physiology and Animal Nutrition*. 57(1–5):204–218.

Veltmann Jr., J., L. Jensen, and G. Rowland. 1987. Partial amelioration of vitamin A toxicosis in the chick and turkey poult by extra dietary vitamin D_3. *Nutrition Reports International*. 35(2):381–392.

Veltmann Jr., J.R., E. Ross, and S.E. Olbrich. 1981. The physiological effects of feeding warfarin to poultry. *Poultry Science*. 60(12):2603–2611.

Vignale, K., E.S. Greene, J.V. Caldas, J.A. England, N. Boonsinchai, P. Sodsee, E.D. Pollock, S. Dridi, and C.N. Coon. 2015. 25-Hydroxycholecalciferol enhances male broiler breast meat yield through the mTOR pathway. *Journal of Nutrition*. 145(5):855–863.

Vigo, C., and D.E. Vance. 1981. Effect of diethylstilboestrol on phosphatidylcholine biosynthesis and choline metabolism in the liver of roosters. *Biochemical Journal*. 200(2):321–326.

Vohra, P., T. Siopes, and W. Wilson. 1979. Egg production and body weight changes of Japanese quail and Leghorn hens following deprivation of either supplementary calcium or vitamin D_3. *Poultry Science*. 58(2):432–440.

Voigt, M.N., R.D. Wyatt, J.C. Ayres, and P.E. Koehler. 1980. Abnormal concentrations of B vitamins and amino acids in plasma, bile, and liver of chicks with aflatoxicosis. *Applied and Environmental Microbiology*. 40(5):870–875.

Walter, E., and L. Jensen. 1964. Serum glutamic-oxaloacetic transaminase levels, muscular dystrophy and certain hematological measurements in chicks and poults as influenced by vitamin E, selenium and methionine. *Poultry Science*. 43(4):919–926.

Ward, N., J. Jones, and D. Maurice. 1985. Inefficacy of propionic acid for depleting laying hens and their progeny of vitamin B_{12}. *Nutrition Reports International*. 32(6):1325–1332.

Ward, N.E., J. Jones, and D.V. Maurice. 1988. Essential role of adenosylcobalamin in leucine synthesis from β-leucine in the domestic chicken. *The Journal of Nutrition*. 118(2):159–164.

Watkins, B., and A. Rogel. 1989. Progression of biotin deficiency and influence of reduced food intake on fatty acids in the chick. *Nutrition Research*. 9(1):57–68.

Watkins, B.A., S.D. Bain, and J.W. Newbrey. 1989. Eicosanoic fatty-acid reduction in the tibiotarsus of biotin-deficient chicks. *Calcified Tissue International*. 45(1):41–46.

Wedegaertner, O., A. Fahrenholz, J. D. Bunod, J. Fontaine, and P. Ferket. 2020. Handling characteristic evaluation of free, standard and lipid microencapsulated vitamin and mineral premixes. *Journal of Agricultural Science and Technology A*. 11:247–256

Wen, Z.G., J. Tang, M. Xie, P.L. Yang, and S.S. Hou. 2016. Effects of dietary methionine levels on choline requirements of starter White Pekin ducks. *Asian-Australasian Journal of Animal Sciences*. 29(12):1742–1747.

West, C.E., S.R. Sijtsma, H.P. Peters, J.H. Rombout, and A.J. Van Der Zijpp. 1992. Production of chickens with marginal vitamin A deficiency. *British Journal of Nutrition*. 68(1):283–291.

White III, H., and A. Merrill Jr. 1988. Riboflavin-binding proteins. *Annual Review of Nutrition*. 8(1):279–299.

White III, H.B., J. Armstrong, and C.C. Whitehead. 1986. Riboflavin-binding protein. Concentration and fractional saturation in chicken eggs as a function of dietary riboflavin. *Biochemical Journal*. 238(3):671–675.

White III, H.B., C.C. Whitehead, and J. Armstrong. 1987. Relationship of biotin deposition in turkey eggs to dietary biotin and biotin-binding proteins. *Poultry Science*. 66(7):1236–1241.

Whitehead, C., and D. Bannister. 1980. Biotin status, blood pyruvate carboxylase (EC 6.4. 1.1) activity and performance in broilers under different conditions of bird husbandry and diet processing. *British Journal of Nutrition*. 43(3):541–549.

Whitehead, C., R.A. Pearson, and K.M. Herron. 1985. Biotin requirements of broiler breeders fed diets of different protein content and effect of insufficient biotin on the viability of progeny. *British Poultry Science*. 26(1):73–82.

Whitehead, C., and W. Siller. 1983. Experimentally induced fatty liver and kidney syndrome in the young turkey. *Research in Veterinary Science*. 34(1):73–76.

Whitehead, C.C., and T. Keller. 2003. An update on ascorbic acid in poultry. *World's Poultry Science Journal*. 2:161–184.

Whitehead, C.C., H.A. McCormack, L. McTeir, and R.H. Fleming. 2004. The maximum legal limit for vitamin D_3 in broiler diets may need to be increased. *British Poultry Science*. 24–26.

Whitehead, C.C., and C.J. Randall. 1982. Interrelationships between biotin, choline and other B-vitamins and the occurrence of fatty liver and kidney syndrome and sudden-death syndrome in broiler-chickens. *British Journal of Nutrition*. 48(1):177–184.

Wideman, R.F. Jr., J. Blankenship, I.Y. Pevzner, and B.J. Turner. 2015. Efficacy of 25-OH vitamin D prophylactic administration for reducing lameness in broilers grown on wire flooring. *Poultry Science*. 94(8):1821–1827.

Wolf, B., G. Heard, J.S. McVoy, and H. Raetz. 1984. Biotinidase deficiency: The possible role of biotinidase in the processing of dietary protein-bound biotin. In *Organic Acidurias*. New York: Springer. 121–122.

Wyatt, R., H. Tung, W. Donaldson, and P. Hamilton. 1973. A new description of riboflavin deficiency syndrome in chickens. *Poultry Science*. 52(1):237–244.

Xie, M., S. Wang, W. Huang, and S. Hou. 2018. Effects of vitamin E on growth performance, tissue α-tocopherol, and lipid peroxidation of starter White Pekin ducks. *Poultry Science*. 97(6):2139–2143.

Xu, G.-L., and A.T. Diplock. 1983. Glutathione peroxidase (EC 1.11. 1.9), glutathione-S-transferase (EC 2.5. 1.13), superoxide dismutase (EC 1.15. 1.1) and catalase (EC 1.11. 1.6) activities in tissues of ducklings deprived of vitamin E and selenium. *British Journal of Nutrition*. 50(2):437–444.

Yacowitz, H., R. Miller, L. Norris, and G. Heuser. 1952. Vitamin B_{12} studies with the hen. *Poultry Science*. 31(1):89–94.

Yang, P., H.K. Wang, M. Zhu, L.X. Li, and Y.X. Ma. 2021. Degradation kinetics of vitamins in premixes for pig: Effects of choline, high concentrations of copper and zinc, and storage time. *Animal Bioscience*. 34(4):701–713.

Yen, J., A. Jensen, and D. Baker. 1977. Assessment of the availability of niacin in corn, soybeans and soybean meal. *Journal of Animal Science*. 45(2):269–278.

Young, R., L. Norris, and G. Heuser. 1955. The chick's requirement for folic acid in the utilization of choline and its precursors betaine and methylaminoethanol. *The Journal of Nutrition.* 55(3):353–362.

Youssef, I.M.I., A. Beineke, K. Rohn, and J. Kamphues. 2012. Influences of increased levels of biotin, zinc or mannan-oligosaccharides in the diet on foot pad dermatitis in growing turkeys housed on dry and wet litter. *Journal of Animal Physiology and Animal Nutrition.* 96(5):747–761.

Yuan, J., A.R. Roshdy, Y. Guo, Y. Wang, and S. Guo. 2014. Effect of dietary vitamin A on reproductive performance and immune response of broiler breeders. *PLoS One.* 9(8):e105677.

Zamierowski, M.M., and C. Wagner. 1977. Effect of folacin deficiency on folacin-binding proteins in the rat. *The Journal of Nutrition.* 107(10):1937–1945.

Zhao, G.P., M.J. Han, M.Q. Zheng, J.P. Zhao, J.L. Chen, and J. Wen. 2011. Effects of dietary vitamin E on immunological stress of layers and their offspring. *Journal of Animal Physiology and Animal Nutrition.* (3):343.

Zhuge, Q., and C.F. Klopfenstein. 1986. Factors affecting storage stability of vitamin A, riboflavin, and niacin in a broiler diet premix. *Poultry Science.* 65(5):987–994.

6

Minerals

INTRODUCTION

In the context of poultry nutrition, minerals are the inorganic part of feeds, feed ingredients, or feed additives that are digested and absorbed from the gastrointestinal tract for storage or utilization within the body. In general, mineral elements are only required at low concentrations in the diet, including a maximum of 4–5 percent for calcium in laying hens but generally no more than 1 percent for each individual mineral. Although all minerals make up a small component of the diet, they are often divided into two categories—major, or macro, and minor, or trace—based on the total amount of mineral required in the diet per species (or category of poultry) and phase of production. Requirements for macro minerals are usually stated as a percentage of the diet, whereas requirements for trace minerals are stated as milligrams per kilogram of diet (i.e., parts per million). (Element concentrations in common mineral sources can be found in Table 6-1.) Minerals are required as components of various compounds with particular functions within the body, as cofactors of enzymes, for the maintenance of osmotic balance within the body of the bird, and for the formation of the skeleton. Minerals are often stored in the body of poultry for use in times of limited feed intake or higher stress. The skeletal system and liver are both major sites for mineral storage within the body for utilization at a later point when required for maintenance but not directly supplied by the gastrointestinal tract in sufficient quantities.

Formulators of poultry diets should be aware of numerous possible mineral interactions and of the potential effects that the chemical form (e.g., cation–anion combination) of mineral sources may have on their utilization by poultry (Allaway, 1986). Mineral salts used as feed supplements are not usually pure compounds but contain variable amounts of other elements.

MACROMINERALS

Requirements for macro minerals are usually stated as a percentage of the diet. Macrominerals include calcium, phosphorus, sodium, potassium, chloride, and magnesium. Although sulfur is a mineral, at least for poultry species, it is only required as a structural component of methionine and cyst(e)ine and will be not be discussed in the context of a mineral requirement.

CALCIUM AND PHOSPHORUS

Calcium is required in poultry due to its regulatory functions important for contracting muscles, conducting nerve impulses, blood clotting, and maintaining a normal heartbeat, and for structural functions in forming and strengthening the skeletal system and eggshell production. A ratio of approximately 2 calcium to 1 nPP (weight/weight) is appropriate for most poultry diets, with the exception of diets for birds that are laying eggs. When poultry are laying eggs, a much higher level of calcium is needed for eggshell formation, and a ratio as high as 12 calcium to 1 nonphytate phosphorus (weight/weight) may be used. But high levels of calcium carbonate (limestone) and calcium phosphates may tend to make the diet unpalatable and dilute the other dietary components. If a calcium source contains a high level of magnesium (as does dolomitic limestone), caution should be used in formulation for poultry diets (Stillmak and Sunde, 1971). An excess of dietary calcium interferes with the availability of other minerals, such as phosphorus (Tamim and Angel, 2003; Hamdi et al., 2015), magnesium (Nugara and Edwards, 1962; Hess and Britton, 1997), manganese (Wedekind and Baker, 1990b), and zinc (Roberson and Schaible, 1960).

Phosphorus is essential as a structural component of nucleic acids, cell membranes, and adenosine 5′-triphosphate (ATP); plays a role in acid–base balance; and is used in combination with calcium for the formation and maintenance of the skeleton as hydroxyapatite, $Ca_5(PO_4)_3(OH)$. Generally, non-phytate forms of phosphorus found in plant feedstuffs can be digested by poultry; however, such digestible forms usually account for only 30–40 percent of the total phosphorus in grains and oilseeds commonly used in poultry

TABLE 6-1 Element Concentrations in Common Mineral Sources (Data on As-Fed Basis)

Entry Number	Feed Name Description	International Feed No.	Calcium (%)	Phosphorous (%)	Sodium (%)	Potassium (%)	Magnesium (%)	Chlorine (%)	Fluorine (%)	Sulfur (%)	Iron (mg/kg)	Copper (mg/kg)	Manganese (mg/kg)	Zinc (mg/kg)
01	Bone meal, steamed	6-00-400	29.8	12.5	0.04	0.2	0.3	–	–	2.4	–	16	30	100
02	Calcium carbonate, $CaCO_3$	6-01-069	38.0	0.0	0.02	0.06	0.05	–	0.00	–	300	24	300	2
03	Calcium phosphate dibasic from defluorinated phosphoric acid	6-01-080	22.0	18.7	0.06	0.1	0.6	0.013	0.18	1.11	10,000	10	300	100
04	Calcium phosphate, mono-dibasic	6-26-137	16.0	21.0	0.06	0.07	0.6	–	0.15	1.2	9,000	15	300	200
05	Calcium sulfate, dihydrate $CaSO_4 \bullet 2H_2O$	6-01-090	22.6	–	–	–	–	–	–	18.1	–	–	–	–
06	Limestone, ground	6-02-632	38.0	–	0.05	0.1	2.1	0.03	<0.0025	–	2,000	–	–	–
07	Magnesium oxide, MgO	6-02-756	3.0	0.03	0.015	0.02	55.0	0.02	0.02	0.04	6,000	10	–	10
08	Meat with bone, meal rendered	5-00-388	10.3	5.1	0.7	1.3	1.1	0.7	–	0.5	490	2	14	93
09	Oyster shells, ground	6-03-481	38.0	0.1	0.2	0.1	0.3	0.01	–	–	500	–	400	–
10	Phosphate, defluorinated	6-01-780	32.0	18.0	4.9	0.1	0.4	–	0.18	–	8,000	20	250	60
11	Phosphate, rock curacao, ground	6-05-586	34.0	14.0	0.2	–	0.8	–	0.53	–	3,500	–	–	–
12	Phosphate, rock, soft	6-03-947	17.0	9.0	0.10	0.30	0.35	0.007	1.25	0.31	15,000	64	39	90
13	Potassium chloride, KCl	6-03-755	0.05	–	1.0	50.5	0.34	47.3	-	0.45	600	7	7	90
14	Potassium and magnesium sulfate	6-06-177	0.06	–	0.76	18.5	11.6	1.25	0.001	22.3	100	2	20	9
15	Potassium sulfate, K2SO4	6-08-098	0.15	–	0.09	41.0	0.6	1.5	–	17.9	700	–	10	–
16	Sodium carbonate, Na_2CO_3	6-12-316	–	–	43.39	–	–	–	–	–	–	–	–	–
17	Sodium bicarbonate, $NaHCO_3$	6-04-272	–	–	27.0	–	–	–	–	–	–	–	–	–
18	Sodium chloride, NaCl (common salt)	6-04-152	0.3	–	39.0	–	0.005	60.0	–	0.2	50	–	–	–
19	Sodium phosphate, dibasic, from furnaced phosphoric acid, Na_2HPO_4	6-04-286	–	20.8	31.0	–	–	–	–	–	–	–	–	–
20	Sodium phosphate, monobasic, $NaH_2PO_4 \bullet H_2O$	6-04-288	–	21.8	16.2	–	–	–	–	–	–	–	–	–
21	Sodium phosphate, decahydrate, $Na_2SO_410H_2O$	6-04-291	–	–	13.8	–	–	–	–	9.7	–	–	–	–
22	Phosphoric acid, H_3PO_4	6-03-707	0.08	23.7	0.05	0.02	0.45	–	0.19	1.1	12,000	10	400	100

NOTES: The mineral supplements used as feed supplements are not chemically pure compounds, and the compositions may vary substantially among sources. The supplier's analysis should be used if it is available. Dashes indicate that no data were available.

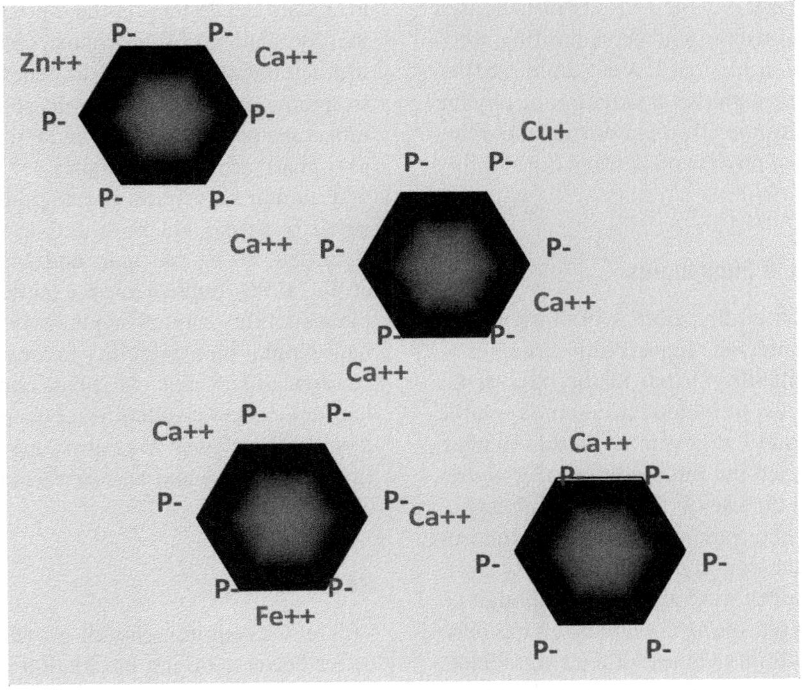

FIGURE 6-1 Association of phytic acid with calcium and other minerals.

production (Eeckhout and De Paepe, 1994). The remaining phosphorus is present mostly as phytate phosphorus and is poorly digested, especially in young birds. The phosphorus from animal products and phosphorus supplements is generally considered to be well utilized.

Phytic acid ($C_6H_{18}O_{24}P_6$), hexakisphosphate, inositol polyphosphate, or phytate (when in salt form) is the phosphate ester of inositol and is a common storage form of phosphorus in the grains and oilseeds commonly used in poultry diets. Under current practical feeding constraints, and without phytase supplementation, only about 10 to 20 percent of the phytate phosphorus is digested by broiler chickens (Nelson, 1976; Sommerfeld et al., 2018). Phytic acid has a strong binding affinity to dietary minerals including calcium, copper, iron, zinc and manganese, thereby inhibiting their absorption and utilization by poultry. It is this interaction among phytic acid and minerals, especially calcium, that has been shown to reduce the availability of phytic acid for digestion (Tamim et al., 2004; Selle et al., 2009). The concept of phytic acid and dietary interactions with minerals, proteins, and enzymes has been extensively reviewed (Singh, 2008; Selle et al., 2009; Woyengo and Nyachoti, 2013).

Most of the recent publications concerning calcium and phosphorus digestion, absorption, metabolism, and utilization within poultry species have included or utilized the addition of exogenously produced phytase (a microbial-derived enzyme that hydrolyzes phosphate groups from phytic acid) that directly liberates phosphorus from phytic acid but also reduces the ability of phytic acid to form phytate complexes with calcium and other minerals. This section will not focus on the direct actions of the phytase enzyme (see Chapter 7, "Non-Nutritive Feed Additives," and Chapter 8, "Digestibility and Bioavailability"), but will use results from these experiments to further the understanding of calcium and phosphorus nutrition.

Historically, purified or semi-purified diets have been used to isolate key nutrients such as minerals to better understand their effects in poultry species. Although this approach has resulted in a better understanding of mineral nutrition, especially deficiency, there are issues with these semi-purified diets in terms of application to commercial poultry production. These types of diets do not contain a typical amount of phytate and fiber (see Chapter 2, "Carbohydrates and Fiber") that can have negative effects on nutrient digestibility and despite reduced anti-nutrients in semi-purified diets, they often result in reduced performance in comparison to diets composed of complex feed ingredients such as corn and soybean meal (Colnago et al., 1984; Shastak et al., 2014). As purified diets without phytate result in lower performance than traditional corn and soybean meal diets, it has been difficult to truly understand the performance effects of phytate in common commercial diets. To this end, high concentrations of phytase have been used in complex diets, resulting in the partial or complete breakdown of phytate in the upper gastro-intestinal track, which results in increased performance when phosphorus has not been a limiting nutrient in the diet (dos Santos et al., 2013; Walk et al., 2014). This increased performance with the destruction of phytate further suggests the anti-nutritive effects of phytate including the negative consequences of phytate on nutrient digestibility and utilization.

Statement on Digestibility of Phosphorus

Generally, the discussion of digestibility of nutrients has been reserved for the chapter on digestibility (Chapter 8, "Digestibility and Bioavailability"), but in the case of Ca and P there is enough interest to include a statement on the rationale of the use of Ca and P digestibility in this chapter. The committee that published the ninth edition of *Requirements of Poultry* settled on the use of "nonphytate phosphorus" (nPP) values for the determination of the phosphorus requirements of growing and egg-laying birds. At that time, phosphorus digestibility values were available. Although no insight on that committee's reasoning is available, it has been noted that although P digestibility values for feed ingredients have been published for 70 years, differences in methodology have resulted in variation and have limited the ability to compare individual feedstuff assays across reports. To address this, a standard assay method was proposed and ring tested using soybean meal (Rodehutscord et al., 2017). Despite using the same diet that contained the same soybean meal and the same laboratory procedures conducted in the same lab for Ca and P analysis, there were still significant differences in determined ileal P and Ca digestibility values across the 17 stations involved in the ring test. There are several compelling arguments to transition to P digestibility, including standardization with amino acids and other nutrients, a refined formulation to reduce over-formulation of P, environmental concern, standardization with other international nutrient requirement publications and easier formulation of diets that contain phytase and other feed additives that alter the digestibly of Ca and P, but the question remains whether a change to a P digestibility system is scientifically justified. At this point, with the results of the most recent well-executed and robust ring test (Rodehutscord et al., 2017), nPP will continue to be used to report phosphorus requirements, although continued research supporting ileal P and Ca digestiblity is warranted, and this system would be an improvement over the current nPP and total Ca requirement systems.

Dietary Sources, Supplementation, and Bioavailability

Calcium and nPP are generally low in the grains (0.02 to 0.06% Ca and 0.08 to 0.15% nPP) and moderate in oil seed meals (0.25 to 0.70% Ca and 0.29 to 0.72% nPP) used commercially in poultry production (Eeckhout and De Paepe, 1994; Anwar et al., 2016). Animal protein meals are generally considered a good source of Ca and nPP, but even those meals are dependent on the amount of bone material that is incorporated into the product (Shirley and Parsons, 2001; Mutucumarana et al., 2015). Therefore, diets are generally supplemented with inorganic sources of Ca (e.g., mined limestone or oyster shells) and P (e.g., mono- and di-calcium phosphate or rock phosphate).

Calcium absorption occurs predominantly in the anterior small intestine via both active and passive mechanisms (DeLuca, 1974; Norman and Means, 1987). Ammerman et al. (1995) published an excellent review chapter on Ca bioavailability, suggesting that several alternative Ca sources have similar bioavailability to limestone with the notable exceptions of soft rock phosphate and steel slag, each of which had reduced bioavailability. Phosphorus bioavailability has been well reviewed in poultry, suggesting tremendous variation both across and among various P sources (Ammerman et al., 1995).

Requirements

The Ca requirements of growing birds and laying hens differ because of egg production in the latter; formation of eggshell increases the amount of dietary calcium needed by three to four times relative to growing meat- and egg-type birds. In contrast, P requirements as a percent of the diet are generally more consistent among bird types but are reduced with increased feed intake as birds grow. In laying hens, additional P is added to the diets above what is needed for growth or egg production, due in part to the feeding of high dietary Ca, which can potentially reduce the utilization of phytate P and total P from the diet (David et al., 2023). Although as laying hens age, P is removed from the diet due to a negative relationship among excess dietary AvP and egg specific gravity and eggshell thickness (Jing et al., 2018). Species- and strain-specific recommendations and the studies they are based on can be found in Chapters 13–19. Requirements are typically based on dietary levels that prevent the expression of deficiency signs or impairments in productivity. Higher levels are typically fed to provide a margin of safety due to interactions with other nutrients, anti-nutritional factors in feedstuffs, potential increases in requirement due to stressful situations and poor bioavailability in some feedstuffs.

Deficiency and Toxicity

Bone formation is highly dependent on the dietary concentrations of calcium and phosphorus (Hart et al., 1922; Dunn, 1924; McGowan and Emslie, 1934). Deficiency of either Ca or P will result in poor feed intake, poor growth, egg production, thin-shelled eggs, and the development of rickets (Klasing and Korver, 2020). Calcium deficiency in pullets and laying hens can result in birds that are unable to stand, appear paralyzed, and have easily broken bones. This condition is technically osteoporosis but is generally termed "cage layer fatigue" (Hester, 2017).

Feeding excessive levels of dietary calcium typically reduces appetite and weight gain, and increases mortality (Urbanyi, 1960; Fangauf et al., 1961; Hester, 2017), while a second common effect is to hinder the intestinal absorption of other minerals in the diet (NRC, 2005), thereby reducing their availability. Nephrosis (kidney damage) and visceral urate deposition (gout) were observed in both growing egg laying pullets and broilers feed high (2–3.68%) concentrations of Ca (Shane et al., 1969; Wideman et al., 1985; Guo et al., 2005; Feizi and Nazeri, 2011). Like calcium, toxicosis from phosphorus is uncommon in food-producing animals (NRC, 2005). However, Harms et al. (1965) reported that feeding laying hens a diet containing an excessive level of total phosphorus reduced egg production and feed conversion (per dozen eggs). Edwards and Veltmann (1983) observed the highest prevalence of tibial dyschondroplasia in chicks fed low calcium and high total phosphorus and concluded that the ratio of calcium to phosphorus in addition to high dietary phosphorus was of importance in the prevalence of this condition.

SODIUM, POTASSIUM, CHLORIDE, AND MAGNESIUM

Sodium, potassium, chloride, and magnesium function with phosphates and bicarbonate to maintain homeostasis of osmotic relationships and pH throughout the body. Active transport of Na^{2+} and K^+ across plasma membranes of intestinal cells accounts for about one-third of the ATP expenditure, hence oxygen consumption, of the whole bird (Park et al., 1998). Higher concentrations of Na^{2+}, Cl^-, and K^+ lead to excessive consumption of water and attendant problems with ventilation control, wet droppings, feet damage (e.g., footpad dermatitis) and impaired absorptive function (Oviedo-Rondon et al., 2001; Mushtaq and Pasha, 2013; Adedokun and Applegate, 2014).

Dietary proportions of sodium, potassium, and chloride are important determinants of acid–base balance (Mongin, 1968; Hurwitz et al., 1973; Hurwitz et al., 1974; Sauveur and Mongin, 1978). Other cations and anions such as calcium, sulfate, and phosphate also contribute to this balance, but with less influence. The appropriate dietary electrolyte balance (DEB) for a particular category of poultry and life-stage is predominantly driven by sodium and potassium, versus chloride, where each element is expressed in milliequivalents (Meq) per kilogram of diet. In this context, sodium and potassium are alkalogenic, whereas chloride is acidogenic. Chloride tends to decrease blood pH and bicarbonate concentration, whereas sodium and potassium tend to increase blood pH and bicarbonate concentration. A dietary balance of sodium, potassium, and chloride is necessary for growth, bone development, eggshell quality, and amino acid utilization (Mongin, 1981). In diets without mineral deficiencies or toxicities, dietary electrolyte balance (i.e., Na^{2+} plus K^+ minus Cl^-) can range from 160 to 320 Meq/kg of feed for layers without adverse effects on eggshell quality (Sauveur and Mongin, 1978), and can range from 150 to 250 Meq/kg of feed for broilers (Mustaq and Pasha, 2013). These ranges in electrolyte balance reflect the concomitant effects of diet composition, feed intake, respiration rate, and mineral sources on optimal electrolyte balance. The interactions of all these factors make it difficult to choose a generally optimal electrolyte balance. The balance of alkalogenic and acidogenic minerals has been reviewed, particularly in relation to counteracting heat stress in broilers (Ahmad and Sarwar, 2006).

Deficiency and Toxicity

Many feedstuffs contain adequate levels of potassium, so deficiencies are relatively rare, and it is not usually necessary to add potassium to practical feed formulations (Klasing and Korver, 2020). However, when a deficiency of potassium occurs, it results in high mortality, reduced feed consumption, decreased plasma and skeletal potassium contents, and reduced growth of chicks, additionally causing reduced egg production and eggshell thickness in laying hens (Ben-Dor, 1941; Gillis, 1948; Rinehart et al., 1969; Leach, 1974). Since potassium is primarily found in intracellular compartments and has an essential role in the maintenance of membrane potential and cellular fluid balance, many of the foregoing symptoms of potassium deficiency are a result of overall muscle weakness, poor intestinal tone and distension, cardiac weakness, and weakness and overall failure of the respiratory muscles (Scott et al., 1982).

In contrast to potassium, many feedstuffs are deficient in sodium and chlorine, so salt (NaCl) is supplemented to most poultry diets (Klasing and Korver, 2020). A deficiency of sodium in chicken diets resulted in poor growth, increased adrenal weight, softening of the bones, corneal keratinization, gonadal inactivity, decreased plasma fluid volume, and decreased egg production (Burns et al., 1952, 1953; Nott and Combs, 1969; Harms et al., 1995; McDowell, 2003). Sodium deficiency in turkeys resulted in decreased serum electrolyte concentrations, body weights, and tibia mineralization/bone density (Jankowski et al., 2012). Signs of chlorine deficiency in chicks (Leach and Nesheim, 1965) and turkeys (Kubicek and Sullivan, 1973) include poor growth, mortality, hemoconcentration, and reduced blood chlorine level. Chlorine-deficient chicks and poults showed nervous conditions resembling tetany and/or balance issues when exposed to a loud, sharp noise (Kubicek and Sullivan, 1973).

Potassium toxicosis in healthy animals is rare due to the body's ability to readily excrete potassium as well as regulate its absorption (NRC, 2005). Teeter and Smith (1986) supplemented both the diet and water with potassium, resulting in decreased growth and increased feed conversion ratio. Increasing dietary potassium in laying hens from 2,300 to 20,000 mg potassium/kg increased water consumption, the water-to-feed ratio, and excreta moisture while decreasing feed intake (Smith et al., 2000).

Early reports of salt (i.e., sodium chloride) toxicity in poultry date back to 1892 and were reviewed by Halpin and colleagues (1936), who demonstrated that feeding Leghorn chicks 2 percent salt and above resulted in increased water intake and wet litter, and 5 percent salt resulted in high mortality and bloated bodies, while in adult hens morality was unaffected, but egg production was reduced (Damron and Kelly, 1987). Administering sodium chloride in the drinking water of chicks above 0.4 percent had adverse effects and above 0.9 percent caused sudden mortality, water retention, and marked renal changes resembling Bright's disease in humans (Selye, 1943; Krista et al., 1961). Egg production in adult laying hens was unaffected by 0.7 percent salt in drinking water but decreased with a 1 percent salt solution (Krista et al., 1961). Feeding young poults diets containing 2 percent sodium chloride or above resulted in increased mortality, edema, and ascites, although one reported this effect with salt at 0.9 percent (Matterson et al., 1946; Scrivner, 1946; Bressler et al., 1951; Morrison et al., 1975). Older poults seemed to tolerate higher dietary salt concentrations as the only reported issues appear to be increased water intake and increased the incidence of pendulous crop (Roberts, 1957; Harper and Arscott, 1962).

Salt in drinking water is comparatively more toxic to poultry than salt in dry feed (NRC, 2005). In this regard, Krakower and Goettsch (1945) equated a dietary level of 3.5 percent salt in a mash diet to 0.9 percent saline solution as drinking water. Poultry are considered to be more sensitive than mammals to excess dietary salt as a result of morphological differences in the renal glomerulus and a more limited effective area of filtration (Krakower and Goettsch, 1945). Cardiac hypertrophy and enlarged, congested kidneys with marked glomerular hypertrophy are among the many pathological changes observed in chicks exhibiting sodium chloride toxicity (Krakower and Goettsch, 1945; Doll et al., 1946; Paver et al., 1953; Mohanty and West, 1969; Sokkar et al., 1983). Young chicks are more susceptible to sodium chloride toxicity than are older chickens (Selye, 1943; Blaxland, 1946; Kare and Biely, 1948; Krista et al., 1961), most likely because their kidneys are too immature to excrete sodium and chloride rapidly when fed high salt diets (NRC, 2005). The balance of sodium and chloride is also of importance to tolerance as Nesheim and coworkers (1964) concluded that excesses of dietary sodium or chloride reduced chick growth and/or increased mortality unless each ion was balanced with equimolar levels of the other.

Normally, adequate magnesium is present in the natural ingredients of practical diets to meet the requirements of poultry (Shastak and Rodehutscord, 2015). However, ingestion of a magnesium-deficient diet negatively affected the growth of chicks and ducks, which were also lethargic, and when disturbed, affected chicks frequently became convulsive and passed into a comatose state (Almquist, 1942; Bird, 1949; Van Reen and Pearson, 1955; Gardiner et al., 1960). In laying hens, magnesium deficiency resulted in a rapid decline in blood magnesium level, withdrawal of magnesium from bone, a decline in egg production, and eventually a comatose state and death (Cox and Sell, 1967). Magnesium content and hatchability of eggs were reduced when hens were fed magnesium-deficient diets (Sell et al., 1967; Hajj and Sell, 1969). As is common among minerals, magnesium interacts with other elements, and increasing either dietary calcium or phosphorus concentrations accentuates the untoward effects of magnesium deficiency (Nugara and Edwards, 1963). Supplementation with excess magnesium in diets naturally adequate in magnesium has been investigated to reduce stress and feather picking in laying hens (Van Krimpen et al., 2005; Shastak and Rodehutscord, 2015).

Feeding chicks magnesium at high concentrations negatively impacted growth, reduced bone ash, increased rickets-like bone lesions and increased mortality (Nugara and Edwards, 1963; Chicco et al., 1967; Lee et al., 1980). In laying hens, high magnesium feeding reduced egg production and egg weights, decreased body weight, and increased feed conversion ratio (McWard, 1967; Stillmak and Sunde, 1971). Excessive dietary magnesium can antagonize calcium metabolism, usually resulting in skeletal abnormalities (Lee et al., 1980; Shastak and Rodehutscord, 2015). Atteh and Leeson (1983) observed reduced eggshell calcium, elevated eggshell magnesium, decreased bone calcium and zinc contents, and elevated bone magnesium with increasing magnesium in laying hen diets.

TRACE MINERALS

Poultry require copper, iodine, iron, manganese, selenium, and zinc at levels between ~0.01 and ~100 mg/kg, and because the levels are low compared to other minerals, they are referred to as trace minerals. Cobalt is also required, but it needs to be supplied as a structural component of vitamin B_{12} in most poultry applications and is not routinely supplemented as inorganic cobalt. As is the case with rodents, it is likely that birds that engage in coprophagy would be able to meet their B_{12} requirement by microbial synthesis when there is sufficient cobalt in the diet, but this is not common in commercial poultry production. A case for a dietary requirement for other trace minerals in vertebrates has been made (Nielsen, 1991; NRC, 1994). The strongest case is for molybdenum because it is a cofactor for several enzymes in vertebrates, including aldehyde oxidase, sulfite oxidase, xanthine oxidase, and xanthine dehydrogenase in some animals (Belaidi and Schwarz, 2017). Supplementation of molybdenum to broilers fed a casein basal diet with 3 mg/kg molybdenum increased the activity of hepatic xanthine oxidase in a dose dependent manner, although weight gain was not affected (Nell and Annison, 1980), but further research is necessary before molybdenum should be considered as nutritionally essential, although potential toxicity will be addressed for molybdenum subsequently. Experimental diets may sometimes be formulated from purified or chemically defined ingredients. Under these conditions, silicon and boron may be inadequate, and biological responses may occur with the addition of these elements to the diet (Carlisle, 1970, 1980; Hunt and Nielsen, 1986).

The requirements for trace minerals are typically based on the minimal level that optimizes production performance (e.g., copper, zinc), the absence of pathologies (e.g., manganese, iodine, iron, and selenium) or the activity of enzymes or functions that are most sensitive to deficiencies (all trace minerals). In practical grain–soybean diets fed to broilers and laying hens, some trace minerals, such as iron, may be present at sufficient levels without supplementation depending on the rate of growth or egg production. However, most of the trace minerals are routinely supplemented to poultry diets for several reasons. First, the concentration of trace minerals in a feedstuff is highly variable relative to other nutrients due to the variability of their content in soils and water used for their production so levels are difficult to predict without routine measurement. Second, the absorption, metabolism, and excretion of a trace mineral can be strongly antagonized by interactions with other dietary components. For these reasons, levels moderately above known mineral requirements are typically fed to provide a margin of safety. Similar to vitamins, the amount of trace minerals added is very small, so they are often diluted in a premix prior to use at a feed mill.

Historically the primary form of supplemented trace minerals were inorganic salts or oxides (NRC, 1994; Leeson, 2009). In the last several decades, organic sources of most of the trace minerals have become available. The organic sources typically consist of minerals chelated or complexed with organic molecules such as amino acids, peptides, proteins, simple carbohydrates, or polysaccharides (Miles and Henry, 2000; AAFCO, 2020). The bioavailabilities of trace mineral supplements are affected by their chemical and physical properties and by their susceptibility to decreased intestinal absorption due to anti-nutrients or high levels of other minerals in diet. The bioavailability of a trace mineral is typically expressed in proportion to a standard, typically a sulfate, chloride, or oxide form of the mineral. The bioavailability of trace mineral salts roughly correlates positively with its solubility in a solution that mimics physiological conditions in the intestinal contents (Henry, 1995), but feeding studies using graded levels of the trace mineral are necessary to precisely define the bioavailability of a mineral source (Chapter 8). Examples of the bioavailability of trace minerals determined by feeding studies are provided subsequently for each trace mineral. Only studies that established slopes using three or more dietary concentrations of the mineral were considered in the following discussion, and studies using levels near the requirement were prioritized over those that use plethoric levels. The rankings of bioavailabilities of mineral sources provided subsequently are estimates from multiple research reports that compare a mineral source to an established standard. However, these studies are often conducted in different experiments at different locations using different husbandry and genetics. For this reason, they are only rough estimates (see Chapter 8). Organic forms of trace minerals are not discussed unless the specific chemical form of the mineral in the organic matrix is known and published in the peer-reviewed literature. For this reason, many of the sources of trace minerals that are in poorly defined organic matrixes were not used by the committee for estimating requirements, bioavailabilities, or toxicities.

The reduction of trace mineral excretion by poultry has become an important area of research due to environmental concerns related to long-term accumulation in soils following application of manures as fertilizers (Dozier et al., 2003; Leeson and Caston, 2008; Spears et al., 2008; Leeson, 2009). Feeding highly bioavailable sources of trace minerals minimizes trace mineral excretion and subsequent manure levels. These concerns have stimulated renewed research interest on trace mineral requirements and the bioavailability of supplemental sources. Additionally, the bioavailability of iron, zinc, copper, and manganese in grains and soybean meal is low due to chelation by phytate; however, feeding phytases results in increased bioavailability and lower rates of excretion (Sebastian et al., 1996; Viveros et al., 2002).

With the exception of zinc and copper, the requirement of poultry for trace minerals are typically based on functional endpoints and the avoidance of pathologies because these endpoints are usually more sensitive than performance criteria. In the case of zinc, and to a lesser extent copper, growth rates are relatively sensitive. Requirement estimates often use semi-purified or partially purified diets that have lower levels of phytate, fiber, and other anti-nutritional factors that decrease the bioavailability of trace minerals. For this reason, the recommendations are minimal requirements and a margin of safety must be added to account for anti-nutritional factors present in other ingredients, as well as differences in rates of performance, genetics, and environmental variations.

Copper

Copper exists as both a metal and as a cation. Its two ionic states are cuprous (+1) and cupric (+2). In its ionic state it is a cofactor for a variety of intracellular and extracellular enzymes that catalyze oxidative reactions, including superoxide dismutase, cytochrome oxidase, lysyl oxidase, ferroxidase, monoamine oxidase, ceruloplasmin, and tyrosinase. Copper-dependent enzymes function in energy metabolism, maturation, and stabilizing the extracellular proteins collagen and elastin, antioxidant defense systems, iron transport and metabolism, and pigmentation as well as other biological processes.

Copper is typically supplemented to the diets of poultry due to the low levels and high variability found in many feedstuffs. Inorganic forms of copper are sometimes added to diets at levels (100–250 mg/kg) far in excess of the requirement because it has growth-enhancing antimicrobial effects (Pesti and Bakalli, 1996; Arias and Koutsos, 2006). Additionally, high levels of dietary copper are used to treat crop mycosis (Harms and Buresh, 1987; Pesti and Bakalli, 1996). However, excess dietary copper results in levels in excreta that are of environmental concern when poultry manure is applied as a fertilizer (Dozier et al., 2003; Leeson, 2009), resulting in maximum dietary levels to be regulated in some jurisdictions.

Dietary Sources, Supplementation, and Bioavailability

Copper levels in plants typically vary from 1 to 50 mg/kg dry matter depending upon the species and soil factors. Cereal grains contain marginal levels of copper (4–8 mg/kg) relative to the requirements of poultry, but seeds of cultivated legumes usually contain higher levels (NRC, 2005). The copper concentrations in fish meals and meat meals are also variable and are often marginal relative to the requirements of poultry. Copper is commonly supplemented to diets fed to poultry from inorganic sources including copper sulfate, copper chloride, tribasic copper chloride ($Cu_2(OH)_3Cl$), copper carbonate, copper oxide, and complexed or chelated organic sources, such as Cu-methionine or Cu-methionine hydroxy analogue.

Copper absorption occurs predominantly in the proventriculus and duodenum (Mertz, 1987; Bao et al., 2009). Because copper is a redox metal, it has the ability to exist in several valence states and interact with other components that inhibit its absorption (Mertz, 1987). Absorption is inhibited by phytate, calcium, oxalate, tannins, fiber, other minerals (including Zn, Mo, and Cd), and dietary reducing agents such as ascorbate and L-Cys (Mertz, 1987; Aoyagi and Baker, 1993a). The apparent digestibility of copper in broiler chicks fed diets near their requirements was reported to be approximately 15 percent in one study (Dozier et al., 2003) and 40 percent in another (Bao et al., 2009), indicating that even when dietary copper levels are low, most of the copper is excreted.

Copper homeostasis is maintained predominantly by controlling the rate of absorption, although copper excretion via the bile increases with increasing dietary copper concentrations (Aoyagi and Baker, 1993b). When dietary copper levels are high, intestinal epithelial cells synthesize metallothionein, a cysteine-rich protein that tightly binds copper and retards its absorption.

The bioavailability of copper in feedstuffs and supplements fed to poultry is typically determined using performance indices, bile copper, or the activity of copper-dependent enzymes, such as lysl oxidase in tendons or plasma ceruloplasmin. $CuSO_4•5H_2O$ is typically used as the standard for comparison (Baker and Ammerman, 1995a). Although copper is often fed at very high levels to increase growth, bioavailability experiments conducted using plethoric levels were not considered for estimating copper bioavailability at physiological levels. Among inorganic reagent grade copper sources added to experimental diets at levels below the requirement, the bioavailability is estimated as CuCl > Cu-citrate > $CuSO_4•5H_2O \approx Cu_2O \approx CuCO_3•Cu(OH)_2 \approx$ Cu-lys ≈ Cu-met > $CuCO_3$, and the bioavailability of CuO is nearly zero, as is that of metallic copper (Zanetti et al., 1991; Aoyagi and Baker, 1993b,c; Baker and Ammerman, 1995a; Pesti and Bakalli, 1996). The bioavailability of copper in feedstuffs fed to poultry are variable and hard to generalize (Aoyagi et al., 1993, 1995; Leeson, 2009). The bioavailability of copper from plant feedstuffs is generally considerably lower than that from animal sources. For example, the bioavailability relative to $CuSO_4•5H_2O$, of soybean meal, cottonseed meal, and corn gluten meal is 50 percent lower than from some animal sources (Aoyagi et al., 1993). However, the bioavailability of copper in some feedstuffs of animal origin, such as pork liver, can be very low (Aoyagi et al., 1995).

Requirements

Poultry have requirements for copper that are less than 15 mg/kg of diet, with the highest being in broiler breeders. The specific recommendations and the studies they are based on can be found in Chapters 13–19. Requirements are typically based on dietary levels that prevent the expression of signs of deficiency or impairments in productivity. Higher levels are typically fed to provide a margin of safety due to interactions among copper and other nutrients, antinutritional factors in feedstuffs, potential increases in requirement due to stressful situations, and poor bioavailability in some feedstuffs.

Deficiency and Toxicity

Copper deficiency signs include microcytic anemia, hemorrhaging, bone deformities and lameness, infertile eggs, and poor feather pigmentation (Scott et al., 1982; NRC, 1994; Klasing and Korver, 2020). Anemia is due to low levels of ceruloplasmin, which reduces iron uptake into tissues. Copper deficiency also results in impaired reproduction (Berwanger et al., 2018), and abnormal embryonic development (Roychoudhury et al., 2016).

Copper is relatively nontoxic to nonruminant animals (NRC, 2005). Poultry are able to tolerate excessive levels of dietary copper. In fact, a dietary copper content of 250 mg/kg from $CuSO_4•5H_2O$ or cupric citrate has been shown to markedly reduce plasma and breast muscle cholesterol concentrations without negatively affecting growth or feed conversion efficiency in broiler chickens (Bakalli et al., 1995; Pesti and Bakalli, 1996). However, other studies with broilers have shown that dietary supplementation of 250 mg copper/kg or 500 mg copper/kg can cause proventricular enlargement (Wideman et al., 1996) and lesions in the oral cavity, tongue, and pharynx (Chiou et al., 1999). In contrast, Pesti and Bakalli (1996) and Ledoux et al. (1991) reported no adverse effects of feeding broilers diets supplemented with 375 or 450 mg copper/kg feed, respectively.

Similar to lowering muscle cholesterol content in broilers, an egg cholesterol-lowering effect of excess dietary copper was observed in laying hens fed up to 250 ppm $CuSO_4 \cdot 5H_2O$ while body weights, feed consumption, and egg weights were generally unaffected (Pesti and Bakalli, 1998).

Waibel et al. (1964) reported that supplemental dietary copper at only 50 mg/kg was highly toxic to turkey poults fed a purified diet (corn starch-isolated soy protein–based) from hatch to 28 d of age, but that poults fed a practical diet (corn-soybean meal–based) supplemented with up to 800 mg copper/kg performed similarly to control birds and "appeared normal." Leeson et al. (1997) reported that feeding 1,000 or 2,000 mg copper/kg diet was without effect on the performance of 6- to 10-week-old poults. High mortality and gastrointestinal tract lesions in 2-d-old turkey poults, resulting from ingestion of bedding material (pine shavings) containing copper sulfate, were reported by Giergiel et al. (2019). The copper sulfate was used as a bedding disinfecting agent but apparently was not applied properly to the pine shavings and, depending upon the building where the birds were housed, averaged 1,404 (Building A) and 3,256 mg copper/kg dry weight (Building B). Supplementation of 500 mg copper/kg diet to ducklings caused a myopathy similar to that observed with selenium-vitamin E deficiency (Van Vleet, 1982), which suggested that ducks may be more sensitive to copper toxicosis than other poultry or that copper may have interfered with selenium utilization.

Iodine

Iodine is a nonmetallic element and an integral part of the thyroid hormones 3, 3′, 5, -triiodothyronine (T_3), and 3, 3′, 5, 5′-tetraiodothyronine (thyroxine, T_4). Iodide is the term used to refer to its ionic form, I, and iodate refers to IO^{3-}. No other metabolic functions other than its requirement for the prohormone T_4 and the hormone T_3 have been described in poultry or other vertebrates. Following absorption, iodine is rapidly taken up by the thyroid gland and converted to organic iodine by combining with tyrosine. Thyroglobulin, a glycoprotein with iodinated tyrosine residues, is the predominant storage form for iodine in the thyroid. Excess iodine is excreted in the urine. Most of the iodine in the body is found in the thyroid. In the egg, most is found in the yolk. Developing follicles in the ovary concentrate iodine, and many studies show that the amount of iodine in the egg reflects that in the diet (Yalçin et al., 2004; Sumaiya et al., 2016; Bakhshalinejad et al., 2018). Enriching eggs in iodine is an important strategy for supplementing the human food supply in deficient areas of the world (Kavtarashvili et al., 2017).

Most iodine in foods is in an inorganic form that is readily absorbed by the gastrointestinal tract as iodide (Miller and Ammerman, 1995; NRC, 2005). Compounds found in some feedstuffs interfere with iodine utilization. Glucosinolates found in *Brassicaceae*, which include rapeseed and cruciferous vegetables, break down to thiocyanates during processing and digestion. Thiocyanates inhibit the uptake and the utilization of iodide in the thyroid gland, blocking the production of thyroid hormones (Willemin and Lumen, 2017). Other foods such as soybeans, linseeds, cassava, peas, lentils, and peanuts also contain goitrogenic substances. Many, but not all, of the effects of goitrogens can be overcome by additional supplementation of iodine (Underwood, 1977).

The iodine concentrations in feedstuffs are highly dependent on geographical location. In general, feedstuffs grown near bodies of waters with high salt content, such as oceans, have high levels of iodine (Zicker and Schoenherr, 2012). Iodine is depleted from soils by leaching from glaciations, flooding, and erosion, and crops grown in these soils are low in iodine (~10 µg/kg). Most poultry diets are supplemented with iodine because supplementation is economically inconsequential relative to analyzing the highly variable levels of iodine in feedstuffs.

Dietary Sources, Supplementation, and Bioavailability

Iodine sources typically used include $Ca(IO_3)_2$, KI, KIO_3, NaI, $NaIO_3$, and organic salts like ethylenediamine dihydroiodide (EDDI). Iodine is supplemented either in a trace mineral mix or as iodized salt used to supplement sodium chloride. Hixon and Rosner (Hixson and Rosner, 1957) compared the addition of 0.6 mg iodine/kg of a corn-based diet from either $Ca(IO_3)_2$ or KI and found similar rates of performance of meat type chicks. Słupczyńska et al. (Słupczyńska et al., 2014) compared KI with KIO_3 at 1, 3, or 5 mg supplemental iodine/kg of a corn- and soybean meal–based diet and found that transfer to egg contents was higher for KI than KIO_3, indicating higher bioavailability for increasing egg iodine to improve human nutrition. They also found that adding rapeseed meal decreased the bioavailability of both sources for increasing egg iodine. Sarlak et al (2020) compared the supplementation of either 2.0, 4.0, or 8.0 mg of iodine/kg of diet on the performance of laying hens fed a diet based on corn and soybean meal. Hens fed EDDI had significantly higher levels of yolk and albumen levels of iodine than those fed $Ca(IO_3)_2$. Addition of 8.0 mg/kg diet of either source significantly decreased egg production with a larger decrease with EDDI, suggesting that its threshold for toxicity is lower.

Requirement

The requirement of poultry for iodine has not been the subject of much investigation, especially recently. In most cases the requirements are rough estimates based on one or a few studies. The estimated iodine requirements of poultry consuming a grain-based diet are less than 0.70 mg/kg dry matter. The specific recommendations and the studies they are based on can be found in Chapters 13–19. The endpoints for estimating the requirement are typically growth, thyroid histology, or hatchability of eggs.

Deficiency and Toxicity

Iodine deficiency results in goiter, which is the enlargement of the thyroid glands due to the release of high amounts of thyroid stimulating hormone from the pituitary, resulting in thyroid hyperplasia (Rogler et al., 1959; Handa and Chiasson, 1980). The gland may increase to many times its usual size. Inadequate production of thyroid hormones results in poor growth, egg production, and egg size. Iodine deficiency in breeders results in low iodine content of the egg and, consequently, decreased hatchability and thyroid enlargement in the embryos.

The continued administration of large doses of iodine greatly reduces iodine uptake by the thyroid gland, thereby causing antithyroidal or goitrogenic effects in many species of domestic and experimental animals (NRC, 2005). As is the case with copper, chickens and swine are more tolerant of excesses of iodine than cattle (NRC, 2005), but iodine toxicoses in poultry are well documented (Christensen and Ort, 1991; Baker et al., 2003; Lewis, 2004) and are typically manifested as decreased egg production, egg weights, and hatchability; low fertility; early embryonic death; enlarged thyroids in hatchlings; and wiry down. Moreover, turkey breeder hens are more sensitive to high iodine diets than chickens (Christensen and Ort, 1991; Baker et al., 2003; Lewis, 2004).

Iron

Iron is found in metals and cationic ionic compounds. Ferrous (+2) and ferric (+3) valence states are nutritionally important because many of their salts have high bioavailability to poultry. Iron is nutritionally unavailable in metals, but these states are commonly found in feedstuffs due to contamination from soil or during processing, making it difficult to rely on the analyzed level of iron in feeds to assess their nutritional efficacy. Iron-containing proteins facilitate the transport, storage, and use of oxygen and also aid in catalyzing enzymatic electron transfer processes. Heme proteins are the primary proteins that facilitate oxygen transport, storage, and use and include hemoglobin, myoglobin, and cytochromes. Iron-containing proteins also have important functions in oxidases and oxygenases. Transferrin facilitates iron transport and is found in the blood plasma. Ferritin and its degradation product, hemosiderin, facilitate iron storage in cells, and the liver, spleen, and bone marrow are the primary tissues for iron storage. Ferritin usually predominates at normal dietary iron levels, and hemosiderin is formed to buffer excess iron. In general, metabolism of absorbed iron in poultry is similar to that in mammals, and it functions to prevent free ionic iron from catalyzing the generation of reactive oxygen molecules via the Fenton reaction and causing damage to tissues. This is done by keeping iron tightly bound to protein during all aspects of iron absorption, transport, and use (Doherty, 2007), resulting in free iron concentrations that are many orders of magnitude lower than those of other required minerals. The extremely low concentrations of free iron in extracellular fluids make it the first limiting nutrient for the growth of many microbes, including pathogens (Doherty, 2007).

In the egg, iron is found in the yolk associated with phosvitin (Romanoff and Romanoff, 1949), and the level increases with dietary iron intake (Bess et al., 2012; Abbasi et al., 2015). Very little iron is found in the albumen; however, it contains large amounts of ovotransferrin (Giansanti et al., 2012). Ovotransferrin complexes free iron and protect the egg from microbial growth. Iron is also an important chromophore in red- and black-pigmented feathers (Davis et al., 1962; Grau et al., 1989).

Dietary Sources, Supplementation, and Bioavailability

Most grains contain 30–85 mg/kg iron; meat meals contain 400–600 mg/kg; and soybean meals contain around 100 mg/kg (Chapter 20). There is high variability in the iron content of feedstuffs from plant origin due to soil attributes, including concentration and chemical form of iron, pH, and levels of other minerals. Calcium phosphate, limestone, and many other commercial mineral supplements contain very high levels of iron (2,000–5,000 mg/kg). Drinking water may also contain nutritionally significant levels of iron (NRC, 2005). Iron is commonly supplemented as ferrous sulfate, ferric chloride, ferric citrate, ferric ammonium citrate, and iron–amino acid complexes (Henry and Miller, 1995).

The absorption of $FeCl_3$ occurs throughout the digestive tract of chickens but is highest in the duodenum (Saiz et al., 1993). The duodenal mucosa was also shown to accumulate more iron than the ileal mucosa when incubated with iron sulfate in vitro (Markovs et al., 2018). Heme iron is absorbed intact, utilizing a transport mechanism that is distinct from that used for ionic iron. Iron absorption is tightly controlled to avoid excess uptake. When body stores of iron are adequate, most of the iron that enters the enterocyte is bound to ferritin and excreted. This mucosal-block is important because iron homeostasis is not achieved by urinary or biliary excretion and only by regulating absorption (Kreuzer and Kirchgessner, 1991).

Many dietary factors influence the absorption of non-heme iron. High levels of dietary ascorbic acid enhance absorption, whereas absorption is decreased by high levels of phytate, polyphenols, fiber, calcium, phosphorus, manganese, copper, and zinc (Henry and Miller, 1995; Kheravii et al., 2018). Experiments examining the bioavailability of iron in feedstuffs or supplements typically use blood hemoglobin concentrations as the response criteria and use $FeSO_4 \cdot 7H_2O$ as a standard for comparison (Henry and Miller, 1995). Reagent grade iron compounds added to experimental diets have the following bioavailabilities: ferric sulfate heptahydrate ≈ ferrous sulfate monohydrate ≈ ferrous chloride ≈ ferric ammonium citrate ≈ ferric ammonium sulfate ≈ ferric choline citrate > ferric chloride ≈ ferric citrate >> ferric pyrophosphate >> ferric oxide ≈ ferric orthophosphate ≈ ferrous carbonate (Henry and Miller, 1995). When compared to highly available mineral salts supplements, iron from plant sources is typically less than 10 percent available, and iron in animal-based feedstuffs is about 30 percent available (Henry and Miller, 1995).

Requirement

The iron requirements of poultry using purified or semi-purified diets have been estimated at between 35 and 120 mg/kg dry matter for all stages of production, though maintenance requirements are likely much lower. The specific recommendations and the studies they are based on can be found in Chapters 13–19. High levels of dietary phytic acid, tannins, fat, zinc, manganese, and copper decrease the absorption of inorganic iron and increase the apparent requirement (Bafundo et al., 1984; Baker and Halpin, 1991; Pimentel et al., 1992), so a margin of safety is necessary for practical diets.

Deficiency and Toxicity

Iron deficiency in chickens and turkeys causes a hypochromic microcytic anemia in which the red blood cells are reduced in size and low in hemoglobin (Davis et al., 1962). In red-feathered chickens, pigmentation does not occur when the diet is deficient in iron (Hill and Matrone, 1961). Egg hatchability is reduced because of embryonic mortality during the late stage of incubation due to anemia (Morck and Austic, 1981).

The characteristic signs of chronic iron toxicosis in poultry include decreased feed intake, weight gain, and efficiency of feed conversion (NRC, 2005). Cao et al. (1996) fed meat-type chicks varying levels of dietary iron from hatch to 21 d of age. Compared to the basal diet containing 188 mg/kg iron, the addition of 400, 600, or 800 mg/kg of additional iron from $FeS04–7H_20$ decreased feed intake and body weight gain. However, there were no significant differences among the three high iron treatments. Delaying feeding the high iron diets until 5 d of age alleviated the depression in feed intake associated with the high iron diets.

Manganese

Manganese can exist in 11 oxidation states, but in animals and their feedstuffs it is usually a divalent cation (Mn^{+2}). However, it is a redox metal, and Mn^{+3} facilitates its function in superoxide dismutase and binding to transferrin. Mn functions as a cofactor in metalloenzymes and also as a regulator of enzymes (Keen et al., 2000; Suttle, 2010). Manganese metalloenzymes include arginase, pyruvate carboxylase, and Mn superoxide dismutase (Mn-SOD). In contrast to the relatively few Mn metalloenzymes, there are many enzymes that are regulated by Mn. Some nutritionally important examples include phosphoenolyruvate carboxykinase, glutamine synthetase, and glycosytransferases. During a deficiency some of the functions of manganese can be accomplished by other minerals, such as magnesium (Keen et al., 2000). Other functions, such as activation of glycosyltransferases, are very sensitive to a deficiency of manganese. These enzymes are important for the synthesis of mucopolysaccharides and glycoproteins involved in bone and cartilage development. Deficiencies of manganese are typically monitored by evaluating declines in joint functions, bone development, and eggshell quality (Leach and Gross, 1983; Liu et al., 1994; Keen et al., 2000).

Manganese absorption occurs throughout the small intestine, but the rate is relatively slow, inefficient, and variable. Chicks fed a semi-purified casein-dextrose diet had an efficiency of absorption of only 2.4 percent, and those fed a corn-soybean meal diet were only 1.7 percent efficient (Halpin et al., 1986). Absorption is negatively affected by the level of other minerals such as phosphorus, phytic acid, and fiber (Halpin et al., 1986; Wedekind et al., 1991). Tissue manganese levels appear to be regulated by the rate of both absorption and excretion, with bile being the primary excretory route. Unlike copper, iron, and zinc, there are no known storage proteins for manganese, but it is incorporated into bone and is released during bone turnover. In growing chickens, body stores of manganese that accumulate during periods of high dietary manganese consumption can supply sufficient manganese to last for about a month when consuming a severely manganese deficient diet (Halpin and Baker, 1986).

Dietary Sources, Supplementation, and Bioavailability

The concentration and bioavailability of manganese in feed ingredients is highly variable. The vegetative parts of plants have high levels of manganese (60–800 mg/kg dry matter) depending on soil and climate conditions. Seeds contain less manganese (10–80 mg/kg). Unlike other trace minerals, manganese is low in meat by-products relative to many plant sources (Chapter 20). Generally, water is a minor source of ingested manganese (NRC, 2005). Diets for poultry based on corn, sorghum, or barley are likely deficient unless they contain manganese-rich ingredients such as wheat bran or wheat middlings. Because of the lack of information on the bioavailability of manganese, the many antagonists of its utilization, and its relatively low toxicity, it is typically supplemented to most poultry diets at levels exceeding the estimated requirement. Chemically characterized sources of manganese include MnO, MnO_2, $MnCO_3$, $MnSO_4$, $MnCl_2$, Mn-methionine, Mn-methionine hydroxy analogue, and Mn-proprionate.

Experiments designed to determine the bioavailability of manganese sources typically use reagent grade $MnSO_4 \bullet H_2O$ as a standard for comparison. The endpoints for bioavailability studies are typically activity of enzymes that use manganese as a cofactor, incidence of perosis, and its concentration in bone (Wedekind and Baker, 1990a; Baker and Oduho, 1994; Keen et al., 2000; Tufarelli and Laudadio, 2017). The following relative bioavailabilities of reagent grade manganese are estimated from multiple studies comparing each source's bioavailability to similar standards in growing chickens: Mn-methionine ≈ Mn-proprionate > $MnSO_4 \bullet H_2O$ ≈ $MnCl_2 \bullet 4H_2O$ > $MnCO_3$ ≈ MnO (Fly et al., 1989; Henry et al., 1989; Henry, 1995; Brooks et al., 2012).

Requirement

The requirement of poultry species for manganese is considerably higher than that of pigs and other domestic mammals, apparently due to inefficient absorption. The manganese requirement of poultry is generally less than 70 mg/kg dry matter. The specific recommendations and the studies they are based on can be found in Chapters 13–19. Performance indices, bone manganese levels, prevention of perosis, and the activity of manganese requiring enzymes or their products are typically used to determine the requirement of poultry. The requirement for birds fed a purified diet is considerably lower due the absence of factors that negatively impact absorption (Halpin et al., 1986). Because requirement studies in poultry use purified or semi-purified diets, a margin of safety should be provided based on the antinutritional effects of other ingredients. In the study by Halpin et al. (1986), the minimal level of safety to account for these factors is an additional 41 percent of added $MnSO_4 \bullet H_2O$.

Deficiency and Toxicity

Manganese deficiencies are relatively common in poultry because the manganese in many feedstuffs is poorly absorbed. Manganese deficiency in chicks and poults results in perosis or slipped tendon (Tufarelli and Laudadio, 2017). Deficiencies of other nutrients, such as riboflavin, niacin, pyridoxine, folate, choline, and biotin, may also be involved in inducing perosis (see Chapter 5, "Vitamins"). The usual signs of perosis are swelling and flattening of the chondyle within the hock joint, with subsequent slipping of the Achilles tendon from its condyles. The tibia and the tarsometatarsus may exhibit bending near the hock joint and lateral rotation. One or both legs may be affected. A shortening and thickening of the long bones of the wings and legs are also observed. The disorder, insofar as manganese is concerned, is aggravated by excess dietary calcium (Wedekind et al., 1991).

In laying and breeding birds, manganese deficiency results in lowered egg production, reduced eggshell strength, poor hatchability, and reduced fertility. Changes in circulating ovarian steroids are one of the causes of the reproductive dysfunction, as manganese is an enzyme cofactor involved in cholesterol biosynthesis, and cholesterol is a precursor of ovarian steroids (Olgun, 2017).

Manganese is considered to be among the least toxic of the essential elements (NRC, 2005). In poultry, decreased growth was not observed until diets were fed to chickens that contained >3,000 mg manganese/kg feed (Southern and Baker, 1983a; Black et al., 1985) or diets were fed to turkeys that contained >4,000 mg manganese/kg feed (Vohra and Kratzer, 1968). The toxicity of excessive manganese appears to be a mineral antagonism manifestation, with iron as the main antagonist (McDowell, 2003). In general, the most common signs of manganese toxicity are depressed iron status and hematological changes (NRC, 2005).

Selenium

Selenium is a metalloid and has chemical properties similar to sulfur, and for that reason some of its metabolic pathways resemble that of sulfur. It has four natural valence states: selenide (–2), elemental selenium (0), selenite (+4), and selenate (+6). Selenium was the last of the trace minerals to be confirmed as nutritionally essential when Patterson et al. reported that it prevented exudative diathesis in chicks (Patterson et al., 1957). Selenium's primary function is as a cofactor in enzymes that catalyze redox reactions. Selenium is unique among the nutrients that serve as cofactors in enzymes because it is incorporated into the functional site of selenium dependent enzymes (selenoenzymes) during their synthesis and not subsequent to their synthesis. Inorganic selenium, usually selenophosphate, is required for the co-translational assembly of selenocysteine at the active site of the enzyme. Dietary selenocysteine or selenomethionine must be catabolized to release inorganic selenium in order to be available for its catalytic role in selenoenzymes.

A quantitative assessment of selenium nutrition for poultry is complicated by its many different chemical forms that differ in their bioavailabilities and toxicities. In feedstuffs selenium is predominately selenomethionine, but inorganic sources are often used as supplements to meet the requirement for selenium. Selenite, selenate, and selenomethionine are readily absorbed in the small intestine. Selenate is likely absorbed by the same transport system used by sulfate, and selenomethionine utilizes the same transporter as methionine (Ha et al., 2019). Selenium absorption is not tightly regulated, and uptake and tissue accumulation are not greatly diminished even when tissue levels are replete (Vendeland et al., 1992). Thus, in comparison to other trace minerals, selenium toxicity can occur at levels that are only moderately higher than the requirement.

The chicken genome predicts the existence of at least 24 selenoenzymes (Zhu et al., 2017; Li et al., 2018), some of which play a role in antioxidant defenses and maintaining redox balance, including the glutathione peroxidases family. The glutathione peroxidases quench peroxides and are important in defending against membrane damage. Vitamin E is also an important anti-oxidant, and glutathione peroxidase functions together with vitamin E to mitigate oxidative damage. This complementary relationship between selenium and vitamin E is the basis for their mutual sparing effect on requirements (Scott, 1966; Combs and Combs, 1986). Selenium also functions in thyroid metabolism because thioredoxin reductase and iodothyronine are selenoenzymes (Sun et al., 2019).

Selenium can be retained in tissues by incorporation into newly synthesized proteins. Selenomethionine is more efficiently stored than inorganic selenium sources because it can be absorbed and directly incorporated into tissue proteins in place of methionine (Surai et al., 2018). During protein catabolism, selenomethionine can be released and used for a supply of selenide that is needed for incorporation into selenoproteins. Muscle contains a large pool of stored selenium, and in the egg, selenium is mostly found in the albumen. The transfer of selenium to the egg is proportional to dietary levels, and supplements that contain selenomethionine result in higher egg levels than equal dietary levels of inorganic sources (Surai, 2000).

Relative to other trace minerals, the amount of selenium added to diets is very small, and the margin between the bird's requirement and the toxic level is narrow, so selenium supplementation must be done carefully. Additionally, high levels of selenium in poultry meat and eggs are a concern for toxicity to humans, and high levels of selenium in manure are a potential environmental hazard, making its supplementation to animal feeds subject to considerable government regulation in many countries (NRC, 2005).

Dietary Sources, Supplementation, and Bioavailability

Most of the selenium in feedstuffs is in the form of selenomethionine, and inorganic forms are a minor contributor. The selenium levels in feedstuffs are extremely variable. Concentrations in plants depend on the plant species and the level of available selenium in the soil and water (NRC, 2005). However, many regions of the world are selenium depleted, and in these regions the plants and their seeds have very low levels of selenium (<0.05 mg/kg dry matter). The amount of selenium in feedstuffs derived from invertebrates and vertebrates is highly dependent upon the amount of selenium in their environment and also their position in the food chain. Supplemental inorganic selenium is commonly added to diets as sodium selenite or sodium selenate (Na_2SeO_4). Metallic selenium (Se^{+0}) is not bioavailable to chickens (Combs et al., 1996). Supplemental organic forms include selenomethionine and Se 2-hydroxy-4-methylselenobutanoic acid (HMSeBA). There are many other organic sources of selenium that are also commercially available for supplementing poultry diets; however, their selenium contents and forms are not well characterized or standardized (Surai et al., 2017). For this reason, they are not included in the committee's consideration for requirement or bioavailability recommendations.

Experiments examining the bioavailability of selenium in feedstuffs or supplements typically use sodium selenite as the standard for comparison (Henry and Ammerman, 1995). The endpoints for estimating bioavailability include prevention of signs of deficiencies, especially exudative diathesis, glutathione peroxidase activity in tissues, or the concentration of selenium in tissues or eggs. As with other microminerals, the bioavailability of selenium varies with the endpoint used for assessment. Tissue concentrations of selenium have been criticized as an endpoint because they are highly dependent on the form of selenium fed and reflect selenium storage and not its use for supplying selenium to the enzymes that require it for their function (Hassan et al., 1993; Henry and Ammerman, 1995). Inorganic selenium is incorporated into selenoenzymes, and selenomethionine or selenocysteine must be catabolized and release inorganic selenium to become available to act as a cofactor in selenoenzymes. The rate of selenium release from selenomethionine so that it can be used as a cofactor in selenoenzymes is dependent on the diet's adequacy in methionine (Sunde et al., 1981). Because the catabolism of ingested methionine is minimal when dietary methionine levels are low but increases at dietary levels above the requirement, the bioavailability of its selenium is correlated with dietary methionine levels. This relationship does not occur for selenite or selenate because their metabolism is not as dependent on the fate of methionine. However, storage levels are important for buffering periods of selenium deficiency (Payne and Southern, 2005) and also for higher transfer to eggs to provision chicks, so those nutritional characteristics can be important (Surai et al., 2017).

The following relative bioavailabilities of reagent grade selenium sources are estimated from multiple studies comparing each source's bioavailability to similar standards in growing chickens fed diets with multiple selenium levels and using the activity of glutathione peroxidase or the prevention of deficiency pathologies: selenomethionine ≈ selenocysteine ≈ selenite (Na_2SeO_3) ≈ selenate (Na_2SeO_4) (Henry and Ammerman, 1995). However, when using the concentration of selenium in tissues or transfer to eggs, selenomethionine and HMSeBA have higher utility than Na_2SeO_3 (Jlali et al., 2013; Surai et al., 2017; Zhao et al., 2017). The bioavailability of selenium in foods of animal origin is generally lower relative to that from plants, a situation that is different than with all other trace minerals (Cantor et al., 1982; Henry and Ammerman, 1995).

Requirement

The selenium requirement of poultry fed purified diets is 0.2 mg/kg or less during all stages of the life cycle. The specific recommendations and the studies they are based on can be found in Chapters 13–19. The endpoints for estimating the requirement include glutathione peroxidase activity and prevention of signs of deficiencies, especially exudative diathesis. Growth rates are not very sensitive to marginal deficiencies of selenium. The requirement for selenium is decreased when dietary vitamin E levels are high (Scott, 1966; Combs and Combs, 1986) and increased when dietary copper or zinc is high (Jensen, 1975).

Deficiency and Toxicity

There is wide variability in the amount and availability of selenium in the soils of different geographic areas (NRC, 2005). Consequently, cereals and plant-derived feedstuffs are variable sources of selenium. Grains from some areas contain sufficient selenium to render them toxic to chicks. Selenium supplementation has a very low margin of error because of its high potential for toxicity and teratogenicity (NRC, 2005). As a result, the amount of supplementary selenium permissible in diets is regulated in the United States and Canada.

Functionally, selenium is closely associated with vitamin E and other antioxidants in practical feed formulation. The principal sign of deficiency in chicks is exudative diathesis (Creech et al., 1957; Patterson et al., 1957; Nesheim and Scott, 1958); myopathies of the gizzard and heart in turkeys (Walter and Jensen, 1963; Scott et al., 1967); pancreatic fibrosis (Thompson and Scott, 1970; Gries and Scott, 1972).

Suchý et al. (2018) stated that in general, inorganic forms of selenium are more toxic than organic forms and listed the following compounds in the order of decreasing toxicity: selenite (most toxic) > selenate > selenocysteine > methylated selenium compounds, resulting in embryonic malformations and decreased hatchability. Heinz and Hoffman (1996) reported that 30 mg/kg seleno-L-methionine, when fed to mallard ducklings from hatch to 14 d of age, was more toxic than seleno-DL-methionine. In addition, biologically incorporated selenium derived from high-selenium wheat was of similar toxicity as the two selenomethionine forms when fed at 15 mg/kg, while selenized yeast was less toxic at each dietary level. The authors did not analyze the chemical forms of selenium in the wheat or yeast product, but based on the work of others who analyzed samples from the same sources, they suggest that they contained primarily seleno-L-methionine. In a reproduction study from the same laboratory, it was demonstrated that the two forms of selenomethionine fed at 10 mg/kg to breeding pairs of mallards were of similar toxicity (decreased hatch of fertile eggs, poorer hatchling survival, and percentage of deformed embryos), and that both compounds were more toxic than selenized yeast (Heinz and Hoffman, 1996).

Zinc

Tissue zinc is usually found in a divalent state, and zinc does not readily exhibit redox chemistry in animals unlike copper, iron, and selenium. Zinc forms complexes with amino acids, peptides, proteins, and nucleotides. It has affinity for thiol and hydroxy groups and for amine electron donors. Approximately 300 enzymes are associated with zinc, and it also has very important structural and regulatory roles. Its catalytic roles are essential to the metabolism of proteins, carbohydrates, nucleic acids, and lipids as well as maintenance of pH (Vallee and Falchuk, 1993). Its structural and regulatory roles include the formation of zinc fingers, which are important structural motifs of small proteins involved in protein-DNA interactions and are important in gene regulation. Zinc is also an important in the pigmentation of feathers. Because zinc is involved in most aspects of metabolism and cellular function, marginal deficiencies result in effects on growth, feed intake, and egg production that are often more pronounced than that of deficiencies of other trace minerals. The immune response is also highly sensitive to zinc deficiency and is sometimes used to assess zinc status (Kidd et al., 1996).

As with copper and selenium, there are environmental concerns with zinc accumulation in soils following application of poultry manure (Mohanna and Nys, 1999; Dozier et al., 2003; Burrell et al., 2004). About half of the zinc consumed by broiler chicks consuming a diet based on corn and soybean meal is excreted when the dietary level is near the requirement and the amount increases at higher levels of dietary zinc (Burrell et al., 2004). However, relatively high levels of zinc in the diets of broilers or laying hens reduces ammonia emissions and increases nitrogen retention in manure by inhibiting microbial uricase activity (Kim and Patterson, 2004, 2005; Hunde et al., 2012).

Dietary Sources, Supplementation, and Bioavailability

Cereal grains typically contain 15 to 30 mg/kg zinc, whereas soybean, peanut, and linseed meals contain 50–70 mg/kg feed, and meat and fish meals typically contain 90–130 mg zinc/kg feed (Chapter 20, "Ingredients"). Drinking water may also provide nutritionally significant levels of zinc. Poultry diets are typically supplemented with zinc due to either low concentrations in feedstuffs or poor bioavailability. Common supplements include $ZnSO_4$, ZnO, $ZnCO_3$, $ZnCl_2$, and complexed or chelated organic sources, such as Zn-methionine, Zn-HMTBa, and other less characterized sources. Water can be an important source of zinc when supplied by galvanized pipe as can galvanized metal wires in cages or fencing (Klasing, 1998).

In broiler chicks, zinc is absorbed by a combination of passive diffusion and carrier-mediated processes and it is absorbed most effectively in the ileum followed by the jejunum and duodenum (Yu et al., 2008). When body stores and dietary zinc are both high, intestinal cells synthesize large amounts of metallothionein, which tightly binds zinc and prevents its transport to the blood (Hempe and Cousins, 1992). Zinc absorption is decreased by high dietary levels of phytic acid, oxalic acid, calcium, phosphorus, copper, cadmium, or chromium. Calcium and phosphate form insoluble precipitates with zinc in the intestinal lumen. Copper, cadmium, copper, and chromium induce metallothionein in the intestinal epithelium, blocking zinc absorption. Zinc complexed to some small ligands such as Ethylenediaminetetraacetic acid (EDTA), citrate, or amino acids (histidine, cysteine, methionine) is less influenced by the negative effects of other dietary minerals or phytic acid (Kratzer and Vohra, 1986; Baker and Ammerman, 1995b). Antagonists of zinc absorption not only decrease tissue stores of zinc but also decrease levels in egg yolk (Skřivan et al., 2005).

Following absorption, the highest uptake of zinc occurs in bone, but the liver, intestines, pancreas, and spleen also accumulate zinc. In these latter tissues, part of this zinc is bound to metallothionein and serves as a storage pool. Bone zinc also contributes zinc to help buffer other tissue pools during a zinc deficiency. However, these pools do not buffer a dietary zinc deficiency for very long. For example, in growing Japanese quail they last for only about four days (Harland et al., 1975) and in growing chickens about six days (Emmert and Baker, 1995) before growth decreases. Thus, of all of the trace minerals, storage pools of zinc are the smallest relative to the requirement, and dietary deficiencies are not well buffered (Harland et al., 1975; Emmert and Baker, 1995).

Experiments designed to determine the bioavailability of zinc sources typically use reagent grade $ZnSO_4 \cdot 7H_2O$ as a standard for comparison. The response criterion is typically growth rate because, unlike other trace minerals, it is sensitive to a marginal deficiency (Harland et al., 1975; Emmert and Baker, 1995). Accumulation of zinc in bone is also sometimes used (Wedekind et al., 1992; Batal et al., 2001), but because it is not a functional measure its utility for estimating bioavailability can be questioned. Reagent-grade zinc compounds added to experimental diets have the following relative bioavailability: Zn-methionine $> ZnSO_4 \cdot 7H_2O \approx ZnSO_4 \cdot H_2O \approx ZnCO_3 \approx ZnCl \approx ZnO \approx Zn_5Cl_2(OH)_8$ (Wedekind et al., 1992; Baker and Ammerman, 1995b; Edwards and Baker, 1999, 2000; Batal et al., 2001). However, feed-grade zinc sources often have lower bioavailability than reagent-grade sources. Wedekind and Baker (Wedekind and Baker, 1990b) found that in Columbian–New Hampshire chicks the bioavailability of feed-grade ZnO produced using the Waelz process relative to ZnSO4•H2O was 61 percent based on weight gain and 44 percent based on tibia Zn concentrations. In a later study using the same type of chicks, Edwards and Baker (1999) examined four feed-grade ZnO sources produced using a variety of manufacturing processes and found that bioavailability relative to reagent grade $ZnSO_4 \cdot 7H_2O$ ranged from 37 percent to 94 percent (average of three experiments). The source with the lowest bioavailability was produced using the Waelz process, and the source with the highest was produced using a hydrosulfide process. Feedstuffs of animal origin generally have higher zinc content and bioavailability than feedstuffs of plant origin (Baker and Ammerman, 1995b). For example, the bioavailability of zinc in beef is indistinguishable from that in $ZnCO_3$ (O'Dell et al., 1972) but the bioavailability in corn, wheat, and soybean meal is 63 percent (O'Dell et al., 1972), 59 percent (O'Dell et al., 1972), and 78 percent (Edwards and Baker, 2000), respectively.

Requirement

Performance indices, bone zinc levels, bone strength, and appetite are typically used to determine the zinc requirement of poultry. The zinc requirements of poultry have been estimated at between 35 and 75 mg/kg dry matter for all stages of production, though maintenance requirements are likely much lower. The zinc requirement is lower for chickens than for other poultry species. The specific recommendations and the studies they are based on can be found in Chapters 13–19.

Deficiency and Toxicity

A marginal deficiency of zinc has a larger impact on growth, feed intake, and egg production than a marginal deficiency of any of the other trace minerals, likely because it is involved as a regulator or cofactor for hundreds of enzymes (Klasing and Korver, 2020). Zinc deficiency causes retarded growth and frayed feathers in chicks and poults (O'Dell et al., 1958; Sullivan, 1961). Layer and breeder diets deficient in zinc have reduced egg production and hatchability (Kienholz et al., 1961).

Zinc is considered to be relatively nontoxic to humans and animals when administered orally (Dean et al., 1991). However, excessive amounts of zinc can accentuate borderline deficiencies of other elements, including copper, iron and selenium (NRC, 2005). As an example, Jensen (Jensen, 1975) reported that dietary zinc levels of 2,100 to 4,100 mg/kg supplemented to a diet containing 0.2 ppm selenium resulted in depressed 4-week body weights, high mortality, exudative diathesis, and muscular dystrophy in broiler chicks, but that when the diet contained 0.5 ppm selenium, growth was still depressed, but the selenium deficiency signs were completely prevented ,and mortality was markedly reduced. In zinc toxicity, pathologic changes first occur in the pancreas, and then in the ventriculus and thyroid (Klasing and Korver, 2020).

TABLE 6-2 Biochemical and Physiological Measurements for Diagnosis of Inorganic Element Deficiencies in Chickens and Turkeys (from NRC, 1994)

Nutrients	Deficiency Signs	References[a]
Calcium	Calcium in hen's blood (but not in chick's unless deficiency is severe); intestinal calcium-binding protein (complicated by D3 metabolites and phosphorus); turkey poults differ from chicks	Bar et al., 1972, 1978a,b; Bar and Hurwitz, 1973
Chlorine	Hemoconcentration; alkalosis	Leach and Nesheim, 1963; Cohen and Hurwitz, 1974; Hamilton and Thompson, 1980
Copper	Plasma ceruloplasmin; lysyl oxidase in aorta, liver, tendon, and bone; erythrocyte superoxide dismutase	Kim and Hill, 1966; Miller and Stake, 1974; Bettger et al., 1979; Opsahl et al., 1982
Iodine	Plasma thyroxine and triiodothyronine	Singh et al., 1968
Iron	Hematocrit; blood hemoglobin concentration; transferrin saturation; anemia with lipemia	Davis et al., 1962; Waddell and Sell, 1964; Planas, 1967
Magnesium	Magnesium concentration in blood	Sell et al., 1967; Hajj and Sell, 1969
Manganese	Chondroitin sulfate in bone; manganese concentration in bone; superoxide dismutase	Leach, 1968; Reid et al., 1973; DeRosa et al., 1980
Phosphorus	Serum inorganic phosphorus; renal calcium-binding protein	Miller and Stake, 1974; Bar et al., 1978a,b
Potassium	Plasma potassium; metabolic acidosis (complicated by sodium)	Burns et al., 1953; Rinehart et al., 1969; Cohen and Hurwitz, 1974
Selenium	Plasma glutathionine peroxidase	Noguchi et al., 1973; Dean and Combs, 1981; Cantor et al., 1982
Sodium	Metabolic acidosis (complicated by potassium)	Nott and Combs, 1969; Cohen and Hurwitz, 1974
Zinc	Plasma and bone zinc; thymidine kinase; alkaline phosphatase and collagenase in bone	Miller and Stake, 1974; Oberleas and Prasad, 1974; Bettger et al., 1979; Starcher et al., 1980

[a]References are listed in NRC. 1994. *Nutrient Requirements of Poultry*, 9th rev. ed. Washington, DC: National Academy Press.

Johnson et al. (1962) reported chickens tolerated 1,000 mg/kg of supplemental dietary zinc (as ZnO) with no adverse effects and that 3,000 mg/kg zinc significantly depressed chick growth. Feeding laying hens 10,000 or 20,000 mg/kg zinc for only 4 d resulted in sharp reductions in feed intake, rapid reduction and cessation of egg production, and lesions in the gizzard and pancreas (Dewar et al., 1983; McCormick and Cunningham, 1984).

Turkeys have been reported to tolerate up to 2,000 ppm zinc (as ZnO) in starter diets without any adverse effects, while a level of 4,000 ppm decreased poult growth (Vohra and Kratzer, 1968). Dean et al. (1991) fed day-old broiler chicks 5,280 ppm zinc for 4 week and observed reduced growth and multiple endocrinopathies, including depressed levels of circulating thyroid hormones, thyroid follicular hyperplasia, and smaller follicles. They suggested that the impaired growth of high-zinc fed birds may be caused, in part, by reduced thyroidal function.

Signs of Nutritional Deficiencies and Toxicities in Poultry

Specific deficiencies and toxicities by mineral are discussed previously, but a summary of known biochemical and physiological measurements for diagnosing each inorganic element deficiency is presented in Table 6-2. Additional information is available in the associated references. Histopathological and biochemical aspects of mineral deficiencies are summarized in a chapter in a frequently updated book targeted at poultry pathologists (Klasing and Korver, 2020).

Inadequate dietary vitamins and minerals in the chicken or turkey hen's diet are likely to reduce the egg contents accordingly and have adverse effects on embryonic development. A summary of inorganic element deficiency symptoms in the embryo are presented in Table 6-3.

In addition to their structural importance, minerals serve as catalysts in enzymes, as regulators of osmotic pressure, acid-base balance, and cell signaling; thus, mineral deficiencies may cause perturbations of one or more of these functions (Klasing and Korver, 2020). Deficiency signs can be similar for different nutrients, and the extent of the inadequacy may change the nature of the symptoms as well as the time when death occurs. Deficiency signs are expressed to a greater extent in growing birds than in adults. As a diagnostic aid, Table 6-4 provides a list of these signs by tissue affected. The table also presents information on these changes such that each can be rationalized in terms of nutrient function. References provided are not complete but are intended to be salient.

Empirical information on the toxic dietary levels of select inorganic elements for poultry is summarized in Table 6-5. Toxicity, as defined here, is any adverse effect on performance. A summary of the maximum tolerable levels (chronic) for poultry of select inorganic elements from the

TABLE 6-3 Signs of Inorganic Element Deficiencies in the Embryo (from NRC, 1994)

Nutrients	Deficiency Signs	References[a]
Copper	Deaths at early blood stage with no malformations	Bird et al., 1963
Iodine	Prolongation of hatching time, reduced thyroid size, and incomplete abdominal closure	Rogler et al., 1959
Iron	Low hematocrit; low blood hemoglobin; poor extraembryonic circulation in candled eggs	Dewar et al., 1974; Morck and Austic, 1981
Manganese	Peak deaths prior to emergence; chondrodystrophy, dwarfism, long bone shortening, head malformations, edema, and abnormal feathering are prominent	Lyons and Insko, 1937
Selenium	High incidence of dead embryos early in incubation	Latshaw et al., 1977
Zinc	Deaths prior to emergence, and the appearance of rumplessness, depletion of vertebral column, eyes underdeveloped, and missing limbs	Kienholz et al., 1961; Turk, 1965

[a]References are listed in NRC. 1994. *Nutrient Requirements of Poultry*, 9th rev. ed. Washington, DC: National Academy Press.

TABLE 6-4 Nutrients Associated with Various Signs of Deficiency in Growing Birds (from NRC, 1994)[a]

Deficiency Signs	Descriptions	Species	Associated Nutrients
Skin lesions	Scaliness on feet	Chick	Zinc, niacin
Feather abnormalities	Frizzled and rough	Chick, poult	Zinc, niacin, pantothenic acid, folic acid, lysine
	Depigmentation	Chick, poult	Copper, iron, folic acid
Nervous disorders	Hyperirritability	Chick, poult, duckling	Magnesium, sodium chloride
	Characteristic fright reaction with tetanic spasms	Chick	Chloride
Blood and vascular system	Microcytic, hypochromic anemia	All poultry	Iron, copper
	Hemorrhage, intramuscular, subcutaneous, internal from aortic rupture	Chick, poult	Vitamin K, copper
	Exudative diathesis	Chick, poult	Selenium, vitamin E
	Enlarged heart	Chick, poult	Copper
Muscle	Muscular dystrophy, white areas of degeneration in skeletal muscle	Chick, duck, poult	Vitamin E, selenium
	Cardiac myopathy	Poult	Vitamin E, selenium
	Gizzard myopathy	Poult	Vitamin E, selenium
Bone disorders	Soft, easily bent bones and beak (rickets)	All poultry	Vitamin D, calcium or phosphorus deficiency or imbalance
	Hock enlargement	Poult, chick, gosling, duckling	Niacin, zinc
	Perosis	Chick, poult	Biotin, choline, vitamin B_{12}, manganese, zinc, folacin
	Shortening and thickening of leg bones	Chick	Zinc, manganese

NOTES: Slow growth and general lack of vigor are generally associated with malnutrition. The signs listed in this table are more specific indications of deficiencies of particular nutrients.

[a]References are listed in NRC. 1994. *Nutrient Requirements of Poultry*, 9th rev. ed. Washington, DC: National Academy Press.

report "Mineral Tolerance of Animals" (NRC, 2005) is presented in Table 6-5. The report defines the maximum tolerable level of an inorganic element as the dietary level that, when fed for a defined period of time, will not impair animal health and/or performance.

Molybdenum Toxicity

Although molybdenum is not addressed as an essential trace mineral due to limited reported information, toxicosis is produced either by high molybdenum intakes alone (>100 mg/kg) or moderately high molybdenum intakes concomitant with low dietary copper levels (<5 mg/kg) or low dietary copper:molybdenum ratios (<2:1) (NRC, 2005). Mild molybdenosis is manifested by increased xanthine oxidase activity or blood uric acid concentrations, while severe cases can be fatal. Arthur et al. (1958) reported that chick growth was depressed with 200 mg molybdenum/kg diet and that the growth depression was partially overcome by supplemental dietary copper, but that supplemental copper was without effect at 500 mg/kg dietary molybdenum. Lepore and Miller (1965) reported that feeding laying hens 1 g molybdenum/kg diet significantly

TABLE 6-5 Toxic Dietary Concentrations of Inorganic Elements and Compounds for Poultry

Element or Compound	Species	Age	Chemical Form	Toxic Concentration (ppm)[a]	Toxic Effects	References[b]
Calcium	Chicken	Immature	$CaCO_3$	>2%	Decreased feed intake and weight gain and increased mortality	Fangauf et al., 1961
Calcium	Chicken	Immature	$CaCO_3$	>2%	Decreased feed intake and weight gain and increased mortality	Urbanyi, 1960
Calcium	Chicken	8 weeks	$CaCO_3$	3% with 0.4% available phosphorus	Visceral urate deposits and renal failure in 10% of the birds fed from 8 to 22 wk of age; 60% developed nephrosis by 16 weeks of age	Shane et al., 1969
Calcium	Chicken	18 weeks	$CaCO_3$	3.25% with 0.4% available phosphorus	14% incidence of urolithiasis in birds fed the diet from 7 to 18 weeks of age	Wideman et al., 1985
Calcium	Chicken	65 d	$CaCO_3$	3.68%	Severe kidney damage in birds fed the diet from 35 to 65 d of age	Guo et al., 2005
Calcium	Chicken	65 d	$CaCO_3$	3.68% plus 24.3% crude protein	Kidney damage and visceral gout in birds fed the diet from 35 to 65 d of age	Guo et al., 2005
Copper	Chicken	Immature	CuO	806	Decreased growth; mortality	Mehring et al., 1960
Copper	Chicken	Immature	$CuSO_4 \cdot 5H_2O$	800	Exudative diathesis; muscular dystrophy	Jensen, 1975
Copper	Chicken	Immature	$CuSO_4 \cdot 5H_2O$	500	Decreased growth; gizzard erosion	Poupoulis and Jensen, 1976
Copper	Chicken	Immature	$CuSO_4 \cdot 5H_2O$	250	Decreased growth; gizzard erosion	Robbins and Baker, 1980
Copper	Chicken	28 d	$CuSO_4$	250–500	Proventriculus enlargement	Wideman et al., 1996
Copper	Chicken	21 d	$CuSO_4$	250–500	Lesions in oral cavity, tongue, and pharynx; gizzard erosion; decreased weight gain and feed intake	Chiou et al., 1999
Copper	Chicken	Laying hen	$CuSO_4$	400	Decreased feed intake and egg production	Jackson and Stevenson, 1981
Copper	Chicken	Laying hen	$CuSO_4$	300–750	Increased gizzard weight; decreased feed intake, egg production, and egg weight	Jackson and Stevenson, 1981
Copper	Turkey	Immature	$CuSO_4 \cdot 5H_2O$	800 (purified diet)	Decreased growth	Supplee, 1964
Copper	Turkey	Immature	$CuCO_3$	50 (purified diet)	Decreased growth	Waibel et al., 1964
Copper	Turkey	Immature	$CuSO_4 \cdot 5H_2O$	676–810	Decreased growth	Vohra and Kratzer, 1968
Copper	Turkey	Immature	$CuSO_4$	500	Decreased body weight and feed intake; gizzard erosion	Christmas and Harms, 1979
Copper	Duck	Immature	$CuSO_4$	500–1,500	Myopathy, characteristic of vitamin E-selenium deficiency; high mortality	Van Vleet, 1982
Iodine	Chicken	Laying hen	KI	312–5,000	Reduced egg production and hatchability; embryonic mortality. Hens resumed egg production 7 d following cessation of supplemental iodine feeding,	Perdomo et al., 1966
Iodine	Chicken	Laying hen	KI	625	Reduced egg production, egg weight, and hatchability	Arrington et al., 1967

TABLE 6-5 Continued

Element or Compound	Species	Age	Chemical Form	Toxic Concentration (ppm)[a]	Toxic Effects	References[b]
Iodine	Chicken	Laying hen	KI	5,000	Reduced egg production and serum calcium	Roland et al., 1977
Iodine	Chicken	Immature	EDDI and KI	600–1,500	Reduced feed intake and body weight; neurological clinical signs	Baker et al., 2003
Iodine	Turkey	Breeder hens	KI	35–350	Reduced egg production, egg weights, and hatchability	Christensen and Ort, 1991
Iron	Chicken	Immature	$Fe_2(SO_4)_3$	4,500	Rickets	Deobald and Elvehjem, 1935
Iron	Chicken	Immature	$Fe_2(SO_4)\cdot 7H_2O$	400–1,600	Decreased body weights	McGhee et al., 1965
Iron	Chicken	Immature	$Fe_2(SO_4)\cdot 7H_2O$	400–800	Decreased feed intake and body weights	Cao et al., 1996
Magnesium	Chicken	Immature	$MgCO_3$	6,400	Decreased growth and bone calcification; mortality	Nugara and Edwards, 1963
Magnesium	Chicken	Immature	$MgCO_3$	7,100	Decreased growth and tibial ash	Chicco et al., 1967
Magnesium	Chicken	Immature	MgO	9,000	Decreased growth; diarrhea; skeletal abnormalities	Lee et al., 1980
Magnesium	Chicken	Immature	MgO	5,700	Decreased eggshell calcium; skeletal development	Atteh and Leeson, 1983
Magnesium	Chicken	Laying hen	$MgSO_4$	19,600	Reduced egg production	McWard, 1967
Magnesium	Chicken	Laying hen	$MgCO_3$	11,200	Reduced egg production and decreased body weight	Stillmak and Sunde, 1971
Manganese	Chicken	Immature	$MnCO_3$	4,779	Mortality	Heller and Penquite, 1937
Manganese	Chicken	Immature	$MnCl_2\cdot 4H_2O$	4,000	Decreased growth, hemoglobin, hematocrit, and liver iron; dietary $MnSO_4.H_2O$, $MnCO_3$, MnO_2 did not affect growth but depressed some or all of the other measures	Southern and Baker, 1983b
Manganese	Chicken	Immature	$MnSO_4\cdot H_2O$	3,000	Decreased feed intake, liver iron, and pancreas iron	Black et al., 1985
Manganese	Turkey	Immature	$MnSO_4\cdot H_2O$	4,800	Decreased growth	Vohra and Kratzer, 1968
Molybdenum	Chicken	Immature	$Na_2MoO4\cdot H_2O$	300	Decreased growth	Kratzer, 1952
Molybdenum	Chicken	Immature	Na_2MoO_4	500	Decreased growth	Motzok et al., 1957
Molybdenum	Chicken	Immature	Na_2MoO_4	200	Decreased growth	Arthur et al., 1958
Molybdenum	Chic(Arthur et al., 1958) ken	Immature	Na_2MoO_4	500	Decreased growth; anemia at 4,000 and 8,000 mg/kg diet; high mortality	Davies et al., 1960
Molybdenum	Chicken	Immature	$Na_2MoO_4\cdot H_2O$	350	Decreased growth	Berg and Martinson, 1972
Molybdenum	Chicken	Laying hen	$Na_2MoO_4\cdot H_2O$	500–2,000	Reduced egg production; decreased body weights; 100% embryonic mortality	Lepore and Miller, 1965
Molybdenum	Turkey	Immature	$NaMoO_4\cdot H_2O$	300	Decreased growth	Kratzer, 1952
Phosphorus	Chicken	Immature	$CaHPO_4$	6,100	Tibial dyschondroplasia (0.70% dietary calcium)	Edwards and Veltmann, 1983
Phosphorus	Chicken	Immature	$CaHPO_4$	8,100–10,900	Tibial dyschondroplasia (0.63–1.67% calcium)	Edwards and Veltmann, 1983

continued

TABLE 6-5 Continued

Element or Compound	Species	Age	Chemical Form	Toxic Concentration (ppm)[a]	Toxic Effects	References[b]
Phosphorus	Chicken	Laying hen	Meat and bone meal	1.15%	Reduced egg production; poorer feed conversion (as kg feed/dozen eggs); diet contained 3.00% calcium	Harms et al., 1965
Phosphorus	Chicken	Laying hen	Defluorinated phosphate	1.28%	Reduced egg production; diet contained 3.30% calcium	Harms et al., 1965
Phosphorus	Chicken	Laying hen	$Ca(H_2PO4)_2.H_2O$	10,200	Reduced egg production; decreased feed intake, body weight, eggshell quality, blood bicarbonate, and plasma calcium. Feeding up to 2.41% dibasic phosphorus ($CaHPO_4$ and $CaHPO4{\cdot}2H_2O$) was without effect on hen performance	Keshavarz, 1994
Potassium	Chicken	Immature	Basal diet + K_2CO_3 in water	7,300 in diet + 1,500 in water	Decreased weight gain and water consumption	Teeter and Smith, 1986
Potassium	Chicken	Laying hen	Basal diet + KCl	5,000–20,000	Linear decrease in feed intake and increase in water intake, weight of water excreted, and excreta moisture as dietary K content increased	Smith et al., 2000
Selenium	Chicken	Immature	Na_2SeO_3	10	Decreased growth; alleviated by supplemental dietary silver or copper	Jensen, 1975
Selenium	Chicken	Immature	Na_2SeO_3	40	Increased mortality; alleviated by supplemental dietary silver or copper	Jensen, 1975
Selenium	Chicken	Immature	Na_2SeO_3	10	Decreased growth and feed efficiency	Jensen and Chang, 1976
Selenium	Chicken	Immature	Na_2SeO_3	5	Decreased growth	Jensen et al., 1986
Selenium	Chicken	Immature	Na_2SeO_3	15	Decreased growth; reduced feed efficiency	Lowry and Baker, 1989
Selenium	Chicken	Immature	Selenomethionine	15	Decreased growth; reduced feed efficiency	Lowry and Baker, 1989
Selenium	Chicken	Laying hen	Se in wheat, barley, and corn	15	Decreased body weight, feed consumption, and egg weight; reduced hatchability; embryonic malformations, including edema of head and neck, twisted neck legs, and missing toes and eyes	Poley et al., 1937
Selenium	Chicken	Laying hen	Se in wheat	10	Reduced hatchability	Moxon and Wilson, 1944
Selenium	Chicken	Laying hen	Na_2SeO_3	5–9	Reduced hatchability; embryonic malformations, including missing limbs and swelling of the head and neck region	Ort and Latshaw, 1978
Selenium	Duck	Immature mallard	Selenized wheat	15	Decreased feed consumption	Heinz and Hoffman, 1996
Selenium	Duck	Immature mallard	Seleno-DL-methionine	30	Decreased growth and feed consumption	Heinz and Hoffman, 1996
Selenium	Duck	Immature mallard	Seleno-L-methionine	30	Decreased growth and feed consumption; increased mortality	Heinz and Hoffman, 1996

TABLE 6-5 Continued

Element or Compound	Species	Age	Chemical Form	Toxic Concentration (ppm)[a]	Toxic Effects	References[b]
Selenium	Duck	Immature mallard	Selenized yease	30	Decreased growth and feed consumption	Heinz and Hoffman, 1996
Selenium	Duck	Laying mallard	Seleno-DL-methionine	10	Decreased hatch of fertile eggs and number of ducklings produced per hen	Heinz and Hoffman, 1996
Selenium	Duck	Laying mallard	Seleno-L-methionine	10	Decreased hatch of fertile eggs and number of ducklings produced per hen	Heinz and Hoffman, 1996
Selenium	Duck	Laying mallard	Selenized yeast	10	Decreased hatch of fertile eggs and number of ducklings produced per hen	Heinz and Hoffman, 1996
Chloride	Chicken	Immature	NaCl; arginine-HCl	0.81%	Decreased growth	Nesheim et al., 1964
Sodium	Chicken	Immature	Na glutamate	8,900[c]	Mortality; decreased growth	Nesheim et al., 1964
Sodium	Chicken	Laying hen	Na_2SO_4	12,000[d]	Reduced egg production	Krista et al., 1961
Sodium chloride	Chicken	Immature	NaCl	5%	Mortality; watery excreta	Halpin et al., 1936
Sodium chloride	Chicken	Immature	NaCl	0.9%[d]	Mortality; nephrosclerosis	Selye, 1943
Sodium chloride	Chicken	Immature	NaCl	7,000[d]	Decreased growth; mortality	Krista et al., 1961
Sodium chloride	Chicken	Laying hen	NaCl	10,000[d]	Reduced egg production	Krista et al., 1961
Sodium chloride	Chicken	Laying hen	NaCl	2–6%	Reduced egg production	Damron and Kelly, 1987
Sodium chloride	Turkey	Immature	NaCl	2.5%	Mortality; edema; ascites	Scrivner, 1946
Sodium chloride	Turkey	Immature	NaCl	4%	Mortality; edema	Matterson et al., 1946
Sodium chloride	Turkey	Immature	NaCl	0.9%	Mortality; ascites	Bressler et al., 1951
Sodium chloride	Turkey	Immature	NaCl	4,000[d]	Decreased growth; mortality	Krista et al., 1961
Sodium chloride	Turkey	Immature	NaCl	40,000	Decreased growth; pendulous crop	Harper and Arscott, 1962
Sodium chloride	Turkey	Immature	NaCl	27,000	Mortality	Morrison et al., 1975
Sodium chloride	Duck	Immature Rouen	NaCl	4,000[d]	Decreased growth	Krista et al., 1961
Sodium chloride	Duck	Immature Rouen	NaCl	10,000[d]	Decreased growth; mortality	Krista et al., 1961
Zinc	Chicken	Immature	$ZnSO_4$, $ZnCO_3$, ZnO	1,500	Decreased growth	Roberson and Schaible, 1960
Zinc	Chicken	Immature	$ZnCO_3$	3,000	Mortality	Roberson and Schaible, 1960
Zinc	Chicken	Immature	ZnO	3,000	Decreased growth	Johnson et al., 1962
Zinc	Chicken	Immature	ZnO	800	Decreased growth; bone ash (sucrose–fish meal diet)	Berg and Martinson, 1972
Zinc	Chicken	Immature	$ZnSO_4$	2,000	Mortality; exudative diathesis; muscular dystrophy (0.2 ppm selenium in the base diet)	Jensen, 1975
Zinc	Chicken	Immature	ZnSO4	2,000	Decreased growth (0.5 ppm selenium in the base diet)	Jensen, 1975
Zinc	Chicken	Immature	ZnO	2,000	Decresed growth; gizzard erosion; pancreatic necrosis	Dewar et al., 1983
Zinc	Chicken	Immature	ZnO	170	Exocrine pancreas damage (purified base diet consisted of sucrose and crystalline amino acids)	Lü and Combs, 1988

continued

TABLE 6-5 Continued

Element or Compound	Species	Age	Chemical Form	Toxic Concentration (ppm)[a]	Toxic Effects	References[b]
Zinc	Chicken	Immature	ZnO	5,280	Decreased growth and multiple endocrinopathies	Dean and Combs, 1981
Zinc	Chicken	Laying hen	ZnO	10,000	Gizzard erosion; pancreatic necrosis	Dewar et al., 1983
Zinc	Chicken	Laying hen	ZnO	20,000	Cessation of egg production after feeding the diet for 5 d	Scott et al., 1967
Zinc	Chicken	Laying hen	ZnO	20,000	Cessation of egg production after feeding the high-zinc diet for 4 d	McCormick and Cunningham, 1984
Zinc	Chicken	Laying hen	ZnO	20,000	High-zinc diet fed for 4 d; 6 d later, ovarian and oviduct weights were reduced by 76% and 58%, respectively	Williams et al., 1989
Zinc	Turkey	Immature	ZnO	4,000	Decreased growth	Vohra and Kratzer, 1968
Zinc	Duck	Immature	$ZnSO_4$	3,000–6,000	Skeletal and cardiac muscle necrosis; smooth muscle necrosis of the small intestine and gizzard; pancreatic necrosis and fibrosis; only the muscle lesions were completely prevented by additional supplemental dietary selenium and vitamin E	Van Vleet, 1982

[a] Dietary concentration of the element in question unless specified.
[b] References are listed in the text of this chapter.
[c] Diet low in Cl^- ion.
[d] In water.

TABLE 6-6 Maximum Tolerable Levels (Chronic) of Select Inorganic Elements for Poultry (NRC, 2005)[a]

Inorganic Element	Maximum Tolerable Level	Species Specified
Calcium	1.5% of diet[b]	Growing poultry (meat birds)
Calcium	5.0% of diet[b]	Laying hen
Copper	250 mg/kg of diet	Chicken and turkey
Copper	100 mg/kg of diet	Duck
Iodine	300 mg/kg of diet	Chicken and turkey
Iron	500 mg/kg of diet	Poultry
Magnesium	0.5% of diet[a]	Growing poultry (meat birds)
Magnesium	0.75% of diet[a]	Laying hens
Manganese	2,000 mg/kg of diet	Chicken
Molybdenum	100 mg/kg of diet	Poultry
Phosphorus	1.0% of diet[c]	Growing poultry (meat birds)
Phosphorus	0.8% of diet[c]	Laying hen
Potassium	10,000 mg/kg of diet	Nonruminants
Selenium	3 mg/kg of diet	Poultry
Sodium chloride	1.5 g/kg BW[d]	Poultry
Zinc	500 mg/kg of diet	Poultry

[a] References are listed in NRC (2005).
[b] When diets contain adequate levels of available phosphorus.
[c] When diets contain adequate levels of calcium.
[d] Body weight.

reduced egg production while feeding hens 2 g molybdenum/kg diet significantly depressed body weights. When they fed hens 500 mg molybdenum/kg of diet, egg molybdenum concentrations of approximately 17 mg/kg were attained, and 100 percent embryonic mortality was observed by day 12. Depending upon the basal diet, Berg and Martinson (1972) observed a growth depression of between 9 percent and 19 percent when chicks were fed a 350 mg molybdenum/kg diet for 14 d.

REFERENCES

AAFCO (Association of American Feed Control Officials). 2020. *Association of American Feed Control Officials*. Official Publication. Champaign, IL: Association of American Feed Control Officials.

Abbasi, M., M. Zaghari, M. Ganjkhanlo, and S. Khalaji. 2015. Is dietary iron requirement of broiler breeder hens at the late stage of production cycle influenced by phytase supplementation? *Journal of Applied Animal Research*. 43:166–176.

Adedokun, S., and T. Applegate. 2014. Dietary electrolyte balance influences ileal endogenous amino acid losses in broiler chickens. *Poultry Science*. 93:935–942.

Ahmad, T., and M. Sarwar. 2006. Dietary electrolyte balance: Implications in heat stressed broilers. *World's Poultry Science Journal*. 62:638–653.

Allaway, W.H. 1986. Soil-plant-animal and human interrelationships in trace element nutrition. In *Trace Elements in Human and Animal Nutrition*, Vol. 2, 5th ed. W. Mertz, editor. Amsterdam, The Netherlands: Academic Press. 465.

Almquist, H.J. 1942. Magnesium requirement of the chick. *Proceedings of the Society for Experimental Biology and Medicine*. 49:544–545.

Ammerman, C.B., D.H. Baker, and A.J. Lewis, eds. 1995. *Bioavilability of Nutrient for Animals: Amino Acids, Minerals and Vitamins*. Amsterdam, The Netherlands: Academic Press.

Anwar, M.N., V. Ravindran, P.C.H. Morel, G. Ravindran, and A.J. Cowieson. 2016. Effect of limestone particle size and calcium to non-phytate phosphorus ratio on true ileal calcium digestibility of limestone for broiler chickens. *British Poultry Science* 57:707–713.

Aoyagi, S., and D.H. Baker. 1993a. Protective effect of copper amino acid complexes against inhibitory effects of L-cysteine and ascorbic acid. *Poultry Science*. 72:82.

Aoyagi, S., and D.H. Baker. 1993b. Bioavailability of copper in analytical-grade and feed-grade inorganic copper sources when fed to provide copper at levels below the chick's requirement. *Poultry Science*. 72:1075–1083.

Aoyagi, S., and D.H. Baker. 1993c. Nutritional evaluation of a copper-methionine complex for chicks. *Poultry Science*. 72:2309–2315.

Aoyagi, S., D.H. Baker, and K.J. Wedekind. 1993. Estimates of copper bioavailability from liver of different animal species and from feed ingredients derived from plants and animals. *Poultry Science*. 72:1746–1755.

Aoyagi, S., K.M. Hiney, and D.H. Baker. 1995. Copper bioavailability in pork liver and in various animal by-products as determined by chick bioassay. *Journal of Animal Science*. 73:799–804.

Arias, V., and E. Koutsos. 2006. Effects of copper source and level on intestinal physiology and growth of broiler chickens. *Poultry Science*. 85:999–1007.

Arrington, L.R., R.A. Santa Cruz, R.H. Harms, and H.R. Wilson. 1967. Effects of excess dietary iodine upon pullets and laying hens. *Journal of Nutrition*. 92:325–330.

Arthur, D., I. Motzok, and H. Branion. 1958. Interaction of dietary copper and molybdenum in rations fed to poultry. *Poultry Science* 37:1181.

Atteh, J.O., and S. Leeson. 1983. Influence of increasing dietary calcium and magnesium levels on performance, mineral metabolism, and egg mineral content of laying hens. *Poultry Science* 62:1261–1268.

Bafundo, K.W., D.H. Baker, and P.R. Fitzgerald. 1984. The iron-zinc interrelationship in the chick as influenced by *Eimeria acervulina* infection. *Journal of Nutrition*. 114:1306–1312.

Bakalli, R.I., G.M. Pesti, W.L. Ragland, and V. Konjufca. 1995. Dietary copper in excess of nutritional requirement reduces plasma and breast muscle cholesterol of chickens. *Poultry Science*. 74:360–365.

Baker, D.H., and C.B. Ammerman. 1995a. Copper bioavailability. In *Bioavailability of Nutrients for Animals*. C.B. Ammerman, D.H. Baker, and A.J. Lewis, editors. Academic Press. 127–156.

Baker, D.H., and C.B. Ammerman. 1995b. Zinc bioavailability. In *Bioavailability of Nutrients for Animals*. C.B. Ammerman, D.H. Baker, and A.J. Lewis, editors. Amsterdam, The Netherlands: Academic Press. 367–398.

Baker, D.H., and K. Halpin, M. 1991. Manganese and iron interrelationship in the chick. *Poultry Science*. 70.

Baker, D.H., and G.W. Oduho. 1994. Manganese utilization in the chick—effects of excess phosphorus on chicks fed manganese-deficient diets. *Poultry Science*. 73:1162–1165.

Baker, D.H., T.M. Parr, and N.R. Augspurger. 2003. Oral iodine toxicity in chicks can be reversed by supplemental bromine. *The Journal of Nutrition*. 133:2309–2312.

Bakhshalinejad, R., A. Hassanabadi, H. Nassiri-Moghaddam, and H. Zarghi. 2018. The effects of dietary calcium iodate on productive performance, egg quality and iodine accumulation in eggs of laying hens. *Journal of Animal Physiology and Animal Nutrition*. 102:746–754.

Bao, Y., M. Choct, P. Iji, and K. Bruerton. 2009. The digestibility of organic trace minerals along the small intestine in broiler chickens. *Asian-Australasian Journal of Animal Sciences*. 23:90–97.

Batal, A.B., T.M. Parr, and D.H. Baker. 2001. Zinc bioavailability in tetrabasic zinc chloride and the dietary zinc requirement of young chicks fed a soy concentrate diet. *Poultry Science*. 80:87–90.

Belaidi, A.A., and G. Schwarz. 2017. Molybdenum cofactor in humans: Health, disease, and treatment. In *Molecular, Genetic, and Nutritional Aspects of Major and Trace Minerals*. J.F. Collins, editor. *Amsterdam, The Netherlands:* Academic Press. 399–410.

Ben-Dor, A. 1941. Requirement of potassium by the chick, *Proceedings of the Society for Experimental Biology and Medicine*. 46:341–343.

Berg, L.R., and R.D. Martinson. 1972. Effect of diet composition of the toxicity of zinc for the chick. *Poultry Science*. 51:1690–1694.

Berwanger, E., S. Vieira, C. Angel, L. Kindlein, A. Mayer, M. Ebbing, and M. Lopes. 2018. Copper requirements of broiler breeder hens. *Poultry Science*. 97:2785–2797.

Bess, F., S.L. Vieira, A. Favero, R.A. Cruz, and P.C. Nascimento. 2012. Dietary iron effects on broiler breeder performance and egg iron contents. *Animal Feed Science and Technology*. 178:67–73.

Bird, F.H. 1949. Magnesium deficiency in the chick; clinical and neuropathologic findings. *The Journal of Nutrition*. 39:13–29.

Black, J., C. Ammerman, P. Henry, and R. Miles. 1985. Effect of dietary manganese and age on tissue trace mineral composition of broiler-type chicks as a bioassay of manganese sources. *Poultry Science*. 64:688–693.

Blaxland, J.D. 1946. The toxicity of sodium chloride for fowls. *The British Veterinary Journal*. 102:157–173.

Bressler, G.O., S. Gordeuck Jr., and G.H. Pretham, 1951. The effect of salt and carbolineum in producing ascites in turkey poults. *Poultry Science*. 30:738–744.

Brooks, M., J. Grimes, K. Lloyd, F. Valdez, and J. Spears. 2012. Relative bioavailability in chicks of manganese from manganese propionate. *Journal of Applied Poultry Research*. 21:126–130.

Burns, C., W. Cravens, and P. Phillips. 1952. The requirement of breeding hens for sodium chloride. *Poultry Science*. 31:302–306.

Burns, C.H., W.W. Cravens, and P.H. Phillips. 1953. The sodium and potassium requirements of the chick and their interrelationship. *The Journal of Nutrition*. 50:317–329.

Burrell, A., W. Dozier, A. Davis, M. Compton, M. Freeman, P. Vendrell, and T. Ward. 2004. Responses of broilers to dietary zinc concentrations and sources in relation to environmental implications. *British Poultry Science*. 45:225–263.

Cantor, A.H., P.D. Moorhead, and M.A. Musser. 1982. Comparative effects of sodium selenite and selenomethionine upon nutritional muscular dystrophy, selenium-dependent glutathione peroxidase and tissue selenium concentrations of turkey poults. *Poultry Science*. 61:478–484.

Cao, J., X. Luo, P. Henry, C. Ammerman, R. Littell, and R. Miles. 1996. Effect of dietary iron concentration, age, and length of iron feeding on feed intake and tissue iron concentration of broiler chicks for use as a bioassay of supplemental iron sources. *Poultry Science*. 75:495–504.

Carlisle, E.M. 1970. Silicon: A possible factor in bone calcification. *Science*. 167:279–280.

Carlisle, E.M. 1980. Biochemical and morphological changes associated with long bone abnormalities in silicon deficiency. *The Journal of Nutrition*. 110:1046–1056.

Chicco, C.F., C.B. Ammerman, P.A. Van Walleghem, P.W. Waldroup, and R.H. Harms. 1967. Effects of varying dietary ratios of magnesium, calcium and phosphorus in growing chicks. *Poultry Science*. 46:368–373.

Chiou, P., C. Chen, K. Chen, and C. Wu. 1999. Effect of high dietary copper on the morphology of gastro-intestinal tract in broiler chickens. *Asian-Australasian Journal of Animal Sciences*. 12:548–553.

Christensen, V.L., and J.F. Ort. 1991. Iodine toxicity in large white turkey breeder hens. *Poultry Science*. 70:2402–2410.

Christmas, R., and R. Harms. 1979. The effect of supplemental copper and methionine on the performance of turkey poults. *Poultry Science*. 58:382–384.

Colnago, G.L., L.S. Jensen, and P.L. Long. 1984. Effect of natural feedstuffs added to a semi-purified diet on *Eimeria tenella* infection. *Poultry Science*. 63:2145–2152.

Combs, G.F., C. Garbisu, B.C. Yee, A. Yee, D.E. Carlson, N.R. Smith, A.C. Magyarosy, T. Leighton, and B.B. Buchanan. 1996. Bioavailability of selenium accumulated by selenite-reducing bacteria. *Biological Trace Element Research*. 52:209–225.

Combs Jr., G., and S.B. Combs. 1986. *The role of selenium in nutrition*. Amsterdam, The Netherlands: Academic Press.

Cox, A.C., and J.L. Sell. 1967. Magnesium deficiency in the laying hen. *Poultry Science*. 46:675–680.

Creech, B., G. Feldman, T. Ferguson, B. Reid, and J. Couch. 1957. Exudative diathesis and vitamin E deficiency in turkey poults. *The Journal of Nutrition*. 62:83–95.

Damron, B.L., and L.S. Kelly. 1987. Short term exposure of laying hens to high dietary sodium chloride levels. *Poultry Science*. 66:825–828.

David, L.S., M.N. Anwar, M.R. Abdollahi, M.R. Bedford, V. Ravindran. 2023. Calcium nutrition of broilers: Current perspectives and challenges. *Animals*. 13:1590.

Davies, R.E., B.L. Reid, A.A. Kurnick, and J.R. Couch. 1960. The effect of sulfate on molybdenum toxicity in the chick. *Journal of Nutrition*. 70:193–198.

Davis, P., L. Norris, and F. Kratzer. 1962. Iron deficiency studies in chicks using treated isolated soybean protein diets. *The Journal of Nutrition*. 78:445–453.

Dean, C.E., B.M. Hargis, and P.S. Hargis. 1991. Effects of zinc toxicity on thyroid function and histology in broiler chicks. *Toxicology Letters*. 57:309–318.

Dean, W., and G. Combs Jr. 1981. Influence of dietary selenium on performance, tissue selenium content, and plasma concentrations of selenium-dependent glutathione peroxidase, vitamin E, and ascorbic acid in ducklings. *Poultry Science*. 60:2655–2663.

DeLuca, H.F. 1974. Vitamin D: The vitamin and the hormone. *Federation Proceedings*. 33:2211–2219.

Deobald, H., and C. Elvehjem. 1935. The effect of feeding high amounts of soluble iron and aluminum salts. *American Journal of Physiology-Legacy Content*. 111:118–123.

Dewar, W.A., P.A. Wight, R.A. Pearson, and M.J. Gentle. 1983. Toxic effects of high concentrations of zinc oxide in the diet of the chick and laying hen. *British Poultry Science*. 24:397–404.

Doherty, C.P. 2007. Host–pathogen interactions: The role of iron. *The Journal of Nutrition*. 137:1341–1344.

Doll, E.R., F.E. Hull, and W.M. Insko. 1946. Toxicity of sodium chloride solution for baby chicks. *Veterinary Medicine*. 41:361–363.

dos Santos, T., S. Saksit, M. Bedford, and C. Walk. 2013. Effect of high phytase inclusion rates on performance of broilers fed diets not severely limited in available phosphorus. *Asian-Australasian Journal of Animal Sciences*. 26:227–232.

Dozier III, W., A. Davis, M. Freeman, and T. Ward. 2003. Early growth and environmental implications of dietary zinc and copper concentrations and sources of broiler chicks. *British Poultry Science*. 44:726–731.

Dunn, L. 1924. The effect of cod live oil in various amounts and forms on the growth of young chickens. *Journal of Biologicial Chemistry*. 61:129–136.

Edwards III, H., and D. Baker. 2000. Zinc bioavailability in soybean meal. *Journal of Animal Science*. 78:1017–1021.

Edwards III, H.M., and D.H. Baker. 1999. Bioavailability of zinc in several sources of zinc oxide, zinc sulfate, and zinc metal. *Journal of Animal Science*. 77:2730–2735.

Edwards Jr., H.M., and J.R. Veltmann Jr. 1983. The role of calcium and phosphorus in the etiology of tibial dyschondroplasia in young chickens. *Journal of Nutrition*. 113:1568–1575.

Eeckhout, W., and M. De Paepe. 1994. Total phosphorus, phytate phosphorus and phytase activity in plant feedstuffs. *Animal Feed Science Technology*. 47:19–29.

Emmert, J.L., and D.H. Baker. 1995. Zinc stores in chickens delay the onset of zinc deficiency symptoms. *Poultry Science*. 74:1011–1021.

Fangauf, R.H., H. Vogt, and W. Penner. 1961. Studies of Ca tolerance in chickens. *Archiv für Geflugelkunde*. 25:82–86.

Feizi, A., and M. Nazeri. 2011. Survey of Newcastle disease effects on broiler breeder performance. *International Journal of Poultry Science*. 10:574–578.

Fly, A.D., O.A. Izquierdo, K.R. Lowry, and D.H. Baker. 1989. Manganese bioavailability in a Mn-methionine chelate. *Nutrition Research*. 9:901–910.

Gardiner, E., J. Rogler, and H. Parker. 1960. Magnesium requirement of the chick. *Poultry Science*. 39:1111–1115.

Giansanti, F., L. Leboffe, G. Pitari, R. Ippoliti, and G. Antonini. 2012. Physiological roles of ovotransferrin. *Biochimica et Biophysica Acta (BBA)—General Subjects*. 1820:218–225.

Giergiel, M., M.M. Durkalec, A. Nawrocka, B. Sell, I. Stolarska, and A. Posyniak. 2019. Ingestion of bedding material as a cause of acute copper sulfate poisoning in turkey poults. *Poultry Science*. 98:707–711.

Gillis, M.B. 1948. Potassium requirement of the chick. *Journal of Nutrition*. 36:351–358.

Grau, C.R., T.E. Roudybush, P. Vohra, F.H. Kratzer, M. Yang, and D. Nearenberg. 1989. Obscure relations of feather melanization and avian nutrition. *World's Poultry Science Journal*. 45:241–246.

Gries, C.L., and M. Scott. 1972. Pathology of selenium deficiency in the chick. *The Journal of Nutrition*. 102:1287–1296.

Guo, X.Q., K.H. Huang, and J.X. Tang. 2005. Clinicopathology of gout in growing layers induced by high calcium and high protein diets. *British Poultry Science*. 465:641–646.

Ha, H.Y., N. Alfulaij, M.J. Berry, and L.A. Seale. 2019. From selenium absorption to selenoprotein degradation. *Biological Trace Element Research*. 192:26–37.

Hajj, R., and J.L. Sell. 1969. Magnesium requirement of the laying hen for reproduction. *Journal of Nutrition*. 97:441–448.

Halpin, J.G., C.E. Holmes, and E.B. Hart, 1936. Salt requirements of poultry. *Poultry Science*. 15:99–103.

Halpin, K.M., and D.H. Baker. 1986. Long-term effects of corn, soybean meal, wheat bran, and fish meal on manganese utilization in the chick. *Poultry Science*. 65:1371–1374.

Halpin, K.M., D.G. Chausow, and D.H. Baker. 1986. Efficiency of manganese absorption in chicks fed corn-soy and casein diets. *The Journal of Nutrition*. 116:1747–1751.

Hamdi, M., S. López-Vergé, E. Garcia Manzanilla, A. Barroeta, and J. Pérez. 2015. Effect of different levels of calcium and phosphorus and their interaction on the performance of young broilers. *Poultry Science*. 94.

Handa, R.J., and R.B. Chiasson. 1980. Comparative effects of three goitrogenic treatments on White Leghorn chickens. *Avian Diseases*. 916–929.

Harland, B.F., M.R. Spivey Fox, and B.E. Fry. 1975. Protection against zinc deficiency by prior excess dietary zinc in young Japanese quail. *Journal of Nutrition*. 105:1509–1518.

Harms, R., K. Kuchinski, D. Sloan, and G. Russell. 1995. Sodium requirement for broiler breeder hens. *Poultry Science*. 74:1311–1316.

Harms, R.H., and R.E. Buresh. 1987. Influence of salinomycin on the performance of broiler chicks. *Poultry Science*, 66:51–54.

Harms R.H., B.L. Damron, and P.W. Waldroup. 1965. Influence of high phosphorus levels in caged layer diets. *Poultry Science*. 44:1249–1253.

Harper, J.A., and G.H. Arscott. 1962. Salt as a stress factor in relation to pendulous crop and aortic rupture in turkeys. *Poultry Science*. 41:497–499.

Hart, E., J. Halpin, H. Steenbock, and A. Black. 1922. The nutritional requirements of baby chicks II. Further study of leg weakness in chickens. *Journal of Biological Chemistry*. 52:379–386.

Hassan, S., J. Hakkarainen, P. Lindberg, and S. Sankari. 1993. Response of whole blood glutathione peroxidase and selenium in chicks fed with sodium selenite, wheat and fish meal. *Animal Feed Science and Technology*. 41:103–111.

Heinz, G.H., and D.J. Hoffman. 1996. Comparison of the effects of seleno-L-methionine, seleno-DL-methionine, and selenized yeast on reproduction of mallards. *Environmental Pollution (Barking, Essex: 1987)*. 91:169–175.

Heller, V., and R. Penquite. 1937. Factors producing and preventing perosis in chickens. *Poultry Science*. 16:243–246.

Hempe, J.M., and R.J. Cousins. 1992. Cysteine-rich intestinal protein and intestinal metallothionein: An inverse relationship as a conceptual model for zinc absorption in rats. *The Journal of Nutrition*. 122:89–95.

Henry, P., C. Ammerman, and R. Miles. 1989. Relative bioavailability of manganese in a manganese-methionine complex for broiler chicks. *Poultry Science*. 68:107–112.

Henry, P.R. 1995. Manganese bioavailability. In *Bioavailability of Nutrients for Animals*. C.B. Ammerman, D.H. Baker, and A.J. Lewis, editors. Amsterdam, The Netherlands: Academic Press. 239–256, 257 ref.

Henry, P.R., and C.B. Ammerman. 1995. Selenium bioavailability. In *Bioavailability of Nutrients for Animals*. C.B. Ammerman, D.H. Baker, and A.J. Lewis, editors. Amsterdam, The Netherlands: Academic Press. 239–256 257 ref.

Henry, P.R., and E.R. Miller. 1995. Iron bioavailability. In *Bioavailability of Nutrients for Animals*. C.B. Ammerman, D.H. Baker, and A.J. Lewis, editors. Amsterdam, The Netherlands: Academic Press. 169–199.

Hess, J., and W. Britton. 1997. Effects of dietary magnesium excess in White Leghorn hens. *Poultry Science*. 76:703–710.

Hester, P.Y. 2017. Chapter 30—Improving egg production and hen health with calcium. In *Egg Innovations and Strategies for Improvements*. P.Y. Hester, editor. Amsterdam, The Netherlands: Academic Press. 319–329.

Hill, C.H., and G. Matrone. 1961. Studies on copper and iron deficiencies in growing chickens. *Journal of Nutrition*. 73:425–431.

Hixson, O., and L. Rosner. 1957. Calcium iodate as a source of iodine in poultry nutrition. *Poultry Science*. 36:712–714.

Hunde, A., P. Patterson, S. Ricke, and W.K. Kim. 2012. Supplementation of poultry feeds with dietary zinc and other minerals and compounds to mitigate nitrogen emissions—a review. *Biological Trace Element Research*. 147:386–394.

Hunt, C.D., and F.H. Nielson. 1986. Dietary boron affects bone calcification in magnesium- and cholecalciferol deficient chick. *Trace Elements in Human and Animal Nutrition*. 275–277.

Hurwitz, S., I. Cohen, A. Bar, and S. Bornstein. 1973. Sodium and chloride requirements of the chick: Relationship to acid-base balance. *Poultry Science*. 52:903–909.

Hurwitz, S., I. Cohen, A. Bar, and U. Minkov. 1974. Sodium and chloride requirements of the 7-9-week-old broiler. *Poultry Science*. 53:326–331.

Jackson, N., and M.H. Stevenson. 1981. A study of the effects of dietary added cupric oxide on the laying, domestic fowl and a comparison with the effects of hydrated copper sulphate. *British Journal of Nutrition*. 45:99–110.

Jankowski, J., K. Lichtorowicz, Z. Zduńczyk, and J. Juśkiewicz. 2012. The effect of different dietary sodium levels on blood mineral concentrations and tibia mineralization in turkeys. *Polish Journal of Veterinary Sciences*. 15:227–232.

Jensen, L.S., G. Colnago, K. Takahashi, and Y. Akiba. 1986. Dietary selenium status and plasma thyroid hormones in chicks. *Biological Trace Element Research*. 10:11–18.

Jensen, L.S. 1975. Precipitation of a selenium deficiency by high dietary levels of copper and zinc. *Proceedings of the Society for Experimental Biology and Medicine*. 149:113–116.

Jensen, L.S., and C.H. Chang. 1976. Fractionation studies on a factor in linseed meal protecting against selenosis in chicks. *Poultry Science*. 55:594–599.

Jing, M., S. Zhao, A. Rogiewicz, B.A. Slominski, and J.D. House. 2018. Assessment of the minimal available phosphorus needs of laying hens: Implications for phosphorus management strategies. *Poultry Science*. 97:2400–2410.

Jlali, M., M. Briens, F. Rouffineau, F. Mercerand, P.-A. Geraert, and Y. Mercier. 2013. Effect of 2-hydroxy-4-methylselenobutanoic acid as a dietary selenium supplement to improve the selenium concentration of table eggs. *Journal of Animal Science*. 91:1745–1752.

Johnson, D., A.L. Mehring, F.X. Savino, and H.W. Titus. 1962. The tolerance of growing chickens for dietary zinc. *Poultry Science*. 41:311–317.

Kare, M.R., and B. Jacob. 1948. The toxicity of sodium chloride and its relation to water intake in baby chicks. 27:751–758.

Kavtarashvili, A.S., I. Stefanova, V. Svitkin, and E. Novotorov. 2017. Functional egg production. II. The roles of selenium, zinc, and iodine. *Agricultural and Biological Chemistry*. 52:700–715.

Keen, C.L., J.L. Ensunsa, and M.S. Clegg. 2000. Manganese metabolism in animals and humans including the toxicity of manganese. *Metal Ions in Biological Systems*. 37:89–121.

Keshavarz, K. 1994. Laying hens respond differently to high dietary levels of phosphorus in monobasic and dibasic calcium phosphate. *Poultry Science*. 73:687–703.

Kheravii, S.K., N.K. Morgan, R.A. Swick, M. Choct, and S.B. Wu. 2018. Roles of dietary fibre and ingredient particle size in broiler nutrition. *World's Poultry Science Journal*. 74:301–316.

Kidd, M., P. Ferket, and M. Qureshi. 1996. Zinc metabolism with special reference to its role in immunity. *World's Poultry Science Journal*. 52:309–324.

Kienholz, E., D. Turk, M. Sunde, and W. Hoekstra. 1961. Effects of zinc deficiency in the diets of hens. *Journal of Nutrition*. 75:211–221.

Kim, W., and P. Patterson. 2004. Effects of dietary zinc supplementation on broiler performance and nitrogen loss from manure. *Poultry Science*. 83:34–38.

Kim, W., and P.H. Patterson. 2005. Effects of dietary zinc supplementation on hen performance, ammonia volatilization, and nitrogen retention in manure. *Journal of Environmental Science and Health Part B*. 40:675–686.

Klasing, K.C. 1998. *Comparative Avian Nutrition*. Wallingford, Oxfordshire, UK: CAB International.

Klasing, K.C., and D.R. Korver. 2020. Nutritional diseases. *Diseases of Poultry*. 1255–1285.

Krakower, C.A., and M. Goettsch. 1945. Effect of excessive ingestion of sodium chloride on the chick, with particular reference to renal changes. *Archives of Pathology*. 40:209–219.

Kratzer, F.H. 1952. Effect of dietary molybdenum upon chicks and poults. Proceedings of the Society for Experimental Biology and Medicine. *Society for Experimental Biology and Medicine*. 80:483–486.

Kratzer, F.H., and P. Vohra. 1986. *Chelates in Nutrition*. Boca Raton, FL: CRC Press.

Kreuzer, M., and M. Kirchgessner. 1991. Endogenous iron excretion. *Biological Trace Element Research*. 29:77–92.

Krista, L.M., C.W. Carlson, and O.E. Olson, 1961. Some effects of saline water on chicks, laying hens, poults and ducklings. *Poultry Science*. 40:938–944.

Kubicek, J.J., and T.W. Sullivan. 1973. Dietary chloride requirement of starting turkeys. *Poultry Science*. 52:1903–1909.

Leach Jr., R., and J. Gross. 1983. The effect of manganese deficiency upon the ultrastructure of the eggshell. *Poultry Science*. 62:499–504.

Leach, R.M. 1974. Studies on the potassium requirement of the laying hen. *Journal of Nutrition*. 104:684–686.

Leach, R., and M. Nesheim. 1965. Nutritional genetic and morphological studies of an abnormal cartilage formation in young chicks. *The Journal of Nutrition*. 86:236–244.

Ledoux, D., P. Henry, C. Ammerman, P. Rao, and R. Miles. 1991. Estimation of the relative bioavailability of inorganic copper sources for chicks using tissue uptake of copper. *Journal of Animal Science*. 69:215–222.

Lee, S.R., W.M. Britton, G.N. Rowland. 1980. Magnesium toxicity: Bone lesions. *Poultry Science*. 59:2403–2411.

Leeson, S. 2009. Copper metabolism and dietary needs. *World's Poultry Science Journal*. 65:353–366.

Leeson, S., and L. Caston. 2008. Using minimal supplements of trace minerals as a method of reducing trace mineral content of poultry manure. *Animal Feed Science and Technology*. 142:339–347.

Leeson, S., A. Zubair, E. Squires, and C. Forsberg. 1997. Influence of dietary levels of fat, fiber, and copper sulfate and fat rancidity on cecal activity in the growing turkey. *Poultry Science*. 76:59–66.

Lepore, P.D., and R.F. Miller. 1965. Embryonic viability as influenced by excess molybdenum in chicken breeder diets. Embryonic viability as influenced by excess molybdenum in chicken breeder diets. *Proceedings of the Society for Experimental Biology and Medicine*. 118:155–157.

Lewis, P.D. 2004. Responses of domestic fowl to excess iodine: A review. *The British Journal of Nutrition*. 91:29–39.

Li, S., F. Gao, J. Huang, Y. Wu, S. Wu, and X.G. Lei. 2018. Regulation and function of avian selenogenome. *Biochimica et Biophysica Acta (BBA)-General Subjects*. 1862:2473–2479.

Liu, A.C.H., B.S. Heinrichs, and R.M. Leach. 1994. Influence of manganese deficiency on the charactersitics of proteoglycans of avian epiphyseal growth plate cartilage. *Poultry Science*. 73:663–669.

Lowry, K.R., and D.H. Baker. 1989. Amelioration of selenium toxicity by arsenicals and cysteine. *Journal of Animal Science*. 67:959–965.

Lü, J., and G.F. Combs Jr. 1988. Effect of excess dietary zinc on pancreatic exocrine function in the chick. *Journal of Nutrition*. 118:681–689.

Lyons, M., and W. Insko Jr. 1937. Chondrodystrophy in the chick embryo produced by a mineral deficiency in the diet of the hen. *Science*. 86.

Markovs, J., A. Galuza, N. Basova, G. Knipse, S. Vasiljeva, and G. Smirnova. 2018. In vitro studies of iron absorption and activity of glutathione peroxidase in intestinal mucosa of the chicken. *European Journal of Biological Research*. 8:168–173.

Matterson, L.D., H.M. Scott, and E. Jungherr, 1946. Salt tolerance of turkeys. *Poultry Science*. 25:539–541.

McCormick, C.C., and D.L. Cunningham. 1984. High dietary zinc and fasting as methods of forced resting: A performance comparison. *Poultry Science*. 63:1201–1206.

McDowell, L.R. 2003. *Minerals in Animal and Human Nutrition*. Amsterdam, The Netherlands: Elsevier Science BV.

McGhee, F., C. Creger, and J. Couch. 1965. Copper and iron toxicity. *Poultry Science*. 44:310–312.

McGowan, J., and A. Emslie. 1934. Rickets in chickens, with special reference to its nature and pathogenesis. *The Biochemical Journal*. 28:1503–1512.

McWard, G.W. 1967. Magnesium tolerance of the growing and laying chicken, *British Poultry Science*. 8:91–99.

Mehring Jr., A.L., J.H. Brumbaugh, A.J. Sutherland, and H.W. Titus. 1960. The tolerance of growing chickens for dietary copper. *Poultry Science*. 39:713–719.

Mertz, W., ed. 1987. *Trace Elements in Human and Animal Nutrition*, 5th ed. Amsterdam, The Netherlands: Academic Press. 301–364.

Miles, R.D., and P.R. Henry. 2000. Relative trace mineral bioavailability. *Brazilian Animal Science* 1:73–93.

Miller, E.R., and C.B. Ammerman. 1995. Iodine bioavailability. In *Bioavailability of Nutrients for Animals*. C.B. Ammerman, D.H. Baker, and A.J. Lewis, editors. Amsterdam, The Netherlands: Academic Press. 169–199.

Mohanna, C., and Y. Nys. 1999. Effect of dietary zinc content and sources on the growth, body zinc deposition and retention, zinc excretion and immune response in chickens. *British Poultry Science*. 40:108–114.

Mohanty, G.C., and J.L. West. 1969. Pathologic features of experimental sodium chloride poisoning in chicks. *Avian Diseases*. 13:762–773.

Mongin P. 1981. Recent advances in dietary anion-cation balance: Applications in poultry. *Proceeding of the Nutrition Society*. 40:285–294.

Mongin, P. 1968. Role of acid-base balance in the physiology of egg shell formation. *World's Poultry Science Journal*. 24:200–230.

Morck, T.A., and R.E. Austic. 1981. Iron requirement of White Leghorn hens. *Poultry Science*. 60:1497–1501.

Morrison, W.D., A.E. Ferguson, J.R. Pettit, and D.C. Cunnigham. 1975. The effects of elevated levels of sodium chloride on ascites and related problems in turkeys. *Poultry Science*. 54:146–154.

Motzok, I., D. Arthur, and H.D. Branion. 1957. Feeding of molybdenum to poultry. *Poultry Science*. 36:1144.

Moxon, A.L., and W.O. Wilson. 1944. Selenium-arsenic antagonism in poultry. *Poultry Science*. 23:149–151.

Mushtaq, M., and T. Pasha. 2013. Electrolytes, dietary electrolyte balance and salts in broilers: An updated review on acid-base balance, blood and carcass characteristics. *World's Poultry Science Journal*. 69:833–852.

Mutucumarana, R., V. Ravindran, G. Ravindran, and A. Cowieson. 2014. Measurement of true ileal digestibility and total tract retention of phosphorus in corn and canola meal for broiler chickens. *Poultry Science*. 93:412–419.

Nell, J., and E. Annison. 1980. Molybdenum requirements of chickens. *British Poultry Science*. 21:183–191.

Nelson, T.S. 1976. The hydrolysis of phytate phosphorus by chicks and laying hens. *Poultry Science*. 55:2262.

Nesheim, M., and M. Scott. 1958. Studies on the nutritive effects of selenium for chicks. *The Journal of Nutrition*. 65:601–618.

Nesheim, M.C., R.M. Leach, T.R. Zeigler, and J.A. Serafín. 1964. Interrelationships between dietary levels of sodium, chlorine and potassium. *The Journal of Nutrition*. 84:361–366.

Nielsen, F.H. 1991. Nutritional requirements for boron, silicon, vanadium, nickel, and arsenic: Current knowledge and speculation. *The FASEB Journal*. 5:2661–2667.

Norman, A., and A. Means. 1987. *Calcium-binding proteins in health and disease*. ScienceDirect.

Nott, H., and G. Combs. 1969. Sodium requirement of the chick. *Poultry Science*. 48:660–665.

NRC (National Research Council). 1994. *Nutrient Requirements of Poultry*, 9th rev. ed. Washington, DC: National Academy Press.

NRC. 2005. *Mineral Tolerance of Animals*. Washington, DC: The National Academies Press.

Nugara, D., and H.M. Edwards Jr. 1963. Influence of dietary Ca and P levels on the Mg requirement of the chick. *The Journal of Nutrition*. 80:181.

O'Dell, B., C. Burpo, and J. Savage. 1972. Evaluation of zinc availability in foodstuffs of plant and animal origin. *The Journal of Nutrition*. 102:653–660.

O'Dell, B., P. Newberne, and J. Savage. 1958. Significance of dietary zinc for the growing chicken. *The Journal of Nutrition*. 65:503–523.

Olgun, O. 2017. Manganese in poultry nutrition and its effect on performance and eggshell quality. *World's Poultry Science Journal*. 73:45–56.

Ort, J.F., and J.D. Latshaw. 1978. The toxic level of sodium selenite in the diet of laying chickens. *The Journal of Nutrition*. 108:1114–1120.

Oviedo-Rondón, E.O., A. Murakami, A. Furlan, I. Moreira, and M. Macari. 2001. Sodium and chloride requirements of young broiler chickens fed corn-soybean diets (one to twenty-one days of age). *Poultry Science*. 80:592–598.

Park, H., B.W. McBride, L.P. Milligan, and L.M. Trouten-Radford. 1998. Acute effects of epidermal growth factor on Na+,K+ ATPase-dependent oxygen consumption and the amount of the enzyme units in enterocytes isolated from the jejunum of chickens. *Canadian Journal of Animal Science*. 78:327–333.

Patterson, E., R. Milstrey, and E.R. Stokstad. 1957. Effect of selenium in preventing exudative diathesis in chicks. *Proceedings of the Society for Experimental Biology and Medicine*. 95:617–620.

Paver, H., A. Robertson, and J.E. Wilson. 1953. Observations on the toxicity of salt for young chickens. *Journal of Comparative Pathology*. 63:31–47.

Payne, R., and L. Southern. 2005. Changes in glutathione peroxidase and tissue selenium concentrations of broilers after consuming a diet adequate in selenium. *Poultry Science*. 84:1268–1276.

Perdomo, J.T., R.H. Harms, and L.R. Arrington. 1966. Effect of dietary iodine upon egg production, fertility and hatchability. *Proceedings of the Society for Experimental Biology and Medicine*. 122:758–760.

Pesti, G.M., and R.I. Bakalli. 1996. Studies on the feeding of cupric sulfate pentahydrate and cupric citrate to broiler chickens. *Poultry Science*. 75:1086–1091.

Pesti, G.M., and R.I. Bakalli. 1998. Studies on the effect of feeding cupric sulfate pentahydrate to laying hens on egg cholesterol content. *Poultry Science*. 77:1540–1545.

Pimentel, J.L., J.L. Greger, M.E. Cook, and J.L. Stahl. 1992. Iron metabolism in chicks fed various levels of zinc and copper. *Journal of Nutritional Biochemistry*. 3:140–145.

Poley, W.E., A.L. Moxon, and K.W. Franke. 1937. Further studies of the effects of selenium poisoning on hatchability. *Poultry Science*. 16:219–225.

Poupoulis, C., and L.S. Jensen. 1976. Effect of high dietary copper on gizzard integrity of the chick. *Poultry Science*. 55:113–121.

Rinehart, K.E., W. Featherston, and J. Rogler. 1969. Influence of dietary potassium on chick growth, food consumption and blood and tissue composition. *Poultry Science*. 48:320–325.

Robbins, K.Y.R., and D.H. Baker. 1980. Effect of sulfur amino acid level and source on the performance of chicks fed high levels of copper. *Poultry Science*. 59:1246–1253.

Roberson, R.H., and P.J. Schaible. 1960. The tolerance of growing chicks for high levels of different forms of Zinc1,2. *Poultry Science*. 39:893–896.

Roberts, R. 1957. Salt tolerance of turkeys. *Poultry Science*. 36:672–673.

Rodehutscord, M., O. Adeola, R. Angel, P. Bikker, E. Delezie, W.A. Dozier, III, M. Umar Faruk, M. Francesch, C. Kwakernaak, A. Narcy, C.M. Nyachoti, O.A. Olukosi, A. Preynat, B. Renouf, A. Saiz del Barrio, K. Schedle, W. Siegert, S. Steenfeldt, M.M. van Krimpen, S.M. Waititu, and M. Witzig. 2017. Results of an international phosphorus digestibility ring test with broiler chickens. *Poultry Science*. 96:1679–1687.

Rogler, J.C., H.E. Parker, F.N. Andrews, and C.W. Carrick. 1959. The effects of an iodine deficiency on embryo development and hatchability. *Poultry Science*. 38:398–405.

Roland Sr., D.A., S.T. McCready, R.H. Stonerock, and R.H. Harms. 1977. Hypercalcemic effect of potassium iodide on serum calcium in domestic fowl. *Poultry Science*. 56:1310–1314.

Romanoff, A.L., and A.J. Romanoff. 1949. *The Avian Egg*. Hoboken, NJ: John Wiley & Sons.

Roychoudhury, S., S. Nath, P. Massanyi, R. Stawarz, M. Kacaniova, and A. Kolesarova. 2016. Copper-induced changes in reproductive functions: In vivo and in vitro effects. *Physiological Research*. 65.

Saiz, M.P., M.T. Marti, M.T. Mitjavila, and J. Planas. 1993. Iron absorption by small intestine of chickens. *Biological Trace Element Research*. 36:7–14.

Sarlak, S., S.A. Tabeidian, M. Toghyani, A.D. Foroozandeh Shahraki, M. Goli, and M. Habibian. 2020. Supplementation of two sources and three levels of iodine in the diet of laying hens: Effects on performance, egg quality, serum and egg yolk lipids, antioxidant status, and iodine accumulation in eggs. *Italian Journal of Animal Science*. 19:974–988.

Sauveur, B., and P. Mongin. 1978. Tibial dyschondroplasia, a cartilage abnormality in poultry. *Annales De Biologie Animale Biochimie Biophysique*. 18:87–98.

Scrivener, L.H. 1946. Experimental edema and ascites in poults. *Journal of the American Veterinary Medical Association*. 108:27–32.

Scott, M. 1966. Studies on the interrelationship of selenium, vitamin E and sulfur amino acids in a nutritional myopathy of the chick. *Annals of the New York Academy of Sciences*. 138:82–89.

Scott, M., G. Olson, L. Krook, and W. Brown. 1967. Selenium-responsive myopathies of myocardium and of smooth muscle in the young poult. *The Journal of Nutrition*. 91:573–583.

Scott, M.L., M.C. Nesheim, and R.J. Young. 1982. *Nutrition of the Chicken*. Ithaca, NY: M.L. Scott & Associates.

Sebastian, S., S.P. Touchburn, E.R. Chavez, and P.C. Lague. 1996. The effects of supplemental microbial phytase on the performance and utilization of dietary calcium, phosphorus, copper, and zinc in broiler chickens fed corn-soybean diets. *Poultry Science*. 75:729–736.

Sell, J.L., R. Hajj, A. Cox, and W. Guenter. 1967. Effect of magnesium deficiency in the hen on egg production and hatchability of eggs. *British Poultry Science*. 8:55–63.

Selle, P., A. Cowieson, and V. Ravindran. 2009. Consequences of calcium interactions with phytate and phytase for poultry and pigs. *Livestock Science*. 124:126–141.

Selye, H. 1943. Production of nephrosclerosis in the fowl by NaCl. *Journal of the American Veterinary Medical Association*. 103:140–143.

Shane, S.M., R.J. Young, and L. Krook. 1969. Renal and parathyroid changes produced by high calcium intake in growing pullets. *Avian Disease*. 13:558–567.

Shastak, Y., E. Zeller, M. Witzig, M. Schollenberge, and M. Rodehutscord. 2014. Effects of the composition of the basal diet on the evaluation of mineral phosphorus sources and interactions with phytate hydrolysis in broilers. *Poultry Science*. 93:2548–2559.

Shastak, Y., and M. Rodehutscord. 2015. A review of the role of magnesium in poultry nutrition. *World's Poultry Science Journal*. 71(1):125–138.

Shirley, R., and C. Parsons. 2001. Effect of ash content on protein quality of meat and bone meal. *Poultry Science*. 80:626–632.

Singh, P. 2008. Significance of phytic acid and supplemental phytase in chicken nutrition: A review. *World's Poultry Science Journal*. 64:553–580.

Skřivan, M., V. Skřivanová, and M. Marounek. 2005. Effects of dietary zinc, iron, and copper in layer feed on distribution of these elements in eggs, liver, excreta, soil, and herbage. *Poultry Science*. 84:1570–1575.

Słupczyńska, M., D. Jamroz, J. Orda, and A. Wiliczkiewicz. 2014. Effect of various sources and levels of iodine, as well as the kind of diet, on the performance of young laying hens, iodine accumulation in eggs, egg characteristics, and morphotic and biochemical indices in blood. *Poultry Science*. 93:2536–2547.

Smith, A., S.P. Rose, R.G. and Wells, and V. Pirgozliev. 2000. Effect of excess dietary sodium, potassium, calcium and phosphorus on excreta moisture of laying hens. *British Poultry Science*. 41:598–607.

Sokkar, S.M., B.M. Hussein, and M.A. Mohammed. 1983. Renal lesions in baby chicks due to sodium chloride poisoning. *Avian Pathology*. 12:277–285.

Sommerfeld, V., M. Schollenberger, I. Kühn, and M. Rodehutscord. 2018. Interactive effects of phosphorus, calcium, and phytase supplements on products of phytate degradation in the digestive tract of broiler chickens. *Poultry Science*. 97:1177–1188.

Southern, L.L., and D.H. Baker. 1983a. Excess manganese ingestion in the chick. *Poultry Science*. 62:642–646.

Southern, L.L., and D.H. Baker. 1983b. Zinc toxicity, zinc deficiency and zinc-copper interrelationship in *Eimeria acervulina*-infected chicks. *Journal of Nutrition*. 113:688–696.

Spears, J., S. Hansen, P. Schlegel, S. Durosoy, and A. Jongbloed. 2008. Bioavailability criteria for trace minerals in monogastrics and ruminants. *Trace Elements in Animal Production Systems*. Wageningen, The Netherlands: Wageningen Academic. 161–177.

Stillmak, S.J., and M.L. Sunde. 1971. The use of high magnesium limestone in the diet of the laying hen. *Poultry Science*. 50:553.

Suchý, P., E. Straková, and I. Herzig. 2018. Selenium in poultry nutrition: A review. *Czech Journal of Animal Science*. 59:495–503.

Sullivan, T. 1961. The zinc requirement of Broad Breasted Bronze poults. *Poultry Science*. 40:334–340.

Sumaiya, S., S. Nayak, R. Baghel, A. Nayak, C. Malapure, and R. Kumar. 2016. Effect of dietary iodine on production of iodine enriched eggs. *Veterinary World*. 9:554.

Sun, L.-H., J.-Q. Huang, J. Deng, and X.G. Lei. 2019. Avian selenogenome: Response to dietary Se and vitamin E deficiency and supplementation. *Poultry Science*. 98:4247–4254.

Sunde, R.A., G.E. Gutzke, and W.G. Hoekstra. 1981. Effect of dietary methionine on the biopotency of selenite and selenomethionine in the rat. *The Journal of Nutrition*. 111:76–86.

Supplee, W. 1964. Observations on the effect of copper additions to purified turkey diets. *Poultry Science*. 43:1599–1600.

Surai, P.F. 2000. Effect of selenium and vitamin E content of the maternal diet on the antioxidant system of the yolk and the developing chick. *British Poultry Science*. 41:235–243.

Surai, P.F., I.I. Kochish, V.I. Fisinin, and O.A. Velichko. 2018. Selenium in poultry nutrition: From sodium selenite to organic selenium sources. *Journal of Poultry Science*. 55:79–93.

Surai, P.F., I.I. Kochish, V.I. Fisinin, and O.A. Velichko. 2017. Selenium in poultry nutrition: From sodium selenite to organic selenium sources. *Journal of Poultry Science*. 55(2):79–93.

Suttle, N. 2010. Manganese. *Mineral Nutrition of Livestock*. 355–376.

Tamim, N.M., and R. Angel. 2003. Phytate phosphorus hydrolysis as influenced by dietary calcium and micro-mineral source in broiler diets. *Journal of Agricultural and Food Chemistry*. 4687–4693.

Tamim, N.M., R. Angel, and M. Christman. 2004. Influence of dietary calcium and phytase on phytate phosphorus hydrolysis in broiler chickens. *Poultry Science*. 83:1358.

Teeter, R.G., and M.O. Smith. 1986. High chronic ambient temperature stress effects on broiler acid-base balance and their response to supplemental ammonium chloride and potassium chloride and potassium carbonate. *Poultry Science*. 65:1777–1781.

Thompson, J., and M. Scott. 1970. Impaired lipid and vitamin E absorption related to atrophy of the pancreas in selenium-deficient chicks. *Journal of Nutrition*. 100:797–809.

Tufarelli, V., and V. Laudadio. 2017. Manganese and its role in poultry nutrition: An overview. *Journal of Experimental Biology and Agricultural Sciences*. 5:749–754.

Urbanyi, L. 1960. Chicken feeding trials with diets containing sufficient phosphorous and increasing amounts of calcium carbonate. *Nutrition Abstract*. 30:691.

Underwood, E. 1977. *Trace Minerals in Human and Animal Nutrition*. New York: Academic Press.

Vallee, B.L., and K.H. Falchuk. 1993. The biochemical basis of zinc physiology. *Physiological Reviews*. 73:79–118.

Van Krimpen, M.M., R.P. Kwakkel, B.F.J. Reuvekamp, C.M.C. Van Der Peet-Schwering, L.A. Den Hartog, and M.W.A. Verstegen. 2005. Impact of feeding management on feather pecking in laying hens. *World's Poultry Science Journal*. 61:663–686.

Van Reen, R., and P.B. Pearson. 1955. Manganese deficiency in the duck. *Journal of Nutrition*. 55:225–234.

Van Vleet, J. 1982. Amounts of twelve elements required to induce selenium-vitamin E deficiency in ducklings. *American Journal of Veterinary Research*. 43:851–857.

Vendeland, S.C., J.A. Butler, and P.D. Whanger. 1992. Intestinal absorption of selenite, selenate, and selenomethionine in the rat. *The Journal of Nutritional Biochemistry*. 3:359–365.

Viveros, A., A. Brenes, I. Arija, and C. Centeno. 2002. Effects of microbial phytase supplementation on mineral utilization and serum enzyme activities in broiler chicks fed different levels of phosphorus. *Poultry Science*. 81:1172–1183.

Vohra, P., and F. Kratzer. 1968. Zinc, copper and manganese toxicities in turkey poults and their alleviation by EDTA. *Poultry Science*. 47:699–704.

Waibel, P., D. Snetsinger, R. Ball, and J. Sautter. 1964. Variation in tolerance of turkeys to dietary copper. *Poultry Science*. 43:504–506.

Walk, C., M. Santos, and M. Bedford. 2014. Influence of superdoses of a novel microbial phytase on growth performance, tibia ash, and gizzard phytate and inositol in young broilers. *Poultry Science*. 93:1172–1177.

Walter, E., and L.S. Jensen. 1963. Effectiveness of selenium and noneffectiveness of sulfur amino acids in preventing muscular dystrophy in the turkey poult. *Journal of Nutrition*. 80:327–331.

Wideman, R.F., J.A. Closser, W.B. Roush, and B.S. Cowen. 1985. Urolithiasis in pullets and laying hens: Role of dietary calcium and phosphorus. *Poultry Science*. 64:2300–2307.

Wedekind, K., and D. Baker. 1990a. Effect of varying calcium and phosphorus level on manganese utilization. *Poultry Science*. 69:1156–1164.

Wedekind, K., and D. Baker. 1990b. Zinc bioavailability in feed-grade sources of zinc. *Journal of Animal Science*. 68:684–689.

Wedekind, K., A. Hortin, and D. Baker. 1992. Methodology for assessing zinc bioavailability: Efficacy estimates for zinc-methionine, zinc sulfate, and zinc oxide. *Journal of Animal Science*. 70:178–187.

Wedekind, K.J., E.C. Titgemeyer, A.R. Twardock, and D.H. Baker. 1991. Phosphorus, but not calcium, affects manganese absorption and turnover in chicks. *Journal of Nutrition*. 121:1776–1786.

Wideman, R.F., Y.K. Kirby, T.L. Barton, D. Clark, G.R. Bayyari, W.E. Huff, P.A. Moore, and P.A. Dunn. 1996. Excess dietary copper triggers enlargement of the proventriculus in broilers. *Journal of Applied Poultry Research*. 5:219–230.

Willemin, M.-E., and A. Lumen. 2017. Thiocyanate: A review and evaluation of the kinetics and the modes of action for thyroid hormone perturbations. *Critical Reviews in Toxicology*. 47:543–569.

Williams, S.N., R.D. Miles, M.D. Ouart, and D.R. Campbell. 1989. Short-term high level zinc feeding and tissue zinc concentration in mature laying hens. *Poultry Science*. 68:539–545.

Woyengo, T.A., and C.M. Nyachoti. 2013. Review: Anti-nutritional effects of phytic acid in diets for pigs and poultry—current knowledge and directions for future research. *Canadian Journal of Animal Science*. 93:21–29.

Yalçin, S., Z. Kahraman, S. Yalcin, S. Yalcin, and H. Dedeoğlu. 2004. Effects of supplementary iodine on the performance and egg traits of laying hens. *British Poultry Science*. 45:499–503.

Yu, Y., L. Lu, X. Luo, and B. Liu. 2008. Kinetics of zinc absorption by in situ ligated intestinal loops of broilers involved in zinc transporters. *Poultry Science*. 87:1146–1155.

Zanetti, M., P. Henry, C. Ammerman, and R. Miles. 1991. Estimation of the relative bioavailability of copper sources in chicks fed on conventional dietary amounts. *British Poultry Science*. 32:583–588.

Zhao, L., L.-H. Sun, J.-Q. Huang, M. Briens, D.-S. Qi, S.-W. Xu, and X.G. Lei. 2017. A novel organic selenium compound exerts unique regulation of selenium speciation, selenogenome, and selenoproteins in broiler chicks. *Journal of Nutrition*. 147:789–797.

Zhu, S.-Y., X.-N. Li, X.-C. Sun, J. Lin, W. Li, C. Zhang, and J.-L. Li. 2017. Biochemical characterization of the selenoproteome in Gallus gallus via bioinformatics analysis: Structure–function relationships and interactions of binding molecules. *Metallomics*. 9:124–131.

Zicker, S., and B. Schoenherr. 2012. The role of iodine in nutrition and metabolism. *Compendium on Continuing Education for the Practising Veternarian*. 34:1–4.

7

Non-Nutritive Feed Additives

INTRODUCTION

Non-nutritive feed additives have numerous important roles in poultry nutrition, ranging from improving ingredient nutrient utilization, protecting feed quality during storage, and improving gut health and human safety issues while handling feed such as dust control. This chapter will provide an overview of the types and functions of the most common feed additives. We realize that certain areas of feed additive research (enzymology) are rapidly changing as new products are being introduced in the poultry feed marketplace and new research is being published annually. So what is reviewed and contained in this chapter is an overview with recognition of the need for a separate publication focused on enzymes, enzyme product combinations, and nutrient release with different combinations in the future.

1. **Enzymes** are biochemical catalysts added to feeds to improve the breakdown of ingredient components to improve availability of those nutrients to poultry species. In effect, the use of exogenous enzymes is intended to supplement the digestive capacity of the bird, thereby providing greater ability for poultry species to utilize otherwise untapped nutrient sources within an array of ingredients. Turkey poults and chickens have limited production of amylase, trypsin, and lipase (Krogdahl and Sell, 1989) during their first few weeks of life (Noy and Skan, 1995). Therefore, incorporation of enzymes serves to lessen the environmental footprint of poultry production by increasing the types of feedstuffs that can be fed to non-ruminant species and decreasing excretion of organic nutrients. Common enzymes added to poultry diets are divided into the categories of carbohydrases, proteases, and phosphatases.

   Use of carbohydrases in poultry diets is intended to increase the digestion of carbohydrate-based energy sources in a vast array of feed ingredients. When the enzymes are successful, they may increase the metabolizable energy available to the bird and reduce untoward effects of anti-nutritive factors, including that of fibrous dietary components as discussed further in Chapter 2, "Carbohydrates." Carbohydrases act to break down NSP and to release more available carbohydrate products for digestion and absorption. They also reduce intestinal viscosity which improves nutrient availability in general (Ravindran, 2013) The general types of carbohydrases used in poultry diets are shown in Table 7-1. Ward (2021) presented a thorough review of the benefits of a combination of core enzymes (cellulases and xylanases) with debranching enzymes (arabinofuranosidase, glucoronsidase, and esterases) to break down glucoronoarabinosylans in cereal grains (see Table 7-2).

   Lipases hydrolyze triglycerides in various fat sources in poultry diets, both of animal and plant origin. Lipases are often of bacterial or mammalian origins. Supplementation of lipases and possible pancreatic products may aid the young chick with fat utilization during times of limited endogenous secretions (Al-Marzooqi and Leeson, 1999).

   Proteases function to hydrolyze proteins of both plant and animal origin. They are most often derived from microbial origins. They may increase nutrient digestibility of proteins and subsequent amino acids in the birds' diet.

   The category of enzyme most predominantly utilized in poultry diets is phosphatases—primarily phytases which hydrolyze plant phytates and thereby primarily increase phosphorus digestibility in the chicken but also secondarily improve trace mineral and amino acid digestibility. Phytase enables the poultry nutritionist to formulate diets with less added phosphate from sources other than corn and soybean meal, therefore decreasing excretion of unavailable phosphorus in poultry manure. A review by Abbasi and

TABLE 7-1 Common Carbohydrases Used in Poultry Diets

Carbohydrase	Substrate	Function
Amylases	Starch (amylose and amylopectin)	Break down starch into more available components
Cellulases	Cellulose (fiber)	Break down cellulose fiber
Galactosidase	Galactosides	Hydrolyze oligosaccharides
Glucanases	Glucans	Hydrolyze beta glucans
Hemicellulases	Hemicellulose (fiber)	Break down hemicellulose
Mannanase	Mannans	Hydrolyzes mannans
Pectinases	Pectin	Breaks down pectins
Xylanases	Xylans	Breaks down xylans

TABLE 7-2 Primary and Debranching Enzymes Associated with Degradation of Glucoronarabinoxylan in Cereal Grains (from Ward, 2021)

Debranching Enzyme	Function
Endo-xylanase	Hydrolyzes mainly interior B-1,4 xylose linkages of the xylan chain
Exo-xylanase	Hydrolyzes mainly B-1,4 xylose linkages releasing xylobiose
B-xylosidase	Releases xylose from xylobiose and short chain oligosaccharides
a-L-arabinofuranosidase	Hydrolyzes terminal nonreducing a-arabinofuransoidase from arabiolxylans
a-Glucouronosidase	Releases glucouronic acid from glucouronxylans
Ferculic acid esterase	Hydrolyzes feruloyl ester bonds in xylans
p-Coumeric acid esterase	Hydrolyzes p-coumaryl ester bonds in xylans

colleagues (2019) shows phytase sources coming from *Escherichia coli*, *Peniphora lycii*, *Aspergillus niger*, and *Ficum*, all having various capacities to improve phosphorus digestibility in poultry diets, and the review thoroughly discusses the environmental benefits of supplementing phytase and subsequent reduction of phosphorus in poultry waste. More than a single dose of phytase enzyme has shown even more benefits in improving the digestibility of other minerals as well as amino acids (Selle and Ranvindran, 2006; Coweison et al., 2011; Selle et al., 2012; Coweison et al., 2016). The efficacy of phytase activity in broilers has also been reported to be affected by over-feeding of calcium and phosphorus in broiler diets (Li et al., 2017, 2018). Slominski (2011) provides an in-depth review of papers studying factors affecting the efficacy of phytase in P excretion of birds fed phytase.

2. **Organic acids**, specifically short chain and medium chain fatty acids, volatile fatty acids, and weak or carboxylic acids (Cherrington et al., 1991), are feed additives utilized in poultry diets to improve gut health and potentially nutrient utilization. Their effects can include improved gut morphology in addition to being an energy source for the gut (Ricke, 2003). Organic acids have been reported to be successful in minimizing *Salmonella* populations in the gut microbiome (Ricke, 2003). It has been proposed that organic acids may interfere with bacterial cell membrane electron transport and ATP production as well as possibly altering epithelial cell gene expression (Darwin and Miller, 1999; Lostroh and Lee, 2001; Ricke, 2003; Van Immerseel et al., 2004), resulting in improved gut health. Butyric acid (C4) and some medium chain fatty acids (C12, lauric acid) have been shown to be effective in improved gut health (microvilli height) and nutrient utilization (Levy et al., 2015; Qaisrani et al., 2015; Kaczmarek et al., 2016; Nguyen et al., 2018).
3. Additives to the diet that protect feed/nutrient quality during storage and feeding include anti-oxidants, mycotoxin binders, and pellet binders. Mycotoxin binders are variably effective against different types of mycotoxins. The most widely researched ones are aluminosilicates and similar absorbents, which have limited absorptive capacity for some aflatoxins, **deoxynivalenol** (DON), and T-2 toxins (Kubena et al., 1990; Huff et al., 1992; Ledoux et al., 1999; Garcia et al., 2003; Binder, 2007). Examples of **pellet binders** include bentonite and lignin sulfonate; their function is to improve pellet quality and durability (Chapter 9, "Feed Processing and Quality"). **Antioxidants** are specifically added to fat sources and high-fat ingredients such as rendered by-product meals and fat-soluble vitamins. Types of **antioxidants** found in poultry rations include, but are not limited to, ethoxyquin, butylated hydroxytoluene (BHT), butylated hydroxyanisole (BHA), tertiary-butylhydoquinone (TBHQ), and certain essential oils. Citric acid may also be supplemented to poultry diets to decrease crop pH and improve galactosidase efficiency (Ao et al., 2009). Probiotics (aka direct-fed microbials) are rapidly gaining use in antibiotic-free poultry feeding programs. The use of live bacteria to modulate the gut microbiome is not a new concept and was first reported in the early 1900s. Probiotics can effectively influence the gut microbiome and inhibit opportunistic pathogens (Patterson and Burkholder, 2003; Park et al., 2013). Probiotics should be nontoxic, nonpathogenic, and normal inhabitants of poultry species, and be able to survive the low pH of the upper gastrointestinal tract to be functional in the small intestine. Bacterial species

found in poultry probiotic products may include those found within the Lactobacillus, *Enterococcus*, *Bacillus*, *Saccharomyces*, *Bifidobacterium*, *Streptoccocus*, *Aspergillus*, and *Candida* genera (Gaggia et al., 2010). Probiotics may also stimulate the immune system due to positive effects on the gut microbiome and host alike (Kabir et al., 2004, 2005; Kabir, 2009).

4. **Prebiotics** are nondigested feed additives, usually complex carbohydrates, that are not digested in the upper gut and are then fermented in the large intestine of the chicken or turkey, where they have positive effects on the bird. Their mode of action is to lower gut pH through lactic acid and other beneficial microbe production and prevent pathogen colonization in the lower gut (Choi et al., 1994; Gibson and Wang, 1994; Patterson and Burkholder, 2003; Gaggia et al., 2010; Hajati and Rezaei, 2010). Prebiotics have been shown to increase production of short chain fatty acids in the lower gut, thereby decreasing cecal pH and stimulating the growth and/or activity of species within the *Bifidobacterium* and *Lactobacillus* genera (Cummings and Mcfarlane, 2002; Donalson, 2005). The most common kinds of prebiotics used in poultry diets are those based on fructooligosaccharides (FOS) (Kim et al., 2011; Park et al., 2013) and mannan oligosaccharides (MOS) (Patterson and Burkholder, 2003; Geier et al., 2009). Spring et al. (2000) showed a reduction of *S. typhimurium* and *S. dublin* colonization in the gut with supplementation of MOS.
5. **Essential oils** are aromatic lipids derived from plant materials (Brenes and Roura, 2010). Essential oils have been proven to elicit antibacterial (Carson et al., 1995; Mourey and Canillac, 2002; Burt, 2004) and antioxidant (Kempaiah and Srinivasan, 2002; Botsoglou et al., 2004; Brenes and Roura, 2010) functions when fed to poultry. The biochemical composition of essential oils is complex and varies according to agronomic conditions of plant growth, with the antimicrobial activity primarily attributed to the presence of phenolic compounds (Cosentino et al., 1999; Burt, 2004). Moreover, the lipophilic nature of essential oils may permit disruption of cell wall membranes of pathogens (Applegate et al., 2010; Brenes and Roura, 2010). Types of essential oils utilized in poultry rations include blends of thymol, carvacrol, eugenol, curcumin, piperin, cinnamaldehyde, and eucalyptus to varying degrees of success against *Clostridium perfringens* (Abildgaard et al., 2010), *Salmonella heidelberg* (Amerah et al., 2012), and *Escherichia coli* (Jang et al., 2007).
6. **Emulsifiers** are also used in poultry feeds to improve nutrient availability, especially digestion of lipids. Types of emulsifiers added to poultry diets include both synthetic and natural lecithin products and other lipid-based additives. A review of emulsifiers is available (Siyal et al., 2017).

TABLE 7-3 List of Authorized EU Carotenoids and Xanthophylls, Adapted from Breithaupt, 2007

Additive	Species	Maximum Content Mg/kg of Feed
Capsanthin	Poultry	80
B-Apo-8-carotenal	Poultry	80
B-Apo-8′-carotenoic acid	Poultry	80
Lutein	Poultry	80
Cryptoxanthin	Poultry	80
Canthaxanthin	Poultry	Laying hens—8 Other Poultry—25
Zeaxanthin	Poultry	80
Citranaxanthin	Poultry	80

7. **Pigmentors (xanthophylls)** are typically carotenoids added to poultry diets to create a desired yellow color in egg yolks and broiler skin color. There are natural carotenoids in feed ingredients such as corn, alfalfa meal, corn gluten, and distillers' products that contribute to egg yolk color and skin color. To achieve desired egg yolk and skin color by some geographical markets, synthetic and natural carotenoids are often supplements to poultry diets (Fefana, 2014). There are also some xanthophylls with biological properties such as canthaxanthin, zeaxanthin, and lutein that have some advantages for human health and adult macular degeneration (Landrum and Bone, 2001; Olmedilla et al., 2001). Xanthophylls are fat soluble and are typically deposited in egg yolk and skin, which increases the yellow color of poultry products. Typically, yellow carotenoids are added to poultry feed to increase overall yellow pigmentation of products, and then red carotenoids such as canthaxanthin are added to intensify the yellow color of poultry products such as eggs and broiler skin color. Both natural and synthetic sources of carotenoids are available worldwide for poultry feeds. Table 7-3 is a list of EU-authorized additives for poultry feeds.

REFERENCES

Abbasi, F., T. Fakhur-un-Nisa, J. Liu, X. Luo, and I.H.R. Abbasi. 2019. *Environmental Science and Pollution Research.* 26:9469–9479.

Abildgaard, L., O. Hojberg, A. Schramm, K.M. Balle, and R.M. Engberg. 2010. The effect of feeding a commercial essential oil product on *Clostridium perfringens* numbers in the intestine of broiler chickens measured by real-time PCR targeting the α-toxin-encoding gene (plc). *Animal Feed Science and Technology.* 157(3–4):181–189.

Al-Marzooqi, W., and S. Leeson. 1999. Evaluation of dietary supplements of lipase, detergent, and crude porcine pancreas on fat utilization by young broiler chicks. *Poultry Science.* 78:1561–1566.

Amaya, E., P. Becquet, S. Carné, S. Peris, and P. Miralles. 2014. Carotenoids in animal nutrition. https://fefana.org/app/uploads/2022/05/2014-12-04_booklet_carotenoids.pdf.

Amerah, A.M., G. Mathis, and C.L. Hofacre. 2012. Effect of xylanase and a blend of essential oils on performance and Salmonella colonization of broiler chickens challenged with *Salmonella Heidelberg*. *Poultry Science*. 91:943–947.

Ao, T., A.H. Cantor, A.J. Pescatore, M.J. Ford, J.L. Pierce, and K.A. Dawson. 2009. Effect of enzyme supplementation and acidification of diets on nutrient digestibility and growth performance of broiler chicks. *Poultry Science*. 88:111–117.

Applegate, T.J., V. Klose, T. Steiner, A. Ganner, and G. Schatzmayr. 2010. Probiotics and phytogenics for poultry: Myth or reality? *Journal of Applied Poultry Research*. 19:194–210.

Binder, E.M. 2007. Managing the risk of mycotoxins in modern feed production. *Animal Feed Science and Technology*. 133:149–166.

Botsoglou, N.A., E. Christaki, P. Floro-Paneri, I. Giannenas, G. Papageorgiou, and A.B. Spais. 2004. The effect of a mixture of herbal essential oils or a α-tocopherylacetate on performance parameters and oxidation of body lipid in broilers. *South African Journal of Animal Science*. 34:52–61.

Breithaupt, D.E. 2007. Modern application of xanthophylls in animal feeding—a review. *Trends in Food Science & Technology*. 18:501–506.

Brenes, A., and E. Roura. 2010. Essential oils in poultry nutrition: Main effects and modes of action. *Animal Feed Science and Technology*. 158:1–14.

Burt, S. 2004. Essential oils: Their antibacterial properties and potential applications in foods: A review. *International Journal of Food Microbiology*. 94:223–253.

Carson, C.F., and T.V. Riley. 1995. Antimicrobial activity of the major components of the essential oil of *Melaleuca alternifolia*. *Journal of Applied Microbiology*. 78:264–269.

Cherrington, C.A., M. Hinton, and I. Chopra. 1990. Effect of short-chain organic acids on macromolecular synthesis in *Escerichia coli*. *Journal of Bacteriology*. 68:69–74.

Choi, K.H., H. Namkung, and I.K. Paik. 1994. Effects of dietary fuctooligosaccharides on suppression of intestinal colonization of *Salmonella typhimurium* in broiler chickens. *Korean Journal of Animal Science*. 36:271–284.

Consentino, S., C.I.G. Tuberoso, B. Pisano, M. Satta, V. Mascia, E. Arzedi, and F. Palmas. 1999. In vitro antimicrobial activity and chemical composition of Sardinian *Thymus* essential oils. *Letters in Applied Microbiology*. 29:130–135.

Cowieson, A.J., J.P. Ruckebusch, I. Knap, P. Guggenbuhl, and F. Fur-Nji. 2016. Phytate-free nutrition: A new paradigm in monogastric animal production. *Animal Feed Science and Technology*. 222:180–189.

Cowieson, A.J., P. Wilcock, and M.R. Bedford. 2011. Super-dosing effects of phytase in poultry and other monogastrics. *World's Poultry Science Journal*. 67:225–235.

Cummings, J.H., and G.T. Macfarlane. 2002. Gastrointestinal effects of prebiotics. *British Journal of Nutrition*. 87(Suppl. 2):S145–S151.

Darwin, K.H., and V.L. Miller. 1999. Molecular basis of the interaction of *Salmonella* with the intestinal mucosa. *Clinical Microbiology Reviews*. 12:405–428.

Donalson, L.M. 2005. The in vivo and in vitro effect of a fructooligosaccharide in prebiotic combined with alfalfa molt diets on egg production and *Salmonella* in laying hens. M.S. Thesis. Texas A&M University.

Gaggia, F., P. Mattarelli, and B. Biavati. 2010. Probiotics and prebiotics in animal feeding for safe food production. *International Journal of Food Microbiology*. 141:S15–S28.

Garcia, A.R., E. Avila, R. Rosiles, and V.M. Petrone. 2003. Evaluation of two mycotoxin binders to reduce toxicity of broiler diets containing ochratoxin A and T-2 toxin contaminated grain. *Avian Diseases*. 47:691–699.

Geier, M.S., R. Butler, and G. Howarth. 2009. Probiotics, prebiotics and synbiotics: A role in chemoprevention for colorectal cancer? *Biological Therapy for Cancer*. 5:1265–1269.

Gibson, G.R., and X. Wang. 1994. Regulatory effects of bifidobacteria on the growth of other colonic bacteria. *Journal of Applied Bacteriology*. 77:412–420.

Hajati, H., and M. Rezaei. 2010. The application of prebiotics in poultry production. *International Journal of Poultry Science*. 9:298–304.

Huff, W.E., L.F. Kubena, R.B. Harvey, and T.D. Phillips. 1992. Efficacy of hydrated sodium calcium aluminosilicate to reduce the individual and combined toxicity of aflatoxin and ochratoxin A. *Poultry Science*. 71:64–69.

Jang, I.S., Y.H. Ko, S.Y. Kang, and C.Y. Lee. 2007. Effect of a commercial essential oil on growth performance, digestive enzyme activity and intestinal microflora population in broiler chickens. *Animal Feed Science and Technology*. 134:304–315.

Kabir, S.M.L. 2009. The role of probiotics in the poultry industry. *International Journal of Molecular Science*. 10:3531–3546.

Kabir, S.M.L., M.M. Rahman, M.B. Rahman, M.M. Rahman, and S.U. Ahmed. 2004. The dynamics of probiotics on growth performance and immune response in broilers. *International Journal of Poultry Science*. 3(5):361–364.

Kabir, S.M.L., M.M. Rahman, M.B. Rahman, M.Z. Hosain, M.S.I. Akand, and S.K. Das. 2005. Viability of probiotics in balancing intestinal flora and effecting histological changes of crop and caecal tissues of broilers. *Biotechnology*. 4:325–330.

Kaczmarek, S.A., A. Barri, M. Hejdsysz, and A. Rutkowski. 2016. Effect of different doses of coated butyric acid on growth performance and energy utilization in broilers. *Poultry Science*. 95:851–859.

Kempaiah, R.K., and K. Srinivasan. 2002. Integrity of erythrocytes of hypercholesterelomic rats during spices treatment. *Molecular and Cellular Biochemistry*. 236:155–161.

Kim, G.-B., Y.M. Seo, C.H. Kim, and I.K. Paik. 2011. Effect of dietary prebiotic supplementation on the performance, intestinal microflora, and immune response of broilers. *Poultry Science*. 90:75–82.

Krogdahl, A., and J.L. Sell. 1989. Influence of age on lipase, amylase and protease activities in pancreatic tissue and intestinal contents of young turkeys. *Poultry Science*. 68:1561–1568.

Kubena, L.F., R.B. Harvey, T.D. Phillips, D.E Corrier, W.E. Huff, and G.E. Rottinghaus. 1990. Efficacy of hydrated sodium calcium aluminosilicate to reduce the toxicity of aflatoxin and T-2 toxin. *Poultry Science*. 78:204–210.

Landrum, J.T., and R.A. Bone. 2001. Lutein, zeaxanthin, and the macular pigment. *Archives of Biochemistry and Biophysics*. 385:28–40.

Ledoux, D.R., G.E. Rottinghaus, A.J. Bermudez, and M. Alonso-Debolt. 1999. Efficacy of a hydrated sodium calcium aluminosilicate to ameliorate the toxic effects of aflatoxin in broiler chicks. *Poultry Science*. 78:204–210.

Levy, A.W., J.W. Kessler, L. Fuller, S. Williams, G.F. Mathis, B. Lumpkins, and F. Valdez. 2015. Effect of feeding an encapsulated source of butyric acid (ButiPEARL) on performance of male Cobb broilers reared to 42 d of age. *Poultry Science*. 94:1864–1870.

Li, W., R. Angel, S.-W. Kim, K. Brady, S. Yu, and P.W. Plumstead. 2017. Impacts of dietary calcium, phytate, and phytase on inositol hexakisphosphate degradation and inositol phosphate release in different segments of the digestive tract of broilers. *Poultry Science*. 96:3626–3637.

Li, W., R. Angel, S.-W. Kim, E. Jimenez-Moreno, M. Proszkowiec-Weglarz, and P.W. Plumstead. 2018. Impacts of age and calcium on phytase efficacy in broiler chickens. *Animal Feed Science and Technology*. 238:9–17.

Lostroh, C.P., and C.A. Lee. 2001. The HilA box and sequences outside it determine the magnitude of HilA-dependent activation of PPGH from *Salmonella* pathogenicity island 1. *Journal of Bacteriology*. 183: 4876–4885.

Mourey, A., and N. Canillac. 2002. Anti-*Listeria monocytogenes* activity of essential oil components of conifers. *Food Control*. 13:289–292.

Nguyen, D.H., K.Y. Lee, M. Mohammeddigheisar, and I.H. Kim. 2018. Evaluation of the blend of organic acids and medium chain fatty acids in matrix coating as antibiotic growth promoter alternative on growth performance, nutrient digestibility, blood profiles, excreta microflora and carcass quality in broilers. *Poultry Science*. 97:4351–4358.

Noy, Y., and D. Skan. 1995. Digestion and absorption in the young chick. *Poultry Science*. 74:366–373.

Olmedilla, B., F. Gradado, I. Blanco, M. Vaquero, and C. Cajigal. 2001. Lutein in patients with cataracts and age related macular degenerations: A long term supplementation study. *Journal of the Science of Food and Agriculture*. 81:904–909.

Park, S.H, I. Hanning, A. Perrota, B.J. Bench, E. Alm, and S.C. Ricke. 2013. Modifying the gastrointestinal ecology in alternatively raised poultry and the potential for molecular and metabolomic assessment. *Poultry Science*. 92:546–561.

Patterson, J.A., and K.M. Burkholder. 2003. Applications of prebiotics and probiotics in poultry production. *Poultry Science*. 82:627–631.

Qaisrani, S.N., M.M. Van Krimpen, R.P. Kwakkel, M.W.A. Verstegen, and W.H. Hendriks. 2015. Diet structure, butyric acid and fermentable carbohydrates influence growth performance, gut morphology, and cecal fermentation characteristics in broilers. *Poultry Science*. 94:2152–2164.

Ravindran, V. 2013. Feed enzymes: The science, practice and metabolic realities. *Journal of Applied Poultry Research*. 22:628–636.

Ricke, S.C. 2003. Perspectives on the use of organic acids and short chain fatty acids as antimicrobials. *Poultry Science*. 82(4):632–639.

Selle, P.H., A.J. Cowieson, N.P. Cowieson, and V. Ravindran. 2012. Protein–phytate interactions in pig and poultry nutrition: A reappraisal. *Nutrition Research Reviews*. 25:1–17.

Selle, P.H., and V. Ravindran. 2006. Microbial phytase in poultry nutrition. *Animal Feed Science and Technology*. 135:1–41.

Siyal, F.A., D. Babazadeh, C. Wang, M.A. Arain, M. Saeed, T. Ayasan, L. Zhang, and T. Wang. 2017. Emulsifiers in the poultry industry. *World's Poultry Science Journal*. 73:611–620.

Slominski, A., 2011. Recent advances in research on enzymes for poultry diets. *Poultry Science*. 90:2013–2023.

Spring, P., C. Wenk, K.A. Dawson, and K.E. Newman. 2000. The effects of dietary mannan-oligosaccharides on cecal parameters and the concentrations of enteric bacteria in the ceca of *Salmonella* challenged broiler chicks. *Poultry Science*. 79:205–211.

Van Immerseel, F., V. Fievez, J. de Buck, F. Pasmans, A. Martel, F. Haesebrouck, and R. Ducatelle. 2004. Microencapsulated short-chain fatty acids in feed modify colonization and invasion early after infection with *Salmonella enteritidis* in young chickens. *Poultry Science*. 83:69–74.

Ward, N.E. 2021. Debranching enzymes in corn/soybean meal-based poultry feeds: A review. *Poultry Science*. 100:765–775.

8

Digestibility and Bioavailability

INTRODUCTION

Nutrient requirements are typically determined using diets containing ingredients that have known nutrient digestibilities and metabolic efficiencies of use for the purpose of meeting their required nutritional functions. Various methods have been developed to assess the nutritional value of commonly used feedstuffs, with measurement of digestibility and bioavailability being the two primary approaches. Digestibility and bioavailability values for a nutrient differ in several important ways that affect their usefulness in poultry nutrition. Digestibility directly measures the disappearance of a nutrient from the digestive tract and attempts to assess the digestive and absorptive phases of nutrient utilization, but it does not directly assess post-absorptive utilization for its required metabolic purposes. Bioavailability measures the value of a nutrient in a feedstuff to support specific required purposes and encompasses digestibility, absorption, and utilization for its required nutritional roles. In addition to the efficiency of digestion and ability to support essential functions, other factors that affect bioavailability include susceptibility to metabolic losses, rates of excretion, and interactions with antagonistic nutrients or antinutritional compounds. Discussions on the comparative bioavailability of different chemical forms of amino acids, vitamins, and minerals can be found in Chapters 3, 5, and 6, respectively.

AMINO ACID DIGESTIBILITY ASSAYS

It has long been recognized that digestible amino acids are a better representation of the amino acid (AA) content of feedstuffs to meet the needs of birds than formulating diets on a total amino acid basis. Until the 1990s, most AA digestibility studies were conducted using an excreta method as described by Kuiken and Lyman in 1948 (Rostagno et al., 1995; Ravindran et al., 1999). This was further developed by Sibbald and Wolynetz (1987) using cecectomized roosters in order to minimize the effects of the hindgut. In this method, birds were tube-fed a predefined quantity of a feed ingredient, and the total excreta voided over a defined time was collected in order to determine the retained AA. However, this method has disadvantages as outlined by Lemme et al. (2004), including using adult roosters rather than growing chicks, considering metabolized AA from uric acid in the calculations, and modifying the feed intake and feeding behavior of the birds. These disadvantages can be mitigated by using ileal digestibility assay as suggested by Payne et al. (1969), which may be conducted by euthanizing birds (Siriwan et al., 1993) or through utilizing a cannula (Raharjo and Farrell, 1984), which is rarely utilized in broilers as compared with larger animals. Ravindran et al. (1999) described an assay for measuring ileal AA digestibility and comparing the results to AA digestibility determined using the excreta collection method. These authors determined that ileal AA digestibility is typically higher than excreta digestibility, likely caused by microbial synthesis of AA in the hindgut, and thus suggested that ileal digestibility is a better representation of the AA available to the bird. The assay for ileal digestibility was further developed in order to standardize for endogenous losses. This was first conducted using fasted birds (Sibbald and Wolynetz, 1987) or with nitrogen-free diets (Adedokun et al., 2007) as well as highly digestible protein diets (Ravindran et al., 2004). Recent work (Ravindran et al., 2017) reported a common methodology for the ileal AA digestibility assay to reduce variation between labs and increase reproducibility of the assay.

Terminology

When discussing AA digestibility, it is important to differentiate between digestibility and bioavailability. Amino acid digestibility may be described as the AA disappearance from the gastrointestinal tract, typically through absorption (Moughan, 2003), or as the AA that are hydrolyzed in the gastrointestinal tract (Fuller, 2003). This differs slightly from bioavailability, which may refer to the AAs that are in a form that an animal can use for normal functions (Batterham, 1992; Ravindran and Bryden, 1999). Digestible amino acid assays are characterized as true, standardized, and apparent digestibility, which are described well in a review by Stein et al. (2007). Apparent ileal amino acid digestibility (AIAAD) is the difference between ingested and ileal digesta or excreted AA without considering any endogenous losses (Stein et al., 2007). Endogenous AA losses represent the nondietary AA losses and are comprised of mucin, sloughed epithelium, digestive enzymes, or other endogenous sources (Nyachoti et al., 1997). These can be divided into basal or specific losses, where basal losses are not related to diet composition (Stein et al., 2007), whereas specific losses are caused by ingredient characteristics such as fiber or antinutritional factors (Schulze et al., 1995). The true precision-fed AA digestibility assay is typically conducted with cecectomized roosters fed a known quantity of an ingredient or test diet and corrects for both specific and basal endogenous losses, while standardized digestibility (SIAAD) is conducted with chicks provided diets for *ad libitum* consumption and only corrects for basal losses (Stein et al., 2007). Standardized and true digestibility assays correct for the effects of feed intake and AA intake, and the ingredient values should be additive in formulation (Stein et al., 2005).

Procedure

Several authors have outlined procedures for a chick bioassay to calculate ileal AA digestibility (Ravindran et al., 1999, 2017; Kadim et al., 2002; Lemme et al., 2004). The procedure described involves feeding chicks a test diet containing an indigestible marker, euthanizing the birds, collecting digesta from the ileum, and then analyzing the proportion of AA and indigestible marker in both the feed and digesta to determine digestibility coefficients. Differences in methodology in the literature include diet design, husbandry, and sample collection and processing (Rodehutscord et al., 2012; Ravindran et al., 2017).

Digestibility values have been shown to be dependent on ingredient quality and diet composition (Adeola and Ileleji, 2009; Rochell et al., 2012), thus, diets should be carefully selected when determining ileal AA digestibility. Ravindran et al. (2017) outlined three methods for calculating AA digestibility: (1) the direct method, where all AAs are from the test ingredient; (2) the difference method, where a basal and a test diet are used; and (3) the regression method, where graded levels of the test ingredient are used. All the methods have their strengths and weaknesses. The choice of method depends on the nutrient and ingredient. The direct method will be described here, as it is most common. Diets are formulated so that all AAs are from the test ingredient, with the remainder of the diet comprised of semi-purified ingredients such as starch or dextrose, as well as oil, vitamins, and minerals (Lemme et al., 2004). Different basal diets should be utilized for grains and oilseed meals, because these feedstuffs have very different nutrient profiles (Ravindran et al., 2017). Even though semi-purified diets are commonly used in this assay, semi-purified diets have been shown to result in underestimation of apparent ileal AA digestibility, possibly through decreasing the retention time of feed in the gastrointestinal tract (Rochell et al., 2012). Additionally, it should be noted that feed form influences digestibility (Barua et al., 2021). Similarly, particle size of feedstuffs should be considered, as a larger particle size can lead to increased digestibility (Carre, 2000; Gabriel et al., 2003).

Experimental design should account for bird number, age, and adaptation to the diet prior to study initiation. Bird number should be chosen to account for the amount of sample needed for analysis. Furthermore, there is considerable bird-to-bird variability in digestion (Cowieson et al., 2020), so increased bird number within each pen should minimize variance. Lemme et al. (2004) reported using a minimum of four replicates per treatment, while Ravindran et al. (2017) suggested a minimum of six replicates. Chicks should be between 21 and 35 d of age (Ravindran et al., 2017). Amino acid digestibility has been shown to be lower at 14 d of age than in older birds (Huang et al., 2005), but age effects tend to disappear after 14 d of age (Batal and Parsons, 2003). Minimal data are available delineating the effects of diet adaptation period on AA digestibility, but Ravindran et al. (2017) suggest providing test diets 5 d prior to digesta collection. Broilers should be provided with a common starter diet until the test diets are provided.

Ileal digestibility assumes that nutrients recovered in the digesta beyond the ileum will not be later absorbed to any appreciable extent by the broiler; therefore, sample location should be considered when conducting a digestibility assay (Rodehutscord et al., 2012). Kluth et al. (2005) observed that amino acids are still absorbed in the proximal third of the ileum, while Rodehutschord et al. (2004) suggested excluding the last 2 cm proximal to the ileocecal junction to prevent contamination from the hindgut. Ileal digesta from each bird should be collected by gently flushing with deionized water the terminal ileum of broilers, which is defined as the terminal one-third of the section between the Meckel's diverticulum and approximately 4 cm anterior from the ileo-cecal junction, using deionized–distilled water (Rodehutscord et al., 2012). Gentle flushing with water is recommended over squeezing or scraping the terminal ileum, as such methods may increase collection of epithelium or mucin (Poureslami et al., 2012; Ravindran et al., 2017). Samples may be dried using either lyophilization or oven-drying as no effects of drying method have been shown on AA digestibility (Olojede et al., 2018). However, these authors favored lyophilization for ease of sample grinding. If titanium is used as the inert marker, the method published by Short et al. (1996) described subsequently can be used. Briefly, 100 mg of dried and ground digesta or feed should be ashed at 580°C for 13 h. The digesta should then be boiled in 10 mL of 7.4 M HCl for approximately 1 h and then combined with

20 mL of hydrogen peroxide (30%) and 70 mL of water. The color of the solution should be allowed to develop for 48 h and then be analyzed on a spectrophotometer at a wavelength of 410 nm. Amino acid concentrations of the feed and digesta should be analyzed using validated High-Performance Liquid Chromatography (HPLC) procedures.

Standardization for basal endogenous losses can be conducted using either a nitrogen-free diet or a diet that has nearly complete amino acid digestibility, with similar results being observed for both diet types (Adedokun et al., 2008). Both methods assume that all AAs in the terminal ileum are of endogenous origin. The semi-purified diet should be provided to chicks in the same manner as the test diets, as standardization is age- and species-dependent (Ravindran et al., 2004; Adedokun et al., 2007, 2008). The characteristics of the diet used for standardization, electrolyte balance, ratio of starch to dextrose, marker type, housing type (floor pens vs. battery cages), and the region of sampling in the gastrointestinal tract, will affect determined endogenous losses and should be considered when determining standardization procedures (Adedokun et al., 2011).

After samples have been analyzed, AIAAD and SIAAD can be calculated by relating the concentrations of the marker and the AA in the diet to the digesta. Apparent digestibility is calculated as follows (Lemme et al., 2004):

$$\text{AIAAD}(\%) = 100 - \left(\frac{M_{\text{diet}} \times AA_{\text{digesta}}}{M_{\text{digesta}} \times AA_{\text{diet}}} \times 100\right)$$

where M_{diet} and $M_{digesta}$ represent the concentration of the marker in the diet and digesta, respectively, while $AA_{digesta}$ and AA_{diet} represent the concentration of the AA being evaluated in the digesta and diet, respectively. The calculation will be conducted individually for each AA being considered. Standardized digestibility is calculated by adding the endogenous flow to the AIAAD, as follows (Adedokun et al., 2008):

$$\text{SIAAD}(\%) = \text{AIADD}(\%) + \left(\frac{\text{IEAA flow, g/kgDMI}}{AA_{\text{ingredient}},\ \text{g/kgDM}} \times 100\right)$$

where IEAA flow is the flow of endogenous AA, $AA_{ingredient}$ is the AA content of the ingredient in question, and DMI is dry matter intake.

Confounding Principles

There is innate variability of AA digestibility due to the digestive capacity of individual birds, and this variability is increased for poorly digestible ingredients (Cowieson et al., 2020). This innate variability is compounded by differing ileal digestibility methodology in the literature and between laboratories (Fontaine et al., 2003; Ravindran et al., 2017), including the factors such as diet formulation, ingredient inclusion, feed form, marker, adaptation period, bird age, collection method, collection location, drying method, and the analytical method employed. Further variability is introduced with the lack of consistent methodology in estimating endogenous losses (Adedokun et al., 2011).

APPARENT METABOLIZABLE ENERGY ASSAYS

An assay for determining metabolizable energy was described by Hill and Anderson (1958) and Hill et al. (1960). The method related the energy in feed consumed to the energy in excreta and is still commonly used today to determine apparent metabolizable energy (AME). This method was expanded on by Sibbald (1975) using predetermined feed allotments and evaluating endogenous losses of energy. Farrell et al. (1978) further modified the apparent metabolizable energy assay using trained cockerels. Interest in a net energy system is not new; Davidson et al. (1968) discussed a method for utilizing comparative slaughter for assessing net energy. Current research on energy assays includes further work on net energy systems (Zuidhof, 2019), assessment of correction for standardized losses (Khalil et al., 2020), and work on improving consistency between trials as well as improving experimental design (Wu et al., 2020). The true metabolizable assay approach developed by Sibbald (1976) has diminished in application because of criticism based on both technical and animal welfare concerns (NRC, 1994).

Terminology

Energy assays are based on the concept of energy partitioning as originally described by Armsby and Fries (1915). Briefly, the gross energy of a feedstuff is determined using direct calorimetry. Digestible energy is gross energy less fecal energy, while metabolizable energy is digestible energy less urinary and gaseous energy. Metabolizable energy is partitioned into the heat increment, or heat generated in the process of digestion, absorption, and metabolism, and then net energy. Net energy is the remaining energy and is useful for maintenance and production. When evaluating studies, care should be taken to understand if the determination method was corrected for nitrogen excretion. One correction unique to energy is the correction for nitrogen excretion, commonly studied as nitrogen-corrected apparent metabolizable energy (AME_n). The purpose of this is to correct the birds to an N equilibrium for comparison (Titus et al., 1959) based on the assumption that oxidation of protein will lead to the production of uric acid, and that this oxidation occurs at a rate that varies with growth (Lopez and Leeson, 2007). A review by Lopez and Leeson (2007) suggested that the value of using AME_n as compared with AME is dependent on bird age and the feed ingredient being evaluated. As expected, the authors found that adult roosters had higher AME_n values than growing broilers, while the converse was found with AME. Additionally, these authors found that AME_n reduced variation of energy in feed ingredients.

Procedure

The most common procedure to determine AME is a balance assay. Test diets are provided to broilers for sufficient time to allow physiological and anatomical acclimation. During the collection period, feed intake and excreta output are quantified and excreta samples are collected. Gross energy of the feed and excreta samples is evaluated on a dry matter basis using bomb calorimetry, and the total energy intake and excretion is calculated using the measured feed intake and excreta weight.

Several experimental designs are available for AME or AME_n balance studies, including the direct, difference, or substitution methods utilizing ingredients of known ME or the regression method (Wu et al., 2019; Khalil et al., 2021). In the direct method, the test ingredient is the only ingredient fed (Sibbald, 1976). Hill and Anderson (1958) suggested using a basal diet with an inclusion of an ingredient with a known ME, such as glucose, and substituting in the test ingredient for the known ingredient. A similar method (Carpenter and Clegg, 1956) suggests feeding diets composed of only ingredients with known ME, such as glucose blended with the test ingredient, which eliminates the need for a basal diet as in the method suggested by Hill and Anderson (1958) and may allow for a more appropriate nutrient profile than the direct method suggested by Sibbald (1976). Another method utilizing a basal diet involves evaluating the energy in a practical corn and soybean-meal basal diet compared to a diet where a percentage of the basal diet was compared with the test ingredient (Farrell et al., 1991). Two final methods are utilization of simple (Dozier et al., 2008; Adeola and Ileleji, 2009) or multiple (Young et al., 1977; Noblet et al., 1993) linear regression analyses to determine metabolizable energy content of ingredients. These methods, as well as other concerns with energy determination assays, are thoroughly discussed in a review by Wu et al. (2020).

Beyond experimental design, diet formulation, including nutrient profile, inclusion rate of test ingredient, and interactions between feed ingredients, as well as feed form, should be considered when determining AME. The maximum test ingredient inclusion rate when using the substitution method is determined based on palatability and nutrient balance (Lopez and Leeson, 2008). Adeola and Ileleji (2009) fed broilers diets varying in inclusion of dried distiller's grains with solubles (DDGS) in either practical corn or soybean-meal diets or in semi-purified diets. They found that DDGS utilization was improved, and AME and AME_n values increased when included in semi-purified diets compared to practical diets. The authors suggested this effect may be due to associative effects of the semi-purified diet and cautioned that diet type may affect determined energy. Balance studies are typically conducted with mash diets (Lee and Kong, 2019), but Khalil et al. (2021) noted that feed form affects determined C with mash diets potentially underestimating AME_n of cereals compared with feeding pelleted diets.

Bird age, adaptation to diet, feeding method, and collection method are other considerations when designing an AME balance study. Yang et al. (2020) observed linear and quadratic increases of AME and AME_n in corn- or wheat-based diets in birds at 1, 2, 3, 4, and 5 weeks of age. As it is well established that bird age affects determined metabolizable energy, the age of the chicks utilized should be representative of the birds for which the research will be utilized (Wu et al., 2020). Chicks should be started on a common starter diet to minimize confounding effects but then should be provided with the test diet with sufficient time for acclimation prior to collection. The adaptation period typically lasts between 3 and 14 d (Kong and Adeola, 2014) but is dependent on species, bird age, and test ingredient (Olukosi et al., 2017). Wu et al. (2020) reviewed multiple energy assays including feed offered for *ad libitum* consumption (Hill and Anderson, 1958), intubation (Sibbald, 1976), training roosters to consume feed (Farrell, 1978), and limited feeding (Fraps et al., 1940). The feeding method is important as feed intake (FI) affects determined AME due to the effect of endogenous losses (Guillaume and Summers, 1970; Sibbald, 1975) as well as due to the potential for spilled feed or measurement error when feeding *ad libitum* (Wu et al., 2020). Finally, the collection method should be considered. While total collection is commonly used for balance studies, utilizing a marker such as titanium dioxide as described by Short et al. (1996) as is common in AA studies can also be utilized to calculate a ME coefficient (Scott and Boldaji, 1997). In the total collection method, care should be taken to avoid collecting feed or feathers as these contaminants will add variation to the results (Wu et al., 2020). Excreta samples and weight should be collected for multiple days to minimize variation.

Using the total collection method, the ME coefficient is calculated according to the following equation:

$$= \frac{\left[\mathrm{FI}(\mathrm{kg,dm}) \times \mathrm{Energy}_{\mathrm{feed}}\left(\mathrm{kcal}/\mathrm{kg}\right)\right] - \left[\mathrm{Excreta\ output}(\mathrm{kg,\ dm}) \times \mathrm{Energy}_{\mathrm{excreta}}\left(\mathrm{kcal}/\mathrm{kg}\right)\right]}{\mathrm{FI}(\mathrm{kg,\ dm}) \times \mathrm{Energy}_{\mathrm{feed}}\left(\mathrm{kcal}/\mathrm{kg}\right)} \times 100$$

where FI is on a dry matter basis, excreta output represents the dry matter total excreta output from the same time period where FI was measured, and the energy of fed and excreta were determined utilizing calorimetry. If the test ingredient is provided directly, the AME of the test ingredient is calculated as follows:

$$\text{AME test ingredient}\left(\text{kcal}/_{\text{kg}}\right) = \frac{\text{ME Coefficient} \times \text{GE test ingredient}\left(\text{kcal}/_{\text{kg}}\right)}{100}$$

If the test ingredient is not provided directly, the AME calculated will represent that of the test diet and should be corrected to the percent inclusion of the test ingredient (Wu et al., 2020).

Confounding Principles

All of the confounding variables related to methodology for AA are also applicable to AME or AME_n. Additionally, the diet composition may have a substantial effect on determined AME due to interactions between ingredients, either additive or anti-nutritive. Further variation may occur in AME due to associated errors in feed disappearance (Guillaume and Summers, 1970) and feathers, feed, or water contamination of excreta (Khalil et al., 2020; Wu et al., 2020).

DEFINING A PHOSPHORUS SYSTEM

Phosphorus (P) and calcium (Ca) are quantitatively the two most important minerals in poultry nutrition regarding absolute body content (predominantly in bone and eggshells), essentiality for cellular metabolism and regulation, and impact on economics, environment, and sustainability of the poultry industry. Inorganic P sources are the third most expensive ingredient in the diet after energy and protein, are nonrenewable, and are of finite accessibility, as current estimates suggest that extraction from the Earth's deposits may be exhausted in 50–100 years (Cordell et al., 2009). Satisfying physiological P and Ca requirements using organic sources (e.g., plant- and animal-based ingredients) or inorganic sources of these minerals is paramount, which necessitates the accurate derivation of mineral availability in feedstuffs and supplements consumed by poultry species. Estimation of available P in any given feedstuff or supplement must account for the concentration and form of P in the raw material, along with the digestive and absorptive capacity of the bird, which varies based on factors including species, genetics, and age. Moreover, the impact of widespread supplementation of poultry diets with exogenous phytase products must also be taken into consideration (Chapter 7).

A process to quantitatively evaluate the relative nutritional value of P in individual ingredients is needed, and ideally, those values should be additive in nature for purposes of practical diet formulation. Various systems have been developed that rely on differing endpoints, which makes it challenging to establish specific recommendations. In this regard, a global effort is currently underway to create a common, accurate, and reliable system for measuring P availability in feedstuffs, with additional efforts focused on the idiosyncrasies of an available Ca system. Since release of the last National Academies of Sciences, Engineering, and Medicine (National Academies) poultry report (NRC, 1994), extensive research focus has been given to P availability evaluation systems. The following section will provide a brief historical context, including discussion of terminology, and provide an overview of current best practices and limitations in the creation of a digestible P system for poultry.

Historical Context and Terminology

Variation in P availability is high among and within feed ingredients (Shastak and Rodehutscord, 2015), which partly stems from the empirical methods used to quantify available P over the past 75 years (Shastak and Rodehutscord, 2013). As with other minerals, bioavailability of P must account for the total content present in dietary ingredients and also for how efficiently that P can be extracted from the intestinal lumen through absorptive processes (Peeler, 1972). Traditional methods of quantifying P bioavailability of raw materials (e.g., plant- or animal-based feedstuffs) involved dose–response relationships between a P-containing ingredient and an inorganic P source used as a standard reference material. As such, the derived estimate is only meaningful relative to that specific reference material. This translates into bioavailability estimates being more qualitative than quantitative, which severely limits the practicality of relative bioavailability values in feed formulation (Coon et al., 2002). Moreover, variation in relative bioavailability values occurs depending on the response criterion used, with outcomes typically related to growth performance, as well as bone (e.g., mineral content, breaking strength) and blood measurements (Shastak and Rodehutscord, 2013).

Inorganic P sources and animal-derived feedstuffs provide phosphate to the bird via solubilization and absorption of P from the gastrointestinal tract. However, plant-based ingredients contain variable proportions of total P content stored as *myo*-inositol, 1,2,3,4,5,6-hexakis (dihydrogen phosphate) (i.e., phytic acid, or phytate when bound to minerals). Whereas some plant-based ingredients may convey measurable activity of phytase (i.e., the hydrolytic enzyme capable of removing phosphate groups from phytic acid), it is commonly held that phytate P is inherently unavailable to the bird due to insufficient endogenous phytase activity (Soares, 1995). Thus, for any feedstuffs containing phytic acid, evaluation of available P must, at a minimum, discriminate between phytate P and non-phytate P, as was established in the previous National Academies poultry report (NRC, 1994). Whether in the form of non-phytate P or phosphates released from phytic acid, it is important to remember that P absorption (i.e., disappearance from the intestinal lumen) must occur for the dietary P to be used for metabolic purposes by the bird. Quantification of non-phytate P occurs strictly through analytical techniques, typically by calculating the difference between total P and phytate P, but this fraction does not account for inefficiencies in extracting P by the bird. Therefore, non-phytate P and available P are *not* synonymous terms, and the most accurate evaluation of available P must also consider its disappearance from the intestinal lumen, which is a process influenced by various confounding factors discussed subsequently.

In broad terms, available P can be defined as the portion of dietary total P, when supplied at a marginal concentration, that can be utilized to meet the physiological P requirement of the bird (Shastak and Rodehutscord, 2015). This definition consolidates total analyzed P content and animal-dependent efficiency of P utilization into a single value that can be applied to any dietary source, which is beneficial for practical feed formulation purposes. Whereas P sources are not technically digested to any extent using endogenous enzymes, the term digestible P is commonly used by swine nutritionists (NRC, 2012) and aligns with the available P definition described here for poultry species. Digestible P technically refers to the proportion of dietary P that is not recovered in feces; however, because birds excrete feces and urine together, colostomy would be required for their independent collection, thus reinforcing the idiosyncratic issues of defining a P system for avian species.

Phosphorus absorption and excretion of endogenous P predominantly occurs in the small intestine, with the large intestine playing an insignificant role in P homeostasis in poultry species. Generally speaking, P digestibility assays rely on luminal contents sampled near the end of the small intestine and thus require the use of an indigestible marker like titanium (IV) oxide (TiO_2) or chromium (III) oxide (Cr_2O_3). Similarly, balance studies are used to quantify P retention, which is defined as the proportion of total dietary P that is not present in the excreta and therefore is assumed to have been retained in the body. Phosphorus digestibility and retention experiments are effectively identical in design and only differ in the site of collection of output samples, which is often demarcated in the literature as ileal versus total tract digestibility. As elaborated upon subsequently, it is most accurate to refer to samples collected near the end of the small intestine as ileal to communicate that the samples have been acted upon by all segments of the gastrointestinal tract up to the terminal ileum. Described subsequently are details of the available P system that quantifies ileal digestible P values as recommended by the current Committee on Nutrient Requirements of Poultry.

Development of a Practical Digestible Phosphorus System

There is clear justification for developing a digestible P system for poultry species, largely following suit with what has occurred for swine. However, as described previously, poultry species present a more challenging technical situation due to emptying of the urinary and digestive tracts into the cloaca, resulting in an inability to separate urine and feces in poultry excreta. Thus, homeostatic mechanisms that drive P and Ca excretion via the urine present a methodological confounder if excreta samples are required to assess the quantity of available P in an ingredient or diet. Chickens have relatively minor hindgut capacity (i.e., cecum plus colon) when compared with swine, and there appears to be no net absorption of P from the hindgut nor net secretion of endogenous P into the colon of either species (Bohlke et al., 2005). The ileal P digestibility response is linear over a wider range of P intake as opposed to use of excreta samples when determining P retention (Rodehutscord et al., 2012), yet both methods produce similar available P values in poultry (Shastak et al., 2012). As such, ileal digestibility serves as an accurate and reliable proxy measure of P availability with the advantage of being less sensitive to dietary P concentration compared with other approaches (Rodehutscord et al., 2012). Moreover, quantifying ileal P digestibility values does not require the use of metabolism cages, meaning such assays can be conducted under varying housing conditions. Therefore, determining ileal disappearance of P negates many caveats associated with P, Ca, and the microbiota beyond the ileo-cecal junction in poultry species, and it offers a way to consolidate P systems for poultry on a global scale.

The Working Group No. 2 (Nutrition) of the European Federation of Branches of the World Poultry Science Association developed a unified protocol for quantification of available P in poultry (WPSA Working Group No. 2, 2013). This protocol recommends using ileal digestible P (idP) determination of available P for feedstuffs containing organic nutrients fed to poultry species, with an emphasis on broiler chickens. In this regard, idP (i.e., g/kg of ingredient) is calculated as the ileal P digestibility (iPd, %; determined using an indigestible marker approach) multiplied by the total P content of that ingredient as defined in the original protocol (WPSA Working Group No. 2, 2013).

Calculations required to estimate available P using this approach involve quantification of ileal P digestibility for each diet and replicated pen of birds based on concentrations of indigestible marker and P in samples of diet and ileal digesta samples:

$$\text{iPd}(\%) = 100 - \left(\frac{M_{\text{diet}} \times P_{\text{digesta}}}{M_{\text{digesta}} \times P_{\text{diet}}} \times 100 \right) \quad \text{[Equation 1]}$$

where M_{Diet} and $M_{Digesta}$ are analyzed marker concentrations (g/kg DM) of diet and digesta samples, respectively, and P_{Diet} and $P_{Digesta}$ are analyzed total P concentrations (g/kg DM) of diet and digesta samples, respectively. Using these calculated ileal P digestibility values, the total content of pcdP in each experimental diet can be derived as:

$$\text{idP (g/kg of diet)} = \text{iPd (\%)} \times \text{total } P_{\text{diet}}/100 \quad \text{[Equation 2]}$$

Ultimately, this available P protocol involves quantification of idP in feed ingredients as tested through a regression analysis. As such, a low-P basal diet is formulated and supplemented with at least two levels of P originating solely from the test ingredient. Differences in experimental diet formulations are required when testing mineral and animal sources of P as opposed to plant P sources, and the reader is referred to the original and amended protocols for complete formulation objectives (Rodehutscord et al., 2017). Regardless of source, the assay sets the highest dietary P concentration below the physiological P requirement of the bird, which means the slope of the linear regression represents the idP content for individual test ingredients. Because a regression analysis is used across graded concentrations of suboptimal P, this procedure inherently negates the need to correct for endogenous P losses, thereby eliciting a practical method by which to assess available P in feed ingredients consumed by poultry.

Complete details, including procedures involving animals, example diet formulations, calculations, and interpretation for the idP assay, can be found in the original publication (WPSA Working Group No. 2, 2013). Refinement of this protocol is ongoing, and recent studies recommend methodological improvements to address the absence of coccidiostats and antibiotics in the common diet; clarification on the use of phytase and Ca, P, and vitamin D_3 concentrations of the experimental diets; and the need for reporting of ingredient composition of diets (Haetinger et al., 2024). Moreover, when testing idP content of plant P sources, diets should not contain any mineral source of P, and a constant Ca:P ratio ranging from 1.3:1 to 1.4:1 should be used. The idP protocol for estimating available P of ingredients containing organic nutrients can also be amended to evaluate phytase efficacy under standardized conditions. As such, the idP content of individual experimental diets is regressed against analyzed supplemental dietary phytase activity to produce a dose–response curve that is inherently nonlinear in shape.

The Committee on Nutrient Requirements of Poultry is recommending a similar digestible P evaluation system for inorganic P sources without specific elements that apply to feedstuffs containing organic nutrients. There are alterations to consider when designing experimental diets for testing P sources of minerals and animal origin (Rodehutscord, 2013; van Harn, 2017). Rodehutscord (2013) outlined a recommended procedure and provided an example of a basal diet when evaluating P sources of mineral and animal origin. Considerations for the basal diet included in the protocol were ingredient selection, available phosphorus concentration (0.15%), total calcium and total P ratio, and limestone source. Supplemental P sources can be supplemented to the basal diet with either two or three concentrations equally spaced, with the maximum supplemental total P concentration being 0.15 percent. The idP concentration is plotted against the total concentration using linear regression to determine idP from the supplemental source. Common intercept multiple linear regression can be used if two or more supplemental sources are evaluated in the same experiment.

Confounding Principles

As with any biological assay, the idP system described previously is vulnerable to a number of factors that may markedly alter outcomes. These serve as the basis for amendments to the original protocol and involve causative factors at the analytical, animal, dietary, and even management methods when performing the assay. After implementing the original pcdP assay described in the previous section at 17 independent research stations around the world, significant variation among stations was observed for the test ingredient (soybean meal), with ileal P digestibility varying from 19 percent to 51 percent (Rodehutscord et al., 2017). However, ileal disappearance of P derived from phytic acid correlated well with estimated ileal P digestibility values, which validated use of the assay as a practical measure of available P. This international cooperative research project on P digestibility in broiler chickens revealed additional factors that warrant additional research to identify their influence on the assay.

Of the factors likely to influence the accuracy of idP estimates, pre-study dietary conditions (i.e., concentrations of dietary P, Ca, vitamin D_3; use of phytase, antibiotics, and coccidiostats) require further standardization. While many pre-study factors have yet to be empirically tested, there is evidence to suggest that bird genetics, sex, handling, housing, and dietary intake each influence idP estimates. Thus, researchers are highly encouraged to report as much information about pre-study experimental conditions as possible when publishing idP estimates for poultry species. Because the idP assay is designed to generate available P values under practical conditions, it is expected that basal diets will include phytate as derived from plant-based ingredients. The presence of phytate remains a significant problem in the poultry industry, hence the almost universal inclusion of exogenous phytase products in commercial formulations. For this reason, transparency in reporting of data from idP assays should include, at a minimum, the formulation and analyzed nutrient composition of the pre-test and experimental diets. Dietary phytate is well known to alter the availability of minerals (especially Ca), amino acids, and other nutrients, which can dramatically alter nutrient availability, energy metabolism, and microbiota composition of poultry species. While it is acceptable and practical to conduct an idP assay using phytate-containing ingredients, controlling the amount of phytate in the basal diet is crucial because it causes underestimation of P digestibility for mineral and animal sources. Additionally, when idP is being estimated in plant-based feedstuffs, the basal diet should not contain any P from mineral sources to avoid the aforementioned interactions.

In addition to the previous P digestibility assay, formulating diets on an analogous digestible calcium basis can be advantageous to optimize nutrient utilization and gut health as excess Ca can reduce P and amino acid utilization, as well as increase the incidence of necrotic enteritis (Walk et al., 2021).

Data on determining Ca digestibility of feed ingredients is limited. Digestibility of Ca in limestone, meat and bone meal, dicalcium phosphate, and monocalcium phosphate sources have been variable with an ileal calcium assay (David et al., 2019). This variance may be due to factors such as differences in methodology, inherent ingredient variability, variations in particle size, influence of Ca:non-phytate P ratios, lack of control for dietary phytate content or phytase inclusion, and influence of vitamin D_3 (Walk et al., 2021). Future research is warranted to refine the methodology of a digestible calcium assay for poultry before reliable calcium digestibility data of feed ingredients can be developed.

BIOAVAILABILITY

The current methodology used to assess bioavailability of a nutrient in a feedstuff calculates its nutritional value relative to that of a standardized reference source. Thus, the bioavailability of a feedstuff is expressed as a relative value (% of a standard) and not an absolute value (g/kg consumed) as is provided by the methods described previously for assessing digestibility (Parsons, 2002). Frequently, the standard reference employed for estimating relative bioavailability is that used in experiments to determine a bird's requirement (Ammerman et al., 1995). For example, L-methionine is the standard for assessing the methionine requirement and also the bioavailability of L-methionine analogues and precursors (Baker and Boebel, 1980). The bioavailability of L-methionine for growth, efficiency of feed utilization, protein accretion, and its various other metabolic roles is defined as 100 percent, even though somewhat less than 100 percent of supplemented L-methionine is actually absorbed and used for its required purposes. For all nutrients, bioavailable values are lower than digestible values due to metabolic inefficiencies of utilizing digested and absorbed nutrients.

Expressing the nutritional value of a nutrient as bioavailability instead of digestibility is typically applied to many vitamins and trace minerals, as well as some chemical forms of amino acids. This is because the utilization of some chemical forms of a nutrient is limited by their metabolic fate once absorbed. Digestibility values do not include inefficiencies that result from poor utilization of absorbed nutrients for their required role in metabolism, but bioavailability values include inefficiencies in both absorption and metabolic utilization. This difference can result in a large divergence between a nutrient's digestibility and bioavailability values. For example, the two enantiomers of lysine, D and L, have very different metabolic fates. Both are absorbed efficiently, but only L-lysine can charge tRNA for use in protein synthesis. D-lysine is absorbed but cannot be converted to L-lysine for use in protein synthesis. Instead, absorbed D-lysine is oxidized or excreted and is thereby of limited nutritional value to the bird. Thus, D- and L-lysine have similar digestibilities but very different bioavailabilities (see Chapter 3, "Protein and Amino Acids"). Similarly, some forms of vitamins and trace minerals are readily absorbed but are less efficient in accomplishing their metabolic functions than other forms (see Chapters 5 and 6). In the case of vitamins, this may be due to the need for conversion to the active form (e.g., conversation of β-carotene to retinal), and for trace minerals this may be due to the chemical form absorbed from the intestine (e.g., selenocysteine vs. selenate). For this reason, the nutritional requirements for many vitamins and minerals are typically expressed as the amount of the reference standard needed to meet the requirement (mg/kg diet), and other sources of that nutrient are adjusted based on their relative bioavailability, which is a ratio of that standard. It is possible that some forms of a nutrient may have a bioavailability above 100 percent. For example, $FeSO_4{\cdot}7H_2O$ is the form used to set requirements and as the standard in bioavailability research in poultry. However, ferric ammonium citrate has a relative bioavailability of 120 percent (Henry and Miller, 1995). Following is an overview of current best practice for measuring nutrient bioavailability.

MEASUREMENTS OF NUTRIENT BIOAVAILABILITY

Experimental Design

As described previously, the bioavailability of a nutrient in a feedstuff indicates the efficiency of its utilization for the same essential purposes that were used to determine its dietary requirement. These endpoints include both economically important indices, such as rates of growth, egg production, and efficiency of feed conversion, as well as nutrient-specific functions in metabolism and health, such as use of co-factors of enzymes and prevention of pathologies (Miles and Henry, 2000; Shen et al., 2015; Klasing and Korver, 2020). Similar to requirement studies, the experimental design of bioavailability studies utilizes a basal diet that is deficient in the nutrient being studied. Additional diets consisting of the basal diet supplemented with graded amounts of the feedstuff under study (i.e., test diets) are fed to establish the test feedstuff's incremental dose–response relationship. Additional groups of birds are fed diets containing equivalent molar amounts of the standard reference nutrient source, which is usually the same form used to establish the requirement. The ratio of the dose response of the test feedstuff to that of the standard reference source gives the relative bioavailability of the test feedstuff. Unlike nutrient requirement studies, bioavailability experiments do not need to establish a breakpoint dietary concentration at which the response variable no longer increases. Thus, they are typically conducted using inclusion concentrations that are below the requirement. At least four dietary concentrations are optimal for establishing the dose–response relationship (i.e., basal diet plus three levels of supplementation).

Relative bioavailability can be calculated based on either the nutrient's dietary concentration (% of the diet) or the amount of the nutrient consumed (absolute ingested quantity per d) versus the change in the response variable. The latter is preferred if the feed intake of the standard and the test ingredient differ. It is important that all diets be chemically analyzed for the nutrient being studied to ensure that the calculated dietary concentrations approximate the actual dietary concentrations. The analytically determined concentrations should be used when analyzing the data. For practical purposes, sometimes the purified nutrient standard is omitted and only commercial sources of an ingredient of interest are compared when the goal is to compare different commercial products.

One criticism of this approach is that formulating a diet that is highly deficient in the nutrient under study often requires the use of purified feed ingredients, and the resulting diets are dissimilar to commercial diets. This is because purified diets typically contain considerably lower concentrations of nutrient antagonists that impair digestive, absorptive, or metabolic processes (e.g., tannins, fiber, phytate). In the case of trace minerals and vitamins that are easily stored in the body, experiments can be designed using a commercial-type basal diet and additional diets containing amounts of the nutrient under study added at concentrations in excess of the requirement and then examine the relative dose responses of deposition into storage pools (Miles and Henry, 2000; Spears and Hansen, 2008). This approach can provide useful information on absorption and retention of a nutrient, but it does not always provide valuable information on the use of the nutrient for its required metabolic role. There are instances where a nutrient source is readily stored but not directly used as efficiently as the reference standard for its metabolic purpose (e.g., organic selenium sources vs. inorganic sources; see Chapter 6). A thorough understanding of the value of a nutrient source for building storage pools versus its value for supporting its required metabolic functions at low dietary concentrations is required in order to use experimental designs that employ plethoric nutrient concentrations.

Data Analysis and Interpretation

Appropriate data analysis and interpretation requires the accurate description of the dose–response relationship between increases in nutrient intake versus increases in biological or economically relevant endpoints. In some cases, the dose–response relationships of a test feedstuff relative to that of the standard reference are sufficiently similar and linear that the resulting calculations are simple and defendable biologically. In these cases, the relative biological value can be calculated by use of linear regression to determine the slopes and dividing the slope of the test feedstuff by that of the standard reference. This method is referred to as the "slope ratio" method (Littell et al., 1995; Miles and Henry, 2000; Kratzer and Littell, 2006). To accommodate the use of the slope ratio method of analysis, experiments are typically designed to provide data on the most linear part of the dose–response relationship so that assumptions regarding linearity, intercepts of the regression lines, and dose–response optima are met (Finney, 1975). If the slopes are nonlinear, transformation of the data may help linearize the response; however, this may call into question the biological validity of the results.

The slope ratio approach has been criticized because the dose response to a nutrient is typically not statistically linear at dietary concentrations near the requirement (Kratzer and Littell, 2006). Thus, bioavailability studies conducted at concentrations considerably below the requirement can produce linear responses but may not be biologically accurate at dietary concentrations closer to the requirement. Additionally, the dose–response relationships of a test feedstuff and a standard reference may be dissimilar at some parts of the full dose–response range. This is because the relative dose response to a nutrient source depends on the aggregated dose responses of its digestion within the digestive tract, absorption from the digestive tract, and metabolic steps required for utilization for required nutritional purposes. Each of these processes typically has multiple steps and each step has its own inherent dose–response relationships, most of which are nonlinear. In some cases, the test feedstuff and the standard differ in the digestive, absorptive, or metabolic pathways needed for utilization. For example, organic forms of selenium follow a different absorptive pathway than the inorganic standard reference, sodium selenite. Following absorption, organic forms must be converted to inorganic selenium to be utilized for its nutritionally required roles. As discussed previously, an analogous situation exists for D-methionine compared with the standard reference L-methionine in that they are absorbed by different pathways, and the D-form requires metabolic conversion to the L-form. For these reasons, multilinear or nonlinear regression techniques, preferably using common plateau asymptotes, may be appropriate for comparing the dose–response relationship of a test feedstuff to a standard (Littell et al., 1995; Lemme et al., 2002; Kratzer and Littell, 2006; Shen et al., 2015), but there is disagreement on the appropriate statistical method for some data sets (Piepho, 2006). Appropriate statistical analysis of dose–response relationships for bioavailability and requirement studies is an area that needs additional research attention.

REFERENCES

Adedokun, S.A., M.S. Lilburn, C.M. Parsons, O. Adeola, and T.J. Applegate. 2007. Endogenous amino acid flow in broiler chicks is affected by the age of birds and method of estimation. *Poultry Science*. 86:2590–2597.

Adedokun, S.A., O. Adeola, C.M. Parsons, M.S. Lilburn, and T.J. Applegate. 2008. Standardized ileal amino acid digestibility of plant feedstuffs in broiler chickens and turkey poults using a nitrogen-free or casein diet. *Poultry Science*. 87:2535–2548.

Adedokun, S.A., O. Adeola, C.M. Parsons, M.S. Lilburn, and T.J. Applegate. 2011. Factors affecting endogenous amino acid flow in chickens and the need for consistency in methodology. *Poultry Science*. 90:1737–1748.

Adeola, O., and K.E. Ileleji. 2009. Comparison of two diet types in the determination of metabolizable energy content of corn distillers dried grains with solubles for broiler chickens by the regression method. *Poultry Science*. 88:579–585.

Ammerman, C.B., D.H. Baker, and A.J. Lewis. 1995. *Bioavailability of Nutrients for Animals*. Amsterdam, The Netherlands: Academic Press.

Armsby, H.P., and J.A. Fries. 1915. Net energy values of feeding stuffs for cattle. *Journal of Agriculture Research*. 3:435–491.

Baker, D.H., and K.P. Boebel. 1980. Utilization of the D- and L-isomers of methionine and methionine hydroxy analogue as determined by chick bioassay. *Journal of Nutrition*. 110:959–964.

Barua, M., M.R. Abdollahi, F. Zaefarian, T.J. Wester, C.K. Girish, and V. Ravindran. 2021. Influence of feed form on the standardized ileal amino acid digestibility of common grains for broiler chickens. *Animal Feed Science and Technology*. 272:114743.

Batal, A.B., and C.M. Parsons. 2003. Utilization of different soy products as affected by age in chicks. *Poultry Science*. 82:454–462.

Batterham, E.S., 1992. Availability and utilization of amino acids for growing pigs. *Nutrition Research Reviews*. 5:1–18.

Bohlke, R.A., R.C. Thaler, and H.H. Stein. 2005. Calcium, phosphorus, and amino acid digestibility in low-phytate corn, normal corn, and soybean meal by growing pigs. *Journal of Animal Science*. 83:2396–2403.

Carpenter, K., and K. Clegg. 1956. The metabolizable energy of poultry feeding stuffs in relation to their chemical composition. *Journal of Science and Food Agriculture*. 7:45–51.

Carré, B. 2000. Effects of feed particle size on the digestive processes in domestic birds. *INRA Animal Production*. 13:131–136.

Coon, C., K. Leske, and S. Seo. 2002. The availability of calcium and phosphorus in feedstuffs. In *Poultry Feedstuffs: Supply, Composition and Nutritive Value*. J.M. McNab and K.N. Boorman, editors. Wallingford, Oxfordshire, UK: CAB International. 151–179.

Cordell, D., J.-O. Drangert, and S. White. 2009. The story of phosphorus: Global food security and food for thought. *Global Environmental Change*. 19(2):292–305.

Cowieson, A.J., M.M. Bhuiyan, J.O.B. Sorbara, G. Pappenberger, M.B. Pendersen, and M. Choct. 2020. Contribution of individual broilers to variation in amino acid digestibility in soybean meal and the efficacy of an exogenous monocomponent protease. *Poultry Science*. 99:1075–1083.

David, L.S., M.R. Abdollahi, G. Ravindran, C.L. Walk, and V. Ravindran. 2019. Studies on the measurement of ileal calcium digestibility of calcium sources in broiler chickens. *Poultry Science*. 98:5582–5589.

Davidson, W., R. Hepburn, J. Mathieson, and J.D. Pullar. 1968. Comparisons of heat loss from young cockerels by direct measurement and by indirect assessment involving body analysis. *British Poultry Science*. 9:93–109.

Dozier, W.A., B.J. Kerr, A. Corzo, M.T. Kidd, T.E. Weber, and K. Bregendahl. 2008. Apparent metabolizable energy of glycerin for broiler chickens. *Poultry Science*. 87:317–322.

Farrell, D. 1978. Rapid determination of metabolisable energy of foods using cockerels. *British Poultry Science*. 19:303–308.

Farrell, D.J., E. Thomson, J.J. Du Preez, and J.P. Hayes. 1991. The estimation of endogenous excreta and the measurement of metabolisable energy in poultry feedstuffs using four feeding systems, four assay methods and four diets. *British Poultry Science*. 32:483–499.

Finney, D.J. 1975. *Statistical Method in Biological Assay*, 3rd ed. London, UK: Charles Griffin & Co.

Fontaine, J., and M. Eudaimon. 2000. Liquid chromatographic determination of lysine, methionine, and threonine in pure amino acids (feed grade) and premixes: Collaborative study. *Journal of the AOAC International*. 83:771–783.

Fraps, G.S. 1928. Digestibility and production coefficients of poultry feeds. *Texas Agricultural Experiment Station Bulletin*. 372:1–24.

Fuller, M. 2003. AA bioavailability—A brief history. In *Digestive Physiology in Pigs. Proceedings 9th International Symposium*. Vol. 1. R.O. Ball, editor. Edmonton, Alberta, Canada: University of Alberta. 183–198.

Gabriel, I., S. Mallet, and M. Leconte. 2003. Differences in the digestive tract characteristics of broiler chickens fed on complete pelleted diet or on whole wheat added to pelleted protein concentrate. *British Poultry Science*. 44:283–290.

Guillaume, J., and J. Summers. 1970. Maintenance energy requirement of the rooster and influence of plane of nutrition on metabolizable energy. *Canadian Journal of Animal Science*. 50:363–369.

Haetinger, V.S., J.Y. Sung, S.A. Adedokun, W.A. Dozier, C.M. Parsons, M. Rodehutscord, and O. Adeola. 2024. Ileal phosphorus digestibility of soybean meal for broiler chickens remains consistent across institutions in a collaborative study regardless of non-phytate phosphorus concentration in the pre-experimental starter diet. *Poultry Science*. 103:103602.

Henry, P.R., and E.R. Miller. 1995. Iron bioavailability. In *Bioavailability of Nutrients for Animals*. C.B. Ammerman, D.H. Baker, and A.J. Lewis, editors. Amsterdam, The Netherlands: Academic Press. 169–199.

Hill, F., and D. Anderson. 1958. Comparison of metabolizable energy and productive energy determinations with growing chicks. *Journal of Nutrition*. 64:587–603.

Hill, F.W., D.L. Anderson, R. Renner, and L.B. Carew Jr. 1960. Studies of the metabolizable energy of grain and grain products for chickens. *Poultry Science*. 39:573–579.

Huang, K.H., V. Ravindran, X. Li, and W.L. Bryden. 2005. Influence of age on the apparent ileal amino acid digestibility of feed ingredients for broiler chickens. *British Poultry Science*. 46:236–245.

Kadim, I.T., P.J. Moughan, and V. Ravindran. 2002. Ileal amino acid digestibility assay for the growing meat chicken—comparison of ileal and excreta amino acid digestibility in the chicken. *British Poultry Science*. 43:588–597.

Khalil, M.M., M.R. Abdollahi, F. Zaefarian, and V. Ravindran. 2020. Measurement of ileal endogenous energy losses and true ileal digestible energy of cereal grains for broiler chickens. *Poultry Science*. 99:6809–6817.

Khalil, M.M., M.R. Abdollahi, F. Zaefarian, and V. Ravindran. 2021. Influence of feed form on the apparent metabolizable energy of feed ingredients for broiler chickens. *Animal Feed Science and Technology*. 271:114754.

Klasing, K.C., and D.R. Korver. 2020. Nutritional diseases. *Diseases of Poultry*. 1255–1285.

Kluth, H., K. Mehlhorn, and M. Rodehutscord. 2005. Studies on the intestine section to be sampled in broiler studies on precaecal amino acid digestibility. *Archives of Animal Nutrition*. 59:271–279.

Kong, C., and O. Adeola. 2014. Evaluation of amino acid and energy utilization in feedstuff for swine and poultry diets. *Asian-Australasian Journal of Animal Science*. 27:917–925.

Kratzer, D., and R. Littell. 2006. Appropriate statistical methods to compare dose responses of methionine sources. *Poultry Science*. 85:947–954.

Kuiken, K.A., and C.M. Lyman, 1948. Availability of amino acids in some foods. *Journal of Nutrition*. 36:359–368.

Lee, J., and C. Kong. 2019. Comparison of energy values estimated by direct and indirect methods for broiler chickens. *Poultry Science*. 18:244–248.

Lemme, A., D. Hoehler, J. Brennan, and P. Mannion. 2002. Relative effectiveness of methionine hydroxy analog compared to DL-methionine in broiler chickens. *Poultry Science*. 81:838–845.

Lemme, A., V. Ravindran, and W.L. Bryden. 2004. Ileal digestibility of amino acid in feed ingredients for broilers. *World's Poultry Science Journal*. 60:423–437.

Littell, R.C., A.J. Lewis, and P.R. Henry. 1995. Statistical evaluation of bioavailability assays. In *Bioavailability of Nutrients for Animals*. Amsterdam, The Netherlands: Academic Press. 5–33.

Lopez, G., and S. Leeson. 2007. Relevance of nitrogen correction for assessment of metabolizable energy with broilers to forty-nine days of age. *Poultry Science*. 86:1696–1704.

Lopez, G., and S. Leeson. 2008. Assessment of the nitrogen correction factor in evaluating metabolizable energy of corn and soy-bean meal in diets for broilers. *Poultry Science* 87:298–306.

Miles, R.D., and P.R. Henry. 2000. Relative trace mineral bioavailability. *Brazilian Animal Science*. 1(2):73–93.

Moughan, P.J. 2003. Amino acid digestibility and availability in food and feedstuffs. In *Digestive Physiology in Pigs. Proceeding 9th International Symposium*. Vol. 1. R.O. Ball, editor. Edmonton, Alberta, Canada: University of Alberta. 199–221.

Noblet, J., H. Fortune, C. Dupire, and S. Dubois. 1993. Digestible, metabolizable and net energy values of 13 feedstuffs for growing pigs: Effect of energy system. *Animal Feed Science Technology*. 42:131–149.

NRC. 1994. *Nutrient Requirements of Poultry*, 9th rev. ed. Washington, DC: National Academy Press.

NRC (National Research Council). 2012. *Nutrient Requirements of Swine*, 11th rev. ed. Washington, DC: The National Academies Press.

Nyachoti, C.M., C.F.M. de Lange, B.W. McBride, and H. Schulze. 1997. Significance of endogenous gut nitrogen losses in the nutrition of growing pigs: A review. *Canadian Journal of Animal Science*. 77:149–163.

Olojede, O.C., M.J. Ford, J.P. Jacob, T. Ao, A.J. Pescatore, and S.A. Adedokun. 2018. The effect of drying method temperature, collection method, and marker type on apparent ileal amino acid digestibility in 21-day-old broiler chickens fed corn-soybean meal-barley based diet. *Poultry Science*. 97:2106–2112.

Olukosi, O.A., S.A. Adedokun, and J.O. Agboola. 2017. Species-dependent response to the influence of adaptation length during assay for metabolisable energy of cereal grains employing the difference method. *Animal Feed Science and Technology*. 231:111–118.

Parsons, C.M. 2002. Digestibility and bioavailability of protein and amino acids. In *Poultry Feedstuffs: Supply, Composition and Nutritive Value*. J.M. McNab and K.N. Boorman, editors. Wallingford, Oxfordshire, UK: CAB International. 115–135.

Payne, W.L., G.F. Combs, R.R. Kifer, and D.G. Snider. 1969. Investigation of protein quality—ileal recovery of amino acids. *Federation Proceedings*. 27:1199–1203.

Peeler, H.T. 1972. Biological availability of nutrients in feeds: Availability of major mineral ions. *Journal of Animal Science*. 35:695–712.

Piepho, H.-P. 2006. A cautionary note on appropriate statistical methods to compare dose responses of methionine sources. *Poultry Science*. 85:1511–1512.

Poureslami, R., A.B. Batal, and B. Jung. 2012. Effect of ileal sub-section and the method of collection of digesta on the determination of apparent ileal digestibility of amino acids in broilers. *Animal Feed Science and Technology*. 177:130–133.

Raharjo, Y., and D.J. Farrell. 1984. A new biological method for determining amino acid digestibility in poultry feedstuffs using a simple cannula, and the influence of dietary fibre on endogenous amino acid output. *Animal Feed Science and Technology*. 12:29–45.

Ravindran, V., and W.L. Bryden. 1999. Amino acid availability in poultry—In vitro and in vivo measurements. *Australian Journal of Agriculture Research*. 50:889–908.

Ravindran, V., P.H. Selle, and W.L. Bryden. 1999. Effects of phytase supplementation, individually and in combination, with glycanase, on the nutritive value of wheat and barley. *Poultry Science*. 78:1588–1595.

Ravindran, V., Y.B. Wu, and W.H. Hendriks. 2004. Effects of sex and dietary phosphorus level on the apparent metabolizable energy and nutrient digestibility in broiler chickens. *Archives of Animal Nutrition*. 58:405–411.

Ravindran, V., O. Adeola, M. Rodehutscord, H. Kluth, J.D. van der Klis, E. van Eerden, and A. Helmbrecht. 2017. Determination of ileal digestibility of amino acids in raw materials for broiler chickens—Results of collaborative studies and assay recommendations. *Animal Feed Science and Technology*. 225:62–72.

Rochell, S.J., T.J. Applegate, E.J. Kim, and W.A. Dozier. 2012. Effects of diet type and ingredient composition on rate of passage and apparent ileal amino acid digestibility in broiler chicks. *Poultry Science*. 91:1647–1653.

Rodehutscord, M., M. Kapocius, R. Timmler, and A. Dieckmann. 2004. Linear regression approach to study amino acid digestibility in broiler chickens. *British Poultry Science*. 45:85–92.

Rodehutscord, M., A. Dieckmann, M. Witzig, and Y. Shastak. 2012. A note on sampling digesta from the ileum of broilers in phosphorus digestibility studies. *Poultry Science* 91:965–971.

Rodehutscord, M. 2013. Determination of phosphorus availability in poultry. *World's Poultry Science Journal*. 69:687–698.

Rodehutscord, M., O. Adeola, R. Angel, P. Bikker, E. Delezie, W.A. Dozier, M. Umar Faruk, M. Francesch, C. Kwakernaak, A. Narcy, C.M. Nyachoti, O.A. Olukosi, A. Preynat, B. Renouf, A. Saiz del Barrio, K. Schedle, W. Siegert, S. Steenfeldt, M.M. van Krimpen, S.M. Waititu, and M. Witzig. 2017. Results of an international phosphorus digestibility ring test with broiler chickens. *Poultry Science*. 96:1679–1687.

Rostagno, H.S., J.M.R. Pupa, and M. Pack. 1995. Diet formulation for broilers based on total versus digestible amino acids. *Journal of Applied Poultry Science*. 4:293–299.

Schulze, H., P. van Leeuwen, M.W.A. Verstegen, and J.W.O. van den Berg. 1995. Dietary level and sources of neutral detergent fiber and ileal endogenous flow in pigs. *Journal of Animal Science*. 73:441–448.

Scott, T.A., and F. Boldaji. 1997. Comparison of inert markers (chromic oxide or insoluble ash (Celite™)) for determining apparent metabolizable energy of wheat- or barley-based broiler diets with or without enzymes. *Poultry Science*. 76:594–598.

Shastak, Y., and M. Rodehutscord. 2013. Determination and estimation of phosphorus availability in growing poultry and their historical development. *World's Poultry Science Journal*. 69:569–586.

Shastak, Y., and M. Rodehutscord. 2015. Recent developments in determination of available phosphorus in poultry. *Journal of Applied Poultry Research*. 24:283–292.

Shastak, Y., M. Witzig, K. Hartung, W. Bessei, and M. Rodehutscord. 2012. Comparison and evaluation of bone measurements for the assessment of mineral phosphorus sources in broilers. *Poultry Science*. 91:2210–2220.

Shen, Y.B., P. Ferket, I. Park, R.D. Malheiros, and S.W. Kim. 2015. Effects of feed grade L-methionine on intestinal redox status, intestinal development, and growth performance of young chickens compared with conventional DL-methionine. *Journal of Animal Science*. 93:2977–2986.

Short, F.J., P. Gorton, J. Wiseman, and K.N. Boorman. 1996. Determination of titanium oxide added as an inert marker in chicken digestibility studies. *Animal Feed Science Technology*. 59:215–221.

Sibbald, I. 1975. The effect of level of feed intake on metabolizable energy values measured with adult roosters. *Poultry Science*. 54:1990–1997.

Sibbald, I.R. 1976. A bioassay for true metabolizable energy in feeding stuffs. *Poultry Science*. 55:303–308.

Sibbald, I.R., and M.S. Wolynetz. 1987. True and apparent metabolisable energy. *British Poultry Science*. 28:782–784.

Siriwan, P., W.L. Bryden, Y. Mollah, and E.F. Annison. 1993. Measurement of endogenous amino acid losses in poultry. *British Poultry Science*. 34:939–949.

Soares, J.H. 1995. Phosphorus bioavailability. In *Bioavailability of Nutrients for Animals: Amino Acids, Minerals, and Vitamins*. C.B. Ammerman, D.H. Baker, and A.J. Lewis, editors. Amsterdam, The Netherlands: Academic Press.

Spears, J., and S. Hansen. 2008. *Bioavailability Criteria for Trace Minerals in Monogastrics and Ruminants: Trace Elements in Animal Production Systems*. Wageningen, The Netherlands: Wageningen Academic.

Stein, H.H., C. Pedersen, A.R. Wirt, and R.A. Bohlke. 2005. Additivity of values for apparent and standardized ileal digestibility of AA in mixed diets fed to growing pigs. *Journal of Animal Science*. 83:2387–2395.

Stein, H.H., B. Seve, M.F. Fuller, P.J. Moughan, and C.F.M. de Lange. 2007. Amino acid bioavailability and digestibility in pig feed ingredients: Terminology and application. *Journal of Animal Science*. 85:172–180.

Titus, H.W., A.L. Mehring Jr., D. Johnson Jr., L.L. Nesbitt, and T. Tomas. 1959. An evaluation of MCF (Micro-Cel-Fat), a new type of fat product. *Poultry Science*. 38:1114–1119.

van Harn, J., J.W. Spek, C.A. van Vuure, and M.M. van Krimpen. 2017. Determination of pre-cecal phosphorus digestibility of inorganic phosphates and bone meal products in broilers. *Poultry Science*. 96:1334–1340.

Walk, C.L., L.F. Romero, and A.J. Cowieson. 2021. Towards a digestible calcium system for broiler chicken nutrition: A review and recommendations for the future. *Animal Feed Science Technology*. 276:114930.

WPSA Working Group No. 2. 2013. Determination of phosphorus availability in poultry. *World's Poultry Science Journal*. 69:687–698.

Wu, S.B., M. Choct, and G. Pesti. 2020. Historical flaws in bioassays used to generate metabolizable energy values for poultry feed formulation: A critical review. *Poultry Science*. 99:385–406.

Wu, S., R. Swick, J. Noblet, N. Rodgers, D. Cadogan, and M. Choct. 2019. Net energy prediction and energy efficiency of feed for broiler chickens. *Poultry Science*. 98:1222–1234

Yang, Z., V.R. Pirgozliev, S.P. Rose, S. Woods, H.M. Yang, Z.Y. Want, and M.R. Bedford. 2020. Effect of age on the relationship between metabolizable energy and digestible energy for broiler chickens. *Poultry Science*. 99:320–330.

Young, L., G. Ashton, and G. Smith. 1977. Estimating the energy value of some feeds for pigs using regression equations. *Journal of Animal Science*. 44:765–771.

Zuidhof, M.J. 2019. A review of dietary metabolizable and net energy: Uncoupling heat production and retained energy. *Journal of Applied Poultry Research*. 28:231–241.

9

Feed Processing

INTRODUCTION

The most widely used technologies for the manufacturing of processed feeds are the extrusion or expansion of individual raw materials and pelleting after grinding and mixing of the dietary ingredients. Extrusion processes consist of the application of high temperatures for short periods of time, with the aim of increasing the digestibility of the starch and protein fractions of the ingredient. During extrusion, the raw material is subjected to intense mechanical shearing through the action of the rotating screws, which disorganizes its original structure. Extrusion cooking might affect dietary fiber, especially for the soluble fraction. The feed is not only compressed as in pelletization, but also cooked, which requires higher levels of moisture, temperature, and pressure. In the case of diets, feed processing refers to any treatment prior to ingestion that alters the physicochemical nature of the ingredients to optimize nutrient utilization and varies from basic mixing of the ingredients to complicated manufacturing operations.

Most ingredients used in poultry diets are ground prior to feed production to reduce particle size and facilitate the mixing process. After grinding and mixing, the feed is offered directly to the birds as mash or can undergo further processes, such as pelleting and crumbling. Pelleting consists of forcing small particles (either individual ingredients or a complete mixed feed) through the holes of a die and typically involves the use of heat, moisture, and pressure to form larger particles. For meat-type birds, ground feedstuffs are commonly pelleted to facilitate feed handling and to improve growth performance. Pellets can be fed intact or crumbled in the case of young broilers and pullets. All of these operations modify the texture and composition of the final feed, thereby affecting feed intake (FI), development and function of the gastrointestinal tract (GIT), rate of feed passage, microbiota profile, fermentation processes, and nutrient digestibility (Engberg et al., 2002; Mateos et al., 2012; Svihus, 2014a). On the other hand, pelleting increases feed manufacturing cost, and consequently the economic benefits of its use must be justified.

FEED PROCESSING FLOW

Most feed manufacturing processes include the following steps: reception of the ingredients, grinding as needed, batching, mixing, hydrothermal processing, pelleting (plus crumbling in certain cases), cooling, and truck loading of the final feed. Upon arrival at the feed mill (as whole grains for cereals and most legumes, coarsely ground for cereal coproducts and protein sources from the oilseed industry, and fine powder for most mineral sources), the ingredients are cleaned as needed and stored as received. Then, most ingredients are ground before or after mixing to reduce particle size.

Hammer mills, roller mills, and multicracker systems are the equipment most commonly used for grinding (Svihus et al., 2004a; Lyu et al., 2020). Hammer mills reduce particle size by impacting slow-moving ingredients with a set of hammers moving at high speed, and they produce spherical-shaped particles with variable size distribution. Roller mills reduce particle size through a compression force between rotating pairs of rollers. Roller mills are more complicated to operate and have higher purchase and maintenance costs than hammer mills. However, rollers have lower specific energy consumption, reduce the proportion of fine particles, produce less noise and moisture, and decrease heat loss during the milling process. In addition, they generate less dust and result in a more uniform distribution of the feed particles than hammer mills (Koch, 1996; Heimann, 2002; Svihus et al., 2004a; Pérez-Bonilla et al., 2014). The multicracker system was developed more recently and is often used for manufacturing mash diets for laying hens in which large particle size, with a limited amount of fines, is the main objective.

After grinding, macroingredients, microingredients, and liquids are weighed and combined, and the resulting batch is mixed. The objective of the mixing process is to produce a mash in which ingredients, additives, and other components of the feed are uniformly distributed. After mixing, the mash can either be fed without further processing or can be pelleted. When pelleted, the mixture is conditioned by steam injection at high temperatures to improve the binding characteristics of the finely ground ingredients and facilitate their adhesion, thereby increasing pellet quality. Once conditioned, the mash passes through the pelleting chamber with a further reduction in particle size (Svihus et al., 2004a) and in the number of feed-borne pathogens (Van Immerseel et al., 2009; Stanley and Bajagay, 2022). Pellets will leave the pellet mill by gravity at a temperature as high as 80–88°C for a short time and a moisture content as high as 17–18 percent. The hot pellets are then cooled by applying a stream of air at ambient temperature that removes any excess of heat and moisture (Zimonja et al., 2007). For proper handling and storage, the final moisture content of the pellets should be less than 12 percent and the temperature about 9°C above atmospheric temperature (California Pellet Mill, 2012). When pellets of smaller size are required (e.g., very young chicks, turkeys, ducks, and pullets), it is desirable to use a crumbling process. A crumble is a pelleted feed that has been broken down between two corrugated rollers, placed usually below the cooler, to reduce its size to better accommodate to the small size of the beak of young birds. From the crumbling equipment the product might pass through a shaker device to separate the undesirable, undersized portions (fines) from the correctly sized material. After cooling, the feed is stored or sent directly to the farms as required. None of the processes described is uniform, and the benefits of pelleting on FI and poultry production vary with the texture and chemical composition of the mixture; friction, heat, and pressure applied during the process; mean particle size; uniformity of the final product; and target species.

GRINDING PROCESS

Grinding is the second most expensive process in feed manufacturing. The majority of the ingredients used in poultry diets require some degree of grinding to improve mixing uniformity, decrease segregation after mixing, facilitate the apprehension of the feed by the bird, and reduce the cost of the pelleting process. In addition, particle size plays a critical role on FI, GIT development and health, and the digestion and absorption of the nutrients (Carré, 2000; Amerah et al., 2007a,b).

Small particles increase the surface area available for the interaction between nutrients and exogenous and endogenous digestive enzymes (Preston et al., 2000; Behnke, 2001; Amerah et al., 2008; Ovi et al., 2021). Fine grinding, together with the friction and the steam and pressure applied during the pelleting process, breaks the cell walls, increasing the accessibility of the enzymes to the encapsulated nutrients. Consequently, the benefits of fine grinding are greater for ingredients with highly protected starch, such as hard wheat and peas, than for ingredients with less protected starch, such as soft wheat and rice (Svihus and Hetland, 2001; Carré et al., 2005; Parera et al., 2010; Abdollahi et al., 2011; Yegani and Korver, 2012). Fine grinding also facilitates the release of the lipids contained in the oil bodies of certain ingredients such as heat-processed soybeans and corn, thereby increasing fat digestibility (Valencia et al., 2009a,b; Abdollahi et al., 2013a). In this respect, the WPSA (1989) recommends to increase the energy content of the processed soybeans by 230 kcal AME_n/kg DM in pelleted diets. However, the higher percentage of small particles that results from very fine grinding impairs the physical characteristics and texture of the mixture, thereby reducing FI when the feed is fed as mash (Herrera et al., 2018a,b; Mateos et al., 2019). Birds usually reject very fine particles when fed mash diets or poor-quality pellets, and consequently large particle are preferentially consumed (Herrera et al., 2018a,b; Ben- Mabrouk et al., 2022, 2023).

The potential benefits of fine grinding of the ingredients on nutrient utilization, is less relevant in poultry than in mammals because, irrespective of the size of the original ingredient, particle size is markedly reduced by the mechanical action of the gizzard (Hetland et al., 2002; Amerah et al., 2007c; Svihus et al., 2011). In fact, the gizzard sorts large particles such as coarse feedstuffs (e.g., grits, whole wheat, and cracked corn) and insoluble, lignified sources of fiber, such as oat hulls, for retention and further processing from very small particles and liquids that are passed along to the small intestine. The large particles that are retained in the gizzard increase the frequency and strength of the contractions of the muscular layers and the intensity of the antiperistaltic movements, improving GIT function (Svihus et al., 2004a, 2005; Amerah et al., 2007a, 2008; Mateos et al., 2012; Lv et al., 2015; Kheravii et al., 2018; Rodrigues and Choct, 2018; Abdollahi et al., 2019; Jiménez-Moreno et al., 2019; Lemons et al., 2019). As a consequence, gizzard pH decreases and pepsin activity, phytate solubilization, and nutrient digestibility increase (Duke, 1986; Angel et al., 2002; Serrano et al., 2013; Jiménez-Moreno et al., 2016; Mateos et al., 2019; Wang-Li et al., 2020). Consequently, under high incidence of digestive disorders, including wet litter problems, the use of coarsely ground diets is recommended. In contrast, fine particles are not retained in the gizzard and the digesta passes rapidly through the upper part of the GIT, which increases FI and BW gain, improving feed efficiency in healthy birds (Amerah et al., 2007a,b) and pullets (Frikha et al., 2009a; Saldaña et al., 2015a,b).

Particle Size and Gastrointestinal Tract Development

Feed processing affects gut refluxes and the size and development of all the organs of the GIT, with more pronounced reducing effects on the weight of gizzard (Amerah et al., 2007b; Zaefarian et al., 2016) and pancreas (Engberg et al., 2002; Röhe, 2013; Abadi et al., 2019; Abd El-Wahab et al., 2020) and on the length of the small intestine (Nir et al.,

1994; Frikha et al., 2009b). A reduction in gizzard weight in birds fed finely ground diets decreases the activity of this organ and increases the passage rate of the ingested feed, which may result in "overeater birds" (Svihus and Hetland, 2001). In addition, finely ground particles, such as those associated with the feeding of high-quality pellets, significantly reduce the relative weight and the activity of the pancreatic enzymes (e.g., amylase, lipase, trypsin, chymotrypsin) in broilers, as compared with coarse particles. All these effects indicate the existence of a feedback mechanism, probably triggered by the intestinal concentration of enzymatically hydrolyzed products or of the respective digestive enzymes (Nir et al., 1994; Engberg et al., 2002).

The gizzard is the organ that controls and regulates the flow of the digesta from the proximal part of the GIT to the duodenum and throughout the entire GIT (Duke, 1992; Svihus, 2011). A well-developed gizzard ensures a more complete grinding of the feed; facilitates the refluxes and the mixing of the chyme and the gastric juices, increasing nutrient digestibility in the small intestine; stabilizes the profile of the intestinal microflora; and favors water reabsorption in the large intestine (Amerah et al., 2007b; Svihus, 2011; Rodrigues and Choct, 2018). Reverse peristaltism possibly serves the bird as a way of prolonging the exposure of the digesta to the action of the enzymes and microbial action and absorption (Sacranie et al., 2007; Rodrigues and Choct, 2018). Structural components, such as coarse particles, potentiate digestive refluxes prolonging the exposure of food to both mechanical and chemical components of the digestion process. Duke (1994) highlighted that reverse peristalsis occurs in three main areas of the GIT: (1) from the gizzard to the proventriculus due to the gizzard contractions, which increases digesta exposure to proventricular enzymes; (2) from the jejunum and duodenum back into the gastric region, which is enhanced in fasted birds; and (3) from the cloaca to the ceca, as a continuous process due to the low amplitude contractions of the intestinal mucosa. The importance of the digestive refluxes in the GIT in poultry was confirmed by Sacranie et al. (2012), who reported the presence of chromium in all the intestinal tract sections, including the crop, when chromium EDTA was introduced via the cloaca.

Whole Wheat

The inclusion of whole wheat (WW) in the diet is a common practice that reduces the cost and improves GIT health in broilers (Bennett et al., 1995, 2002; Ferket, 2000; Svihus et al., 2004b; Gabriel et al., 2008; Singh, 2013, 2019; Liu et al., 2015; Abdollahi et al., 2019). The benefits of WW feeding are also apparent, but less studied in turkeys (Jankoswki et al., 2016a,b; Ahmed at al., 2018), pullets (Foster and Taha, 1978), and laying hens (Kermanshahi and Classen, 2001; Hetland et al., 2003; MacIsaac and Anderson, 2007; Koçer et al., 2016; Mtei et al., 2019). Whole wheat is retained for longer in the gizzard than finely ground wheat, increasing transit time of the digesta through the upper part of the GIT and increasing gizzard function, HCl production, and villus height-to-crypt depth ratio (VH:CD) (Qaisrani et al., 2022). As a result, WW supplementation often modifies the profile of the microbiota of the small intestine and the ceca, thereby improving litter quality. In contrast, WW supplementation reduces diet uniformity and increases the selection of coarse particles by the bird, which might affect FI. Consequently, WW supplementation requires precise control of the mixing process (e.g., distribution of pelleted feed and grain in the mixture) and of the risk of microbial contamination of the final feed, caused by the native bacterial and fungi content of the unprocessed grain. Basically, as shown previously, WW supplements the diet with very coarse particles, and consequently, provides to the birds the same benefits (and challenges) as the use of large particles.

The main strategy used to incorporate WW into the diet consists of the substitution, on a weight-to-weight basis, of part of the finely ground wheat of the pellet diet by WW added post-pelleting. However, in order to reduce feed cost, the WW and the pelleted diet might be mixed before transport to the farm or added directly on top of the pellets, diluting the nutrient content of the original feed. Under this last strategy, when substantial amounts of WW are added as a supplement of the original diet, growth performance might decrease because of a deficiency in nutrients such as Lys and other essential amino acids, P, and Ca (Amerah and Ravindran, 2008; Kadardar et al., 2022). This could be circumvented by formulation of the diet so that the consumption of the supplemental WW is accounted for. The benefits of WW supplementation on broiler production depend on factors such as (1) age, with young birds preferring pellets to wheat grain; (2) inclusion level of WW, especially when the grain is added on top, in diets with small safety margins for key limiting nutrients; (3) learning procedure, with better results when the amount of supplemental WW increases with age; (4) health status of the birds, with greater benefits in commercial farms under poor management conditions; and (5) poor mixing of pellets and WW at the feed mill, during transport, or at the farm (Bennett et al., 2002; Engberg et al., 2004; Singh and Ravindran, 2019). In addition, when WW is fed continuously to slaughter, the risk of carcass contamination during processing increases, because whole grain is retained for longer in the gizzard than pelleted diet (Qaisrani et al., 2022).

In addition to WW, cracked corn and cracked or whole barley grains can also be used to stimulate gizzard function and reduce feed cost (Svihus et al., 1997; Bennett et al., 2002; Parsons et al., 2006; Singh and Ravindran, 2019; Ovi et al., 2020; Wang-Li et al., 2020). In this respect, Clark et al. (2009) reported same performance of broilers fed diets with up to 25 percent cracked corn than of broilers fed ground corn. Similar or even beneficial results have been reported by Bennet et al. (2002), Singh et al. (2014), Jelveh et al. (2020), and Tari et al. (2022) with the use of whole barley and cracked or coarsely ground corn. Under most circumstances, large particles are retained for longer in the proximal part of the GIT, slowing rate of passage and thus improving AME and N retention at the expense of a reduction in feed intake.

In summary, the inclusion of WW (or cracked corn) in the diet is a sound strategy that helps to improve the development of the GIT and the productivity of commercial poultry integrations. A poor uniformity of the final diet (a mixture of WW and pelleted diet), especially in young birds, and the lack of training might reduce FI in healthy birds, a problem that is less relevant in flocks affected by digestive disorders.

STEAM CONDITIONING AND HYDROTHERMAL PROCESSING

Hydrothermal processing (HTP), including expansion, extrusion, and pelleting, impacts GIT function and nutrient availability while reducing bacterial load and the profile of the feed-borne microbiota (Kiarie and Mills, 2019; Córdova-Noboa et al., 2020). The final effects of HTP on bird production depend on factors such as (1) the physicochemical characteristics of the cereals and other ingredients of the diet, including the presence of anti-nutritional factors in the feed; (2) the nutrient considered (e.g., protein, starch, lipid, and fiber fractions); (3) the initial microbial load of the feed; (4) the amount of heat applied during the process (e.g., balance among the steam, moisture, and mechanical pressure), and (5) the type of bird, including species, age, and GIT health (Plavnik and Sklan, 1995; Abdollahi et al., 2010a,b, 2013b; Frikha et al., 2013; Goodarzi Boroojeni et al., 2016).

Heat processing of cereals at temperatures above 90°C is a common practice to increase nutrient digestibility and feed efficiency in piglets (Medel et al., 2004). However, the benefits of the use of cooked cereals or the application of extra heat to the feed prior to pelleting, in terms of nutrient availability and growth performance, are not well documented in poultry. Many factors, such as the cereal tested, the gradient of temperatures, the time and moisture applied, and the methodology used to process the cereal (e.g., expansion vs. extrusion vs. micronization), affect the response of the bird to HTP (Gracia et al., 2003; Abdollahi et al., 2010b; Kianfar et al., 2013). In fact, the potential benefits of high temperatures (e.g., extrusion and expansion processes) on the ileal digestibility of the protein and starch fractions of ingredients and diets are elusive in broilers (Mateos et al., 2002; Svihus et al., 2004a; Moritz et al., 2005; González-Alvarado et al., 2007; Jiménez-Moreno et al., 2009; Bergeron et al., 2018; Zentek and Goodarzi Boroojeni, 2020) and turkey poults (Plavnik et al., 1997) after the first 2 to 4 weeks of life. Positive effects of heat processing of the cereal on nutrient retention are sometimes reported in young birds fed corn (González-Alvarado et al., 2007; Gracia et al., 2009; Jiménez-Moreno et al., 2009) whereas no effects or even a reduction in nutrient digestibility is observed when fed wheat (Abdollahi et al., 2011), barley (Herstad and McNab, 1975), sorghum (Abdollahi et al., 2010b), or rice (González-Alvarado et al., 2007; Jiménez-Moreno et al., 2009). Heat processing of cereals with a high content of β-glucans and xylans, such as barley and wheat, increases digesta viscosity and thus will reduce nutrient digestibility. The negative effect of an excess of heat might be compensated by the addition of adequate levels of exogenous enzymes (Vukic Vranjes et al., 1995).

The changes that occur in the structure of the starch during the pelleting process have been extensively studied (Moritz et al., 2005; Svihus et al., 2005; Abdollahi et al., 2011; Muramatsu et al., 2015). Gelatinization markedly increases the susceptibility of the starch to amylolytic degradation due to the loss of the crystalline structure. During steam conditioning and pelleting, however, the gelatinization process affects only to 10 to 20 percent of the starch present (Svihus et al., 2004a, 2014b), which implies that the pelleting process might not have a marked effect on the final digestibility of the dietary starch. In fact, under most practical situations, heat processing of the cereals results in little effect on growth performance, and in most cases the benefits disappear with age (Gracia et al., 2003, 2009; González-Alvarado et al., 2007; Kianfar et al., 2013). As a general rule, severe HTP conditions for a prolonged time reduce nutrient digestibility, as documented subsequently.

Protein

Under moderate HTP conditions, the spatial structure of the protein chains is denatured, which facilitates the access of proteolytic enzymes to nutrients, increasing amino acid digestibility. In contrast, an excess of heat increases the incidence of Maillard reactions which reduces the availability of Lys and other essential amino acids (Abdollahi et al., 2020).

Starch

Hydrothermal processing of ingredients and feeds modifies the secondary structure of the glucose chains, deactivates potential amylose inhibitors present in the feed, and disrupts the protein/lipid matrix that protects the starch granules from degradation (Grant et al., 1995). The effects of HTP on swelling, gelatinization, and digestibility of the starch are greater for corn than for wheat and rice and greater for peas and other legume grains than for cereals (Abd El-Khalek and Janssons, 2010). For corn, HTP disrupts existing lipid/amylose complexes and bonds between the starch and the protein matrix, increasing starch digestibility. The digestibility of rice starch in its native state, however, is almost complete, and therefore no increases in starch digestibility due to HTP are expected (González-Alvarado et al., 2007, 2008; Vicente et al., 2008). On the other hand, an excess of heat applied to diets based on wheat or barley increases the viscosity of the digesta, reducing starch digestibility especially when the diets are not supplemented with carbohydrases. In this respect, Abdollahi et al. (2011) reported that starch digestibility in wheat was reduced from 98 percent to 91 percent as the temperature of the heating process increased from 60°C to 90°C.

Lipids

Heat application and the mechanical pressure are used disrupt the structure of the cell walls and release the lipids in ingredients such as corn, rapeseeds, and soybeans, increasing their digestibility (Shen et al., 1983; Kan et al., 1988; Abdollahi et al., 2013a). A moderate excess of heat during the pelleting process, however, does not affect the stability and digestibility of the lipid fraction of most commonly used feed ingredients (Lehtinen et al., 2003; Parkash, 2015; Ali et al., 2017).

Fiber

Heating modifies the structure and molecular weight of the soluble fiber fraction of the diet, thereby increasing digesta viscosity and reducing nutrient digestibility. Consequently, an excess of heat is more deleterious in diets based on cereals with a high non-starch polysaccharides content (e.g., β-glucans and xylans), such as rye, wheat, and barley, than in diets based on corn or sorghum (Lázaro et al., 2003; Goodarzi Boorojeni et al., 2016). Starch and saturated lipids are the ingredients most affected by an increase in digesta viscosity, with effects being less evident and even absent when adequate exogenous non-starch polysaccharide-degrading enzymes are supplemented (Dänicke et al., 1997; Amerah et al., 2008; Kianfar et al., 2013).

Macrominerals

Naderinejad et al. (2016) reported that pelleting reduces the apparent ileal digestibility of Ca and P but increases that of Na in broilers. The mechanism for these contrasting effects is unknown but might be related to the decrease in HCl production by the proventriculus and the increase in gizzard pH caused by pelleting. The effects of heat processing on mineral availability depends on numerous factors, including the ingredient considered and the conditions of the process applied. In this respect, Abdollahi et al. (2013b) reported that HTP increased the ileal digestibility of Ca and P in corn-based diets but not in wheat-based diets, a difference that was probably related to the variable effects of the heat applied on digesta viscosity. Phosphorus bioavailability is higher for the distiller's dry grain solubles than for the original cereal grain, an effect that might have been caused by the fermentation processes that take place as well as by the high temperature applied during the ethanol production process (Salim et al., 2010).

Vitamins and Additives

Feed processing at high temperature reduces the stability and the activity of many vitamins, such as vitamin A, thiamine, pyridoxine, folic acid, and cyanocobalamine, and of additives such as enzymes and probiotics (Coelho, 2002; Riaz et al., 2009; Abdollahi et al., 2013). When unprotected, the vitamins most affected by the high humidity and the oxidative stress caused the pelleting process are vitamin A, D, K, and C and thiamine. In practice, current protection methods (e.g., microencapsulation and chemical form) reduce to a minimum the loss in activity of most vitamins in commercial feeds (Yang et al., 2020). Most of the losses in vitamins' and additives' stability takes place after feed manufacturing because the pressure, friction, and steam applied during the pelleting process might damage the matrix that protects the vitamins, exposing them to a loss of activity as high as 30–50 percent (Coelho, 2002; McDowell and Ward, 2008; Goodarzi Boroojeni et al., 2016).

Anti-Nutritional Factors

Overheating inactivates some of the thermo-labile, anti-nutritional factors present in ingredients (e.g., lectins and trypsin inhibitors), which interfere with metabolic processes and nutrient digestion (Goodarzi Boroojeni et al., 2016; Bergeron et al., 2018). However, the heat applied during the pelleting process is usually below 85°C, a temperature that under most practical situations will have little impact on the degradation of most anti-nutritional factors present in poultry feeds.

In summary, heat processing is expensive because it requires high levels of energy and the use of sophisticated equipment, which increases cost. Altogether, the benefits of HTP on nutrient digestibility might not justify the cost under most practical situations. Consequently, the use of heat processed ingredients in poultry diets should be carefully evaluated.

PELLETING

Pelleted feeds have been defined as "agglomerated feeds formed by extruding a mixture of ingredients by compacting and forcing through die openings by any mechanical process." The main purpose of pelleting is to take a finely ground, sometimes dusty, unpalatable, and difficult to handle feed material, to form large particles (California Pellet Mill, 2012). Pellets are formed as the feed mash passes through a metal die that compacts the mixture, usually of finely ground ingredients, under the combined effects of heat, moisture, and pressure. Pelleting affects the size, development, and function of the GIT and compared to mash diets increases BW gain and feed efficiency in broilers under most circumstances (Svihus et al., 2005; Serrano et al., 2012; Abdollahi et al., 2013a, 2014; Jiménez-Moreno et al., 2016; Naderinejad et al., 2016), turkeys (Proudfoot and Hulan, 1982; Favero et al., 2009; Flores et al., 2020), and young pullets (Frikha et al., 2009a,b; Guzmán et al., 2015; Saldaña et al., 2015a; Bozkurt et al., 2019). The benefits of pelleting on poultry performance are less evident in laying hens, especially after reaching the peak of egg production (Hamilton and Proudfoot, 1995; Röhe et al., 2014; Wan et al., 2021) and of limited interest in pullets from 5 to 17 weeks of age (Saldaña et al., 2016). Pelleting might help to reduce the microbial load of the feed, but in feed restricted pullets and broiler breeders, it intensifies the competition for feed among birds, reducing BW uniformity during the rearing and laying phases (Zuidhof et al., 2015; Gous and Danisman, 2016).

The increase in BW gain and feed efficiency observed in broilers, turkeys, and other poultry species when fed pellets are explained by its positive effects on (1) voluntary FI (Nir et al., 1994; Serrano et al., 2012; Jiménez-Moreno et al., 2016; Abdollahi et al., 2018), (2) reduction of feed wastage between feed manufacturing and the beak of the chick (Serrano et al., 2013; Saldaña et al., 2015b; Jiménez-Moreno et al., 2016), and (3) less energy required for physical activity and feed procurement (Jensen, 2000; Skinner-Noble et al., 2005). Other benefits of pellet feeding are (1) improved feed hygiene, (2) less fines and better flowability of the diets, (3) preventing the choosing between ingredients and the selection of coarse particles by the bird, (4) improvement of the physicochemical characteristics of the feed, (5) easier access of the enzymes present in the GIT to the nutrients of the digesta, and (6) increased bulk density, which results in better handling of the feed at the feed mill, improving transport, and a reduction in the cost of the farm logistic process.

Feeding crumbles rather than pellets might be appropriate in broilers, turkeys, ducks, and pullets for the first 2 to 3 weeks of age. Serrano et al. (2012, 2013) compared the growth performance of broilers fed micro-pellets (2 mm Ø), crumbles, or mash feeds in two trials (floor pen and battery). The data show a clear advantage of the pellet and crumble feeds over the mash feed for FI and BW gain. Also, crumbling and pelleting improved feed efficiency, a benefit that was more pronounced in the trial conducted on floor pens than in the trial conducted in battery cages because in the first experiment feed wastage was not recorded.

Pelleting and Digestive Tract Function

The GIT of the birds, including the gizzard, is immature at birth but develops fast when diets with a high insoluble fiber content in mash form are fed (Choi et al., 1986). As described above, the gizzard plays an important role in bird physiology, reducing particle size and controlling the flow of the digesta to the small intestine, which aids in the digestion of most nutrients. In contrast, HCl production is reduced in broilers with poorly developed gizzards with a decrease in the retention time of the digesta in the foregut and less opportunities for exogenous and endogenous enzymes action (Engberg et al., 2002; Amerah et al., 2007b; Svihus, 2011, 2014a).

The potential benefits of using pellets versus mash diets in poultry depend on the step of the digestive process. From feed manufacturing to the beak of the bird, feed form is more important than particle size. In the feed acquisition step, birds eat more and faster, rest for longer, and waste less feed with pellets than with mash diets (Serrano et al., 2012, 2013). Consequently, pellet feeding is more advantageous than mash feeding during this period. In the second step (from the beak of the bird to the cloaca), once the bird ingests the feed, pellets disintegrate and consequently only the particle size of the digesta is of relevance (Mateos et al., 2012). The fine particles of the pelleted diets pass more quickly from the gizzard to the small intestine than coarse particles. Consequently, the birds feel hungrier when fed pellets than when fed mash, which results in an increase in FI (Preston et al., 2000; Mateos et al., 2012; Svihus, 2014b). As described subsequently, the effects of particle size on the development of the gizzard and the GIT, change quickly soon after the feed texture is modified. The proximal part of the GIT adapts quickly to an increase of the coarseness of the diet by increasing gizzard size and the production of HCl in the proventriculus (Svihus, 2011; Saldaña et al., 2015b), with an opposite effect observed when the percentage of large feed particles is reduced (Saldaña et al., 2015a).

Pelleting and Nutrient Digestibility

Research conducted after the last edition of the *Nutrient Requirements of Poultry* (NRC, 1994) has shown little benefit and even a reduction in the digestibility of starch, protein, and other nutrients in broilers fed diets in pellet form (Svihus, 2001, 2011, 2014b; Svihus et al., 2005; Zimonja et al., 2007; Abdollahi et al., 2011, 2013, 2014; Mateos et al., 2012; Serrano et al., 2013; Jiménez-Moreno et al., 2016, 2019). Most research available shows that the improvement in feed efficiency observed in broilers fed pellets is not a consequence of better nutrient digestibility but of an increase in FI caused by the fast emptying of the gizzard, a reduction in energy requirements for feed procurement, and a decrease in feed wastage. In this respect, Plavnik et al. (1997) compared feeding pelleted or mash diets to broilers and turkey poults. The authors reported an advantage in growth performance in both species when the diets were pelleted. However, the grinding of the pellets completely abolished the growth and feed efficiency response observed when the original physical form was preserved.

Pellet Quality

Pellet quality is defined as the ability to resist fragmentation and abrasion during handling without breaking up, and to reach the feeders with absence of fines (Briggs et al., 1999; Muramatsu et al., 2015). Physical quality of the pellets, with a texture adequate to the sensibility of the mechanical receptors present in the beak, is critical to optimize feed efficiency and growth response in broilers and turkey poults (Proudfoot and Hulan, 1982; Behnke, 2001; Flores et al., 2021). Fine particles improve pellet quality and consequently, fine grinding of the ingredients is usually preferred when the diets will be fed as pellets, unless GIT health is an issue. Durability and hardness are the two main physical indicators of the physical quality of the pellets. Durability measures indirectly the amount of fines produced during transport from the feed mill to the farm and during the distribution of the feed to the birds. Pellet durability index (method 52694; ASAE, 2003) is the main method used by the industry to determine pellet quality, as it indicates the percentage of pellets that remain intact after being submitted to mechanical forces. Pellet hardness, expressed as Newtons, measures the resistance of intact pellets, without any visible deformation, to breakage when submitted to external pressure (Thomas and Van der Poel, 1996).

Because of the friction created as the feed passes through the die holes, the pelleting process further reduces the particle size of the mixture, increasing the presence of fine particles in the foregut, which might affect GIT function (Amerah et al., 2007b; Naderinejad et al., 2016). Other adverse situations in relation to pellet quality of diets fed to young broilers are (1) pellets not optimized in diameter and length to the size of the chick's beak (Cerrate et al., 2009; Serrano et al., 2013) and (2) manufacturing of hard pellets based on wheat with a low moisture content. In practice, pellet durability, which improves with the inclusion of wheat in the diet, the reduction in particle size of the ingredients, and the increase in the temperature of the steam used in the process, increases the performance of meat-type birds. These three factors, however, contribute also to increase the hardness of the pellets, which might negatively affect FI and feed wastage (Abdollahi et al., 2010b).

Pellet quality is relevant not only because of its effects on broiler growth and feed efficiency but also for its important role in feed marketing. High-quality pellets are obtained often at the expense of nutritional optimization. In fact, an increase in steam temperature usually improves the physical quality of the pellets but reduces nutrient digestibility, resulting in variable end effects on broiler performance (Abdollahi et al., 2010b). Under most commercial practices, farm managers give preference to the quality of the pellets over the health status of the birds (Amerah et al., 2007c; Naderinejad et al., 2016). However, increased physical quality of the pellets does not always justify the potential losses in GIT tract health and the cost of the process. Consequently, a compromise between pellet quality and GIT function and health, nutrient availability, and marketing issues, is needed.

FEED FORM AND OTHER AREAS OF INTEREST

The effects of feed pelleting on the control of feed-borne microbial contamination, villus development and mucosa integrity, water intake and litter quality, bird mortality, and interactions between feed form and energy density of the diet, are relevant for the poultry industry, and are summarized in the following sections.

Pelleting and Feed-Borne Microbial Contamination

Pellet feeding affects the microbial load and profile of the digesta, effects that vary according to the hygienic conditions of the farm, the health status of the birds, the physicochemical characteristics of the diet, and the conditions applied during the pelleting process (Kiarie and Mills, 2019). Hydrothermal heat processing has a considerable impact on feed hygiene by reducing microbial load in the GIT and increasing the concentration of *Lactobacilli* (and lactate production) in the proximal part of the GIT, particularly in the crop (Goodzari Boorojeni et al., 2016). These authors reported that HTP (e.g., expansion, extrusion, and pelleting) improved the hygiene status of the feed with positive effects on the morphology and microbiota profile of the GIT. Engberg et al. (2002) reported that pellet-fed birds have larger numbers of coliform bacteria in the ileum and a reduced number of *Clostridium perfringens* in the distal end of the large intestine (caeca and rectum), with an increase in the concentration of VFA in the cecum.

Strategies available to control microbial contamination along the production chain, include coarse grinding of ingredients independently of feed form (Engberg et al., 2012), WW supplementation (Bjerrum et al., 2005), HTP of ingredients and diets (Goodarzi Boroojeni et al., 2016; Abadi et al., 2019), inclusion in the feed of inert fiber sources in adequate proportions (Mateos et al., 2012), and the supplementation of the diet with organic acids (Khan et al., 2022). A well-developed gizzard, a consequence of feeding diets based on coarse particles or of the inclusion of inert fiber in the diet, increases the intensity of the antiperistaltic movements of the GIT and reduces digesta pH, thereby preventing the entry of pathogenic bacteria to the small intestine and their adherence to the digestive mucosa. Moreover, the colonies of *Clostridium perfringens* and *Enterobacteriacea* spp. in the intestines are reduced (Engberg et al., 2004; Abadi et al., 2019) while microbiota diversity improves, and as a consequence the risk of enteric disorders decreases (Engberg, 2002; Bjerrum et al., 2005). In this respect, Bjerrum et al. (2005) reported that WW feeding reduced significantly the number of *Clostridium perfringens* in the intestinal tract of the birds. Moreover, Engberg et al. (2004) and Singh (2013) reported an increase in caecal counts of beneficial bacteria such as *Lactobacilli* spp. and *Bifidobacteria* spp. and a decreased count of *Clostridium* spp. and other pathogens with WW supplementation of broilers' diets.

Contamination of the feed, digestive tract, litter, eggs, and poultry meat by *Salmonella* spp. and other pathogens is a major concern for the poultry industry (Jones et al., 1991; Jones, 2011). Large feed particles promote the development of the GIT and reduce the concentration of pathogens such as *Salmonella* spp. and *Clostridium* spp. in the GIT of the birds (Bjerrum et al., 2005; Yegani and Korver, 2008). Bjerrum et al. (2005), in an experiment with broilers infected with a rifampicyn-resistant *Salmonella typhimurium* strain, reported that WW feeding influenced the course of infection, which was probably related to changes in the pH of the gizzard contents, which decreased as the amount of WW in the diet increased over time. Following infection, lower numbers of *Salmonella typhimurium* were found in the gizzard and ileum of birds when the pelleted diet was supplemented with WW.

Time, temperature, pressure, and moisture applied during the conditioning process, and the resistance to heat of the species present, affect the growth of *Salmonella* spp. Pelleting above 80–85°C reduces *Salmonella* colonies in feed by 82–93 percent (Jones et al., 1991; Veldman et al., 1995; Van Immerseel et al., 2009; Jones, 2011; Goodarzi Boorojeni et al., 2016) but do not eliminate microbial contamination of the feed (Goodarzi Boorojeni et al., 2016). When the objective is the complete elimination of feed-borne pathogens, including *Salmonella* spp., *Clostridium perfringens*, and spore-forming bacteria, a minimum temperature of 90°C is needed. However, high temperatures are detrimental for nutrient availability, especially when the birds are fed wheat-based diets.

Post-pelleting recontamination of the feed during the cooling, transportation, delivery, and storage processes is frequent, especially when the coolers are dirty with presence of dust and condensed water inside the equipment (Jones and Richardson, 2004; Jones, 2011). Cleaning and the use of adequate levels of organic acids reduce the incidence of recontamination of the pellets, especially of non-spore-forming bacteria (Khan et al., 2022).

In summary, the interest to achieve a high physical quality of the pellets to improve the acceptance of the feed by the farmers and to reduce feed-borne contamination by pathogens such as *Salmonella* spp. and *Clostridium* spp. has led to the use of high steam conditioning temperatures in poultry diets. However, the effects of thermal processing of the feeds during the pelleting process on nutrient availability are equivocal, with minimal impact on starch and protein digestibility and on the inactivation of anti-nutritional factors. On the other hand, when high temperatures are applied, the stability of heat-sensitive additives (e.g., vitamins and feed enzymes) and the digestibility of many nutrients will decrease. The use of coarse particles in conveniently heated mash or pellet diets improves GIT function and reduces the adherence of pathogens to the mucosa of the intestines, decreasing microbial contamination and improving the health status of the GIT.

Particle Size, Pelleting, and Mucosa Development

The intestinal mucus plays an important role in protecting the epithelial cells against pathogens and might be affected by particle size and HTP of the ingredients and by pelleting of the diet. During the digestion process the sloughed mucus can be utilized by the bacteria in the distal part of the GIT, increasing the endogenous losses of protein and energy. Duangnumsawang et al. (2021) reported that HTP of the feed might reduce this loss by decreasing mucus shedding into the lumen.

Villus height to crypt depth ratio (VH:CD) of the intestinal mucosa of broilers usually increases when (1) the diet is fed as mash rather than as pellets, (2) the ingredients are coarsely ground, and (3) the main cereal of the diet is heat processed (Gracia et al., 2003; Naderinejad et al., 2016; Mtei et al., 2019; Wan et al., 2021). Qaisrani et al. (2015) studied the effects of feed particle size on the development of the GIT and villi characteristics in broilers. An increase in diet coarseness increased villus height by 19 percent and decreased crypt depth by 18 percent, and as a consequence, the VH:CD was 48 percent greater in chickens fed the coarsely ground diet than in chickens fed the fine diet. Sogunle et al. (2013) reported 89 percent greater villus height in the duodenum of broilers with an increase in the feed particle size of the diet from 1 to 2 mm. Ravindran et al. (2006), Gabriel et al. (2008), Liu et al. (2015), and Jankoswki et al. (2016a) reported that VH:CD and the surface of the duodenum available for nutrient absorption of meat-type birds increased when the diet was supplemented with WW. Abadi et al. (2019) reported greater villus height and deeper crypt depth in the small intestine in broilers fed a pellet diet than in broilers fed a mash diet.

In laying hens, Röhe et al. (2014) reported an increase in the duodenal VH:CD in hens fed mash feeds compared to hens fed expanded feed. Wan et al. (2021) studied the effects of feed form on the intestinal morphology of two strains of laying hens. Compared to mash diets, pelleting increased villus height and crypt depth in the small intestine and VH:CD in the duodenum in both strains, but no effects were observed in the jejunum or in the ileum. The authors suggested that the increased development of the villi was a consequence of the greater FI of hens fed pellets as compared with hens fed mash.

Pelleting and Water Intake

An increase in water intake reduces litter quality, animal welfare, incidence of diseases in the flock (e.g., *Staphylococcus* spp.), and carcass condemnations at the slaughterhouse (e.g., footpad dermatitis and breast blisters) (Francesch and Brufau, 2004). Under moderate weather conditions, birds drink twice as much as the feed they consume, with pelleting increasing the water intake to FI intake (WI:FI) ratio as compared with mash feeding, resulting in wet litter problems and in an excess of ammonia production (Serrano et al., 2013; Van der Klis and de Lange, 2013; Jiménez-Moreno et al., 2016). Under hot weather conditions, the increase in water intake together with the reduction in FI, results in an increase in the WI:FI ratio, an effect that increases with age (Jiménez-Moreno et al., 2016).

Laying hens fed pellets dedicate less time to eat, drink more, and show a more aggressive behavior than hens fed mash, probably because of the extra time available, which is dedicated to feather pecking, increasing mortality and reducing flock profitability (Aerni et al., 2000; El-Lethei et al., 2000). In contrast, diets based on coarse particles increase gizzard function in broilers, turkeys, and laying hens, thereby increasing water reabsorption at the large intestine and improving litter quality and animal welfare.

Pelleting and Mortality

Pelleting tends to increase mortality in broilers, a consequence of the greater FI and growth rate and the concomitant increase in the incidence of metabolic problems (Proudfoot and Hulan, 1982; Jensen, 2000; Bennett et al., 2002; Bricket et al., 2007; Baghbanzadeh and Decuypere, 2008; Serrano et al., 2012). An increase of the amount of feed ingested, caused by pellet feeding, increases the incidence of ascites, right heart failure, flip-overs, and skeletal abnormalities (Decuypere et al., 2000; Bennett et al., 2002). Also, over-eating of pellets manufactured using finely ground ingredients might cause dilation of the proventriculus and other metabolic disorders (Serrano et al., 2013; Liermann et al., 2019), thereby increasing mortality and reducing the advantages of this feed form. In addition, compared with coarsely ground feed, the presence of fine particles in the GIT of young chicks, a consequence of feeding high-quality pellets, may increase mortality associated with necrotic enteritis and coccidiosis (Branton et al., 1987). The effects of pellet feeding on mortality are more evident in young birds with a very fast rate of growth (e.g., broilers vs. pullets) and in heavy BW birds for the last week before slaughter (Xin et al., 1994; Serrano et al., 2013).

Interaction Between Feed Form and Energy Content of the Diets

The interactions between feed form and energy content of the diet and their effects on growth performance have been studied in broilers (McNaughton and Reece, 1984; Brickett et al., 2007; Abdollahi et al., 2018) and pullets (Frikha et al., 2009b; Saldaña et al., 2015a). When the energy content of the diet is very low, FI is limited by the physical capacity of the GIT rather than by the need of the birds to satisfy their energy requirements. Consequently, pelleting is often more effective at increasing broiler growth and feed efficiency in flocks fed low-energy diets than in flocks fed high energy diets (Abdollahi et al., 2018). In contrast, the increase in fat content associated with high-energy diets impairs pellet quality, which might reduce FI. Frikha et al. (2009b), in a study conducted with young pullets from 1 to 6 week of age, did not find any interaction between energy concentration and feed form of the diet for any of the variables studied. However, FI and BW were greater in pullets fed pellets than in pullets fed mash and also when fed the high- than when fed the low-energy diet (3,025 vs. 2,735 kcal AME_n/kg). The authors reported also that the relative weights of all the segments of the GIT, including the gizzard, were the lowest in pullets fed the high-energy diet in pellet form.

In summary, pellet feeding increases FI and reduces energy requirements for maintenance and feed wastage, improving broiler performance. The benefits of pelleting on growth performance are more relevant when low- rather than high-energy density diets are used. An increase in pellet durability increases broiler performance but might result in harder pellets, especially when the diets are based on wheat. On the other hand, mash feeding reduces FI but increases gizzard development, gut health, and nutrient digestibility (Mateos et al., 2012; Svihus, 2014a,b; Abdollahi et al., 2019). Consequently, the decision on the best feed form for broilers depends on the main objective of the integration: increase growth and feed efficiency or improve GIT health and litter quality.

FEED FORM AND PARTICLE SIZE RECOMMENDATION IN POULTRY SPECIES

Feed form and particle size of the diet affect growth and performance of poultry in different ways depending on the age and the species considered. While pelleting or crumbling is standard for meat-type birds, the high cost and the increase in energy intake associated with pellet feeding, is a disadvantage, under most situations, for broiler breeders and commercial layers, because it might result in overconsumption and unfavorable BW gain (Gous and Danisman, 2016).

Current practices, based on a literature review, suggest that a good physical quality of pellets and crumbles is essential to increase poultry performance, especially in ducks, turkeys, young broilers, and pullets. Consequently, very fine grinding of the ingredients is preferred when the diet will be fed as pellets. The recommended size of the sieve to grind the cereals (and other ingredients) destined for pellet feeds depends on the main objective of the producer (pellet quality vs. GIT health). If the objective is to improve pellet quality, fine grinding (sieve diameter $\leq$ 2.5 mm) is recommended. However, if the objective is to optimize the health status of the GIT, an increase of the diameter of the sieve to 3.0–3.5 mm might be more appropriate. Svihus (2011) indicated that in order to improve GIT physiology, at least 20–30 percent of the feed particles should have a diameter above 1.0 mm. Consequently, to reduce the incidence of wet litter and of other problems related to GIT function, particle size of the feed should be increased, to the detriment of pellet quality. In all cases the absence of fines and the uniformity of feed particles should be a priority (Serrano et al., 2012, 2013).

The use of crumbles or micro-pellets for the prestarter period (0 to 7 to 12 d of age for broilers and 0 to 21 to 28 d for turkeys and pullets) might be preferred to pellets in spite of the higher cost. The micro-pellets commercially available have a diameter of 2.0–2.3 mm and a length below 5.0 mm to better adapt the feed form to the size of the beak. After this age, the recommended diameter of the pellet increases gradually to 3.5–4.0 mm or even 4.5 mm in the finisher period. Crumbles or pellets, not adapted in size to the beak of the bird, will reduce FI and increase feed wastage (Brickett et al., 2007; Serrano et al., 2012). The durability of the pellets at the feeder level should be as high as possible (over 80–85%) with turkeys being more sensitive to poor quality pellets than broilers.

In pullets, the use of pellets increases FI and BW gain and improves flock uniformity from 0 to 4 weeks of age (Frikha et al., 2009a; Guzmán et al., 2015). When young pullets are fed mash diets, the percentage of fine particles should be maintained as low as possible. From 5 to 17 weeks of age, pellet feeding reduces the anatomical development and the physical capacity of the GIT of the pullets, which affects voluntary FI at the start of egg production cycle (Saldaña et al., 2015a,b). In contrast, coarsely ground mash diets fed during the late phase of the rearing phase and at the initiation of the egg cycle favor the development of the GIT in a period in which the requirements for growth and egg production of the young hens increase sharply (Saldaña et al., 2015a).

The main objective in laying hen operations is to produce uniform feeds with the percentage of fine (<800 μm) and large (<3,200 μm) particles as low as possible (Portella et al., 1988a; Herrera et al., 2018a). In most cases, hens are fed mash diets coarsely ground, with reduced amounts of fines, during the laying period. The sieve diameter most commonly used to grind the cereal fraction of these diets varies between 8 and 12 mm. Within the limits indicated, screen size has limited impact on hen production, provided that the proportion of very fine particles remains under control (Pérez-Bonilla et al., 2014; Herrera et al., 2018a). When the sieve diameter is reduced below 6 mm, the proportion of fines increases and FI decreases, affecting hen production (MacIsaac and Anderson, 2007; Herrera et al., 2018a). Strategies available to reduce the percentage of fines in poultry diets include (1) grinding of the ingredients using roller mills or multicracker equipment rather than hammer mills; (2) increasing the percentage of coarse limestone in the diet at the expenses of fine limestone (2 to 4 mm vs. 1 mm of diameter); (3) bypassing the grinding step by including some ingredients already ground, such as soybean meal, directly into the mixer; and (4) agglomerating the fine particles by supplementing the diet with extra amounts of fat (2–3%).

Under special circumstances, such as the use of high energy density diets, low in inert fiber, from 10 to 17 weeks of age, feeding crumbles during the pre-peak phase of the laying period might be recommended (Wan et al., 2021). Low fiber diets reduce the physical capacity of the GIT, which affects FI at the onset of egg production. However, the benefits are less pronounced or even absent after the peak of egg production (Hamilton and Proudfoot, 1995; Röhe, 2013). In broiler breeders, the use of HTP-mash diets is often recommended to reduce microbial load and improve the hygiene of the feed. Crumble diets, however, might alter FI of the smaller birds because when fed under a feed restriction program the competition among hens for the feed will reduce BW uniformity of the birds, flock performance, and farm profitability.

REFERENCES

Abadi, M.H.M.G., H. Moravej, M. Shivazad, M.A.K. Torshizi, and W.K. Kim. 2019. Effects of feed form and particle size, and pellet binder on performance, digestive tract parameters, intestinal morphology, and cecal microflora populations in broilers. *Poultry Science.* 98:1432–1440.

Abd El-Khalek, E., and G.P.J. Janssens. 2010. Effect of extrusion processing on starch gelatinization and performance in poultry. *World's Poultry Science Journal.* 66:53–64.

Abd El-Wahab, A., J.P. Kriewitz, J. Hankel, B. Chuppava, C. Ratert, V. Taube, C. Visscher, and J. Kamphues. 2020. The effects of feed particle size and floor type on the growth performance, GIT development, and pododermatitis in broiler chickens. *Animals.* 10:1256.

Abdollahi, M.R., V. Ravindran, T.J. Wester, G. Ravindran, and D.V. Thomas. 2010a. Influence of conditioning temperature on the performance, nutrient utilisation and digestive tract development of broilers fed on maize- and wheat-based diets. *British Poultry Science.* 51:648–657.

Abdollahi, M.R., V. Ravindran, T.J. Wester, G. Ravindran, and D.V. Thomas. 2010b. Influence of conditioning temperature on performance, apparent metabolisable energy, ileal digestibility of starch and nitrogen and the quality of pellets, in broiler starters fed maize- and sorghum-based diets. *Animal Feed Science and Technology.* 162:106–115.

Abdollahi, M.R., V. Ravindran, T.J. Wester, G. Ravindran, and D.V. Thomas. 2011. Influence of feed form and conditioning temperature on performance, apparent metabolisable energy and ileal digestibility of starch and nitrogen in broiler starters fed wheat-based diet. *Animal Feed Science and Technology.* 168:88–99.

Abdollahi, M.R., V. Ravindran, and B. Svihus. 2013a. Pelleting of broiler diets: An overview with emphasis on pellet quality and nutritional value. *Animal Feed Science and Technology.* 179:1–23.

Abdollahi, M.R., V. Ravindran, and B. Svihus. 2013b. Influence of grain type and feed form on performance, apparent metabolisable energy and ileal digestibility of nitrogen, starch, fat, calcium and phosphorous in broilers. *Animal Feed Science and Technology.* 186:193–203.

Abdollahi, M.R., V. Ravindran, and B. Svihus. 2014. Influence of feed form on growth performance, ileal nutrient digestibility, and energy utilization in broiler starters fed a sorghum-based diet. *Livestock Science.* 165:80–86.

Abdollahi, M.R., F. Zaefarian, V. Ravindran, and P.H. Selle. 2018. The interactive influence of dietary nutrient density and feed form on the performance of broiler chickens. *Animal Feed Science and Technology.* 239:33–43.

Abdollahi, M.R., F. Zaefarian, H. Hunt, M.N. Anwar, D.G. Thomas, and V. Ravindran. 2019. Wheat particle size, insoluble fibre sources and whole wheat feeding influence gizzard musculature and nutrient utilisation to different extents in broiler chickens. *Journal of Animal Physiology and Animal Nutrition.* 103:146–161.

Abdollahi, M.R., F. Zaefarian, L. Hall, and J.A. Jendza. 2020. Feed acidification and steam-conditioning temperature influence nutrient utilization in broiler chickens fed wheat-based diets. *Poultry Science.* 99:5037–5046.

Aerni, V., H. El-Lethei, and B. Wechster. 2000. Effect of foraging material and feed form on feather pecking in laying hens. *British Poultry Science.* 41:16–21.

Ahmed, R., D.T., Juniper, A. Tonks, and C. Rymer. 2018. The effect of incremental inclusion of whole grain wheat in the diet of growing turkeys on growth performance, feed conversion ratio, cecal health, and digesta characteristics. *Livestock Science.* 214:36–41.

Ali, M.A., M.A. Islam, N.H. Othman, and A.M. Noor. 2017. Effect of heating on oxidation stability and fatty acid composition of microwave roasted groundnut seed oil. *Journal of Food Science and Technology.* 54:4335–4343.

Amerah, A.M., V. Ravindran, R.G. Lentle, and D.G. Thomas. 2007a. Feed particle size: Implications on the digestion and performance of poultry. *World's Poultry Science Journal.* 63:439–455.

Amerah, A., V. Ravindran, R. Lentle, and D. Thomas. 2007b. Influence of feed particle size and feed form on the performance energy utilization, digestive tract development, and digesta parameters of broiler starters. *Poultry Science.* 86:2615–2623.

Amerah, A.M., R.G. Lentle, and V. Ravindran. 2007c. Influence of feed form on gizzard morphology and particle size spectra of duodenal digesta in broiler chickens. *Poultry Science.* 44:175–181.

Amerah, A.M., and V. Ravindran, 2008. Influence of method of whole-wheat feeding on performance, digestive tract development and carcass traits of broilers chickens. *Animal Feed Science and Technology.* 147:326–339.

Amerah, A.M., V. Ravindran, R.G. Lentle, and D.G. Thomas. 2008. Influence of particle size and xylanase supplementation on the performance, energy utilisation, digestive tract parameters and digesta viscosity of broiler starters. *British Poultry Science.* 49:455–462.

Angel, R., N.M. Tamim, T.J. Applegate, A.S. Dhandu, and L.E. Ellestad. 2002. Phytic acid chemistry: Influence on phytin-phosphorous availability and phytase efficacy. *Journal of Applied Poultry Research.* 11:471–480.

ASAE (Agriculture Engineers Yearbook of Standards). 2003. Cubes, pellets, and crumbles—Definitions and methods for determining density, durability, and moisture content. American Society of Agricultural Engineers. Standard S269.4. In *Agriculture Engineers Yearbook of Standards.* St. Joseph, MI: ASAE.

Baghbanzadeh, A., and E. Decuypere. 2008. Ascities syndrome in broilers: Physiological and nutritional perspectives. *Avian Pathology.* 37:117–126.

Behnke, K.C. 2001. Factors influencing pellet quality. *Feed Technology.* 5:19–22.

Ben-Mabrouk, J., G.G. Mateos, A.F. de Juan, L. Aguirre, and L. Cámara. 2022. Effect of beak trimming at hatch and the inclusion of oat hulls in the diet on growth performance, feed preference, exploratory pecking behavior, and gastrointestinal tract traits of brown-egg pullets from hatch to 15 weeks of age. *Poultry Science.* 101:102044.

Ben-Mabrouk, J., G.G. Mateos, A.F. de Juan, L. Aguirre, and L. Cámara. 2023. Effect of beak trimming at hatch and oat hulls inclusion in the diet on growth performance, preference behavior for coarse particles, and gastrointestinal tract traits of brown-egg pullets from 0 to 5 weeks of age. *Animal Feed Science and Technology.* 101:1–11.

Bennett, C.D., H.L. Classen, and C. Riddell. 1995. Live performance and health of broiler chickens fed diets diluted with whole or crumbled wheat. *Canadian Journal of Animal Science.* 75:611–614.

Bennett, C.D., H.L. Classen, and C. Riddell. 2002. Feeding broiler chickens wheat and barley diets containing whole, ground, and pelleted grain. *Poultry Science.* 81:995–1003.

Bergeron, A.N., J.W. Boney, and J.S. Moritz. 2018. The effects of diet formulation and thermal processes associated with pelleting on 18-day broiler performance and digestible amino acid concentration. *Journal of Applied Poultry Research.* 27:540–549.

Bjerrum, L., A.B. Pedersen, and R.M. Engberg. 2005. The influence of whole wheat feeding on *Salmonella* infection and gut flora composition in broilers. *Avian Diseases.* 49:9–15.

Bozkurt, M., B. Koçer, G. Ege, A.E. Tüzüm, H. Biyik, and E. Poyrazoglu. 2019. Influence of the particle size and form of feed on growth performance, digestive tract traits, and nutrient digestibility of white egg-laying pullets from 1 to 112 d of age. *Poultry Science.* 98:4016–4029.

Branton, S.L., F.N. Reece, and W.M. Hagler. 1987. Influence of a wheat diet on mortality of broiler chickens associated with necrotic enteritis. *Poultry Science.* 66:1326–1330.

Brickett, K.E., J.P. Dahiya, H.L. Classen, and S. Gomis. 2007. Influence of dietary nutrient density, feed form, and lighting on growth and meat yield of broiler chickens. *Poultry Science.* 86:2172–2181.

Briggs, J.L., D.E. Maier, B.A. Wakins, and K.C. Behnke. 1999. Effect of ingredients and processing parameters on pellet quality. *Poultry Science.* 78:1464–1471.

California Pellet Mill. 2012. The pelleting process. http://ww1.prweb.com/prfiles/2012/01/09/9090113/Animal%20Feed%20Pelleting.

Carré, B. 2000. Effets de la taille des particules alimentaires sur les processus digestifs chez les oiseaux d'élevage. *INRA Production Animals.* 13:131–136.

Carré, B., N. Muley, J. Gómez, F.X. Oury, E. Laffite, D. Guillou, and C. Signoret. 2005. Soft wheat instead of hard wheat in pelleted diets results in high starch digestibility in broiler chickens. *British Poultry Science.* 46:66–74.

Cerrate, S., Z. Wang, C. Coto, F. Yan, and P.W. Waldroup. 2009. Effect of pellet diameter in broiler prestarter diets on subsequent performance. *Journal of Applied Poultry Research.* 18:590–597.

Choi, J.H., B.S. So, K.S. Ryu, and S.L. Kang. 1986. Effects of pelleted or crumbled diets on the performance and the development of the digestive organs of broilers. *Poultry Science.* 65:594–597.

Clark, P.M., K.C. Behnke, and A.C. Fahrenholz. 2009. Effects of feeding cracked corn and concentrate protein pellets on broiler growth performance. *Journal of Applied Poultry Research.* 18:259–268.

Coelho, M. 2002. Vitamin stability in premixes and feeds. A practical approach in ruminant diets. *Proceedings. 13th Annual Florida Ruminant Nutrition Symposium*, Gainesville, FL. 127–145.

Córdova-Noboa, H.A., E.O. Oviedo-Rondón, A. Ortiz, Y. Matta, S. Hoyos, G.D. Buitrago, J.D. Martínez, J. Yanquen, L. Peñuela, J.O.B. Sorbara, and A.J. Cowieson. 2020. Corn drying temperature, particle size, and amylase supplementation influence growth performance, digestive tract development, and nutrient utilization of broilers. *Poultry Science.* 99:5681–5696.

Dänicke, S., O. Simon, H. Jeroch, and M. Bedford. 1997. Interactions between dietary fat type and xylanase supplementation when rye-based diets are fed to broiler chickens. 2. Performance, nutrient digestibility and the fat-soluble vitamin status of livers. *British Poultry Science.* 38:546–556.

Decuypere, E., J. Buyse, and N. Buys. 2000. Ascites in broiler chickens: Exogenous and endogenous structural and functional causal factors. *World's Poultry Science Journal.* 56:367–377.

Duangnumsawang, Y., J. Zentek, and F. Goodazri Boroojeni. 2021. Development and functional properties of intestinal mucus layer in poultry. *Frontiers of Immunology.* 12:745849.

Duke, G.E. 1986. Alimentary canal: anatomy, regulation of feeding and mortality. In *Avian Physiology*, 4th ed. P.D. Sturkie, editor. Berlin, Germany: Springer-Verlag. 269–288.

Duke, G.E. 1992. Recent studies on regulation of gastric motility in turkeys. *Poultry Science.* 71:1–8.

Duke, G.E. 1994. Physiology of digestion and metabolism. *Proceedings, 43th Western Poultry Disease Conference.* 29–31.

El-Lethei, H., V. Aerni, T.W. Jungi, and R. Wechster. 2000. Stress and feather pecking in laying hens in relation to housing conditions. *British Poultry Science.* 41:22–28.

Engberg, R.M., M.S. Hedemann, and B.B. Jensen. 2002. The influence of grinding and pelleting of feed on the microbial composition and activity in the digestive tract of broiler chickens. *British Poultry Science.* 43:569–579.

Engberg, R.M., M.S. Hedemann, S. Steenfeldt, and B.B. Jensen. 2004. Influence of whole wheat and xylanase on broiler performance and microbial composition and activity in the digestive tract. *Poultry Science.* 83:925–938.

Favero, A., A. Maiorka, F. Dahlke, R.F.P. Meurer, R.S. Oliveira, and R.F. Sens. 2009. Influence of feed form and corn particle size on the live performance and digestive tract development of turkeys. *Journal of Applied Poultry Research.* 18:772–779.

Ferket, P. 2000. Feeding whole grains to poultry improves gut health. *Feedstuffs.* 72:12–16.

Flores, K.R., J.L. Grimes, and A. Fahrenholz. 2020. Effect of feed form, soybean meal protein content, and Rovabio® advance on poult live performance to 3 weeks of age. *Poultry Science.* 99:6705–6714.

Flores, K.R., A. Fahrenholz, and J.L. Grimes. 2021. Effect of pellet quality and biochar litter amendment on male turkey performance. *Poultry Science*. 100:101002.

Foster, W.H., and Z.T. Taha. 1978. Whole grain diets for pullets. *British Poultry Science*. 19:233–242.

Francesch, M., and J. Brufau. 2004. Nutritional factors affecting excreta/litter moisture and quality. *World's Poultry Science Journal*. 60:64–75.

Frikha, M., H.M. Safaa, M.P. Serrano, X. Arbe, and G.G. Mateos. 2009a. Influence of the main cereal and feed form of the diet on performance and digestive tract traits of brown-egg laying pullets. *Poultry Science*. 88:994–1002.

Frikha, M., H.M. Safaa, E. Jiménez-Moreno, R. Lázaro, and G.G. Mateos. 2009b. Influence of energy concentration and feed form of the diet on growth performance and digestive traits of brown-egg laying pullets from 1 to 120 days of age. *Animal Feed Science and Technology*. 153:292–302.

Frikha, M., D.G. Valencia, A. de Coca-Sinova, R. Lázaro, and G.G. Mateos. 2013. Ileal digestibility of amino acids of unheated and autoclaved pea protein concentrate in broilers. *Poultry Science*. 92:1848–1857.

Gabriel, I., S. Mallet, and M. Leconte. 2003. Differences in the digestive tract characteristics of broiler chickens fed on complete pelleted diet or on whole wheat added to pelleted protein concentrate. *British Poultry Science*. 44:283–290.

Gabriel, I., S. Mallet, M. Leconte, A. Travel, and J. Lalles. 2008. Effects of whole wheat feeding on the development of the digestive tract of broiler chickens. *Animal Feed Science and Technology*. 142:144–162.

González-Alvarado, J.M., E. Jiménez-Moreno, R. Lázaro, and G.G. Mateos. 2007. Effects of type of cereal, heat processing of the cereal, and inclusion of fiber in the diet on productive performance and digestive traits of broilers. *Poultry Science*. 86:1705–1715.

González-Alvarado, J.M., E. Jiménez-Moreno, R. Lázaro, and G.G. Mateos. 2008. Effects of fiber source and heat processing of the cereal on the development and pH of the gastrointestinal tract of broilers fed diets based on corn or rice. *Poultry Science*. 87:1779–1795.

Goodarzi Boroojeni, F., B. Svihus, H.G. von Reichenbach, and J. Zentek. 2016. The effects of hydrothermal processing on feed hygiene, nutrient availability, intestinal microbiota and morphology in poultry—A review. *Animal Feed Science and Technology*. 220:187–215.

Gous, R.M., and R. Danisman. 2016. Age, lighting treatment, feed allocation, and feed form influence broiler breeder feeding time. *South African Journal of Animal Science*. 46:29–34.

Gracia, M.I., M.A. Latorre, M. García, R. Lázaro, and G.G. Mateos. 2003. Heat processing of barley and enzyme supplementation of diets for broilers. *Poultry Science*. 82:1281–1291.

Gracia, M.I., R. Lázaro, M.A. Latorre, P. Medel, M.J. Aranibar, E. Jiménez-Moreno, and G.G. Mateos. 2009. Influence of enzyme supplementation of diets and cooking-flaking of maize on digestive traits and growth performance of broilers from 1 to 21 days of age. *Animal Feed Science and Technology*. 150:303–315.

Grant, G., J.E. Edwards, and A. Pusztai. 1995. α-Amylose inhibitor levels in seeds generally available in Europe. *Journal of Science Food and Agriculture*. 67:235–238.

Guzmán, P., H.A.B. Saldaña, A. Pérez-Bonilla, R. Lázaro, and G.G. Mateos. 2015. Productive performance of brown-egg laying pullets from hatching to 5 weeks of age as affected by fiber inclusion, feed form, and energy concentration of the diet. *Poultry Science*. 94:249–261.

Hamilton, R.M.G., and F.G. Proudfoot. 1995. Effects of ingredient particle size and feed form on the performance of Leghorn hens. *Canadian Journal of Animal Science*. 75:109–114.

Heimann, M. 2002. The "bottom line" of grinding. *Feed International*. 23(5):32–34.

Herrera, J., B. Saldaña, L. Cámara, J.D. Berrocoso, and G.G. Mateos. 2018a. Influence of grinding size of the main cereal of the diet on egg production and eggs quality of brown egg laying hens from 33 to 65 weeks of age. *Poultry Science*. 97:2506–2515.

Herrera, J., B. Saldaña, P. Guzmán, M.A. Ibáñez, H. Mandalawi, L. Cámara, and G.G. Mateos. 2018b. Particle size affects short-term preference behavior of brown-egg laying hens fed diets based on corn or barley. *Poultry Science*. 97:1324–1333.

Herstad, O., and J.M. McNab. 1975. The effect of heat treatment and enzyme supplementation on the nutritive value of barley for broiler chicks. *British Poultry Science*. 16:1–8.

Hetland, H., B. Svihus, and V. Olaisen. 2002. Effect of feeding whole cereals on performance, starch digestibility, and duodenal particle size distribution in broiler chickens. *British Poultry Science*. 43:416–423.

Hetland, H., B. Svihus, and A. Krogdahl. 2003. Effects of oat hulls and wood shavings on digestion in broilers and layers fed diets based on whole or ground wheat. *British Poultry Science*. 44:275–282.

Jankoswki, J., Z. Zdunczyk, and J. Juskiewicz. 2016a. Whole grain in turkey nutrition. Part 1: Gastrointestinal development and function. *World's Poultry Science Journal*. 72:521–530.

Jankoswki, J., Z. Zdunczyk, and J. Juskiewicz. 2016b. Whole grain in turkey nutrition. Part 2: Production results in different feeding systems. *World's Poultry Science Journal*. 72:563–572.

Jelveh, K., B. Rasouli, A. Seidavi, P. Makovicky, E. Hatzigiannakis, and C. Milis. 2020. Effect of dilution of broiler chicken diet with cracked maize on performance and intestinal morphology. *Journal of Animal and Feed Science*. 29:250–257.

Jensen, L.S. 2000. Influence of pelleting on the nutritional needs of poultry. *Asian-Australian Journal of Animal Science*. 13:35–46.

Jiménez-Moreno, E., J.M. González-Alvarado, R. Lázaro, and G G. Mateos. 2009. Effects of type of cereal, heat processing of the cereal, and fiber inclusion in the diet on gizzard pH and nutrient utilization in broilers at different ages. *Poultry Science*. 88:1925–1933.

Jiménez-Moreno, E., A. de Coca-Sinova, J.M. González-Alvarado, and G.G. Mateos. 2016. Insoluble fiber sources in mash or pellet diets for broilers. 1. Effects on growth performance and water intake. *Poultry Science*. 95:41–52.

Jiménez-Moreno, E., J.M. González-Alvarado, A. de Coca-Sinova, R.P. Lázaro, L. Cámara, and G.G. Mateos. 2019. Insoluble fiber sources in mash or pellets diets for young broilers. 2. Effects on gastrointestinal tract development and nutrient digestibility. *Poultry Science*. 98:2531–2547.

Jones, F.T., R.C. Axtell, D.V. Rives, S.E. Scheideler, F.R. Tarver Jr., R.L. Walker, and M.J. Wineland. 1991. A survey of *Salmonella* contamination in modern broiler production. *Journal of Food Protection*. 54:502–507.

Jones, F.T., and K.E. Richardson. 2004. *Salmonella* in commercial manufactured feeds. *Poultry Science*. 83:384–391.

Jones, F.T. 2011. A review of practical *Salmonella* control measures in animal feed. *Journal of Applied Poultry Research*. 20:102–113.

Kadardar, H., L. Aguirre, G. Talegón, R. Lázaro, L. Cámara, and G.G. Mateos. 2022. Effects of the nutritional density of the diet and whole wheat supplementation on growth performance and gizzard development of broilers from 0 to 40 days of age. *Poultry Science*. 101:116427.

Kan, C.A., C.W. Scheele, and W.M.M.A. Janssen. 1988. The energy content of full-fat soya beans in meal and pelleted feeds for adult cocks and broilers. *Animal Feed Science and Technology*. 19:97–104.

Kermanshahi, H., and H.L. Classen. 2001. Feeding whole wheat with or without a dietary enzyme or grit to laying hens. *Journal of Agricultural Science and Technology*. 3:193–198.

Khan, R.U., S. Naz, F. Raziq, Q. Qudratullah, N.A. Khan, V. Laudadio, V. Tufarelli, and M. Ragui. 2022. Prospects of organix acids as safe alternative to antibiotics in broiler chicken diets. *Environmental Science and Poultry Research*. 29:32594–32604.

Kheravii, S.K., N.K. Morgan, R.A. Swick, M. Choct, and S.B. Wu. 2018. Roles of dietary fibre and ingredients particle size in broiler nutrition. *World's Poultry Science Journal*. 74:301–316.

Kianfar, R., H. Moravej, M. Shivazad, M. Taghinejad-Roudbaneh, and M.A. Shahrasb. 2013. The effects of dry heat processing, autoclaving, and enzyme supplementation on the nutritive value of wheat for growing Japanese quails. *Journal of Applied Animal Research*. 41:93–102.

Kiarie, E.G., and A. Mills. 2019. Role of feed processing on gut health and function in pigs and poultry: conundrum of optimal particle size and hydrothermal regimens. *Frontiers in Veterinary Science*. 6:19.

Koçer, B., M. Bozkurt, K. Küçükyilmaz, G. Ege, H. Akşit, A. Orojpour, S. Topbaş, A.E. Tüzün, E. Bintaş, and K. Seyrek. 2016. Effects of particle sizes and physical form of the diet on performance, egg quality and size of the digestive organs in laying hens. *European Poultry Science*. 80:1–41.

Koch, K. 1996. Hammer mills and roller mills. MF-2048 Feed manufacturing. Manhattan, KS: Department of Grain Science, Indiana Kansas State University.

Lázaro, R., M. García, M.J. Aranibar, and G.G. Mateos. 2003. Effect of enzyme addition to wheat-, barley-, and rice-based diets on nutrient digestibility and performance of laying hens. *British Poultry Science*. 44:256–265.

Lehtinen, P., K. Kiiliäinen, I. Lehtomäki, and S. Laakso. 2003. Effects of heat treatment on lipid stability in processed oats. *Journal of Cereal Science*. 37:215–221.

Lemons, M.E., C.D. McDaniel, J.S. Moritz, and K.G.S. Wamsley. 2019. Increasing average feed particle size during the starter period maximizes Ross × Ross 708 male broiler performance. *Journal Applied Poultry Research*. 28:420–434.

Liermann, W., M. Bochnia, A. Berk, V. Böschen, L. Hüther, A. Zeyner, and S. Dänicke. 2019. Effects of feed particle size and hydro-thermal processing methods on starch modification, nutrient digestibility, and the performance and the gastrointestinal tract of broilers. *Animals*. 9:294.

Liu, S.Y., H.H. Truong, and P.H. Selle. 2015. Whole-grain feeding for chicken meat production: Possible mechanisms driving enhanced energy utilization and feed conversion. *Animal Production Science*. 55:559–572.

Lv, M., L. Yan, Z. Wang, S. An, M. Wu, and Z. Lv. 2015. Effects of feed form and feed particle size on growth performance, carcass characteristics and digestive tract development of broilers. *Animal Nutrition*. 1:252–256.

Lyu, F., M. Thomas, W.H. Hendriks, and A.F.B. van der Poel. 2020. Size reduction in feed technology and methods for determining, expressing, and predicting particle size: A review. *Animal Feed Science and Technology*. 261:114347.

MacIsaac, J.L., and D.M. Anderson. 2007. Effect of whole wheat, enzyme supplementation, and grain texture on the production performance of laying hens. *Canadian Journal of Animal Science*. 87:579–589.

Mateos, G.G., R. Lázaro, and M.I. Gracia. 2002. The feasibility of using nutritional modifications to replace drugs in poultry feeds. *Journal of Applied Poultry Research*. 11:437–452.

Mateos, G.G., E. Jiménez-Moreno, M.P. Serrano, and R.P. Lázaro. 2012. Poultry response to high levels of dietary fiber sources varying in physical and chemical characteristics. *Journal of Applied Poultry Research*. 21:156–174.

Mateos, G.G., L. Cámara, G. Fondevila, and R.P. Lázaro. 2019. Critical review of the procedures used for estimation of the energy content of diets and ingredients in poultry. *Journal of Applied Poultry Research*. 28:506–525.

McDowell, L.R., and N.E. Ward. 2008. Optimum vitamin nutrition for poultry. *International Poultry Production*. 16(4):27–34.

McNaughton, J.L., and F.N. Reece. 1984. Factor affecting pelleting response. 1. Influence of dietary energy in broiler starter diets. *Poultry Science*. 63:682–685.

Medel, P., M.A. Latorre, C. de Blas, R. Lázaro, and G.G. Mateos. 2004. Heat processing of cereals in mash or pellet diets for young pigs. *Animal Feed Science and Technology*. 113:127–140.

Moritz, J.S., A.S. Parsons, N.P. Buchanan, W.B. Calvalcanti, K.R. Cramer, and R.S. Beyer. 2005. Effect of gelatinizing dietary starch through feed processing on zero to three-week broiler performance and metabolism. *Journal of Applied Poultry Research*. 14:47–54.

Mtei, A.W., R. Abdollahi, and V. Ravindran. 2019. Impact of corn particle size on nutrient digestibility varies depending on bird type. *Poultry Science*. 98:5504–5513.

Muramatsu, K., A. Massuquetto, F. Dahlke, and A. Maiorka. 2015. Factors that affect pellet quality: A review. *Journal of Agriculture Science and Technology*. A.5:717–722

Naderinejad, S., F. Zaefarian, M.R. Abdollahi, A. Hassanabadi, H. Kermanshahi, and V. Ravindran. 2016. Influence of feed form and particle size on performance, nutrient utilisation, and gastrointestinal tract development and morphometry in broiler starters fed maize-based diets. *Animal Feed Science and Technology*. 215:92–104.

Nir, I., Y. Twina, E. Grossman, and Z. Nitsan. 1994. Quantitative effects of pelleting on performance, gastrointestinal tract and behaviour of meat-type chickens. *British Poultry Science*. 35:589–602.

NRC (National Research Council). 1994. *Nutrient Requirements of Poultry*, 9th rev. ed. Washington, DC: National Academy Press.

Ovi, F.K., C. Bortoluzzi, T.J. Applegate, C.R. Starkey, K. Macklin, A. Morey, and W. Pacheco. 2020. Effects of pre-pelleting whole corn inclusion on broiler performance, intestinal microbiota, and carcass characteristics. *Journal of Applied Poultry Research*. 30:100114.

Parera, N., R.P. Lázaro, M.P. Serrano, D.G. Valencia, and G.G. Mateos. 2010. Influence of the inclusion of cooked cereals and pea starch in diets based on soy or pea protein concentrate on nutrient digestibility and performance of young pigs. *Journal of Animal Science*. 88:671–679.

Parkash, S. 2015. Effects of processing on oil quality. In Functional Dietary Lipids. T. Sanders, editor. Amsterdam, The Netherlands: Woodhead Publishing, Elsevier.

Parsons, A.S., N.P. Buchanan, K.P. Blemings, M.E. Wilson, and J.S. Mortiz. 2006. Effect of corn particle size and pellet texture on broiler performance in the growing phase. *Journal of Applied Poultry Research*. 15:245–255.

Pérez-Bonilla, A., M. Frikha, R.P. Lázaro, and G.G. Mateos. 2014. Type of grinding of the main cereal of the diet affects production of Brown egg-laying hens. *Animal Feed Science and Technology*. 194:121–130.

Plavnik, I., and D. Sklan. 1995. Nutritional effects of expansion and short time extrusion on feeds for broilers. *Animal Feed Science and Technology*. 55:247–251.

Plavnik, I., E. Wax, D. Sklan, and S. Hurwitz. 1997. The response of broiler chickens and turkey poults to steam-pelleted diets supplemented with fat or carbohydrates. *Poultry Science*. 76:1006–1013.

Portella, F.J., L.J. Caston, and S. Leeson. 1988a. Apparent feed particle size preference by laying hens. *Canadian Journal Animal Science*. 68:915–922.

Preston, C.M., K.J. McCracken, and A. McAllister. 2000. Effect of diet form and enzyme supplementation on growth, efficiency, and energy utilization of wheat-based diets for broilers. *British Poultry Science*. 41:324–331.

Proudfoot, F.G., and H.W. Hulan. 1982. Feed texture effects on the performance of turkey broilers. *Poultry Science*. 61:327–330.

Qaisrani, S.N., M.M. van Krimpen, R.P. Kwakkel, M.W.A. Verstegen, and W.H. Hendriks. 2015. Diet structure, butyric acid, and fermentable carbohydrates influence growth performance, gut morphology, and cecal fermentation characteristics in broilers. *Poultry Science*. 94:2152–2164.

Qaisrani, S.N., A.I. Hussain, S. Naveed, F. Bibi, C.A. Akram, T.N. Pasha, M. Asif, I. Irshad, and R.M. Bilal. 2022. Effects of protein source, whole wheat, and butyric acid on live performance, gut health, and amino acid digestibility in broiler chickens. *Metabolites*. 12:989.

Ravindran, V., Y.B. Wu, D.G. Thomas, and P.C.H. Morel. 2006. Influence of whole wheat feeding on the development of gastrointestinal tract and performance of broiler chickens. *Australian Journal of Agriculture Research*. 57:21–26.

Riaz, M.N., M. Asif, and R. Ali. 2009. Stability of vitamins during extrusion. *Critical Reviews in Food Science and Nutrition*. 49:361–368.

Rodrigues, I., and M. Choct. 2017. The foregut and its manipulation via feeding practices in the chicken. *Poultry Science*. 97:3188–3206.

Röhe, I. 2013. Effects of feed structure on animal performance, gastrointestinal morphology, gut-associated lymphoid tissue and jejunal glucose transport in laying hens. Ph.D. Thesis. Berlin University, Germany.

Röhe, I., I. Ruhnke, F. Knorr, A. Mader, F. Goodarzi Boroojeni, R. Löwe, and J. Zentek. 2014. Effects of grinding method, particle size, and physical form of the diet on gastrointestinal morphology and jejunal glucose transport in laying hens. *Poultry Science*. 93:2060–2068.

Sacranie, A., P. Iji, L.L. Mikkelsen, and M. Choct. 2007. Occurrence of reverse peristalsis in broiler chicken. *Proceedings of Australian Poultry Science Symposium*. 19:161–164.

Sacranie, A., B. Svihus, V. Denstadli, B. Moen, P.A. Iji, and M. Choct. 2012. The effect of insoluble fiber and intermittent feeding on gizzard development, gut motility, and performance of broiler chickens. *Poultry Science*. 91:693–700.

Saldaña, B., P. Guzmán, L. Cámara, J. Garcia, and G.G. Mateos. 2015a. Feed form and energy concentration of the diet affect growth performance and digestive tract traits of brown-egg laying pullets from hatching to 17 weeks of age. *Poultry Science*. 94:1879–1893.

Saldaña, B., P. Guzmán, H.M. Safaa, R. Harzalli, and G.G. Mateos. 2015b. Influence of the main cereal and feed form of the rearing phase diets on performance, digestive tract, and body traits of brown-egg laying pullets from hatch to 17 weeks of age. *Poultry Science*. 94:2650–2661.

Saldaña, B., C.E. Gewehr, P. Guzmán, J. García, and G.G. Mateos. 2016. Influence of feed form and energy concentration on the rearing phase diets on productivity, digestive tract development and body measurements of brown-egg laying hens fed diets varying in energy concentration from 17 to 46 wk of age. *Animal Feed Science and Technology*. 221:87–100.

Salim, H.M., Z.A. Kruk, and B.D. Lee. 2010. Nutritive value of corn distillers dried grains with solubles as an ingredient of poultry diets: A review. *World's Poultry Science*. 66:411–432.

Serrano, M.P., D.G. Valencia, J. Mendez, and G.G. Mateos. 2012. Influence of feed form and source of soybean meal of the diet on growth performance of broilers from 1 to 42 days of age. 1. Floor pen study. *Poultry Science*. 91:2838–2844.

Serrano, M.P., M. Frikha, J. Corchero, and G.G. Mateos. 2013. Influence of feed form and source of soybean meal on growth performance, nutrient retention, and digestive organ size of broilers. 2. Battery study. *Poultry Science*. 92:693–708.

Shen, H., J.D. Summers, and S. Leeson. 1983. The influence of steam pelleting and grinding on the nutritive value of canola rapeseed for poultry. *Animal Feed Science and Technology*. 8:303–311.

Singh, Y. 2013. Whole grain inclusion in poultry diets: Effects on performance, nutrient utilization, gut development, caecal microflora profile and coccidiosis challenge. Ph.D. Thesis. Massey University, Palmerston North, New Zealand.

Singh, Y., V. Ravindran, T.J. Wester, A.L. Moland, and G. Ravindran. 2014. Influence of feeding coarse corn on performance, nutrient utilization, digestive tract measurements, carcass characteristics, and cecal microflora counts of broilers. *Poultry Science*. 93:607–616.

Singh, Y., and V. Ravindran. 2019. Influence of feeding whole maize, differing in endosperm hardness, on the performance, nutrient utilization and digestive tract development of broiler starters. *Journal of Applied Animal Nutrition*. 7:10.1017.

Skinner-Noble, D.O., L.J. Mckinney, and R.G. Teester. 2005. Predicting effective caloric value of non-nutritive factors: III. Feed form affects broiler performance by modifying behavior patterns. *Poultry Science*. 84:403–411.

Sogunle, O.M., B.B. Olatoye, M. Agric, L.T. Egbeyale, A.V. Jegede, O.A. Adeyemi, D.A. Ekunseitan, and K.O. Bello. 2013. Feed forms of different particle sizes: Effects on growth performance, carcass characteristics, and intestinal villus morphology of cockerel chickens. *Pacific Journal of Science and Technology*. 14:405–415.

Stanley, D., and Y.S. Bajagay. 2022. Feed safety and development of poultry intestinal microbiota. *Animals*. 12:103390.

Svihus, B. 2001. Research note: A consistent low starch digestibility observed in pelleted broiler chicken diets containing high levels of different wheat varieties. *Animal Feed Science and Technology*. 92:45–49.

Svihus, B. 2011. The gizzard: Function, influence of diet structure and effects on nutrient availability. *World's Poultry Science Journal*. 67:207–224.

Svihus, B. 2014a. Function of the digestive system. *Journal Applied Poultry Research*. 23:306–314.

Svihus, B. 2014b. Starch digestion capacity of poultry. *Poultry Science*. 93:2394–2399.

Svihus, B., O. Herstad, C.W. Newman, and R.K. Newman. 1997. Comparison of performance and intestinal characteristics of broiler chickens fed on diets containing whole, rolled or ground barley. *British Poultry Science*. 38:524–529.

Svihus, B., and H. Hetland. 2001. Ileal starch digestibility in growing broiler chickens fed on a wheat-based diet is improved by mash feeding, dilution with cellulose or whole wheat inclusion. *British Poultry Science*. 42:633–637.

Svihus, B., K.H., Kløvstad, V. Pérez, O. Zimonja, S. Sahlström, R.B. Schüller, W.K. Jeksrud, and E. Prestløkken. 2004a. Physical and nutritional effects of pelleting of broiler chicken diets made from wheat ground to different coarsenesses by the use of roller mill and hammer mill. *Animal Feed Science and Technology*. 117:281–293.

Svihus, B., E. Juvik, H. Hetland, and Å. Krogdahl. 2004b. Causes for improvement in nutritive value of broiler chicken diets with whole wheat instead of ground wheat. *British Poultry Science*. 45:55–60.

Svihus, B., A.K. Uhlen, and O.M. Harstad. 2005. Effect of starch granule structure, associated components, and processing on nutritive value of cereal starch: A review. *Animal Feed Science and Technology*. 122:303–320.

Tari, L.M., W.N.U. Perera, F. Zaefarian, M.R. Abdollahi, A.J. Cowieson, and V. Ravindran. 2022. Influence of barley inclusion method and protease supplementation on growth performance, nutrient utilization, and gastrointestinal tract development in broiler starters. *Animal Nutrition*. 8:61–70.

Thomas, M., and A.F.B. Van der Poel. 1996. Physical quality of pelleted animal feed 1. Criteria for pellet quality. *Animal Feed Science and Technology*. 61:89–112.

Valencia, D.G., M.P. Serrano, E. Jiménez-Moreno, R. Lázaro, and G.G. Mateos. 2009a. Ileal digestibility of amino acids of pea protein concentrate and soya protein sources in broiler chicks. *Livestock Science*. 121:21–27.

Valencia, D.G., M.P. Serrano, R. Lázaro, E. Jiménez-Moreno, and G.G. Mateos. 2009b. Influence of micronization (fine grinding) of soya bean meal and full-fat soya bean on the ileal digestibility of amino acids for broilers. *Animal Feed Science and Technology*. 150:238–248.

Van der Klis, J.D., and L. de Lange. 2013. Water intake of poultry. *Proceedings, 19th European Poultry Nutrition Symposium*, Potsdam, Germany. 102–107.

Van Immerseel, F., L. De Zutter, K. Houf, F. Pasmans, F. Haesebrouck, and R. Ducatelle. 2009. Strategies to control *Salmonella* in the broiler production chain. *World's Poultry Science Journal*. 65:367–392.

Veldman, A., H.A. Vahl, G.J. Borggreve, and D.C. Fuller. 1995. A survey of the incidence of Salmonella species and Enterobacteriaceae in poultry feeds and feed components. *Veterinary Records*. 136:169–172.

Vicente, B., D.G. Valencia, M. Pérez-Serrano, R.P. Lázaro, and G.G. Mateos. 2008. The effects of feeding rice in substitution of corn and the degree of starch gelatinization of rice on the digestibility of dietary components and productive performance of young pigs. *Journal of Animal Science*. 86:119–126.

Vukic Vranješ, M., H.P. Pfirter, and C. Wenk. 1994. Influence of processing treatment and type of cereal on the effect of dietary enzymes in broiler diets. *Animal Feed Science Technology*. 46:261–270.

Wan, Y., R. Ma, A. Khalid, L. Chai, R. Qi, W. Liu, J. Li, Y. Li, and K. Zhan. 2021. Effect of the pellet and mash feed forms on the productive performance, egg quality, nutrient metabolism, and intestinal morphology of two laying hen breeds. *Animals*. 11:701.

Wang-Li, L., Y. Xu, A.P. Shivkumar, M. Williams, and J. Brake. 2020. Effect of dietary coarse corn inclusion on broiler live performance, litter characteristics, and ammonia emission. *Poultry Science*. 99:869–878.

WPSA (World's Poultry Science Association). 1989. *European Table of Energy Values for Poultry Feedstuffs*, 3rd ed. Subcommittee of Energy of the Working Group no 2. Nutrition of the European Federation of Branches of the World's Poultry Science Association, Spelderholt Centre for Poultry Research, Beekbergen, The Netherlands.

Xin, H., I.L. Berri, T.L. Barton, and G.T. Tabler. 1994. Feed and water consuption, growth, and mortality of male broilers. *Poultry Science*. 73:610–616.

Yang, P., H. Wang, M. Zhu, and Y. Ma. 2020. Evaluation of extrusion temperatures, pelleting parameters, and vitamin forms on vitamin stability in feed. *Animals*. 10:84.

Yegani, M., and D.R. Korver. 2008. Factors affecting intestinal health in poultry. *Poultry Science*. 87:2052–2063.

Yegani, M., D.R. Korver. 2012. Review: Prediction of variation in energetic value of wheat for poultry. *Canadian Journal of Animal Science*. 92:261–273.

Zaefarian, F., M.R. Abdollahi, and V. Ravindran. 2016. Particle size and feed form in broiler diets: Impact on gastrointestinal tract development and gut health. *World's Poultry Science Journal*. 72:277–290.

Zentek, J., and F. Goodarzi Boroojeni. 2020. (Bio) Technological processing of poultry and pig feed: Impact on the composition, digestibility, antinutritional factors, and hygiene. *Animal Feed Science and Technology*. 268:114576.

Zimonja, O., A. Stevnebø, and B. Svihus. 2007. Nutritional value of diets for broiler chickens as affected by fat source, amylose level and diet processing. *Canadian Journal of Animal Science*. 87:553–562.

Zuidhof, M.J., D.E. Holm, R.A. Renema, M.A. Jalal, and F.E. Robinson. 2015. Effects of broiler breeder management on pullet body weight and carcass uniformity. *Poultry Science*. 94:1389–1397.

10

Nutrient Excretion and the Environment

INTRODUCTION

For many decades, commercial poultry production has expanded and concentrated in specific geographical areas driven by economic considerations that maximize production, efficiency, and competitiveness with other foods. While these benefits have resulted in some positive impacts on environmental sustainability, they have also resulted in some negative environmental challenges that must be mitigated. In particular, concentration of poultry production in some geographical areas has resulted in nutrient emissions into the local environment that exceed agronomic use and impact environmental quality. Feedstuffs are transported from crop production regions to the areas of intensive livestock production. Due to inefficiencies in conversion of feed into poultry meat or eggs, dietary nutrients are excreted, and the manure is often used to fertilize local landscapes. Land application of poultry manure can result in unintended environmental losses that may impact the quality of surface and ground waters as well as excess concentrations of nutrients in the soil. Consequently, increased regulatory pressures have been imposed on animal agriculture to mitigate these problems. A variety of state and federal laws regulate environmental pollutants from point sources and often compel reductions over time (Rosov et al., 2020). Depending on the geographical area, regulations on the disposal of animal manure may target N, several minerals, and total organic matter. Among minerals, the disposal of P, Cu, Zn Se, As, and total salts are regulated depending on regional soil, topography, and rainfall characteristics.

Environmental problems become especially serious when nutrients applied to the land exceed plant utilization and nutrients migrate to surface or ground waters and accelerate eutrophication (N and P) or if they are emitted into the atmosphere (N and S) and cause acidification of terrestrial environments or contribute to climate change. At the time of this report, several countries and some U.S. states have initiated the regulation of greenhouse gases from animal production activities. Currently, mitigation of the contributions of poultry production to greenhouse gas emissions has become an emphasis in nutrition research, given that the production of feedstuffs fed to poultry and the excretion of some dietary nutrients into the environment are both accounted for in lifecycle assessment and greenhouse gas emission models.

The extent of nutrient emission from poultry production is the difference between that which is consumed by animals and what is retained in meat or eggs destined for the human food supply. There are five basic ways to reduce environmental problems resulting from nutrient excretion during poultry production (Ferket et al., 2002; Wiliams et al., 1999): (1) reduce poultry production; (2) collect and transport poultry manures out of nutrient excessive areas to areas where the soil can be improved by the nutrients; (3) collect the poultry manures and recycle the nutrients into ruminant feeds for consumption in other locations, depending on regulatory constraints; (4) generate energy from the waste; and (5) increase the efficiency of nutrient utilization by animals, thereby converting more of the input nutrients into meat and eggs. The last of these potential solutions can be addressed by poultry nutritionists (Nahm, 2002). Following is an overview of the nutritional approaches to reduce the environmental impact from poultry production. Higher levels of detail on the mechanistic and nutritional basis can be found in the other chapters of this report. Additionally, nutritional solutions that reduce atmospheric gases emitted due to poultry production are briefly summarized. This is a developing area of research, especially in poultry nutrition, and much of the research is based on outdated standards for predicting their impact on climate change. This should certainly be an area for additional research and for coverage in future editions of this report. Additionally, some of the nutritional solutions for reducing greenhouse gases are like those for mitigating environmental impacts from N excretion, which are summarized subsequently.

MINIMIZING NUTRIENT EXCRETION BY IMPROVING THE EFFICIENCY OF NUTRIENT UTILIZATION

Nutritional approaches to limiting nutrient excretion can have significant impacts by maximizing the conversion of consumed nutrients into meat and eggs. The two primary approaches toward this goal are (1) aligning the concentration of bioavailable nutrients in the diet to meet the bird's requirements as closely as economically practical; and (2) increasing the digestibility and bioavailability of nutrients in the diet.

Aligning Bioavailable Nutrient Concentrations with Requirements

Levels of bioavailable nutrients in the diet that are above or below the animal's requirement will increase nutrient excretion. When a nutrient is fed at a level above its dietary requirement, the excess is not utilized for its nutritionally required roles and is excreted. In some cases, a dietary excess of one nutrient can also increase the excretion of another nutrient. For example, dietary calcium in excess of the requirement results in excretion of the excess calcium, but it also antagonizes the absorption of P and several trace minerals, which concomitantly increases their excretion. When a nutrient is supplied at a level below the animal's requirement, it is usually efficiently utilized, but excretion of other dietary nutrients is usually increased due to lower rates of tissue deposition for growth or egg production. For example, when a required amino acid is deficient, the rate of protein accretion decreases, resulting in a large portion of all other amino acids being catabolized and their N excreted. Additionally, the lower rate of tissue accretion results in increased excretion of minerals that would normally be used for growth.

In the preceding examples, optimizing nutrient balance decreases nutrient excretion while often simultaneously decreasing feed costs. Matching bioavailable nutrient levels in the diet to the bird's requirements is highly dependent on having accurate estimates of the bioavailable dietary nutrient concentrations and the specific requirements of the bird given its age, genetic strain, sex, rate of production, and management conditions. Estimates of nutrient requirements of poultry and the concentrations, digestibilities, bioavailabilities, and the many factors that impact them are detailed in various chapters throughout this report.

Because a bird's nutrient requirements change daily due to changes in rates of growth or egg production, aligning nutrient input with needs requires continual adjustments of the composition of the bird's diet. Currently, this is typically done by phase feeding, which means supplying a series of feeds formulated to approximate the requirement of the bird during segmented periods of its life (Ferket et al., 2002). For example, during the growth phase, the requirement (mg/kg diet) for each nutrient generally decreases with age, and a series of diets with declining nutrient levels can be fed for specific periods of time. To minimize nutrient excretion, the time that birds consume a specific diet should be as short as economically practical (Ferket et al., 2002; Nahm, 2002). When birds are individually fed, the emerging strategy of matching nutrient delivery to individual birds in real time through the use of sensor and precision feeding technologies further decreases nutrient excretion.

Accurate analytical nutrient compositional data of each feedstuff (Chapter 20), expressed on an available nutrient basis, are crucial to meeting nutrient requirements with minimal excesses (Ferket et al., 2002). This is particularly difficult when there is significant batch-to-batch and day-to-day nutrient variability in feedstuffs (Chapter 9). Accurate diet formulation requires frequent analysis of each feed ingredient, which may include real-time assessment using nondestructive sensors (e.g., near-infrared spectroscopy) and adjustment of formulations at the time of mixing. Additionally, nutrient analysis of finished feed is necessary to identify problems in the feed mixing and manufacturing processes. It is common practice to add safety margins above nutrient requirement targets, in part to remedy the adverse effects of nutrient variability on the productive performance of animals. Safety margins are typically broadened with increasing nutrient variability, and this relationship inherently increases the probability of higher nutrient excretion. Therefore, accurately estimating nutrient requirements, minimizing feedstuff variation, and frequent and accurate feedstuff analysis are necessary to minimize nutrient excretion. Unfortunately, in many real-world commercial situations, it is not possible to obtain a complete nutrient analysis of the feed before it is fed to the birds, so if there are performance problems or excessive excretion of nutrients due to dietary deficiencies, excesses, or toxicities, they are usually determined after the fact via analysis of archived feed samples.

Increasing the Digestibility of Nutrients in the Diet

Nutrients that are not digested and absorbed are quickly excreted. Nutrients that are absorbed but are not bioavailable to be used for their required metabolic role also contribute to nutrient excretion (Chapter 10). Maximizing the digestibility and bioavailability of nutrients in poultry feed contributes markedly to minimizing nutrient excretion. Individual nutrients in feedstuffs have different degrees of digestibility, and the digestibility of any given nutrient can vary greatly between and within feedstuffs. Factors that affect digestibility include the nutrient's chemical form in the feedstuff, excesses or deficiencies of other nutrients, and the presence of a wide range of anti-nutritional factors. Optimizing nutrient digestibility requires selecting feedstuffs with high inherent digestibilities due to low levels of fiber, anti-nutrients, and nutrient antagonists. Nutrient digestibilities in feedstuffs can also be modified by post-harvest processing methods that remove poorly digestible components, such as by milling to remove husks and by mitigating the activity of anti-nutrients by a wide variety of techniques that include heating or solvent extraction. Supplementing diets with feed additives that increase digestion are effective and often economically feasible. Enzymes (e.g., phytase, carbohydrases, and proteases) have been especially useful for increasing the digestibility of nutrients and mitigating the negative effects of anti-nutrients like phytate and indigestible oligosaccharides and polysaccharides (e.g., β-glucans and arabinoxylans). Improvement can also be accomplished by optimizing the particle size of the ingredients and adjustments in pelleting or extrusion conditions (Chapters 2 and 9).

NITROGEN EXCRETION

Excessive N emissions from animal production can occur when the application of manure for fertilizing crops and forages is excessive or if animal density exceeds approved nutrient management guidelines. This can adversely influence water quality (both surface and groundwater) and contribute to acidification of soils via both direct deposition and acidified precipitation in the form of rain or snow. In geographical regions where poultry production is highly concentrated relative to available land for manure application this is a problem that requires mitigation.

Manure is a valuable source of organic N for crops and is an important co-product of poultry production in many locations. Poultry manure is typically land-applied as a dry litter. The N excreted by poultry is mostly in uric acid; microorganisms in the manure, or in soil following land application, degrade uric acid and other nitrogenous compounds, resulting in the release of ammonia. Uric acid is metabolized to ammonia by microbes at a slightly slower rate than urea excreted by mammalian species. Ammonia can enter the environment as a volatile emission or be converted to nitrate by nitrification or to gaseous N by denitrification. Ammonia release to the atmosphere contributes to acidic precipitation that acidifies the terrestrial environment. Ammonia converted to nitrate contributes to ground water contamination.

Poultry nutritionists can have a large impact on this problem by formulating diets that minimize the amount of N excreted (Kebreab et al., 2016; Belloir et al., 2017; Tallentire et al., 2017; Cappelaere et al., 2021). Research toward lowering dietary levels of protein and total N has been in progress since the early 1940s (Kidd et al., 2021) and has mostly been driven by regulatory and economic factors, with environmental factors becoming a more recent incentive. As with all nutrients, minimizing excretion requires the use of highly digestible feedstuffs and providing them at levels that align dietary nutrient concentrations with the bird's requirements. Important developments since the last report (NRC, 1994) include advances in the ideal protein ratio approach (Baker, 1997; Kidd et al., 2013; Van Harn et al., 2019; Brown et al., 2021) and the use of the digestible rather than total amino acid concentrations of ingredients (Chapter 8) when formulating diets. Applying the ideal protein ratio approach to diet formulation is accomplished by formulating the diet to provide a low level of crude protein and supplementing chemically or biologically synthesized amino acids (e.g., methionine, lysine, threonine) to meet the bird's essential amino acid requirements (Chapter 3). This approach results in dietary amino acid concentrations that are similar to the bird's requirements and results in low levels of excess amino acids that would be deaminated and excreted. The use of supplemental amino acids reduces N excretion by more than 20 percent in most poultry diets as compared to those containing only intact protein, without adversely affecting growth performance (Nahm, 2002). One caveat is that the bioavailability of supplemental synthetic amino acids is impacted by the feeding regimen (Baker and Izquierdo, 1985), likely because the kinetics of their absorption differs from that of intact protein. The rate of absorption of purified amino acids is faster than amino acids from intact proteins and may result in increased oxidation and N excretion when infrequent meal feeding is employed instead of frequent meals or provision for ad libitum consumption.

Supplementing protease, carbohydrase, and phytase enzymes (Ravindran and Son, 2011; Lee et al., 2018; Borda-Molina et al., 2019; Cowieson et al., 2019), and optimizing physical aspects of the diet, such as particle size of the ingredients and manufacturing techniques, increases the digestibility of dietary protein (Chapters 8 and 9). The relationship between the amount of N excretion that occurs with each percentage unit decrease in dietary crude protein is highly variable (Kerr, 1995) and recent review estimated the reduction to be about 10 percent per 1 percent decrease (Belloir et al., 2017). Application of linear programing and other formulation approaches to minimizing the impact of diet composition on N excretion and ammonia emissions has been a recent area of research (Méda et al., 2019; Cappelaere et al., 2021). Integrating approaches of N reduction with the sometimes-competing goals of reducing P excretion and greenhouse gas emissions is important for minimizing the total impact of poultry nutrition on the environment (Nguyen et al., 2012; Tallentire et al., 2017; Méda et al., 2019; Cappelaere et al., 2021).

MINERAL EXCRETION

Use of poultry manure to supply N to crops may result in the application of more minerals than can be utilized and removed by crops, resulting in their buildup in soils and, in the case of P, runoff into surface waters (Sharpley et al., 1994). Along with N, P is a limiting nutrient for algae and other aquatic plant growth that, when excessive, can cause environmental and economic harm and potential human health problems (Kotak, 1993).

Minimizing P excretion can be accomplished by feeding levels of bioavailable P that closely match the bird's requirement and using a source of P that has a high digestibility

(Maguire et al., 2005). The P requirements of poultry (Chapters 13 to 19) and the uncertainties that might require the addition of a margin of safety are detailed in this report (Chapters 6 and 8). Briefly, ingredients based on plants typically have low P digestibility relative to that from ingredients from animals or phosphate supplements. About two-thirds of the P in plants is covalently bound to inositol in phytate (Ravindran, 1996). Poultry have very low levels of the intestinal enzyme phytase, which releases P from phytate. In the absence of phytase supplemented to the diet, most of the P in phytate is unavailable and excreted. An important development in poultry nutrition since the last report (NRC, 1994) is that the inclusion of phytase in a diet is typically economically justifiable and the bioavailability challenge caused by phytate has been greatly mitigated (Angel et al., 2005, 2006; Kidd et al., 2016). It is important that diets supplemented with phytate account for the additional P released by the enzyme in the matrix values used in feed formulation (Shelton et al., 2004; Cowieson et al., 2016; Dersjant-Li et al., 2019). Additionally, low-phytate crops have been developed by genetic selection and have applications in some circumstances (Raboy, 2020). Furthermore, nutritionists should be cognizant of the mechanism of action of phytase. Phytase does not liberate additional P from feed ingredients such as inorganic (i.e., rock) phosphate sources, nor from animal-based sources such as meat and bone meal. Because phytase does not itself directly provide available P, the increase in available P made possible is dependent on the presence of phytate phosphorus. Certain plant-based co-products, such as distiller's dried grains with solubles, may have already had a substantial proportion of phytate-bound P released via the fermentation process and may be less responsive to dietary phytase addition.

The efficiency of P absorption and utilization can be diminished by excess Ca, and this remains a challenge due to difficulties of estimating digestible Ca and P requirements of the bird (Chapter 6) and the digestibility of dietary sources of Ca and P (Chapter 8). Additionally, maximizing P absorption requires including vitamin D or its metabolites (Chapter 5) at dietary levels higher than the requirement for preventing signs of vitamin D deficiency (Angel et al., 2006). Methodological approaches for minimizing these issues are discussed in the chapters referenced previously. The effectiveness of nutritional strategies for mitigating the environmental impacts of dietary P requires measuring their impact on both total P and water-soluble P excretion (Maguire et al., 2005). Water soluble P in manures causes dissolved reactive P losses in runoff immediately following application to land.

Fertilization of soils with poultry manures is an important source of trace minerals in some geographical regions and increases crop yields and trace mineral concentrations in human and animal foodstuffs. The increased levels in human foods are important to human health in many underdeveloped countries. However, repeated applications of manures may eventually cause accumulation of excessive levels of trace minerals in soils and decreased plant growth (Nys et al., 2018). Unlike excess land application of N and P, most trace minerals do not migrate to water supplies but remain bound to soil. In some geological regions excess trace minerals accumulation of Cu and Zn are especially problematic. Poultry are sometimes fed diets containing more than 10 times the NRC recommended levels for Cu or Zn, presumably because high levels are antimicrobial and replace in-feed antibiotics (Chapter 11). The use of high levels of these trace minerals is often prohibited by legislation in areas prone to environmental degradation due to these minerals (Nys et al., 2018). Natural Se levels in the soil are highly variable and fertilization with animal manures can be helpful in those areas that are deficient but can exacerbate toxicities in regions that have high soil Se levels. For this reason, the concentration of Se in poultry diets is regulated in many countries, including the United States. Arsenic-containing dietary additives used in the diet (e.g., some coccidiostats) may result in excessive accumulate of As in soils (NRC, 2005), but most of these additives are no longer permitted for use in feed and have been removed from the market.

GREENHOUSE GAS EMISSIONS

Greenhouse gases and their associated global warming potential (GWP) are emitted at all stages of production (Kebreab et al., 2016; Martinelli et al., 2020; Costantini et al., 2021). These stages include (1) producing feedstuffs, including growing commodities of plant origin (including fertilizer), transporting those commodities, and the changes in land use that result; (2) processing commodities into feedstuffs appropriate for poultry and incorporation of those feedstuffs into poultry feeds via manufacturing and transport; (3) bird production including reproduction, grow-out, housing, and manure disposal; and (4) processing poultry products and transporting products to retail stores.

Research examining the lifecycle GWP of poultry production has lagged behind that of other livestock species. This is likely due to poultry's higher efficiency of incorporating dietary nutrients into tissues used as human foods, and due to the lower emission of methane from enteric and manure sources (Roma et al., 2015; Seidavi et al., 2019). Producing crops such as grains and legumes for feedstuffs is the largest contributor to GWP of the previous four stages of poultry production and GWP of producing high protein feedstuffs are especially great (Tallentire et al., 2017; Méda et al., 2019).

Poultry nutritionists can contribute substantially toward mitigating GWP emissions by formulating diets that (1) maximize the efficiency of feed conversion (Kebreab et al., 2016; Tallentire et al., 2017; Heidari et al., 2021); (2) incorporate protein sources with lower embedded GWP and by using purified amino acid supplements and complementary approaches to minimize dietary protein levels (Kebreab et al., 2016; Cappelaere et al., 2021); and (3) include an estimate of the GWP for the production of each specific ingredient in the matrix used for feed formulation and include GWP of the diet as a formulation constraint based on the value provided by marketing and governmental incentives (Tallentire et al., 2017; Garcia-Launay et al., 2018; Méda et al., 2019).

Currently, the contribution of some important sources to GWP is controversial and require refinement to be reliable. In particular, the contributions from land use change due to crop production (e.g., soybean meal) and due to the use of co-products from human food (e.g., meat meals) and energy production (e.g., distillers' dried grains with solubles) are highly debated (NRC, 2011; Martinelli et al., 2020). For example, a lifecycle assessment study by Tallentire et al. (2017) found least-cost diets that minimize GWP of broiler production in the United States, without impinging on production rates or efficiencies, provide a reduction of GWP of 7 percent and increase the cost of the diet by 25 percent. For comparison, in the United Kingdom, formulating to minimize GWP decreased GWP by 37 percent and increased cost by 17 percent. This study also found that the UK broiler production system had a much greater GWP than the U.S. system, largely due to the higher GWP of soybean meal imported from South America and the legal prohibition of incorporating animal co-products (e.g., meat meal) into poultry diets. Because the U.S. poultry market predominantly feeds domestically produced soybean meal, the associated GWP due to land use change is markedly lower. However, attribution of GWP due to land use change is controversial, and the consensus continually evolves (NRC, 2011; Castanheira and Freire, 2013; Martinelli et al., 2020). Continued developments in estimating the GWP of feedstuffs grown in different regions will be necessary for accurate predictions on the impact of animal production on climate change.

Diets optimized to minimize GWP are often not helpful at simultaneously reducing other environmentally important emissions, such as ammonia and P, and in many cases actually increases them. Such "pollution swapping" has been observed across several different feed formulation objectives (Nguyen et al., 2012; Garcia-Launay et al., 2018). Future research should work to solve the methodological challenge of accounting for the aggregated environmental impacts caused by the relative environmental impacts of N, P, and GWP (Mackenzie et al., 2016; Tallentire et al., 2017; Méda et al., 2021).

The environmental impacts of specific feed ingredients fed to poultry have been tabulated by several research groups and used to formulate diets using traditional least-cost linear programming, goal programming, and optimization approaches (Méda et al., 2019; Heidari et al., 2021). At this time the geographical area considered for the above estimates of environmental and sustainability impacts are at the country level and do not capture the large regional differences in important agricultural practices (Nguyen et al., 2012; Pelletier, 2017; Heidari et al., 2021). Resolving these information gaps should be an important area for future research.

REFERENCES

Angel, C.R., W.J. Powers, T.J. Applegate, N.M. Tamim, and M.C. Christman. 2005. Influence of phytase on water-soluble phosphorus in poultry and swine manure. *Journal of Environmental Quality*. 34:563–571.

Angel, R., W. Saylor, A. Mitchell, W. Powers, and T. Applegate. 2006. Effect of dietary phosphorus, phytase, and 25-hydroxycholecalciferol on broiler chicken bone mineralization, litter phosphorus, and processing yields. *Poultry Science*. 85:1200–1211.

Baker, D.H. 1997. *Ideal Amino Acid Profiles for Swine and Poultry and Their Applications in Feed Formulation*. Cape Girardeau, MO: Biokyowa.

Baker, D.H., and O.A. Izquierdo. 1985. Effect of meal frequency and spaced crystalline lysine ingestion on the utilization of dietary lysine by chickens. *Nutrition Research*. 5:1103–1112.

Belloir, P., B. Méda, W. Lambert, E. Corrent, H. Juin, M. Lessire, and S. Tesseraud. 2017. Reducing the CP content in broiler feeds: Impact on animal performance, meat quality and nitrogen utilization. *Animal*. 11:1881–1889.

Borda-Molina, D., T. Zuber, W. Siegert, A. Camarinha-Silva, D. Feuerstein, and M. Rodehutscord. 2019. Effects of protease and phytase supplements on small intestinal microbiota and amino acid digestibility in broiler chickens. *Poultry Science*. 98:2906–2918.

Brown, A., J. Lee, R. Adhikari, K. Haydon, and K. Wamsley. 2021. Determining the optimal digestible isoleucine to lysine ratio of Ross 708 × Ross YP male broilers from 14 to 28 days of age. *Journal of Applied Poultry Research*. 30:100192.

Cappelaere, L., J. Le Cour Grandmaison, N. Martin, and W. Lambert. 2021. Amino acid supplementation to reduce environmental impacts of broiler and pig production: A review. *Frontiers in Veterinary Science*. 799.

Castanheira, É.G., and F. Freire. 2013. Greenhouse gas assessment of soybean production: Implications of land use change and different cultivation systems. *Journal of Cleaner Production*. 54:49–60.

Costantini, M., V. Ferrante, M. Guarino, and J. Bacenetti. 2021. Environmental sustainability assessment of poultry productions through life cycle approaches: A critical review. *Trends in Food Science & Technology*. 110:201–212.

Cowieson, A., J. Ruckebusch, I. Knap, P. Guggenbuhl, and F. Fru-Nji. 2016. Phytate-free nutrition: A new paradigm in monogastric animal production. *Animal Feed Science and Technology*. 222:180–189.

Cowieson, A., M. Toghyani, S. Kheravii, S. Wu, L. Romero, and M. Choct. 2019. A mono-component microbial protease improves performance, net energy, and digestibility of amino acids and starch, and upregulates jejunal expression of genes responsible for peptide transport in broilers fed corn/wheat-based diets supplemented with xylanase and phytase. *Poultry Science*. 98:1321–1332.

Dersjant-Li, Y., M. Hruby, C. Evans, and R. Greiner. 2019. A critical review of methods used to determine phosphorus and digestible amino acid matrices when using phytase in poultry and pig diets. *Journal of Applied Animal Nutrition*. 7.

Ferket, P., E. Van Heugten, T. Van Kempen, and R. Angel. 2002. Nutritional strategies to reduce environmental emissions from nonruminants. *Journal of Animal Science*. 80:E168–E182.

Garcia-Launay, F., L. Dusart, S. Espagnol, S. Laisse-Redoux, D. Gaudré, B. Méda, and A. Wilfart. 2018. Multiobjective formulation is an effective method to reduce environmental impacts of livestock feeds. *British Journal of Nutrition*. 120:1298–1309.

Heidari, M.D., S. Gandasasmita, E. Li, and N. Pelletier. 2021. Proposing a framework for sustainable feed formulation for laying hens: A systematic review of recent developments and future directions. *Journal of Cleaner Production*. 288:125585.

Kebreab, E., A. Liedke, D. Caro, S. Deimling, M. Binder, and M. Finkbeiner. 2016. Environmental impact of using specialty feed ingredients in swine and poultry production: A life cycle assessment. *Journal of Animal Science*. 94:2664–2681.

Kerr, B. 1995. Nutritional strategies for waste reduction management: Nitrogen. *Proceedings of New Horizons in Animal Nutrition and Health.* Raleigh, NC.

Kidd, M., C. Maynard, and G. Mullenix. 2021. Progress of amino acid nutrition for diet protein reduction in poultry. *Journal of Animal Science and Biotechnology.* 12:1–9.

Kidd, M., T. Nelson, R. Brister, and M. Donohue. 2016. Chapter 7, Current and future amino acid formulation trends–phytate and phytase history and implications. In *Phytate Destruction-Consequences for Precision Animal Nutrition.* Wageningen, The Netherlands: Wageningen Academic Publishers. 1298–1304.

Kidd, M., P. Tillman, P. Waldroup, and W. Holder. 2013. Feed-grade amino acid use in the United States: The synergetic inclusion history with linear programming. *Journal of Applied Poultry Research.* 22:583–590.

Kotak, B. 1993. Occurrence and toxicological evaluation of cyanobacterial toxins in Alberta lakes and farm dugouts. *Water Research.* 27:495–506.

Lee, S., M. Bedford, and C. Walk. 2018. Meta-analysis: Explicit value of mono-component proteases in monogastric diets. *Poultry Science.* 97:2078–2085.

Mackenzie, S., I. Leinonen, N. Ferguson, and I. Kyriazakis. 2016. Towards a methodology to formulate sustainable diets for livestock: Accounting for environmental impact in diet formulation. *British Journal of Nutrition.* 115:1860–1874.

Maguire, R., Z. Dou, J. Sims, J. Brake, and B. Joern. 2005. Dietary strategies for reduced phosphorus excretion and improved water quality. *Journal of Environmental Quality.* 34:2093–2103.

Martinelli, G., E. Vogel, M. Decian, M.J.U.S. Farinha, L.V.M. Bernardo, J.A.R. Borges, R.M.T. Gimenes, R.G. Garcia, and C.F. Ruviaro. 2020. Assessing the eco-efficiency of different poultry production systems: An approach using life cycle assessment and economic value added. *Sustainable Production and Consumption.* 24:181–193.

Méda, B., P. Belloir, A. Narcy, and A. Wilfart. 2019. Improving environmental sustainability of poultry production using innovative feeding strategies. In *22nd European Symposium on Poultry Nutrition.*

Méda, B., F. Garcia-Launay, L. Dusart, P. Ponchant, S. Espagnol, and A. Wilfart. 2021. Reducing environmental impacts of feed using multi-objective formulation: What benefits at the farm gate for pig and broiler production? *Animal.* 15:100024.

Nahm, K. 2002. Efficient feed nutrient utilization to reduce pollutants in poultry and swine manure. *Critical Reviews in Environmental Science and Technology.* 32:1–16.

Nguyen, T.T.H., I. Bouvarel, P. Ponchant, and H.M. van der Werf. 2012. Using environmental constraints to formulate low-impact poultry feeds. *Journal of Cleaner Production.* 28:215–224.

NRC (National Research Center). 1994. *Nutrient Requirements of Poultry,* 9th rev. ed. Washington, DC: National Academy Press.

NRC. 2005. *Mineral Tolerance of Animals,* 2nd rev. ed. Washington, DC: National Academy Press.

NRC. 2011. *Renewable Fuel Standard: Potential Economic and Environmental Effects of U.S. Biofuel Policy.* Washington, DC: The National Academies Press.

Nys, Y., P. Schlegel, S. Durosoy, C. Jondreville, and A. Narcy. 2018. Adapting trace mineral nutrition of birds for optimising the environment and poultry product quality. *World's Poultry Science Journal.* 74:225–238.

Pelletier, N. 2017. Life cycle assessment of Canadian egg products, with differentiation by hen housing system type. *Journal of Cleaner Production.* 152:167–180.

Raboy, V. 2020. Low phytic acid crops: Observations based on four decades of research. *Plants.* 9:140.

Ravindran, V. 1996. Occurrence of phytic acid in plant feed ingredients. *Phytase in Animal Nutrition and Waste Management. A BASF Reference Manual.* M.B. Coelho and E.T. Kornegay, editors. Mount Olive, NJ: BASF Corporation.

Ravindran, V., and J.-H. Son. 2011. Feed enzyme technology: Present status and future developments. *Recent Patents on Food, Nutrition & Agriculture.* 3:102–109.

Roma, R., S. Corrado, A. De Boni, M.B. Forleo, V. Fantin, M. Moretti, N. Palmieri, A. Vitali, and D.C. Camillo. 2015. Life cycle assessment in the livestock and derived edible products sector. In *Life Cycle Assessment in the Agri-Food Sector.* New York: Springer. 251–332.

Rosov, K.A., M.A. Mallin, and L.B. Cahoon. 2020. Waste nutrients from U.S. animal feeding operations: Regulations are inconsistent across states and inadequately assess nutrient export risk. *Journal of Environmental Management.* 269:110738.

Seidavi, A., H. Zaker-Esteghamati, and C. Scanes. 2019. Present and potential impacts of waste from poultry production on the environment. *World's Poultry Science Journal.* 75:29–42.

Sharpley, A.N., S. Chapra, R. Wedepohl, and J.T. Sims. 1994. Managing agricultural phosphorus for protection of surface waters: Issues and options. *Journal of Environmental Quality.* 23:437–451.

Shelton, J., L. Southern, L. Gaston, and A. Foster. 2004. Evaluation of the nutrient matrix values for phytase in broilers. *Journal of Applied Poultry Research.* 13:213–221.

Tallentire, C., S. Mackenzie, and I. Kyriazakis. 2017. Environmental impact trade-offs in diet formulation for broiler production systems in the UK and USA. *Agricultural Systems.* 154:145–156.

Van Harn, J., M. Dijkslag, and M. Van Krimpen. 2019. Effect of low protein diets supplemented with free amino acids on growth performance, slaughter yield, litter quality, and footpad lesions of male broilers. *Poultry Science.* 98:4868–4877.

Wiliams, C., J. Barker, and J. Sims. 1999. Management and utilization of poultry wastes. *Review of Environmental Contamination and Toxicology.* 105–157.

11

Nutritional Implications of Environmental Stressors

INTRODUCTION

A bird's nutrient requirements arise from its genetics and physiological state. The ability of a diet to meet the nutrient requirements of a bird depends on the amounts and availability of nutrients present in the diet. The bird's nutrient requirements can be altered by various stresses that alter the bird's physiology in ways that cause maldigestion/malabsorption, decrease nutrient utilization, or increase cell turnover or nutrient turnover/loss (Combs, 2012). Stress responses can be broadly defined as host defense mechanisms against environmental stimuli (i.e., stressors) designed to restore homeostatic conditions in the bird (Siegel, 1995). Regardless of whether in a commercial production setting or backyard environment, the ability of birds to counteract stressors largely depends on the type and severity of the collective stressors being imposed when mounting a stress response. Hormesis is a dose-dependent stress response phenomenon characterized by stimulation of performance by low-dose stress and compromised performance with high-dose stress (Calabrese and Baldwin, 2002). Poultry species are exposed to a wide variety of environmental stressors that represent both direct and indirect threats to the health and well-being of the bird, including stressors related to climatic, ambient environmental, nutrition, physical, social, psychological, microbiological, and pathological factors (Thaxton and Siegel, 1970; Davison and Flack, 1981; Gross and Siegel, 1983; McFarlane et al., 1989; Maxwell et al., 1992; Rosales, 1994; Scheele, 1997; Puvadolpirod and Thaxton, 2000; Post et al., 2003; Cheng and Muir, 2004; Mashaly et al., 2004; Hangalapura et al., 2005; Odihambo Mumma et al., 2006; Shini et al., 2008; Freeman, 2013). Importantly, some stressors act by directly altering nutrient requirements of the bird, while others (e.g., heat stress) act by suppressing appetite and thus the ingestion of feed and the nutrients it contains. Birds in either situation may benefit from an increase in the nutrient density of feed, but for different reasons (Siegel, 1995). Some stress responses can be altered through selective breeding strategies (Brown and Nestor, 1973; Gross and Siegel, 1985; Muir and Craig, 1998; Jones and Hocking, 1999). Additionally, emerging evidence suggests that a stress response can be shaped by host–microbe interactions (Pan and Yu, 2014; Allaire et al., 2018), further complicating interpretation of diet matrix effects on nutrient requirements versus nutrient availability.

Short-term stressors (e.g., a transient change in lighting) typically have minor impacts in poultry production, but long-term or chronic exposure to stressors elicits a coordinated stress response that will negatively affect the economics of production (Virden and Kidd, 2009). Such regulatory responses may be specific to a stressor and thereby influence a particular organ or targeted physiological pathway, or result in a nonspecific response, thereby culminating in a generalized and pervasive state of stress experienced by the bird (Siegel, 1995). Regardless of the stressor type and duration, however, the combined neuroendocrine and immune systems serve to integrate responses through consolidated pathways. Birds experiencing stress beyond the level of hormesis manifest a diverse array of physiologic, metabolic, behavioral, and immune effects that often include negative impacts on digestive health, nutrient partitioning to muscle and other organs, and alterations of nutrient requirements of the bird.

Upon experiencing an acute stressor, activation of the sympathetic–adrenomedullary system causes release of catecholamines (i.e., epinephrine and norepinephrine) as part of the "fight-or-flight" response. This results in marked increases in blood pressure, muscle tone, nerve sensitivity, circulating glucose concentrations, and respiration (Siegel, 1980; Virden and Kidd, 2009). Continued exposure to the stressor(s) in the absence of habituation results in subsequent activation of the hypothalamic–pituitary–adrenal (HPA) axis and release of glucocorticoids, including secretion of corticotrophin-releasing factor from the hypothalamus and adrenocorticotrophic hormone (ACTH) from the pituitary, and ultimately the systemic release of corticosterone from the adrenal cortex of the bird (Shini et al., 2008; Virden and

Kidd, 2009). As a hallmark indicator of the stress response in birds, sustained elevations in circulating corticosterone concentrations elicit a myriad of physiological effects, including alterations in glucose, energy, and mineral metabolism (e.g., inhibition of skeletal calcification), detriments to immune system function (e.g., regression of the thymus, spleen, and bursa of Fabricius), disruption of intestinal integrity and barrier function, and shifts in nutrient partitioning between tissues (e.g., increased lipogenesis and decreased muscle protein accretion) (Siegel, 1995). Collectively, the physiological effects of chronic stress result in losses in productivity due to decreased feed intake and detriments to productive and reproductive outcomes, and the issue is potentially compounded by negative changes to the composition and yield of saleable poultry products.

Whereas physiological consequences of chronic stress are relatively well-characterized, especially in the context of animal well-being or welfare, there remain many gaps in knowledge about the nutritional implications of stress for poultry species. This chapter will provide an overview of available and objective evidence of where dietary nutrient requirements were quantified in birds exposed to defined environmental stressors. Most research on nutrition and health has involved thermal stressors (heat or cold stress) and their influence on intestinal barrier integrity and function, so there are natural connections with outcomes including efficiency of nutrient utilization and metabolism. However, relatively less is known about how nutrient requirements are affected by other stressors, including the physical environment associated with different management systems, form and quality of diets, or items related to husbandry, handling, and transport. Moreover, evidence of how interactions among the microbiota, host, and immune system influence nutrient requirements are lacking, largely due to the additional variation in responses of different populations of birds to a defined immune activation protocol. Current evidence strongly supports the notion that broad overlap exists between the microbiota and host (Klasing et al., 1987; Klasing, 1988; Klasing and Leshchinsky, 1999; Klasing, 2004), especially with regard to intestinal health, which is an important and burgeoning area of research for establishing nutrient requirements.

The effects of chronic stress are completely dependent on environmental circumstances, and as a result it is imperative to realize that no single set of nutrient recommendations can be universally applied, even within a category of poultry or at a given location. Thus, variations in genetics, microbiological/immune status, and management styles among production settings will ostensibly require different approaches to understand the implications of stress on nutritional needs. Additionally, evidence suggests that when occurring simultaneously, individual stressors may elicit additive or synergistic effects, and the cumulative stress response experienced by the bird is also shaped by the magnitude and duration of the stressors. Most of the available evidence presented in the available literature on the interaction of stress and nutrition involves chickens (laying hens, broiler breeders, and broilers) and to a lesser extent Japanese quail. Finally, where applicable, reference to other chapters within this report will be made with respect to the effects of stress on either specific poultry categories, classes of nutrients, or energy metabolism.

ENVIRONMENTAL STRESSORS RELATED TO HOUSING CONDITIONS

A variety of poultry species are raised for meat, egg, and feather production; game hunting; embryo production for the vaccine industry; and exhibition purposes, resulting in inherent variation in bird management. Management practices and housing conditions may induce physiologic, metabolic, nutritional, or even behavioral stressors to a bird and thereby influence nutrient requirement recommendations.

Housing Systems and Stocking Density

Housing type may influence activity levels, energy expenditure, and exposure to oxidative and pathogenic stressors, as well as impart physical strain on the skeletal system of poultry species (e.g., keel bone perturbations in laying hens raised in aviaries). Each of these variables has the potential to alter nutrient requirements of the bird. Non-cage systems have been shown to increase feed intake, decrease body weight gain, and impair feed efficiency. These effects were attributed to increased activity levels and the resultant increase in energy expenditure (Lebas, 2001; Elson and Croxall, 2006; Bogosavljević-Bošković et al., 2012; Leenstra et al., 2014). However, there is no consensus on whether enriched (e.g., furnished) or conventional caging systems yield better, more desirable production outcomes. Some studies suggest similar production outcomes among the different types of caging systems (Valkonen et al., 2008) while other reports suggest lower feed intake (Elson and Croxall, 2006) and higher utilization of Ca and P with the enriched cages (Neijat et al., 2011). Differences in energy expenditure and mineral requirements may exist between housing systems, which is why research is warranted to identify specific changes in nutrient requirements.

A variable feature of the physical rearing environment is stocking density (i.e., the number of birds housed in a given area of floor space), which influences the natural behaviors of birds. Several studies reported that increased stocking density decreases feed intake and weight gain (Cengiz et al., 2015) as attributed to a restriction of movement (Sørensen et al., 2000; Simitzis et al., 2012). One study provided evidence for a poorer feed conversion ratio (Cengiz et al., 2015), while others found no effect, when stocking density was increased (Puron et al., 1995; Simitzis et al., 2012; Gopinger et al., 2015). While there are conflicting reports, any factors that influence the ability of birds to access feed and water will concomitantly alter productive or reproductive outcomes, and possibly lead to increased mortality. In general, there is insufficient evidence to suggest specific changes in nutrient intake or requirements due to the multitude of variations that exist within housing environments used today. This is a management variable where focused research is needed.

Ambient Environment (Temperature, Humidity, and Ventilation)

Even small alterations in ambient conditions can greatly influence physiologic systems within the bird, which is why the impact of this variable on nutrient requirements has received a great deal of attention. As an example, the ideal temperature for optimizing lean tissue accretion is not the same as the ideal temperature for optimizing feed efficiency or egg weight (Daghir, 2008). Non-cage housing may subject birds to colder temperatures in the winter, potentially due to lower stocking density and/or outdoor access (Miao et al., 2005; Green and Xin, 2009; Xin et al., 2011); exposure to lower temperatures may increase energy expenditure and feed intake to maintain body temperature (Miao et al., 2005; Xin et al., 2011; Cândido et al., 2016). Thus, cold stress alone is known to increase feed intake and so will influence the concentrations and patterns of nutrients required to manage birds with the goal of achieving production targets.

Heat stress has a more pronounced effect in breeds with high growth potential, such as commercial broilers compared to slower-growing breeds (Cahaner and Leenstra, 1992; Cahaner et al., 1995; Yunis and Cahaner, 1999). For every 1.0°C increase in the temperature range of 22–32°C, feed intake may be reduced by 1.2 percent, and for every 1.0°C increase in the temperature range of 32–38°C, feed intake may be reduced by as much as 5 percent (Naga Raja Kumari and Narendra Nath, 2017). Moreover, metabolizable energy requirements decrease with temperatures above 21°C, reaching a minimum at approximately 27°C, with increased energy requirements being recognized for temperatures above 27°C and up to 34°C (Daghir, 2008). Nutrients involved in temperature regulation and energy expenditure would be expected to change as birds experience heat stress. Moreover, differences in heat increment associated with each nutrient class should be considered when formulating diets based on the ambient temperature and humidity of a given rearing context, as explained in detail in Chapter 1 (Energy).

The relationship between ambient temperature and feed intake in poultry makes it crucial to realize that a change in the consumption of an absolute mass of feed has a direct impact on formulated nutrient profiles. Thus, when formulated nutrients are expressed in proportion to the total amount of feed (e.g., percent of the diet), that relative inclusion level must increase to permit the bird to ingest the same absolute nutrient mass (e.g., mg of nutrient per day) to meet its physiological requirements. In this regard, several classes of nutrients are best expressed relative to the energy density of the diet (e.g., amino acids) as the bird will effectively adjust consumption patterns to satisfy its energy intake.

The impact of ambient temperature on the stress experienced by birds is further affected by humidity and ventilation. Elevated humidity is known to exacerbate the effect of high temperatures, because high humidity increases the workload of dissipating excess body heat (Balnave, 2004). As such, evaporative cooling is a critical regulatory mechanism to maintain body temperature when ambient temperature is above 26°C. Evaporative cooling occurs by conversion of liquid water into vapor using the thermal energy in the air, resulting in a lower air temperature at the evaporate surface, and this is why the phenomenon is largely driven by body surface area. In birds, the respiratory tract is the primary site of evaporative cooling via panting, gular fluttering, and in some species, tissue swelling to increase evaporative surface area (Weathers and Schoenbaechler, 1976; Gaunt, 1980; St-Laurent and Larochelle, 1994). Species vary in their heat tolerance; for example, the maximum growth rate of broilers raised at 35°C can be achieved at a higher relative humidity than the maximum growth rate of turkeys (Yahav et al., 1995). However, temperature and humidity are integrally linked to ventilation rate, and inadequate ventilation can cause air stagnation zones of higher temperatures and ammonia accumulation, especially for caged birds housed closer to the floor (Tong, 2019). Preventing these zones by increasing ventilation can decrease heat stress and increase growth performance and egg production (Wang, 2019), but caution must be exercised as birds housed closer to exhaust units may yield undesirable production outcomes (Tong, 2019). Overall, there are known links between ambient environmental conditions and physiologic and metabolic stress in birds, with implications toward nutrient intake linked to feed and energy intake.

Ammonia

Exposure to ammonia causes irritation of mucous membranes of the respiratory tract and eyes (Kristensen and Wathes, 2007). As related to ventilation rates described previously, there is consensus that ammonia levels at or above 50 ppm decrease bird performance due to decreased feed intake and poorer feed conversion that results in reduced weight gain (Charles and Payne, 1966; Quarles and Kling, 1974; Reece et al., 1980; Miles et al., 2004; Kristensen and Wathes, 2007). Moreover, some studies suggest that ammonia as low as 25 ppm may reduce weight gain of broilers (Reece et al., 1980). The decrease in live weight gain due to relatively high environmental ammonia concentrations may also delay the onset of maturity in pullets (Charles and Payne, 1966). With regard to the link between dietary intake and ammonia, most research emphasis has been placed on nutritional strategies to reduce the excretion of uric acid, which is readily converted by ammonia by microbial enzymes. Litter moisture management is an important non-dietary variable to limit ammonia production at different ambient temperatures (Miles et al., 2011). Whereas there is evidence to link inhalation of ammonia and reduced productive performance of birds, no studies currently exist on whether ammonia exposure elicits changes in dietary nutrient requirements.

Lighting

Lighting conditions are known to influence reproduction and growth of birds, which makes this an important tool in management of the production and well-being of poultry species. However, there exist conflicting results with regard to ideal lighting conditions for optimal production outcomes, and this point is further complicated by the variety of housing types used in the poultry industry. In general, establishing ambient artificial lighting conditions involves manipulation of duration, wavelength, and intensity of lighting (Deep et al., 2010). The effects of lighting are most clearly interpreted as changes in feed intake patterns, which is how this environmental factor may lead to changes in nutrient intake and ability to achieve maximal performance. With regard to photoperiod, some studies suggest that nearly continuous lighting may elicit optimal feed efficiency and body weight gain of broilers (Abreu et al., 2011), while other evidence suggests that intermittent lighting may reduce feed intake to yield better feed efficiency (Buyse et al., 1996). Additionally, fluorescent bulbs were shown to cause lower body weights when compared with incandescent lighting (Rogers et al., 2015; Morrissey et al., 2017). Light from light-emitting diode (LED) bulbs of various colors may elicit the same response (Santana et al., 2014) or higher body weight gain (Morrissey et al., 2017) compared with fluorescent bulbs.

Higher light intensity has also been associated with increased activity level in laying hens (Boshouwers and Nicaise, 1987), which may cause an increase in energy expenditure and thereby require changes in formulated dietary nutrient concentrations. Moreover, increased light intensity has also been associated with decreased feed intake in pullets (Lewis and Morris, 2007). In broilers, variation in lighting intensity ranging from 1 to 40 lux (i.e., common intensity levels used within commercial broiler operations) did not affect broiler performance but did change carcass characteristics, including decreases in carcass, thigh, and drum yield as a percentage of live weight (Deep et al., 2010). Altered carcass composition could warrant altered dietary nutrient intake in situations where the changes are desirable. Overall, a relationship between ambient lighting and performance likely exists, but no specific nutrient recommendations can be ascribed to a particular lighting condition until further research is conducted.

DIETARY AND NUTRITIONAL STRESSORS

Feed can also be an environmental stressor through its nutritional composition, as well as through the physical and chemical properties of feed ingredients used to formulate poultry diets. In this way, the diet can contain anti-nutritional factors, feed deterioration products (e.g., peroxyradicals of oxidized dietary oils and fats), and mycotoxins. Additionally, immunogens, compounds that stimulate an immune reaction, may be derived from the diet as either nutrients per se, non-nutritive compounds (e.g., soluble β-mannans), or microbiological substrates (e.g., of pathogenic or nonpathogenic origin; discussed later in this chapter). Finally, diet form (e.g., mash, pellet, and crumbles) in various feeding phases may constitute a stressor through impacts on intestinal morphology, health, and overall productivity of the bird (Chapter 8: Digestibility and Bioavailability).

Mycotoxins

Feedstuffs contaminated with mycotoxins have adverse effects on poultry productivity and health (Binder et al., 2007; Guerre, 2016). Molds of the genera *Aspergillus*, *Penicillium*, *Fusarium*, *Claviceps*, and *Alternaria* are commonly encountered in feedstuffs used in poultry production, and these molds can produce several toxins including aflatoxins, zearalenone, T2-toxin, deoxynivalenol, ochratoxin A, fumonisins, ergot alkaloids, and patulin (Hussein and Brasel, 2001; Dänicke, 2015, 2016, 2017). Chronic ingestion of low mycotoxin concentrations via contaminated feed is associated with poor production performance (Hamilton et al., 1982). However, an acute mycotoxicosis can cause organ dysfunction [e.g., liver damage in acute aflatoxicoses and fumonisin toxicoses (Weibking et al., 1993; Devegowda et al., 2005) and kidney failure in ochratoxicosis (Glahn et al., 1988)], immune suppression (Leeson et al., 1995), and high mortality rates (Hamilton et al., 1982). Several environmental factors including optimum moisture, temperature, and a slightly elevated CO_2 concentration favor growth of mycogenic fungal species in feedstuffs (Magan et al., 2011). Feed contaminated with a combination of different mycotoxins has severe toxigenic potential due to synergism/additive effects (CAST Task Force, 2003). Therefore, exercising caution during feed storage is critical to minimize the incidence of mycotoxicoses in a production setting. The global nature of the poultry industry includes the movement of feedstuffs between diverse climates, which may cause the spread of new genera of mycotoxin-producing fungi (Bryden, 2012). Whereas ingestion of mycotoxins causes oxidative stress and disruption of intestinal barrier function, the predominant effect is an overall reduction in feed intake (Antonissen et al., 2014; Chen et al., 2016; Liu et al., 2020). However, further research is warranted to elucidate how nutrient requirements change in poultry exposed to mycotoxins.

Alkaloids and Phytosterols

Pyrrolizidine, tropane, and steroid alkaloids are naturally occurring and found in some plant species (Elkin, 2017). In poultry, feed contaminated with seeds from *Crotalaria* spp.

are thought to be the main source of pyrrolizidine alkaloids, which are hepatotoxic (Gaul et al., 1993) and can cause poor growth in broilers (Hess and Mosjidis, 2008). Chronic ingestion of pyrrolizidine alkaloid-contaminated feed can cause decreased egg production, increased mortality, and liver damage with internal hemorrhage and ascites in laying hens (Schmittle et al., 1959). Major tropane alkaloids include atropine, scopolamine, and hyoscyamine, which are found mainly in plants of the *Datura* genera. Contamination of soybean and wheat products with seeds from *Datura* spp. is a common occurrence (Bucher and Meszaros, 1989). Generally, tropane toxicity causes respiratory and cardiac disturbances in species other than poultry (Kingsbury, 1964). However, growth restriction without any respiratory and cardiac symptoms was observed in broilers fed *Datura*-derived feedstuffs at a level of 150 mg alkaloid/kg feed (Kovatsis et al., 1993). Phytosterols are present as a minor constituent in vegetable oils, cereals, nuts, legumes, and vegetables. Phytosterols are considered as a useful dietary component as they reportedly lower serum cholesterol concentrations in humans (Kritchevsky and Chen, 2005). However, dietary bioavailable soy phytosterols have been shown to be without effect on either plasma or egg yolk cholesterol (expressed as mg per yolk) content when fed to laying hens (Elkin and Lorenz, 2009; Liu et al., 2010).

Glycosides

Plant glycosides are composed of a sugar moiety bound to another functional group (e.g., phenol or sterol) via a glycosidic bond (Elkin, 2017). Goitrogenic glycosides are found mostly in plants of the *Crucifera* family but also in oilseed meals derived from canola (rapeseed) and crambe. These substances can cause altered synthesis of thyroid hormones, thereby affecting temperature and metabolic regulation (Divi et al., 1997), but data on toxicity of goitrogenic glycosides in poultry species are lacking. Cyanogenic glycosides are more problematic in grazing animals than poultry, and the main source of cyanogenic glycosides in poultry would likely be cassava root. Chronic cassava supplementation in poultry lowers egg quality and production (Jalaludin and Leong, 1973) and reduces hatchability (Ngoka et al., 1982). In broilers, a total cyanogen (HCN) content of 140 mg/kg of diet reduced growth (Panigrahi et al., 1992).

Saponins are triterpenes or steroids linked to one or more sugar groups, and these compounds possess detergent-like properties. They are found in common feed ingredients including products derived from soybeans, peas, sugar beet, sunflower, oats, capsicum, yam, yucca, alliums, tea, and others (Fenwick et al., 1991). Soybean saponins failed to impart growth depression when provided at up to five times the normal level of soybean inclusion (Ishaaya et al., 1969). However, experimental feeding of tea saponins negatively affected growth performance of broilers (Ueda and Tanoue, 2000). The negative effects of saponins on growth appear mainly due to reduced feed intake (Jacobsen et al., 1997), intestinal motility (Klita et al., 1996), and protein digestibility (Shimoyamada et al., 1998). Faba beans (*Vicia faba*) are used as a substitute for soybean in many parts of the world, and this feedstuff contains high concentrations of vicine/convicine, which is a glucosidic aminopyrimidine derivative. High inclusion of unprocessed (i.e., unheated) faba beans reduced growth performance and feed efficiency of broilers, increased mass of the liver and pancreas (Marquardt et al., 1974), and decreased egg production and egg weight in layers (Robblee et al., 1977; Lessire et al., 2017).

Protease Inhibitors

Protease inhibitors occur mainly in legumes, potatoes, and cereal grains. The quantity of inhibitor depends on the variety and physiological status of the plant, as well as the level of insect infestation and damage during the growing season (Richardson, 1977). Trypsin and chymotrypsin inhibitory proteins have been characterized from a variety of legume seeds including soybeans (Liu and Markakis, 1991), red kidney beans (Wu and Whitaker, 1990), and pigeon peas (Godbole et al., 1994). Protease inhibitors are classified as part of either the Bowman-Birk (10 kDa; inhibits only trypsin) or Kunitz (21 kDa; inhibits both trypsin and chymotrypsin) families of inhibitors (Xavier-Filho and Campos, 1989). Upon ingestion, protease inhibitors reduce the availability and efficiency of active protein digestive enzymes by forming direct, nondissociable complexes. This causes pancreatic hypertrophy and hyperplasia due to the continual need for increased enzyme secretion to aid digestion. The negative implications are further exacerbated by the passage of undigested dietary protein into the ceca and colon, where microbial fermentation of proteins occurs to produce noxious compounds including biogenic amines. As such, ingestion of feedstuffs containing protease inhibitors reduces feed intake and impairs digestion, absorption, and metabolism of nutrients to negatively affect the overall health and productivity of birds, including by increasing the incidence of necrotic enteritis (Liener, 1994; Clarke and Wiseman, 2005; Palliyeguru et al., 2010, 2011). The effects of protease inhibitors can be counteracted by increasing dietary amino acid concentrations, but this approach inherently increases excretion of nitrogenous compounds into the environment (Liener, 1994). Thus, mitigation strategies that include ingredient processing methods to reduce or remove activity of the native protease inhibitors (e.g., toasting of soybeans in the production of soybean meal) enhance environmental sustainability.

Lectins

Lectins are carbohydrate-binding proteins found mainly in legumes. Because they are resistant to digestive processes, lectins can bind to glycoprotein receptors on enterocytes within the gastrointestinal tract (GIT) of poultry and thereby reduce growth performance by interfering with nutrient absorption (Pusztai et al., 1990). Lectin binding can cause disruption of brush border integrity and reduce epithelial cell viability, thus leading to hyperplasia within crypts and increasing intestinal weight (Liener, 1994). Collectively, these negative effects will alter feed intake patterns and increase maintenance requirements of the bird. However, insufficient evidence exists to support specific nutrient recommendations due to lectin-mediated stress.

Phenolic Compounds

Phenolic compounds contain one or more aromatic rings with one or more hydroxyl groups attached (Elkin, 2017). Major phenolic compounds that may be found in poultry diets include condensed tannins, gossypol, and sinapine. Condensed tannins are found mainly in sorghum, canola (rapeseed), faba beans, barley, and field peas. Feeding sorghum-derived tannins to broilers causes reduced growth (Kumar et al., 2005), reduced protein and amino acid digestibility (Kumar et al., 2007; Mansoori and Acamovic, 2009), and leg abnormalities (Elkin et al., 1978; Gualtieri and Rapaccini, 1990). Evidence on the negative effects of feeding sorghum tannins in laying hens is limited (Sell et al., 1983; Sell and Rogler, 1984). Interestingly, it has been shown that condensed tannins are only partially responsible for variations in nutrient digestibilities of sorghum grain cultivars (Elkin et al., 1996), and subsequent work (Elkin et al., 2002) suggested that the kafirin protein body structure, as well as the starch content/carbohydrate profile, may also be important in this regard.

Gossypol is the major anti-nutritional factor present in cotton seed. Gossypol reacts with amino acids and minerals due to the presence of reactive phenolic hydroxyl and aldehyde groups (Cheeke, 1998). Gossypol causes olive discoloration of egg yolks (Schaible et al., 1934) and firm, elastic cooked egg yolk texture (Qi et al., 2017). In broiler chicks, ingestion of 800 mg of gossypol/kg body weight for 21 days caused poor growth and feed efficiency, reduced plasma iron concentration, and resulted in liver damage and hepatic cholestasis (Henry et al., 2001). Canola (rapeseed) is the major source of sinapine, a choline ester of sinapic acid, in poultry diets, and it is known to impart fishy taint in eggs (Hobson-Frohock et al., 1977) in some genetic lines of laying hens (Ward et al., 2009). Sinapine alone is not known to cause growth depression in broilers (Qiao and Classen, 2003).

Amino Acid Derivatives

One of the well-studied nutritional stressors in nonruminant species is the reduced availability of amino acids due to formation of by-product compounds during feed processing. The best known example involves the ability for the epsilon amino group of lysine (Lys) to react with a reducing sugar (among other nutrients) via the Maillard reaction (Maillard, 1912; Hodge, 1953). Regardless of whether at an early (i.e., Lys is irreversibly bound) or advanced (i.e., Lys is destroyed) stage of product formation, the Maillard reaction effectively reduces the bioavailable content of Lys in an over-processed ingredient. As such, heat processing and unfavorable storage conditions of feed ingredients may reduce dietary bioavailable Lys concentrations to a greater extent than detectable through analytical means (Moughan et al., 2014). This is a classical example of a nutrient that can be quantified using analytical chemistry procedures (e.g., gross concentration), yet it effectively provides little to no nutritive value to the bird (i.e., net concentration). Bioassays involving poultry species can be conducted to empirically determine the reduction in Lys availability for feedstuffs subjected to given processing methods (Moughan and Rutherfurd, 1996; Moughan, 2005), and such information is required to enable quantifiable increases in dietary Lys concentrations to meet physiological requirements of the bird. Presently, less attention has been given to quantitating the changes in available Lys concentrations, and instead the impetus is for ingredient suppliers to carefully control processing methods to ensure a quality feedstuff that has minimal formation of Maillard reaction products.

Lathyrogens constitute free, non-protein amino acids (β-N-oxalylamino-l-alanine) and nitriles (β-aminopropionitrile) found mainly in *Lathyrus* and *Vicia* spp. (Roy and Spencer, 1989). Specific health disturbances caused by lathyrogen ingestion include neurolathyrism and oteolathyrism. Neurolathyrism is rare in broilers and layers, but osteolathyrism caused by β-aminopropionitrile manifests as structural and functional deformities of the aorta and bones (Roy and Spencer, 1989). A higher inclusion (100–200 mg/kg body weight) of β-aminopropionitrile is needed to impart toxicity in laying hens (Chowdhury, 1990) and is rare in commercial poultry production. Toxicity effects of β-aminopropionitrile include decreased egg production and eggshell abnormalities (Chowdhury, 1988). In broilers, consumption of raw seeds of *Latyrus sativus* cause reduced growth and feed intake (Latif et al., 1976), whereas ingestion of lathyrogens by laying hens cause soft-shelled eggs due to interference with collagen cross-linking during eggshell membrane formation (Elkin, 2017).

Lipids

Erucic acid and cyclopropenoid fatty acids are the most common fatty acid toxicoses in poultry (Elkin, 2017). Canola (rapeseed) and its oil are the main sources of erucic acid (C22:1 n−13) in poultry feed (Busch et al., 1994; Lu et al., 2020) and is associated with decreased egg production and egg size. In severe cases, consumption of erucic acid can increase mortality due to liver and kidney disorders (March and Soong, 1976). In broilers, adverse effects include reduced feed intake, growth, and digestibility of lipids and individual fatty acids (Leeson and Summers, 2001). However, due to genetic improvement of commodity crops, erucic acid comprises less than 2 percent of total fatty acids in newer canola cultivars (Rzehak et al., 2011) and essentially no longer poses a threat to poultry. Cyclopropenoid fatty acid consumption from plant sources of the *Malvaceae*, *Sterculiaceae*, *Tilliaceae*, and *Bombacaceae* families (e.g., cotton seed and velvetweed) decreases feed intake and weight gain; resulting in poor feed conversion in broilers (Mukhtar, 2007), and it is associated with "pink white" deterioration in stored eggs (Lorenz, 1939; Keshavarz, 1993).

Feed deterioration due to poor handling and storage of poultry feed is a concern for the industry. Vegetable oils added to feed are highly digestible and contain high amounts of polyunsaturated fatty acids. However, oils and fats included in the diet may oxidize during diet manufacture and storage, which may cause poor growth and health of birds upon consumption. Oxidation of polyunsaturated fatty acids is greatly accelerated by heat; the presence of divalent cations, especially iron; or low levels of dietary antioxidants. Conflicting results have been reported regarding the influence of oxidized oil on growth. Inclusion of 0.055 percent sunflower oil (% of diet by weight of oil) [peroxide value (PV) = 400 mEq/kg of oil] (Lin et al., 1989), 11 percent canola plus soy oil (PV = 156 mEq/kg of oil) (Engberg et al., 1996), or 4 percent poultry fat (PV = 175 mEq/kg of fat) (Cabel et al., 1988) caused a clear reduction in growth of broilers. However, this effect was not apparent with addition of 1.5 percent fish oil (PV = 520 mEq/kg of oil) (Carpenter and L'estrange, 1966) or 0.6 percent sunflower oil (PV = 148 mEq/kg of oil) (Açıkgöz et al., 2011). It is important to note that PV and thiobarbituric acid reactive substances (TBARS) are the most commonly reported assays due to their simplicity and low cost, but neither are accurate as individual metrics when assessing oil peroxidation status (Hung et al., 2017). Moreover, the range of PV in commercially available fat and oil sources can range from 0.1 to 181 mEq/kg of lipid (Shurson et al., 2015), and the level of oil or fat inclusion must be taken into account when estimating the influence on dietary components. Nevertheless, consumption of oxidized oils and fats causes oxidative stress indicated by elevated concentrations of oxidative products within the liver (Zhang et al., 2010) and signs of oxidative stress in the intestinal epithelium (Dibner et al., 1996). Moreover, ingestion of oxidized oil products also negatively affects egg production and quality in laying hens (Saki, 2016) and meat quality in broilers (Lin et al., 1989).

Indigestible Carbohydrates

Plant-derived non-starch oligosaccharides and polysaccharides are resistant to enzymatic and chemical digestion by the bird (see Chapter 2, "Carbohydrates and Fiber"). Better known as dietary fibers, the diverse array of compounds in this category serve as a source of nutrients for microbes resident in the GIT of every category of poultry. As such, inclusion of dietary fiber effectively serves to dilute the energy or caloric density of a dietary formulation (Coon et al., 1990). Variable amounts of energy can be salvaged from microbial end-products of fermentation (i.e., from short-chain fatty acids), but this is thought to be a minor process in poultry (see Chapter 1, "Energy").

Non-starch polysaccharides may be in the form of cellulose polymers, non-cellulose polymers (e.g., pentosans, arabinoxylans, xylans, and β-glucans), or pectic polysaccharides (e.g., glucomannans, galactomannans, arabinans, xyloglucans, and galactans). Cellulose polymers are insoluble in nature and minimally fermented by microbes, if at all, because the residence time in the hindgut of poultry is limited (de Vries et al., 2012). Some soluble non-starch polysaccharides increase the viscosity of digesta, thereby causing profound changes in gut physiology and microbial profiles (Angkanaporn et al., 1994). These effects negatively influence nutrient digestibility (Choct and Annison, 1990) depending on the dietary source and inclusion level (Lemme et al., 2004; Mateos et al., 2012). Stachyose, raffinose, and verbascose are the major oligosaccharides present in grains and legumes. Broiler diets enriched with stachyose, raffinose, and verbascose were shown to lower feed intake and weight gain, increase digesta passage rate due to fluid retention, and cause impaired nutrient absorption (Irish and Balnave, 1993).

Diet Form

Physical form of poultry feed has profound impacts on the growth and productivity of birds (Dozier et al., 2010). Depending on the age, feed may be offered in mash, pellet, or crumble forms. Pellets and crumbles have an advantage over mash diets in terms of growth (McKinney and Teeter, 2004; Chewning et al., 2012) and nutrient digestibility (Wahlström et al., 1999). This improvement in growth and nutrient utilization may be partly due to adequate nutrient intake by alleviating selective feeding, improving palatability and digestibility, and potentially through a reduction in pathogenic exposure, assuming sufficient heat and moisture were used during the pelleting process. Because pelleting and extrusion processes require relatively extreme processing conditions involving heat and moisture, care must be given to labile nutrients (especially oils, vitamins, and exogenous enzymes) to ensure sufficient concentrations of undamaged forms are available in the final diet presented to the bird to avoid nutrient deficiencies and oxidative stress. If the mash form is considered as a baseline, increasing feed intake and nutrient digestibility by pelleting, crumbling, or even extruding processes is certainly an advantage. However, there is currently no evidence to suggest that specific nutrient requirements are altered when the form of the diet is changed, and diet form can be more accurately attributed to changes in nutrient bioavailability.

ENVIRONMENTAL STRESSORS RELATED TO IMMUNE ACTIVATION

The interaction of nutrition and health requires a basic understanding of how the bird's immune components are organized, and this section will emphasize how intestinal immunity relates to dietary nutrient requirements. An overview of how the poultry immune system is organized will provide the means to discuss nutrient requirements related to immunity. Considering progress toward judicious use of in-feed antibiotics or removal of in-feed antibiotics altogether in the last decade, studying the influence of diet on the composition and function of the microbiota, and by extension the host, is of high priority in poultry nutrition. Knowledge in this realm is rapidly evolving as we gain a better understanding of host–microbe interactions as a component of an emerging field known as integrative physiology (Leulier et al., 2017).

Components of Intestinal Health

The GIT is the primary interface between the external environment (e.g., nutrients, microbes, antigens, and toxins derived via ingestion) and the host, which is why a considerable proportion of all immune cells in the body can be found within close proximity of the GIT organs and tissues (Celi et al., 2019). Intestinal leukocytes cells are predominantly found in three distinct GIT compartments, including the gut-associated lymphoid tissue (GALT), mucosal lamina propria, and surface epithelium (Celi et al., 2019). The bursa of Fabricius, cecal tonsils, and Peyer's patches are all found within the GALT. Additionally, physical and chemical barriers of the GIT play important roles in maintaining gut barrier integrity that prevents the passive entry of pathogens and toxins. The physical barrier includes epithelial cells (a single cellular layer covering the length of the GIT), their interdigitating tight junction proteins, and the overlaying mucus layer. Chemical barrier components include a relatively low pH (acidic conditions) in the proventriculus and gizzard, and antimicrobial proteins including immunoglobulins, lysozymes, and other proteolytic proteins. Immune modulation within the GIT of poultry occurs primarily via three pathways: (1) sampling by M-cells in Peyer's patches, (2) sampling by macrophages and dendritic cells, and (3) direct interactions between luminal contents and epithelial cells or leukocytes. Overall, the immune system represents a robust and redundant system that works to carefully maintain a balance between tolerance and response to antigens by distinguishing between nutrients, beneficial microbes, and harmful pathogens. Without such defenses, desired productive outcomes and health of birds can be compromised if prolonged or inappropriate immune responses occur. Additional details about the various components of intestinal health follow.

1. **Peyer's patches** are dome-like structures that can be found scattered along the intestinal mucosa, with their distribution in poultry species occurring sporadically throughout the duodenum and ileum (Makala et al., 2002). In mammals, they are particularly enriched in lymphoid tissues, making them key sites for coordinating immune responses to pathogens while promoting tolerance to harmless microbes (Chassaing et al., 2014). Typically, lymphatics from Peyer's patches and villi drain into mesenteric lymph nodes as a conduit to connect with other immune organs. However, it should be noted that of the categories of avian species discussed in this report, chickens have relatively poorly developed lymph nodes and generally lack lymphatic drainage of intestinal tissues without a central lacteal (Noyan et al., 1964; Moran, 1982). The key function of Peyer's patches is to constantly sample antigens from the lumenal environment for evaluation by leukocytes. As such, Peyer's patches contain specialized phagocytic cells called M-cells (described subsequently), which can transport materials across the epithelial barrier via a process called transcytosis, though the number and distribution of M-cells is known to differ between mammalian and poultry species.
2. The **bursa of Fabricius** is located at the proctodeal region of bird's cloaca and is responsible for development of B cells. Unlike the thymus, which facilitates development and maturation of T-lymphocytes associated with cell-mediated immunity, the bursa assists in the development of B-cells associated with humoral immunity (Kaiser, 2010). As a chick grows, the bursa transitions from a primary immune tissue for B-lymphocyte development to a secondary immune tissue where immune responses occur, and eventually it regresses and becomes vestigial in adults.
3. **Cecal tonsils** are located on both ceca at the place where this organ joins the ileum. The ceca are morphologically similar to mammalian Peyer's patches in that they contain both T and B cells (Wang et al., 2019). Considering most of the microbiota present in the GIT of poultry species resides in the ceca, cecal tonsils are particularly important for functions including surveillance, cytotoxic killing, and tolerance with respect to environmental microbes.
4. **Goblet cells, mucins, and the mucus layer** collectively serve to protect the epithelial surface of the GIT. A layer of mucus covers the GIT epithelial surface, which acts as both a lubricant and mechanism for trapping ingested components. Goblet cells that are scattered throughout the intestinal epithelium secrete mucin (i.e., high molecular weight glycoproteins), which is the major component of the mucus layer (Celi et al., 2019). There are two types of mucins: gel-forming and membrane-bound. In chickens, nine mucin genes have been identified, including *MUC 1*, *2*, *4*, *5ac*, *5b*, *6*, *13*, and *16*. *MUC 2*, *5ac*, *5b*, and *6* genes are translated to gel-forming mucins, while the remainder of the genes

are translated to membrane-bound mucins (Kaiser, 2010).

5. **Microfold cells (M-cells)** are located on the surface of Peyer's patches and isolated lymphoid follicles. These cells are specialized epithelia cells with shorter microvilli that import lumenal antigens from pathogens and microbes and funnel them to lymphoid tissues for evaluation by the host immune system (Chassaing et al., 2014).
6. **Dendritic cells (DCs)** are primary antigen-presenting cells found within the GALT that are able to perform endocytosis, exocytosis, antigen processing and presentation to naïve T and B cells, production of regulatory cytokines, and initiation of innate and adaptive immune responses (Vu Manh et al., 2014; Zmrhal and Slama, 2020). As such, DCs extend their dendrites between epithelial cells to capture antigens from the intestinal lumen and distribute them to naïve $CD4^+$ T cells in lymphoid organs, including Peyer's patches and cecal tonsils in poultry. In mammals, the capture of antigens by DCs activates tolerogenic processes upon presentation to T cells, which involves T regulatory cells, but similar processes have yet to be fully elucidated in poultry.
7. **Paneth cells** are columnar-shaped and contain large, apical-facing, lysozyme-rich secretory granules that are released via exocytosis (i.e., merocrine secretion) in response to bacterial invasion but not fungal or protozoal infection. Typically located within intestinal crypts at the base of villi in mammalian species, Paneth cells have also been characterized in the small intestine of chickens, as confirmed by phloxine-tartrazine staining, microscopy, and expression of lysozyme-related genes (Wang et al., 2016). However, much remains unknown of the development and responsivity of Paneth cells in poultry as related to effects on nutritional requirements.
8. **Tight junction proteins** provide the ability for intestinal epithelial cells to form a continuous barrier against lumenal contents. Tight junction proteins have been more completely described in mammalian species and are grouped into two categories: cytoplasmic and membrane-bound. In mammals, zona occludens are cytoplasmic proteins, whereas claudins, occludin, tricellulin, and marvelD3 are membrane-bound proteins (Runkle and Mu, 2013). However, in poultry, elucidation of tight junction proteins homologous to those found in mammals is ongoing, and presently, only zona occludens, occludin, and claudins have been characterized (Denker and Nigam, 1998; Kearsey et al., 2004). Weakening of tight junction proteins creates a gap between epithelial cells, causing nonspecific intestinal permeability and access for microbes and other exogenous substances to infiltrate host tissues, thereby leading to immune system stimulation and potentially inflammation. However, much remains unknown about the expression of tight junction proteins in poultry species.
9. **Pattern recognition receptors (PRRs)** function as sensors on innate immune cells that recognize pathogen-associated molecular patterns (PAMPs), which may be more appropriately referred to as microbe-associated molecular patterns (MAMPs). Examples of PRRs include C-type lectin receptors and toll-like receptors (TLRs), with 10 TLR each recognizing distinct molecules as identified in poultry species (Temperley et al., 2008; Keestra et al., 2013). TLR 1, 2, 4, and 6 recognize bacterial lipopolysaccharides, whereas TLR 3, 7, 8 and 9 are involved in the recognition of nucleic acids (Kaiser, 2010). Recognition of pathogens or toxins by these receptors results in the production of pro-inflammatory cytokines and initiates recruitment of immune cells from the bloodstream and into tissues at the site of infection. Similar to PRR, nucleotide-binding oligomerization domain-like (NOD) receptors assist in regulating the inflammatory response toward resident microbes and thereby help to preserve epithelial barrier integrity.
10. **Defensins** are cysteine-rich cationic peptides that have broad activity against bacteria, fungi, and viruses. Their antimicrobial activities include binding to microbial cell membranes and forming pores in the cell wall that ultimately cause death of the microbe. In mammals, there are three classes of defensins, including α, β, and θ, but poultry species only possess the β-defensin family, which is primarily expressed by leukocytes (heterophils) and epithelial cells (van Dijk et al., 2008).

Innate Immunity

Newly hatched chicks possess maternal antibodies transmitted through the yolk that convey some degree of pathogen resistance via innate immune mechanisms (e.g., PRR, TLR) and effector cells (macrophages, dendritic cells, heterophils, and natural killer cells) (Kogut, 2009). The innate immune system does not have memory (i.e., it does not develop greater efficiency with repeated stimulation), but this arm of the immune system is rapid and vital in removing 99 percent of pathogenic organisms. The innate immune system consists of physical, humoral, and cellular barriers to infection that collectively neutralize most infections before they spread beyond the intestine.

A number of cell types are associated with the innate immune system. Heterophils comprise 40–60 percent of the total white blood cell population in the body. Macrophages are present throughout the body but are especially important within the lamina propria of intestinal tissues. Like dendritic cells, macrophages contain PRRs that detect antigens and constitute another antigen-presenting cell type that serves to influence the development and function of T cells (Chassaing et al., 2014). Like other innate immune cells, natural killer (NK) cells recognize invading pathogens, but instead of destroying the external threat, NK cells eliminate the compromised host cells. Finally, mast cells are present in tissues associated with nerves and blood vessels, and release granules containing histamine and heparin during times of allergy (Chassaing et al., 2014). There exist sparse reports that histamine-induced anaphylaxis mediated by mast cells can occur in avian species (Gershwin, 2015), but this remains an understudied and controversial area, and it is likely that other immune cell types (e.g., basophils) also play a contributing role in the bird.

Adaptive Immunity

The adaptive immune system includes cells that increase specificity with repeated exposure to antigens, and the T and B cells elicit a more efficient attack with the development of memory each time a pathogen is encountered. Adaptive immune cells in poultry species typically take three or more weeks after hatch to develop and mature (Beal et al., 2006). B cells are versatile cells associated with multiple functions including antigen presentation, antibody production, and cytokine generation. In mammals, B cells mature in bone marrow, but in poultry species, B cells mature in the bursa of Fabricius (Kaiser, 2010). Activation of memory B cells occurs when target antigens had been previously encountered by parent B cells. Whereas some B cells are T cell-independent, most B cells require interactions with T cells for activation (Ziegler-Heitbrock and Ulevitch, 1993).

T cells are lymphocytes with cell-surface receptors that are developed in the thymus. Precursors are derived from bone marrow and develop into several phenotypes within the thymus based on cell-surface markers, including CD4 alone (helper T cells), CD8 alone (killer/cytotoxic T cells), and the combination of CD4 and CD8 (regulatory or suppressor T cells). Helper T cells are required for the development of other lymphocytes, including maturation of B cells into antibody-secreting plasma cells and activation of killer T cells (Ganeshan and Chawla, 2014).

Disease-Related Stress

The nutritional status of an animal is influenced by infectious diseases, and the magnitude of the effect is roughly proportional to the extent of tissue damage and functional losses inflicted by the pathogen, along with the nutritional costs of mounting a protective immune response. Tissue-specific pathology resulting from a pathogen can cause important functional changes that impact nutrient absorption, metabolism, and excretion in the bird. For example, the negative effects of *Eimeria acervulina* or necrotic enteritis due to *Clostridium perfringens* on gain and efficiency of feed conversion is partially attributed to damage to the intestinal mucosa and impaired nutrient absorption, even in mild or subclinical cases (Waldenstedt et al., 2000; Van Immerseel et al., 2009; Castro et al., 2020). The immune response also has costs related to the expansion of leukocyte populations and secretion of protective proteins by leukocytes and the liver. Additionally, the immune system initiates a generalized stress response that orchestrates systemic changes of nutritionally important processes (Johnson, 1997; Humphrey and Klasing, 2004; Smirnov et al., 2005). These changes follow a similar pattern regardless of the causative pathogen because the early responding phagocytes secrete inflammatory cytokines that influence virtually all physiological and developmental processes, including a downregulation of muscle protein synthesis (i.e., the major component of body weight gain) and an increase in skeletal muscle breakdown. Detailed subsequently is evidence pertaining to specific influences of pro-inflammatory cytokines that are released during many types of avian infection. These proteinaceous pro-inflammatory cytokines orchestrate the classical acute phase response that may include decreased feed intake, fever, hepatic production of acute phase proteins, impaired digestibility of nutrients, and changes in tissue priorities for uptake and use of nutrients. Overall, such effects elicit impairment of anabolic processes, including growth and reproduction, by diversion of nutrients to the tissues involved in immunity, including lymphoid organs and the liver.

Effects on Nutrient Digestibility and Nutrient Utilization

Jeurissen et al. (2002) defined intestinal integrity as "the cells and products constituting the barrier against leakage or translocation of feed components, microbial toxins, and microorganisms from the lumen to the body." As described previously, integrity of the intestinal epithelium relies on the continuous single layer of epithelial cells that are tightly held together by adhesion molecules, the layer of mucous on the apical surface of those epithelial cells, and the protective proteins and peptides secreted by the epithelia into the mucous layer. The amount of adhesion between epithelial cells and the amount of mucin is affected by the microbial milieu within the intestinal lumen and by the diet (Smirnov et al., 2004, 2005; Amerah et al., 2009; Bortoluzzi et al., 2020). Epithelial damage and inflammation diminish intestinal integrity, which allows the influx of more microorganisms and their toxins, thereby amplifying the inflammatory response. Damage can be caused directly by a pathogen or indirectly by the host immune response to the pathogen.

Even mild infections of the intestinal epithelium result in marked impairments in nutrient digestion and absorption (Tomar, 2001). This impairment is due to bolstered defensive processes, such as infiltration of lymphocytes into the epithelium, accelerated turnover of epithelial cells, decreased villi length and surface area, increased rate of peristalsis (i.e., decreased intestinal transit time), and profuse mucus secretion (Nusrat et al., 2001; Tomar, 2001; Schiffrin and Blum,

2002). In general, the absorption of micronutrients, especially fat-soluble vitamins and iron, is impacted to a greater degree than that of macronutrients. Of the macronutrients, fat absorption is typically impacted the most (Sharma and Fernando, 1975; Adams et al., 1996; Gautier et al., 2020), with three general mechanisms mediating the response. First, increased endogenous losses and decreased nutrient absorption can be a direct result of pathological changes in the integrity and function of the intestinal epithelium. This is especially evident with infections that inflict direct pathological damage to the absorptive surface within the small intestine (e.g., *Eimeria* species), which causes decreased amino acid digestibility (Persia et al., 2006; Adedokun et al., 2016; Rochell et al., 2016). Second, malabsorption can result from an increase in the rate of passage of digesta through the intestines, which diminishes the time available for digestion and absorption. Third, immune responses via the release of pro-inflammatory cytokines decrease absorption of many nutrients even in the absence of intestinal pathology. This is illustrated by changes in the absorption of iron, vitamin A, carotenoids, and fat due to inflammation in locations distant from the intestines (Klasing, 1984; Koutsos et al., 2003). The changes in nutrient absorption that are mediated by the immune system appear to have protective value for the host in parallel with sequestration of nutrients required by microbes for proliferation (e.g., iron) or for defenses against the immune system's effector mechanisms (e.g., antioxidants such as vitamin E and carotenoids). In many enteric infections, such as coccidiosis and necrotic enteritis, damage to the intestinal epithelium triggers a large immune response, and this combination results in a severe reduction in absorption of vitamin A, fat, carotenoids, and iron (Marusich et al., 1973; Sharma and Fernando, 1975; Turk, 1981; Allen, 1992; West et al., 1992).

Decreased growth and egg production is a classic consequence of most enteric challenges, including coccidiosis, necrotic enteritis, spirochetosis, rotavirus infection, campylobacteriosis, clostridial ulcerative enteritis, and viral malabsorption syndrome (Swayne, 2020). Undesired production outcomes are caused by sub-clinical infections such as those induced by poor sanitation (Roura et al., 1992; Gautier et al., 2020) or dietary ingredients that induce intestinal inflammation (Gewirtz et al., 2002). The inflammatory effects of high-rye diets are likely due to high levels of non-starch polysaccharides that may cause a shift in microbiota composition. Additionally, ingestion of high levels of dietary rye has been shown to cause intestinal villus fusion, a thinner tunica muscularis, increased enterocyte apoptosis, T-lymphocyte infiltration into the epithelial layer, and increased number and size of goblet cells, all of which are indicative of compromised intestinal integrity (Teirlynck et al., 2009). Although impaired digestion explains part of the undesired production outcomes of poultry during infections, there are many other important changes orchestrated by the immune response that also contribute. For example, nutritional costs associated with the activation of the systemic (non-mucosal) immune system include decreased feed intake, fever, and the redistribution of nutritional resources from anabolic processes (e.g., growth and reproduction) to meet the needs of the immune system and the liver for their protective effector functions (Johnson, 1997; Klasing, 2004; Chen et al., 2018).

Effect of Immune Responses on Nutrient Requirements of Poultry

The nutritional costs of immunity can be itemized into three areas: (1) those needed for development of the immune system including the selection and maturation of lymphocytes, (2) those needed to maintain a mature system of sufficient size and capacity to have an effective response to a pathogen, and (3) the additional costs of using the system to thwart a challenge (Klasing, 2004). Unfortunately, these costs have not been objectively tabulated for most nutrients in any species due to the complexity and diffuse location of the leukocytes in the immune system. Upon examining the cells and protective proteins of the systemic (non-mucosal) component of the immune system in healthy adult Leghorns at maintenance, Iseri and Klasing (2014) quantified that lysine associated with these resources constituted 0.38 percent of the bird's total lysine content. During the acute phase of a response to killed *Escherichia coli*, the amount of additional lysine needed for the expansion of leukocyte populations and production of protective proteins was equivalent to the amount of lysine in 17 percent of an egg or 5.5 percent of the lysine in the pectoralis muscle of an adult chicken. However, the largest need for lysine during the acute phase response was due to the hypertrophy of the liver to facilitate its switch from a primary homeostatic responsibility to one that supports the immune response. Including the lysine needed for hepatic hypertrophy gives a proportion of lysine needed for a robust acute phase response to an amount equal to that contained in 52 percent of an egg, which is a nutritionally and economically significant quantity. Interestingly, the amount of lysine needed for adaptive immunity, including expansion of antigen-responding lymphocytes and secretion of immunoglobulins, was equal to only 1.3 percent of an egg and could easily be supplied by the catabolism of hepatic tissue and acute phase proteins as the acute phase response declines. The challenge model used in the preceding experiment elicited a response that was more similar to vaccination than a clinically important infection; however, it does put potential nutritional costs into perspective. The challenge used was transient, so the metabolic costs diminished quickly, whereas metabolic costs associated with an active infectious challenge would presumably remain elevated throughout the infection and recovery phases. Clearly, the diverse types of challenges (e.g., bacterial, viral, protozoan), types of immune response, durations of challenge, and amounts of immunopathology create a situation in which the quantitative effects of infection on nutrient requirements would be expected to be highly variable. In general, this remains an important and evolving area of study, which requires thinking that goes beyond a reductionist approach to attribute changes in nutrient requirements to individual pathogenic challenge contexts.

Several studies have examined the requirement of growing chicks for various nutrients while undergoing an immune response relative to that of healthy chicks. Most often the immune system was stimulated with an inflammatory agent such as lipopolysaccharide (LPS), killed bacteria, or an adjuvant. This is because an inflammatory response usually has greater negative effects on growth, feed intake, and body composition relative to responses mediated by lymphocytes. The use of a non-infectious inflammatory challenges (e.g., LPS administration) eliminates pathogen-specific pathology and allows for broader interpretation because an acute phase response accompanies most natural pathogen challenges regardless of the tissues affected. Broiler chickens challenged by LPS had decreased growth and impaired feed conversion but did not have a detectable change in the arginine requirement, expressed as percent of the diet, to maximize growth or the efficiency of feed conversion (Tan et al., 2014). Similarly, multiple vaccinations of broiler chicks decreased growth and feed efficiency (Rubin et al., 2007) but did not affect the optimal dietary methionine concentration. An acute phase response in broilers induced by LPS or killed *S. aureus* decreased the both the methionine and lysine requirements, expressed as percent of the diet or mg per day, for growth and efficiency of feed conversion (Klasing and Barnes, 1988), likely due to the marked decrease in growth and muscle deposition that is caused by pro-inflammatory cytokines. Using slower growing Colombian × New Hampshire chicks, Webel et al. (1998a,b) also observed lower absolute requirements (expressed as mg per day) for lysine and arginine to maximize growth and rate of protein accretion following an LPS-induced inflammatory response; however, when expressed as percent of the diet, no significant changes were noted. They also showed that the efficiency of lysine and arginine deposition in the body was not affected by LPS. Taken together, amino acid requirements due to an inflammatory response to a non-infectious challenge are either decreased or are unchanged. This is apparently due to a reduced capacity to accrete protein (i.e., alterations in body protein turnover rates) and is not due to a change in the efficiency of amino acid utilization for protein accretion. Supplying surfeit levels of lysine or methionine does not typically improve growth and feed efficiency of immunologically stressed poultry in the absence of intestinal pathology and the resulting malabsorption.

Although an immune response decreases body stores of many micronutrients, especially iron, zinc, carotenoids, retinol, and vitamin E, quantitative estimates of changes in the dietary requirement are largely unknown. For example, Koh et al. (1996) found that copper requirements are increased as a result of its use as a cofactor for ceruloplasmin during the acute phase of an innate immune response, but they were unable to establish the magnitude of the change. Distinct from macronutrients, the role of micronutrients is generally less structural and more supportive in a metabolic sense, since many are cofactors for enzymes and other biologically important molecules (e.g., hemoglobin), but that does not lessen the importance of micronutrient status in the event of immune stimulation. Considering the diverse functions of micronutrients with regard to immune function, this remains a needed area of research in poultry nutrition.

Quantitative changes in nutrient requirements due to true pathogens are very difficult to establish using methodology typically used in healthy poultry. This is because the responses of individual birds to an infectious challenge differ greatly in onset, magnitude, and duration due to many factors that are difficult to control and to account for including genetics, such as the major histocompatibility complex (MHC) haplotype. Consequently, this research requires relatively high numbers of animals maintained in highly controlled environments, which exceeds the standard experimental conditions typically used with titration studies designed to estimate nutrient requirements of healthy poultry. In general, use of *Eimeria* challenges provides better experimental control and less variation than many other infectious challenge models. Also, studies using different levels of sanitation are promising for investigating the effects of sub-clinical disease stress on desired production outcomes and nutrient requirements (Roura et al., 1992; Corzo et al., 2007; Macelline et al., 2020).

At this time, there is not sufficient evidence to estimate requirements for specific nutrients affected by immune activation, and this is mainly because the replication used (i.e., variability estimates) does not meet our inclusion criteria for generating objective requirement estimates (see Chapters 13 through 19 for species-specific inclusion criteria for requirement studies). However, several studies have determined the directional changes in the requirement due to experimental infection (Bortoluzzi et al., 2020). For example, broilers challenged with *E. maxima* and *C. perfringens* had positive production responses to increased dietary threonine-to-lysine ratios (Star et al., 2012) relative to that of controls. Two studies showed no beneficial effects of added methionine during an *Eimeria* infection even though intestinal pathology indicative of impaired nutrient absorption was observed (Castro et al., 2020; Ren et al., 2020). Studies designed to titrate nutrient requirements using standardized disease challenge models are needed to make definitive recommendations on the changes in nutrient requirements due to the stress of economically important infectious diseases.

Implications of Decreased Use of Antibiotics

The prophylactic use of antibiotics as a diet supplement to improve the health and productivity of poultry was first reported in 1946 (Moore et al., 1946). This discovery was rapidly adopted by industry because subtherapeutic levels of antibiotics optimized the health and productivity of poultry and were cost-effective. In the past decades, use of prophylactic in-feed antibiotics has decreased considerably due to consumer, retailer, and legislative factors. As such, the U.S. Food and Drug Administration (FDA) enacted the Veterinary Feed Directive (VFD) final rule on January 1, 2017, as progress toward the judicious use of medically important antibiotics when included in feed or water for food-producing animals (Anonymous, 2015). Because prophylactic antibiotics increase the health, welfare, and productivity of poultry, their removal compels mitigation by developing and adopting alternative management strategies.

A solid understanding of the mechanism by which feeding subtherapeutic levels of antibiotics improve poultry production and health is not known with confidence, and this situation has impaired the development of alternatives. The dominant hypothesis is that antibiotics act by affecting the population dynamics and the metabolome of the microbiome, resulting in a healthier intestinal epithelium (Dibner and Richards, 2005; Brown et al., 2017). The hypothesized underlying reasons include decreased microbial production of toxins including nitrogenous compounds (Visek, 1978), decreased competition between the microbiome and the host for nutrients, and fewer subclinical pathogen challenges, thereby resulting in lower levels of inflammatory cells and inflammatory cytokines, and a greater villi surface area in the intestinal epithelium (Roura et al., 1992; Smirnov et al., 2005; Brown et al., 2017; Gadde et al., 2017). A more direct, but not mutually exclusive, hypothesis is that antibiotics may have intrinsic anti-inflammatory effects on the epithelium that are unrelated to their antimicrobial effects, thus causing diminished losses in productivity as a result of immune responses to nonpathogenic enteric challenges (Niewold, 2007). Regardless of the exact mechanism, antibiotics increase growth, efficiency, and welfare of poultry raised in conventional production environments, though they are much less effective in highly sanitized environments (Roura et al., 1992). Because they do not increase desired production outcomes beyond a bird's genetic potential, they should not technically be referred to as growth promoters and are more accurately described as growth supportive or growth permitting.

Research on replacing antibiotics so that the desired production outcomes and welfare of poultry is not diminished has generally approached the problem by testing candidates with similar potential mechanisms and activities as described previously for antibiotics (Yang et al., 2009; Gadde et al., 2017; Suresh et al., 2018; Cowieson and Kluenter, 2019). For example, antimicrobial candidates that are unrelated to antibiotics include bacteriophages, peptides, proteins such as lysozyme, metals such as copper and zinc, and plant antimicrobials such as thymol and other phytogenics. Candidates that modify the intestinal immune system and decrease inflammation act by many mechanisms, such as by binding to nuclear receptors in leukocytes, and they include nutrients such as long chain polyunsaturated fatty acids and vitamins A, D, and E. Candidates that optimize the microbiota diverge into those that are substrates that nourish the growth of specific microbes (prebiotics), enzymes that hydrolyze fiber for the same purposes, and those that supply specific strains of microbes that are enriched in the enteric environment of relatively high-performing poultry (probiotics, also referred to as direct-fed microbials).

It is often difficult to judge if a replacement candidate truly replaces antibiotics or if it has growth permitting properties that differ from antibiotics that nevertheless helps fill the void that occurs in diets without antibiotics. A true replacement for antibiotics would be expected to act through a similar mechanism, have similar effects on the intestinal phenotype, and its effects would not be predicted to be additive with those of antibiotics. In other words, an exact replacement for antibiotics should be more effective in the absence than in the presence of an antibiotic. Those that are growth permitting but have mechanisms independent of antibiotics might be expected to have similar effectiveness regardless of the presence or absence of dietary antibiotics. Commercial approaches to antibiotic replacements often combine several candidate compounds, resulting in a mixture that has some candidates that act by mechanisms similar to those of antibiotics and partially replace them, and others that are additive because they operate through independent mechanisms.

REFERENCES

Abreu, V.M.N., P.G.d. Abreu, A. Coldebella, F.R.F. Jaenisch, J.I.d. Santos Filho, and D.P.d. Paiva. 2011. Curtain color and lighting program in broiler production: I—general performance. *Revista Brasileira de Zootecnia.* 40:2026–2034.

Açıkgöz, Z., H. Bayraktar, Ö. Altan, S.T. Akhisaroglu, F. Kırkpınar, and Z. Altun. 2011. The effects of moderately oxidised dietary oil with or without vitamin E supplementation on performance, nutrient digestibility, some blood traits, lipid peroxidation and antioxidant defence of male broilers. *Journal of the Science of Food and Agriculture.* 91 (7):1277–1282.

Adams, C., H. Vahl, and A. Veldman. 1996. Interaction between nutrition and *Eimeria acervulina* infection in broiler chickens: Diet compositions that improve fat digestion during *Eimeria acervulina* infection. *British Journal of Nutrition.* 75(6):875–880.

Adedokun, S., A. Helmbrecht, and T. Applegate. 2016. Investigation of the effect of coccidial vaccine challenge on apparent and standardized ileal amino acid digestibility in grower and finisher broilers and its evaluation in 21-day-old broilers. *Poultry Science.* 95(8):1825–1835.

Allaire, J.M., V. Morampudi, S.M. Crowley, M. Stahl, H. Yu, K. Bhullar, L.A. Knodler, B. Bressler, K. Jacobson, and B.A. Vallance. 2018. Frontline defenders: Goblet cell mediators dictate host–microbe interactions in the intestinal tract during health and disease. *American Journal of Physiology—Gastrointestinal and Liver Physiology.* 314(3): G360–G377.

Allen, P.C. 1992. Effect of coccidiosis on the distribution of dietary lutein in the chick. *Poultry Science.* 71(9):1457–1463.

Amerah, A.M., V. Ravindran, and R.G. Lentle. 2009. Influence of insoluble fibre and whole wheat inclusion on the performance, digestive tract development and ileal microbiota profile of broiler chickens. *British Poultry Science.* 50(3):366–375.

Angkanaporn, K., M. Choct, W.L. Bryden, E.F. Annison, and G. Annison. 1994. Effects of wheat pentosans on endogenous amino acid losses in chickens. *Journal of the Science of Food and Agriculture*. 66(3):399–404.

Antonissen, G., F. Van Immerseel, F. Pasmans, R. Ducatelle, F. Haesebrouck, L. Timbermont, M. Verlinden, G.P. Janssens, V. Eeckhaut, M. Eeckhout, S. De Saeger, S. Hessenberger, A. Martel, and S. Croubels. 2014. The mycotoxin deoxynivalenol predisposes for the development of *Clostridium perfringens*-induced necrotic enteritis in broiler chickens. *PLoS One*. 9(9):e108775.

Balnave, D. 2004. Challenges of accurately defining the nutrient requirements of heat-stressed poultry. *Poultry Science*. 83(1):5–14.

Beal, R.K., C. Powers, T.F. Davison, and A.L. Smith. 2006. Immunological development of the avian gut. In *Avian Gut Function in Health and Disease*. G.C. Perry, editor. Wallingford, Oxfordshire, UK: CAB International. 85–103.

Binder, E.M., L.M. Tan, L.J. Chin, J. Handl, and J. Richard. 2007. Worldwide occurrence of mycotoxins in commodities, feeds and feed ingredients. *Animal Feed Science and Technology*. 137(3):265–282.

Bogosavljević-Bošković, S., S. Rakonjac, V. Dosković, and M.D. Petrović. 2012. Broiler rearing systems: A review of major fattening results and meat quality traits. *World's Poultry Science Journal*. 68(2):217–228.

Bortoluzzi, C., J. Fernandes, K. Doranalli, and T. Applegate. 2020. Effects of dietary amino acids in ameliorating intestinal function during enteric challenges in broiler chickens. *Animal Feed Science and Technology*. 262:114383.

Boshouwers, F.M.G., and E. Nicaise. 1987. Physical activity and energy expenditure of laying hens as affected by light intensity. *British Poultry Science*. 28(1):155–163.

Brown, K., R.R. Uwiera, M.L. Kalmokoff, S.P. Brooks, and G.D. Inglis. 2017. Antimicrobial growth promoter use in livestock: A requirement to understand their modes of action to develop effective alternatives. *International Journal of Antimicrobial Agents*. 49(1):12–24.

Brown, K.I., and K.E. Nestor. 1973. Some physiological responses of turkeys selected for high and low adrenal response to cold stress. *Poultry Science*. 52(5):1948–1954.

Bryden, W.L. 2012. Mycotoxin contamination of the feed supply chain: Implications for animal productivity and feed security. *Animal Feed Science and Technology*. 173(1):134–158.

Bucher, E., and L. Meszaros. 1989. Stechapfelsamen (*Datura* spec.) und deren alkaloide in futtermitteln. *Kraftfutter*. 3(89):76–82.

Busch, L., V. Gunter, T. Mentele, M. Tachikawa, and K. Tanaka. 1994. Socializing nature: Technoscience and the transformation of rapeseed into canola. *Crop Science*. 34(3):607–614.

Buyse, J., E.R. Kuhn, and E. Decuypere. 1996. The use of intermittent lighting in broiler raising. 1. Effect on broiler performance and efficiency of nitrogen retention. *Poultry Science*. 75(5):589–594.

Cabel, M.C., P.W. Waldroup, W.D. Shermer, and D.F. Calabotta. 1988. Effects of ethoxyquin feed preservative and peroxide level on broiler performance. *Poultry Science*. 67(12):1725–1730.

Cahaner, A., and F. Leenstra. 1992. Effects of high temperature on growth and efficiency of male and female broilers from lines selected for high weight gain, favorable feed conversion, and high or low fat content. *Poultry Science*. 71(8):1237–1250.

Cahaner, A., Y. Pinchasov, I. Nir, and Z. Nitsan. 1995. Effects of dietary protein under high ambient temperature on body weight, breast meat yield, and abdominal fat deposition of broiler stocks differing in growth rate and fatness. *Poultry Science*. 74(6):968–975.

Calabrese, E.J., and L.A. Baldwin. 2002. Defining hormesis. *Human and Experimental Toxicology*. 21(2):91–97.

Cândido, M.G.L., I.F.F. Tinôco, F.A.C. Pinto, N.T. Santos, and R.P. Roberti. 2016. Determination of thermal comfort zone for early-stage broilers. *Engenharia Agrícola*. 36:760–767.

Carpenter, K.J., and J.L. L'estrange. 1966. Effects of moderate levels of oxidized fat in animal diets under controlled conditions. *Proceedings of the Nutrition Society*. 25(1):25–31.

CAST (Council for Agricultural Science and Technology) Task Force. 2003. *Mycotoxins: Risks in Plant, Animal, and Human Systems*. Vol. 139. Ames, IA: Council for Agricultural Science and Technology.

Castro, F., Y. Tompkins, R. Pazdro, and W. Kim. 2020. The effects of total sulfur amino acids on the intestinal health status of broilers challenged with *Eimeria* spp. *Poultry Science*. 99(10):5027–5036.

Celi, P., V. Verlhac, E. Pérez Calvo, J. Schmeisser, and A.-M. Kluenter. 2019. Biomarkers of gastrointestinal functionality in animal nutrition and health. *Animal Feed Science and Technology*. 250:9–31.

Cengiz, O., B.H. Koksal, O. Tatli, O. Sevim, U. Ahsan, A.G. Uner, P.A. Ulutas, D. Beyaz, S. Buyukyoruk, A. Yakan, and A.G. Onol. 2015. Effect of dietary probiotic and high stocking density on the performance, carcass yield, gut microflora, and stress indicators of broilers. *Poultry Science*. 94(10):2395–2403.

Charles, D.R., and C.G. Payne. 1966. The influence of graded levels of atmospheric ammonia on chickens. *British Poultry Science*. 7(3):177–187.

Chassaing, B., M. Kumar, M.T. Baker, V. Singh, and M. Vijay-Kumar. 2014. Mammalian gut immunity. *Biomedical Journal*. 37(5):246–258.

Cheeke, P.R. 1998. *Natural Toxicants in Feeds, Forages, and Poisonous Plants*, 2nd ed. Danville, IL: Interstate Publishers.

Chen, X., K. Naehrer, and T.J. Applegate. 2016. Interactive effects of dietary protein concentration and aflatoxin B1 on performance, nutrient digestibility, and gut health in broiler chicks. *Poultry Science*. 95(6):1312–1325.

Chen, Y., H. Zhang, Y. Cheng, Y. Li, C. Wen, and Y. Zhou. 2018. Dietary l-threonine supplementation attenuates lipopolysaccharide-induced inflammatory responses and intestinal barrier damage of broiler chickens at an early age. *British Journal of Nutrition*. 119(11):1254–1262.

Cheng, H.W., and W.M. Muir. 2004. Chronic social stress differentially regulates neuroendocrine responses in laying hens: II. Genetic basis of adrenal responses under three different social conditions. *Psychoneuroendocrinology*. 29(7):961–971.

Chewning, C.G., C.R. Stark, and J. Brake. 2012. Effects of particle size and feed form on broiler performance. *Journal of Applied Poultry Research*. 21(4):830–837.

Choct, M., and G. Annison. 1990. Anti-nutritive activity of wheat pentosans in broiler diets. *British Poultry Science*. 31(4):811–821.

Chowdhury, S.D. 1988. Lathyrism in poultry—A review. *World's Poultry Science Journal*. 44(1):7–16.

Chowdhury, S.D. 1990. Effects of low and high dietary levels of beta-aminopropionitrile (BAPN) on the performance of laying chickens. *Journal of the Science of Food and Agriculture*. 52(3):315–329.

Clarke, E., and J. Wiseman. 2005. Effects of variability in trypsin inhibitor content of soya bean meals on true and apparent ileal digestibility of amino acids and pancreas size in broiler chicks. *Animal Feed Science and Technology*. 121(1–2):125–138.

Combs, G.F. 2012. *The Vitamins: Fundamental Aspects in Nutrition and Health*, 4th ed. Amsterdam, The Netherlands: Elsevier/AP.

Coon, C.N., K. Leske, O. Akavanichan, and T. Cheng. 1990. Effect of oligosaccharide-free soybean meal on true metabolizable energy and fiber digestion in adult roosters. *Poultry Science*. 69(5):787–793.

Corzo, A., M. Kidd, W. Dozier III, G. Pharr, and E. Koutsos. 2007. Dietary threonine needs for growth and immunity of broilers raised under different litter conditions. *Journal of Applied Poultry Research*. 16(4):574–582.

Cowieson, A., and A. Kluenter. 2019. Contribution of exogenous enzymes to potentiate the removal of antibiotic growth promoters in poultry production. *Animal Feed Science and Technology*. 250:81–92.

Daghir, N.J. 2008. Nutrient requirements of poultry at high temperatures. In *Poultry Production in Hot Climates*, 2nd ed. N.J. Daghir, editor. Trowbridge, UK: Cromwell Press. 132–159.

Dänicke, S. 2015. Ergot alkaloids in feed for Pekin ducks: Toxic effects, metabolism and carry over into edible tissues. *Toxins*. 7(6):2006–2023.

Dänicke, S. 2016. Toxic effects, metabolism, and carry-over of ergot alkaloids in laying hens, with a special focus on changes of the alkaloid isomeric ratio in feed caused by hydrothermal treatment. *Mycotoxin Research*. 32(1):37–52.

Dänicke, S. 2017. Ergot alkaloids in fattening chickens (broilers): Toxic effects and carry over depending on dietary fat proportion and supplementation with non-starch-polysaccharide (NSP) hydrolyzing enzymes. *Toxins*. 9(4):118.

Davison, T.F., and I.H. Flack. 1981. Changes in the peripheral blood leucocyte populations following an injection of corticotrophin in the immature chicken. *Research in Veterinary Science*. 30(1):79–82.

de Vries, S., A.M. Pustjens, H.A. Schols, W.H. Hendriks, and W.J.J. Gerrits. 2012. Improving digestive utilization of fiber-rich feedstuffs in pigs and poultry by processing and enzyme technologies: A review. *Animal Feed Science and Technology*. 178(3):123–138.

Deep, A., K. Schwean-Lardner, T.G. Crowe, B.I. Fancher, and H.L. Classen. 2010. Effect of light intensity on broiler production, processing characteristics, and welfare. *Poultry Science*. 89(11):2326–2333.

Denker, B.M., and S.K. Nigam. 1998. Molecular structure and assembly of the tight junction. *American Journal of Physiology—Renal Physiology*. 274(1):F1–F9.

Devegowda, G., T. Murthy, and D. Diaz. 2005. *Mycotoxins: Their Effects in Poultry and Some Practical Solutions*. D. Diaz, editor. Nottingham, UK: Nottingham University Press. *The Mycotoxin Blue Book 1*.

Dibner, J., and J. Richards. 2005. Antibiotic growth promoters in agriculture: History and mode of action. *Poultry Science*. 84(4):634–643.

Dibner, J.J., C.A. Atwell, M.L. Kitchell, W.D. Shermer, and F.J. Ivey. 1996. Feeding of oxidized fats to broilers and swine: Effects on enterocyte turnover, hepatocyte proliferation and the gut associated lymphoid tissue. *Animal Feed Science and Technology*. 62(1):1–13.

Divi, R.L., H.C. Chang, and D.R. Doerge. 1997. Anti-thyroid isoflavones from soybean: Isolation, characterization, and mechanisms of action. *Biochemical Pharmacology*. 54(10):1087–1096.

Dozier, W.A., K.C. Behnke, C.K. Gehring, and S.L. Branton. 2010. Effects of feed form on growth performance and processing yields of broiler chickens during a 42-day production period. *Journal of Applied Poultry Research*. 19(3):219–226.

Elkin, R., W. Featherston, and J. Rogler. 1978. Investigations of leg abnormalities in chicks consuming high tannin sorghum grain diets. *Poultry Science*. 57(3):757–762.

Elkin, R.G. 2017. Influence of plant toxins on laying hen performance and egg quality. In *Egg Innovation and Strategies for Improvement*. P.Y. Hester, editor. Amsterdam, The Netherlands: Academic Press. 499–512.

Elkin, R.G., E. Arthur, B.R. Hamaker, J.D. Axtell, M.W. Douglas, and C.M. Parsons. 2002. Nutritional value of a highly digestible sorghum cultivar for meat-type chickens. *Journal of Agricultural and Food Chemistry*. 50(14):4146–4150.

Elkin, R.G., M.B. Freed, B.R. Hamaker, Y. Zhang, and C.M. Parsons. 1996. Condensed tannins are only partially responsible for variations in nutrient digestibilities of sorghum grain cultivars. *Journal of Agricultural and Food Chemistry*. 44(3):848–853.

Elkin, R.G., and E.S. Lorenz. 2009. Feeding laying hens a bioavailable soy sterol mixture fails to enrich their eggs with phytosterols or elicit egg yolk compositional changes. *Poultry Science*. 88(1):152–158.

Elson, H.A., and R. Croxall. 2006. European study on the comparative welfare of laying hens in cage and non-cage systems. *Archiv für Geflügelkunde*. 70:194–198.

Engberg, R.M., C. Lauridsen, S.K. Jensen, and K. Jakobsen. 1996. Inclusion of oxidized vegetable oil in broiler diets. Its influence on nutrient balance and on the antioxidative status of broilers. *Poultry Science*. 75(8):1003–1011.

Fenwick, G.R., K.R. Price, C. Tsukamoto, and K. Okubo. 1991. Saponins in toxic substances in crop plants. In Saponins. J.F. D'Mello, C.M. Duffus, and J.H. Duffus, editors. London, UK: The Royal Society of Chemistry.

Freeman, B.M. 2013. The stress syndrome. *World's Poultry Science Journal*. 43(1):15–19.

Gadde, U., W. Kim, S. Oh, and H.S. Lillehoj. 2017. Alternatives to antibiotics for maximizing growth performance and feed efficiency in poultry: A review. *Animal Health Research Reviews*. 18(1):26–45.

Ganeshan, K., and A. Chawla. 2014. Metabolic regulation of immune responses. *Annual Review of Immunology*. 32(1):609–634.

Gaul, K., P. Gallagher, D. Reyes, S. Stasi, and J. Edgar. 1993. Poisoning of pigs and poultry by stock feed contaminated with heliotrope seed. *Proceedings of the 4th International Symposium on Poisonous Plants (ISOPP4)*.

Gaunt, S.L. 1980. Thermoregulation in doves (*Columbidae*): A novel esophageal heat exchanger. *Science*. 210(4468):445–447.

Gautier, A.E., J.D. Latorre, P.L. Matsler, and S.J. Rochell. 2020. Longitudinal characterization of coccidiosis control methods on live performance and nutrient utilization in broilers. *Frontiers in Veterinary Science*. 6: 468.

Gershwin, L.J. 2015. Comparative immunology of allergic responses. *Annual Review of Animal Biosciences*. 3(1):327–346.

Gewirtz, A.T., A.S. Neish, and J.L. Madara. 2002. Mechanisms of active intestinal inflammation and potential down-regulation via lipoxins. In *Eicosanoids and Other Bioactive Lipids in Cancer, Inflammation, and Radiation Injury, 5*. K.V. Honn, L.J. Marnett, S. Nigam, E. Dennis and C. Serhan, editors. New York: Springer US. 229–235.

Glahn, R.P., R.F. Wideman, J.W. Evangelisti, and W.E. Huff. 1988. Effects of ochratoxin a alone and in combination with citrinin on kidney function of single comb White Leghorn pullets. *Poultry Science*. 67(7):1034–1042.

Godbole, S.A., T.G. Krishna, and C.R. Bhatia. 1994. Further characterisation of protease inhibitors from pigeon pea (*Cajanus cajan* (l) millsp) seeds. *Journal of the Science of Food and Agriculture*. 64(3):331–335.

Gopinger, E., V.S.d. Avila, D. Perondi, A.A.d.S. Catalan, E.L. Krabbe, and V.F.B. Roll. 2015. Performance, carcass characteristics and litter moisture in broilers housed at two densities. *Acta Scientiarum Animal Sciences*. 37:35–39.

Green, A., and H. Xin. 2009. Effects of stocking density and group size on heat and moisture production of laying hens under thermoneutral and heat-challenging conditions. *Transactions of the ASABE*. 52(6): 2027–2032.

Gross, W.B., and H.S. Siegel. 1983. Evaluation of the heterophil/lymphocyte ratio as a measure of stress in chickens. *Avian Diseases*. 27(4):972–979.

Gross, W.B., and P.B. Siegel. 1985. Selective breeding of chickens for corticosterone response to social stress. *Poultry Science*. 64(12):2230–2233.

Gualtieri, M., and S. Rapaccini. 1990. Sorghum grain in poultry feeding. *World's Poultry Science Journal*. 46(3):246–254.

Guerre, P. 2016. Worldwide mycotoxins exposure in pig and poultry feed formulations. *Toxins*. 8(12):350.

Hamilton, P.B., W.E. Huff, J.R. Harris, and R.D. Wyatt. 1982. Natural occurrences of ochratoxicosis in poultry. *Poultry Science*. 61(9):1832–1841.

Hangalapura, B.N., M.G. Nieuwland, G. De Vries Reilingh, J. Buyse, H. Van Den Brand, B. Kemp, and H.K. Parmentier. 2005. Severe feed restriction enhances innate immunity but suppresses cellular immunity in chicken lines divergently selected for antibody responses. *Poultry Science*. 84(10):1520–1529.

Henry, M.H., G.M. Pesti, and T.P. Brown. 2001. Pathology and histopathology of gossypol toxicity in broiler chicks. *Avian Diseases*. 45:598–604.

Hess, J.B., and J.A. Mosjidis. 2008. Effect of sunn hemp seed inclusion in broiler starter diets on live performance attributes. *Journal of Applied Animal Research*. 33(2):105–108.

Hobson-Frohock, A., G. Fenwick, R. Heaney, D. Land, and R. Curtis. 1977. Rapeseed meal and egg taint: Association with sinapine. *British Poultry Science*. 18(5):539–541.

Hodge, J.E. 1953. Dehydrated foods, chemistry of browning reactions in model systems. *Journal of Agricultural and Food Chemistry*. 1(15):928–943.

Humphrey, B., and K. Klasing. 2004. Modulation of nutrient metabolism and homeostasis by the immune system. *World's Poultry Science Journal*. 60(1):90–100.

Hung, Y.T., A.R. Hanson, G.C. Shurson, and P.E. Urriola. 2017. Peroxidized lipids reduce growth performance of poultry and swine: A meta-analysis. *Animal Feed Science and Technology*. 231:47–58.

Hussein, H.S., and J.M. Brasel. 2001. Toxicity, metabolism, and impact of mycotoxins on humans and animals. *Toxicology*. 167(2):101–134.

Irish, G.G., and D. Balnave. 1993. Non-starch polysaccharides and broiler performance on diets containing soyabean meal as the sole protein concentrate. *Australian Journal of Agricultural Research.* 44(7):1483–1499.

Iseri, V., and K. Klasing. 2014. Changes in the amount of lysine in protective proteins and immune cells after a systemic response to dead *Escherichia coli*: Implications for the nutritional costs of immunity. *Integrative and Comparative Biology.* 54:922–930.

Ishaaya, I., Y. Birk, A. Bondi, and Y. Tencer. 1969. Soyabean saponins IX—Studies of their effect on birds, mammals and cold-blooded organisms. *Journal of the Science of Food and Agriculture.* 20(7):433–436.

Jacobsen, E., B. Skadhauge, and S.-E. Jacobsen. 1997. Effect of dietary inclusion of quinoa on broiler growth performance. *Animal Feed Science and Technology.* 65(1–4):5–14.

Jalaludin, S.Y., and S.K. Leong. 1973. Response of laying hens to low and high levels of tapioca meal. *Malays Agriculture Research.* 2(1):47–51.

Jeurissen, S.H., F. Lewis, J.D. van der Klis, Z. Mroz, J.M. Rebel, and A.A. ter Huurne. 2002. Parameters and techniques to determine intestinal health of poultry as constituted by immunity, integrity, and functionality. *Current Issues in Intestinal Microbiology.* 3(1):1–14.

Johnson, R.W. 1997. Inhibition of growth by pro-inflammatory cytokines: An integrated view. *Journal of Animal Science.* 75(5):1244–1255.

Jones, R., and P. Hocking. 1999. Genetic selection for poultry behaviour: Big bad wolf or friend in need? *Animal Welfare.* 8(4):343–359.

Kaiser, P. 2010. Advances in avian immunology—prospects for disease control: A review. *Avian Pathology.* 39(5):309–324.

Kearsey, J., S. Petit, C. De Oliveira, and F. Schweighoffer. 2004. A novel four transmembrane spanning protein, CLP24. *European Journal of Biochemistry.* 271(13):2584–2592.

Keestra, A.M., M.R. de Zoete, L.I. Bouwman, M.M. Vaezirad, and J.P.M. van Putten. 2013. Unique features of chicken Toll-like receptors. *Developmental & Comparative Immunology.* 41(3):316–323.

Keshavarz, K. 1993. Effect of corn contaminated with velvetweed seeds on eggs. *Journal of Applied Poultry Research.* 2(3):232–238.

Kingsbury, J.M. 1964. Poisonous plants of the United States and Canada. *Soil Science.* 98(5):349.

Klasing, K.C. 1984. Effect of inflammatory agents and interleukin 1 on iron and zinc metabolism. *American Journal of Physiology—Regulatory, Integrative and Comparative Physiology.* 247(5):R901–R904.

Klasing, K.C. 1988. Nutritional aspects of leukocytic cytokines. *Journal of Nutrition.* 118(12):1436–1446.

Klasing, K.C. 2004. The costs of immunity. *Acta Zoologica Sinica.* 50(6): 961–969.

Klasing, K.C., and D.M. Barnes. 1988. Decreased amino acid requirements of growing chicks due to immunologic stress. *The Journal of Nutrition.* 118(9):1158–1164.

Klasing, K.C., D.E. Laurin, R.K. Peng, and D.M. Fry. 1987. Immunologically mediated growth depression in chicks: Influence of feed intake, corticosterone and interleukin-1. *Journal of Nutrition.* 117(9):1629–1637.

Klasing, K.C., and T.V. Leshchinsky. 1999. Interactions between nutrition and immunity. In *Nutrition and Immunology: Principles and Practice.* M.E. Gershwin, J.B. German, and C.L. Keen, editors. Totowa, NJ: Humana Press. 363–373.

Klita, P., G. Mathison, T. Fenton, and R. Hardin. 1996. Effects of alfalfa root saponins on digestive function in sheep. *Journal of Animal Science.* 74(5):1144–1156.

Kogut, M.H. 2009. Impact of nutrition on the innate immune response to infection in poultry. *Journal of Applied Poultry Research.* 18:111–124.

Koh, T., R. Peng, and K. Klasing. 1996. Dietary copper level affects copper metabolism during lipopolysaccharide-induced immunological stress in chicks. *Poultry Science.* 75(7):867–872.

Koutsos, E.A., C.C. Calvert, and K.C. Klasing. 2003. The effect of an acute phase response on tissue carotenoid levels of growing chickens (*Gallus gallus domesticus*). *Comparative Biochemistry and Physiology Part A: Molecular & Integrative Physiology.* 135(4):635–646.

Kovatsis, A., J. Flaskos, E. Nikolaidis, V.P. Kotsaki-Kovatsi, N. Papaioannou, and F. Tsafaris. 1993. Toxicity study of the main alkaloids of *Datura ferox* in broilers. *Food and Chemical Toxicology.* 31(11):841–845.

Kristensen, H.H., and C.M. Wathes. 2007. Ammonia and poultry welfare: A review. *World's Poultry Science Journal.* 56(3):235–245.

Kritchevsky, D., and S.C. Chen. 2005. Phytosterols—health benefits and potential concerns: A review. *Nutrition Research.* 25(5):413–428.

Kumar, V., A.V. Elangovan, and A.B. Mandal. 2005. Utilization of reconstituted high-tannin sorghum in the diets of broiler chickens. *Asian-Australasian Journal of Animal Sciences.* 18:538–544.

Kumar, V., A.V. Elangovan, A.B. Mandal, P.K. Tyagi, S.K. Bhanja, and B.B. Dash. 2007. Effects of feeding raw or reconstituted high tannin red sorghum on nutrient utilisation and certain welfare parameters of broiler chickens. *British Poultry Science.* 48(2):198–204.

Latif, M.A., T.R. Morris, and D.J. Jayne-Williams. 1976. Use of khesari (*Lathyrus sativus*) in chick diets. *British Poultry Science.* 17(5): 539–546.

Lebas, F. 2001. Fattening farms: Advantages and disadvantages. *Cuniculture.* 28:163–168.

Leenstra, F., V. Maurer, F. Galea, M. Bestman, Z. Amsler-Kepalaite, J. Visscher, I. Vermeij, and M. van Krimpen. 2014. Laying performance in different production systems; why do they differ and how close to the gap? Results of discussions with groups of farmers in The Netherlands, Switzerland and France, benchmarking and model calculations. *European Poultry Science.* 78.

Leeson, S., G. Diaz, and J. Summers. 1995. *Poultry Metabolic Disorders and Mycotoxins.* Guelph, Ontario, Canada: University Books.

Leeson, S., and J.D. Summers. 2001. *Nutrition of the Chicken*, 4th ed. Guelph, Ontario, Canada: University Books.

Lemme, A., V. Ravindran, and W.L. Bryden. 2004. Ileal digestibility of amino acids in feed ingredients for broilers. *World's Poultry Science Journal.* 60(4):423–438.

Lessire, M., V. Gallo, M. Prato, O. Akide-Ndunge, G. Mandili, P. Marget, P. Arese, and G. Duc. 2017. Effects of faba beans with different concentrations of vicine and convicine on egg production, egg quality and red blood cells in laying hens. *Animal.* 11(8):1270–1278.

Leulier, F., L.T. MacNeil, W.J. Lee, J.F. Rawls, P.D. Cani, M. Schwarzer, L. Zhao, and S.J. Simpson. 2017. Integrative physiology: At the crossroads of nutrition, microbiota, animal physiology, and human health. *Cell Metabolism.* 25(3):522–534.

Lewis, P.D., and T.R. Morris. 2007. Light intensity and performance of domestic pullets. *World's Poultry Science Journal.* 55(3):241–250.

Liener, I.E. 1994. Implications of antinutritional components in soybean foods. *Critical Reviews in Food Science & Nutrition.* 34(1):31–67.

Lin, C.F., A. Asghar, J.I. Gray, D.J. Buckley, A.M. Booren, R.L. Crackel, and C.J. Flegal. 1989. Effects of oxidised dietary oil and antioxidant supplementation on broiler growth and meat stability. *British Poultry Science.* 30(4):855–864.

Liu, J.D., B. Doupovec, D. Schatzmayr, G.R. Murugesan, C. Bortoluzzi, A.M. Villegas, and T.J. Applegate. 2020. The impact of deoxynivalenol, fumonisins, and their combination on performance, nutrient, and energy digestibility in broiler chickens. *Poultry Science.* 99(1):272–279.

Liu, K., and P. Markakis. 1991. Aqueous ethanol extraction of soybean trypsin inhibitors and characterization of a calcium-sensitive fraction. *Journal of Food Biochemistry.* 15(3):159–168.

Liu, X., H. Zhao, S. Thiessen, J. House, and P. Jones. 2010. Effect of plant sterol-enriched diets on plasma and egg yolk cholesterol concentrations and cholesterol metabolism in laying hens. *Poultry Science.* 89(2):270–275.

Lorenz, F.W. 1939. Egg deterioration due to ingestion by hens of malvaceous materials. *Poultry Science.* 18(4):295–300.

Lu, S., M. Aziz, D. Sturtevant, K.D. Chapman, and L. Guo. 2020. Heterogeneous distribution of erucic acid in *Brassica napus* seeds. *Frontiers in Plant Science.* 10:1744.

Macelline, S.P., S.S. Wickramasuriya, H.M. Cho, E. Kim, T.K. Shin, J.S. Hong, J.C. Kim, J.R. Pluske, H.J. Choi, and Y.G. Hong. 2020. Broilers fed a low protein diet supplemented with synthetic amino acids maintained growth performance and retained intestinal integrity while reducing nitrogen excretion when raised under poor sanitary conditions. *Poultry Science.* 99(2):949–958.

Magan, N., A. Medina, and D. Aldred. 2011. Possible climate-change effects on mycotoxin contamination of food crops pre- and postharvest. *Plant Pathology*. 60(1):150–163.

Maillard, L.C. 1912. Action des acides aminés sur les sucres; formation des mélanoïdines par voie methodique. *Comptes rendus de l'Académie des Sciences*. 154:66–68.

Makala, L.H.C., N. Suzuki, and H. Nagasawa. 2002. Peyer's patches: Organized lymphoid structures for the induction of mucosal immune responses in the intestine. *Pathobiology*. 70(2):55–68.

Mansoori, B., and T. Acamovic. 2009. Influence of tannic acid and polyethylene glycol on the excretion and digestibility of amino acids in gelatin-fed broilers. *British Poultry Science*. 50(2):199–206.

March, B.E., and R. Soong. 1976. Mortality and production characteristics of laying chickens fed high- and low-erucic acid rapeseed oils. *Poultry Science*. 55(4):1557–1560.

Marquardt, R.R., L.D. Campbell, S.C. Stothers, and J.A. McKirdy. 1974. Growth responses of chicks and rats fed diets containing four cultivars of raw or autoclaved faba beans. *Canadian Journal of Animal Science*. 54(2):177–182.

Marusich, W.L., E.F. Ogrins, E. Schildknecht, P.R. Brown, and M. Mitrovic. 1973. The effect of subclinical infections with *Eimeria praecox* and *Eimeria tenella* on pigmentation and vitamin A absorption in broilers. *British Poultry Science*. 14(6):541–546.

Mashaly, M.M., R.G.L. Hendricks, M.A. Kalama, A.E. Gehad, A.O. Abbas, and P.H. Patterson. 2004. Effect of heat stress on production parameters and immune responses of commercial laying hens. *Poultry Science*. 83(6):889–894.

Mateos, G.G., E. Jiménez-Moreno, M.P. Serrano, and R.P. Lázaro. 2012. Poultry response to high levels of dietary fiber sources varying in physical and chemical characteristics. *Journal of Applied Poultry Research*. 21(1):156–174.

Maxwell, M.H., P.M. Hocking, and G.W. Robertson. 1992. Differential leucocyte responses to various degrees of food restriction in broilers, turkeys and ducks. *British Poultry Science*. 33(1):177–187.

McFarlane, J.M., S.E. Curtis, R.D. Shanks, and S.G. Carmer. 1989. Multiple concurrent stressors in chicks. 1. Effect on weight gain, feed intake, and behavior. *Poultry Science*. 68(4):501–509.

McKinney, L.J., and R.G. Teeter. 2004. Predicting effective caloric value of nonnutritive factors: I. Pellet quality and II. Prediction of consequential formulation dead zones. *Poultry Science*. 83(7):1165–1174.

Miao, Z.H., P.C. Glatz, and Y.J. Ru. 2005. Free-range poultry production—A review. *Asian-Australasian Journal of Animal Science*. 18(1):113–132.

Miles, D.M., S.L. Branton, and B.D. Lott. 2004. Atmospheric ammonia is detrimental to the performance of modern commercial broilers. *Poultry Science*. 83(10):1650–1654.

Miles, D.M., D.E. Rowe, and T.C. Cathcart. 2011. High litter moisture content suppresses litter ammonia volatilization. *Poultry Science*. 90(7):1397–1405.

Moore, P., A. Evenson, T. Luckey, E. McCoy, C. Elvehjem, and E. Hart. 1946. Use of sulfasuxidine, streptothricin, and streptomycin in nutritional studies with the chick. *Journal of Biological Chemistry*. 165:437–441.

Moran, E.T. 1982. *Comparative Nutrition of Fowl & Swine: The Gastrointestinal Systems*. Guelph, Ontario, Canada: Office for Educational Practice, University of Guelph.

Morrissey, S.M., E.R. Benson, R.L. Alphin, and D.P. Hougentogler. 2017. Compatibility of dimmers and alternative technology lamps for commercial broiler houses. *The Journal of Applied Poultry Research*. 26(4):548–557.

Moughan, P.J. 2005. Absorption of chemically unmodified lysine from proteins in foods that have sustained damage during processing or storage. *Journal of AOAC International*. 88(3):949–954.

Moughan, P.J., V. Ravindran, and J.O.B. Sorbara. 2014. Dietary protein and amino acids—Consideration of the undigestible fraction. *Poultry Science*. 93(9):2400–2410.

Moughan, P.J., and S.M. Rutherfurd. 1996. A new method for determining digestible reactive lysine in foods. *Journal of Agricultural and Food Chemistry*. 44(8):2202–2209.

Muir, W.M., and J.V. Craig. 1998. Improving animal well-being through genetic selection. *Poultry Science*. 77(12):1781–1788.

Mukhtar, M.A. 2007. The effect of feeding rosella (*Hibiscus sabdariffa*) seed on broiler chicks performance. *Research Journal of Animal and Veterinary Sciences*. 2(13):21–23.

Naga Raja Kumari, K., and D. Narendra Nath. 2017. Ameliorative measures to counter heat stress in poultry. *World's Poultry Science Journal*. 74(1):117–130.

Neijat, M., J.D. House, W. Guenter, and E. Kebreab. 2011. Calcium and phosphorus dynamics in commercial laying hens housed in conventional or enriched cage systems. *Poultry Science*. 90(10):2383–2396.

Ngoka, D., E. Chike, A. Awoniyi, T. Enyinnia, and S. Odurukwe. 1982. Effect of the cassava meal on the hatchability of chickens. *Proceedings of the Proceeding Triennial Symposium of International E: For Tropical Root Crops*.117.

Niewold, T.A. 2007. The nonantibiotic anti-inflammatory effect of antimicrobial growth promoters, the real mode of action? A hypothesis. *Poultry Science*. 86(4):605–609.

Noyan, A., W.J. Lossow, N. Brot, and I.L. Chaikoff. 1964. Pathway and form of absorption of palmitic acid in the chicken. *Journal of Lipid Research*. 5(4):538–541.

Nusrat, A., S.V. Sitaraman, and A. Neish. 2001. Interaction of bacteria and bacterial toxins with intestinal epithelial cells. *Current Gastroenterology Reports*. 3(5):392–398.

Odihambo Mumma, J., J.P. Thaxton, Y. Vizzier-Thaxton, and W.L. Dodson. 2006. Physiological stress in laying hens. *Poultry Science*. 85(4):761–769.

Palliyeguru, M.W.C.D., S.P. Rose, and A.M. Mackenzie. 2010. Effect of dietary protein concentrates on the incidence of subclinical necrotic enteritis and growth performance of broiler chickens. *Poultry Science*. 89(1):34–43.

Palliyeguru, M.W.C.D. 2011. Effect of trypsin inhibitor activity in soya bean on growth performance, protein digestibility and incidence of sub-clinical necrotic enteritis in broiler chicken flocks. *British Poultry Science*. 52(3):359–367.

Pan, D., and Z. Yu. 2014. Intestinal microbiome of poultry and its interaction with host and diet. *Gut Microbes*. 5(1):108–119.

Panigrahi, S., J. Rickard, G. O'Brien, and C. Gay. 1992. Effects of different rates of drying cassava root on its toxicity to broiler chicks. *British Poultry Science*. 33(5):1025–1041.

Persia, M., E. Young, P. Utterback, and C. Parsons. 2006. Effects of dietary ingredients and *Eimeria acervulina* infection on chick performance, apparent metabolizable energy, and amino acid digestibility. *Poultry Science*. 85(1):48–55.

Post, J., J.M. Rebel, and A.A. ter Huurne. 2003. Physiological effects of elevated plasma corticosterone concentrations in broiler chickens: An alternative means by which to assess the physiological effects of stress. *Poultry Science*. 82(8):1313–1318.

Puron, D., R. Santamaria, J.C. Segura, and J.L. Alamilla. 1995. Broiler performance at different stocking densities. *Journal of Applied Poultry Research*. 4(1):55–60.

Pusztai, A., S.W. Ewen, G. Grant, W.J. Peumans, E.J. van Damme, L. Rubio, and S. Bardocz. 1990. Relationship between survival and binding of plant lectins during small intestinal passage and their effectiveness as growth factors. *Digestion*. 2:308–316.

Puvadolpirod, S., and J.P. Thaxton. 2000. Model of physiological stress in chickens 1. Response parameters. *Poultry Science*. 79(3):363–369.

Qi, M., L.h. Sun, J. Guo, J. Liu, Y.Z. Duan, X. Wang, C.P. Zeng, D.S. Qi, and N.Y. Zhang. 2017. Abnormal characteristics of eggs laid by hens fed with high levels of cottonseed meal. *Journal of Applied Poultry Research*. 26(1):122–129.

Qiao, H., and H.L. Classen. 2003. Nutritional and physiological effects of rapeseed meal sinapine in broiler chickens and its metabolism in the digestive tract. *Journal of the Science of Food and Agriculture*. 83(14): 1430–1438.

Quarles, C.L., and H.F. Kling. 1974. Evaluation of ammonia and infectious bronchitis vaccination stress on broiler performance and carcass quality. *Poultry Science*. 53(4):1592–1596.

Reece, F.N., B.D. Lott, and J.W. Deaton. 1980. Ammonia in the atmosphere during brooding affects performance of broiler chickens. *Poultry Science*. 59(3):486–488.

Ren, Z., D.E. Bütz, R. Whelan, V. Naranjo, M.K. Arendt, M.D. Ramuta, X. Yang, T.D. Crenshaw, and M.E. Cook. 2020. Effects of dietary methionine plus cysteine levels on growth performance and intestinal antibody production in broilers during *Eimeria* challenge. *Poultry Science*. 99(1):374–384.

Richardson, M. 1977. The proteinase inhibitors of plants and microorganisms. *Phytochemistry*. 16(2):159–169.

Robblee, A.R., D.R. Clandinin, R.T. Hardin, G.R. Milne, and K. Darlington. 1977. Studies on the use of faba beans in rations for laying hens. *Canadian Journal of Animal Science*. 57(3):421–425.

Rochell, S., C. Parsons, and R. Dilger. 2016. Effects of *Eimeria acervulina* infection severity on growth performance, apparent ileal amino acid digestibility, and plasma concentrations of amino acids, carotenoids, and α1-acid glycoprotein in broilers. *Poultry Science*. 95(7):1573–1581.

Rogers, A.G., E.M. Pritchett, R.L. Alphin, E.M. Brannick, and E.R. Benson. 2015. II. Evaluation of the impact of alternative light technology on male broiler chicken stress. *Poultry Science*. 94(3):331–337.

Rosales, A.G. 1994. Managing stress in broiler breeders: A review. *The Journal of Applied Poultry Research*. 3(2):199–207.

Roura, E., J. Homedes, and K.C. Klasing. 1992. Prevention of immunologic stress contributes to the growth-permitting ability of dietary antibiotics in chicks. *The Journal of Nutrition*. 122(12):2383–2390.

Roy, D.N., and P.S. Spencer. 1989. Lathyrogens: Toxicants of plant origin. In *Proteins and Amino Acid*. P.R. Cheeke, editor. Boca Raton, FL: CRC Press. 170–201.

Rubin, L.L., C.W. Canal, A. Ribeiro, A. Kessler, I. Silva, L. Trevizan, T. Viola, M. Raber, T.A. Gonçalves, and R. Krás. 2007. Effects of methionine and arginine dietary levels on the immunity of broiler chickens submitted to immunological stimuli. *Brazilian Journal of Poultry Science*. 9(4):241–247.

Runkle, E.A., and D. Mu. 2013. Tight junction proteins: From barrier to tumorigenesis. *Cancer Letters*. 337(1):41–48.

Rzehak, P., S. Koletzko, B. Koletzko, S. Sausenthaler, D. Reinhardt, A. Grübl, C.P. Bauer, U. Krämer, C. Bollrath, and A. von Berg. 2011. Growth of infants fed formula rich in canola oil (low erucic acid rapeseed oil). *Clinical Nutrition*. 30(3):339–345.

Saki, A.A. 2016. Effects of various levels of oxidized oil on performance, egg quality and some blood metabolites in laying hens. *Poultry Science Journal*. 4(1):13–18.

Santana, M.R.d., R.G. Garcia, I.d.A. Naas, I.C.d.L.A. Paz, F.R. Caldara, and B. Barreto. 2014. Light emitting diode (LED) use in artificial lighting for broiler chicken production. *Engenharia Agrícola*. 34:422–427.

Schaible, P.J., L.A. Moore, and J.M. Moore. 1934. Gossypol, a cause of discoloration in egg yolks. *Science*. 79: 372.

Scheele, C.W. 1997. Pathological changes in metabolism of poultry related to increasing production levels. *Veterinary Quarterly*. 19(3):127–130.

Schiffrin, E.J., and S. Blum. 2002. Interactions between the microbiota and the intestinal mucosa. *European Journal of Clinical Nutrition*. 56(Suppl. 3):S60–S64.

Schmittle, S.C., D.J. Richey, and J.T. Tumlin. 1959. Toxicity of crotalaria-spectabilis seed in poultry. *Poultry Science*. 38(5):1244–1245.

Sell, D.R., and J.C. Rogler. 1984. The effects of sorghum tannin and methionine level on the performance of laying hens maintained in two temperature environments. *Poultry Science*. 63(1):109–116.

Sell, D.R., J.C. Rogler, and W.R. Featherston. 1983. The effects of sorghum tannin and protein level on the performance of laying hens maintained in two temperature environments. *Poultry Science*. 62(12):2420–2428.

Sharma, V., and M. Fernando. 1975. Effect of *Eimeria acervulina* infection on nutrient retention with special reference to fat malabsorption in chickens. *Canadian Journal of Comparative Medicine*. 39(2):146.

Shimoyamada, M., S. Ikedo, R. Ootsubo, and K. Watanabe. 1998. Effects of soybean saponins on chymotryptic hydrolyses of soybean proteins. *Journal of Agricultural and Food Chemistry*. 46(12):4793–4797.

Shini, S., P. Kaiser, A. Shini, and W.L. Bryden. 2008. Biological response of chickens (*Gallus gallus domesticus*) induced by corticosterone and a bacterial endotoxin. *Comparative Biochemistry and Physiology Part B: Biochemistry and Molecular Biology*. 149(2):324–333.

Shurson, G.C., B.J. Kerr, and A.R. Hanson. 2015. Evaluating the quality of feed fats and oils and their effects on pig growth performance. *Journal of Animal Science and Biotechnology*. 6(1):10.

Siegel, H.S. 1980. Physiological stress in birds. *BioScience*. 30(8):529–534.

Siegel, H.S. 1995. Gordon Memorial Lecture. Stress, strains and resistance. *British Poultry Science*. 36(1):3–22.

Simitzis, P.E., E. Kalogeraki, M. Goliomytis, M.A. Charismiadou, K. Triantaphyllopoulos, A. Ayoutanti, K. Niforou, A.L. Hager-Theodorides, and S.G. Deligeorgis. 2012. Impact of stocking density on broiler growth performance, meat characteristics, behavioural components and indicators of physiological and oxidative stress. *British Poultry Science*. 53(6):721–730.

Smirnov, A., R. Perez, E. Amit-Romach, D. Sklan, and Z. Uni. 2005. Mucin dynamics and microbial populations in chicken small intestine are changed by dietary probiotic and antibiotic growth promoter supplementation. *Journal of Nutrition*. 135(2):187–192.

Smirnov, A., D. Sklan, and Z. Uni. 2004. Mucin dynamics in the chick small intestine are altered by starvation. *Journal of Nutrition*. 134(4):736–742.

Sørensen, P., G. Su, and S.C. Kestin. 2000. Effects of age and stocking density on leg weakness in broiler chickens. *Poultry Science*. 79(6):864–870.

St-Laurent, R., and J. Larochelle. 1994. The cooling power of the pigeon head. *Journal of Experimental Biology*. 194(1):329–339.

Star, L., M. Rovers, E. Corrent, and J. Van der Klis. 2012. Threonine requirement of broiler chickens during subclinical intestinal *Clostridium* infection. *Poultry Science*. 91(3):643–652.

Suresh, G., R.K. Das, S. Kaur Brar, T. Rouissi, A. Avalos Ramirez, Y. Chorfi, and S. Godbout. 2018. Alternatives to antibiotics in poultry feed: Molecular perspectives. *Critical Reviews in Microbiology*. 44(3):318–335.

Swayne, D. 2020. *Diseases of Poultry*, 14th ed. Hoboken, NJ: Wiley-Blackwell.

Tan, J., S. Liu, Y. Guo, T.J. Applegate, and S.D. Eicher. 2014. Dietary L-arginine supplementation attenuates lipopolysaccharide-induced inflammatory response in broiler chickens. *British Journal of Nutrition*. 111(8):1394–1404.

Teirlynck, E., L. Bjerrum, V. Eeckhaut, G. Huygebaert, F. Pasmans, F. Haesebrouck, J. Dewulf, R. Ducatelle, and F. Van Immerseel. 2009. The cereal type in feed influences gut wall morphology and intestinal immune cell infiltration in broiler chickens. *British Journal of Nutrition*. 102(10):1453–1461.

Temperley, N.D., S. Berlin, I.R. Paton, D.K. Griffin, and D.W. Burt. 2008. Evolution of the chicken Toll-like receptor gene family: A story of gene gain and gene loss. *BMC Genomics*. 9(1):62.

Thaxton, P., and H.S. Siegel. 1970. Immunodepression in young chickens by high environmental temperature. *Poultry Science*. 49(1):202–205.

Tomar, B.S. 2001. Intestinal infections. *Indian Journal of Pediatrics*. 68(Suppl. 3):S8–S18.

Tong, X. 2019. CFD modeling of airflow, thermal environment, and ammonia concentration distribution in a commercial manure-belt layer house with mixed ventilation systems. *Computers and Electronics in Agriculture V*. 162:281–299.

Turk, D.E. 1981. Coccidial infections and iron absorption. *Poultry Science*. 60(2):323–326.

Ueda, H., and K. Tanoue. 2000. Growth-depressing and cholesterol-lowering effects of *Quillaja* and tea saponins in chicks as influenced by diet composition. *Nihon Chikusan Gakkaiho*. 71(4):393–399.

Valkonen, E., E. Venalainen, L. Rossow, and J. Valaja. 2008. Effects of dietary energy content on the performance of laying hens in furnished and conventional cages. *Poultry Science*. 87(5):844–852.

van Dijk, A., E.J.A. Veldhuizen, and H.P. Haagsman. 2008. Avian defensins. *Veterinary Immunology and Immunopathology*. 124(1):1–18.

Van Immerseel, F., J.I. Rood, R.J. Moore, and R.W. Titball. 2009. Rethinking our understanding of the pathogenesis of necrotic enteritis in chickens. *Trends in Microbiology*. 17(1):32–36.

Virden, W.S., and M.T. Kidd. 2009. Physiological stress in broilers: Ramifications on nutrient digestibility and responses. *The Journal of Applied Poultry Research*. 18(2):338–347.

Visek, W. 1978. The mode of growth promotion by antibiotics. *Journal of Animal Science*. 46(5):1447–1469.

Vu Manh, T.-P., H. Marty, P. Sibille, Y. Le Vern, B. Kaspers, M. Dalod, I. Schwartz-Cornil, and P. Quéré. 2014. Existence of conventional dendritic cells in *Gallus gallus* revealed by comparative gene expression profiling. *The Journal of Immunology*. 192(10):4510–4517.

Wahlström, A., K. Elwinger, and S. Thomke. 1999. Total tract and ileal nutrient digestibility of a diet fed as mash or crumbled pellets to two laying hybrids. *Animal Feed Science and Technology*. 77(3–4):229–239.

Waldenstedt, L., K. Elwinger, A. Lunden, P. Thebo, M.R. Bedford, and A. Uggla. 2000. Intestinal digesta viscosity decreases during coccidial infection in broilers. *British Poultry Science*. 41(4):459–464.

Wang, F., Z. Zuo, K. Chen, X. Peng, J. Fang, H. Cui, G. Shu, M. He, and L. Tang. 2019. Selenium rescues aflatoxin B1-inhibited T cell subsets and cytokine levels in cecal tonsil of chickens. *Biological Trace Element Research*. 188(2):461–467.

Wang, L., J. Li, J. Li, R.X. Li, C.F. Lv, S. Li, Y.L. Mi, and C.Q. Zhang. 2016. Identification of the Paneth cells in chicken small intestine. *Poultry Science*. 95(7):1631–1635.

Wang, Y. 2019. A new ventilation system to reduce temperature fluctuations in laying hen housing in continental climate. *Biosystems Engineering V*. 181:52–62.

Ward, A.K., H.L. Classen, and F.C. Buchanan. 2009. Fishy-egg tainting is recessively inherited when brown-shelled layers are fed canola meal. *Poultry Science*. 88(4):714–721.

Weathers, W.W., and D.C. Schoenbaechler. 1976. Contribution of gular flutter to evaporative cooling in Japanese quail. *Journal of Applied Physiology*. 40(4):521–524.

Webel, D.M., R.W. Johnson, and D.H. Baker. 1998a. Lipopolysaccharide-induced reductions in body weight gain and feed intake do not reduce the efficiency of arginine utilization for whole-body protein accretion in the chick. *Poultry Science*. 77(12):1893–1898.

Webel, D.M., R.W. Johnson, and D.H. Baker. 1998b. Lipopolysaccharide-induced reductions in food intake do not decrease the efficiency of lysine and threonine utilization for protein accretion in chickens. *The Journal of Nutrition*. 128(10):1760–1766.

Weibking, T.S., D.R. Ledoux, A.J. Bermudez, J.R. Turk, G.E. Rottinghaus, E. Wang, and A.H. Merrill Jr. 1993. Effects of feeding *Fusarium moniliforme* culture material, containing known levels of fumonisin B1, on the young broiler chick. *Poultry Science*. 72(3):456–466.

West, C.E., S.R. Sijtsma, B. Kouwenhoven, J.H. Rombout, and A.J. van der Zijpp. 1992. Epithelia-damaging virus infections affect vitamin A status in chickens. *The Journal of Nutrition*. 122(2):333–339.

Wu, C., and J.R. Whitaker. 1990. Purification and partial characterization of four trypsin/chymotrypsin inhibitors from red kidney beans (*Phaseolus vulgaris*, var. Linden). *Journal of Agricultural and Food Chemistry*. 38(7):1523–1529.

Xavier-Filho, J., and F.A.P. Campos. 1989. *Protease Inhibitors in Toxicants of Plant Origin, Volume III, Proteins and Amino Acids*. P.R. Cheeke, editor. Boca Raton, FL: CRC Press.

Xin, H., R.S. Gates, A.R. Green, F.M. Mitloehner, P.A. Moore Jr., and C.M. Wathes. 2011. Environmental impacts and sustainability of egg production systems. *Poultry Science*. 90(1):263–277.

Yahav, S., S. Goldfeld, I. Plavnik, and S. Hurwitz. 1995. Physiological responses of chickens and turkeys to relative humidity during exposure to high ambient temperature. *Journal of Thermal Biology*. 20(3):245–253.

Yang, Y., P. Iji, and M. Choct. 2009. Dietary modulation of gut microflora in broiler chickens: A review of the role of six kinds of alternatives to in-feed antibiotics. *World's Poultry Science Journal*. 65(1):97–114.

Yunis, R., and A. Cahaner. 1999. The effects of the naked neck (Na) and frizzle (F) genes on growth and meat yield of broilers and their interactions with ambient temperatures and potential growth rate. *Poultry Science*. 78(10):1347–1352.

Zhang, W., S. Xiao, E.J. Lee, and D.U. Ahn. 2010. Consumption of oxidized oil increases oxidative stress in broilers and affects the quality of breast meat. *Journal of Agricultural and Food Chemistry*. 59(3):969–974.

Ziegler-Heitbrock, H.W., and R.J. Ulevitch. 1993. CD14: Cell surface receptor and differentiation marker. *Immunology Today*. 14(3):121–125.

Zmrhal, V., and P. Slama. 2020. Immunomodulation of avian dendritic cells under the induction of prebiotics. *Animals*. 10(4):698.

12

Nutritional Qualities of Products

INTRODUCTION

Poultry meat and eggs are the most widely consumed animal proteins among human diets. Both poultry and eggs are starting ingredients for myriad food products. The environmental impact of poultry production is the least of all animal proteins, and production systems exist that accommodate small home flocks as well as large commercial complexes capable of raising millions of birds at a time. Poultry meat is less expensive than other animal proteins and bird genetics offer further optimization for local conditions and customs. Flexible, economical, and sustainable; poultry production provides the means and methods to improve human health and diet quality in even the most underdeveloped countries.

Documenting the variability in nutritional contributions that food products containing poultry and eggs make to the diet is beyond the scope of this report. For this reason, this section will address the nutritional content of raw food materials derived from poultry and eggs. General compositional information on poultry and eggs is available in national and international food composition databases. One of the most comprehensive sources, FoodData Central is maintained by the U.S. Department of Agriculture (USDA, 2025) Agricultural Research Service (ARS) and should be consulted for details of specific products and cookery effects. Other limitations to the scope of this section include the omission of nutritional benefits derived from poultry by-products such as manure and feathers when used as fertilizer for other crops (Moore et al., 1995) or used to feed other animals (Papadopoulos et al., 1985; Fakhfakh et al., 2011; Ghaly and MacDonald, 2012). The use of components isolated from poultry tissues or eggs that provide nutritional or health benefit to humans as described in more specialized texts will also not be addressed here, but these certainly represent the diversified value of poultry products (Walzem, 2004; Sim and Sunwoo, 2006; Watson and De Meester, 2015).

Poultry and Poultry Product Use

The Food and Agriculture Organization (FAO) noted that in 2017, chickens accounted for 92 percent of the world's poultry population (2022). Birds in addition to chickens, ducks, and turkeys contribute to poultry meat and egg production (Table 12-1, FAO, 2022), but globally, some 89 percent of poultry meat and 92 percent of eggs are produced from chickens. Most of the published data describing variation in the nutritional qualities of poultry and eggs come from studies with chickens and form the basis of statements in this section. However, where possible, outcomes at variance to those in chickens are noted.

Edible Components

The primary edible components of poultry include specific large muscle groups on the breast and legs of birds as

TABLE 12-1 Global Sources of Edible Poultry[a] and Eggs

Species/Common Name	Global Poultry Population, %	Global Poultry Meat Production, %	Global Egg Production by Species, %[c]
Chicken[b]	92.0	89.0	92
Duck[b]	5.0	4.0	Varies by region[c]
Turkey[b]	2.0	5.0	
Geese and guinea fowl[b]	<1.0	2.0	
Other	<1.0	<1.0	

[a] Values from 2017 FAOSTAT—live animals, http://www.fao.org/faostat/en/#data/QA. This database defines poultry as indigenous and commercial breeds of chickens, Muscovy ducks, mallard ducks, turkeys, guinea fowl, geese, quail, pigeons, ostriches, and pheasants.

[b] In 2017, the global number of chickens was >22 billion; ducks were 1.15 billion; turkeys numbered 459 million, with geese and guinea fowls together being 371 million birds. In 2017, the United States processed 26.6 million ducks into 134 million pounds (60.8 million kg) of meat, exporting 7.0 million pounds (3.2 million kg) to 32 destination countries (Sheats, 2018).

[c] Regionally, non-chicken egg production accounts for 12.0, 2.0, 3.0, 0.6, 0.8% and nil in Asia, North America, South America, Oceania, Europe, and Africa, respectively.

well as eggs. Wings and feet (paws) are also widely consumed in certain regional markets. A variety of meat portions are available; trade descriptions establish voluntary trade standards for ready-to-cook chicken and turkey products produced in the United States. USDA trade descriptions provide standardized codes to specify not only the cut but also the species, methods of rearing, slaughtering, post-slaughter processing, piece weights, and product options followed by packaging options (USDA, 2000a). Similar specifications are available for turkey (USDA, 2000b). The Agricultural Marketing Service website also provides egg and egg product specifications (USDA AMS, 2015, 2017).

CONTRIBUTIONS TO HUMAN NUTRITION

The importance of these poultry products in the human diet are reflected in their designation as key foods within the USDA's National Food and Nutrient Analysis Program (NFNAP) (Haytowitz, 2015). The premise of the key foods approach is that a relatively small number of foods contribute most of the intake of a nutrient or component for the United States population. The initial 666 foods selected as key foods was primarily based on their consumption pattern as reported in the Continuing Survey of Food Intakes by Individuals 1994–1996 (CSFII) and by the U.S. Department of Health and Human Services in the National Health and Nutrition Examination Survey (NHANES). The approach supported creation of readily updated "evergreen" information databases that underpin nutrition policy development (Holden et al., 2006). The methodological adequacy of nutrient determinations within the databases are routinely evaluated for the ability to satisfy qualitative, quantitative, and economic criteria, in much the same way as Association of American Feed Control Officials (AAFCO) numbers mentioned in Chapters 4 and 20 are, as they have regulatory and policy-setting applications. In the most recent update (Haytowitz, 2015), eggs, fresh or frozen, were one of the top 10 foods for protein, Fe, P, Zn, cholesterol, and monounsaturated fatty acids (MUFAs). While poultry meat per se did not make the top 10 list for single ingredient/simple foods, chicken-containing foods or mixed dishes appeared in 13 different forms or cookery methods (Haytowitz, 2015). Notably, using data from the NHANES for 2007–2010, poultry meat accounted for 22.0 percent of total protein intake of U.S. adults aged 19 years and older; eggs and omelet accounted for an additional 7.2 percent of protein intake (Pasiakos et al., 2015). Overall, animal proteins accounted for 46 percent of protein intake while dairy and plant proteins accounted for 16 percent and 30 percent, respectively. The source of the remaining 8 percent of protein intake was not specified. Poultry and eggs are core foods with regards to dietary vitamin intake (Combs and McClung, 2017; NIOH ODS, 2020). Eggs provide significant amounts of vitamins A, B_{12}, and D and thiamin, as well as riboflavin, biotin, pantothenic acid, folate, and choline. Poultry meat is an important source for vitamins B_6 and B_{12}, thiamin, niacin, and pantothenic acid.

PROTEIN QUALITY CONSIDERATIONS

The nutrient of highest concentration in poultry and egg products is protein, along with its constituent amino acids. Protein quality for human nutrition is primarily determined by a protein's amino acid composition and secondarily by its digestibility (IOM, 2003, 2005; FAO, 2011). The method of *Protein Digestibility–Corrected Amino Acid Score* (PDCAAS) has served in food regulatory settings since 1989 to describe protein content or make claims (IOM, 2003, 2005). The reference pattern for this index is based on the essential amino acid requirements for preschool children aged 1–3 years because they are the population group with high, growth-related amino acid requirements (IOM, 2005). The amino acid score is defined as the content of the first limiting indispensable amino acid in the test protein divided by the content of the same amino acid in a reference protein (Schaafsma, 2012). To list protein on the Nutrition Facts panel, the protein must have a PDCAAS of 1.0 (100%). The PDCAAS index assumes that ileal digestibility of all amino acids can be predicted from the total tract digestibility of crude protein. This scoring system caps values at 1.0, which is considered indicative of meeting human requirements. In a 2014 FAO expert consultation (Gidrewicz and Fenton, 2014), the PDCAAS method was criticized for not crediting the extra nutritional value of high-quality proteins and overestimating nutritional quality of proteins containing antinutritional factors. The bases for critique included failing to adequately account for the bioavailability of amino acids and overestimating the quality of poorly digestible proteins supplemented with limiting amino acids and of proteins that are co-limiting in more than one amino acid. That report advocated adoption of the *Digestible Indispensable Amino Acid Score* (DIAAS) for regulatory purposes. The DIAAS calculation differs from PDCAAS in that it is not capped and uses true ileal digestibility of each amino acid in its calculations. Detailed examples of the calculation for single foods and multiple ingredient dishes and diets can be found in the FAO report number 92 that recommended adoption of DIAAS as the regulatory norm (FAO, 2011). Table 12-2 provides examples of differences in apparent protein quality calculated using both PDCAAS and DIAAS approaches.

The amino acid composition of protein in poultry parts or eggs is not modified by nutritional intervention. All poultry-related protein sources reflect the aggregate composition of individual proteins that comprise, in the case of muscle or organ sources, the intact tissue or, in the case of the egg, the cell. Modification of lipid concentration and composition are possible through nutritional intervention as addressed in a later section.

TABLE 12-2 Differences in Apparent Protein Quality Using PDCAAS and DIAAS Calculations[a]

Protein Source	Wheat[b]	Pea Protein Concentrate[b]	Isolated Soy Protein[b]	Chicken Meat[c]	Whole Egg[c]
PDCAAS	51	84	100	100	100
DIAAS	54	73	98	108	116

[a] $\text{PDCAAS}(\%) = \frac{\text{mg of limiting amino acid in 1g of the dietary protein}}{\text{mg of same amino acid in 1g of the reference protein}} \times \text{true fecal digestibility}(\%) \times 100$

[b] $\text{DIASS}(\%) = \frac{\text{mg of digestible indispensible amino acid in 1g of the dietary protein}}{\text{mg of the same indispensable amino acid in 1g of the reference protein}}$

[c] Values from Mathai et al. (2017).

[d] Values from Ertl et al. (2016).

TABLE 12-3 Energy and Macronutrient Content of 100 g Boneless, Skinless Meat[a]

Meat Type	FDC ID[b]	Energy, kcal	Protein, g	Lipid, g	PUFA/SAT[a]	Iron, mg
Chicken, breast	171077	120	22.5	2.62	0.75	0.37
Turkey, breast	174515	114	23.3	2.33	1.86	0.76
Duck, breast (non-domestic)	174469	123	19.9	4.25	0.43	4.51
Ruffed grouse (breast)	172831	112	25.9	0.88	1.02	0.58
Duck, meat (domestic)	172410	135	18.3	5.95	0.32	2.40
Goose, meat (domestic)	172413	161	22.8	7.13	0.32	2.57
Quail, breast	172829	123	22.6	2.99	0.89	2.31
Chicken, thigh	173627	121	19.6	4.12	0.86	0.81
Turkey, thigh	174518	116	20.6	3.69	0.92	1.42
Beef, ribeye[c]	174699	187	19.5	11.4	0.10	2.64

[a] Polyunsaturated (PUFA) to saturated fat (SAT) ratio. Other common abbreviations for saturated fat include SFA, used in other sections of this monograph. SAT is the abbreviation used in FoodData Central (USDA, 2025).

[b] Values taken from FoodData Central (USDA, 2025), individual products are identified by a numeric code abbreviated as FDC ID.

[c] Ribeye, cap steak, boneless, separable lean only, trimmed to 0 in. fat, choice.

Muscle

The breast muscle of chicken, when served cooked, is commonly referred to as white meat, while cooked muscles from the leg muscle are referred to as dark meat, as related to the perception that muscle fiber type influences color of the cooked meat (Scanes, 2015). White meat is comprised of fast-twitch glycolytic fibers (Type IIb) that are physiologically useful for brief, powerful bursts of activity. This muscle fiber type relies on carbohydrate for fuel and so contains little intramuscular fat. In contrast, dark meat is comprised of slow-twitch oxidative fibers (Type I) that are useful for prolonged activities such as sustained flying or running. This muscle fiber type also uses carbohydrate to some extent, but it primarily relies on fatty acid oxidation to sustain prolonged vigorous activity. Oxidative metabolism in the muscle requires myoglobin, which is the protein that gives the muscle its red color due to its heme-iron content. Genetic selection in both chickens and turkeys aims to increase the proportion of breast meat comprised of Type IIb fibers on each carcass. Poultry that remain capable of flying, such as ducks, pheasants, geese, and wild turkeys, have breast muscle comprised of Type 1 fibers, which is why breast meat is dark in those species. The distinctive physiology of these muscle fiber types results in fundamental differences in the nutritional characteristics of light and dark meat, including the higher proportion of lipid and highly available heme iron in dark meat cuts (Table 12-3).

Values in Table 12-3 show that intramuscular lipid is relatively low in poultry meat, regardless of muscle fiber type. The dominant adipose depot in poultry is found in the abdomen and is readily separable at slaughter. Additional adipose depots within and under the skin (subcutaneous) can modify the total lipid content of chicken pieces. Subcutaneous adipose is more abundant around Type 1 fiber muscle groups; in chicken, this fiber type is found in the leg and thigh portions. Energy content of poultry pieces will vary with the presence of skin as well as amounts of subcutaneous adipose left on the part (Table 12-4).

Table 12-4 shows that retaining the skin and subcutaneous adipose on poultry meat will significantly increase the energy content of carcass pieces. Total caloric values can increase by 40–80 percent. Protein content is similar in pieces with and without skin. Note that subcutaneous adipose is not readily separable when skin is left on the piece.

Eggs

The egg's high nutrient content is consistent with its physiologic role as the complete nutrient supply for the developing embryo. Chicken eggs are a mainstream, cost-effective protein source that are considered the single most

TABLE 12-4 Differences in Protein and Fat Content of Chicken Pieces with and without Skin and Subcutaneous Adipose on Lipid and Energy Content of 100 g of Raw Boneless Poultry Meat Pieces[a]

Portion	FDC ID[a]	Protein, g	Lipid, g	Energy, kcal	Increase in kcal with Skin, %
Breast, skinless	171077	22.5	2.62	120	—
Breast, with skin	171474	20.9	9.25	172	43.3
Drumstick, skinless	173614	19.4	3.71	116	—
Drumstick, with skin	172373	18.1	9.2	161	38.8
Thigh, skinless	173627	19.7	4.12	121	—
Thigh, with skin	172385	16.5	16.6	221	82.6
Skin, with subcutaneous adipose	172855	9.58	44.23	440	—

[a] Values taken from FoodData Central (USDA, 2025); individual products are identified by a numeric code abbreviated as FDC ID.

versatile ingredient available for food preparation and consumption (Ruhlman, 2014). This perception is based on the functional properties of both whole and separated egg components that are extensively used in food production. Eggs and their components are key to the production of a wide variety of food products commonly purchased by the consumer; everything from mayonnaise to meringue. These properties include functionality for binding, emulsification, coagulation, and adhesion (Stadelman et al., 2017). Eggs act as a binder for cakes and breads and add a desired richness to soft fillings and fondants through their rheological properties (Stadelman et al., 2017). The functional properties of eggs and egg components also benefit food products such as ice cream, cream sauces, yogurts, probiotic, and frozen microwaveable dishes (Ruhlman, 2014; Stadelman et al., 2017). Eggs and egg components are available in a variety of forms such as pasteurized liquid egg products and spray-dried egg powders, each of which are more stable than raw eggs. The functional properties of these stabilized forms differ from those of raw forms and benefit production of meat items such as sausages, hamburgers, and hams (dry albumen or "whites") and high-temperature tolerant emulsions (yolk powders) (Uysal et al., 2017).

In whole egg, shell comprises 9–12 percent of total egg weight, with albumin or egg white comprising 60 percent of total egg weight and yolk 30–33 percent of total egg weight (Stadelman et al., 2017). The proportion of egg white to yolk varies with species, hen age, and energy intake (Ahn et al., 1997; Chen et al., 2006; Nys and Guyot, 2011). Proximate analysis shows that whole egg is 75 percent water, 12 percent protein, 12 percent lipids, and 1 percent carbohydrates plus minerals. The protein content is equally distributed between white and yolk (Williams et al., 2001), but lipids, vitamins, and minerals are concentrated in the yolk (Réhault-Godbert et al., 2019). Whole egg composition found in FoodData Central does not include shell components (Table 12-5) but shows that eggs from different types of poultry are similar in composition. Egg white mainly consists of water (88%) and protein (11%) with the remaining components being trace amounts of minerals and lipids (1%) (USDA, 2025). A comprehensive inventory of proteins in the white revealed the presence of 78 different proteins (Mann, 2007).

Egg yolk is physically separated from the white by the vitelline membrane (Bellairs et al., 1963). The vitelline membrane is a multilayered proteinaceous structure containing as many as 137 proteins. The strength of this membrane is enhanced in hens fed supplemental vitamin E and selenium (Scheideler et al., 2010). Membrane strength is an important variable in breaker egg operations (Kirunda and McKee, 2000). Chicken egg yolk comprises 52 percent water, 31–35 percent lipids, 15–17 percent protein, and ~1 percent carbohydrate (Table 12-5). The yolk also contains 1 percent carotenoids; their composition can be adjusted through altering dietary levels to produce variation in yolk color (Hencken, 1992; Hernandez, 2005; Olson et al., 2008). (See also Chapter 4, "Lipids.") Proteomic studies of yolk found 119 different protein species (Mann and Mann, 2008). Historically, yolk proteins were broadly classified as lipovitellins (36%), livetins (38%), phosvitin (8%), and low-density lipoproteins (17%) (Stadelman et al., 2017).

Other

Secondary edible components of poultry include chicken feet (referred to in industry as "paws"), wings, heart, gizzard, and liver (the latter three organs are collectively referred to as "giblets"), head, neck, testicles, yolk follicles, blood, and carcass. Often, these components are part of specific ethnic dishes (Table 12-6).

NUTRIENTS MODIFIABLE THROUGH POULTRY DIETS

Lipids

Birds reared with nutritionally adequate diets produce highly nutritious meat and eggs. Feed is the greatest expense in poultry production, and producers work diligently to achieve best production with least-cost, wholesome feed ingredients. For most producers in the United States, diets are comprised of mixtures of corn and soybean meal to deliver macronutrients. The protein components of the diet support lean tissue growth, while carbohydrate and fat serve as the major energy sources. Other cereal and fat sources are also used in least-cost formulations (Ahiwe et al., 2018). Poultry

TABLE 12-5 Energy and Macronutrient Content of 100 g of Egg or Egg Component[a]

Amounts per 100 g	Unit	Chicken	Turkey	Duck	Goose	Quail	Chicken Egg White	Chicken Egg Yolk
FDC ID		1711287	172192	172189	172190	172191	339017	339018
Typical egg weight, g		50	79	70	144	9	N/A	N/A
Water	g	76.15	72.5	70.83	70.43	74.35	87.57	52.31
Energy	kcal	143	171	185	185	158	52	322
Protein	g	12.56	13.68	12.81	13.87	13.05	10.9	15.86
Total lipid (fat)	g	9.51	11.88	13.77	13.27	11.09	0.17	26.54
Carbohydrate, by difference	g	0.72	1.15	1.45	1.35	0.41	0.73	3.59
Calcium, Ca	mg	56.0	99.0	64.0	60.0	64.0	7.0	129.0
Iron, Fe	mg	1.75	4.1	3.85	3.64	3.65	0.08	2.73
Magnesium, Mg	mg	12	13	17	16	13	11	5
Phosphorus, P	mg	198	170	220	208	226	15	390
Potassium, K	mg	138	142	222	210	132	163	109
Sodium, Na	mg	142	151	146	138	141	166	48
Zinc, Zn	mg	1.29	1.58	1.41	1.33	1.47	0.03	2.3
Copper, Cu	mg	0.072	0.062	0.062	0.062	0.062	0.023	0.077
Manganese, Mn	mg	0.028	0.038	0.038	0.038	0.038	NR	NR
Selenium, Se	µg	30.7	34.3	36.4	36.9	32	20	56
Thiamin	mg	0.04	0.11	0.156	0.147	0.13	0.004	0.176
Riboflavin	mg	0.457	0.47	0.404	0.382	0.79	0.439	0.528
Niacin	mg	0.075	0.024	0.2	0.189	0.15	0.105	0.024
Pantothenic acid	mg	1.533	1.889	1.862	1.759	1.761	NR	NR
Vitamin B_6[b]	mg	0.17	0.131	0.25	0.236	0.15	0.005	0.35
Folate, total	µg	47	71	80	76	66	4	146
Choline, total	mg	293.8	NR	263.4	263.4	263.4	1.1	820.2
Vitamin B_{12}[b]	µg	0.89	1.69	5.4	5.1	1.58	0.09	1.95
Vitamin A, RAE[b]	µg	160	166	194	187	156	0	381
Retinol	µg	160	166	192	185	155	0	371
Lutein + zeaxanthin	µg	503	NR	459	442	369	0	1094
Vitamin E (alpha-tocopherol)	mg	1.05	NR	1.34	1.29	1.08	0	2.58
Vitamin D ($D_2 + D_3$)[c]	µg	2	NR	1.7	1.7	1.4	0	5.4
Vitamin K (phylloquinone)	µg	0.3	NR	0.4	0.4	0.3	0	0.7
Total SAT[b]	g	3.126	3.632	3.681	3.595	3.557	0	9.551
16:0	g	2.231	NR	NR	NR	NR	NR	NR
18:0	g	0.811	0.877	0.632	0.696	0.838	0	2.417
Total MUFA[b]	g	3.658	4.571	6.525	5.747	4.324	0	11.738
16:1 n-7	g	0.201	0.672	0.441	0.394	0.473	0	0.918
18:1 n-9	g	3.411	3.899	6.084	5.353	3.851	0	10.701
Total PUFA[b]	g	1.911	1.658	1.223	1.672	1.324	0	4.204
18:2 n-6 (LNA)	g	1.531	1.166	0.558	0.68	0.94	0	3.538
18:3 n-3 (ALA)	g	0.036	0.079	0.102	0.554	0.044	0	0.103
20:4 n-6 (ARA)	g	0.188	NR	0.319	0.279	0.122	0	0.438
20:5 n-3 (EPA)	g	0	NR	0	0	0	0	0.011
22:5 n-3 (DPA)	g	0.007	NR	0	0	0	0	0
22:6 n-3 (DHA)	g	0.058	NR	0	0	0	0	0.114
Cholesterol	mg	372	933	884	852	844	0	1085

[a] Values taken from FoodData Central (USDA, 2025); individual products are identified by a numeric code abbreviated as FDC ID. NR: not reported.

[b] Terms used in FoodData Central (USDA, 2025): vitamin B_6, pyridoxine forms; vitamin B_{12}, cobalamin; RAE, retinol activity equivalents; SAT, saturated fatty acids; MUFA, monounsaturated fatty acids; PUFA, polyunsaturated fatty acids; LNA, linoleic acid, 18:2 n-6; ALA, α-linolenic acid, 18:3 n-3; ARA, arachidonic acid, 20:4 n-6; EPA, eicosapentaenoic acid, 20:5 n-3; DPA; docosapentaenoic acid, 22:5 n-3; DHA, docosahexaenoic acid, 22:6 n-3.

[c] Vitamin D ($D_2 + D_3$): While ergocalciferol (vitamin D_2) excretion is rapid in birds, humans readily use this vitamin form, which can be present in feed and resultant eggs (Mattila et al., 2004). FoodData Central (USDA, 2025) did not provide reports for the different vitamin forms in all cases.

diets are typically low in fat (less than 6% by weight, or 17% of energy in a 3,100 kcal/kg diet) while providing at least 17 percent of preformed fatty acids as linoleic acid (LNA, 18:2 n-6). Typical poultry diets are rich in carbohydrates and promote de novo lipogenesis that, in poultry, produces palmitic acid (16:0 n-9) and oleic acid (18:1 n-9) as the dominant SAT and MUFA products, respectively. Linoleic acid and its elongation product arachidonic acid (ARA, 20:4, n-9) are the dominant PUFA and LC-PUFA, respectively (Komprda et al., 2005), in poultry products because LNA is the predominant

TABLE 12-6 Secondary Edible Portions of Poultry[a]

Body Part	Nutritional Features
Feet/paws	Skin and cartilage: protein, lipid, some carotenoids
Head	Brain: protein, lipid, iron, choline
Neck	Skin used as casing for fillings that combine neck meat with grains and vegetables. Recipes vary: protein, lipid, and some carotenoids.
Carcass	Broth: protein, lipid, Na, K, P, and small amounts of Ca
Refined adipose (e.g., schmaltz)	Lipid, choline, vitamin D
Blood[b]	Protein, heme iron, Mg, Cu, Zn, Mn
Heart[c]	Protein, heme iron, P, K, Zn, niacin, B_{12}
Gizzard[c]	Protein, heme iron, P, K, Zn, niacin, B_{12}
Liver[c]	Protein, fat (high if pate), Fe, P, K, Zn, niacin, pantothenate, B_{12}, folate, choline, vitamin A

[a] Nutritional features taken from FoodData Central (USDA, 2025) or primary literature as needed (Sorapukdee and Narunatsopanon, 2017).
[b] Chicken and duck blood had similar nutritional characteristics (Sorapukdee and Narunatsopanon, 2017).
[c] Similar characteristics for chicken, turkey, and duck in FoodData Central (USDA, 2025). Liver values taken for non-fattened liver. Selenium was present but varied with diet content, and so calculated values appeared in FoodData Central in some instances.

PUFA in most poultry feed. Linoleic acid is an essential fatty acid and must be provided preformed in the diet. In acylglycerol lipids, PUFA and LC-PUFA are found primarily in the 2-position (see Chapter 4, Lipids), while MUFA and SAT primarily occupy the 1- and 3-positions. Lean tissue contains little triacylglycerol (TAG) and mostly contains phospholipids whose 1- and 2-positions are occupied by SAT/MUFA and PUFA, respectively, and a phosphate group in the 3-position. Chickens (Naber and Biggert, 1989; Desert et al., 2018), like other monogastric vertebrates (Horton et al., 2002), reduce de novo lipogenesis rates to utilize preformed dietary fatty acids when fed diets containing greater amounts of lipid. In chickens, a change from 2 percent to 8 percent lipid by weight in isocaloric, isonitrogenous diets reduced the transcriptional program for de novo lipogenesis (Desert et al., 2018). Fatty acids themselves cause additional effects (Sanz, 1999; Selvaraj et al., 2010) largely through transcriptional changes (Jump et al., 2013), including increased and decreased de novo lipogenesis of SAT and long chain n-3 PUFA, respectively (Sanz, 1999; Jump et al., 2013). Activation of peroxisome proliferator-activated receptor-α (PPAR-α) by n-3 fatty acids likely causes changes in adipose biology, albeit effects differ with strain, sex, and dietary ingredient selection (Hargis et al., 1991; Selvaraj et al., 2010; Royan et al., 2011; González-Ortiz et al., 2013; Torchon et al., 2017). These features of lipid metabolism make poultry highly responsive to dietary interventions that modify the amount and composition of lipids present in meat (Sirri and Meluzzi, 2011; Poureslami and Batal, 2012; Lilburn, 2017) and eggs (Van Elswyk, 1997; Nys and Guyot, 2011; Sirri and Meluzzi, 2011; Cherian, 2012).

The most commonly sought diet-induced modification to tissue and egg fatty acid composition is a reduction in n-6 fatty acids and increase in n-3 fatty acids (Cherian, 2012). This goal is based on the rationale that reduced tissue n-6/n-3 reduces inflammatory tone via adoption of gene expression and bioactive lipid patterns in the body that reduce diet-dependent chronic disease development (Simopoulos, 2008). The rationale is not accepted by all (Willett, 2007), but if true, by changing the fatty acid composition of poultry tissues and eggs, health benefits are presumed to accrue for both birds and consumers of poultry products and eggs.

There are multiple sources of dietary n-3 fatty acids including flax, high in ALA, and fish oil or microalgae that are high in long chain n-3 PUFAs (n-3 LC-PUFA) such as eicosapentaenoic acid (EPA, 20:5 n-3) and docosahexaenoic acid (DHA, 22:6 n-3) (Van Elswyk, 1997; Rymer and Givens, 2005, 2006; Ao et al., 2015; Park et al., 2015; Lee et al., 2019; Manor et al., 2019). Wild fish rich in LC-PUFA (n-3) obtain these fatty acids from small aquatic plants, microalgae, that are consumed at the bottom of the marine food chain (Nichols et al., 2010). Apart from fishmeal or fish oil from wild fish, sustainable sources of n-3 LC-PUFA are increasingly being developed due to declining fishery stocks (Nichols et al., 2010; Wells et al., 2017; Remize et al., 2021; Saini et al., 2021). The purported health benefits of n-3 fatty acids differ, with the effects of n-3 LC-PUFA (i.e., those with chain lengths greater than 20) being more pronounced than those of n-3 PUFA with chain lengths less than 20 carbons, that is, ALA (Rymer and Givens, 2005; Cherian, 2012). Altering the fatty acid composition of edible poultry tissue or eggs to increase unsaturated fatty acid content increases the need for antioxidants (Raederstorff et al., 2015) to prevent oxidative damage to organoleptic and storage stability of poultry products (Rymer and Givens, 2005; Sirri and Meluzzi, 2011). Antioxidants are included in diets and/or at further processing steps to prevent oxidative damage and maintain product quality as well as prevent undesirable product flavor and odor profiles. Despite adequate antioxidant protection, there are limits to consumer acceptance of the flavors arising from EPA and DHA enrichment (Van Elswyk, 1997; Ribeiro et al., 2013). Table 12-7 shows the substantial degree of enrichment in poultry meat that can occur with the addition of preformed n-3 LC-PUFA to the diets of both chickens and turkeys. Note that the fishmeal addition did not appreciably alter meat ALA content because EPA and DHA are the dominant n-3 fatty acids in marine fish (Shireman, 2003). A meta-analysis

TABLE 12-7 Effect of Long Chain n-3 Fatty Acid Supplementation on Fatty Acid Composition of Poultry Meat

	White Meat				Dark Meat			
Feed	Typical Chicken[a]	Supplemented Chicken[a,b]	Typical Turkey[c]	Supplemented Turkey[a,b]	Typical Chicken[a]	Supplemented Chicken[a,b]	Typical Turkey[c]	Supplemented Turkey[a,b]
Lipid, % fresh tissue	0.90	0.95	0.81	0.70	2.2	1.95	2.25	2.23
% total fatty acids								
ALA[d]	0.50	0.45	1.10	0.95	0.70	0.60	1.29	1.85
EPA + DHA[d]	2.70	4.95	0.61	6.85	1.50	2.60	0.62	3.45
mg/100 g edible tissue								
ALA	4.50	4.30	9.00	6.65	15.5	11.7	29.0	42.90
EPA + DHA	22.50	46.80	5.00	47.90	35.30	50.75	14.0	78.90

[a] Adapted from Rymer and Givens (2005).
[b] Birds were fed diets containing 40–50 g fishmeal/kg diet.
[c] Calculated from data in FoodData Central (USDA, 2025), FDC ID171098 and 174518.
[d] ALA, α-linolenic acid, 18:3 n-3; EPA, eicosapentaenoic acid, 20:5 n-3; DHA, docosahexaenoic acid, 22:6 n-3.

showed that feeding 100 g ALA per kg feed resulted in approximately 136 mg of DHA per egg (Irawan et al., 2022). Such outcomes highlight the limited conversion of ALA to EPA or DHA in vertebrates (Rymer and Givens, 2005; Sirri and Meluzzi, 2011; Cherian, 2012; Poureslami and Batal, 2012). Poultry diets supplemented with ALA cause fewer adverse flavor effects on products than comparable levels of LC-PUFA EPA and DHA supplementation, but the majority of health benefits are associated with those latter LC-PUFA. It is currently undetermined whether ALA supplementation and associated n-3 fatty acid enrichment result in significant health benefits (Abdelhamid et al., 2020). Nevertheless, as Table 12-8 shows, eggs from hens fed ALA, EPA, and DHA have substantially greater amounts of these fatty acids in the egg compared to hens fed diets containing the fatty acid profile typical of corn/soy diets.

Antioxidant additions to poultry diets include vitamins C and E and phytochemicals, as well as the trace minerals selenium, iron, copper, zinc, and manganese, which serve as cofactors of enzymes in the antioxidant defense system (O'Dell and Sunde, 1997; Combs, 2007). These additions may alter the vitamin and mineral content of meat and egg products depending on the level of supplemental addition and amounts in the basal diet (Schiavone and Barroeta, 2011). Cholesterol has been another target for modification (Clarke, 1997). Reduced tissue concentrations are reported for meat and other products, but most work has sought to reduce egg cholesterol content. Thought to be immutable in eggs (Griffin, 1992), modest (5–10%) reductions are possible through dietary intervention (c.f. Table 12-8). Some pharmacologic interventions with 3-hydroxy 3 methylglutaryl coenzyme A (HMG-CoA) reductase inhibitors can nearly halve yolk cholesterol concentrations (Elkin et al., 1999). In that study, treatment with HMG-CoA reductase inhibitors did not impair egg production during the 5-week trial.

Vitamins and Minerals

As noted previously, concentrations of many vitamins and minerals in tissues can be increased by diet (Naber, 1993; Schiavone and Barroeta, 2011), albeit there are biologic limits due to homeostatic mechanisms, competition and, for some nutrients, toxicity (O'Dell and Sunde, 1997; Combs, 2007). Diet components, supplement formulations, bird strain, and age create a large degree of variability in possible outcomes that preclude meaningful predictive specifics using currently available data. In general, eggs are a frequent target of nutritional enhancement. The terms "enriched" and "fortified" have specific definitions in U.S. food regulations (IOM, 2003). Both terms mean that nutrients have been added to make the food more nutritious. "Enriched" involves restoration of nutrients that were lost during food processing steps. "Fortified" means vitamins or minerals have been added to a food where they were not originally present. The method of nutrient increase in the case of unprocessed poultry and eggs is through the diet, and so neither term is well suited for general descriptions. Relative content claims (IOM, 2003), such as "good" or "excellent" source, are commonly used to indicate elevated nutrient contents relative to a standard of nutrient requirements.

Unlike meat or other components that must be weighed to calculate nutrient content, eggs in the shell provide a ready means to communicate nutrition content information for each portion as purchased when expressed on a "per egg" basis. Egg standards include defined weights for eggs of different sizes (USDA AMS, 2017). For example, that regulatory document specifies content weights for medium, large, and jumbo eggs are 44 g, 50 g, and 63 g, respectively. Food recipes are usually based on the large egg size, and nutrition facts are based on the established serving size of one egg (in grams) (USDA AMS, 2017). Eggs act as a unique nutrient delivery system that generally improves the bioavailability

TABLE 12-8 Change in Yolk Composition with Diet Reformulation

	Unit	Conventional Egg	Specialty Egg	Compared to Conventional Egg Content
Amounts per 100 g	FDC ID	1711287[a]	N/A[b]	%
Energy	kcal	143	120	83.9%
Protein	g	12.56	12	95.5%
Total lipid (fat)	g	9.51	8	84.1%
Iron, Fe	mg	1.75	1.8	102.9%
Sodium, Na	mg	142	130	91.5%
Zinc, Zn	mg	1.29	1.4	108.5%
Iodine, I	µg	56	120	214.3%
Selenium, Se	µg	30.7	44	143.3%
Riboflavin	mg	0.457	0.6	131.3%
Pantothenic acid	mg	1.533	2	130.5%
Folate, total	µg	47	90	191.5%
Choline, total	mg	293.8	300	102.1%
Vitamin B_{12}[c]	µg	0.89	2	224.7%
Vitamin A, RAE	µg	160	160	100.0%
Lutein + zeaxanthin	µg	290	400	137.9%
Vitamin E (α-tocopherol)	mg	1.05	10	952.4%
Vitamin D (D_2+D_3)[d]	µg	2	12	600.0%
Total SAT[c]	g	3.126	2	64.0%
Total MUFA[c]	g	3.658	4	109.3%
Total PUFA[c]	g	1.911	2	104.7%
18:2 n-6 (LNA)[c]	g	1.531	1.247	81.5%
18:3 n-3 (ALA)[c]	g	0.036	0.12	333.3%
20:4 n-6 (ARA)[c]	g	0.188	0.153	81.4%
20:5 n-3 (EPA)[c]	g	0.00	0.004	NC
22:6 n-3 (DHA)[c]	g	0.058	0.126	217.2%
Cholesterol	mg	372	340	91.4%

[a] Values taken from FoodData Central (USDA, 2025). NC: not calculated due to 0.0 value in content reported for FDC ID 1711287.

[b] Information provided by food brand owners is label data. Brand owners are responsible for descriptions, nutrient data, and ingredient information. USDA calculates values per 100 g or 100 ml from values per serving. Values calculated from % daily value (%DV) use current standard for daily values; those for an adult consuming a 2,000-calorie diet (21 CFR 101.9(c)).

[c] Term used in FoodData Central (USDA, 2025): vitamin B_{12}, cobalamin; SAT, saturated fatty acids; MUFA, monounsaturated fatty acids; PUFA, polyunsaturated fatty acids; LNA, linoleic acid, 18:2 n-6; ALA, α-linolenic acid, 18:3 n-3; ARA, arachidonic acid, 20:4 n-6; EPA, eicosapentaenoic acid, 20:5 n-3; DPA; docosapentaenoic acid, 22:5 n-3; DHA, docosahexaenoic acid, 22:6 n-3.

[d] Vitamin D (D_2+D_3): While ergocalciferol (vitamin D_2) excretion is rapid in birds, humans readily use this vitamin form, which can be present in feed and resultant eggs (Mattila et al., 2004). FoodData Central did not provide reports for the different vitamin forms in all cases but provided the combined value most often.

(Réhault-Godbert et al., 2019) of nutrients through the biological processes that serve to incorporate them into tissues (Walzem, 2004). Diet reformulation to increase egg nutrients are successful to varying degrees (calculated as percentage difference from standard egg) depending on initial concentrations in the standard egg. For example, it would be difficult to increase the amount of the essential fatty acid LNA in hens fed LNA-rich corn/soy diets, whereas dietary DHA supplementation elicits impressive gains owing to the low starting concentrations of this fatty acid in standard eggs laid by hens fed corn/soy diets.

Fat-soluble, as well as many water-soluble vitamins, are concentrated in yolk (Schiavone and Barroeta, 2011). Several of the fat-soluble and water-soluble vitamins can be increased several-fold with dietary supplementation (Nys and Guyot, 2011; Schiavone and Barroeta, 2011; Réhault-Godbert et al., 2019) (Table 12-8). The comparator-enriched egg composition shown in Table 12-8 is drawn from product information for eggs from a large commercial n-3 specialty egg producer (Michella and Slaugh, 2000). This comparison shows typical alteration patterns of various nutrients or properties (i.e., energy) that are possible with supplemented diets. Note the greater than 500 percent increase in vitamins D and E and the doubling of vitamin B_{12}, iodine, and folate, with 25–50 percent increases in selenium, riboflavin, pantothenic acid, and lutein + zeaxanthin. In other settings vitamins A and K, as well as thiamin and biotin, can also be elevated.

REFERENCES

Abdelhamid, A.S., T.J. Brown, J.S. Brainard, P. Biswas, G.C. Thorpe, H.J. Moore, K.H.O. Deane, C.D. Summerbell, H.V. Worthington, F. Song, et al. 2020. Omega-3 fatty acids for the primary and secondary prevention of cardiovascular disease. *Cochrane Database of Systematic Reviews*. 3:CD003177.

Ahiwe, E., A.A. Omede, M.B. Abdallh, and P.A. Iji. 2018. Managing dietary energy intake by broiler chickens to reduce production costs and improve product quality. In *Animal Husbandry and Nutrition*. B. Yücel and T. Taşkin, editors. London, UK: IntechOpen.

Ahn, D.U., S.M. Kim, and H. Shu. 1997. Effect of egg size and strain and age of hens on the solids content of chicken eggs. *Poultry Science*. 76(6):914–919.

Ao, T., L.M. Macalintal, M.A. Paul, A.J. Pescatore, A.H. Cantor, M.J. Ford, B. Timmons, and K.A. Dawson. 2015. Effects of supplementing microalgae in laying hen diets on productive performance, fatty-acid profile, and oxidative stability of eggs. *Journal of Applied Poultry Research*. 24(3):394–400.

Bellairs, R., M. Harkness, and R.D. Harkness. 1963. The vitelline membrane of the hen's egg: A chemical and electron microscopical study. *Journal of Ultrastructure Research*. 8(3):339–359.

Chen, S.E., J.P. McMurtry, and R.L. Walzem. 2006. Overfeeding-induced ovarian dysfunction in broiler breeder hens is associated with lipotoxicity. *Poultry Science*. 85(1):70–81.

Cherian, G. 2012. Modifying egg lipids for enhancing human health. *Fats and Fatty Acids in Poultry Nutrition and Health*. G. Cherian and R. Poureslami, editors. Packington, England: Context Products.

Clarke, A.D. 1997. Reduction of cholesterol levels in meat, poultry and fish products. In *Production and Processing of Healthy Meat, Poultry and Fish Products*. A.M. Pearson and T.R. Dutson, editors. Princeton, NJ: Springer US.

Combs, G.F. 2007. *The Vitamins: Fundamental Aspects in Nutrition and Health*, 3rd ed. Amsterdam, The Netherlands: Academic Press.

Combs, G.F., and J.P. McClung. 2017. *The Vitamins: Fundamental Aspects in Nutrition and Health*, 5th ed. Amsterdam, The Netherlands: Academic Press.

Desert, C., E. Baéza, M. Aite, M. Boutin, A. Le Cam, J. Montfort, M. Houee-Bigot, Y. Blum, P.F. Roux, C. Hennequet-Antier, C. Berri, S. Metayer-Coustard, A. Collin, S. Allais, E. Le Bihan, D. Causeur, F. Gondret, M.J. Duclos, and S. Lagarrigue. 2018. Multi-tissue transcriptomic study reveals the main role of liver in the chicken adaptive response to a switch in dietary energy source through the transcriptional regulation of lipogenesis. *BMC Genomics*. 19(1):187.

Elkin, R.G., Z. Yan, Y. Zhong, S.S. Donkin, K.K. Buhman, J.A. Story, J.J. Turek, R.E. Porter Jr., M. Anderson, R. Homan, and R.S. Newton. 1999. Select 3-hydroxy-3-methylglutaryl-coenzyme A reductase inhibitors vary in their ability to reduce egg yolk cholesterol levels in laying hens through alteration of hepatic cholesterol biosynthesis and plasma VLDL composition. *Journal of Nutrition*. 129(5):1010–1019.

Ertl, P., W. Knaus, and W. Zollitsch. 2016. An approach to including protein quality when assessing the net contribution of livestock to human food supply. *Animal*. 10(11):1883–1889.

Fakhfakh, N., N. Ktari, A. Haddar, I.H. Mnif, I. Dahmen, and M. Nasri. 2011. Total solubilisation of the chicken feathers by fermentation with a keratinolytic bacterium, *Bacillus pumilus* A1, and the production of protein hydrolysate with high antioxidative activity. *Process Biochemistry*. 46(9):1731–1737.

FAO (Food and Agriculture Organization of the United Nations). 2011. *Dietary Protein Quality Evaluation in Human Nutrition*. Rome, Italy: FAO.

FAO. 2022. *Gateway to Poultry Production and Products*. Rome, Italy: FAO.

Ghaly, A.E., and K.N. MacDonald. 2012. Drying of poultry manure for use as animal feed. *American Journal of Agricultural and Biological Sciences*. 7(3):239–254.

Gidrewicz, D.A., and T.R. Fenton. 2014. A systematic review and meta-analysis of the nutrient content of preterm and term breast milk. *BMC Pediatrics*. 14(1):1–14.

González-Ortiz, G., R. Sala, E. Cánovas, N. Abed, and A.C. Barroeta. 2013. Consumption of dietary n-3 fatty acids decreases fat deposition and adipocyte size, but increases oxidative susceptibility in broiler chickens. *Lipids*. 48(7):705–717.

Griffin, H.D. 1992. Manipulation of egg-yolk cholesterol—A physiologist's view. *World's Poultry Science Journal*. 48(2):101–112.

Hargis, P.S., M.E. Van Elswyk, and B.M. Hargis. 1991. Dietary modification of yolk lipid with menhaden oil. *Poultry Science*. 70(4):874–883.

Haytowitz, D.B. 2015. Updating USDA's key foods list for what we eat in America, NHANES 2011–12. *Procedia Food Science*. 4:71–78.

Hencken, H. 1992. Chemical and physiological behavior of feed carotenoids and their effects on pigmentation. *Poultry Science*. 71(4):711–717.

Hernandez, J.M. 2005. European consumer surveys about egg quality: How to improve the nutritional value. *Proceedings of the XIth European Symposium on the Quality of Eggs and Egg Products, Doorwerth, The Netherlands* (CD Symposium Proceedings).

Holden, J.M., J.M. Harnly, and G.R. Beecher. 2006. Food composition. In *Present knowledge in nutrition*. M. Traber, B. Bowman, and R. Russell, eds.: Washington, DC: ILSI Press.

Horton, J.D., J.L. Goldstein, and M.S. Brown. 2002. SREBPs: Activators of the complete program of cholesterol and fatty acid synthesis in the liver. *The Journal of Clinical Investigation*. 109(9):1125–1131.

IOM (Institute of Medicine). 2003. *Dietary Reference Intakes: Guiding Principles for Nutrition Labeling and Fortification*. Washington, DC: The National Academies Press.

IOM. 2005. *Dietary Reference Intakes for Energy, Carbohydrate, Fiber, Fat, Fatty Acids, Cholesterol, Protein, and Amino Acids*. Washington, DC: The National Academies Press.

Irawan, A., N. Ningsih, Hafizuddin, R.K. Rusli, W.P.S. Suprayogi, N. Akhirini, R.F. Hadi, W. Setyono, and A. Jayanegara. 2022. Supplementary n-3 fatty acids sources on performance and formation of omega-3 in egg of laying hens: A meta-analysis. *Poultry Science*. 101(1):101566.

Jump, D.B., S. Tripathy, and C.M. Depner. 2013. Fatty acid–regulated transcription factors in the liver. *Annual Review of Nutrition*. 33:249.

Kirunda, D.F.K., and S.R. McKee. 2000. Relating quality characteristics of aged eggs and fresh eggs to vitelline membrane strength as determined by a texture analyzer. *Poultry Science*. 79(8):1189–1193.

Komprda, T., J. Zelenka, E. Fajmonova, M. Fialova, and D. Kladroba. 2005. Arachidonic acid and long-chain n–3 polyunsaturated fatty acid contents in meat of selected poultry and fish species in relation to dietary fat sources. *Journal of Agricultural and Food Chemistry*. 53(17):6804–6812.

Lee, S.A., N. Whenham, and M.R. Bedford. 2019. Review on docosahexaenoic acid in poultry and swine nutrition: Consequence of enriched animal products on performance and health characteristics. *Animal Nutrition*. 5(1):11–21.

Lilburn, M.S. 2017. Enhancing the nutritional quality of poultry meat. In *Achieving Sustainable Production of Poultry Meat Volume 1*. Sawston, UK: Burleigh Dodds Science Publishing.

Mann, K. 2007. The chicken egg white proteome. *Proteomics*. 7(19): 3558–3568.

Mann, K., and M. Mann. 2008. The chicken egg yolk plasma and granule proteomes. *Proteomics*. 8(1):178–191.

Manor, M.L., T.J. Derksen, A.D. Magnuson, F. Raza, and X.G. Lei. 2019. Inclusion of dietary defatted microalgae dose-dependently enriches ω-3 fatty acids in egg yolk and tissues of laying hens. *The Journal of Nutrition*. 149(6):942–950.

Mathai, J.K., Y. Liu, and H.H. Stein. 2017. Values for digestible indispensable amino acid scores (DIAAS) for some dairy and plant proteins may better describe protein quality than values calculated using the concept for protein digestibility-corrected amino acid scores (PDCAAS). *British Journal of Nutrition*. 117(4):490–499.

Mattila, P., J. Valaja, L. Rossow, E. Venäläinen, and T. Tupasela. 2004. Effect of vitamin D2- and D3-enriched diets on egg vitamin D content, production, and bird condition during an entire production period. *Poultry Science*. 83(3):433–440.

Michella, S.M., and B.T. Slaugh. 2000. Producing and marketing a specialty egg. *Poultry Science*. 79(7):975–976.

Moore, P.A., T.C. Daniel, A.N. Sharpley, and C.W. Wood. 1995. Poultry manure management: Environmentally sound options. *Journal of Soil and Water Conservation*. 50(3):321.

Naber, E.C. 1993. Modifying vitamin composition of eggs: A review. *Journal of Applied Poultry Research*. 2(4):385–393.

Naber, E.C., and M.D. Biggert. 1989. Patterns of lipogenesis in laying hens fed a high fat diet containing safflower oil. *Journal of Nutrition*. 119(5):690–695.

Nichols, P.D., J. Petrie, and S. Singh. 2010. Long-chain omega-3 oils: An update on sustainable sources. *Nutrients*. 2(6):572–585.

NIH ODS (National Institutes of Health Office of Dietary Supplements). 2020. *Supplements. Nutrient Recommendations and Databases*. Bethesda, MD: NIH ODS.

Nys, Y., and N. Guyot. 2011. 6—Egg formation and chemistry. In *Improving the Safety and Quality of Eggs and Egg Products*. Y. Nys, M. Bain, and F. Van Immerseel, editors. Sawston, UK: Woodhead Publishing. 83–132.

O'Dell, B.L., and R.A. Sunde. 1997. *Handbook of Nutritionally Essential Mineral Elements,* 1st ed. New York: Marcel Dekker.

Olson, J.B., N.E. Ward, and E.A. Koutsos. 2008. Lycopene incorporation into egg yolk and effects on laying hen immune function. *Poultry Science*. 87(12):2573–2580.

Papadopoulos, M.C., A.R. El Boushy, and E.H. Ketelaars. 1985. Effect of different processing conditions on amino acid digestibility of feather meal determined by chicken assay. *Poultry Science*. 64(9):1729–1741.

Park, J.H., S.D. Upadhaya, and I.H. Kim. 2015. Effect of dietary marine microalgae (*Schizochytrium*) powder on egg production, blood lipid profiles, egg quality, and fatty acid composition of egg yolk in layers. *Asian-Australasian Journal of Animal Sciences*. 28(3):391.

Pasiakos, S.M., S. Agarwal, H.R. Lieberman, and V.L. Fulgoni III. 2015. Sources and amounts of animal, dairy, and plant protein intake of US adults in 2007–2010. *Nutrients*. 7(8):7058–7069.

Poureslami, R., and A. Batal. 2012. Modifying meat lipids for human health. In *Fats and Fatty Acids in Poultry Nutrition and Health*. G. Cherian and R. Poureslami, editors. Packington, England: Context Products.

Raederstorff, D., A. Wyss, P.C. Calder, P. Weber, and M. Eggersdorfer. 2015. Vitamin E function and requirements in relation to PUFA. *British Journal of Nutrition*. 114(8):1113–1122.

Réhault-Godbert, S., N. Guyot, and Y. Nys. 2019. The golden egg: Nutritional value, bioactivities, and emerging benefits for human health. *Nutrients*. 11(3):684.

Remize, M., Y. Brunel, J.L. Silva, J.Y. Berthon, and E. Filaire. 2021. Microalgae n-3 PUFAs production and use in food and feed industries. *Marine Drugs*. 19(2):113.

Ribeiro, T., M.M. Lordelo, S.P. Alves, R.J.B. Bessa, P. Costa, J.P.C. Lemos, L.M.A. Ferreira, C.M.G.A. Fontes, and J.A.M. Prates. 2013. Direct supplementation of diet is the most efficient way of enriching broiler meat with n-3 long-chain polyunsaturated fatty acids. *British Poultry Science*. 54(6):753–765.

Royan, M., G.Y. Meng, F. Othman, A.Q. Sazili, and B. Navidshad. 2011. Effects of conjugated linoleic acid, fish oil and soybean oil on PPARs (α & γ) mRNA expression in broiler chickens and their relation to body fat deposits. *International Journal of Molecular Sciences*. 12(12):8581–8595.

Ruhlman, M. 2014. *Egg: A Culinary Exploration of the World's Most Versatile Ingredient*. New York: Little Brown & Co.

Rymer, C., and D.I. Givens. 2005. n-3 fatty acid enrichment of edible tissue of poultry: A review. *Lipids*. 40(2):121–130.

Rymer, C., and D.I. Givens. 2006. Effect of species and genotype on the efficiency of enrichment of poultry meat with n-3 polyunsaturated fatty acids. *Lipids*. 41(5):445–451.

Saini, R.K., P. Prasad, R.V. Sreedhar, K. Akhilender Naidu, X. Shang, and Y.-S. Keum. 2021. Omega-3 polyunsaturated fatty acids (PUFAs): Emerging plant and microbial sources, oxidative stability, bioavailability, and health benefits—A review. *Antioxidants*. 10(10):1627.

Sanz, M. 1999. Higher lipid accumulation in broilers fed on saturated fats than in those fed on unsaturated fats. *British Poultry Science*. 40(1):95–101.

Scanes, C.G. 2015. Protein metabolism. In *Sturkie's Avian Physiology*, 6th ed. C.G. Scanes and P.D. Sturkie, editors. Amsterdam, The Netherlands: Academic Press.

Schaafsma, G. 2012. Advantages and limitations of the protein digestibility-corrected amino acid score (PDCAAS) as a method for evaluating protein quality in human diets. *British Journal of Nutrition*. 108(S2): S333–S336.

Scheideler, S.E., P. Weber, and D. Monsalve. 2010. Supplemental vitamin E and selenium effects on egg production, egg quality, and egg deposition of α-tocopherol and selenium. *Journal of Applied Poultry Research*. 19(4):354–360.

Schiavone, A., and A.C. Barroeta. 2011. 15—Egg enrichment with vitamins and trace minerals. In *Improving the Safety and Quality of Eggs and Egg Products*. F. Van Immerseel, Y. Nys, and M. Bain, editors. Sawston, UK: Woodhead Publishing.

Selvaraj, R.K., R. Shanmugasundaram, and K.C. Klasing. 2010. Effects of dietary lutein and PUFA on PPAR and RXR isomer expression in chickens during an inflammatory response. *Comparative Biochemistry and Physiology Part A: Molecular & Integrative Physiology*. 157(3):198–203.

Sheats, M. 2018. Wanna buy a duck? *Economic Landscape Report* (March 2018).

Shireman, R. 2003. Essential fatty acids. In *Encyclopedia of Food Sciences and Nutrition,* 2nd ed. B. Caballero, editor. Amsterdam, The Netherlands: Academic Press.

Sim, J.S., and H.H. Sunwoo. 2006. *The Amazing Egg: Nature's Perfect Functional Food for Health Promotion*. Department of Agricultural, Food and Nutritional Science, Edmonton, Canada: University of Alberta.

Simopoulos, A.P. 2008. The importance of the omega-6/omega-3 fatty acid ratio in cardiovascular disease and other chronic diseases. *Experimental Biology and Medicine*. 233(6):674–688.

Sirri, F., and A. Meluzzi. 2011. Modifying egg lipids for human health. In *Improving the Safety and Quality of Eggs and Egg Products*. Amsterdam, The Netherlands: Elsevier. 272–288.

Sorapukdee, S., and S. Narunatsopanon. 2017. Comparative study on compositions and functional properties of porcine, chicken and duck blood. *Korean Journal for Food Science of Animal Resources*. 37(2):228.

Stadelman, W.J., D. Newkirk, and L. Newby. 2017. Egg science and technology. Boca Raton, FL: CRC Press.

Torchon, E.T., S. Das, R.C. Beckford, and B.H. Voy. 2017. Enriching the starter diet in n-3 polyunsaturated fatty acids reduces adipocyte size in broiler chicks. *Current Developments in Nutrition*. 1(11):e001644.

USDA (U.S. Department of Agriculture). 2000a. U.S. trade descriptions for poultry. Marketing and Regulatory Programs. AMS 71. USDA.

USDA. 2000b. U.S. trade descriptions for poultry, turkey. Marketing and Regulatory Programs. AMS 71. USDA.

USDA. 2025. *FoodData Central*. Agricultural Research Service. https://fdc.nal.usda.gov.

USDA AMS (Agriculture Marketing Service). 2015. Federal purchase program specification (FPPS) for shell eggs. Washington, DC: USDA AMS.

USDA AMS. 2017. Specifications for shell eggs. Washington, DC: USDA AMS.

Uysal, R.S., İ.H. Boyacı, E.A. Soykut, and N. Ertaş. 2017. Effects of heat treatment parameters on liquid whole egg proteins. *Food Chemistry*. 216:201–208.

Van Elswyk, M.E. 1997. Comparison of n-3 fatty acid sources in laying hen rations for improvement of whole egg nutritional quality: A review. *British Journal of Nutrition*. 78(1):S61–S69.

Walzem, R.L. 2004. Poultry, eggs, and biotechnology. *Bioprocesses and Biotechnology for Functional Foods and Nutraceuticals*. Boca Raton, FL: CRC Press.

Watson, R.R., and F. De Meester. 2015. *Handbook of Eggs in Human Function*. Vol. 9. Wageningen, The Netherlands: Wageningen Academic Publishers.

Wells, M.L., P. Potin, J.S. Craigie, J.A. Raven, S.S. Merchant, K.E. Helliwell, A.G. Smith, M.E. Camire, and S.H. Brawley. 2017. Algae as nutritional and functional food sources: Revisiting our understanding. *Journal of Applied Phycology*. 29(2):949–982.

Willett, W.C. 2007. The role of dietary n-6 fatty acids in the prevention of cardiovascular disease. *Journal of Cardiovascular Medicine*. 8(Suppl. 1): S42–S45.

Williams, T.D., W.L. Reed, and R.L. Walzem. 2001. Egg size variation: Mechanisms and hormonal control. In *Avian Endocrinology*. New Delhi, India: Narosa Publishing House. 205–217.

13

Nutrient Requirements of Laying Hens

WORLD EGG PRODUCTION, THE MAJOR TABLE EGG BREEDS, AND HOUSING SYSTEMS

Domestication of the chicken occurred in Southeast Asia approximately 3,500 years ago according to most recent estimates (Peters et al., 2022). Due to its ability to adapt to a wide variety of environments as well as its genetic diversity, the domestic fowl became distributed globally (Yamada, 1988). It is estimated that it initially arrived in Europe by approximately 800 BCE (Peters et al., 2022) and subsequently was introduced to the New World in the mid-16th century (Yamada, 1988). Maintained initially for cockfighting or religious rituals, the chicken eventually was used for food (especially egg production) during the Roman Empire (Chambers et al., 2017) and became a subject for scientific research during the 16th century (Yamada, 1988). During the 19th century, chicken breeds with novel traits were developed for exhibitions, initially throughout Europe and later in the United States, with some breeds eventually contributing to today's commercial strains (Yamada, 1988; Chambers et al., 2017).

With the development of the broiler industry in the United States in the early 1940s, a clear differentiation was established between poultry meat and commercial egg production, which gave rise to the highly specialized sectors of the modern poultry industry (Yamada, 1988). Practically all commercial white-egg lines of chickens are Single Comb White Leghorns, while breeders have developed special lines and crosses with Rhode Island Red stock for the production of pullets that lay brown-shelled eggs (Bell, 2002). In most cases, brown egg lines lay larger eggs than those produced by white egg lines, with only minor differences in shell and internal egg quality (Bell, 2002). Recently, due to animal well-being concerns associated with the culling of layer-type male chicks, there has been a renewed interest in dual-purpose chicken strains in which the females are used to produce eggs and the males are fattened (Mueller et al., 2018). Traditional dual-purpose breeds include New Hampshire, Plymouth Rock, and Wyandotte, while new commercial lines that are crosses between layer and broiler genetics also have been developed (Mueller et al., 2018). However, dual-purpose chickens typically underperform compared to commercial layer or fast-growing broiler strains (Mueller et al., 2018).

According to FAOSTAT (2020), the global production of chicken eggs totaled approximately 1.6 trillion in number and exceeded 83 million tonnes (Mt) in 2019. China, which has been the world's leading producer of eggs for over 30 years (Yang et al., 2018), accounted for over 36 percent of all eggs produced and approximately 34 percent of global egg tonnage in 2019 (569 billion eggs and 28.4 Mt, respectively). The United States ranked a distant second, with 7.2 percent and 8.0 percent of the world's egg production and tonnage, respectively (113.3 billion eggs produced and 6.7 Mt). India, Indonesia, Brazil, Mexico, Japan, the Russian Federation, Turkey, and Pakistan (egg numbers) or Ukraine (tonnes) rounded out the top 10 in that order. Collectively, the ten leading countries accounted for 71.7 percent of the global egg numbers and 70.8 percent of the global egg tonnage (FAOSTAT, 2020).

The commercial laying hen sector has experienced many dramatic challenges in the past three decades while advancing its genetic potential from 230–240 eggs/hen/year produced in the 1990s to 300–310 per year in 2021. Challenges included control of *Salmonella enteriditis* contamination and consumer concerns over egg cholesterol content. The most dramatic challenge at present is the consumer demand for cage-free egg production systems, which move the laying hen out of traditional battery cages into colony cages, aviaries, or free-run or free-range environments. As the hen moves to these new systems, the impending challenge is to maintain a healthy chicken that can lay optimum numbers of eggs with low overall maintenance inputs. Several hybrid breeds are currently used for egg production, with white egg layer crosses and brown egg crosses being predominant, followed by more traditional breeds, such as the Rhode Island

Red. The White Leghorn is the most common white egg–type chicken breed in the world; the majority of nutrient requirement studies reported and utilized for this report were conducted using this strain.

Hen housing systems are currently undergoing rapid changes due to consumer demand for eggs produced in cage-free egg systems. This has culminated with over 27 percent of the egg production in the United States being in cage-free production systems in 2021, as compared to less than 5 percent in 2006 (Egg Industry Center, 2025). This trend is likely to continue over the next 5 to 10 years, partly driven by restaurants, grocery chains, and other egg-utilizing entities that have committed to selling cage-free eggs. However, the rate of cage-free systems expansion will be limited by both available financing and producer willingness to convert conventional cage houses or undertake the construction of new facilities.

The housing system can have a marked effect on the hens' environment and potentially on their nutrient requirements for maintenance. Confinement facilities are highly automated with computer-controlled environments. The array of housing types prevalent in global egg production systems include conventional (battery) cage systems, colony cages, aviary systems, floor rearing, and free-range systems that include pasture access. Battery cage systems were widely adopted decades ago when the global egg industry first industrialized. They provide hens with an environment safe from predators and parasites and a clean place to lay their eggs, since their manure falls through the wire bottom of the cage, either into a pit or onto a belt, which removes it from the house. Following oviposition, eggs typically roll out onto an automated belt, which moves them into a central collecting area or directly into a processing facility. Water and feed are provided for ad libitum consumption, with the latter automatically delivered via auger systems.

New housing systems are becoming more popular and prevalent on several continents due to consumer pressure to improve the well-being of the laying hen. There are pros and cons of all housing systems for laying hens. One of the cons of the traditional cage systems has been the inability of the hen to exhibit instinctual behaviors, such as dust bathing, perching, and nesting. Newer housing systems, such as furnished colony cages and aviary systems, have been designed to include amendments for these behaviors. Colony cages are much larger than the traditional battery cage and allow the hen access to a scratch area, nesting area, and perches, all on one horizontal level. Like battery cages, eggs are laid on a wire mesh and kept separate from manure and litter, and feed and water are delivered by automated systems. Aviaries are designed with horizontal and vertical space for hens to move about while also having floor litter areas for dustbathing. Floor housing systems, which resemble broiler breeder houses, have one horizontal layer of floor-housed hens with nest boxes and automated egg collection, feeding, and watering equipment. Activity and maintenance energy requirements are likely increased for hens housed in such systems versus battery or colony cages. Floor eggs can be more prevalent in aviary and floor systems, which may present some food safety challenges. Other challenges in such open systems include aggressive feather pecking, which can also be addressed with some nutrient modifications such as increased dietary fiber (Debruslais et al., 2021).

Free-range systems provide outdoor pen access and opportunities to consume forage, weather permitting. It is very difficult to assess the nutrient contributions of outdoor forage access for laying hens. At best, it is a partial supplement of fiber and some carbohydrates, xanthophylls, polyunsaturated fatty acids, and limited protein for the hen and is dependent on the relative presence of edible plant material. Hens housed in a free-range environment will still need access to a balanced ration to meet their amino acid, vitamin, mineral, and energy needs on a daily basis. Organic egg production in the United States and many other countries has regulations mandating outdoor access.

Non-Fasting Molt Diets

Many layer flocks in the United States are not molted and continue into production until 90–100 weeks of age without molting. However, non-fasting molts are still a management tool utilized in the United States to lengthen the laying hens' productive life. Approximately 16 percent of the U.S. flock is currently molted (Egg Industry Center, 2025), and due to welfare concerns, all molts are non-fasting. A non-fasting molt diet is typically low in calories and nutrient density, resulting in the cessation of lay, and thus allowing hens to replenish skeletal and nutrient reserves for a determined period of time after which they will be brought back into lay with light stimulation and a post-molt type of ration. Good sources of low-density feed ingredients for molting diets include alfalfa meal, soy hulls, dried distillers grains, or wheat middlings (Donalson et al., 2005; Mejia et al., 2010). In addition to being a low energy feed, molt and post-molt diets need to have a minimal level of amino acids and other essential nutrients to meet the non-laying hen's basic nutrient needs.

NUTRIENT RECOMMENDATIONS

A comprehensive literature review of nutrient requirements was conducted for each of the essential amino acids (AAs), minerals, and vitamins and essential fatty acids. Peer-reviewed articles published since September 1991, which was the date that the review of literature for the previous report (NRC, 1994) was completed, were considered. Specifically, published studies were excluded from consideration if the following conditions occurred: (1) insufficient dietary levels for establishing a dose–response relationship were employed; (2) the study failed to report analyzed dietary contents for the nutrient(s) under study; and (3) the study failed to provide these details or inadequate experimental design, housing, diet nutrient levels, analytical methodologies, or statistical procedures were used. Many of the recommended nutrient requirements of laying hens have not changed from the prior report (NRC, 1994). However, the committee has chosen to present them based on egg weights and masses during the peaking, peak, and post-peak egg production periods, rather than as phase feeding over set time periods as per the pullet recommendations. This acknowledges the physiologic and anatomical changes that are occurring during each production period, which may differ temporally depending on the strain of hen. When multiple empirical studies were available for a particular nutrient and time period of production, the values were averaged to estimate the minimum nutrient recommendations as presented in Tables 13-1 to 13-4. Nutrient requirements for pullets have been expressed in terms of units per kg of feed, which is consistent with the other chapters. This approach is based on relatively predictable feed intakes with increasing age. Nutrient requirements for laying hens have been expressed in terms of units of intake per hen per day.

Water

Water is quantitatively the most important nutrient needed for the growth, maintenance, and production of eggs of pullets and layers. Access to a clean source of water free of chemical and microbial contamination is absolutely necessary for the laying hen. Under thermoneutral conditions, adult laying hens will typically consume 200–275 ml of water/d (Nasr et al., 2013; Riek et al., 2020). The lower end would be for smaller white egg layers and the greater consumption for brown egg layers in free-range conditions. Drinking water or dietary sodium content (e.g., from bakery meals), environmental temperature, and disease may also markedly affect water intake as well. A measurement of water intake per flock is an important management tool along with feed intake to ascertain bird health.

Amino Acids

The majority of AA requirement studies were conducted in White Leghorn hens and centered on the first two limiting AAs in corn-soybean meal–based diets, namely methionine and lysine. When empirical data were insufficient, guidance levels in Tables 13-1 to 13-4 were estimated based on the ideal AA ratios to lysine (Lys) reported by Bregendahl et al. (2008), who conducted seven experiments with Hy-Line W-36 White Leghorn hens from 28 to 34 weeks of age to determine the ideal ratio of arginine (Arg), isoleucine (Ile), methionine (Met), methionine + cyst(e)ine (Met + Cys; total sulfur amino acids [TSAA]), threonine (Thr), tryptophan (Trp), and valine (Val) to Lys for maximal egg mass. Specifically, the ratios to Lys (as 1.00) based on true digestible requirements for maximal egg mass were as follows: methionine, 0.47; TSAA, 0.94; threonine, 0.77; tryptophan, 0.22; arginine, 1.07 or less; and isoleucine (Ile), 0.79 (Bregendahl et al., 2008). These values were similar to those reported by CVB (1996), which were based on digestible amino acid requirements, and Leeson and Summers (2005), which were based on total amino acid requirements. In the case where no specific AA ratio was available and requirement studies were lacking, the values recommended in NRC (1994) were utilized and reported in shaded areas in Tables 13-1 to 13-4. Model recommendations from Sakomura et al. (2015) were simulated in 2017 by Sakomura and coworkers and are also shown in Tables 13-1 to 13-4 for white- and brown-egg laying pullets. The genetic lines used to estimate required digestible lysine intake were Lohman LSL white egg layer and Lohman Brown egg layer (Bendezu et al., 2018; Alves et al., 2019). The factorial model for metabolizable energy recommendation (Sakomura, 2004) considers the energy requirements for maintenance and for gain based on lipid and protein depositions. The model to estimate digestible lysine recommendation has a similar concept, as it describes the necessity of lysine for maintenance and body gain. However, the body gain factor accounts for protein deposition in the body and an additional amount of lysine deposition for the reproductive tract, ovaries, and oviduct. The inclusion of ovary and oviduct growth in the factorial model allows for additional lysine required after 15 weeks of age. In Table 13-5, the values of digestible lysine decrease from the first phase (0 to 4 weeks) until the third phase (11 to 15 weeks) reported, after which the recommendation for digestible lysine increases. A similar pattern is observed in the digestible lysine recommendation found in the literature review.

For laying hens, the recommendation of digestible lysine estimated using a current adaptation of the Reading model (Fisher et al., 1973; Sakomura et al., 2015) and literature review is presented in Table 13-6. An appealing trait from the Reading model is the possibility to predict amino acid requirements to meet the requirement of a desired population. According to Hauschild et al. (2010), the amount of amino acid estimated by the factorial model to meet the needs of approximately 80 percent of the population gives a similar value to what is found using an empiric function. Although this behavior was observed in swine, for reasons of simplicity, the Reading model was similarly applied in this chapter. Requirements in mg/d are based on body weight, egg output, and their respective standard deviations, representing AA requirements to meet the needs of 85 percent of the population. The standard deviations used in this exercise are assumed to be 15 percent of body weight and 25 percent of egg output, and the correlation between body weight and egg output was assumed to be 0.

Lysine

Lysine was the most frequently researched AA in the context of daily nutrient requirements of pullets and laying hens.

TABLE 13-1 Nutrient Requirements of Immature White Egg-Laying Pullets as Percentages or Units per Kilogram of Diet (90% Dry Matter)[a,b]

Nutrient	Unit	Starter (0–4 Weeks)		Grower (5–10 Weeks)		Developer (11–15 Weeks)		Pre-Lay (16–17 Weeks)	
ME_n	Kcal/kg	2,940		2,820		2,700		2,780	
Crude protein[c]			20.00		17.5		16.0		16.5
		Model	Minimum	Model	Minimum	Model	Minimum	Model	Minimum
Amino acids (digestible)									
Arginine	%	1.24	0.94	1.01	0.78	0.72	0.62	0.92	0.72
Histidine	%	0.43	0.25	0.35	*0.21*	0.25	0.16	0.32	0.18
Isoleucine	%	0.81	***0.81***	0.66	***0.58***	0.47	***0.40***	0.60	***0.63***
Leucine	%	1.30	***1.10***	1.06	***0.85***	0.76	***0.70***	0.96	***0.80***
Lysine	%	0.97	0.97	0.70	0.70	0.48	0.48	0.75	0.75
Methionine	%		***0.45***		***0.33***		***0.23***		***0.35***
Methionine + cyst(e)ine	%	0.84	***0.91***	0.68	***0.66***	0.49	***0.45***	0.62	***0.66***
Phenylalanine + tyrosine	%	1.39	**0.94**	1.13	**0.78**	0.81	**0.63**	1.03	**0.70**
Threonine		0.77	***0.79***	0.63	***0.54***	0.45	***0.37***	0.57	***0.58***
Tryptophan	%	0.21	***0.19***	0.17	***0.15***	0.12	***0.15***	0.15	***0.17***
Valine	%	0.89	***0.89***	0.72	***0.65***	0.52	***0.45***	0.66	***0.70***
Fat	%								
Linoleic acid	%		1.00		1.00		1.00		1.00
Macrominerals									
Calcium	%		0.90		0.80		0.80		2.00
Nonphytate phosphorus	%		0.30		0.275		0.25		0.275
Potassium	%		***0.25***		***0.25***		***0.25***		***0.25***
Sodium	%		0.15		0.15		0.15		0.15
Chlorine	%		0.15		***0.12***		***0.12***		***0.15***
Magnesium	mg		600		***500***		***400***		***400***
Trace minerals									
Manganese	mg		60		***30***		***30***		***30***
Zinc	mg		40		***35***		***35***		***35***
Iron	mg		80		***60***		***60***		***60***
Copper	mg		***5.0***		***4.0***		***4.0***		***4.0***
Iodine	mg		0.35		***0.35***		***0.35***		***0.35***
Selenium	mg		0.15		***0.10***		***0.10***		***0.10***
Fat-soluble vitamins									
A	IU		1,500		1,500		1,500		1,500
D	IU		200		200		200		300
E	IU		***10***		***5***		***5***		***5***
K	mg		0.5		***0.5***		***0.5***		***0.5***
Water-soluble vitamins									
Thiamine	mg		***1***		***1***		***0.8***		***0.8***
Riboflavin	mg		3.6		1.8		1.8		2.2
Niacin	mg		27		11		11		11
Pantothenic acid	mg		***10***		***10***		***10***		***10***
Pyridoxine	mg		3		***3***		***3***		***3***
Folacin	mg		***0.55***		***0.25***		***0.25***		***0.25***
Biotin	mg		***0.15***		***0.10***		***0.10***		***0.10***
B_{12}	mg		0.009		0.003		0.003		0.004
Choline	mg		1,300		900		500		500

[a] Values in normal font are minimum nutrient requirement recommendations based on empirical studies summarized in Table 13-7.

[b] Values in shaded areas are NRC (1994) recommendations for which no new studies are available.

[c] Chickens do not have a requirement for crude protein per se. However, there should be sufficient crude protein to ensure an adequate amino nitrogen supply for synthesis of nonessential amino acids. Suggested requirements for crude protein are typical of those derived from corn-soybean meal–based diets, and levels can be reduced somewhat when synthetic amino acids are used.

TABLE 13-2 Nutrient Requirements of Immature Brown Egg-Laying Pullets as Percentages or Units per Kilogram of Diet (90% Dry Matter)[a,b]

Nutrient	Unit	Starter (0–4 Weeks)		Grower (5–10 Weeks)		Developer (11–15 Weeks)		Pre-Lay (16–17 Weeks)	
ME_n	Kcal/kg	2,940		2,820		2,700		2,730	
Crude protein[c]	%		20.0		17.5		15.0		16.5
		Model	Minimum	Model	Minimum	Model	Minimum	Model	Minimum
Amino acids (Digestible)									
Arginine	%	1.28	1.00	1.03	0.83	0.76	0.67	0.84	0.75
Histidine	%	0.44	0.26	0.35	*0.22*	0.26	0.17	0.29	0.20
Isoleucine	%	0.83	***0.88***	0.76	***0.68***	0.49	***0.59***	0.54	***0.63***
Leucine	%	1.33	***0.80***	1.08	***0.56***	0.79	***0.42***	0.88	***0.49***
Lysine	%	1.05	1.05	0.82	0.82	0.70	0.70	0.75	0.75
Methionine	%		***0.49***		***0.49***		***0.33***		***0.35***
Methionine + cyst(e)ine	%	0.86	***0.96***	0.70	***0.96***	0.51	***0.66***	0.57	***0.71***
Phenylalanine + tyrosine	%	1.43	1.00	1.16	***0.83***	0.85	0.67	0.94	0.75
Threonine		0.79	***0.81***	0.64	***0.81***	0.47	***0.54***	0.52	***0.54***
Tryptophan	%	0.21	***0.19***	0.17	***0.18***	0.13	***0.15***	0.14	***0.17***
Valine	%	0.91	***0.98***	0.74	***0.98***	0.54	***0.65***	0.60	***0.70***
Fat									
Linoleic acid	%		***1.00***		***1.00***		***1.00***		***1.00***
Macrominerals									
Calcium	%		***0.90***		***0.80***		***0.80***		***1.80***
Nonphytate phosphorus	%		0.30		0.275		0.25		0.275
Potassium	%		***0.25***		***0.25***		***0.25***		***0.25***
Sodium	%		***0.15***		***0.15***		***0.15***		***0.15***
Chlorine	%		***0.12***		***0.11***		***0.11***		***0.11***
Magnesium	mg		***570***		***470***		***370***		***370***
Trace minerals									
Manganese	mg		***56***		***28***		***28***		***28***
Zinc	mg		***38***		***33***		***33***		***33***
Iron	mg		***75***		***56***		***56***		***56***
Copper	mg		***5.0***		***4.0***		***4.0***		***4.0***
Iodine	mg		***0.33***		***0.33***		***0.33***		***0.33***
Selenium	mg		***0.14***		***0.10***		***0.10***		***0.10***
Fat-soluble vitamins									
A	IU		***1,420***		***1,420***		***1,420***		***1,420***
D	IU		***190***		***190***		***190***		***280***
E	IU		***9.5***		***4.7***		***4.7***		***4.7***
K	mg		***0.47***		***0.47***		***0.47***		***0.47***
Water-soluble vitamins									
Thiamine	mg		***1***		***1***		***0.8***		***0.8***
Riboflavin	mg		***3.4***		***1.7***		***1.7***		***1.7***
Niacin	mg		***26***		***10.3***		***10.3***		***10.3***
Pantothenic acid	mg		***9.4***		***9.4***		***9.4***		***9.4***
Pyridoxine	mg		***2.8***		***2.8***		***2.8***		***2.8***
Folacin	mg		***0.52***		***0.23***		***0.23***		***0.23***
Biotin	mg		***0.14***		***0.09***		***0.09***		***0.09***
B_{12}	mg		***0.009***		***0.003***		***0.003***		***0.003***
Choline	mg		***1,225***		***850***		***470***		***470***

[a] Values in normal font are minimum nutrient requirement recommendations based on empirical studies summarized in Table 13-7.

[b] Values in shaded areas are NRC (1994) recommendations for which no new studies are available. Guidance recommendations are signified by bold and italicized font.

[c] Chickens do not have a requirement for crude protein per se. However, there should be sufficient crude protein to ensure an adequate amino nitrogen supply for synthesis of nonessential amino acids. Suggested requirements for crude protein are typical of those derived from corn-soybean meal–based diets, and levels can be reduced somewhat when synthetic amino acids are used.

TABLE 13-3 Nutrient Requirements of White Egg-Layers as Percentages or Units per Day[a,b]

	Unit	Peaking		Peak		Post-Peak	
Body weight	g	1,520		1,560		1,580	
Egg weight	g	58.0		62.0		63.0	
Daily egg mass	g	54.5		55.8		52.0	
Nutrient							
ME_n	kcal/kg diet	2,780		2,780		2,730	
Crude protein[c]	%		17		16.5		15.5
		Model	Minimum	Model	Minimum	Model	Minimum
Amino acids (digestible)							
Arginine	mg/day	***744***	700	***762***	1,000	***714***	700
Histidine	mg/day	***207***	170	***213***	**170**	***199***	170
Isoleucine	mg/day	***558***	***643***	***572***	***600***	***536***	***585***
Leucine	mg/day	***865***	820	***887***	820	***831***	820
Lysine	mg/day	715	770	733	718	687	700
Methionine	mg/day		340		300		280
Methionine + cyst(e)ine	mg/day	***586***	790	***601***	778	***563***	700
Phenylalanine			470		470		470
Threonine	mg/day	***493***	560	***506***	544	***474***	462
Tryptophan	mg/day	***150***	195	***154***	195	***144***	195
Valine	mg/day	***651***	***716***	***667***	***725***	***625***	***651***
Fat							
Linoleic acid	mg/day		***1,000***		***1,000***		***1,000***
Macrominerals							
Calcium[d]	mg/day		3,600		4,000		4,800
Nonphytate phosphorus	mg/day		300		300		300
Potassium	mg/day		130		130		130
Sodium	mg/day		130		130		130
Chlorine	mg/day		110		110		110
Magnesium	mg/day		50		50		50
Trace minerals							
Manganese	mg/day		2.0		2.0		2.0
Zinc	mg/day		3.5		3.5		3.5
Iron	mg/day		4.5		4.5		4.5
Copper	mg/day		—		—		—
Iodine	mg/day		0.004		0.004		0.004
Selenium	mg/day		0.006		0.006		0.006
Fat-soluble vitamins							
A	IU/day		300		300		300
D	IU/day		30		30		30
E	IU/day		0.5		0.5		0.5
K	mg/day		**0.05**		**0.05**		**0.05**
Water-soluble vitamins							
Thiamin	mg/day		***0.07***		***0.07***		***0.07***
Riboflavin	mg/day		0.25		0.25		0.25
Niacin	mg/day		1.0		1.0		1.0
Pantothenic acid	mg/day		0.20		0.20		0.20
Pyridoxine	mg/day		0.25		0.25		0.25
Folacin	mg/day		0.025		0.025		0.025
Biotin	mg/day		***0.01***		***0.01***		***0.01***
B_{12}	mg/day		***0.0004***		***0.0004***		***0.0004***
Choline	mg/day		105		105		105

[a] Values in normal font are minimum nutrient requirement recommendations based on empirical studies summarized in Table 13-7.

[b] Values in shaded areas are NRC (1994) recommendations for which no new studies are available. Guidance recommendations are signified by bold and italicized font.

[c] Chickens do not have a requirement for crude protein per se. However, there should be sufficient crude protein to ensure an adequate amino nitrogen supply for synthesis of nonessential amino acids. Suggested requirements for crude protein are typical of those derived from corn-soybean meal–based diets, and levels can be reduced somewhat when synthetic amino acids are used.

[d] Ca is typically provided as a ratio of fine (0–2 mm diameter) and coarse (2–4 mm diameter) particles, with an increasing proportion of large particle Ca as the hens age.

TABLE 13-4 Nutrient Requirements of Brown Egg-Layers as Percentages or Units per Kilogram of Diet (90% Dry Matter)[a,b]

	Unit	Peaking		Peak		Post-Peak	
Body weight	g	1,700		1,850		1,900	
Egg weight	g	58.5		61.5		63.0	
Daily egg mass	g	56.0		57.2		53.5	
Nutrient							
ME_n	kcal/kg diet	2,730		2,730		2,700	
Crude protein[c]	mg/day		17.0		16.5		15.5
		Model	Minimum	Model	Minimum	Model	Minimum
Amino acids							
Arginine	mg/day	***770***	770	***790***	770	***746***	770
Histidine	mg/day		190		190		190
Isoleucine	mg/day	***577***	***655***	***593***	***630***	***559***	***614***
Leucine	mg/day	***895***	900	***920***	900	***868***	900
Lysine	mg/day	740	784	760	755	717	735
Methionine	mg/day		***368***		***355***		***345***
Methionine + cyst(e)ine	mg/day	***607***	***679***	***623***	***679***	***588***	***626***
Phenylalanine			520		520		520
Threonine	mg/day	***511***	566	***524***	506	***495***	491
Tryptophan	mg/day	***155***	215	***160***	215	***151***	215
Valine	mg/day	***673***	***753***	***692***	***725***	***652***	***705***
Fat							
Linoleic acid	mg/day		***1,100***		***1,100***		***1,100***
Macrominerals							
Calcium[d]	mg/day		3,600		4,000		4,800
Nonphytate phosphorus	mg/day		300		300		300
Potassium	mg/day		***165***		***165***		***165***
Sodium	mg/day		***165***		***165***		***165***
Chlorine	mg/day		***145***		***145***		***145***
Magnesium	mg/day		***55***		***55***		***55***
Trace minerals							
Manganese	mg/day		***2.2***		***2.2***		***2.2***
Zinc	mg/day		***3.9***		***3.9***		***3.9***
Iron	mg/day		***5.0***		***5.0***		***5.0***
Copper	mg/day		—		—		—
Iodine	mg/day		***0.004***		***0.004***		***0.004***
Selenium	mg/day		***0.006***		***0.006***		***0.006***
Fat-soluble vitamins							
A	IU/day		***330***		***330***		***330***
D	IU/day		***33***		***33***		***33***
E	IU/day		***0.55***		***0.55***		***0.55***
K	mg/day		***0.055***		***0.055***		***0.055***
Water-soluble vitamins							
Thiamine	mg/day		***0.08***		***0.08***		***0.08***
Riboflavin	mg/day		0.28		0.28		0.28
Niacin	mg/day		***1.1***		***1.1***		***1.1***
Pantothenic acid	mg/day		***0.22***		***0.22***		***0.22***
Pyridoxine	mg/day		***0.28***		***0.28***		***0.28***
Folacin	mg/day		***0.028***		***0.028***		***0.028***
Biotin	mg/day		***0.011***		***0.011***		***0.011***
B_{12}	mg/day		***0.0004***		***0.0004***		***0.0004***
Choline	mg/day		***115***		***115***		***115***

[a] Values in normal font are minimum nutrient requirement recommendations based on empirical studies summarized in Table 13-7.

[b] Values in shaded areas are NRC (1994) recommendations for which no new studies are available. Guidance recommendations are signified by bold and italicized font.

[c] Chickens do not have a requirement for crude protein per se. However, there should be sufficient crude protein to ensure an adequate amino nitrogen supply for synthesis of nonessential amino acids. Suggested requirements for crude protein are typical of those derived from corn-soybean meal–based diets, and levels can be reduced somewhat when synthetic amino acids are used.

[d] Ca is typically provided as a ratio of fine (0–2 mm diameter) and coarse (2–4 mm diameter) particles, with an increasing proportion of large particle Ca as the hens age.

TABLE 13-5 Pullet Requirements for Digestible Lysine Estimated by Factorial Model and Literature Review

	Phase	Metabolizable Energy[b]	Feed Intake[c]	Digestible Lysine Requirement (%)[d]	
Pullet[a]	Weeks	kcal/bird/day	g//bird/day	Model	Review
White egg laying pullets	0–4	36.7	12.7	0.941	0.966
	5–10	94.2	32.5	0.772	0.700
	11–15	116	40.1	0.520	0.480
	16–17	116	40.0	0.759	0.750
Brown egg laying pullets	0–4	44.2	15.2	0.966	1.05
	5–10	127	43.7	0.788	0.819
	11–15	160	55.2	0.544	0.700
	16–17	159	54.8	0.686	0.750

[a] Pullet simulated for factorial model was Lohman LSL and Lohman Brown, as published by Alves et al. (2018) and Silva et al. (2014).

[b] Metabolizable energy requirement $(ME) = BW^{0.75} \times 92.4 + 13.5 \times Ld + 12.3 \times Pd$.

[c] Feed intake calculated from ratio between metabolizable energy (ME) requirement and ME of diet (Mcal/kg of feed). The ME in the diets were 3.00, 3.11, 3.16, 3.19, and 3.21 kcal/kg of feed from 1 to 14, 15 to 28, 29 to 35, 36 to 46, and 47 to 56 days of age, respectively.

[d] $Lys = [(AAm \times BP^{0.73} \times u) + (FL \times FP \times AA f] + [(AAb \times BPD + AA f \times FPD)/k + (AAova \times OvaPD)/k + (AAovi \times OviPD)/k]$.

TABLE 13-6 Laying Hen Requirements for Digestible Lysine Estimated by the Reading Model and Literature Review

		Body Weight	Average Daily Egg Mass	Metabolizable Energy[b]	Digestible Lysine Requirement (mg/day)[c]	
Laying hen[a]	Phase	g	g	kcal/hen/day	Model	Review
White egg layers	Peaking	1,520	54.5	270	715	770
	Peak	1,560	55.8	276	733	718
	Post-Peak	1,580	52.0	268	687	700
Brown egg layers	Peaking	1,700	56.0	285	740	784
	Peak	1,850	57.2	297	760	755
	Post-Peak	1,900	53.5	291	717	735

[a] The parameter used in the reading model is published by Sakomura et al. (2015).

[b] Metabolizable energy requirement $= BW^{0.75} \times 94.2 + 6.68 \times BWG + 2.40 x EM$.

[c] IAA (mg/hen/day) $= [(a \times EOmax) + (b \times BW)] + Z \times \sqrt{a^2 \times \sigma^2_{EO_{max}} + b^2 \times \sigma^2_{BW}}$. Where, EO is egg output, BW is body weight, Z is the value under normal curve that represents 85% of the population, $\sigma^2_{EO_{max}}$ and σ^2_{BW} are the square of standard deviations observed for maximum egg output and body weight.

As previously noted, requirements for all of the other essential AAs were ratioed to Lys if empirical data were not available (Bregendahl, 2008). Lysine recommendations for pullets in the model were based on research by Silva et al. (2000a,b,c). The recommended daily intakes of Lys for white egg laying hens are 770 mg/hen, 718 mg/hen, and 700 mg/hen at peaking, peak, and post-peak egg production (Table 13-3) and are based on an average from the reports of Faria et al. (2003), Novak et al. (2004), Sá et al. (2007), Schmidt et al. (2009), Jardim et al. (2010), Silva et al. (2010), Figueiredo et al. (2012), and Pastore et al. (2015, 2018). The recommended daily intakes of Lys are 784 mg/hen, 755 mg/hen, and 735 mg/hen for brown egg layers at peaking, peak, and post-peak egg production and are based on the reports of Sá et al. (2007), Schmidt et al. (2009), and Scappaticcio et al. (2021).

Methionine

No empirical requirement studies for Met were found for white egg or brown egg laying pullets. In those instances, recommendations were based on the ideal Met:Lys ratio of 0.48 recommended by Bregendahl et al. (2008). In contrast, numerous Met requirement studies employing white egg layers or brown egg layers have been published since the previous edition of this report (NRC, 1994). For pullets, the primary response variable used to evaluate dietary requirements was growth (body weight); for layers, the primary response variables used to evaluate dietary requirements were egg production and egg weights. Several Met requirement studies were reported for white egg layers at varying ages. The daily Met intake recommendations made in Table 13-3 for white egg layers at peaking, peak, and post-peak production are 340 mg/hen based on Saki et al. (2012) during peaking, 300 mg/hen based on Harms et al. (2003) at peak, and 280 mg/hen based on Gomez et al. (2016) at post-peak.

Total Sulfur Amino Acids

The daily TSAA intake recommendations made in Table 13-3 for white egg layers at peaking, peak, and

TABLE 13-7 Documentation of Nutrient Requirements of Pullets and Laying Hens[a]

Nutrient and Estimated Requirement (%)	Age (Weeks) or Period	Response Criteria	Breed	Reference
Arginine				
1.27	33–45	Egg production	Brown Asian Breed	Yuan et al. (2015)
Lysine				
0.99	1–3	Weight gain	Hy-Line W-36 White Leghorn pullets	Silva et al. (2015)
0.94	4–6	Weight gain	Hy-Line W-36 White Leghorn pullets	Silva et al. (2015)
0.70	7–12	Weight gain	White pullets	Silva et al. (2000b)
0.48	13–20	Weight gain	White pullets	Silva et al. (2000c)
0.80	0–6	Weight gain	Brown pullets	Silva et al. (2000a)
0.70	7–12	Weight gain	Brown pullets	Silva et al. (2000b)
0.48	13–20	Weight gain	Brown pullets	Silva et al. (2000c)
1.05	1–3	Weight gain	Hy-Line Brown Leghorn pullets	Silva et al. (2015)
0.82	4–6	Weight gain	Hy-Line Brown Leghorn pullets	Silva et al. (2015)
0.90	24	Egg mass, egg production	Hy-Line W-36 White Leghorn	Pastore et al. (2018)
0.77	24–40	Egg production, egg mass	Hy-Line W-36 White Leghorn	Rocha et al. (2009)
0.68	34–50	Egg production, egg mass	Lohman White	Sá et al. (2007)
0.75	42–58	Egg mass	Hy-Line W-36 White Leghorn	Figueiredo et al. (2012)
0.68	44–58	Egg production	Hy-Line W-36 White Leghorn	Faria et al. (2003)
0.84	60	Egg mass	Hy-Line W-36 White Leghorn	Pastore et al. (2015)
0.66	60	Egg production	Hy-Line W-36 White Leghorn	Pastore et al. (2015)
0.88	68	Egg mass	Hisex White	Silva et al. (2010)
0.68	68	Egg production	Hisex White	Silva et al. (2010)
0.76	79–95	Egg production, egg mass	Lohmann White	Schmidt et al. (2009)
0.68	34–50	Egg production, egg mass	Lohmann Brown	Sá et al. (2007)
0.73	34–50	Egg production, egg mass	Lohmann Brown	Sá et al. (2007)
0.71	79–95	Egg mass	Lohmann Brown	Schmidt et al. (2009)
0.68	19–59	Egg production	Lohmann Brown	Scappaticcio et al. (2021)
Methionine				
0.34	22–36	Egg mass	White Leghorn	Saki et al. (2012)
0.30	45	Egg production	Hy-Line W-36 White Leghorn	Harms and Russell (2003)
0.51	68	Egg production, egg mass	Hisex White Leghorn	Silva et al. (2010)
0.28	68–75	Egg production, egg mass	Hy-Line W-36 White Leghorn	Gomez and Angeles (2016)
0.45	2nd cycle	Egg production, egg mass	Hy-Line W-36 White Leghorn	Gomez and Angeles (2009)
Methionine + cyst(e)ine				
0.90	24–40	Egg production, egg mass	Hy-Line W-36 White Leghorn	Brumano et al. (2010a)
0.68	34–50	Egg production, egg mass	Lohmann White	Sá et al. (2007)
0.90	42–58	Egg production, egg mass	Hy-Line W-36 White Leghorn	Brumano et al. (2010b)
0.70	54–70	Egg mass	Lohmann White	Cupertino et al. (2009)

post-peak egg production were 790 mg/hen based on Sá et al. (2007) and Brumano et al. (2010a), 778 mg/hen based on Barbosa et al. (1999), Cupertino et al. (2009), and Brumano et al. (2010b), and 700 mg/hen based on Schmidt et al. (2011) and Gomez et al. (2016). For peaking and peak brown egg layers, the daily TSAA intake recommendations made in Table 13-4 were 679 mg/hen based on Sá et al. (2007) and 626 mg/hen during post-peak according to Cupertino et al. (2009), Barbosa et al. (1999), and Schmidt et al. (2009). No white or brown laying pullet reports were found that met the committee's criteria for inclusion, so the TSAA pullet recommendations in Tables 13-1 and 13-2 were estimated based on the TSAA:Lys ratio of 0.94 recommended by Bregendahl et al. (2008).

Threonine

No reports on pullet Thr requirements that met the committee's criteria for inclusion were found; therefore, recommendations included in Tables 13-1 and 13-2 are estimated based on the modeling equations. In contrast, several reports on the Thr requirements of white and brown egg layers were found. For peaking white egg layers, a recommendation of 560 mg/hen/d was based on the reports of Faria et al. (2002), Schmidt et al. (2009), and Figueiredo et al. (2012); at peak egg production, 544 mg/hen/day was recommended based on the reports of Martinez-Amezcua et al. (1999) and Cardoso et al. (2014); and for the post-peak period, a recommendation of 462 mg/hen/day was based on Cupertino et al. (2010) and

Schmidt et al. (2011) (Table 13-3). For brown egg layers, the daily Thr intake recommendations for peaking, peak, and post-peak egg production are 566 mg/hen, 506 mg/hen, and 491 mg/hen, respectively, based on the publications of Schmidt et al. (2011) and Nunes et al. (2015) (Table 13-4).

Tryptophan

Tryptophan recommendations are based on a limited number of reports. For starter white egg pullets, Harms et al. (2000) recommended a dietary level of 0.20 percent. Based on the reports of Russell and Harms (1999) and Wen et al. (2019), this value was lowered to 0.19 percent, and to 0.15 percent for all grower and developer diets, with a slight increase to 0.17 percent for pre-lay diets (Table 13-1). Similar recommendations are made for brown egg laying pullets, with the exception of 0.18 percent Trp during the grower phase (Table 13-2). The daily recommended intake of Trp for white egg layers (Table 13-3) is 195 mg/hen for all three phases of egg production, which was based on the reports of Russell and Harms (1999), Wen et al. (2019), and Harms et al. (2000); and 215 mg/hen during all three phases for brown egg layers (Table 13-4), based on the reports of Peganova et al. (2003) and Schmidt et al. (2009).

Arginine

Because of a lack of Arg requirement studies for pullets and laying hens, the dietary requirements were not changed from NRC (1994), with the exception of a recommended daily intake of 1,000 mg/hen/d for white egg layers at peak production (Table 13-3). Although Bregendahl et al. (2008) did not include an Arg:Lys recommendation in their ideal amino acid profile for laying hens, the referenced model used a ratio estimation of 1.08 for pullets and 1.04 for laying hens in order to calculate Arg recommendations. Yuan et al. (2015) recommended 1,270 mg/day intake for brown egg layers, which is much higher than the NRC (1994) recommendation and much higher than the peak or post-peak diet recommendations, so the Arg recommendations are kept at 770 mg/hen/day for all three phases of egg production in Table 13-4.

Valine

A daily Val intake recommendation of 725 mg/hen for white egg layers at peak production (Table 13-3) was based on one study (Wen et al., 2019). Lelis et al. (2014) recommended a Val:Lys ratio of 0.96 for peaking brown egg layers, and this was used to calculate their recommended daily Val intake of 753 mg/hen (Table 13-4).

Isoleucine

Although Rocha et al. (2013) reported an ideal Ile:Lys ratio of 0.88, the Ile requirements of white and brown egg layers were not changed from NRC (1994) because isoleucine studies were too few to establish a requirement.

Minerals

Calcium

Reproducing female chickens (and other poultry species) have a large requirement for calcium (Ca) during egg production. This large demand for Ca to support eggshell formation, and the subsequent use of bone reserves in addition to dietary Ca to meet this need, leads to unique aspects of Ca and P metabolism in laying hens. Until approximately 2 weeks to 10 days before the onset of lay (i.e., sexual maturity; Taylor and Dacke, 1984), egg-type pullets deposit bone tissue in the form of cortical (the outer structural shell) and trabecular (inner struts) bone (Kim et al., 2012). This arrangement provides strength while minimizing the weight of the skeleton as an adaptation for flight in most species of birds. As estrogen levels increase with the impending onset of lay, deposition of structural bone ceases and medullary bone is deposited (Whitehead, 2004). Medullary bone is a labile, nonstructural bone tissue that can be deposited and mobilized on a daily basis (Kerschnitzki et al., 2014). It is formed when calcium supply exceeds demand (i.e., when Ca is being absorbed from the digestive tract and no shell is being formed) and is mobilized when demand exceeds the supply (i.e., when an eggshell is being formed and insufficient Ca is being absorbed from the digestive tract). This latter situation occurs particularly during the night, when the bird is not eating and the feed ingested prior to the dark period has passed through the digestive tract (Kerschnitzki et al., 2014). In persistent egg layers such as the commercial laying hen, medullary bone deposition and mobilization cycle on a daily basis and contributes 35–40 percent of the calcium of each egg shell (Comar and Driggers, 1949; Mueller et al., 1964). Over time, there is a gradual loss of structural bone as well, but unlike medullary bone, structural bone is not replaced as long as the hen is in lay.

Overall, the Ca requirement of the laying hen is extremely high, both in absolute amount and in proportion to non-phytate phosphorus (nPP), because the eggshell, which accounts for ~10 percent of the egg weight, consists of ~95 percent Ca carbonate ($CaCO_3$; Nys et al., 2004). Since $CaCO_3$ contains ~40 percent Ca, a 60 g egg containing 6 g of shell would contain approximately 2.4 g of Ca. Commercial strains of white-egg and brown-egg hens lay at extremely high rates (>95% at peak lay) and therefore need to produce an eggshell almost every day during their productive life, thereby necessitating a high dietary Ca intake and robust metabolic mechanisms to mobilize and replenish bone Ca. de Juan et al. (2023) published a paper investigating the influence of calcium in pre-lay brown egg laying hens and reported that increasing dietary Ca from 2.5 to 3.8 percent from 16 to 19 weeks of age improved eggshell quality but not tibia mineralization during their 16- to 63-week-old subsequent lay period.

As the number of eggs produced per hen continues to increase via extension of the laying cycle, resulting from advances in genetic selection, Ca nutrition will continue to be a critical nutrient for laying hen health and welfare. In addition to Ca content, other dietary factors including P content, phytate content, the presence of exogenous phytase, and supplemental vitamin D (or metabolites) can alter the hens' Ca homeostasis. Due to the complex nature of dietary Ca availability and homeostasis, it is unfortunate that an available or digestible Ca method has not been validated and that total Ca requirements are still used in feed formulation. This reliance on diet formulation on a total Ca basis may result in slightly higher requirement values based on feed ingredients in common use. Additionally, Ca requirements have been determined using diets without either supplemental phytase or higher-than-required levels of vitamin D. Ca particle size has also not been directly accounted for in most of the Ca requirement studies referenced, especially the feeding of larger particle size and more slowly digested limestone, which has been shown to increase egg shell quality (Lichovnikova, 2007; Safaa et al., 2008; Swiatkiewicz et al., 2015) and typically accounts for half of the dietary Ca provided.

Although an attempt was made to identify a difference in the Ca requirement of white and brown egg laying chickens, too few peer-reviewed articles were available to make this assessment. Moreover, it was difficult to justify changing the Ca requirement of laying hens for peak or post-peak egg production. Feeding between 3.3 and 5.88 g of Ca per day did not significantly affect feed intake, egg production, egg mass, plasma Ca, egg specific gravity, or eggshell weight (both as total g of shell and as a percent of the egg), or egg shell breaking strength during peak or post-peak production (Keshavarz and Nakajima, 1993; Bar et al., 2002; Castillo et al., 2004; Pelicia et al., 2009; Swiatkiewicz et al., 2015; An et al., 2016). A limitation of the literature is that almost no experiments quantified the effects of early Ca nutrition on eggshell quality of the hen later in the production phase, with the exception of Safaa and coworkers (2008), who reported that feeding 4.8 g Ca daily increased egg production and egg mass, and decreased feed conversion ratio, as compared to 4.1 g of Ca daily from 58 to 73 weeks of age. However, several researchers (Bar et al., 2002; Safaa et al., 2008; Swiatkiewicz et al., 2015) observed that dietary Ca intake affected eggshell quality, and at post-peak egg production, and multiple reports suggest that higher feeding of Ca is required for optimum eggshell quality. Results of two studies (Bar et al., 2002; An et al., 2016) suggested that feeding dietary Ca above 4.7 g per day was required, although these reports did not contain sufficient dietary levels to allow for calculation of a true requirement.

Supplemental calcium may be provided. Inorganic sources, such as limestone and oyster shell, provide the largest amounts of Ca in most laying hen and pullet diets. Calcium source particle size is also an important element in balancing a laying hens' Ca needs (Eusebio-Balzacar et al., 2018). Ca is typically initially provided as a 50:50 ratio mixture of coarse (2 to 4 mm diameter) and fine (0 to 2 mm diameter) particles, with an increasing proportion of large particle Ca as the hens age (up to 70:30). Large particles of calcium sources are fed to slow the rate of digestion during feeding hours allowing an extended digestion of calcium and thereby supply of calcium to the hen during her dark, non-feeding hours of the day.

Phosphorus

This is a sometimes forgotten nutrient for egg production because the shell contains very small amounts of P. However, P plays an important role in medullary bone and cortical bone formation, but it is reliant on the storage and use pattern of Ca as a component of hydroxyapatite ($Ca_{10}[PO_4]_6[OH]_2$). Any P in excess of metabolic needs is excreted in the urine via the kidney.

Little direct P requirement work has been completed with table egg laying pullets since the last report (NRC, 1994). However, several studies have been published using multiple concentrations of dietary nPP, both with and without phytase supplementation, and those articles were used to validate and update the recommended pullet P requirements. The major drawback to this approach is that after the starter phase of production, all comparisons are confounded by differences in previous dietary phosphorus nutrition, so caution is needed in extrapolation of grower, developer, and pre-lay recommendations. As with laying hens, the nPP requirements are determined in studies using diets without feed additives like phytase, vitamin D levels and metabolites, or other potential modifiers. When feeding laying hens, the nutritional implications of those feed additives would need to be considered. In the case of pullets, the requirement is based on the P needed to maximize body weight and not necessarily skeletal health. Although skeletal responses were noted in the cited reports, body weight was equally or more sensitive to dietary P than skeletal measurements. Additionally, when skeletal measurements were performed at the end of the pullet phase (approximately 17 to 18 weeks of age), no differences due to dietary P level were noted among the treatments regardless of previous responses (Keshavarz, 2000a; Toledo, 2007; Jing et al., 2018).

The only publication to estimate a P requirement suggested that caged white egg pullet chicks from hatch to 42 d of age require 0.28 percent available P (AvP) based on body weight gain (Toledo et al., 2007). When white egg laying pullet chicks were fed 0.2, 0.3, 0.4, and 0.5 percent AvP from hatch to 42 d of age, both body weight and feed efficiency were maximized between 0.2 and 0.3 percent AvP (Jing et al., 2018). Keshavarz (2000b) fed 0.2, 0.3, and 0.4 percent nPP diets to White Leghorn pullet chicks from 0 to 42 d of age and observed that body weights were not different between chicks fed 0.3 or 0.4 percent nPP. These reports suggest or support the idea that the requirement for egg laying pullets for nPP is between 0.2 and 0.3 percent nPP. One report did suggest a lower requirement for nPP in growing White Leghorn chicks fed 0.13, 0.23, 0.33, and 0.43 percent nPP, where no increase in body weight was noted after 0.23 percent nPP diets were fed (Punna and Roland, 2000). These reports suggest that the requirement for nPP in starting egg laying chicks is lower than previous NRC estimates

As noted previously, the reports for grower phase nPP are confounded by previous dietary nPP status. The starter chicks used by Toledo and coworkers (2007) were continued on diets that contained 0.15, 0.20, 0.25, 0.30, and 0.35 percent AvP without difference in body weight gain, tibia P content, or breaking strength through 84 d of feeding. Additional experiments that support a lower requirement include increases in body weight that occurred in pullets fed between 0.1 and 0.2 percent nPP (Punna and Roland, 2000); 0.2 and 0.3 percent nPP (Keshavarz, 2003a); 0.175 and 0.275 percent AvP (Jing et al., 2018), without any further increase in body weight at higher levels of nPP.

As with the data generated for the grower phase, developer phase data are also confounded with the previous dietary treatments and again need to be viewed within this context. Toledo and coworkers (2007) fed White Leghorn pullets 0.15, 0.20, 0.25, 0.30, and 0.35 percent AvP from 84 to 126 d of age without observing any difference in pullet body weight, tibia P content, or tibia breaking strength. Additional experiments that support a lower requirement include increases in body weight that occurred in pullets fed between 0.1 and 0.2 percent nPP (Punna and Roland, 2000); 0.2 and 0.3 percent nPP (Keshavarz, 2003a); and 0.175 and 0.275 percent AvP (Jing et al., 2018), without any further increase in body weight at higher levels of P.

Limited data were available to generate an update to requirements during the pre-lay period, so although the requirement has been lowered, this is primarily due to the data reported for the pullet starter, grower, and developer phases being reduced and was not based on data from empirical studies. An increased nPP recommendation as compared to the developer phase diet was included, and this was based on the increased recommendation for Ca and the associated reduction in P utilization as the pullet readies for the initiation of egg production by increasing deposition of medullary bone.

The pullet nPP recommendations have each been reduced based on pullet body weight and skeletal structure responses reported in the literature. One area where data are limited are actual pullet nPP feeding studies and subsequent effects on egg production of laying hens once they reach maturity. Only one report was found to explore this question, in which egg production and egg specific gravity up to 30 weeks of age was not altered by pullet dietary nPP ranging from 0.2 to 0.4 percent nPP (Keshavarz, 2000a), suggesting that short-term body weight is an acceptable response variable to approximate egg production responses, but longer-term and end-of-lay data have not been generated.

A thorough search of the literature yielded few peer-reviewed articles that addressed the laying hens' dietary needs for P alone. An attempt was made to justify a difference in P requirement between white and brown egg laying chickens, but too few peer-reviewed articles were available for individual assessment. Additionally, the nPP requirement for laying hens is reported in diets without feed additives like phytase, and higher-than-requirement feeding of vitamin D and its metabolites, or other potential modifiers. Impacts of those on P utilization would need to be considered when feeding laying hens. Moreover, when looking at egg production alone, no changes—or even a possible reduction—in the dietary nPP requirements of laying hens could be justified on the basis of egg production responses (Keshavarz, 2000b; Sohail and Roland, 2002; Snow et al., 2005; Panda et al., 2005). Three of these reports included measurements of bone responses, either tibia ash or breaking strength, and two of the three articles recommended a higher dietary nPP requirement for bone health than for egg production. Panda and coworkers (2005) did not calculate a requirement but did show that tibia ash and breaking strength were greatly increased when hens consumed 323 mg of nPP/day as compared to 287 mg (or less) of nPP/per day. Sohail and Roland (2002) reported a quadratic response in tibia breaking strength to dietary nPP in laying hens, which was estimated to be maximized at an intake of 280 mg nPP/ per day; mortality was estimated to be minimized at an intake of 325 mg nPP/day. These reports suggested that the dietary requirement of nPP for laying hens is greater for tibia structural integrity than for egg production alone.

The report of Keshavarz (2000b) illustrates the complex nature of the hens' response to dietary nPP. Reduced nPP diets of 119, 144, 199, 245, 295, and 351 mg/d were fed to White Leghorn hens from 30 to 66 weeks of age. There were no differences in performance or tibia ash measurements after 12 weeks in the 119, 144, and 199 mg/day fed hens for egg production and in the 144 and 199 mg/day fed hens for tibia ash. The complexity of the issue is highlighted by the finding that hens fed 119 mg of nPP/day had higher tibia ash contents than hens fed 144 mg of nPP/day. This was most likely due to the large decrease in egg production, which reduced the hens' need for dietary Ca for eggshell formation and allowed the dietary Ca and nPP to be diverted to replenish tibia supplies of these minerals, thereby increasing tibia ash over time. The study by Keshavarz (2000b) highlights the need for a holistic approach to understanding the effects of dietary nPP on hen performance and health.

Although consideration of the effect of dietary nPP levels on tibia ash has resulted in an increased recommended dietary requirement for nPP in the present report, there are several experiments that have shown a negative effect of high dietary nPP levels on eggshell quality. Keshavarz (2003b) fed four strains of White Leghorn hens either 191, 240, or 437 mg nPP/day and observed that specific gravity of eggs from hens fed the highest level was decreased. Sohail and Roland

(2002) fed white egg layers 13 diets that provided between 77 and 588 mg of nPP/day and observed a linear decrease in egg specific gravity with increasing dietary nPP levels. In an experiment when hens were exposed to heat stress, the heat stress did not affect eggshell breaking strength or interact with dietary nPP, but increasing dietary nPP intakes from 304 to 472 to 620 mg/day resulted in progressively reduced eggshell breaking strength (Usayran and Balnave, 1995).

A more recently published review of minimum nPP levels in laying hens was reported by Rodehutscord et al. (2023), which summarized that date supported a minimum requirement of only 2.2 g nPP/kg of feed for laying hens.

Trace Minerals and Electrolytes

No new research on pullet requirements for trace minerals or electrolytes that meets the criteria set forth for inclusion in this report has been published since the last report (NRC, 1994). Therefore, the trace mineral and electrolyte requirements of pullets were not changed from those provided in the previous report (NRC, 1994).

Kuchiniski et al. (1999) examined the sodium (Na) requirements of a commercial strain of white egg laying hens from 28 to 46 weeks of age and determined that the requirements for maximal egg production and egg solids were 113 mg/day (0.12% of the diet) and 102 mg/day (0.11% of the diet), respectively. In a second experiment using 23- to 48-week-old white egg layers, they determined that a daily intake of 99 mg of Na (0.10% of the diet) was necessary to maximize egg production. These levels of dietary Na were also sufficient to maintain maximal fertility and hatchability. No other research on trace minerals or electrolyte requirements of laying hens that met the inclusion criteria was published since the last report (NRC, 1994), which recommended 0.15 percent sodium for hens consuming 100 g of feed per day, which is equivalent to 150 mg/day. The NRC (1994) recommendation was based on a study by Reid (1977), who reported that Single Comb White Leghorn hens had a Na requirement of 0.126 percent of the diet (141 mg/day) in order to maximize egg production during the early phase of lay. Based on these three experiments, the recommended Na requirement of white egg laying hens has been reduced to 0.12 percent (130 mg/day) during the early phase of egg production (Table 13-3), whereas the requirement for brown egg layers remains at 165 mg/day (Table 13-4).

Vitamins

Due to a paucity of both pullet and laying hen studies meeting the criteria for inclusion, few changes from the previous report (NRC, 1994) have been made to the vitamin requirements of either pullets or laying hens. Pullet nutrient requirements are estimated based on current growth rates and mature body weight goals for White Leghorn and Brown Leghorn strains, with the recognition that Brown Leghorn strains have a more robust growth curve than White Leghorns. The recommended nutrient requirement tables provided for white- and brown-egg strain pullets and laying hens with the realization that other breeds, such as the more robust Rhode Island Red, would have different nutrient needs than the smaller commercial-strain chickens. The nutrient requirements of larger dual-purpose breeds may actually be closer to those of the broiler chickens as described in Chapter 14. Despite this lack of work on daily vitamin requirements per se, several studies have reported benefits of added vitamin E during heat stress. Puthpongsiriporn et al. (2001) and Kirunda et al. (2001) both reported benefits of supplementing 50 to 60 IU of vitamin E/kg diet to laying hens on egg production, egg weight, and egg quality measurements, such as vitelline membrane strength, during heat stress. Because of its antioxidant properties, Scheideler and Froning (1996) also reported benefits of supplementing 50 IU vitamin E/kg diet to flax-based diets fed to increase omega 3 fatty acids in the egg. Flax is rich in linolenic acid (18:3 n-3). Puthpongsiriporn et al. (2001) also reported limited benefits of dietary vitamin C supplementation (1,000 ppm provided in the drinking water) on immune function in heat-stressed laying hens.

REFERENCES

Alves, W.J., E.B. Malheiros, N.K. Sakomura, E.P. da Silva, G. da Silva Viana, M. de Paula Reis, C.A. Gonçalves, and R.M. Suzuki. 2018. In vivo description of body growth and chemical components of egg-laying pullets. *Livestock Science*. 220:221–229.

An, S.H., D.W. Kim, and B.K. An. 2016. Effects of dietary calcium levels on productive performance, eggshell quality and overall calcium status in aged laying hen. *Asian-Australasian Journal of Animal Science*. 29(10):1477–1482.

Bar, A., V. Razaphkovsky, and E. Vax. 2002. Re-evaluation of calcium and phosphorus requirements in aged laying hens. *British Poultry Science*. 43(2):261–269.

Barbosa, B.A.C., P.R. Soares, H.S. Rostagno, M. de Almeida Silva, L.F.T. Albino, and A.S. das Graças. 1999. Nutritional requirements of methionine + cystine by white-egg and brown-egg layer hens in the second production cycle: 1. productive traits. *Revista Brasileira de Zootecnia*. 28:526–533.

Bell, D.D. 2002. Modern breeds of chickens. In *Commercial Chicken Meat and Egg Production*, 5th ed. D.D. Bell and W.D. Weaver, editors. Dordrecht, The Netherlands: Kluwer Academic Publishers.

Bendezu, H.P., N.K. Sakomura, E.B. Malheiros, J.C. De Paula Dorigam, E.P. Da Silva, and M. De Paula Reis. 2018. Estimation of desired feed intake for growth and reproductive organ development in pre-laying hens. *Animal Production Science*. 59(7):1228–1234.

Bregendahl, K., S.A. Roberts, B. Kerr, and D. Hoehler. 2008. Ideal ratios of isoleucine, methionine, methionine plus cysteine, threonine, tryptophan, and valine relative to lysine for White Leghorn type laying hens of twenty-eight to thirty-four weeks of age. *Poultry Science*. 87:744–758.

Brumano, G., P.C. Gomes, J.L. Donzele, H.S. Rostagno, T.C. da Rocha, and R.L. de Almeida. 2010a. Levels of methionine + digestible cystine in light laying diets in the 24 to 40 week period. *Revista Brasileira de Zootecnia*. 39(6):1228–1236.

Brumano, G., P. Gomes, J.L. Donzele, H.S. Rostagno, T.C. da Rocha, H. H. de Carvalho Mello. 2010b. Methionine + digestible cystine levels for light laying hens from 42 to 58 weeks of age. *Revista Brasileira de Zootecnia*. 39(9):1984–1992.

Cardoso, A.S., F.G.P. Costa, M.R. de Lima, E.T. Nogueira, C.S. Santos, R.B. de Sousa, R.C. Lima, and D.V.G. Vieira. 2014. Nutritional requirement of digestible threonine for white egg layers of 60 to 76 weeks of age. *Journal of Applied Poultry Research*. 23:724–728.

Castillo, C., M. Cuca, A. Pro, M. González, and E. Morales. 2004. Biological and economic optimum level of calcium in White Leghorn laying hens. *Poultry Science*. 83:868–872.

Chambers, J.R., K. Zaheer, H. Akhtar, and E.S.M. Adbel-Aal. 2017. Chicken eggs. In *Egg Innovations and Strategies for Improvements*. P.Y. Hester, editor. Amsterdam, The Netherlands: Elsevier. 3–11.

Comar, C.L., and J.C. Driggers. 1949. Secretion of radioactive calcium in the hen's egg. *Science*. 109:282.

Cupertino, E.S., P.C. Gomes, H.S. Rostagno, J.L. Donzele, M. Schmidt, and H.H. de C. Mello. 2009. Nutritional requirement of digestible methionine + cystine for hens from 54 to 70 weeks of age. *Revista Brasileira de Zootecnia*. 38(7):1238–1246.

CVB (Centraal Veevoederbureau). 1996. Aminozurenbehoefte van leghennen en vleeskuikens [Amino acid requirements for laying hens and broiler chickens]. Documentation Report nr. 18 (in Dutch), Centraal Veevoederbureau, Lelystad, The Netherlands.

de Juan, A.F., R. Scappaticcio, L. Aguirre, G. Fondevila, J. Garcia, L. Camara, and G.G. Mateos. 2023. Influence of the calcium and nutrient content of the prelay diet on egg production, egg quality, and tibiae mineralization of brown egg-laying hens from 16–63 wk of age. *Poultry Science*. 102:102491.

Desbruslais, A., A. Wealleans, D. Gonzalez-Sanchez, and M. di Benedetto. 2021. Dietary fibre in laying hens: a review of effects on performance, gut health and feather pecking. *World's Poultry Science Journal*. 77(4): 797–823.

Donalson, L.M., W.K. Kim, C.L. Woodward, P. Herrera, L.F. Kubena, D.J. Nisbet, and S.C. Ricke. 2005. *Poultry Science*. 84:362–369.

Egg Industry Center. 2025. Industry analysis. https://www.eggindustrycenter.org/industry-analysis/categories/15de6a876f6348ef8fe28e9ed5827163.

Eusebio-Balcazar, P.E., S. Purdum, K. Hanford, and M.M. Beck. 2018. Limestone particle size fed to pullets influences subsequent bone integrity of hens. *Poultry Science*. 9:1471–1483.

FAOSTAT (Food and Agriculture Organization of the United Nations). 2020. FAOSTAT database. http://faostat.fao.org/site/291/default.aspx.

Faria, D.E., R.H. Harms, R.S. Antar, and G.B. Russell. 2003. Re-evaluation of the lysine requirement of the commercial laying hen in a corn-soybean meal diet. *Journal of Applied Animal Research*. 23(2):161–174.

Faria, D.E., R.H. Harms, and G.B. Russell. 2002. Threonine requirement of commercial laying hens fed a corn-soybean meal diet. *Poultry Science*. 81:809–814.

Figueiredo, G.O., A.G. Bertechini, E.J. Fassani, P.B. Rodrigues, J.Á.G. Brito, and S.F. Castro. 2012. Performance and egg quality of laying hens fed with dietary levels of digestible lysine and threonine. *Arquivo Brasileiro de Medicina Veterinária e Zootecnia*. 64(3):743–750.

Fisher, C., T.R. Morris, and R.C. Jennings. 1973. A model for the description and prediction of the response of laying hens to amino acid intake. *British Poultry Science*. 14:469–484.

Gomez, R.S., and M.L. Angels. 2016. Requirement of digestible sulfur amino acids in laying hens fed sorghum- and soybean meal-based diets. *Brazilian Journal of Poultry Science*. 18(2):231.

Harms, R.H., and G.B. Russell. 2000. Evaluation of tryptophan requirement of the commercial layer by using a corn-soybean meal basal diet. *Poultry Science*. 79:740–742.

Harms, R.H., and G.B. Russell. 2003. Performance of commercial laying hens fed diets with various levels of methionine. *Journal of Applied Poultry Research*. 12:449–455.

Hauschild, L., C. Pomar, and P.A. Lovatto. 2010. Systematic comparison of the empirical and factorial methods used to estimate the nutrient requirements of growing pigs. *Animal*. 4(5):714–723.

Jardim, F.R.d.M., J.H. Stringhini, M.A. Andrade, M.B. Café, N.S.M. Leandro, and F.B. de Carvalho. 2010. Levels of digestible lysine for Hy-Line W-36 hens in production period. *Revista Brasileira de Zootecnia*. 39(4):787–795.

Jing, M., S. Zhao, A. Rogiewicz, B.A. Slominski, and J.D. House. 2018. Assessment of the minimal available phosphorus needs of pullets during the pre-laying period. *Poultry Science*. 97:557–567.

Kerschnitzki, M., T. Zander, P. Zaslansky, P. Fratzl, R. Shahar, and W. Wagermaier. 2014. Rapid alterations of avian medullary bone material during the daily egg-laying cycle. *Bone*. 69:109–117.

Keshavarz, K. 2000a. Nonphytate phosphorus requirement of laying hens with and without phytase on a phase feeding program. *Poultry Science*. 79:748–763.

Keshavarz, K. 2000b. Reevaluation of nonphytate phosphorus requirement of growing pullets with and without phytase. *Poultry Science*. 79:1143–1153.

Keshavarz, K. 2003a. Effects of continuous feeding of low-phosphorus diets with and without phytase during the growing and laying periods on performance of two strains of Leghorns. *Poultry Science*. 82:1444–1456.

Keshavarz, K. 2003b. The effect of different concentrations of nonphytate phosphorus with and without phytase on the performance of four strains of laying hens. *Poultry Science*. 82:71–91.

Keshavarz, K., and S. Nakajima. 1993. Re-evaluation of calcium and phosphorus requirements of laying hens for optimum performance and eggshell quality. *Poultry Science*. 72:144–153.

Kim, W.K., S.A. Bloomfield, T. Sugiyama, and S.C. Ricke. 2012. Concepts and methods for understanding bone metabolism in laying hens. World's Poultry Science Association. *World's Poultry Science Journal*. 68.

Kirunda, D.F.K., S.E. Scheideler, and S.R. McKee. 2001. The efficacy of vitamin E (DL-α-tocopheryl acetate) supplementation in hen diets to alleviate egg quality deterioration associated with high temperature exposure. *Poultry Science*. 80:1378–1383.

Kuchinski, K.K., R.H. Harms, H.R. Wilson, G.B. Russell, and L.R. McDowell. 1999. Re-evaluation of sodium requirement of commercial laying hen. *Journal of Applied Animal Research*. 15(1):25–34.

Leeson, S., and J.D. Summers. 2005. *Commercial Poultry Production*, 3rd ed. Guelph, Ontario, Canada: University Books.

Lelis, G.R., L.F.T. Albino, F.C. Tavernari, A.A. Calderano, H.S. Rostagno, V.R.S.M. Barros, and R.C. Maia. 2014. Digestible valine-to-digestible lysine ratios in brown commercial layer diets. *Journal of Applied Poultry Research*. 23:683–669.

Lichovnikova, M. 2007. The effect of dietary calcium source, concentration and particle size on calcium retention, eggshell quality and overall calcium requirement in laying hens. *British Poultry Science*. 48(1):71–75.

Martínez-Amezcua, C., J.L. Laparra-Vega, E. Avila-Gonzalez, F. Fuente, T. Jinez, and M.T. Kidd. 1999. Dietary L-threonine responses in laying hens. *Journal of Applied Poultry Research*. 8(2):236–241.

Mejia, L., E.T. Meyer, P.L. Utterback, C.W. Utterback, C.M. Parsons, and K.W. Koelkebeck. 2010. Evaluation of limit feeding corn and distillers dried grains with solubles in non-feed-withdrawal molt programs for laying hens. *Poultry Science*. 89:386–392.

Mueller, S., M. Kreuzer, M. Siegrist, K. Mannale, R.E. Messikommer, and I.D.M. Gangnat. 2018. Carcass and meat quality of dual-purpose chickens (Lohmann Dual, Belgian Malines, Schwiezerhuhn) in comparison to broiler and layer chicken types. *Poultry Science*. 97:3325–3336.

Mueller, W.J., R. Schraer, and H. Schraer. 1964. Calcium metabolism and skeletal dynamics of laying pullets. *Journal of Nutrition*. 84:20–26.

Nasr, M.A.F., J. Murrell, and C.J. Nicol. 2013. The effect of keel fractures on egg production, feed and water consumption in individual laying hens. *British Poultry Science*. 54:165–170.

Novak, C., H. Yakout, and S. Scheideler. 2004. The combined effects of dietary lysine and total sulfur amino acid level on egg production parameters and egg components in Dekalb Delta laying hens. *Poultry Science*. 83:97–984.

NRC (National Research Council). 1994. *Nutrient Requirements of Poultry*, 9th rev. ed. Washington, DC: National Academy Press.

Nunes, R.V., S.E. Schneider, C. Souza, C.P. Sangali, C. Polese, R.S. Bueno, and F.M. Vieites. 2015. Digestible lysine requirement for semi-heavy laying hens 50 to 66 weeks of age. *Arquivo Brasileiro de Medicina Veterinária e Zootecnia*. 67(6):1675–1683.

Nys, Y., J. Gautron, J.M. Garcia-Ruiz, and M.T. Hincke. 2004. Avian eggshell mineralization: Biochemical and functional characterization of matrix proteins. *Comptes Rendus Palevol.* 3:549–562.

Panda, A.K., S.V. Rama Rao, M.V.L.N. Raju, and S.K. Bhanja. 2005. Effects of dietary non-phytate phosphorus levels on egg production, shell quality and nutrient retention in White Leghorn layers. *Asian-Australasian Journal of Animal Science.* 18:1171–1175.

Pastore, S.M., P.C. Gomes, S.L.T. Barreto, G. da Silva, V.E.A. da Silva, R.L. de Almeida, L.V.S. Barbosa, and W.P. de Oliveira. 2015. Nutritional requirement of digestible lysine for white-egg laying hens in production. *Ciência Rural, Santa Maria.* 45(8):1496–1503.

Pastore, S.M., P.C. Gomes, G. Da Silva Viana, E.A. Da Silva, W.P. De Oliveira, L.V.S. Barbosa, A.Z. Fraga, and W.J. Alves. 2018. Standardized ileal digestible lysine requirement of white commercial layers in peak egg production. *Bioscience Journal.* 34(1):186–193.

Peganova, S., F. Hirche, and K. Eder. 2003. Requirement of tryptophan in relation to the supply of large neutral amino acids in the laying hen. *Poultry Science.* 82:815–822.

Pelicia K., E.A. Garcia, A.B.G. Faitarone, A.P. Silva, D.A. Berto, A.B. Molino, and F. Vercese. 2009. Calcium and available phosphorus levels for laying hens in second production cycle. *Revista Brasileira de Ciencia Avicola.* 11(1):39–49.

Peters, J., O. Lebrasseur, E.K. Irving-Pease, and G. Larson. 2022. The biocultural origins and dispersal of domestic chickens. *Proceedings of the National Academy of Sciences.* 119:e2121978119.

Punna, S., and D.A. Roland. 2000. Dietary supplementation of pullets fed varying levels of dietary phosphorus from day 1 through 19 wk. *Journal of Applied Poultry Research.* 9:35–42.

Puthpongsiriporn, U., S.E. Scheideler, J.L. Sell, and M.M. Beck. 2001. Effects of vitamin E and C supplementation on performance, in vitro lymphocyte proliferation, and antioxidant status of laying hens during heat stress. *Poultry Science.* 80:1190–1191.

Reid, B.L. 1977. Dietary sodium for laying hens. *Poultry Science.* 56:373–374.

Riek, A., S. Petow, J.R. Speakman, and L. Schrader. 2020. Daily energy expenditure and water turnover in two breeds of laying hens kept in floor housing. *Animal: The International Journal of Animal Biosciences.* 15:100047.

Rocha, T.A., P.C. Gomez, and J.L. Donele. 2009. Digestible lysine levels in feed for 24 to 40 week old laying hens. *Revista Brasileira de Zootecnia.* 38(9):1726–1731.

Rocha, T.C.d., J.L. Donzele, P.C. Gomes, H.H.d. Carvalho Mello, A.R. Troni, and G.d. Silva Viana. 2013. Ideal digestible isoleucine: Digestible lysine ratio in diets for laying hens aged 24–40 weeks. *Revista Brasileira de Zootecnia.* 42(11):780–784.

Rodehutscord, M., V. Sommerfeld, C.R. Angel, and D.R. Korver. 2023. Minimum phosphorus requirements for laying hen feed formulations. *Poultry Science.* 102:102344.

Russell, G.B., and R.H. Harms. 1999. Tryptophan requirement of the commercial hen. *Poultry Science.* 78:1283–1285.

Sá, L. M., P.C. Gomes, L.F.T. Albino, H.S. Rostagno, and C.C. Cardoso Nascif. 2007. Nutritional requirement of digestible methionine + cystine for hens from 34 to 50 weeks of age. *Revista Brasileira de Zootecnia.* 36(6):1837–1845.

Safaa, H.M., M.P. Serrano, D.G. Valencia, M. Frikha, E. Jiménez-Moreno, and G.G. Mateos. 2008. Productive performance and egg quality of brown egg-laying hens in the late phase of production as influenced by level and source of calcium in the diet. *Poultry Science.* 87:2043–2051.

Saki, A.A., R. Naseri Harsini, M.M. Tabatabaei, P. Zamani, and M. Haghight. 2012. Estimates of methionine and sulfur amino acid requirements for laying hens using different models. *Revista Brasileira De Ciencia.* 14(3):209–216.

Sakomura, N.K. 2004. Modeling energy utilization in broiler breeders, laying hens and broilers. *Brazilian Journal of Poultry Science.* 6:1–11.

Sakomura, N.K., E.P. Silva, J.C.P. Burigam, R.M. Gous, and N. St. Pierre. 2015. Modeling amino acid requirements of poultry. *The Journal of Applied Poultry Research.* 24:267–282.

Scappaticcio, R., J. Garcia, G. Fondevila, A.F. de Juan, L. Camara, and G.G. Mateos. 2021. Influence of the energy and digestible lysine contents of the diet on performance and egg qualify traits of brown egg laying hens from 19–59 weeks of age. *Poultry Science.* 100:101211.

Scheideler, S.E., and G. Froning. 1996. The combined influence of dietary flaxseed variety, level, form, and storage conditions on egg production and composition among vitamin E-supplemented hens. *Poultry Science.* 75:1221–1226.

Schmidt, M., P.C. Gomes, H.S. Rostagno, L.F. Teixeira Albino, R.V. Nunes, and G. Brumano. 2009. Nutritional levels of methionine + digestible cystine for low-fat in the second production cycle. *Revista Brasileira de Zootecnia.* 38(10):1962–1968.

Schmidt, M., P.C. Gomes, H.S. Rostagno, L.F.T. Albino, R.V. Nunes, and H.H.d. Carvalho Mello. 2011. Nutritional levels of methionine + digestible cystine for light layers in the second production cycle. *Revista Brasileira de Zootecnia.* 40(1):142–147.

Silva, J.H.V.d., L.F.T. Albino, H.S. Rostagno, P.C. Gomes, and R.F. Euclydes. 2000a. Lysine requirement for replacement birds 0 to 6 weeks old. *Revista Brasileira de Zootecnia.* 29(6):1777–1785.

Silva, J.H.V.d., L.F.T. Albmeino, H.S. Rostagno, P.C. Gomes, and R.F. Euclydes. 2000b. Lysine requirement for replacement birds 7–12 weeks old. *Revista Brasileira de Zootecnia.* 29(6):1786–1794.

Silva, J H.V.d., L.F.T. Albino, H.S. Rostagno, P.C. Gomes, and R.F. Euclydes. 2000c. Lysine requirement for replacement birds from 13 to 20 weeks. *Revista Brasileira de Zootecnia.* 29(6):1795–1802

Silva, M.F.R., D.E.d. Faria, P.W. Rizzoli, A.L. dos Santos, M.I. Sakamoto, and H.R.B.d. Souza. 2010. Performance and egg quality of commercial laying hens fed diets containing different levels of methionine and lysine. *Revista Brasileira de Zootecnia.* 39(10):2246–2252.

Silva, J.H.V. 2014. Response of pullets to digestible lysine intake. *Czech Journal of Animal Science.* 59:208–218.

Silva, E.P., E.B Malheiros, N.K. Sakomura, K.S. Venturini, L. Hauschild, J.C.P. Dorigam, and J.B.K., Fernandes. 2015. Lysine requirements of laying hens. *Livestock Science.* 173:69–77.

Snow, J.L., K.A. Rafacz, P.L. Utterback, C.W. Utterback, R.W. Leeper, and C.M. Parsons. 2005. Hy-Line W-36 and Hy-Line W-98 laying hens respond similarly to dietary phosphorus levels. *Poultry Science.* 84(5): 757–763.

Sohail, S.S., and D.A. Roland Sr. 2002. Influence of dietary phosphorus on performance of Hy-line W36 hens. *Poultry Science.* 81:75–83.

Swiatkiewicz, S., A. Arczewska-Włosek, J. Krawczyk, M. Puchała, and D. Józefiak. 2015. Effects on performance and eggshell quality of particle size of calcium sources in laying hens' diets with different Ca concentrations. *Archives Animal Breeding.* 58:301–307.

Taylor, T.G., and C.G. Dacke. 1984. Calcium metabolism and its regulation. In *Physiology and Biochemistry of the Domestic Fowl.* 5:125–170. London: Academic Press.

Toledo, Y.V., M.C. García, A.P. Martíneza, C.M. Becerril Péreza, J.L. Figueroa Velascoa, and M. Del Rosario Tovar Gómez. 2007. Apparent available phosphorus levels for Leghorn white pullets during starter, growing, and development stages, and the beginning of laying period. *Técnica Pecuaria en México.* 45:161–175.

Usayran, N., and D. Balnave. 1995. Phosphorus requirements of laying hens fed on wheat-based diets. *British Poultry Science.* 36:285–301.

Wen, J., A. Helmbrecht, M.A. Elliot, J. Thomson, and M.E. Persia. 2019. Evaluation of the tryptophan requirement of small-framed first cycle laying hens. *Poultry Science.* 98:1263–1271.

Whitehead, C.C. 2004. Overview of bone biology in the egg-laying hen. *Poultry Science.* 83:193–199.

Yamada, Y. 1988. The contribution of poultry science to society. *World's Poultry Science Journal.* 44:172–178.

Yang, Z., S.P. Rose, H.M. Yang, V. Pirgozliev, and Z.Y. Wang. 2018. Egg production in China. *World's Poultry Science Journal.* 74(3):417–426.

Yuan, C., J.M. Li, Y. Ding, Q. He, H.X. Yan, J.J. Lu, and X.T. Zou. 2015. Estimation of L-arginine requirement for Xinyang Black laying hens from 33 to 45 weeks of age. *Journal of Applied Poultry Research.* 24:463–469.

14

Nutrient Requirements of Broiler Chickens

COMMERCIAL CHICKEN MEAT PRODUCTION: A CHANGING AND GLOBAL INDUSTRY

Domestication of the chicken in Southeast Asia is summarized in Chapter 13, "Nutrient Requirements of Laying Hens," but the history of the modern broiler chicken is less than 100 years old (Siegel, 2014; Peters et al., 2022). Historically, poultry enterprises included raising both male and female chicks until the two sexes could be identified and separated based on secondary sexual characteristics, which was about 8 weeks of age (Griffin and Goddard, 1994). In that model, female chicks were raised to adulthood for egg production while male chicks were raised until about 16 weeks of age for meat production. Methods to sex chicks at day of hatch were developed in Japan and subsequently translated and brought to North America in the early 1930s (Gibbs, 1934). This technique allowed male and female chicks to be separated early in life and facilitated the development of distinct lines of chickens selected for egg and meat production (Griffin and Goddard, 1994). In this context, it is important to define the term "modern broiler" to include a meat-type chicken raised for marketing at an early age of approximately 5 to 7 weeks (Hurnik et al., 1995). Historic and current definitions of broiler chickens have not included body weight estimates due to the annual changes in growth rates and market conditions (Havenstein et al., 2003; Schmidt et al., 2009; Zuidhof et al., 2014). Broiler selection began on a larger and more organized scale in the United States with the "Chicken of Tomorrow" contest held from 1946 to 1948 (Elfick, 2011). Initial selection criteria included growth rate, feed conversion efficiency, and the amount of meat on breasts and drumsticks. As the broiler industry grew and developed, the cost of feed dictated the importance of selection of broilers for feed conversion (Emmerson, 1997). Thus, research interest on nutrient requirements became a complementary goal. It soon became apparent that reproduction and growth were generally inversely correlated; therefore, reproductive traits became additional selection criteria. As the chicken meat market evolved, broiler selection also changed with a strong focus on parts yield, and specifically in the early 1980s, increased breast meat yield was a primary selection target. More recently, primary breeders have additionally applied substantial selection pressure for traits related to sustainability, welfare, and product quality (Katanbaf and Hardiman, 2010; Tallentire et al., 2016; Castro et al., 2023). This sustained genetic selection has resulted in a broiler that has changed dramatically in rate of body weight (BW) gain (BWG), feed conversion, meat yield, and relative organ weights over time, as outlined by experiments that have compared modern broilers to historic benchmarks (Havenstein et al., 2003; Schmidt et al., 2009; Zuidhof et al., 2014). In the context of this chapter, the term *broiler* will be further categorized into early-developing broilers, selected for rapid and efficient BWG earlier in life, and late-developing broilers or higher yield broilers that have been selected for increased breast meat yield at a heavier BW.

The global production of chicken meat totaled 99.079 million metric tonnes on a ready-to-cook basis in 2020 (USDA ERS, 2022). The United States produced 20.4 percent of that total, while other leading chicken meat-producing countries included Brazil, China, and the European Union with each accounting for more than 10 percent of world production. Russia, Mexico, Thailand, Argentina, Turkey, and Japan rounded out the top 10 chicken meat producing countries. Overall, these 10 countries accounted for 78.2 percent of worldwide chicken meat production in 2020. Chicken meat per capita consumption has continued to steadily increase in the United States over the past 10 years (2012 to 2021) while beef, pork and turkey consumption has increased only modestly or even declined (USDA ERS, 2022). Worldwide broiler meat consumption is harder to determine, but worldwide chicken meat production has continued to increase (FAO, 2021). The continued genetic improvement of broiler chickens along with the increased production of broiler meat worldwide highlight the importance of maintaining up-to-date information on nutrient requirements of broiler chickens.

NUTRIENT REQUIREMENT CRITERIA

A comprehensive literature review of nutrient requirements was conducted for each of the essential nutrients for broiler meat production Although several of the nutrient requirements for broiler meat production were not changed from the prior report (NRC, 1994), the committee adapted values to be presented based on BW ranges rather than a general age range as in previous reports, in acknowledgment of the continued annual changes in broiler feed intake and growth rates. The BW ranges selected for this report correspond with traditional feeding phases, including a starter period in which birds weighing 0 to 600 g are included, a grower range for birds weighing from 601 to 1,600 g, a finisher range for birds weighing from 1,601 to 2,800 g, a second finisher range for birds weighing from 2,801 to 3,600 g, and a final finisher range for birds weighing from 3,600 to 4,300 g. The three finisher BW ranges correspond to the current span of final target BWs for market broilers. These final live BWs correspond to approximately 1,800 g for fast food chicken cuts, to 2,700 to 3,200 g for tray pack, rotisserie, and whole birds, and up to 4,300 g for deboned and cutup birds. General feed conversion estimates will not be provided in this document as this number is constantly in flux with genetic selection, dietary nutrient and energy density, and management changes. The most recent and relevant data can be estimated from annual updates of breeder expectations and commercial data. Such data and expectations can be found on primary breeder websites.

Water

Water is crucial to life because it facilitates whole body temperature regulation, lubricates and cushions joints, allows enzymes and other proteins in the body to function, and rids the body of waste through urine and feces. The water requirements of broiler chickens are not only met through feed and "metabolic water" but through supplemental sources, such as wells or cisterns, and delivered to the birds through water lines that stretch the length of the broiler house. Access to water in water lines is provided by devices such as nipple valves whose numbers and spacing are designed to provide ready access to all birds present in the broiler house. Local water sources vary in quality, and so a number of water quality variables such as turbidity, hardness, oxidation-reduction potential, iron, manganese, nitrate, pH, alkalinity, total dissolved solids, toxic compounds, dissolved oxygen, and bacteria are monitored to minimize potential water quality issues (Fairchild and Ritz, 2015; Maharjan et al., 2017). Microbial growth and biofilm generation occurs more readily in low-pressure water lines and can cause performance and health issues, requiring proper water line management and sanitation (Hancock et al., 2007). Water consumption generally increases with bird age and BW; typical daily consumption was estimated at 5.28 g per bird per day of age (i.e., a 5 d old bird would require 5.28 g/d × 5 d = 26.4 g water for that day) in the 1980s, but current estimates are 10.26 g per bird per day of age, reflecting the need for additional water with increased growth rates (Pesti et al., 1985; Hiltz et al., 2021). Perhaps a more consistent method to represent water consumption would be as a ratio to BWG or to feed intake, although these measurements are also subject to variation due to broiler age. Environmental temperature plays a role in water consumption; birds in cold conditions will reduce water intake and birds in hot conditions will increase water intake (Pesti et al., 1985; Fairchild and Ritz, 2015). Daily water consumption on a house basis should increase with age, but daily water intake can vary; if variation lasts for more than a single day, further investigation should be completed to ensure proper waterline management (water line height, water availability, water quality) or a potential disease response (Watkins and Tabler, 2009). Feed form has been associated with water intake and water to feed consumption ratio, and further details can be found in Chapter 9, "Feed Processing."

Amino Acids

Previous publications of the *Nutrient Requirements of Poultry* have summarized empirical amino acid studies based on bird age published in the scientific literature. This chapter presents requirements based on defined body weight ranges. Moreover, requirements are also expressed as digestible amino acid percentages, and on an intake (mg/d) basis to account for discrepancy in feed intake among published research. This should allow estimates of requirements for specific amino acids at defined body weight ranges to be more reflective of the modern poultry strains used for meat production. In addition to reliance on empirical data, a major change from previous *Nutrient Requirements of Poultry* is that a model will be provided to estimate the lysine requirements of broilers. Digestible amino acid requirements of broilers were estimated by utilizing factorial modeling and empirical methods. Chickens do not have a requirement for crude protein per se. However, there should be sufficient crude protein to ensure an adequate amino nitrogen supply for synthesis of nonessential amino acids. Estimated digestible Lys requirements of early-developing male and female broilers, selected for accelerated growth rate from to 5 to 6 weeks of age, were determined using a factorial model (Table 14-1). Broiler growth rates were estimated from 1 to 56 d of age using the Gompertz equation (Gompertz, 1825; Wellock, 2004). Based on these estimated digestible Lys requirements, ideal ratios shown in Table 14-2 were used to determine the requirements of other essential amino acids (Baker et al., 2002). Accurate estimates of feed intake are central to formulation of diets with appropriate amino acid contents because feed intake may be influenced by environmental conditions (Ipek and Sahan, 2006; Steenfeldt et al., 2019; Liu et al., 2020) and bulk capacity of the intestinal

TABLE 14-1 Estimated Digestible Lys Requirements of Broilers Expressed on Intake (mg/d) and In-Feed (% of Diet) Bases[a]

	Male					Female				
Age	BW[b]	ME[c]	FI[c]	SID Lys[a]		BW[b]	ME[c]	FI[c]	SID Lys[a]	
(d)	(g)	(kcal/bird/d)	(g)	(% of diet)	(mg/d)	(g)	(kcal/bird/d)	(g)	(% of diet)	(mg/d)
1	55	40	13	1.27	172	56	37	12	1.29	160
2	67	50	17	1.20	204	68	45	15	1.23	188
3	82	60	20	1.20	241	82	53	17	1.22	218
4	99	70	23	1.19	281	98	62	20	1.21	252
5	118	82	27	1.19	326	116	72	24	1.20	289
6	141	95	31	1.18	375	137	82	27	1.19	328
7	166	110	36	1.17	427	161	94	31	1.18	370
8	195	125	41	1.16	484	187	107	35	1.17	414
9	227	141	47	1.15	544	217	120	40	1.15	461
10	263	159	53	1.14	608	250	134	44	1.14	510
11	303	178	59	1.14	674	286	149	49	1.13	560
12	347	198	66	1.13	744	325	165	55	1.11	612
13	395	219	73	1.12	815	368	181	60	1.10	666
14	447	241	80	1.15	889	414	199	66	1.13	720
15	504	264	84	1.14	964	464	216	69	1.11	774
16	565	287	92	1.13	1,040	517	234	75	1.10	829
17	630	311	100	1.12	1,116	574	253	81	1.08	883
18	700	335	108	1.11	1,192	635	272	87	1.07	936
19	774	360	116	1.09	1,268	699	291	93	1.06	989
20	853	385	124	1.08	1,342	766	310	99	1.04	1,040
21	936	410	132	1.07	1,415	837	329	106	1.03	1,090
22	1,023	435	140	1.06	1,486	911	348	112	1.02	1,138
23	1,114	460	148	1.05	1,555	988	367	118	1.00	1,183
24	1,208	484	156	1.04	1,621	1,068	386	124	0.98	1,227
25	1,307	508	163	1.03	1,684	1,151	404	130	0.97	1,268
26	1,409	531	171	1.02	1,743	1,236	422	136	0.96	1,306
27	1,514	554	178	1.01	1,799	1,324	439	141	0.94	1,341
28	1,622	576	185	1.02	1,851	1,414	456	147	0.95	1,373
29	1,733	597	189	1.00	1,898	1,506	472	149	0.93	1,402
30	1,847	617	195	0.99	1,942	1,600	488	154	0.92	1,428
31	1,963	636	201	0.98	1,981	1,696	502	159	0.91	1,451
32	2,081	655	207	0.97	2,015	1,793	516	163	0.89	1,470
33	2,200	672	213	0.96	2,046	1,891	530	168	0.88	1,487
34	2,322	688	218	0.95	2,071	1,991	542	172	0.87	1,500
35	2,444	703	223	0.94	2,093	2,091	554	175	0.87	1,510
36	2,568	717	225	0.93	2,110	2,192	565	177	0.85	1,518
37	2,692	730	229	0.92	2,123	2,293	575	180	0.84	1,522
38	2,817	742	233	0.91	2,132	2,394	584	183	0.83	1,524
39	2,942	753	236	0.90	2,136	2,496	592	186	0.82	1,523
40	3,068	762	239	0.89	2,137	2,597	600	188	0.80	1,519
41	3,193	771	242	0.88	2,134	2,698	607	190	0.79	1,513
42	3,318	778	244	0.87	2,128	2,798	613	192	0.78	1,505
43	3,442	785	246	0.86	2,119	2,898	618	194	0.77	1,495
44	3,565	790	248	0.85	2,106	2,997	623	196	0.76	1,483
45	3,688	795	250	0.84	2,091	3,096	627	197	0.75	1,469
46	3,809	799	251	0.83	2,072	3,193	630	198	0.74	1,453
47	3,930	802	252	0.82	2,052	3,289	633	199	0.72	1,436
48	4,049	804	251	0.81	2,029	3,384	635	198	0.71	1,417
49	4,166	805	251	0.79	2,004	3,477	636	198	0.70	1,397
50	4,282	806	251	0.78	1,977	3,569	637	199	0.69	1,376
51	4,396	806	251	0.77	1,948	3,660	638	199	0.68	1,354
52	4,508	806	251	0.76	1,917	3,749	638	199	0.67	1,331
53	4,618	805	251	0.75	1,886	3,836	638	199	0.65	1,307
54	4,726	803	250	0.74	1,853	3,922	637	198	0.64	1,283
55	4,832	801	250	0.72	1,819	4,006	636	198	0.63	1,258
56	4,937	798	249	0.71	1,785	4,088	634	198	0.62	1,232

[a] Early-developing broilers are selected for rapid and efficient BW gain early in life.

[b] Standardized ileal digestible Lys values, determined according to the factorial equation $Lys = [(151 \times BPm^{0.73} \times (BP/BPm)) + (0.01 \times FP \times 18)] + [(75 \times BPD + 18 \times FPD)/ELU]$, where Lys is Lys requirement (mg/bird/d); BPm, BP, and BPD are body protein at maturity (kg), current body protein (kg), and body protein deposition (g/d), respectively; FP and FPD are, respectively, feather protein (kg) and feather protein deposition (g/d); and ELU is efficiency of Lys utilization, estimated as 0.77 (Siqueira et al., 2013). The body weight (BW) and the contents of protein and lipid, including feather protein content, are calculated based on Gonçalves et al. (2020). This model was based on data from Cobb 500 genetic stock, and interpretation of these values should be limited to this strain.

[c] Feed intake calculated from ratio between metabolizable energy (ME) requirement (kcal/bird/d) and ME of diet (Mcal/kg of feed). The ME requirement is determined as $[BW^{0.75} \times 113 + 13.5 \times Ld + 12.3 \times Pd]$, where BW is BW (kg), Ld is lipid deposition (g/d) and Pd is protein deposition (g/d). The ME in the diets were assumed to be 3.00, 3.11, 3.16, 3.19, and 3.21 kcal/kg of feed from 1 to 14, 15 to 28, 29 to 35, 36 to 46, and 47 to 56 days of age, respectively.

TABLE 14-2 Minimum Digestible Amino Acid Values as Ratios to Lys for Determination of Broiler Amino Acid Requirement[a]

Amino Acid	Ratio
Methionine + Cyst(e)ine	0.76[b]
Methionine	0.41
Threonine	0.68[c]
Valine	0.77[d]
Isoleucine	0.69[e]
Arginine	1.07
Tryptophan	0.17
Leucine	1.07
Histidine	0.37
Phenylalanine	0.63

[a] When not noted, the basis of ideal amino acid ratios were obtained from Baker et al. (2002) and Rostagno et al. (2011).

[b] From Dozier and Mercier (2013).

[c] From Corzo et al. (2009) and Dozier et al. (2015, 2016).

[d] From Corzo et al. (2007, 2008), Rostagno et al. (2011), Tavernari et al. (2013), and Maynard et al. (2021).

[e] From Rostagno et al. (2011) and Tavernari et al. (2012).

tract (Nascimento et al., 2020; Taylor et al., 2021). The factorial model estimates the amino acid needs necessary to meet maintenance and body protein accumulation, but it is important to keep in mind that feed disappearance is not accounted for in this approach.

Data used to estimate empirical digestible amino acid requirements across various BW ranges of broilers were identified using the following selection criteria: (1) study designs in which the test diet was deficient in the amino acid of interest but adequate in other essential amino acids; (2) a minimum of six treatments to establish at least three points below the purported requirement, one point near the purported requirement, and at least two points beyond the purported requirement to allow application of breakpoint methodology; and (3) a minimum of six replicate pens per treatment. Due to variability in ingredient composition, nutrient contents, dietary analyses, and digestibility ratios within the published literature, the following procedures were implemented when using published data sets adhering to the selection criteria listed previously to estimate the amino acid composition of the test diet from each data set. Total amino acid content of ingredients reported in the swine NRC (2012) publication were used across all data sets unless the total amino acid analyses of the ingredients were provided in the manuscript. The total amino acid content of ingredients was multiplied by their dietary inclusion rate and by a common set of digestible amino acid coefficients of ingredients. The dose titrations were calculated by adding amino acid inclusion to the calculation of the test diet. The total amino acid content of ingredients was multiplied by their dietary inclusion rate and by a common set of digestible amino acid coefficients of ingredients. Amino acid digestibility coefficients used in formulation were averages from a summary of peer-review articles and AMINO DAT 5.0 and were as follows: for Lys grain sorghum, 87.47 percent; corn, 83.58 percent; hard wheat, 78.62 percent; triticale, 85.00 percent; maize gluten meal, 86.08 percent; peanut meal, 76.00 percent; soybean meal, 90.90 percent; soybean meal—dehulled, 89.13; casein, 97.00 percent; fish meal, 83.54 percent; meat and bone meal, 72.99 percent; pet-food grade poultry by-product meal, 72.00 percent; Pro-Plus (animal protein blend), 71.95 percent; and L-Lys, 100.00 percent.

Minimum digestible amino acid requirements of broilers post-hatching until 4,300 g are presented in Table 14-3. Due to the genetic increases in broiler growth rate and meat yield, the committee only considered data published from 2007 to 2021 to be relevant to modern broiler nutrition. This corresponds to a 10-year period before the time that the literature review was initiated for the current National Academies publication (i.e., in 2017). Due to significant knowledge gaps with various amino acids across the BW ranges reported in Table 14-3, digestible amino acid requirements were estimated by multiplying the digestible Lys requirements by digestible amino acid ratios presented in Table 14-2. The optimum digestible Lys requirement was estimated using linear and quadratic broken-line regression (Garcia-Neto and Perri, 2015) on the dose–response diets among 13 data sets from 10 experiments using Ross and Cobb broilers based on the dietary calculations determined from previously described methods. For each data set, the midpoint of the BW range was used to regress digestible Lys requirements on a percentage of diet and intake (mg/d) basis to determine quadratic equations to predict digestible Lys requirements. The midpoint BW ranged from 108 to 4,306 g for the 13 data sets used to estimate digestible Lys requirements. Each digestible Lys requirement was regressed with the midpoint BW to estimate the requirements for the BW range from 108 to 4,306 g in one-g increments. Requirements determined for the separate studies were averaged among response criteria within each data set. Data from the regression equations for digestible Lys requirements expressed as a percentage of the diet and the Lys requirement on an intake basis were averaged from 0 to 600, 601 to 1,600, 1,601 to 2,800, 2,801 to 3,600, and 3,601 to 4,300 g. The quadratic equation developed to estimate the Lys requirement as percent of the diet was $1.265 - 0.000167(BW) + 0.0000000178(BW \times BW)$, and the quadratic equation developed to estimate the Lys requirement on an intake basis was $328.1 + 0.979(BW) - 0.000133(BW \times BW)$. Documentation of the references used to estimate the digestible Lys requirement is presented in Table 14-6.

Vitamins

The minimum requirements, estimated requirements, and guidance levels for vitamins (Table 14-5) are expressed in terms of BW ranges instead of bird age used in the previous report (NRC, 1994). Few studies conducted under conditions that would allow for a confident revision of vitamin requirements were published since completion of the literature review of the

TABLE 14-3 Minimum Digestible Amino Acid Requirements of Broilers Weighing from 0 to 4,300 g[a,b]

	0 to 600 g		601 to 1,600 g		1,601 to 2,800 g		2,801 to 3,600 g		3,601 to 4,300 g	
	SID[c]		SID[c]		SID[c]		SID[c]		SID[c]	
Amino Acid	(%)[d]	(mg/d)[e]	(%)[d]	(mg/d)[e]	(%)[d]	(mg/d)[e]	(%)[d]	(mg/d)[e]	(%)[d]	(mg/d)[e]
Sulfur amino acids	0.91	495	0.83	955	0.74	1,411	0.69	1,621	0.68	1,626
Methionine	0.49	267	0.45	515	0.40	761	0.37	875	0.36	877
Lys	1.20	651	1.09	1,257	0.97	1,857	0.91	2,133	0.89	2,139
Threonine	0.82	443	0.74	855	0.66	1,263	0.62	1,450	0.61	1,455
Valine	0.92	501	0.84	968	0.75	1,430	0.70	1,642	0.69	1,647
Isoleucine	0.83	449	0.75	867	0.67	1,281	0.63	1,472	0.61	1,476
Arginine	1.28	697	1.17	1,345	1.04	1,987	0.97	2,282	0.95	2,289
Tryptophan	0.20	111	0.19	214	0.16	316	0.15	363	0.15	364
Leucine	1.28	697	1.17	1,345	1.04	1,987	0.97	2,282	0.95	2,289
Histidine	0.44	241	0.40	465	0.36	687	0.34	789	0.33	791
Phenylalanine	0.76	410	0.69	792	0.61	1,170	0.57	1,344	0.56	1,348

[a] Lys requirements were based on regression equations developed from 13 data sets published from 2007 to 2017. The 13 data sets were from 10 experiments using male broilers. Seven experiments utilized Ross broilers whereas six originated from Cobb broilers. The other essential amino acid requirements were based on ratios to Lys. Ratios used were 0.76, 0.41, 0.68, 0.77, 0.69, 1.07, 0.17, 1.07, 0.37, and 0.63, respectively, for sulfur amino acids, methionine, threonine, valine, isoleucine, arginine, tryptophan, leucine, histidine, and phenylalanine (Table 14-3). In order to convert these digestible amino acid requirements to a total basis, divide the digestible Lys requirement percentage by 0.89 based on corn and soybean meal as the primary ingredients and using feed-grade methionine, Lys, and threonine in the formulation. The other essential amino acids for a total basis could be estimated using the ratios.

[b] The average metabolizable energy values used in diet formulation with the published research to estimate amino acid requirements within various weight categories were 3,032, 3,125, 3,164, 3,168, and 3,220 kcal/kg from 0 to 600, 601 to 1,600, 1,601 to 2,800, 2,801 to 3,600, and 3,601 to 4,300 g, respectively. A change in metabolizable energy as well as other factors can influence amino acid consumption within the various weight categories.

[c] Standardized ileal amino acid digestibility values were determined using total amino acid concentrations of ingredients from the reviewed papers and using the amino acid composition of each feedstuff as listed in the swine NRC (2012). The SID amino acid digestibility coefficients of each feed ingredient were summarized from peer-reviewed articles in poultry-based manuscripts (Chapter 3). Lys requirements were then determined using a broken-line analysis for each requirement data set reviewed. For each requirement data set reviewed, the midpoint of the BW range was used to regress digestible Lys requirements on a percentage and mg/d basis. Overall, the midpoint BW ranged from 108 to 4,306 grams for the 13 data sets used.

[d] Quadratic equation developed to estimate the Lys requirement as percent of the diet: $1.265 - 0.000167(BW) + 0.0000000178(BW \times BW)$.

[e] Quadratic equation developed to estimate the Lys requirement on an intake basis: $328.1 + 0.979(BW) - 0.000133(BW \times BW)$.

previous report (NRC, 1994) in September 1991. In almost all cases, minimum requirement and estimated requirement values have been carried forward from the previous report. For several of the vitamins, only a few studies were reported in the literature that confirm, or at least do not contradict, the previous estimates but were not designed to determine requirements.

Fat-Soluble Vitamins

Vitamin A

No vitamin A requirement studies published after September 1991 were found that met the inclusion criteria for the current report, so the estimated vitamin A requirement values from the previous report (NRC, 1994) have been retained. Various studies involving pharmacologic levels of vitamin A supplementation (Sklan et al., 1994; Aburto and Britton, 1998) do not provide justification for changing the previous estimated requirements.

Vitamin D_3 (Cholecalciferol)

The vitamin that has received the most attention since the previous report has been vitamin D_3 (cholecalciferol), focusing mainly on the effects of various vitamers of vitamin D_3. Most of those studies conducted since September 1991 tested relatively few levels of supplementation, thus rendering them unsuitable for requirement determination. The vitamers include 25-hydroxycholecalciferol (Fritts and Waldroup, 2003; Ledwaba and Roberson, 2003; Chou et al., 2009; Leyva-Jimenez et al., 2019a; Prokoski et al., 2021), 1α-hydroxycholecalciferol (Han et al., 2017; Kheiri and Landy, 2019; Yang et al., 2020), and 1,25-dihydroxycholecalciferol (Souza et al., 2013; Vieites et al., 2018). Each of these vitamers (25-hydroxycholecalciferol, 1α-hydroxycholecalciferol) possesses vitamin D activity through conversion to the active form in the body of the bird or by the supplementation of the active form itself (1,25-dihydroxycholecalciferol). However, none of the published studies allowed for the determination of a vitamin D_3 requirement from any of these vitamers. Most of the studies using cholecalciferol supplements did not report analyzed vitamin D_3 values for experimental diets and therefore cannot be used to confidently determine the vitamin D_3 requirement of broilers. Of the few studies that have reported analyzed values for vitamin D_3 content of all experimental diets, most have had inadequate numbers of dietary levels of vitamin D_3 above and below the requirement to confidently determine a plateau in response and have resulted in a wide range in proposed vitamin D_3 optima. For BWG up to 600 g, positive responses were

TABLE 14-4 Documentation of Macromineral Requirements of Growing Broilers

Age (wk)	BW Range or Stage of Production	Response Criteria	Basal Diet Level	Requirement % or Other	Notes (Strain, Diet Type, Diet Form, Vitamin Form)	Reference
Vitamin A						
0–37 d		BW Antibody production to *Mycobacterium tuberculosis* or β-casein; T lymphocyte proliferation	Not reported	BW: lowest for 0 IU/kg; no difference among other treatments Antibody: maximum response in 22,178 IU Proliferation: maximum response in 22,178 IU	0, 0.8325,1.50, 3.33, 6.66, and 13.2 µg/g retinol equivalents (0, 2,772, 4,995, 11,089, 22,178, and 43,956 IU/kg)	Sklan et al., 1994
		BW	Not reported; no diet vitamin A measurement reported	Regression: levels above 40,000 reduced BW and G:F	Vitamin A (5,000, 10,000, 20,000, 40,000, 80,000, and 160,000 IU/kg—supplemented levels). Study focused on interaction (interference) of vitamin A with vitamin D (dietary and UV light exposure). Ross × Ross	Aburto and Britton, 1998
Vitamin D						
0–16 d		BWG, G:F, bone traits, rickets incidence	Not analyzed	Experiment 3: 275 ± 126 ICU/kg for 16 d BW, 186 ± 49,153 ICU/kg for gain:feed ratio, 904 ± 91 ICU/kg for rickets, 503 ± 51 ICU/kg for bone ash, 552 ± 144 ICU/kg for plasma Ca Experiment 2: Results consistent with Experiment 3, although highest level was 400 ICU/kg	Male broilers (strain not mentioned) Presence vs. absence of UV light Experiment 2: 0, 50, 100, 200, and 400 ICU/kg Experiment 3: 0, 200, 400, 800, and 1,600 ICU/kg UV light blocked	Edwards et al., 1994
0–21 d		Mean BW, bone density, bone Ca, and bone phosphorus	Diets not analyzed for vitamin D	20 µg/kg diet (800 IU/kg); note that 15 µg/kg treatment excluded from analysis because of mixing error	Cornish giant, Peterson Arbor Acre Corn–soy diet 5, 10, 15, 20, 25, 37.5, or 75 µg/kg diet (200, 400, 600, 800, 1,000, 1,500, or 3,000 IU/kg); 600 IU group excluded from analysis because of mixing error.	Goff et al., 1995
1–16 d		Growth, bone ash	Cholecalciferol-free premix; no diet analysis	BW increased @ 5 µg/kg Linear increase in bone ash up to 20 µg/k diet	High TD line; low TD line cholecalciferol at 0, 5, 10, 20, or 40 µg/kg (0, 200, 400, 800, 1,600 IU/kg).	Mitchell et al., 1997
7–19 d		Performance, bone traits	Cholecalciferol-free premix; no diet analysis	Linear BWG response up to 1,250 in Ca- and P-deficient diet, inflection at 12.5; linear increase in bone ash, no inflection. No toxic effect of highest dose.	New Hampshire × Columbian No diet vitamin D analyses Deficient in Ca, aP (Exp. 1) ANOVA Supplemental cholecalciferol: 0, 3.3, 6.25, 12.5, 25.0, 37.5, or 1,250 µg/kg (0, 132, 250, 500, 1,000, 1,500, or 50,000 IU/kg).	Baker et al., 1998

7–19 d	Performance, bone traits	Cholecalciferol-free premix; no diet analysis	Linear increase in BWG, bone ash to 5 μg/kg. No toxic effect of highest dose. 200 IU/kg	Ross × Hubbard No diet vitamin D analyses Adequate in P Slightly deficient in Ca (Exp. 2) ANOVA Supplemental cholecalciferol: 0, 2.5, 5, 10, 20, 40, 1,250 μg/kg (0, 100, 200, 400, 800, 1,600, 50,000 IU/kg)	Baker et al., 1998
8–19 d	Performance, bone traits	Cholecalciferol-free premix; no diet analysis	Linear increase in BWG, bone ash to 5 μg/kg, further increase in bone ash w/ 1,250 μg/kg. No toxic effect of highest dose. 200 IU/kg for growth; no plateau for bone ash	Ross × Hubbard No diet vitamin D analyses Adequate in P Adequate in Ca (exp. 3) ANOVA Supplemental cholecalciferol: 0, 2.5, 5, 10, 20, 30, 1,250 μg/kg (0, 100, 400, 800, 1,200, 50,000 IU/kg)	Baker et al., 1998
1–16 d	Bone ash		For the three products: 843 ± 85, 911 ± 106, and 986 ± 131 IU/kg of diet as compared with 915 ± 82 IU/kg (Sigma Reference Standard)	Mixed sex Peterson × Hubbard 150, 300, 600, and 1,200 ICU/kg; three different cholecalciferol products.	Kasim and Edwards, 2000
	BW, bone ash	Diets not analyzed	5 to 10 mg/kg (200 to 400 IU) of the D source were sufficient to ensure maximal BW and bone ash	Cobb Varying levels of $CaCO_3$ and ground limestone (trials 1–3) or Ca carbonate (trials 4 and 5) and $CaH_2PO_4{\cdot}2H_2O$ Vitamin D vs. 25-OHD3	Bar et al., 2003
0–21 d	BW, FCR, bone ash, TD severity	Diets analyzed; fairly close to calculated values	> 500 IU/kg for BW regardless of source. 2,000 IU/kg to maximize bone ash (with D3)	125, 250, 500, 1,000, 2,000, or 4,000 IU/kg fed as D3 or 25-OHD3	Fritts and Waldroup, 2003
21–42 d	BW, FCR, bone ash, TD severity	Diets analyzed; fairly close to calculated values	1,000–2,000 IU/kg when D3 fed; no level effect with 25OHD 2,000 IU/kg to maximize bone ash (with D3)	125, 250, 500, 1,000, 2,000, or 4,000 IU/kg fed as D3 or 25-OHD3	Fritts and Waldroup, 2003
1–17 d	BW, G:F, rickets, tibial dyschondroplasia incidence		Supplementation increased BW10 μg/kg increased bone ash and reduced rickets, 70 μg/kg reduced severe TD lesions. 10 μg/kg supplemental 25-OHD3 for bone quality; performance data not shown	Ross × Ross Deficient Ca diet; dietary 25-(OH)D3 was added at0, 10, or 70 μg/kg, no supplemental cholecalciferol. UV light exposure vs filtered light. Experiment 1. No feed vitamin D analysis. ANOVA	Ledwaba and Roberson, 2003
1–17 d	BW, G:F, tibial dyschondroplasia incidence		10 μg/kg supplemental 25-OHD3 for bone quality	Ross × Arbor Acres 25-(OH)D3 was added at 0, 10, 40, or 70 μg/kg to a TD-inducing diet containing 27.5 μg/kg (1,100 IU) added cholecalciferol. Experiment 2. No feed vitamin D analysis. ANOVA	Ledwaba and Roberson, 2003

continued

TABLE 14-4 Continued

Age (wk)	BW Range or Stage of Production	Response Criteria	Basal Diet Level	Requirement % or Other	Notes (Strain, Diet Type, Diet Form, Vitamin Form)	Reference
Vitamin D						
1–20 d		Tibial dyschondroplasia incidence		Linear decrease in TD. Addition of 70 μg/kg of 25-(OH)D3 was needed to significantly reduce TD incidence to 5%. However, the effect of feeding 40 μg/kg of 25-(OH)D3 on TD incidence was not significantly different from 70 μg/kg of 25-(OH)D3. 70 μg/kg needed to reduce severe lesions.	Ross × Ross Marginal Ca deficiency; 25-(OH)D3 was added at 0, 10, 40, or 70 μg/kg to a TD-inducing diet containing 55 μg/kg (2,200) added cholecalciferol. Experiment 3. No feed vitamin D analysis. ANOVA, linear regression.	Ledwaba and Roberson, 2003
1–20 d		Tibial dyschondroplasia incidence		No treatment effects in either experiment	Ross × Ross In experiments 4 and 5, 25-(OH)D3 was added at 0, 18, 36, 54, 72, or 90 μg/kg to a diet marginal in Ca. Not clear if cholecalciferol added. No feed vitamin D analysis. ANOVA, linear regression.	Ledwaba and Roberson, 2003
0–21 d 22–42 d		Various immune parameters	Basal diet not supplemented; Diets analyzed for vitamin D	No treatment effects on most parameters, inconsistent effects on WBC proportions	Commercial broiler strain males. 125, 250, 500, 1,000, 2,000, or 4,000 IU/kg as either vitamin D or 25OHD.	Fritts et al., 2004
0–42 d		BW, biomechanics, bone ash, blood ionized Ca	Diets not analyzed	BW (14 d): 250 μg =greatest (no plateau) Bone: 250 μg =greatest (no plateau) BW (42 d) 20 μg/kg (substantially lower)	Experiment 1: 5, 20, 125, and 250 μg cholecalciferol/kg) at P (10g Ca, 4.5 g avP/kg; 8 g Ca, 3.5 g avP/kg) and vitamin A (2.4 and 4.5 mg retinol/kg). Experiment 2: 5, 20, 125, and 250 μg cholecalciferol/kg) at P (13g Ca, 5 g avP/kg; 8 g Ca, 3.5 g avP/kg)	Whitehead et al., 2004
			Diets not analyzed	125–25 ug/kg (5,000 to 10,000) Higher requirement with lower Ca/aP levels	Conference paper, not a lot of details (may be the basis for the other Whitehead et al., 2004 paper). 2 × 4 factorial 2 concentrations of Ca and available P (10 g Ca, 4.5 g avP/kg; 8 g Ca, 3.5 g avP/kg) and 4 concentrations of vitamin D3 (5, 20, 125, and 250 μg vitamin D3/kg)	Whitehead et al., 2004
0–16 d		Broiler performance, bone characteristics	Not measured; assumptions stated regarding the lack of D_3 in ingredients	BW maximized when 2,125 IU/kg in chick diet when maternal diet contained 2,907 IU/kg (predicted)	Ross × Ross Breeder diets: 0, 125, 250, 500 1,000, 2,000 IU/kg diet (hens receiving 0 IU/kg switched to 4,000 IU/kg @ 36 wk of age Broiler diets: 0, 200, 400, 800, 1,600, 3,200 IU/kg feed. *No effect of maternal diet on broiler response at 27 weeks of breeder age or younger.* Linear increase without plateau with increasing vitamin D in broiler diets (up to 3,200 IU/kg) Vitamin D activity in diet not measured.	Atencio et al., 2005

0–3 wk 3–6 wk 6–7 wk	BW, bone traits	Analyzed vitamin D; good agreement with calculated values	0–21: no effect of D level 22–42: BW of 4,000 IU chicks > 1,000 and 2,000 chicks 43–49: BW of 4,000 IU chicks > 1,000 chicks No effect on tibia ash	Low and high P diets 1,000, 2,000, or 4,000 IU/kg as vitamin D or 25OHD ANOVA	Fritts and Waldroup, 2005
1–21 d 2–42 d	BW, bone traits	Diets not analyzed	BW: 200 IU > 1,800 IU	Two levels of vitamin D 200 IU/kg, 1,800 IU/kg	Lohakare et al., 2005
		Diets not analyzed for vitamin D	In low Ca and P diets BWG: 2,400 increased vs. lower levels FE, bone abnormalities, tibia ash: 3,600 In reference diet (200 IU w/ adequate Ca and aP), often best performance	Cobb 100 200 vs. 1,200, 2,400, and 3,600 ICU/kg supplemental control, two levels of reduced aP Increasing D3 in a Ca and aP reduced diet increased BW and bone ash Adequate Ca and P only fed in a diet containing 200 IU/kg	Rao et al., 2006
2–42 d	BWG, FCR, leg abnormality	Diets not analyzed for D3		Vencob broiler female chicks Two levels of Ca and P deficiency, <u>no adequate control</u> 200, 1,200, 2,400, or 3,600 ICU/kg Cages Not suitable for requirement study	Rao et al., 2006
		Diets not analyzed for D3		*Vencobb* female broiler chicks Two levels of Ca and P deficiency, <u>no adequate control</u> Not suitable for requirement study	Rao et al., 2007
				Not suitable for requirement study	Rao et al., 2008
				D3 vs. 25OHD Not suitable for requirement study	Chou et al., 2009
0–21 d 22–42 d		Analyzed values close to calculated	1,500 IU increased BW at both ages relative to control, highest BW w/ 3,500 Progressive increase in bone ash with increasing D3	Hubbard straight-run 200 (control), 1,500, 2,500, or 3,500 IU/kg No plateau	Khan et al., 2010
				Substitution of 25OHD for D3 Not suitable for requirement study	Michalczuk et al., 2010
				Varying Ca:total P levels 0 vs. 600 IU D3/kg Not suitable for requirement study	Singh, 2010
				Multiple non-vitamin D factors, 90, 125, or 250 μg/kg diet Not suitable for requirement study	Kim et al., 2011
				Comparison of different vitamin D forms at a common level of inclusion Not suitable for requirement study Six levels of 1,25(OH)2D3	Garcia et al., 2013

continued

TABLE 14-4 Continued

Age (wk)	BW Range or Stage of Production	Response Criteria	Basal Diet Level	Requirement % or Other	Notes (Strain, Diet Type, Diet Form, Vitamin Form)	Reference
Vitamin D						
					Positive effect on BW of two lowest levels (vs 0); reduction thereafter Not suitable for requirement study	Souza et al., 2013
1–21 d 22–35 d 36–42 d		BWG, FI, FCR, walking ability	Diets not analyzed	Final BW, carcass composition reduced by 100 μg/kg; no effect on growth overall	Ross 308 5, 50, and 100 μg/kg Two stocking densities ANOVA	Sun et al., 2017
					Various levels of D3 with/without various levels of 25OHD3 Not suitable for requirement study	Gómez-Verduzco et al., 2013
					Basal vitamin D (3,500 IU) plus various levels of 25OHD Not suitable for requirement study	Colet et al., 2015
					Thesis, see published papers	Leyva-Jimenez, 2015
					1-alpha D3 vs. 25OHD3 at various levels Not suitable for requirement study	Han et al., 2017
					Review article—vitamin D metabolites Summary of various studies that report a higher (than 200 IU) requirement.	Świątkiewicz et al., 2017
1–21 d		Growth, bone traits, humoral, and cellular immune responses	Diets not analyzed	5,000 IU increased growth and tibia ash relative to 200 IU	Two levels of vitamin D (200, 5,000 IU) × 2 sources (vitamin D, 25OHD)	Vazquez et al., 2018
10–17 d		Relative bioavailability of beadlet to powdered (as standard)			Basal diet with 0 (NC), 62.5, 125, 250, 500, or 1,000 IU D3/kg of feed added Two sources of vitamin D (powdered, beadlet) Relative bioavailability of beadlet to powdered (as standard) Not suitable for requirement study	Leyva-Jimenez et al., 2018
					1,25-dihydroxycholecalciferol from *Solanum glaucophyllum*	Vieites et al., 2018
					1.25(OH)2D3 as a substitute of vitamin D3 (cholecalciferol)	Alves et al., 2019
					1a-hydroxycholecalciferol (1a-OH-D3) in Ca- and P-deficient diets Not suitable for requirements	Ghasemi et al., 2019
			Diets analyzed		1 alpha-hydroxycholecalciferol (1-α(OH)D3) individually or in combination with different concentrations of cholecalciferol Not suitable for requirements	Kheiri and Landy, 2019

1–21 d	Performance, cocci lesion scores, intestinal morphology	Diets analyzed	Increasing from 5 to 50 μg decreased FCR at 7, 14 and 21 d; increased BWG 7–14 and 14–21 d, BW at 21 d	Partial replacement of D3 with 25OHD3 in coccidiosis-vaccinated broilers 5 = 200 IU, 50 = 2,000 IU, 100 = 4,000 IU, 200 = 8,000 IU/kg 5, 50, 100, 200 μg D3/kg of feed Or 25 + 25, 50 + 50, 75 + 75 or 100 + 100 μg D3 in combination with 25-OH-D3 per kg of feed Not a requirement study.	Leyva-Jimenez et al., 2019a
10–21 d	Bone mineralization, performance	Diets analyzed	BWG:185 IU/kg FE: 116 IU/kg (both of the above had low R2; no requirement confidently estimated) Bone mineralization: 285 IU/kg	Cobb 700, males Ca and NPP of 0.90 and 0.45%, respectively, 0, 50, 100, 200, 400, 800, 1,600, and 3,200 IU D3/kg 9 d maternal depletion, 12 d feeding period One-slope straight broken-line regression model Two-slope broken line Quadratic broken line Results variable depending on model used, higher for bone traits than for growth	Leyva-Jimenez et al., 2019b
	Performance, bone mineralization, cytokines		200 ug/kg increased performance	25 vs. 100 μg/kg as D3 or 25OHD Eimeria challenge (+/–)	Sakkas et al., 2019a
0–10 d 11–24 d 25–38 d	Performance, bone traits	Diets not analyzed	Performance not affected Gait score decreased (improved) with increasing supplementation, transient effects on bone traits (increased with increasing levels)	Ross 308, Ross 708 1,000, 4,000, or 7,000 IU/kg of D3 1,000 IU D3 + 3,000 IU as 25OHD No strain × diet interaction	Sakkas et al., 2019b
1–14 d 15–31 d	Performance, bone traits, phosphate transporters	Diets not analyzed	BWG 3,125 > 200 @ 14 d, no effect at other ages FI same pattern as BWG @ 15–31 and 0–31 d 14 d: toe ash increased @ 3,125 and above, tibia ash increased @ 7,813 IU 31 d: highest w/ 7813 IU	Ross 308 80, 200, 500, 1,250, 3,125, or 7,813 IU/kg	Cho et al., 2020
1–42 d	Performance tibia traits, VDR, Na-P cotransporter Growth, carcass traits, bone strength, mTOR	Stock solution analyzed, but not diets	Growth, bone ash maximized wth 2 to 5 μg/kg	Ross 308 1α-OHD3 0, 1.25, 2.5, 5, and 10 μg/kg in experiment 1 0.625, 1.25, 2.5, 5, 7.5, and 10 μg/kg in experiment 2 2, 2.5, 3, 3.5, 4, 4.5, and 5 μg/kg in experiment 3 D3 + 25OHD3 Not suitable for requirement, 25OHD3 increased breast meat yield, no effect on performance	Yang et al., 2020 Prokoski et al., 2021

continued

TABLE 14-4 Continued

Age (wk)	BW Range or Stage of Production	Response Criteria	Basal Diet Level	Requirement % or Other	Notes (Strain, Diet Type, Diet Form, Vitamin Form)	Reference
Vitamin E						
0–6 d		Antibodies to naturally occurring *E. coli*, NDV vaccine, turkey pox vaccine	"Determination of vitamin E in the experimental diets confirmed the calculated values (data not shown)"		0, 50, and 150 mg/kg added vitamin E (DL-a-tocopheryl acetate). Basal diet contained no added vitamin E. No effect on BW gain, FCR Significantly higher responses were measured in chicks that received 0 and 10 mg/kg added vitamin E, whereas in chicks receiving 30 and 150 mg/kg, antibody production was significantly lower	Friedman et al., 1998
					Two levels of vitamin E: 10 IU and 10,000 IU/kg Not suitable for requirement study	Aburto and Britton, 1998
		Humoral, cell-mediated immune response, innate immune response	Basal diet 10.2 IU/kg Diet levels confirmed by HPLC	25–50 IU most immunomodulatory Difficult to set a requirement based on immune responses	Male broiler chicks 0, 10, 17.5, 25, 37.5, 50, 100, and 200 IU/kg added	Leshchinsky and Klasing, 2001
		BW, pathology following *Eimeria* infection	Basal diets analyzed (13.2 IU)/kg	BW maximum at 39.8 IU/kg High levels did not protect against pathology from coccidiosis	13.2, and calculated levels 25, 50, 100, and 200 mg/kg	Allen and Fetterer, 2002
		BWG, FCR, IgA antibody titres	Basal diets not analyzed	No effect on performance 250 mg increased % TH cells and Ia+ cells	Male Cobb broiler chicks 25, 250, 2,500, or 5,000 mg dl-a-tocopherol acetate/kg	Muir et al., 2002
		Carcass composition			All-rac-α-tocopheryl acetate (70 and 140 mg/kg); two levels of fish oil, two levels of Zn supplementation Not suitable for requirement study	Bou et al., 2004
		Phagocytic activity			16 (control), 110, or 220 mg of VE/kg of feed Not suitable for requirement study	Konjufca et al., 2004
		Fat absorption, tocopherol incorporation	No effect on performance		0, 100, 200, and 400 mg/kg Variable degree of dietary unsaturation	Villaverde et al., 2004
		BW, meat oxidation products	No effect on performance		Female broilers 20, 40, or 200 mg/kg for 35 d All levels above NRC 1994 requirements	Eder et al., 2005
0–21 d		BW,	No effect on performance		0, 50, 75 IU/kg Interaction with vitamin C	Nameghi et al., 2007
22–42 d		BW, antibody production	Increase BWG with 75 IU/kg		0, 50, 75 IU/kg Interaction with vitamin C	Nameghi et al., 2007
		BW, antibody production, H:L ratio	BW, antibody max with 65 mg/kg		Male Ross 308 30, 65, or 100 mg/kg of vitamin E Birds vaccinated or not against coccidiosis	da Silva et al., 2009

1–42 d	Performance, cell mediated, humoral immunity		0, 100, or 200 mg/kg Thermoneutral or heat stress No effect on BWG, FI Higher FCR @ 100 IU	Niu et al., 2009
	Antibody response		40, 80, or 200 mg/kg Various levels of arginine	Ruiz-Feria and Abdukalykova, 2009
	Performance, immunity	No effect on performance	30 mg/kg of all-rac-α-tocopherol acetate vs. RRR-α-tocopherol succinate at 10, 30, and 50 mg/kg	Zhang et al., 2009
21–42 d	Meat composition, consumer acceptance		100, 150, or 200 mg LD-α-tocopherol acetate kg Various levels of fish oil	Rymer and Givens, 2010
	Cell-mediated immunity vs. coccidiosis vaccine		30, 65, and 100 mg/kg	da Silva et al., 2011
0–21 d	Growth, leg meat enrichment, antioxidant	BW—no effect at 21 d; max BW at 35 d with 50 IU/kg	Ross 308	Englmaierova et al., 2011
22–35 d	Oxidative stress, oxidative stability		0, 50, and 100 mg/kg Two levels of lycopene 0 vs. 85 or 200 IU of vitamin E as all-rac-α-tocopherol and 85 IU as RRR-α-tocopherol	Voljč et al., 2011
0–56 d	Performance, antibody response	250, 500 greatest BWG at 4 and 8 weeks of age	250 or 500 mg/kg VE with or without 250 or 500 mg/kg vitamin C	Beura et al., 2012
0–49 d	Performance, antibody responses, H:L ratios	No effect on growth; FCR reduced with 125 mg/kg	0, 125 and 250 mg/kg Three levels of Se TN or heat stress	Habibian et al., 2014
	Fatty acids, antioxidants, immune function	Performance not measured	10 vs. 200 IU/kg, various antioxidants and pro-oxidant conditions	Lu et al., 2014
		No effect on performance	0, 150, and 300 mg/kg Three levels of Zn	Kakhki et al., 2016
	Performance, meat quality and enrichment, oxidation	No effect on performance	0, 150, and 300 mg/kg Three levels of Zn	Kakhki et al., 2017
	Meat quality, antioxidant, enrichment	No performance data		Niu et al., 2017

continued

TABLE 14-4 Continued

Age (wk)	BW Range or Stage of Production	Response Criteria	Basal Diet Level	Requirement % or Other	Notes (Strain, Diet Type, Diet Form, Vitamin Form)	Reference
Vitamin E						
					Meta-analysis "Growth performance is not linked to the supplementation level of vitamin E" "Some studies used a basal level of vitamin E, which can be considered to be adequate or to marginally exceed the minimum requirements of broiler chickens as recommended by NRC (1994)." "There is no evidence of dietary vitamin E supplementation benefits for the growth performance of male broiler chickens, indicating that the basal vitamin levels contained in the principal dietetic ingredients such as corn and soybean meal may be sufficient to ensure performance"	Pompeu et al., 2018
		Growth performance, meat lipid oxidation, gut pro- and anti-inflammatory gut genes	No effects on growth performance		0, 33, 65, or 100 IU/kg as water-miscible vitamin E (WVE) or lipid-soluble vitamin E (LVE)	Pitargue et al., 2019
			No effect on performance vs. comparable control		0, 50, 100, or 150 mg/kg supplemental vitamin E	Khatun et al., 2020
Vitamin K						
0–40 d		BW, BWG, feed intake FCR; bone quality	1.6 mg/kg (calculated)	<1.6 mg/kg	Cobb-Vantress 11.6, 21.6, 41.6, 61.6, and 81.6 mg/kg diet	Duarte et al., 2014
0–21 d	BW = 1,008 g (control males), 926 g control females	Gain, feed intake, bone characteristics	0.02 mg/kg diet (analyzed)		Arbor Acres male and female—sex-separate	Guo et al., 2020
22–42 d	BW = 2,818 g (control males), 2,436 g control females				0, 0.5, and 4.0 mg/kg menadione (experimental diets not analyzed). Increase in several traits going from 0 to 0.5 mg/kg supplemental, no further increase	

Age (wk)	BW Range or Stage of Production	Response Criteria	Basal Diet Level	Requirement % or Other	Notes (Strain, Diet Type, Diet Form, Vitamin Form)	Reference
Thiamin						
1–21 d 22–40 d		BWG, FI, mortality, tissue thiamin	1.8 mg/kg diet, measured	≤ 1.8 mg/kg	Added thiamine to basal diet at 0, 2, 8, or 32 mg/kg diet No response in BWG, FI, FCR, mortality Suggested a requirement of 10 to 12 mg/kg thiamin to saturate thiamin in blood, liver, and brain Heart may have a preferential requirement for higher levels of thiamin. Strain: not specified.	Olkowski and Classen, 1996
0–21 d	~750 g BW	BWG, G:F	Not reported	< 8 mg/kg; 32 mg/kg decreased BWG vs. 0, 2, 8 mg/kg	Hubbard breeders supplemented thaimine at 0, 2, 8, and 32 mg/kg Maternal diet had no effect on broiler growth. From each maternal treatment, broilers supplemented thaimine at 0, 2, 8, and 32 mg/kg	Olkowski and Classen, 1999
0–21 d	787 to 805 g BWG	FI, BWG, FCR	Not reported Analyzed values 1.63 to 2.0 mg thiamin/kg feed for experimental diets	No effect on performance	No deficient control group. Different forms of thiamin added to a deficient (not analyzed) purified basal diet. A thiamin-deficient diet was supplemented with either 1.8 or 1.5 mg/kg thiamin equivalent as water-soluble salts, or with 1.5 and 1.2 mg/kg thiamin equivalent as benfotiamin, respectively, and fed to three replicate groups/treatment for 21 days No dose response design; can't determine whether requirements met or not No effect on performance Strain: Sena-hybrid broilers	Geyer et al., 2000
Riboflavin						
0–5 d	1,750 g	BWG, FI, FCR	0.65 mg/kg	3 mg/kg (supplemental)	Ross 0, 1, 2, 3, 4, and 8 mg/kg feed (supplemental) In German, summary in English Not suitable for requirement determination; consistent with NRC 1994.	Roth-Maier and Kirchgessner, 1997
0–21 d		Acquired tomaculous neuropathy	Not analyzed	Deficient vs. sufficient	Cobb 500 Riboflavin-deficient diet (1.8 mg/kg riboflavin) or a conventional diet (5.0 mg/kg riboflavin).	Cai et al., 2006a
0–31 d		Fibroblastic onion bulb-like structures in peripheral nerves	Not analyzed	Deficient vs. sufficient	Cobb 500 Riboflavin-deficient diet (1.8 mg/kg riboflavin) or a conventional diet (5.0 mg/kg riboflavin)	Cai et al., 2006b

continued

TABLE 14-4 Continued

Age (wk)	BW Range or Stage of Production	Response Criteria	Basal Diet Level	Requirement % or Other	Notes (Strain, Diet Type, Diet Form, Vitamin Form)	Reference
Riboflavin						
0–36 d		Intestinal BW, tract BW, cocci infection	Not analyzed	No effect of supplementation on performance	Ross 708. 0.75, 6.6, and 20 ppm/kg supplemented on basal diet ANOVA. Treatments were arranged in a 3 (riboflavin) × 2 (*B. subtilis*) × 2 (coccidial challenge) factorial arrangement in a randomized complete block design	Poudel et al., 2021
Niacin						
8 to ~22 d				> 12 mg/kg nicotinamide (supplemental) for maximum growth	New Hampshire × Columbian Efficacy of various sources of niacin for chicks Increasing levels of tryptophan (0, 100, 200, 300, 400, and 500 mg/kg), nicotinic acid (0, 5, and 10 mg/kg), nicotinamide (0, 6, 12, and 120 mg/kg of added nicotinamide) (separate experiments)	Oduho and Baker, 1993
0–2 wk		BWG, FI	24.5 mg/kg—not clear if this is analyzed or calculated	> 9 kg/kg supplemental; > 33.5 mg/kg total	Arbor Acres Efficacy of tryptophan conversion to niacin in chicks and ducks 0, 3, 6, 9 mg/kg supplemental niacin At 9 mg/kg, some chicks still showed niacin deficiency signs	Chen, 1996
1–42 d	One dietary phase	Leg deformities, dermatitis, growth, FI, carcass yield	14.8 mg/kg (analyzed)	44.8 mg/kg total for growth	Broiler chicks (strain not identified) Guinea corn–palm kernelmeal based diet supplemented with 0.0, 7.5, 15.0, 22.5, 30.0, 37.5, and 45.0 mg niacin/kg feed 15.0 mg/kg supplemental to prevent dermatitis, 22.5 mg/kg to prevent leg deformities 30 mg/kg for growth, carcass BW, total edible meat, total bone BW, and abdominal fat expressed as percent of carcass BW. ~2.7 kg @ 8 wk of age (Arbor Acres)—a bit small	Oloyo, 1997
0–42 d		Gain, FI, feed intake, apparent utilization of nitrogen, ME, Ca, and P and carcass characteristics, incidence of dermatitis and leg deformities	15 mg/kg (analyzed); 0.25% tryptophan (calculated); complete diets not analyzed	37.5 mg/kg niacin (total) for growth; 30.0 mg/kg niacin (total) to prevent dermatitis and support bone growth	Commercial broiler chicks (strain not specified) Supplemented up to 15.0, 22.5, 30.0, 37.5, 45.0, 52.5, and 60.0 mg of niacin/kg of feed Broilers were pretty small at 42 d of age	Oloyo and Ogunmodede, 2000

8–21 d			4 mg/kg (calculated); tryptophan-adequate diet (0.16%); diet not analyzed	19.5 mg bioavailable niacin/kg diet	New Hampshire × Columbian Plymouth Rock Supplemental niacin in a Trp-adequate diet (0, 4, 8, 12, 16, 20 mg/kg) The requirement for supplemental niacin was estimated by fitting the data to a single-slope broken-line least-squares model	Augspurger and Baker, 2007
0–4 d 5–8 d 9–16 d		BWG, FI, FE, carcass characteristics, meat quality	26.17 mg/kg 25.93 24.02 Analyzed values Tryptophan appeared adequate (0.25, 0.22, 0.19% of the diet)		Arbor Acres (0–8 wk) Beijing-You (0–16 wk) low IM fat Beijing-You (0–16 wk) high IM fat Supplemented with 0, 30, 60, and 120 mg NA/kg Ignore Bejing-You results; compared at 16 wk vs. 8 wk for Arbor Acres ADFI, BW, AFW, PAF, and SFS up to 60 mg NA/kg and then decreased for chickens fed the diet supplemented with 120 mg NA/kg Arbor Acres ~2.5 kg at 8 wk	Jiang et al., 2011
0–4 d 5–8 d 9–16 d		BW, ADFI Fat metabolism	26.17 mg/kg 25.93 24.02 Analyzed values Tryptophan appeared adequate (0.25, 0.22, 0.19% of the diet)	81–84 mg NA/kg diet.	Arbor Acres (0–8 wk) female Beijing-You (0–16 wk) female 0, 30, 60, and 120 mg nicotinic acid /kg supplemental For both breeds, supplemental NA increased (quadratic, $p < 0.01$) ADFI, BW, AFW, PAF, and SFS up to 60 mg NA/kg and then decreased for chickens fed the diet supplemented with 120 mg NA/kg Ignore Bejing-You results; compared at 16 wk vs. 8 wk for Arbor Acres Arbor Acres ~2.7 kg at 8 wk	Jiang et al., 2014
Pantothenic Acid						
0–28 d	Starter	BWG, FCR	5.5 mg/kg in basal (analyzed); complete diets not analyzed	10 mg/kg is adequate	Vantress × Arbor Acres 10, 20, 40 mg/kg no response to increasing levels, unclear whether levels were final levels or supplemental amounts added to basal diet	Deyhim et al., 1992
29–56 d	Grower	BWG, FCR	5.0 mg/kg in basal (analyzed); complete diets not analyzed	10 mg/kg is adequate	Vantress × Arbor Acres 10, 20, 40 mg/kg no response to increasing levels, unclear whether levels were final levels or supplemental amounts added to basal diet	Deyhim et al., 1992
0–21 d		BWG, FI, FCR	4.74 mg/kg (based on analyzed ingredient composition)	4.74 mg/kg in basal diet, no response to supplementation	Arbor Acres—mixed sex 0, 0.3, 0.6, 2.4, 4.8, or 14.4 mg of supplemental pantothenic acid/kg of diet	Harms and Nelson, 1992

continued

TABLE 14-4 Continued

Age (wk)	BW Range or Stage of Production	Response Criteria	Basal Diet Level	Requirement % or Other	Notes (Strain, Diet Type, Diet Form, Vitamin Form)	Reference
Pyridoxine						
8–20 d		BW gain	Cited a previous paper stating the basal diet was devoid of pyridoxine	0.73 mg/kg (when diet was adequate in methionine 0.2 g/kg) 1.05 mg/kg (when diet had excess methionine 12 g/kg)	Chicks crosses of New Hampshire males and Columbian females Semi-purified diet, with 0.2, 0.4, 0.6, 0.8, 1.0, 1.2, and 1.4 mg/kg pyridoxine (note—all levels below NRC 1994 recommendation) Linear and quadratic responses, one-slope broken line model	Scherer and Baker, 2000
Folacin						
0 to 18 d	Adequate diet BWG=537 g	BW, FE	0.465%	N/A	Experiment 1: 0.465% (analyzed) vs. 0.665% (not analyzed) Addition of 2 mg/kg feed folic acid increased BWG and FE to 18 d Two levels only Folic acid levels not measured in test diets Basal diet lacked vitamin B_{12} Strain: not stated Low levels of choline and methionine can make a corn–soy started diet deficient in folate	Pesti et al., 1991
0 to 18 d	Methionine trt BWG=431 g	BW, FE, bone traits, blood hemoglobin, liver BW, liver lipids, PCV	0.465%	N/A	Experiment 2: 0.465% (analyzed) vs. 0.565% (not analyzed) Addition of 1 mg/kg feed folic acid increased growth but not FE to 18 d vs. basal Addition of 1,000 mg/kg choline increased BWG and FE relative to basal, increased FE relative to 1 mg/kg feed folic acid Addition of 0.25% additional DL-methionine (0.38% in basal) greater BWG response than folic acid, greater FE response than folic acid or choline Similar pattern of responses to supplementation for bone traits and blood hemoglobin Strain: not stated	Pesti et al., 1991
0 to 18 d		BW, FE	0.465%	N/A	Experiment 4: 0.465% (analyzed) vs. 0.665% (not analyzed) Addition of 2 mg/kg of feed folic acid increased BWG, FE relative to basal. Addition of (200 mg/kg DL-methionine + 1000 mg/kg choline further increased BWG and FE No additional response in BWG or FE to (2 mg folic acid + 200 mg DL-methionine + 1000 mg choline)/kg feed Strain: not stated	Pesti et al., 1991

0–14 d	Choline + methionine trt BW = 383 g	BW, FE, blood hemoglobin, liver BW, PCV	0.465%	N/A	Experiment 3: 0.465% (analyzed) vs. 0.665% (not analyzed) Addition of 2 mg/kg feed folic acid increased BWG relative to basal, but not FE Addition of (1,000 mg choline + 0.2% DL-methionine) to basal increased BWG and FE relative to folic acid diet Strain: not stated	Pesti et al., 1991
0–42 d		BW, FE, blood hemoglobin, liver BW, PCV	0.465%	N/A	Experiment 3: 0.465% (analyzed) vs. 0.665% (not analyzed) No significant differences in either treatment diet relative to basal Both experimental diets reduced blood hemoglobin relative to basal at 42 d One dietary phase throughout the study Strain: not stated	Pesti et al., 1991
0–18 d		Growth FCR	Basal and experimental diets analyzed (0.09 mg/kg)	1.79 and 1.47 mg/kg for 18 d growth, and 1.62 and 0.92 mg/kg for feed efficiency in Experiments 1 and 2, respectively 0.24 mg/kg (0.17 mg/kg analyzed) if diets adequate in methionine	Arbor Acres × Arbor Acres 0.24, 0.54, 1.14, and 2.34 mg/kg	Ryu et al., 1995a
0–18 d		Growth FCR	Basal and experimental diets analyzed (0.7 mg/kg—Experiments 1 and 2; 0.6 mg/kg—Experiments 3–5), but not experimental diets Diets low in methionine (8.74 g/kg vs. requirement of 9.3 g/kg)	Maximum growth and feed efficiencies from 1.45 mg/kg diet Chicks fed practical ingredient-based diets require 1.3 mg folic acid/kg diet with low levels of choline, but only 1.2 mg folic acid/kg when choline is offered near the NRC recommended level of 1,300 mg/kg of choline	Arbor Acres × Arbor Acres 0.70 to 2.20 mg/kg in increments of 0.25 mg/kg	Ryu et al., 1995b
0–3 d	Starter		Basal diet analyzed for folate (0.39 mg total folate/kg), supplemented with 0, 0.5, 1.0, or 2.0 mg folic acid/kg Met + cys appeared to be adequate	1.7 to 2 mg/kg	Cobb Three experiments; supplemented 0 to 2 mg/kg Experiment 1: 0, 0.5, 1.0, and 2.0 Experiment 2: 0, 0.75, 1.5, and 2.0 Experiment 3: 0, 0.75, 1.5, and 2.0 Wheat-based and corn-based diets ANOVA.	Whitehead et al., 1995

continued

TABLE 14-4 Continued

Age (wk)	BW Range or Stage of Production	Response Criteria	Basal Diet Level	Requirement % or Other	Notes (Strain, Diet Type, Diet Form, Vitamin Form)	Reference
Folacin						
3–6 d	Finisher		Basal diet analyzed for folate (estimated 0.2 mg total folate/kg [wheat-based] or 0.5 mg total folate/kg [corn-based]), supplemented with 0, 0.5, 1.0, or 2.0 mg folic acid/kg-met+cys appeared to be adequate	1 to 1.5 mg/kg	Cobb Three experiments; supplemented 0 to 2 mg/kg Experiment 1: 0, 0.5, 1.0, and 2.0 Experiment 2: 0, 0.75, 1.5, and 2.0 Experiment 3: 0, 0.75, 1.5, and 2.0 Wheat-based (Experiment 1) and corn-based (Experiment 2) diets ANOVA	Whitehead et al., 1995
Biotin						
20–48 d	BW, FI, deficiency signs		Not reported; not analyzed	120 μg/kg diet with 2% palm kernel oil; 160 μg/kg diet with 2% palm kernel oil to prevent deficiency symptoms. FI, BW 120 μg/kg diet regardless of palm kernel oil level	40, 80, 120, 160, 200, and 240 μg/kg diet Two replicates of 20 chicks per treatment	Oloyo and Ogunmodede, 2000
0–21 d 22–43 d 44–61 d		BW, bone size, and mineralization	Not analyzed	< 100 μg (0–21) biotin supplementation decreased growth Between 100–200 (22–61 d) μg 200, 300, and 400 μg/kg of feed increased growth rate during the second and third period of growth, FCR was improved throughout the productive cycle with the higher dose of biotin at 400 μg/kg of feed	Ross 508 0, 100, 200, 300, 400, and 500 μg/kg feed (supplemental); diets not analyzed 200 mg and beyond increased BMD relative to 0 and 100 only at 61 d Birds only 3.2 kg @ 61 days	Quarantelli et al., 2007
Expt 1: 1–21 d 22–42 d 43–56 d Expt 2: 1–15 d 16–29 d 30–43 d 44–57 d		BW, FCR, footpad dermatitis	Calculated biotin for basal diets: 0.00016 mg/kg (starter), 0.00014 mg/kg (grower), 0.00013 mg/kg (finisher), and 0.00012 mg/kg (withdrawal)	No effect of biotin on any parameter	Ross 708 0.05, 0.1, 0.2, 0.4 mg/kg supplemental biotin Diets not analyzed ANOVA	Cengiz et al., 2012

		Performance, FPD	Basal, experimental diets not analyzed Biotin-free vitamin premix	No effect on performance High biotin reduced footpad dermatitis at 21 d and beyond (Experiment. 1) and 14, 35, and 40 d (Experiment. 2)		Ross 708 2×2 factorial (two biotin levels—300 or 2,000 μg/kg; two zinc levels) FPD challenge model	Abd El-Wahab et al., 2013
22–42 d		Performance, footpad dermatitis	Basal: 140 μg/kg Experimental diets 155 and 1,521 μg/kg	High biotin increased BW gain in wet litter challenge model, at both high and low density High biotin reduced FPD and hock burn scores		Ross 308 Experimental diets 155 and 1,521 μg/kg	Sun et al., 2017
Vitamin B_{12}							
1–14 d 15–21 d 22–35 d	BW of 10 μg/kg trt 330 g 661 g 1893 g	BWG, FI, FCR, carcass composition	Analyzed—4 μg vitamin B_{12}	No response observed beyond 24 μg B_{12}/kg diet (0.024 mg)		Ross 308 male Supplemental: 0, 10, 20, 40 μg vitamin B_{12}, fed with or without 0.65 mg/kg cobalt Insufficient levels to determine requirement; ANOVA, interaction with Co level	Halle et al., 2011
Choline							
0–42 d		BWG, FI, FCR	1,300 mg/kg (not analyzed)	≤ 1,300 mg/kg		SML (synthetic male line) × ONBP (open nucleus broiler population) cross 0.0, 1,000, and 2,000 mg/kg supplemental No effect on performance parameters	Swain and Johri, 2000
1–7 d 1–14 d 1–21 d		BWG	Semi-purified diet; analyzed choline values (choline + betaine), used to determine choline intake Ingredients, not complete diets, analyzed.	1–7 = 1,228 1–14 = 1,111 1–21 = 1,042 mg/kg dietary choline for the requirement level of digestible methionine and 1–7 = 1,276 1–14 = 1,046 1–21 = 1,010 mg/kg dietary choline for the broilers receiving the diet with the 25% reduction in digestible methionine	1–7 = 27,013 1–14 = 44,458 1–21 = 62,535 mg/bird/day for the requirement level of digestible methionine and 1–7 = 26,796 1–14 = 41,820 1–21 = 56,578 mg/bird/day for the broilers receiving the diet with the 25% reduction in digestible methionine	Cobb 715, 1,040, 1,365, 1,690, and 2,015 mg/kg supplemented as choline chloride (62.5%). The BW gain responses were fitted using quadratic polynomial (QP) and broken-line (BL) models. Tested in adequate and methionine-reduced diets. Study 1: adequate digestible methionine Study 2: 25% reduced digestible methionine	de Lima et al., 2018

continued

TABLE 14-4 Continued

Age (wk)	BW Range or Stage of Production	Response Criteria	Basal Diet Level	Requirement % or Other	Requirement (Units/Day)	Notes (Strain, Diet Type, Diet Form, Vitamin Form)	Reference
Choline							
1–21 d			24 (analyzed)	1,335 to 1,372 mg/kg for BWG, carcass BW, breast BW		Ross 308 Encapsulated choline chloride 24, 323, 634, 930, 1,233, 1,290 (commercial choline chloride); 1,520, 1,810 mg/kg (analyzed) Linear broken-line analysis	Azghadi et al., 2020
Expt. 1 4–21 d 22–28 d Expt. 2 1–21 d 22–35 d 36–42 d		BWG	Diets analyzed	Expt. 2: 778 mg/kg 1–7 d 632 mg/kg 1–35 d 645 mg/kg 1–42 d		Cobb 500 males both experiments Experiment 1—bioequivalence of two choline sources Bio-choline™ levels (0, 100, 200, or 300 mg/kg) and choline (200, 400, or 600 mg/kg) Experiment 2—choline requirement 0, 200, 400, 600, or 800 mg/kg choline as choline chloride Not suitable for requirement studies—requirement is only slightly below the highest level of inclusion	Farina et al., 2017
1–7 d 8–21 d	170 g (1,427 mg/kg trt group) 775 g (1,427 mg/kg trt group)		727 (basal) Analyzed diet levels in agreement with formulated	Average requirements: 2,516 ppm for BWG 2,533 ppm for FCR 2,202 ppm to minimize locomotor disorders		Cobb × Cobb 500, males Dietary choline supplemented at 0, 700, 1,400, 2,100, and 2,800 ppm from choline Cl, totaling 727, 1,427, 2,127, 2,827, and 3,527 ppm, respectively	Santiago et al., 2020

TABLE 14-5 Daily Vitamin and Mineral Requirements of Broiler Chicken Strains Based on BW[a,b]

		BW (g)				
Vitamin	Units	0 to 600	601 to 1,600	1,601 to 2,800	2,801 to 3,600	> 3,600
Vitamin A	IU	***1,500***	***1,500***	***1,500***	***1,500***	***1,500***
Vitamin D	IU	285	200	200	200	***200***
Vitamin E	IU	***10***	***10***	***10***	***10***	***10***
Vitamin K	mg	0.5	0.5	0.5	0.5	***0.5***
Thiamin	mg	***1.8***	***1.8***	***1.8***	***1.8***	***1.8***
Riboflavin	mg	3.6	3.6	3.6	3.0	***3.0***
Niacin	mg	35	30	30	25	***25***
Pantothenic acid	mg	***10***	***10***	***10***	***10***	***10***
Pyridoxine	mg	3.5	3.5	***3.0***	***3.0***	***3.0***
Folate	mg	0.55	0.55	***0.50***	***0.50***	***0.50***
Biotin	mg	0.15	0.15	0.15	***0.12***	***0.12***
Vitamin B_{12}	mg	0.01	***0.01***	***0.01***	***0.01***	***0.01***
Choline	mg	1,300	***1,000***	***750***	***750***	***750***
Macromineral						
Nonphytate P	%	0.45	0.35	0.30	0.25	***0.25***
Ca	%	***0.90***	***0.80***	***0.70***	***0.70***	***0.70***
Potassium	%	0.30	0.30	0.30	0.30	0.30
Sodium	%	0.20	0.15	0.15	0.12	0.12
Chloride	%	0.20	0.15	0.15	0.12	0.12
Magnesium	mg	600	600	600	600	600
Trace mineral						
Copper	mg	***6***	***6***	***6***	***6***	***6***
Iodine	mg	***0.5***	***0.5***	***0.5***	***0.5***	***0.5***
Iron	mg	80	45	45	45	45
Manganese	mg	45	***45***	***45***	***45***	***45***
Se	mg	0.15	***0.15***	***0.15***	***0.15***	***0.15***
Zinc	mg	35	***22***	***22***	***22***	***22***

[a] Values in normal font are minimum nutrient requirement recommendations based on empirical studies summarized in Tables 14-6 and 14-7. Values in bold italics are guidance levels based on estimates due to insufficient experimental data (see front matter for details). Values in shaded areas are the NRC (1994) recommendations, which remained unchanged.

[b] Nutrient requirements are presented on a percentage of the diet or an IU or mg per kg of feed provided.

observed up to 500 IU/kg feed (Fritts and Waldroup, 2003) and increasing from 200 to 2,000 IU/kg feed (Leyva-Jimenez et al., 2019a). The most appropriately designed study available reported requirement for BWG to 21 d of age of 185 IU/kg of feed (Leyva-Jimenez et al., 2019b), which is close to the previous recommendation of 200 IU/kg feed (NRC, 1994). Although dietary vitamin D was not analyzed, 200 IU/kg feed resulted in equivalent performance to levels up to 3,600 IU/kg feed when the diets were adequate in Ca and available P (Rao et al., 2006). For BWG from 601 to 2800 g, increasing responses were observed up to 1,000 (Fritts and Waldroup, 2003), 3,500 (Khan et al., 2010) or 4,000 IU/kg feed (Fritts and Waldroup, 2005). Above 2,800 g BW, increasing responses were observed up to the tested maximum dietary level of 4,000 IU/kg feed (Fritts and Waldroup, 2005). Although the requirement for BWG from 0 to 600 g was 185 IU/kg feed, the requirement for bone mineralization during this same age was 285 IU/kg feed (Leyva-Jimenez et al., 2019b). In general, the dietary vitamin D_3 requirement for maximal bone mineralization (Mitchell et al., 1997; Whitehead et al., 2004; Khan et al., 2010; Vazquez et al., 2018), was greater than that for BWG (Sattar Khan et al., 2000; Whitehead et al., 2004; Atencio et al., 2005; Vazquez et al., 2018; Cho et al., 2020), meat yield (Vignale et al., 2015) and immune function (Vazquez et al., 2018). Based on the requirement for bone mineralization (Leyva-Jimenez et al., 2019b), the requirement value for vitamin D_3 from 0 to 600 g has been increased to 285 IU/kg feed. Due to a lack of appropriate studies, the requirement values for vitamin D have not changed for heavier BW, but it is likely that the actual requirements for vitamin D are higher than indicated for all stages of growth. Future studies are required to determine the vitamin D_3 requirements of modern broiler genotypes fed diets adequate in all other nutrients.

Vitamin E

No suitable vitamin E requirement studies have been conducted since the publication of the previous NRC (1994) report. Based on a meta-analysis conducted by Pompeu et al. (2018), it appears that intrinsic levels of vitamin E in practical diets based mostly on corn and soybean meal are sufficient to meet the requirements of broiler chickens, as there was a lack of a growth response to vitamin E supplementation of basal diets (Corrales et al., 2023). Supplementation of vitamin E above the current estimated requirement for growth can increase various aspects of immune function (da Silva et al., 2011), although at higher levels, immune function can be negatively affected (Leshchinsky and Klasing, 2001). Increasing vitamin E supplementation can also increase muscle vitamin E enrichment and antioxidant protection (Pitargue et al., 2019). Estimated requirements for vitamin E have not been changed from the previous report, although broilers fed a diet high in highly unsaturated fatty acids may have an increased vitamin E requirement to maintain tissue vitamin E levels (Barraoeta, 2007; Voljč et al., 2011).

Vitamin K

Vitamin K in broiler diets has received very little recent attention. Only two studies (Duarte et al., 2014; Guo et al., 2020) investigating this vitamin in broiler diets have been published since September 1991, and although neither is a requirement study, the data presented do not contradict the current minimum requirement value of 0.5 mg/kg.

Water-Soluble Vitamins

Due to a lack of studies conducted since September 1991 that met the committee's criteria for inclusion, the NRC (1994) dietary requirement recommendations for thiamin, riboflavin, niacin, pantothenic acid, or pyridoxine remain unchanged. However, several well-designed experiments that did not meet the committee's requirements, due mostly to a limited number of treatments for regression, are described in the following sections for these vitamins. Neither riboflavin nor niacin are detailed as no new relevant experiments were identified to expand current knowledge.

Thiamin

In a study by Olkowski and Classen (1999), no positive growth responses were observed when a basal diet containing 1.8 mg thiamin/kg was supplemented with up to 32 mg thiamin/kg diet; however, supplementation of 32 mg thiamin/kg diet significantly reduced broiler growth rate. In an earlier study from this laboratory (Olkowski and Classen, 1996), adding 2 or 8 mg of supplemental thiamin/kg diet to a basal diet containing 5.6 to 6.5 mg thiamine/kg was without effect on broiler growth performance.

Pantothenic Acid

Two studies investigating pantothenic acid supplementation in broiler diets have been published since the last NRC (1994) report. Supplementing 20 or 40 mg/kg pantothenic acid/kg diet in broilers to 4 weeks of age (930 g BW) or 8 weeks of age (2,760 g BW), respectively, resulted in similar growth performance to broilers fed 10 mg pantothenic acid/kg diet (Deyhim et al., 1992; Harms and Nelson, 1992). Similarly, feeding broilers a basal diet containing 4.74 mg pantothenic acid/kg (analyzed) from 1 to 21 d of age resulted in equivalent growth, feed intakes, and feed conversion ratios to those of birds fed various levels of pantothenic acid up to 19.14 mg/kg diet (Harms and Nelson, 1992).

Pyridoxine

Scherer and Baker (2000) estimated that the pyridoxine requirement of New Hampshire crossbred chicks was 0.73 mg/kg diet when the diet was adequate in methionine, but 1.05 mg/kg when the diet contained excess methionine.

Folic Acid

A small number of studies with limited folacin concentrations (Whitehead et al., 1995) or confounding factors, such as low dietary methionine or choline content (Ryu et al., 1995a,b), indicated that the requirement for folate may be higher than was indicated in the NRC (1994) report, particularly when other methyl donor nutrients are limiting. However, the lack of studies meeting the inclusion criteria for requirement determination prevented any change in the recommended folacin requirement.

Biotin

Supplementation of a biotin-deficient basal starter (0 to 21 d), grower (22 to 42 d) and finisher (43 to 56 d) diets containing calculated biotin levels of 0.00016, 0.00014, and 0.00013 mg/kg, respectively, with 0.05, 0.1, 0.2, or 0.4 mg of biotin/kg diet had no significant effect on broiler growth, feed conversion ratio, nor incidence and severity of footpad dermatitis (Cengiz et al., 2012). Surprisingly, from 0 to 21 d of age, birds fed an unsupplemented basal diet (biotin concentration not analyzed) had a significantly heavier BW (663 g) than broilers fed the basal diet supplemented with 0.1 to 0.5 mg of biotin/kg of diet (BW range 618 to 642 g) (Quarantelli et al., 2007). However, in the same report, at 41 d and 61 d, the basal diet supplemented with 0.2, 0.3, or 0.4 mg biotin/kg feed resulted in significantly greater BW than those of birds fed the unsupplemented diet or the diets supplemented with 0.1 or 0.5 mg biotin/kg diet. In a third study, there was a positive response (indicated by increased BW, reduced footpad dermatitis, and hock burn scores) from 22 to 42 d of age to increasing dietary biotin from 0.155 mg to 1.521 mg/kg diet in a wet litter challenge model (Sun et al., 2017), suggesting a higher requirement than previous studies. However, further research is needed to confirm their observations.

Vitamin B_{12}

Halle et al. (2011) reported that adding 0.04 mg vitamin B_{12}/kg to a basal diet containing 0.004 mg vitamin B_{12}/kg diet increased 35 d BW relative to broilers fed the same basal diet that was either unsupplemented or supplemented with 0.01 mg vitamin B_{12}/kg diet. When 0.02 mg supplemental vitamin B_{12}/kg was fed, it resulted in an intermediate 35 d BW response (Halle et al., 2011). Although the previous requirement of 0.01 mg/kg diet was not changed, this study suggests that it may be higher, although confirmation is needed.

Choline

de Lima et al. (2018) reported choline requirements of 1,128, 1,111, and 1,042 mg/kg diet for broiler BW gains from 1 to 7, 1 to 14, and 1 to 21 d of age, respectively, indicating a decrease with age even during this early period. Slightly higher values of 1,335 to 1,372 mg choline/kg diet were reported to maximize BW gain, carcass weight, and breast muscle weight in broilers from 1 to 21 d of age (Azghadi et al., 2020). These values agree with the recommendation in the previous report (NRC, 1994) of 1,300 mg choline/kg feed. Choline requirements for BW gain to 21 d of age were estimated to be 2,936 mg/kg using a quadratic polynomial regression, and 2,077 mg/kg using a quadratic broken line regression model (Santiago et al., 2020). However, Farina et al. (2017) reported lower requirements for BW gain from 1 to 7, 1 to 35, and 1 to 42 d of age (778, 632, and 645 mg choline/kg diet, respectively), although the determined requirement fell between the two highest levels of supplementation. The studies used to estimate the choline requirement in the previous NRC (1994) report either used noncommercial strains of birds, were published in the 1980s or prior, or both. Because three of the four studies published since the previous report (NRC, 1994) suggested that the choline requirement of broilers was lower, the value has been changed to an estimate (bold, italics) until these discrepancies can be resolved.

Minerals

Research published since September 1991 led to the refinement or confirmation of many of the NRC (1994) recommended dietary mineral requirements of broilers. In reviewing available literature, it was clear that the amount of research conducted on various minerals was often in direct proportion to either economic value or to likelihood of dietary deficiency in practical diets. Additionally, substantial research was conducted using new commercial products that provide minerals in various forms or that alter the digestibly or availability of the mineral. One general issue with this research, in the context of requirement generation, is that the commercial products containing novel mineral formulations require additional controls, thus increasing the total number of treatments needed to correctly determine the ability of different mineral forms to meet requirement and distinguish requirements from bioavailability. A lack of suitably controlled studies eliminated much of this literature from consideration by the committee. Moreover, there is a great deal of literature concerning Ca and P and some of the trace minerals, but this research was not designed to establish requirements. When possible, these experiments were examined for their agreement with the NRC (1994) recommendations or with other recently published data that resulted in a requirement update. However, such experiments were not used as a sole source to update any requirements. Experiments using minerals in an uncharacterized organic matrix, complexed with proteinates, or in microbes such as yeast were not considered because manufacturing techniques are not standardized across the various sources. Both macromineral and micromineral requirements are presented in Table 14-6.

Macrominerals

Phosphorus

Requirements for P are expressed in terms of nonphytate P (nPP). For further discussion on the opportunities and limitations of the use of a digestible or available P concept in poultry the reader is referred to Chapter 6, "Minerals," and Chapter 8, "Bioavailability and Digestibility"). Although phytases are now ubiquitous in commercial broiler diets and increase the availability of phosphorus from phytate provided by plant components of the diet, Ca and P or nPP requirements are expressed in this report in the absence of phytase supplementation. This is because increases in phosphorus availability due to phytase-mediated phosphorous release does not change the bird's nutritional requirement for phosphorus; it increases the ability of a phytate-containing diet to meet that requirement.

The nPP requirement of 0.45 percent for the 0 to 600 g broiler chick remains unchanged from the starter phase requirement reported (NRC, 1994). Following are brief summaries of recent experiments that were evaluated but did not result in a change to the overall requirement. Waldroup et al. (2000) conducted a large experiment that investigated nPP requirement of broilers. They employed diets consisting of yellow dent corn or a high-available P cultivar with or without phytase supplementation and reported that 1 to 21 d old broilers required 0.39 percent nPP (0.35 to 0.44% nPP—95% confidence interval) based on tibia ash when fed the yellow dent corn diet without phytase supplementation. The committee only considered the yellow dent corn diet without phytase for the purpose of requirement evaluation. Dietary treatments included 0.20, 0.25, 0.30, 0.35, 0.40, 0.45, and 0.50 percent nPP. Li et al. (2017) fed diets that contained 0.18, 0.23, 0.28, 0.33, 0.38, 0.43, 0.48, 0.53, and 0.58 percent nPP to 1 to 21 d old broiler chicks and reported a 0.39 percent nPP requirement (0.37 to 0.40% nPP—95% confidence interval) based on tibia breaking strength. Two strains of broiler chickens were

TABLE 14-6 Documentation of Digestible Lys Requirements of Broilers[a]

Estimated Requirements		Age Period	Gender	BW Gain	Response Criteria	Strain	References	Comments
%	mg/d	Days		g/d				
1.30	312	1 to 7	Females	21.7	BWG	Ross × Ross 708	(Dozier and Payne, 2012)	Exp 1
1.33	319	1 to 7	Females	21.7	Feed conversion	Ross × Ross 708	(Dozier and Payne, 2012)	Exp 1
1.22	488	1 to 14	Females	32.7	BWG	Ross × Ross 708	(Dozier and Payne, 2012)	Exp 1
1.21	472	1 to 14	Females	32.7	Feed conversion	Ross × Ross 708	(Dozier and Payne, 2012)	Exp 1
1.23	280	1 to 7	Females	21.8	BWG	Hubbard × Cobb 500	(Dozier and Payne, 2012)	Exp 2
1.08	244	1 to 7	Females	21.8	Feed conversion	Hubbard × Cobb 500	(Dozier and Payne, 2012)	Exp 2
1.14	445	1 to 14	Females	32.7	BWG	Hubbard × Cobb 500	(Dozier and Payne, 2012)	Exp 2
1.22	476	1 to 14	Females	32.7	Feed conversion	Hubbard × Cobb 500	(Dozier and Payne, 2012)	Exp 2
1.17	783	8 to 22	Males	36.1	BWG	Cobb × Cobb 500	(Siqueira et al., 2013)	
1.06	709	8 to 22	Males	36.1	Feed conversion	Cobb × Cobb 500	(Siqueira et al., 2013)	
1.23	309	1 to 11	Males	29.5	BWG	Ross	(Todelo et al., 2011)	Exp 1
1.24	310	1 to 11	Males	29.5	Feed conversion	Ross	(Todelo et al., 2011)	Exp 1
1.18	892	12 to 22	Males	29.4	BWG	Ross	(Todelo et al., 2011)	Exp 2
1.16	449	1 to 12	Males	29.4	BWG	Cobb 500	(Cemin et al., 2017)	Exp 1
1.07	449	1 to 12	Males	53.8	Feed conversion	Cobb 500	(Cemin et al., 2017)	Exp 1
1.26	1,030	8 to 21	Males	53.8	BWG	Cobb	(de Oliveria et al., 2013)	
1.43	1,173	8 to 21	Males	53.8	Feed conversion	Cobb	(de Oliveria et al., 2013)	
1.13	1,375	14 to 28	Males	84.3	BWG	Ross × TP16	(Dozier et al., 2009a)	Exp 1
1.20	1,469	14 to 28	Males	84.3	Feed conversion	Ross × TP16	(Dozier et al., 2009a)	Exp 1
0.99	1,083	14 to 28	Females	68.4	BWG	Ross × TP16	(Dozier et al., 2009a)	Exp 2
1.02	1,122	14 to 28	Females	68.4	Feed conversion	Ross × TP16	(Dozier et al., 2009a)	Exp 2
1.14	1,283	14 to 28	Males	75.9	BWG	Ross × TP16	(Dozier et al., 2009b)	Exp 2
1.19	1,336	14 to 28	Males	75.9	Feed conversion	Ross × TP16	(Dozier et al., 2009b)	Exp 2
1.08	1,208	15 to 28	Mixed	68.0	BWG	Ross 308	(Mehri et al., 2010)	
1.13	1,182	15 to 28	Mixed	68.0	Feed conversion	Ross 308	(Mehri et al., 2010)	
0.95	1.155	12 to 28	Males	80.1	BWG	Cobb 500	(Cemin et al., 2017)	Exp 2
1.04	1,240	12 to 28	Males	80.1	Feed conversion	Cobb 500	(Cemin et al., 2017)	Exp 2
0.97	1,712	28 to 42	Males	94.3	BWG	Ross × TP16	(Dozier et al., 2010)	Exp 1
1.04	1,908	28 to 42	Males	94.3	Feed conversion	Ross × TP16	(Dozier et al., 2010)	Exp 1
0.88	1,662	28 to 42	Males	94.3	Breast meat yield	Ross × TP16	(Dozier et al., 2010)	Exp 1
1.01	1,883	28 to 42	Males	105.6	BWG	Cobb × Cobb 700	(Dozier et al., 2010)	Exp 2
1.03	1,915	28 to 42	Males	105.6	Feed conversion	Cobb × Cobb 700	(Dozier et al., 2010)	Exp 2
1.03	1,911	28 to 42	Males	105.6	Breast meat yield	Cobb × Cobb 700	(Dozier et al., 2010)	Exp 2
0.96	2,087	28 to 42	Males	91.6	BWG	Cobb × Cobb 500	(Cemin et al., 2017)	Exp 2
0.99	2,157	28 to 42	Males	91.6	Feed conversion	Cobb × Cobb 500	(Cemin et al., 2017)	Exp 2
0.95	2,012	28 to 42	Males	91.6	Breast meat yield	Cobb × Cobb 500	(Cemin et al., 2017)	Exp 2
1.07	2,226	35 to 49	Males	107.8	BWG	Ross 708	(Mejia et al., 2012)	Exp 1
1.10	2,289	35 to 49	Males	107.8	Feed conversion	Ross 708	(Mejia et al., 2012)	Exp 1
0.87	2,112	49 to 63	Males	86.7	BWG	Ross 708	(Dozier et al., 2008)	Exp 2
0.88	2,137	49 to 63	Males	86.7	Feed conversion	Ross 708	(Dozier et al., 2008)	Exp 2
0.86	2,088	49 to 63	Males	86.7	Breast meat yield	Ross 708	(Dozier et al., 2008)	Exp 2

[a] Only references published from 2007 to 2021 were considered by the committee as being relevant to modern broiler strains.

fed 0.15, 0.23, 0.31, 0.39, 047, and 0.55 percent nPP diets from 5 to 23 d of age with and without phytase treatment to the diet (Persia and Saylor, 2006). Broilers had a higher nPP requirement for maximal skeletal structure, including tibia ash percent, tibia ash weight, and tibia breaking strength, rather than BWG. Therefore, the committee used the estimated requirements, based on tibia ash percentage, from birds fed diets without phytase resulting in an estimate requirement of 0.44 and 0.48 percent nPP for the two strains.

The nPP recommendations for broilers with BW of 601 to 1,600 g and 1,601 to 2,800 g remain unchanged from the NRC (1994) grower and finisher phase recommendations at 0.35 and 0.30 percent nPP, respectively. Two experiments support the 0.35 percent nPP requirement for broilers weighing between 601 to 1,600 g. Yan and coworkers (2001) determined the nPP requirement of 21 to 42 d old broilers to be 0.33 percent. In their study, starting BWs were not reported, but BWG over the experimental period was 1,400 g. In that experiment, nPP treatments included diets with 0.10, 0.15, 0.20, 0.25, 0.30, 0.35, 0.40, and 0.45 percent nPP fed with and without phytase supplementation. Again, for the purpose of NRC requirement generation, only treatments that were free from phytase treatment were considered for requirement generation. A second study by Li and coworkers (2016) did not include enough treatments to determine nPP requirements and failed to report a 21 d starting BW for broilers reared from 21 to 42 d. However, their reported 1,170 g BWG was considered appropriate and supportive of the finding that 0.30 percent nPP maximized tibia ash response in broilers in this BW range. No experiments could be found to directly examine the nPP requirements of broilers from 1,601 to 2,800 g BW, but Dhandu and Angel (2003) fed broilers diets that contained 0.15, 0.19, 0.26, and 0.31 percent nPP, with the 0.19 percent nPP diet resulting in maximized bone ash responses without differences in performance over 32 to 42 d of age in which BW increased from 540 and 2,200 g, respectively. This single experiment with an inadequate number of treatments to generate a nPP requirement is not sufficient to change the requirement but should encourage research to determine whether a reduction in dietary nPP concentration is appropriate.

The BW category 2,801 to 3,600 g corresponds to what the previous report (NRC, 1994) referred to as the finisher phase. A single publication by Dhandu and Angel (2003) that evaluated two dietary concentrations of nPP supported a reduction in broiler nPP estimate to 0.25 percent nPP for broilers weighing 2,801 to 3,600 g. In a follow-up experiment that crossed BW categories, they estimated a nPP requirement for broilers based on tibia ash responses of 0.16 to 0.18 percent nPP for 42 d old broilers weighing 2,400 g to 49 d old broilers weighing 3,008 g. In that report, the birds were fed diets containing 0.10, 0.13, 0.22, 0.27, and 0.31 percent nPP, a design that only provides two data points below the plateau region. while the requirements of broilers weighing more than 3,601 g was not directly addressed in the previous report (NRC, 1994). No reports relevant to nPP needs of broilers over 3,601 g were found; therefore, the requirement value was set to 0.25 percent nPP, consistent with that of birds in the 2,801 to 3,600 g range.

Calcium

The committee chose to report broiler Ca requirements on a total Ca basis despite recent publications citing advantages in expressing requirements on a digestible Ca basis (David et al., 2021; Walk et al., 2021, 2022). The committee agrees that the movement to a digestible Ca basis for expressing requirements could prove beneficial but that the underlying literature is currently insufficient to implement such a change in requirement assessment. The Ca estimate of 0 to 600 g broilers has been reduced to 0.90 percent. Driver et al. (2005) fed diets that contained 0.325, 0.400, 0.475, 0.550, 0.625, and 0.900 percent Ca using both male (550 g) and female (520 g) broiler chicks from 1 to 16 d of age. A Ca requirement of 0.72 percent Ca was generated by regression analysis based on tibia ash, but only the 0.90 percent diet was above the requirement level, suggesting caution in interpretation of these data. Although limited in the number of dietary Ca treatments and not suitable for actual Ca requirement determination, a study by Mitchell and Edwards (1996) reported the results of two experiments support a reduced Ca requirement when four diets with 0.77, 0.87, 0.97, and 1.07 percent Ca diets were fed to broilers, and each experiment showed no additional tibia ash responses of 1 to 21 d old broiler chicks beyond 0.77 and 0.87 percent Ca, respectively.

One requirement experiment was reported in 19 to 42 d old broiler chickens that suggests that tibia ash was maximized by the feeding of 0.625 percent dietary Ca, but only two values fell within the plateau region of the response (Driver et al., 2005). Although an initial BW was not reported for these birds, 19 d old broilers would be expected to fall within 601–1,600 g BW. Based on this assumption, these data were used as support to lower the Ca estimate of 601–1,600 g broiler chicks, but reduction to 0.625 percent Ca was not possible based on one report with a limited number of treatments in the plateau region. The Ca estimate for 601–1,600 g broiler chicks was set at 0.80 percent Ca to be consistent with the reduced Ca requirement noted in NRC (1994) between the starter and grower dietary phases. No new requirement data were found for broiler chickens with BW over 1,600 g. Therefore, the reduction in Ca estimate between 601–1,600 and 1,601–2,800 g was based on the reduction between the grower and finisher dietary phases reported in NRC (1994). This reduction resulted in a dietary Ca estimate of 0.70 percent for 1,601–2,800 g broilers. The lack of data for broilers above 2,800 g precludes the ability to further reduce the Ca requirement at heavier BW ranges. One experiment by Skinner et al. (1992) supported the general reduction in dietary Ca for 42 to 56 d old broilers that were fed 0.48 and 0.60 percent Ca diets and performed similarly in all skeletal measurements compared to control broilers fed 0.80 percent Ca.

Potassium, Sodium, Chlorine, and Magnesium

No changes are proposed for potassium, sodium, chlorine, or magnesium requirements. Tabular requirement values for these minerals were translated from an age basis to a BW basis. No experiments meeting the committee's criteria were reported for potassium, chlorine, or magnesium. Several experiments were conducted with limited ranges of dietary sodium concentrations, and those experiments were consistent with historic sodium requirements for broiler chickens (Murakami et al., 1997, 2001; Ravindran et al., 2008; Akter et al., 2019). As the sodium requirements reported agreed with or were lower than the previous NRC (1994) requirements, no changes were recommended outside of the translation from day of age basis to a BW range basis. Feeding broilers dietary sodium concentrations above accepted requirement levels compromised their environment due to increased water consumption and excretion, resulting in increased litter moisture content or litter scores (Murakami et al., 1997; Ravindran et al., 2008; Akter et al., 2019).

Trace Minerals

Similar to vitamins, deficiencies of trace minerals are sometimes recognized by the expression of pathologies that impact health or product quality at levels lower than those that impact growth (Klasing and Korver, 2020). In those cases where pathology is the most sensitive endpoint it is the preferred response criteria for establishing a requirement. However, pathology is a subjective endpoint, and in some cases, an important functional endpoint that requires the mineral and whose mineral dependent change also highly correlates with an important pathology is a useful quantitative proxy. Depending on the mineral, the response criteria used to set requirement recommendations were based on these considerations. This lack of reliance on BW data makes translation of certain micro-minerals from a diet phase basis to a broiler BW basis difficult; in these cases, the committee estimated the changes as best as possible with limited data.

Requirement studies for trace minerals have used a variety of diet types ranging from those based on purified sources of protein and energy to those based on common feed ingredients, mostly of plant origin. In the former, a control diet based on purified ingredients, such as isolated soybean protein and starch, which typically has low levels of the mineral under study, is supplemented with a mineral source of high availability that is added in relatively large amounts to provide a dose response. In studies using practical ingredients, such as corn and soybean meal, the unsupplemented control diet typically contains higher levels of the mineral under investigation while lower amounts of the high bioavailability standard are added to establish a dose response. In studies using purified ingredients, the determined requirement approximates the true requirement for the standard reference mineral source, which is usually highly available. When studies use common feed ingredients, estimates reflect amounts of a standard reference source needed in addition to that provided by the practical ingredients, which is typically of lower bioavailability. The requirement values in Table 14-6 assumed that broilers were fed diets containing corn and soybean meal as the primary ingredients. Mineral sulfates or chlorides were used as the standard reference materials in most trace mineral requirement studies, with the exception of studies on selenium (Se) requirements. See Table 14-7 for the specific form used in each study. When other mineral sources are used, the amount added should be adjusted for the weight percentage of the mineral of interest in the source and its bioavailability relative to those used in studies for establishing requirements (Table 14-6). The recommended trace mineral requirements are for diets without phytase. Because phytase releases divalent cations bound to phytate, lower levels of supplementation are appropriate when exogenous phytase is supplemented to the diet; however, specific studies would be required to determine by exactly how much the level of supplementation can be reduced.

Copper

The recommendations in the previous report (NRC, 1994) for hatchlings through slaughter was based on a single report (McNaughton and Day, 1979) that examined the requirement in broilers from hatch to d 21. More recently, five studies examined the copper requirement of broilers fed diets based on corn and either soybean meal or isolated soybean protein. Three studies examined the copper requirement during the early production period, and each estimated the requirement for growth and feed efficiency to be lower than the requirement of 8 mg Cu/kg diet recommended previously. Du et al. (1996) did not observe a significant response by adding incremental amounts of $CuSO_4$ to a basal diet with 4.6 mg Cu/kg when BWG and feed efficiency were the response criteria. Unfortunately, they did not measure copper concentration in water nor specify the source of the water; thus, their results might be confounded by consumption of copper from the water supply. Schmidt et al. (2005b) reported an even lower requirement of <1.5 mg/kg diet using the same response criteria as well as tibia copper concentration, and this could not be attributed to copper in the water, which was quantified. Li et al. (2017) examined the requirement in a slow-growing yellow-feathered broiler breed and found that the requirement was <2.8 mg/kg diet for BWG and feed efficiency and for minimizing malondialdehyde, a by-product of lipid oxidation. Based on maximal liver CuZn-superoxide dismutase (SOD) activity, the copper requirement was reported to be between 6.8 and 10.8 mg/kg diet. For this reason, a tentative recommendation for the starter phase of broilers based on the average of the higher end of the estimates in these three studies is 6 mg Cu/kg in a corn-soybean meal–based diet.

Schmidt et al. (2005a) examined the copper requirement of broilers from 22 to 42 d of age and found that additions of $CuSO_4$ to the basal corn–soybean diet containing 1.4 mg Cu/kg (average for grower and finisher diets) did not increase BWG, feed efficiency, or tibia Cu. However, Li et al. (2014) found that the requirement during the late finishing phase (1.28–2.08 kg BW) was 19.9 mg copper/kg diet for BW gain and 38.3 to maximize Cu-Zn SOD, but <2.7 mg/kg diet for maximal feed efficiency using a slow-growing broiler strain (Lingnan yellow-feathered). The previous report (NRC, 1994) listed a tentative recommendation of 8 mg Cu/kg based on the study using hatchling broilers because there were no studies using broilers during the grower or finisher phases. Because of the disparity of these two more recent studies, the recommendation for the grower and finisher diets of fast-growing broiler strains fed corn-soybean meal-based diets is 6 mg/kg diet based on the previous estimate for the early phase and is tentative until further studies are done.

Iodine

The concentration of iodine (I) in feedstuffs is greatly variable with high levels typically found in crops grown near

TABLE 14-7 Documentation of Macromineral Requirements of Growing Broilers

BW and Bird Age	Estimated Requirement	Response Criteria	Strain, Sex	Dietary Treatments and Mineral Form	Comments	Reference
(g, d)	(%)			(range, %)		
Non-phytate Phosphorus (nPP)						
Hatch–740 (1–21)	0.34 0.39 0.37 0.36 0.39	BWG Tibia breaking Tibia ash Toe ash Serum P	Arbor Acres, males	0.18, 0.23, 0.28, 0.33, 0.38, 0.43, 0.48, 0.53, 0.58 nPP	All diets contained 1% Ca	Liu et al., 2017
72–734 (5–23)	0.41 0.41	BWG Tibia ash	Aviagen 308, males	0.15, 0.23, 0.31, 0.39, 0.47, 0.55 nPP	Experiment 2 All diets contained 1% Ca	Persia and Saylor, 2006
71–697 (5–23)	0.44 0.48	BWG Tibia ash	Aviagen 708 males	0.15, 0.23, 0.31, 0.39, 0.47, 0.55 nPP	Experiment 2 All diets contained 1% Ca	Persia and Saylor, 2006
Hatch–687 (1–21)	0.32 0.39	BW Tibia ash	Commercial strain, males	0.20, 0.25, 0.30, 0.35, 0.40, 0.45, 0.50 nPP	The yellow dent corn without phytase diet was considered All diets contained 1% Ca	Waldroup et al., 2000
1,170 gain (21–42)	ND 0.30 to 0.35	BWG Tibia ash	Arbor Acres Plus, males	0.25, 0.30, 0.35, 0.40 nPP	Response was a main effect across a high and low nutrient plane without interactions Basal diets contained 0.56% Ca	Li et al., 2016
1,400 gain (21–42)	0.19 0.33 0.16	BWG Tibia ash FCR	Cobb 500, males	0.10, 0.15, 0.20, 0.25, 0.30, 0.35, 0.40, 0.45 nPP	The without phytase diet was considered All diets contained 0.9% Ca	Yan et al., 2001
1,541–2,195 (32–42)	0.20 0.21	Tibia ash BW Tibia ash percent	Ross 308, males	0.15, 0.19, 0.26, 0.31 nPP	Experiment 1 All diets contained 0.7% Ca	Dhandu and Angel, 2003
2,396–3,800 (42–49)	0.16 0.18	Tibia ash BW Tibia ash percent	Ross 308, males	0.10, 0.13, 0.22, 0.27, 0.31 nPP	Experiment 2 Diets contained Ca dependent on nPP ranging from 0.7 to 0.8% Ca	Dhandu and Angel, 2003
Ca						
Hatch–650 (1–16)	0.625 to 0.90	Tibia ash percent	Cobb × Cobb male and female	0.325, 0.400, 0.475, 0.550, 0.625, 0.90 Ca	Experiment 2 Response was a main effect across male and female chicks without interactions All diets contained 0.64% P	Driver et al., 2005
Hatch–679 (1–21)	≤0.77 ≤0.77	BW Tibia ash	Commercial broiler, male	0.77, 0.87, 0.97, 1.07 Ca	Experiment 1 P concentrations of the diet were varied, but only treatments that contained at least 0.55% P were considered	Mitchell and Edwards, 1996
Hatch-624 (1–21)	0.75 to 0.87 0.75 to 0.87	BW Tibia ash	Commercial broiler, male	0.63, 0.75, 0.87, 0.99 Ca	Experiment 2 P concentrations of the diet were varied, but only treatments that contained at least 0.55% P were considered	Mitchell and Edwards, 1996

continued

TABLE 14-7 Continued

BW and Bird Age	Estimated Requirement	Response Criteria	Strain, Sex	Dietary Treatments and Mineral Form	Comments	Reference
1,798 gain (19–42)	0.55 to 0.625	Tibia ash percent	Cobb × Cobb male and female	0.325, 0.400, 0.475, 0.550, 0.625, 0.90 Ca	Experiment 1 Response was a main effect across high and low CP diets and male and female chicks without interactions	Driver et al., 2005
Sodium						
Hatch–249 (1–10)	0.15 to 0.24	FCR	Ross 308, male	0.15, 0.24, 0.35 Na bicarbonate	Response was a main effect across phytase treatment without interactions	Akter et al., 2019
1,543 gain (21–42)	≤0.10 ≤0.10	BWG FCR	Cobb, male	0.10, 0.15, 0.20, 0.25, 0.30, 0.35 Na bicarbonate	Experiment 1	Murakami et al., 2001
Hatch–920 (1–21)	≤0.15 ≤0.15	BWG FCR	Ross 308, male	0.15, 0.18, 0.35, 0.52 Na bicarbonate	Response was a main effect across phytase treatment without interactions.	Ravindran et al., 2008

Documentation of Trace Mineral Requirements of Growing Broilers

BW, Start/End of Study, kg (Age d)	Estimated Require-ment (mg/kg)	Response Criteria	Strain, Sex	Number of Levels, (Range, mg/kg) and Mineral Form	Comments	Reference
Copper						
Hatch–0.440 (1–14)	<4.6 <4.6 <4.6	BW gain Feed efficiency Leg pathology	Arbor Acres, male and female	5 (4.6–12.6) $CuSO_4$	The corn–soybean meal basal diet contained 4.6 mg/kg copper Data were analyzed by a one-factor ANOVA and means comparisons	Du et al., 1996
0.041–0.329	<2.8 <2.8 <2.8 <2.8 6.8–18.8 6.8–10.8 <2.8 <2.8	BW gain Feed efficiency Serum Cu Serum MDA[a] Serum CuZn-SOD[a] Liver CuZn-SOD[a] Liver MDA Serum cerulo-plasmin	Lingnan yellow-feathered, male	6 (2.8–66.8)	The corn starch, glucose, soybean protein basal diet contained 2.8 mg/kg copper Data were analyzed by a one-factor ANOVA, broken line regression, and quadratic models	Li et al., 2016
(8–21)	<1.5 <1.5 <1.5 <1.5	BW gain Feed efficiency Tibia Cu Leg pathology	Avian Farms, male and female	6 (1.5–19.0) $CuSO_4 \cdot 5H_2O$	Level of Cu in the corn-isolated soybean meal diet was 1.5 mg/kg Birds were fed a diet with adequate Cu prior to d 8 Data were analyzed by SAEG program and by linear response plateau and quadratic models There were no sex by copper level interactions	Schmidt et al., 2005b

TABLE 14-7 Continued

BW and Bird Age	Estimated Requirement	Response Criteria	Strain, Sex	Dietary Treatments and Mineral Form	Comments	Reference
(22–42)	<1.4 <1.4 <1.4	BW gain Feed efficiency Tibia Cu	Avian Farms, males and females	6 (1.4–18.9) $CuSO_4 \cdot 5H_2O$	Levels of Cu in the grower and finisher corn-isolated soybean meal diets were 1.4 and 1.33 mg/kg Diets fed prior to the experiment are not reported Data were analyzed by SAEG program and by linear response plateau and quadratic models There were no sex by copper level interactions	Schmidt et al., 2005a
1.28–2.08 (43–63)	19.9 <2.7 38.3	BW gain Feed efficiency Liver CuZn-SOD[a]	Lingnan yellow-feathered, male	6 (2.7–66.7) $CuSO_4$	The corn starch, glucose, soybean protein basal diet contained 2.7 mg/kg copper Prior to the start of the experiment, chicks were fed a diet with 2.8 mg/kg Cu These diets did not support normal growth, likely due to the semi-purified diet with low copper fed from hatch Data were analyzed by a one-factor ANOVA, broken line regression, and quadratic models	Li et al., 2014
Iodine						
(1–35)	<0.5 0.5–5	BW gain Feed efficiency	Ross 308, male	4 (0.5–5.5) KI	The corn–wheat–soybean meal basal diet contained 0.5 mg/kg iodine Data were analyzed by ANOVA and means comparison	Roettger et al., 2011
Iron						
Hatch–0.644 (1–21)	67–107 <67 <67 97 125 129 104 121	BW gain Hemoglobin Hematocrit Tibia bone Fe Liver SDH[a] Liver COX[a] Liver CAT[a] Heart SDH	Arbor Acres, male	7 (67–177) $FeSO_4 \cdot 7H_2O$	The corn–soybean meal basal diet contained 67 mg/kg iron Data were analyzed by a one-factor ANOVA and by broken-line, quadratic, and asymptotic models; the model with the best fit was used to estimate the requirement	Ma et al., 2016
8–21 d	85 86.5	Hemoglobin Serum iron	Columbian–New Hampshire cross, female	7 (47–97) $FeSO_4 \cdot 7H_2O$	The semi-purified basal diet contained 46.5 mg/kg iron, most of which was from phytate in soy protein concentrate The iron in the basal diet was judged to be mostly not available Hatchlings were fed an iron-deficient diet prior to the experimental period Data were analyzed by broken line regression	Aoyagi and Baker, 1995

continued

TABLE 14-7 Continued

BW and Bird Age	Estimated Requirement	Response Criteria	Strain, Sex	Dietary Treatments and Mineral Form	Comments	Reference
Hatch–0.374 (1–21) 0.374–1.21 (22–42) 1.21–2.26 kg (43–63)	<48 <48 <48 <49 <49 <49 <49 <49 <49 89	BW gain Feed efficiency Hematocrit Carcass drip loss	Lingnan yellow-feather, male	6 (48–148) 6 (49–149) 6 (49–149)	The corn–soy protein basal diets for the starter, grower, and finisher had 48, 49, and 49 mg/kg Fe The birds used to examine the requirements from d 22–42 and from d 43–63 were fed a diet from hatch to the start of these separate experiments containing 80 mg/kg Fe Data were analyzed by a one-factor ANOVA and means comparisons	Lin et al., 2020
22–42 d	<47 <47 <47 <47 104	BW gain Hemoglobin Hematocrit Tibia bone Fe Heart COX	Arbor Acres, male	6 (47–142) $FeSO_4{\cdot}7H_2O$	The corn–soybean meal basal diet contained 47 mg/kg iron Data were analyzed by a one-factor ANOVA and by broken-line, quadratic, and asymptotic models; the model with the best fit was used to estimate the requirement. However, the authors accepted P values < 0.10 as significant. Those > 0.05 are not considered in this report.	Liao et al., 2017
Manganese						
0.221–0.731 (10–21)	<45 <45 <45	BW gain Food intake Mn-SOD	Cobb 500, male	4 (45–184) $MnSO_4$	The corn–soybean meal basal diet contained 45 mg/kg Mn Data were analyzed by ANOVA and means comparisons	Conly et al., 2012
Hatch–0.75 (1–21)	<19 <19 38–98 144	BW gain Feed efficiency Leg abnormality Heart MnSOD[a] activity	Arbor Acres, male	8 (19–158) $MnSO_4{\cdot}H_2O$	The corn–soybean meal basal diet contained 10 mg/kg Mn Data were analyzed by a one-factor ANOVA and by quadratic and asymptotic models	Li et al., 2011
Hatch–0.894 (1–21) 0.894–2.393 (22–42)	< 33 < 33 < 32 < 32	BW gain Feed efficiency	Cobb 500, male	5 (33–182) Mn—methionine 5 (32–153)	The corn–soybean meal basal diet contained 33 mg/kg Zn in the starter diet (d 1–21) and 32 mg/kg in the finisher diet (d 22–42) Data were analyzed by ANOVA	Pacheco et al., 2017
Hatch–0.461 (d 1–14)	<45 <45	BW gain Feed efficiency	Ross 708	5 (41–201) MnHCl—tribasic	The corn–soybean meal basal diet contained 45, 40, 39, 34, and 38 mg/kg Mn in the five diets fed (starter, grower, and three finisher diets) The values for d 1–55 are an average across the five diets fed	Jasek et al., 2019
Hatch–4.30 (d 1–55)	<39 39–119 39–119	Tibia Mn		5 (37–197)	Data were analyzed by one-factor ANOVA	

TABLE 14-7 Continued

BW and Bird Age	Estimated Requirement	Response Criteria	Strain, Sex	Dietary Treatments and Mineral Form	Comments	Reference
Se						
Hatch–0.490 (1–21)	<0.05 <0.05 0.08–0.23 0.13	BW gain Feed efficiency Plasma MDA[a] Liver MDA	Lingnan yellow feather, sex unknown	5 (0.04–0.34) Na_2SeO_3	The corn–soybean meal basal diet contained 0.04 mg/kg Se Data were analyzed by a one-factor ANOVA and by a quadratic model Rate of gain and feed efficiency were reduced at Se 0.23–0.34	Li et al., 2017
Hatch–1.21 (1–29)	0.03 0.32 0.35	BW gain Pancreas GPX1[a] activity Pancreas GPX4 activity	Cornish cross, male	10 (0.005–1.005) Na_2SeO_3	The torula yeast–sucrose basal diet contained 0.005 mg/kg Se and was supplemented with sodium selenite Data analyzed by ANOVA and by hyperbolic regression analysis	Li and Sunde, 2016
Zinc						
Hatch–0.712 (1–21)	28–48 <28 <28 59 62	BW gain Feed efficiency Liver CuZnSOD[a] Pancreas Zn Bone Zn	Arbor Acres, male	8 (28–168) $ZnSO_4 \cdot 7H_2O$	The corn–soybean meal basal diet contained 28 mg/kg Zn Data were analyzed by a one-factor ANOVA and by quadratic and asymptotic models	Huang et al., 2007
Hatch–0.894 (1–21) 0.894–2.393 (22–42)	<33 <33 <33 <32 <32 <32	BW gain Feed efficiency Mortality BW gain Feed efficiency Mortality	Cobb 500, male	5 (33–182) Zn—methionine 5 (32–153)	The corn–soybean meal basal diet contained 33 mg/kg Zn in the starter diet (d 1–21) and 32 mg/kg in the finisher diet (d 22–42) Data were analyzed by ANOVA	Pacheco et al., 2017
0.039–0.425 (1–21)	85 45–105 93	BW gain Feed efficiency Tibia Zn	Lingnan yellow feather, male	6 (25–145) $ZnSO_4 \cdot 7H_2O$	The corn starch–soybean meal basal diet contained 25 mg/kg Zn Data were analyzed by a one-factor ANOVA and by nonlinear regression and broken line regression	Jiang et al., 2010a
8–22 d	27	BW gain	New Hampshire × Columbian, female	5 (14–32) $ZnSO_4 \cdot 7H_2O$	The soy concentrate–dextrose diet contained 14 mg/kg Zn and was fed for 7 d post-hatch to deplete zinc stores The amount of bioavailable zinc in the basal diet was 9 mg/kg Data were analyzed by broken line regression	Batal et al., 2001
Hatch–0.710 (1–21) 0.710–2.450 (22–42)	<35 <35 51–84 <28 <28 <28	BW gain Feed efficiency Tibia Zn BW gain Feed efficiency Tibia Zn	Ross 308	7 (35–145) $ZnSO_4 \cdot 7H_2O$ 9 (28–131)	The corn–soybean meal basal diet contained 35 and 28 mg/kg in the basal starter and grower diets, respectively Data were analyzed by ANOVA and means comparisons	Zhang et al., 2018

continued

TABLE 14-7 Continued

BW and Bird Age	Estimated Requirement	Response Criteria	Strain, Sex	Dietary Treatments and Mineral Form	Comments	Reference
d 8–0.600 g (d 8–21)	<12 <12 37–62 12–37	BW gain Feed efficiency Tibia Zn (males) Tibia Zn (females)	Avian Farms, .ales and females	6 (12–137) ZnO	The corn–soybean meal basal diet contained 12 mg/kg Zn Supplemental Zn was added as ZnO from a commercial sourced and purity unknown Data were analyzed by ANOVA	Gomes et al., 2008
(22–42)	<28 <28 62	BW gain Feed efficiency Tibia Zn	Arbor Acres, male	8 (28–166) $ZnSO_4{\cdot}7H_2O$	The corn starch–soybean meal basal diet contained 28 mg/kg Zn From hatch to the beginning of the experiment, all chicks were fed diets with 86 mg/kg Zn Data were analyzed by a one-factor ANOVA and by broken-line, quadratic, and asymptotic models	Liao et al., 2013
1.070–2.060 (43–63)	<18 <1860	BW gain Feed efficiency Tibia Zn	Lingnan yellow feather, male	6 (18–138) $ZnSO_4{\cdot}7H_2O$	The corn starch–soybean meal basal diet contained 18 mg/kg Zn From hatch to the beginning of the experiment, all chicks were fed diets with 40 mg/kg Zn Data were analyzed by a one-factor ANOVA and by nonlinear regression and broken-line regression	Jiang et al., 2010b

[a] Activity of succinate dehydrogenase (SDH); cytochrome C oxidase (COX); catalase (CAT); Mn superoxide dismutase (MnSOD), (MDA), glutathione peroxidase (GPX); and Cu-Zn superoxide dismutase (CuZnSOD).

oceans and other saline waters and low levels in crops grown in intracontinental regions. The previous report (NRC, 1994) tentatively recommended that 0.35 mg/kg diet of iodine was required, and this was based on a single study (Creek et al., 1957) examining the histology of the thyroid in response to potassium iodide supplementation in historical dual-purpose breeds and a laying breed. More recently, Roettger et al. (2011) examined the iodine requirement in broilers and found that a corn–wheat–soybean meal diet containing 0.5 mg/kg diet was adequate for BWG but between 0.5 and 5 mg I/kg were needed to optimize feed efficiency from hatch to d 35. Based on this study the tentative recommended requirement for iodine is increased to 0.5 mg I/kg in a cereal grain and soybean meal diet.

Iron

In the previous report (NRC, 1994), the recommended iron (Fe) requirement for broilers from hatch to slaughter was 80 mg Fe/kg diet based on two studies. Since that report, there have been two new Fe requirement studies (Table 14-6) in the early phase of production and two studies in later stages using broilers fed corn-soybean meal–based diets. Ma et al. (2016) estimated the requirement of Arbor Acres broilers from hatch to d 21 to be between 67 and 107 mg Fe/kg diet based on BWG (0.63 kg BW) and 97 mg Fe/kg based on tibia ash, but <67 mg Fe/kg for maximal plasma hemoglobin and hematocrit values. Lin et al. (2020) found that adding $FeSO_4{\cdot}H_2O$ to a basal corn–soybean meal diet with 48 mg Fe/kg did not increase BWG, feed efficiency, or hematocrit levels of Lingnan yellow-feather broilers from hatch to 374 g BW. The differing results between these two breeds might be related to the much slower growth rate of the Lingnan strain relative to most other broiler strains. Aoyagi and Baker (1995) examined the Fe requirements of Columbian × New Hampshire chicks and estimated it to be 85 mg/kg diet based on maximizing plasma hemoglobin from 8 to 21 d. Based on these studies and the previous report by McNaughton and Day (1979), the recommended requirement for starting broiler chicks fed corn-soybean meal–based diets remains unchanged at 80 mg/kg diet.

In the growing phase (22 to 42 d), Liao et al. (2017) reported that the iron requirement was <47 mg Fe/kg diet, and Lin et al. (2020) found that it was <49 mg Fe/kg to optimize BWG, feed efficiency, and hematocrit. Based on these two studies, the recommended requirements for Fe in the growing and finishing stages were 50 mg Fe/kg.

Manganese

The previous report (NRC, 1994) provided a tentative recommendation of 60 mg/kg diet of manganese (Mn) across all age groups. Because of the lack of studies using modern broilers, this was based on an old study (Gallup and Norris, 1939) using New Hampshire chicks fed a corn-dried milk diet. Since then, four studies (Li et al., 2011; Conly et al., 2012; Pacheco et al., 2017; Jasek et al., 2019) examined the Mn requirement of broilers in the early stage of production and none of them found an increase in BWG or feed efficiency as a result of adding additional Mn to the unsupplemented corn–soybean meal diets in which Mn concentrations ranged from 19 to 45 mg mg/kg. Li et al. (2011) found adding $MnSO_4{\cdot}H_2O$ to a basal diet containing 19 mg Mn/kg reduced the incidence of leg abnormalities at a level between 38 and 98 mg Mn/kg. Using quadratic modeling, they found that it took approximately 144 mg Mn/kg diet to maximize Mn-SOD activity. However, the activity of this enzyme has not been directly implicated in signs of manganese pathology, likely due to the high redundancy in tissue antioxidant enzymes. Additionally, Conly et al. (2012) did not observe increased Mn-SOD activity after adding Mn to the basal diet containing 45 mg Mn/kg from day 10 to 21 (221–731 g BW). Based on these studies, the recommended requirement for Mn is reduced to 45 mg Mn/kg diet in in the early stage of production when using diets based on corn and soybean meal.

Two studies (Pacheco et al., 2017; Jasek et al., 2019) examined the broiler's requirement for Mn from hatch to slaughter. Pacehco et al. (2017) did not observe an increase in BWG nor feed efficiency with Mn added to a corn–soybean meal diet containing 45 mg/kg in the starter diet and 32 mg/kg in the finisher diet. Jasek et al. (2019) did not observe an increase in BWG, but they did find that the requirement was 39 and 119 mg/kg for feed efficiency and tibial manganese content, respectively. There were no studies that depleted birds and then tested the requirements during the growing or finishing phases. Based on these two studies and those in hatchling broilers, the tentative recommended requirement is 45 mg Mn/kg diet for growing and finishing broilers.

Selenium

For most species, including chickens, the requirement for selenium (Se) is typically based on the prevention of pathology because pathology is frequently apparent at levels lower than those that maximize growth or feed efficiency in studies of short duration. The activity of Se-dependent enzymes that correlate with the expression of pathology are often used as a proxy for qualitative pathology scores, especially glutathione peroxidase (GPX). Accumulation of Se in tissues was not considered to be an appropriate endpoint because the Se in tissues is present as a substitution for sulfur in methionine or cysteine. In addition, only ionic Se is also available to act as a cofactor for seleno-enzymes (Sunde, 1997). In the previous report (NRC, 1994) the committee recommended 0.15 mg Se/kg diet during all phases of broiler production based primarily on the study of Jensen et al. (1986). Since then, there have been two additional studies using broilers that meet the inclusion requirements described previously. Li and Sunde (2016) used a purified diet containing 0.005 mg Se/kg and found that BWG was maximized at 0.03 mg Se/kg diet, while all isoforms of GPX in the liver, plasma, and gizzard were maximized at 13 mg Se/kg diet or less from hatch to 1,210 g BW. Pancreas GPX1 and GPX4 activity were maximized at 0.31 and 0.34 mg Se/kg diet, respectively, but no gross signs of pathology were reported in any of the treatment groups, indicating that the activity of these two isoforms of GPX were not correlated with gross pathology or BWG. In the later stage of production, Li et al. (2017) found that the basal corn–soybean meal diet containing 0.04 mg/kg Se was sufficient to optimize BWG and feed efficiency from hatch to 0.49 kg BW and that 0.13 mg Se/kg diet minimized liver levels of malondialdehyde, which is a by-product of lipid oxidation. Based on these studies, the recommended Se requirement remains at 0.15 mg/kg diet for all stages of production.

Zinc

In the previous report (NRC, 1994), the committee recommended 40 mg zinc (Zn)/kg diet for all stages of production based on nine studies conducted in the early stage of production. Only one study used a broiler strain, and the rest used historical dual-purpose breeds. In the broiler study, Dewar and Downie (1984) reported a requirement of 18 mg Zn/kg diet using a purified diet whereas most of the studies using historical dual-purpose breeds concluded that the requirement was closer to 40 mg Zn/kg diet using a variety of diet types. Since then, there have been many additional studies that meet the committee's inclusion criteria described previously. Most of these studies used modern broiler strains fed a corn-soybean meal–based diet with $ZnSO_4$ as the Zn source (Table 14-7).

Six studies examined Zn requirements in the early phase of production. Pacheco et al. (2017) found that the Zn requirement of Cobb broilers fed a basal corn–soybean meal diet with 33 mg Zn/kg diet was less than that that previously recommended because adding supplemental Zn did not improve indices of performance or mortality from hatch to 894 g BW. Zhang et al. (2018) found that the requirement for Ross broilers to be <35 mg Zn/kg diet based on BW gain and feed efficiency and 51 to 84 mg Zn/kg diet to maximize tibia Zn content from hatch to 710 g BW. Gomes et al. (2008) found a requirement of <12 mg Zn/kg diet for BWG and feed efficiency in Avian broilers, but maximizing tibia Zn required 37 to 62 mg/kg diet from hatch to 0–710 g BW. Huang et al. (2007) estimated a Zn requirement between 28 and 48 mg/kg diet for maximum BW gain from hatch to 712 kg BW, an outcome consistent with the NRC (1994) recommendation. Jiang et al. (2010a) used a slower-growing Lingnan yellow feather breed and found that 85 mg Zn/kg diet was required to maximize BWG from hatch to 425 g BW. Batal et al. (2001) estimated the requirement of a historical dual-purpose breed to be 27 mg Zn/kg diet for BWG. Based on these results, the recommended requirement for the Zn using a corn–soybean meal diet was decreased to 35 mg/kg diet; however, less heavily selected breeds might have a higher requirement, and maximizing storage in bone also requires higher levels.

Four studies (Gomes et al., 2008; Jiang et al., 2010b; Liao et al., 2013; Zhang et al., 2018) examined the Zn requirement in the later stages of production. In each of these studies the addition of Zn to the basal diets that had Zn concentrations ranging between 12 and 28 mg Zn/kg diet did not increase BWG or feed efficiency, although higher levels were needed to increase tibia Zn content. The average of the concentration of Zn in the basal diets of these studies was 22 mg/kg diet. Based on these studies, an adequate amount of Zn for grower and finisher broilers fed a corn–soybean meal diet is 22 mg/kg diet, but the requirement has yet to be determined. Phytase releases Zn from phytate and suggests that diets treated with phytase can be supplemented with lower concentrations of Zn due to the increased bioavailability of Zn in the plant-based ingredients (Zaghari et al., 2015).

REFERENCES

Abd El-Wahab, A., D. Radko, and J. Kamphues. 2013. High dietary levels of biotin and zinc to improve health of foot pads in broilers exposed experimentally to litter with critical moisture content. *Poultry Science*. 92:1774–1782.

Aburto, A., and W.M. Britton. 1998. Effects of different levels of vitamins A and E on the utilization of cholecalciferol by broiler chickens. *Poultry Science*. 77(4):570–577.

Akter, M., H. Graham, and P.A. Iji. 2019. Response of broiler chickens to diets containing different levels of sodium with or without microbial phytase supplementation. *Journal of Animal Science and Technology*. 61:87–97.

Allen, P.C., and R. Fetterer. 2002. Recent advances in biology and immunobiology of Eimeria species and in diagnosis and control of infection with these coccidian parasites of poultry. *Clinical Microbiology Reviews*. 15:58–65.

Alves, O.D.S., L.F.L. Calixto, A.H.B. Araujo, K.A.A. Torres-Cordido, T.L. Reis, and A.A. Calderano. 2018. Decreased levels of vitamin D3 and supplementation with 1, 25-dihydroxyvitamin D3-glycoside on performance, carcass yield and bone quality in broilers. *Ciencia Rural*. 48.

Aoyagi, S., and D.H. Baker. 1995. Iron requirement of chicks fed a semipurified diet based on casein and soy protein concentrate. *Poultry Science*. 74:412–415.

Augspurger, N.R., and D.H. Baker. 2007. Excess dietary lysine increases growth of chicks fed niacin-deficient diets, but dietary quinolinic acid has no niacin-sparing activity. *Poultry Science*. 86:349–355.

Atencio, A., H.M. Edwards Jr., and G. Pesti. 2005. Effects of vitamin D3 dietary supplementation of broiler breeder hens on the performance and bone abnormalities of the progeny. *Poultry Science*. 84:1058–1068.

Azghadi, M.A., H. Kermanshahi, A. Golian, R. Kadkhodaee, and A.R. Vakili. 2020. Effect of encapsulated choline chloride on performance, carcass characteristics and some blood parameters in broiler chicks. *Poultry Science Journal*. 8:51–58.

Baker, D.H., A.B. Batal, T.M. Parr, N.R. Augspurger, and C.M. Parsons. 2002. Ideal ratio (relative to Lys) of tryptophan, threonine, isoleucine, and valine for chicks during the second and third wk posthatch. *Poultry Science*. 81:485–494.

Baker, D.H., R.R. Biehl, and J.L. Emmert. 1998. Vitamin D3 requirement of young chicks receiving diets varying in calcium and available phosphorus. *British Poultry Science*. 39(3):413–417.

Bar, A., D. Shinder, S. Yosefi, E. Vax, and I. Plavnik. 2003. Metabolism and requirements for calcium and phosphorus in the fast growing chicken as affected by age. *British Journal of Nutrition*. 89:51–60.

Barroeta, A.C. 2007. Nutritive value of poultry meat: Relationship between vitamin E and PUFA. *World's Poultry Science Journal*. 63:277–284.

Batal, A.B., T.M. Parr, and D.H. Baker. 2001. Zinc bioavailability in tetrabasic zinc chloride and the dietary zinc requirement of young chicks fed a soy concentrate diet. *Poultry Science*. 80:87–90.

Beura, T.K., N. Panda, P. Mishra, B. Panigrahi, H.K. Panda, and P.K. Pati. 2012. Effect of vitamin E and C on the growth and immunocompetence of coloured birds during summer. *Animal Nutrition and Feed Technology*. 12:13–24.

Bou, R., S. Grimpa, F. Guardiola, A. Barroeta, and R. Codony. 2006. Effects of various fat sources, tocopheryl acetate, and ascorbic acid supplements on fatty acid composition and tocopherol content in raw and vacuum-packed, cooked dark chicken meat. *Poultry Science*. 85:1472–1481.

Cai, Z., J.W. Finnie, P.C. Blumbergs, J. Manavis, M.N. Ghabriel, and P.D. Thompson. 2006a. Early paranodal myelin swellings (tomacula) in an avian riboflavin deficiency model of demyelinating neuropathy. *Experimental Neurology*. 198:65–71.

Cai, Z., J.W. Finnie, and P.C. Blumbergs. 2006b. Avian riboflavin deficiency: An acquired tomaculous neuropathy. *Veterinary Pathology*. 43:780–781.

Castro F.L.S., L. Chai, J. Arango, C.M. Owens, P.A. Smith, S. Reichelt, C. DuBois, and A. Menconi. 2023. Poultry industry paradigms: Connecting the dots. *Journal of Applied Poultry Research*. 32:100310.

Cemin, H.S., S.L. Vieira, C. Stefanello, M. Kipper, L. Kindlein, and A. Helmbrecht. 2017. Digestible Lys requirements of male broilers from 1 to 42 days of age. *PloS One*. 12:e0179665.

Cengiz, O., J.B. Hess, and S.F. Bilgili. 2012. Dietary biotin supplementation does not alleviate the development of footpad dermatitis in broiler chickens. *Journal of Applied Poultry Research*. 21:764–769.

Chen, B., T.F. Shen, and R. E. Austic. 1996. Efficiency of tryptophan-niacin conversion in chickens and ducks. *Nutrition Research*. 16:91–104.

Cho, T.Z.A., M.B. Sadiq, P. Srichana, and A.K. Anal. 2020. Vitamin D3 enhanced intestinal phosphate cotransporter genes in young and growing broilers. *Poultry Science*. 99:2041–2047.

Chou, S.H., T.K. Chung, and B. Yu. 2009. Effects of supplemental 25-hydroxycholecalciferol on growth performance, small intestinal morphology, and immune response of broiler chickens. *Poultry Science*. 88:2333–2341.

Colet, S., R. Garcia, I.C.L. Paz, F. Caldara, R. Borille, A.F.B. Royer, I.A. Nääs, and S. Sgavioli. 2015. Bone Characteristics of Broilers Supplemented with Vitamin D. *Revista Brasileira de Ciência Avícola*. 17:325–332.

Conly, A., R. Poureslami, E. Koutsos, A. Batal, B. Jung, R. Beckstead, and D. Peterson. 2012. Tolerance and efficacy of tribasic manganese chloride in growing broiler chickens. *Poultry Science*. 91:1633–1640.

Corrales, N.L., F. Sevillano, R. Escudero, G.G. Mateos, and D. Menoyo. 2023. Replacement of vitamin E by an extract from an olive oil by-product, rich in hydroxytyrosol, in broiler diets: Effects on growth performance and breast meat quality. *Antioxidants*. 12:1940.

Corzo, A., M.T. Kidd, W.A. Dozier III, and S.L. Vieira. 2007. Marginality and needs of dietary valine for broilers fed certain all-vegetable diets. *Journal of Applied Poultry Research*. 16:546–554.

Corzo, A., W.A. Dozier III, and M.T. Kidd. 2008. Valine nutrient recommendations for Ross × Ross 308 broilers. *Poultry Science*. 87:335–338.

Corzo, A., R. Loar, and M. Kidd. 2009. Limitations of dietary isoleucine and valine in broiler chick diets. *Poultry Science*. 88:1934–1938.

Creek, R., H. Parker, S. Hauge, F. Andrews, and C. Carrick. 1957. The iodine requirements of young chickens. *Poultry Science*. 36:1360–1365.

da Silva, I.C.M, A.M. Leal Ribeiro, C. Wageck Canal, C.C. Pinheiro, M. de Moraes Vieira, T.A. Gonçalves, R. Alves Pereira, and L. Lacerda. 2009. Broiler chicken responses to immunological stimuli as mediated by different levels of vitamin E in the diet. *Journal of Applied Poultry Research*. 18:752–760.

da Silva, I.C.M., A.M.L. Ribeiro, C.W. Canal, M.M. Vieira, C.C. Pinheiro, T. Goncalves, M.L. de Moraes, and V.S. Ledur. 2011. Effect of vitamin E levels on the cell-mediated immunity of broilers vaccinated against *Coccidiosis*. *Brazilian Journal of Poultry Science*. 13:53–56.

de Lima, M.B., E.P. da Silva, R. Pereira, G.G. Romano, L.W. de Freitas, C.T.S. Dias, and J.F.M. Menten. 2018. Estimate of choline nutritional requirements for chicks from 1 to 21 days of age. *Journal of Animal Physiology and Animal Nutrition*. 102:780–788.

David, L.S., M.R. Abdollahi, M.R. Bedford, and V. Ravindran. 2021. Requirement of digestible Ca at different dietary concentrations of digestible P for broiler chickens. 1. Broiler starters (d 1 to 10 post-hatch). *Poultry Science*. 100:101439.

Dewar, W., and J. Downie. 1984. The zinc requirements of broiler chicks and turkey poults fed on purified diets. *British Journal of Nutrition*. 51:467–477.

Deyhim, F., T. Belay, and R.G. Teeter. 1992. An evaluation of dietary pantothenic acid needs of broilers through eight wk posthatching. *Nutrition Research*. 12:1549–1554.

Dhandu, A.S., and R. Angel. 2003. Broiler nonphytin phosphorus requirement in the finisher and withdrawal phases of a four phase feeding program. *Poultry Science*. 82:1257–1265.

Dozier III, W.A., A. Corzo, M.T. Kidd, and M.W. Schilling. 2008. Dietary digestible Lys requirements of male and female broilers from forty-nine to sixty-three days of age. *Poultry Science*. 87:1385–1391.

Dozier III, W.A., and Y. Mercier. 2013. Ratio of digestible total sulfur amino acids to Lys of broiler chicks from 1 to 15 days of age. *Journal of Applied Poultry Research*. 22:862–871.

Dozier III, W.A., A. Corzo, M.T. Kidd, P.B. Tillman, and S.L. Branton. 2009a. Digestible Lys requirements of male and female broilers from fourteen to twenty-eight days of age. *Poultry Science*. 88:1676–1682.

Dozier III, W.A., A. Corzo, M.T. Kidd, P.B. Tillman, J.L. Purswell, and B.J. Kerr. 2009b. Dietary Lys responses of male broilers from fourteen to twenty-eight days of age subjected to different environmental conditions. *Journal of Applied Poultry Research*. 18:690–698.

Dozier III, W.A., A. Corzo, M.T. Kidd, P.B. Tillman, J.P. McMurtry, and S.L. Branton. 2010. Digestible Lys requirements of male broilers from 28 to 42 days of age. *Poultry Science*. 89:2173–2182.

Dozier III, W.A., and R.L. Payne. 2012. Digestible Lys requirement of female broilers from 14 to 28 days of age. *Journal of Applied Poultry Research*. 21:348–357.

Dozier III, W.A., K.J. Meloche, P.B. Tillman, and Z. Jiang. 2015. Growth performance of male broilers fed diets varying in digestible threonine to Lys ratio from 1 to 14 days of age. *Journal of Applied Poultry Research*. 24:457–462.

Dozier III, W.A., P.B. Tillman, and Z. Jiang. 2016. Growth performance and carcass characteristics of male broilers fed diets varying in digestible threonine to Lys ratio from 21 to 49 days of age. *Journal of Applied Poultry Research*. 25:571–580.

Driver, J.P., G.M. Pesti, R.I. Bakalli, and H.M. Edwards Jr. 2005. Phytase and 1α-hydroxycholecalciferol supplementation of broiler chickens during the starting and growing/finishing phases. *Poultry Science*. 84:1616–1628.

Du, Z., G. Zhou, and Y. Han. 1996. Study on copper requirement for broilers. *Journal Acta Zoonutrimenta Sinica*. 4.

Duarte, C., F. Bratti, A. Murakami, J. Fernandes, I. Ospina-Rojas, and A. Furlan. 2014. Effects of vitamin K-3 supplementation on growth performance and bone quality of broiler chickens. *Archivos de Medicina Veterinaria*. 46:305–313.

Eder, K., G. Grünthal, H. Kluge, F. Hirche, J. Spilke, and C. Brandsch. 2005. Concentrations of cholesterol oxidation products in raw, heat-processed and frozen-stored meat of broiler chickens fed diets differing in the type of fat and vitamin E concentrations. *British Journal of Nutrition*. 93:633–643.

Edwards Jr., H.M., M.A. Elliot, S. Soonchareryinying, and W.M. Britton. 1994. Quantitative requirement for cholecalciferol in the absence of ultraviolet light. *Poultry Science*. 73:288–294.

Elfick, D. 2011. A brief history of broiler selection: How chicken became a global food phenomenon in 50 yr. Huntsville, AL: Aviagen.

Emmerson, D.A. 1997. Commercial approaches to genetic selection for growth and feed conversion in domestic poultry. *Poultry Science*. 76:1121–1125.

Englmaierová, M., I. Bubancovà, T. Vit, and M. Skøivan. 2011. The effect of lycopene and vitamin E on growth performance, quality and oxidative stability of chicken leg meat. *Czech Journal of Animal Science*. 56:536–543.

Fairchild, B.D., and C.W. Ritz. 2015. *Poultry Drinking Water Primer*. Bulletin 1301. Athens, GA: University of Georgia Extension.

FAO (UN Food and Agriculture Organization). 2021. *World Food and Agriculture—Statistical Yearbook 2021*. Rome, Italy: FAO.

Farina, G., A.D.M. Kessler, P.D. Ebling, F.R. Marx, R. César, and A.M.L. Ribeiro. 2017. Performance of broilers fed different dietary choline sources and levels. *Ciência Animal Brasileira*. 18:e-37633.

Friedman, A., I. Bartov, and D. Sklan. 1998. Humoral immune response impairment following excess vitamin E nutrition in the chick and turkey. *Poultry Science*. 77:956–962.

Fritts, C.A., and P.W. Waldroup. 2003. Effect of source and level of vitamin D on live performance and bone development in growing broilers. *Journal of Applied Poultry Research*. 12:45–52.

Fritts, C., and P. Waldroup. 2005. Comparison of cholecalciferol and 25-hydroxycholecalciferol in broiler diets designed to minimize phosphorus excretion. *Journal of Applied Poultry Research*. 14(1):156–166.

Gallup, W.D., and L. Norris. 1939. The amount of manganese required to prevent perosis in the chick. *Poultry Science*. 18:76–82.

Garcia, A.F., A.E. Murakami, C.R. Duarte, I.C. Rojas, K.P. Picoli, and M.M. Puzotti. 2013. Use of vitamin D3 and its metabolites in broiler chicken feed on performance, bone parameters, and meat quality. *Asian-Australasian Journal of Animal Science*. 26:408–415.

Garcia-Neto, M., and S.H.V. Perri. 2015. *PPM: Programa prático de modelagem*. Disponivelem.

Geyer, J., M. Netzel, I. Bitsch, T. Frank, R. Bitsch, K. Krämer, and P.P. Hoppe. 2000. Bioavailability of water- and lipid-soluble thiamin compounds in broiler chickens. *International Journal for Vitamin and Nutrition Research. Internationale Zeitschrift für Vitamin- und Ernährungsforschung. Journal International de Vitaminologie et de Nutrition*. 70:311–316.

Ghasemi, P., M. Toghyani, and N. Landy. 2019. Effects of dietary 1 alpha-hydroxycholecalciferol in calcium- and phosphorous-deficient diets on growth performance, tibia related indices and immune responses in broiler chickens. *Animal Nutrition*. 5:134–139.

Gibbs, C.S. 1934. Sexing baby chicks. *Poultry Science*. 8:208–211.

Goff, J.P., T.A. Reinhardt, and R.L. Horst. 1995. Milk fever and dietary cation-anion balance effects on concentration of vitamin D receptor in tissue of periparturient dairy cows. *Journal of Dairy Science*. 78(11): 2388–2394.

Gomes, P.C., D.C.M. Rigueira, H.S. Rostagno, L.F.T. Albino, G. Brumano, and M. Schmidt. 2008. Zinc requirements for male and female broilers in the initial phase. *Revista Brasileira de Zootecnia*. 37:79–83.

Gómez-Verduzco, G., R. Morales-Lopez, and E. Avila-Gozalez. 2013. Use of 25-hydroxycholecalciferol in diets of broiler chickens: Effects on growth performance, immunity and bone calcification. *Journal of Poultry Science*. 50:60–64.

Gompertz, B. 1825. On the nature of the function expressive of the law of human mortality, and on a new mode of determining the value of life contingencie. *Philosophical Transactions of the Royal Society of London*. 115:513–583.

Gonçalves, C.A., N.K. Sakomura, M.F. Sarcinelli, L.G. Pacheco, L. Soares, M.C. Melaré, W.J. Alves, and R.M. Gous. 2020. In vivo assessment of body composition and growth potential of modern broiler using dual-energy X-ray absorptiometry. *Animal Production Science*. 60:1959–1968.

Griffin, H.D., and C. Goddard. 1994. Rapidly growing broiler (meat-type chickens): Their origin and use for comparative studies of the regulation of growth. *International Journal of Biochemistry*. 26:19–28.

Guo, S., J. Xv, Y. Li, Y. Bi, Y. Hou, and B. Ding. 2020. Interactive effects of dietary vitamin K3 and *Bacillus subtilis* PB6 on the growth performance and tibia quality of broiler chickens with sex separate rearing. *Animal*. 14:1–9.

Habibian, M., S. Ghazi, M. Moeini, and A. Abdolmohammadi. 2014. Effects of dietary selenium and vitamin E on immune response and biological blood parameters of broilers reared under thermoneutral or heat stress conditions. *International Journal of Biometeorology*. 58:741–752.

Halle, I., M. Henning, and P. Kohler. 2011. Influence of vitamin B-12 and cobalt on growth of broiler chickens and Pekin ducks. *Landbauforschung*. 61:299–306.

Han, J.C., G.H. Chen, J.L. Zhang, J.G. Wang, H.X. Qu, Y.F. Yan, X.J. Yang, and Y.H. Cheng. 2017. Relative biological value of 1α-hydroxycholecalciferol to 25-hydroxycholecalciferol in broiler chicken diets. *Poultry Science*. 96:2330–2335.

Hancock, A., J. Hughes, and S. Watkins. 2007. In search of the ideal water line cleaner. *Avian Advice*. 9:1–4. Fayetteville, AR: University of Arkansas Cooperative Extension Service.

Harms, R.H., and D.S. Nelson. 1992. A lack of response to pantothenic-acid supplementation to a corn and soybean-meal broiler diet. *Poultry Science*. 71:1952–1954.

Havenstein, G.B., P.R. Ferket, and Qureshi M.A. 2003. Growth, livability, and feed conversion of 1957 versus 2001 broilers when fed representative 1957 and 2001 broiler diets. *Poultry Science*. 82:1500–1508.

Hiltz, J.Z., S.K. Orlowski, L.N. Harrington, C.W. Maynard, T.W. Tabler, and N.B. Anthony. 2021. Applied research note: Development of a novel low flow water monitoring system in poultry/agriculture systems. *Journal of Applied Poultry Research*. 30:100151.

Huang, Y.L., L. Lu, X.G. Luo, and B. Liu. 2007. An optimal dietary zinc level of broiler chicks fed a corn-soybean meal diet. *Poultry Science*. 86:2582–2589.

Hurnik, J.F., S.P. Webster, and P.B. Siegel. 1995. *Dictionary of Farm Animal Behavior*. Ames, IA: Iowa State University Press.

Ipek, A.Y.D.I.N., and U. Sahan. 2006. Effects of cold stress on broiler performance and ascites susceptibility. *Asian-Australasian Journal of Animal Science*. 19:734–738.

Jasek, A., C.D. Coufal, T.M. Parr, and J.T. Lee. 2019. Evaluation of increasing manganese hydroxychloride level on male broiler growth performance and tibia strength. *Journal of Applied Poultry Research*. 28:1039–1047.

Jensen, L., G. Colnago, K. Takahashi, and Y. Akiba. 1986. Dietary Se status and plasma thyroid hormones in chicks. *Biological Trace Element Research*. 10:11–18.

Jiang, R.R., G.P. Zhao, J.L. Chen, M.Q. Zheng, J.P. Zhao, P. Li, J. Hu, and J. Wen. 2011. Effect of dietary supplemental nicotinic acid on growth performance, carcass characteristics and meat quality in three genotypes of chicken. *Journal of Animal Physiology and Animal Nutrition*. 95:137–145.

Jiang, Z., X. Liu, S. Jiang, G. Zhou, Y. Lin, and F. Chen. 2010a. Zinc requirement of yellow broilers aged from one to twenty-one days. *Chinese Journal of Animal Nutrition*. 22:301–309.

Jiang, Z., X. Liu, S. Jiang, G. Zhou, Y. Lin, F. Chen, and X. Ma. 2010b. Zinc requirement of yellow broilers from forty-three to sixty-three days of age. *Scientia Agricultura Sinica*. 43:4295–4302.

Jiang, R.R., G.P. Zhao, J.P. Zhao, J.L. Chen, M.Q. Zheng, R.R. Liu, and J. Wen. 2014. Influence of dietary nicotinic acid supplementation on lipid metabolism and related gene expression in two distinct broiler breeds of female chickens. *Journal of Animal Physiology and Animal Nutrition*. 98:822–829.

Kakhki, R.A.M., R. Bakhshalinejad, and M. Shafiee. 2016. Effect of dietary zinc and α-tocopheryl acetate on broiler performance, immune responses, antioxidant enzyme activities, minerals and vitamin concentration in blood and tissues of broilers. *Animal Feed Science and Technology*. 221:12–26.

Kakhki, R.A.M., R. Bakhshalinejad, A. Hassanabadi, and P. Ferket. 2017. Effects of dietary organic zinc and α-tocopheryl acetate supplements on growth performance, meat quality, tissues minerals, and α-tocopherol deposition in broiler chickens. *Poultry Science*. 96:1257–1267.

Kasim, A.B., and H.M. Edwards Jr. 2000. Evaluation of cholecalciferol sources using broiler chick bioassays. *Poultry Science*. 79:1617–1622.

Katanbaf, M.N., and J.W. Hardiman. 2010. Primary broiler breeding—Striking a balance between economic and well-being traits. *Poultry Science*. 89:822–824.

Khan, S.H., R. Shahid, A.A. Mian, R. Sardar, and M.A. Anjum. 2010. Effect of the level of cholecalciferol supplementation of broiler diets on the performance and tibial dyschondroplasia. *Journal of Animal Physiology and Animal Nutrition*. 94:584–593.

Khatun, J., T. Loh, H.L. Foo, H. Akit, R. Mohamad, and N. Shazali. 2020. Effects of vitamin E, an oil blend and L-Arginine on breast meat from broiler chickens. *South African Journal of Animal Science*. 50:745–757.

Kheiri, F., and N. Landy. 2019. The effects of dietary 1-alpha-hydroxycholecalciferol individually or in combination with different levels of cholecalciferol on growth performance and tibia criteria in broiler chickens. *Livestock Science*. 221:172–176.

Kim, W.K., S.A. Bloomfield, and S.C. Ricke. 2011. Effects of age, vitamin D3, and fructooligosaccharides on bone growth and skeletal integrity of broiler chicks. *Poultry Science*. 90:2425–2432.

Klasing, K.C., and D.R. Korver. 2020. Nutritional diseases. *Diseases of Poultry*. 1255–1285.

Konjufca, V.K., W.G. Bottje, T.K. Bersi, and G.F. Erf. 2004. Influence of dietary vitamin E on phagocytic functions of macrophages in broilers. *Poultry Science*. 83:1530–1534.

Ledwaba, M.F., and K.D. Roberson. 2003. Effectiveness of twenty-five-hydroxycholecalciferol in the prevention of tibial dyschondroplasia in Ross cockerels depends on dietary Ca level. *Poultry Science*. 82:1769–1777.

Leshchinsky, T.V., and K.C. Klasing. 2001. Relationship between the level of dietary vitamin E and the immune response of broiler chickens. *Poultry Science*. 80:1590–1599.

Leyva-Jimenez, H. 2015. Evaluation of the dietary vitamin D3 requirement of modern broiler chickens and partial replacement of vitamin D3 with 25-hydroxycholecalciferol. Dissertation. *Texas A&M*.

Leyva-Jimenez, H., K. Gardner, Y. Al-Jumaa, J. Padgett, and C. Bailey. 2019a. Partial replacement of dietary cholecalciferol with 25-hydroxycholecalciferol on broiler chickens subjected to *a Coccidiosis* vaccine challenge. *Journal of Applied Poultry Research*. 28:743–754.

Leyva-Jimenez, H., M. Khan, K. Gardner, R.A. Abdaljaleel, Y. Al-Jumaa, A.M. Alsadwi, and C.A. Bailey. 2019b. Developing a novel oral vitamin D3 intake bioassay to re-evaluate the vitamin D3 requirement for modern broiler chickens. *Poultry Science*. 98:3770–3776.

Li, J., and R.A. Sunde. 2016. Selenoprotein transcript level and enzyme activity as biomarkers for Se status and Se requirements of chickens (*Gallus gallus*). *PLoS One*. 11.

Li, L., S. Jiang, C. Zheng, Z. Gou, F. Chen, D. Ruan, and X. Lin. 2014. Copper requirement of male yellow-feathered broilers aged from 43 to 63 days. *Chinese Journal of Animal Nutrition*. 26:3266–3275.

Li, L., S. Jiang, C. Zheng, Z. Gou, F. Chen, D. Ruan, and D. Yu. 2015. Copper requirement of yellow-feathered broilers aged from 1 to 21 days. *Chinese Journal of Animal Nutrition*. 27:578–587.

Li, L., S. Jiang, C. Zheng, Z. Gou, F. Chen, Y. Xiong, Y. Qiu, and X. Xie. 2017. Optimal dietary Se allowance for Lingnan yellow-feathered broilers aged from 1 to 21 days. *Chinese Journal of Animal Nutrition*. 29:1141–1147.

Li, S., Y. Lin, L. Lu, L. Xi, Z. Wang, S. Hao, L. Zhang, K. Li, and X. Luo. 2011. An estimation of the manganese requirement for broilers from 1 to 21 days of age. *Biological Trace Element Research*. 143:939–948.

Li J., J. Yuan, Z. Miao, Z. Song, Y. Yang, W. Tian. 2016. Effect of dietary nutrient density on small intestinal phosphate transport and bone mineralization of broilers during the growing period. *PLoS One*. 11(4): e0153859.

Liao, X., A. Li, L. Lu, S. Liu, S. Li, L. Zhang, G. Wang, and X. Luo. 2013. Optimal dietary zinc levels of broiler chicks fed a corn–soybean meal diet from 22 to 42 days of age. *Animal Production Science*. 53:388–394.

Liao, X., C. Ma, L. Lu, L. Zhang, and X. Luo. 2017. Determination of dietary iron requirements by full expression of iron-containing cytochrome c oxidase in the heart of broilers from 22 to 42 d of age. *British Journal of Nutrition*. 118:493–499.

Lin, X., Z. Gou, Y. Wang, L. Li, Q. Fan, F. Ding, C. Zheng, and S. Jiang. 2020. Effects of dietary iron level on growth performance, immune organ indices and meat quality in Chinese Yellow broilers. *Animals*. 10(4):670.

Liu, L., M. Ren, K. Ren, Y. Jin, and M. Yan. 2020. Heat stress impacts on broiler performance: A systematic review and meta-analysis. *Poultry Science*. 99:6205–6211.

Liu, S.B., X.D. Liao, L. Lu, S.F. Li, L. Wang, L.Y. Zhang, Y. Jiang, and X.G. Luo. 2017. Dietary non-phytate phosphorus requirement of broilers fed a conventional corn-soybean meal diet from 1 to 21 d of age. *Poultry Science*. 96:151–159.

Liu, X., J.A. Byrd, M. Farnell, and C.A. Ruiz-Feria. 2014. Arginine and vitamin E improve the immune response after a Salmonella challenge in broiler chicks. *Poultry Science*. 93(4):882–890.

Lu, T., A.F. Harper, J. Zhao, and R.A. Dalloul. 2014. Effects of a dietary antioxidant blend and vitamin E on growth performance, oxidative status, and meat quality in broiler chickens fed a diet high in oxidants. *Poultry Science*. 93:1649–1657.

Ma, X., X. Liao, L. Lu, S. Li, L. Zhang, and X. Luo. 2016. Determination of dietary iron requirements by full expression of iron-containing enzymes in various tissues of broilers. *Journal of Nutrition*. 146:2267–2273.

Maharjan, P., T. Clark, C. Kuenzel, M.K. Foy, and S. Watkins. 2017. On farm monitoring of the impact of water system sanitation on microbial levels in broiler house water supplies. *Journal of Applied Poultry Research*. 25:266–271.

Maynard, C.W., S.Y. Liu, J.T. Lee, J.V. Caldas, J.J.E. Diehl, S.J. Rochell, S. Dridi, and M.T. Kidd. 2021. Determination of digestible valine requirements in male and female Cobb 500 broilers. *Journal of Animal Feed Science and Technology*. 275:114847.

McNaughton, J.L., and E.J. Day. 1979. Effect of dietary Fe to Cu ratios on hematological and growth responses of broiler chickens. *The Journal of Nutrition*. 109:559–564.

Mehri, M., H. Nassiri-Moghaddam, H. Kermanshahi, and M. Danesh-Mesgaran. 2010. Digestible Lys requirements of straight-run broiler chickens from fifteen to twenty-eight days of age. *Journal of Animal Veterinary Advances*. 9:2321–2324.

Mejia, L., C.D. McDaniel, and A. Corzo. 2012. Dietary influence of digestible Lys concentration on Cobb 500 hen broiler breeder reproductive performance. *Poultry Science*. 91:426–431.

Michalczuk, M., J. Niemiec, D. Pietrzak, and J., Mroczek. 2010. Effectiveness of vitamin D3 and calcidiol (25-OH-D3) application in feeding broiler chickens: Production performance and meat quality. *Polish Journal of Food and Nutrition Sciences*. 60:121–126.

Mitchell, R.D., and H.M. Edwards. 1996. Additive effects of 1,25-dihydroxycholecalciferol and phytase on phytate phosphorus utilization and related parameters in broiler chickens. *Poultry Science*. 75:111–119.

Mitchell, R.D., H.M. Edwards Jr., and G.R. McDaniel. 1997. The effects of ultraviolet light and cholecalciferol and its metabolites on the development of leg abnormalities in chickens genetically selected for a high and low incidence of tibial dyschondroplasia. *Poultry Science*. 76:346–354.

Muir, W.I., A.J. Husband, and W.L. Bryden. 2002. *Vitamin E Modulation of Avian Intestinal Immunity*. 11th International Congress of Mucosal Immunology (ICMI), Orlando, Florida, USA. Bethesda, MD: Society for Mucosal Immunology (SMI).

Murakami, A.E., E.O. Oviedo-Rondon, E.N. Martins, M.S. Pereira, and C. Scapinello. 2001. Sodium and chloride requirements of growing broiler chickens (twenty-one to forty-two days of age) fed corn-soybean diets. *Poultry Science*. 80:289–294.

Murakami, A.E., S.E. Watkins, E.A. Saleh, J.A. England, and P.W. Waldroup. 1997. Estimation of the sodium and chloride requirements for the young broiler chick. *Journal of Applied Poultry Research*. 6:155–162.

Nameghi, A.H., H.N. Moghaddam, J.T. Afshari, and H. Kermanshahi. 2007. Effects of vitamin E and C supplementation on performance and immune response of broiler chicks. *Journal of Animal and Veterinary Advances*. 6:1060–1069.

Nascimento, M.Q.D., R.M. Gous, M.D.P. Reis, J.B.K. Fernandes, and N. Sakomura. 2020. Prediction of maximum scaled feed intake in broiler chickens based on physical properties of bulky feeds. *British Poultry Science*. 61:676–683.

Niu, Z.Y., F.Z. Liu, Q.L. Yan, and W.C. Li. 2009. Effects of different levels of vitamin E on growth performance and immune responses of broilers under heat stress. *Poultry Science*. 88:2101–2107.

Niu, Z., Y. Min, and F. Liu. 2017. Dietary vitamin E improves meat quality and antioxidant capacity in broilers by upregulating the expression of antioxidant enzyme genes. *Journal of Applied Animal Research*. 46:1–5.

NRC (National Research Council). 1994. *Nutrient Requirements of Poultry*, 9th rev. ed. Washington, DC: National Academy Press.

NRC. 2012. *Nutrient Requirements of Swine*, 11th rev. ed. Washington, DC: The National Academies Press.

Oduho, G.W., and D.H. Baker. 1993. Quantitative efficacy of niacin sources for chicks: Nicotinic acid, nicotinamide, NAD and tryptophan. *The Journal of Nutrition*. 123:2201–2206.

Oliveira, W.P.d., R.F.M.d. Oliveira, J.L. Donzele, L.F.T. Albino, P.H. Reis, F. Campos, E.M. Balbino, A.P.d.A. Maia, and S.M. Pastore. 2013. Lysine levels in diets for broilers from 8 to 21 days of age. *Revista Brasileira de Zootecnia*. 42:869–878.

Olkowski, A.A., and H.L. Classen. 1996. The study of thiamine requirement in broiler chickens. *International Journal of Vitamin and Nutrient Research*. 66:332–341.

Olkowski, A.A., and H.L. Classen. 1999. The effects of maternal thiamine nutrition on thiamine status of the offspring in broiler chickens. *International Journal for Vitamin and Nutrition Research*. 1:32.

Oloyo, R.A. 1997. Effect of guineacorn palm kernel-meal based diet with niacin supplementation on the performance of broilers. *Indian Journal of Animal Sciences*. 67:422–425.

Oloyo, R.A. 2000. Niacin requirement of broilers fed on a diet based on maize-palm kernel meal. *Pertanika Journal of Tropical Agricultural Science*. 23:37–41.

Pacheco, B.H.C., V.S. Nakagi, E.H. Kobashigawa, A.R.M. Caniatto, D.E. Faria, and D.E. Faria Filho. 2017. Dietary levels of zinc and manganese on the performance of broilers between 1 to 42 days of age. *Brazilian Journal of Poultry Science*. 19:171–178.

Persia, M.E., and W.W. Saylor. 2006. Effects of broiler strain, dietary non-phytate phosphorus, and phytase supplementation on chick performance and tibia ash. *Journal of Applied Poultry Research*. 15:72–81.

Pesti, G.M., S.V. Amato, and L.R. Minear. 1985. Water consumption of broiler chickens under commercial conditions. *Poultry Science*. 64:803–808.

Pesti, G.M., G.N. Rowland, and K.S. Ryu. 1991. Folate deficiency in chicks fed diets containing practical ingredients. *Poultry Science*. 70:600–604.

Peters, J., O. Lebrasseur, E.K. Irving-Pease, P.D. Paxinos, J. Best, R. Smallman, C. Callou, A. Gardeisen, S. Trixl, L. Frantz, N. Sykes, D.Q. Fuller, and G. Larsond. 2022. The biocultural origins and dispersal of domestic chickens. *PNAS*. 119:24.

Pitargue, F.M., J.H. Kim, D. Goo, J.B. Delos Reyes, and D.Y. Kil. 2019. Effect of vitamin E sources and inclusion levels in diets on growth performance, meat quality, alpha-tocopherol retention, and intestinal inflammatory cytokine expression in broiler chickens. *Poultry Science*. 98:4584–4594.

Pompeu, M.A., L.F.L. Cavalcanti, and F.L.B. Toral. 2018. Effect of vitamin E supplementation on growth performance, meat quality, and immune response of male broiler chickens: A meta-analysis. *Livestock Science*. 208:5–13.

Poudel, S., L. Zhang, G.T. Tabler, J. Lin, and W. Zhai. 2021. Effects of riboflavin and *Bacillus subtilis* on internal organ development and intestinal health of Ross 708 male broilers with or without coccidial challenge. *Poultry Science*. 100:100973.

Prokoski, K., L.C. Bittencourt, L.V. Teixeira, C. Bortoluzzi, E. Vanroo, S. Palma, and J.I.M. Fernandes. 2021. Classic and non-classic effects of the duration of supplementation of 25-hydroxicholecalciferol in broiler chicken diets. *Animals*. 11:2971.

Quarantelli, A., A. Cacchioli, S. Romanelli, F. Righi, I. Alpigiani, and C. Gabbi. 2007. Effects of different levels of dietary biotin on the performance and bone structure of broilers. *Italian Journal of Animal Science*. 6:5–17.

Rao, S.V.R., M.V.L.N. Raju, A.K. Panda, G.S. Sunder, and R.P. Sharma. 2006. Effect of high concentrations of cholecalciferol on growth, bone mineralization, and mineral retention in broiler chicks fed suboptimal concentrations of Ca and nonphytate phosphorus. *Journal of Applied Poultry Research*. 15:493–501.

Rao, S.V.R., M.V.L.N. Raju, and M.R. Reddy. 2007. Performance of broiler chicks fed high levels of cholecalciferol in diets containing sub-optimal levels of calcium and non-phytate phosphorus. *Animal Feed Science and Technology*. 134:77–88.

Ravindran, V., A.J. Cowieson, and P.H. Selle. 2008. Influence of dietary electrolyte balance and microbial phytase on growth performance, nutrient utilization, and excreta quality of broiler chickens. *Poultry Science*. 87:677–688.

Roettger, A.S., I. Halle, H. Wagner, G. Breves, and G. Flachowsky. 2011. The effect of various iodine supplementations and two different iodine sources on performance and iodine concentrations in different tissues of broilers. *British Poultry Science*. 52:115–123.

Rostagno, H.S., L.F.T. Albino, J.L. Donzele, P.C. Gomes, R.F.M. Oliveira, D.C. Lopes, A.S. Ferreira, and S.L.T. Barreto. 2011. *Brazilian Tables for Poultry and Swine—Composition of Feedstuffs and Nutritional Requirements*, 3rd ed. Viçosa, Minas Gerais, Brazil: Viçosa.

Roth-Maier, D.A., and M. Kirchgessner. 1997. Investigations on riboflavin requirement of fattening chickens. *Archiv fuer Gefluegelkunde*.

Ruiz-Feria, C.A., and S.T. Abdukalykova. 2009. Arginine and vitamin E improve the antibody responses to infectious bursal disease virus (IBDV) and sheep red blood cells in broiler chickens. *British Poultry Science*. 50:291–297.

Rymer, C., and D.I. Givens. 2005. n-3 Fatty Acid Enrichment of Edible Tissue of Poultry: A Review. *Lipids*. 40:121–130.

Ryu, K.S., G.M. Pesti, K.D. Roberson, H.M. Edwards, and R.R. Eitenmiller. 1995a. The folic-acid requirements of starting broiler chicks fed diets based on practical ingredients. 2. Interrelationships with dietary methionine. *Poultry Science*. 74:1456–1462.

Ryu, K.S., K.D. Roberson, G.M. Pesti, and R.R. Eitenmiller. 1995b. The folic-acid requirements of starting broiler chicks fed diets based on practical ingredients. 1. Interrelationships with dietary choline. *Poultry Science*. 74:1447–1455.

Sakkas, P., I. Oikeh, D.P. Blake, S. Smith, and I. Kyriazakis. 2019a. Dietary vitamin D improves performance and bone mineralisation, but increases parasite replication and compromises gut health in *Eimeria*-infected broilers. *The British Journal of Nutrition*. 122:676–688.

Sakkas, P., S. Smith, T.R. Hill, and I. Kyriazakis. 2019b. A reassessment of the vitamin D requirements of modern broiler genotypes. *Poultry Science*. 98:330–340.

Santiago, G.S., S.L. Vieira, C. Stefanello, C.T. Simoes, L. Kindlein, D.D. Maria, and P. Ibairro. 2020. Dietary choline affects field performance and broiler leg deviations. *Livestock Science*. 240:104127.

Sattar Khan, M., S. Nakamura, M. Ogawa, E. Akita, H. Azakami, and A. Kato. 2000. Bactericidal action of egg yolk phosvitin against *Escherichia coli* under thermal stress. *Journal of Agricultural Food Chemistry*. 48:1503–1506.

Scherer, C.S., and D.H. Baker. 2000. Excess dietary methionine markedly increases the vitamin B-6 requirement of young chicks. *Journal of Nutrition*. 130:3055–3058.

Schmidt, C.J., M.E. Persia, E. Feierstein, B. Kingham, and W.W. Saylor. 2009. Comparison of a modern broiler and a heritage line unselected since the 1950s. *Poultry Science*. 88:2610–2619.

Schmidt, M., P.C. Gomes, L.F.T. Albino, H.S. Rostagno, P.R. Cecon, and E.S. Cupertino. 2005a. Copper nutritional levels for male and female broilers in the growing and finishing phases. *Revista Brasileira de Zootecnia*. 34(3):890–899.

Schmidt, M., P.C. Gomes, H.S. Rostagno, L.F.T. Albino, and E.S. Cupertino. 2005b. Copper nutritional levels for male and female broilers in the initial phases. *Revista Brasileira de Zootecnia*. 34(5):1599–1605.

Siegel, P.B. 2014. Evolution of the modern broiler and feed efficiency. *Annual Review of Animal Biosciences*. 2:375–385.

Singh, J. 2010. Effect of supplementation of vitamin D3 at varying dietary Ca:tP ratio in diets supplemented with phytase on growth performance of broilers chicks. *Animal Nutrition and Feed Technology*. 10:245–253.

Siqueira, J.C., N.K. Sakomura, L.R.B. Dourado, J.M.B. Ezequiel, N.A.A. Barbosa, and J.B.K. Fernandes. 2013. Diet formulation techniques and Lys requirements of 1- to 22-day-old broilers. *Revista Brasileira de Ciência Avícola*. 15:123–134.

Skinner, J.T., M.H. Adams, S.E. Watkins, and P.W. Waldroup. 1992. Effects of calcium and nonphytate phosphorus levels fed during 42 to 56 days of age on performance and bone strength of male broilers. *Journal of Applied Poultry Research*. 1:167–171.

Sklan, D., D. Melamed, and A. Friedman. 1994. The effect of varying levels of dietary vitamin A on immune response in the chick. *Poultry Science*. 73:843–847.

Souza, C.S., F.M. Vieites, C.H.F. Vasconcellos, A.A. Calderano, R.V. Nunes, C.M. Ferreira, T.V.S. Pereira, and G.H.K. Moraes. 2013. Supplement of 1,25-dihydroxycholecalciferol and reduction of Ca and available phosphorus in broiler chickens. *Arquivo Brasileiro De Medicina Veterinaria E Zootecnia*. 65:519–525.

Steenfeldt, S., P. Sørensen, and B.I. Nielsen. 2019. Effects of choice feeding and lower ambient temperature on feed intake, growth, foot health, and panting of fast- and slow-growing broiler strains. *Poultry Science*. 98:503–513.

Sun, Z.W., Q.H. Fan, X.X. Wang, Y.M. Guo, H.J. Wang, and X. Dong. 2017. High dietary biotin levels affect the footpad and hock health of broiler chickens reared at different stocking densities and litter conditions. *Journal of Animal Physiology and Nutrition*. 101:521–530.

Sunde, R.A. 1997. Selenium. In *Handbook of Nutritionally Essential Mineral Elements*. B.L. O'Dell and R.A. Sunde, editors. New York: Marcel Dekker. 493–556.

Swain, B.K., and T.S. Johri. 2000. Effect of supplemental methionine, choline and their combinations on the performance and immune response of broilers. *British Poultry Science*. 41:83–88.

Świątkiewicz, S., A. Arczewska-Włosek, D. Bederska-Lojewska, and D. Józefiak. 2017. Efficacy of dietary vitamin D and its metabolites in poultry: Review and implications of the recent studies. *World's Poultry Science Journal*. 73:57–68.

Tallentire, C.W., I. Leinonen, and I. Kyriazakis. 2016. Breeding for efficiency in the broiler chicken: A review. *Agronomy and Sustainable Development*. 36:66.

Tavernari, F.D.C., G.R. Lelis, P.R.D.O. Carneiro, R.A. Vieira, R.C. Polveiro, J.A.P. Luengas, H.S. Rostagno, and L.F.T. Albino. 2012. Effect of different digestible isoleucine/Lys ratios for broiler chickens. *Revista Brasileira de Zootecnia*. 41:1699–1705.

Tavernari, F.C., G.R. Lelis, R.A. Vieira, H.S. Rostagno, L.F. Albino, and A.R. Oliveira Neto. 2013. Valine needs in starting and growing Cobb (500) broilers. *Poultry Science*. 92:151–157.

Taylor, J., P. Sakkas, and I. Kyriazakis. 2021. What are the limits to feed intake of broilers on bulky feeds? *Poultry Science*. 100:100825.

Toledo, R., R. Horacio, A. Luiz, D. Marli. C. Débora, and N. Eduardo. 2011. Lysine nutritional requirements of broilers reared in clean and dirty environments during the pre-starter and starter phases. *Revista Brasileira de Zootecnia*. 40:2205–2210.

USDA ERS (U.S. Department of Agriculture Economic Research Service). 2022. Per capita red meat and poultry consumption expected to decrease modestly in 2022. https://www.ers.usda.gov/data-products/charts-of-note/chart-detail?chartId=103767.

Vazquez, J.R., G.V. Gomez, C.C. Lopez, A.C. Cortes, A.C. Diaz, S.R.T. Fernandez, E.M. Rosales, and A.G. Avila. 2018. Effects of 25-hydroxycholecalciferol with two D-3 vitamin levels on production and immunity parameters in broiler chickens. *Journal of Animal Physiology and Animal Nutrition*. 102:E493–E497.

Vieites, F.M., E. Brusamarelo, B.S. Vieira, F.G. da Silva, C.S. Souza, G.d.S.S. Corrêa, J.G.C. Júnior, and G.H.K. Moraes. 2018. 1, 25-dihydroxycholecalciferol from *Solanum glaucophyllum* supports normal growth and reduces the negative effects of Ca and phosphorus restriction on broilers' bone tissue. *Semina: Ciências Agrárias*. 39:2205–2214.

Vignale, K., E.S. Greene, J.V. Caldas, J.A. England, N. Boonsinchai, P. Sodsee, E.D. Pollock, S. Dridi, and C.N. Coon. 2015. 25-Hydroxycholecalciferol enhances male broiler breast meat yield through the mTOR pathway. *Journal of Nutrition*. 145:855–863.

Villaverde, C., L. Cortinas, A.C. Barroeta, S.M. Martín-Orúe, and M.D. Baucells. 2004. Relationship between dietary unsaturation and vitamin E in poultry. *Journal of Animal Physiology and Animal Nutrition*. 88:143–149.

Voljč, M., T. Frankič, A. Levart, M. Nemec, and J. Salobir. 2011. Evaluation of different vitamin E recommendations and bioactivity of a-tocopherol isomers in broiler nutrition by measuring oxidative stress in vivo and the oxidative stability of meat. *Poultry Science*. 90:1478–1488.

Waldroup, P.W., J.H. Kersey, E.A. Saleh, C.A. Fritts, F. Yan, H.L. Stilborn, R.C. Crum Jr., and V. Raboy. 2000. Nonphytate phosphorus requirement and phosphorus excretion of broiler chicks fed diets composed of normal or high available phosphate corn with and without microbial phytase. *Poultry Science*. 79:1451–1459.

Walk, C.L., Z. Wang, S. Wang, J.O.B. Sorbara, and J. Zhang. 2022. Determination of the standardized ileal digestible Ca requirement of male Arbor Acres Plus broilers from day 11 to 24 post-hatch. *Poultry Science*. 101:101836.

Walk, C.L., Z. Wang, S. Wang, J. Wu, J.O.B. Sorbara, and J. Zhang. 2021. Determination of the standardized ileal digestible Ca requirement of male Arbor Acres Plus broilers from hatch to day 10 post-hatch. *Poultry Science*. 100:101364.

Watkins, S., and G.T. Tabler. 2009. Broiler water consumption. *Avian Advice*. 11(2):11–12. Fayetteville, AR: University of Arkansas Cooperative Extension.

Wellock, I.J., G.C. Emmans, and I. Kyriazakis. 2004. Modeling the effects of stressors on the performance of populations of pigs. *Journal of Animal Science*. 82:2442–2450.

Whitehead, C.C., H.A. McCormack, J.S. Rennie, and M. Frigg. 1995. Folic acid requirements of broilers. *British Poultry Science*. 36:113–121.

Whitehead, C.C., H.A. McCormack, L. McTeir, and R.H. Fleming. 2004. High vitamin D3 requirements in broilers for bone quality and prevention of tibial dyschondroplasia and interactions with dietary Ca, available phosphorus and vitamin A. *British Poultry Science*. 45:425–436.

Yan, F., J.H. Kersey, and P.W. Waldrop. 2001. Phosphorus requirements of broiler chicks three to six wk of age as influenced by phytase supplementation. *Poultry Science*. 80:455–459.

Yang, X., N. Zhang, X. Wang, H. Qu, J. Zhang, Y. Yan, Y. Cheng, and J. Han. 2020. Optimal dietary levels of 1 alpha-hydroxycholecalciferol in broiler chickens from 1 to 42 days of age. *Journal of Poultry Science*. 57:124–130.

Zaghari, M., M. Avazkhanllo, and M. Ganjkhanlou. 2015. Reevaluation of male broiler zinc requirement by dose-response trial using practical diet with added exogenous phytase. *Journal of Agricultural Science and Technology*. 17:333–343.

Zhang, X.H., X. Zhong, Y.M. Zhou, H.M. Du, and T. Wang. 2009. Effect of RRR-alpha-tocopherol succinate on the growth and immunity in broilers. *Poultry Science*. 88:959–966.

Zhang, T.Y., J.L. Liu, J.L. Zhang, N. Zhang, X. Yang, H.X. Qu, L. Xi, and J.C. Han. 2018. Effects of dietary zinc levels on the growth performance, organ zinc content, and zinc retention in broiler chickens. *Brazilian Journal of Poultry Science*. 20:127–132.

Zhang, M., J. Li, Y. Zhu, Q. Wu, Y. Li, D. Huang, Z. Gan, L. Zhong, J. Huang, H. Li, W. Xu, M. Zhang, Q. Hung, and Y. Sun. 2021. Effect of Vitamin E Supplementation on Deposition and Gene Expression Profiling of Abdominal Fat in Broiler Chickens. *The Journal of Poultry Science*. 58:40–50.

Zuidhof, M.J., B.L. Schneider, V. L. Carney, D.R. Korver, and F.E. Robinson. 2014. Growth, efficiency, and yield of commercial broilers from 1957, 1978, and 2005. *Poultry Science*. 93:2970–2982.

15

Nutrient Requirements of Broiler Breeder Pullets, Hens, and Males

INTRODUCTION

As noted in Chapter 14, poultry meat was the most consumed form of animal protein in the world in 2022 and is predicted to remain so for the future (OECD et al., 2021). Most poultry meat comes from strains of chickens selected for early rapid growth, increased feed efficiency, and breast meat yield (Muir et al., 2008; Zuidhof et al., 2014). Chicken meat production relies upon hatching egg production from growth-selected, "meat-type" breeder hens, also referred to as broiler breeder hens. Aggressive selection for early rapid growth and leanness appears to have compromised fertile egg production (Cash et al., 2025), which can be partially mitigated by rigorous nutritional management.

FEED ALLOCATION PROGRAMS

Properly managed broiler breeder hens can be expected to lay up to 180 eggs during a 44-week production cycle; in contrast, hens genetically selected for egg production lay ~364 eggs during a 60-week production cycle (Walzem and Chen, 2014). However, it has long been recognized that when broiler breeder hens are improperly managed by being allowed to consume feed to appetite and realize their full growth potential, they become obese, and settable egg production is markedly reduced (Yu et al., 1992; Renema et al., 1999; Chen et al., 2006; Walzem and Chen, 2014; Weaver and Ramachandran, 2023). Furthermore, improperly managed growth rates and feed intakes result in increased morbidity (Richards, 2003; Renema and Robinson, 2004) and mortality primarily due to heart failure (Lin et al., 2019b). For these reasons, feed allocation programs are key to the proper juvenile growth, sexual maturation, and adult reproductive performance and liveability in broiler breeders. Growth profiles of pullets can be highly controlled, and research continues to identify optimal growth profiles for adult reproductive performance (Noetzold et al., 2024).

The degree of feed restriction applied to broiler breeders during rearing (juvenile growth) strongly influences their rate of sexual development (Lewis et al., 2007), and management techniques that produce pullet flocks with highly uniform body weights and rates of gain are desirable as a means to synchronize reproductive development. Uniformity is promoted through the use of diets with higher protein content starting at the day of place to 3 weeks of age before limiting growth by reformulating diets to have reduced proportions of dietary crude protein and limiting both protein and energy intake by limiting feed intake. Modest increases in dietary crude protein are needed to support puberty and egg production (Table 15-1). The degree of growth restriction required for adequate egg production is most severe from 8 to 16 weeks of age during the juvenile growth phase of breeder development (Renema et al., 2007). With the onset of sexual maturation, feed allocations are increased to support adequate adipose, muscle, and non-muscle tissue development necessary for the onset of the energy-, mineral-, and amino acid-demanding process of egg production. In addition to supporting egg production, feed allocation programs to optimize adult reproductive performance also must take into account the energy and amino acid demands for tissue maintenance (van Emous et al., 2020).

Broiler breeder males also require careful nutritional management to control body weight gain to maintain fertility and breeding performance (Renema et al., 2007), albeit their specific requirements for growth and adult performance are far less studied compared to females. Juvenile males are commonly fed the same feed as females, although differing in amounts to maintain a different growth pattern. Use of sex-specific feed formulations is an emerging practice within the United States (Hudson, 2022). A major advantage of sex-separate feeding is improved control of body weight gain and subsequent improvement in fertility and mating ability (Moyle et al., 2012). The benefits in bird growth and performance come with new associated facility costs and personnel protocols to ensure effective implementation of separate sex feeding. Facility costs include separate feed bin and hopper setups in the breeder house as well as capacity within the feed mill for separate feed production streams, storage, and transport. Personnel protocols must be implemented that ensure that drivers deliver the correct feed to the correct feed bin at the breeder farm.

In the breeder house, males and females are housed together, and because males represent a small proportion of the total population, they have often been fed the female diet. This reduces costs because of the need for fewer feed deliveries and for additional feed storage on-farm. Adult male breeders are taller and have larger head sizes than adult females; these differences have long been used to allocate different amounts of feed to roosters and hens within the same barn environment (van Emous et al., 2020). Sex-specific growth and adult body weight gain profiles could be used in conjunction with sex-specific diets to maximize reproductive performance. At present, sex-specific feeding facilitates

TABLE 15-1 Minimum Nutrient Requirements of Broiler Breeder Pullets as Percentages or Units per Kilogram of Diet (90% Dry Matter) unless Otherwise Indicated[a,b]

Nutrient	Unit	Starter	Grower	Developer	Pre-breeder	Breeder I, Peak[d]	Breeder II, Post Peak[d]	Breeder III, Post Peak[d]
		Rearing Period				*Laying Period*		
		1 to 21 d	22 to 42 d	43 to 105 d	106 to 5% production	5% production to 252 d	259 to 322 d	329 to 455 d
Metabolizable energy, ME_n[c]	Kcal/kg diet	***2,800***	***2,800***	***2,800***	***2,800***	2,800	2,800	2,800
Feed intake, g/day		***20–49***	***27–63***	***40–74***	***56–112***	154	149.5	145
Crude protein[e]	%	***19.0***	***14.5***	***14.5***	***14.5***	15.0	14.5	14.0
Amino Acids[f]								
Arginine	%, (mg/day)	***1.00***	***0.63***	***0.61***	***0.61***	***0.69 (1,069)***	***0.65 (968)***	***0.60 (877)***
Histidine	% (mg/day)	***0.40***	***0.25***	***0.24***	***0.24***			
Isoleucine	%, (mg/day)	***0.64***	***0.42***	***0.41***	***0.41***	***0.56 (869)***	***0.53 (786)***	***0.49 (713)***
Leucine	%, (mg/day)	***1.05***	***0.66***	***0.64***	***0.64***	—	—	—
Lysine	%, (mg/day)	***0.95***	***0.60***	***0.58***	***0.58***	***0.62 (955)***	***0.58 (864)***	***0.54 (783)***
Methionine	%, (mg/day)	***0.47***	***0.33***	***0.32***	***0.32***	***0.29 (439)***	***0.27 (398)***	***0.25 (360)***
Methionine + cyst(e)ine	%, (mg/day)	***0.84***	***0.54***	***0.53***	***0.53***	***0.61 (945)***	***0.57 (856)***	***0.53 (775)***
Phenylalanine	%, (mg/day)	***0.79***	***0.50***	***0.49***	***0.49***	***0.47 (726)***	***0.44 (657)***	***0.41 (595)***
Phenylalanine + tyrosine	%, (mg/day)					***0.67 (1,031)***	***0.63 (933)***	***0.58 (846)***
Threonine	%, (mg/day)	***0.61***	***0.38***	***0.37***	***0.37***	***0.42 (647)***	***0.39 (583)***	***0.36 (522)***
Tryptophan	%, (mg/day)	***0.18***	***0.15***	***0.14***	***0.14***	***0.17 (267)***	***0.16 (242)***	***0.15 (219)***
Valine	%, (mg/day)	*0.72*	*0.45*	***0.44***	***0.44***	***0.54 (832)***	***0.51 (754)***	***0.47 (682)***
Fat								
Linoleic acid	%	1.0	1.0	1.0	1.0	1.0	1.0	1.0
Macrominerals								
Calcium	%, (g/day)	0.90	***0.80***	***0.70***	*(1.2–1.5)*	*(4.0)*	*(4.4)*	*(4.0)*
Chloride	%, (mg/day)	0.20	0.15	0.15	***0.12***	*(185)*	*(185)*	*(185)*
Magnesium	mg	600	***600***	***600***	***600***	*(81)*	*(81)*	*(81)*
Nonphytate phosphorus	%, (mg/day)	0.45	0.35	0.30	***0.35–0.45***	*(350)*	*(350)*	*(350)*
Potassium	%	0.30	0.30	0.30	***0.30***	*(227)*	*(227)*	*(227)*
Sodium	mg/d	0.20	0.15	0.12	***0.12***	*(150)*	*(150)*	*(150)*
Trace Minerals								
Copper	mg (mg/day)	8	*8*	*8*	*6*	*13.5 (2.08)*	*(2.0)*	*(1.96)*
Iodine	mg	***0.5***	***0.5***	***0.5***	***0.5***	*(0.006)*	*(0.006)*	*(0.006)*
Iron	mg (mg/day)	80	***45***	***45***	***45***	*92 (13.2)*	*65 (9.7)*	*(9.1)*
Manganese	mg	45	***45***	***45***	***45***	*(3.23)*	*(3.23)*	*(3.23)*
Selenium	mg	0.15	0.15	0.15	***0.15***	*(0.009)*	*(0.009)*	*(0.009)*
Zinc	mg (mg/day)	40	***22***	***22***	***22***	***81.0 (12.4)***	***71. (13.9)***	—
Fat-Soluble Vitamins								
A	IU	***1,500***	***1,500***	***1,500***	***1,500***			
D_3	IU	285	200	200	200	***1,400 (216)***	***2,800 (419)***	***2,800 (406)***
E	IU	***10***	***10***	***10***	***10***	*(1.54)*	*(1.54)*	*(1.54)*
K	mg	0.50	0.50	0.50	***0.50***	*(0.08)*	*(0.08)*	*(0.08)*

(*continued*)

TABLE 15-1 Continued

Nutrient	Unit	Starter	Grower	Developer	Pre-breeder	Breeder I, Peak[d]	Breeder II, Post Peak[d]	Breeder III, Post Peak[d]
Water-Soluble Vitamins								
B_{12}	mg	0.01	***0.01***	***0.007***	***0.007***	***(0.002)***	***(0.002)***	***(0.002)***
Biotin	µg	0.15	0.15	0.15	0.12	***16***	***(18)***	***(18)***
Choline	mg	1,300	***1,000***	***750***	750	***(169)***	***(169)***	***(169)***
Niacin	mg	35	30	25	25	***(3.85)***	***(3.85)***	***(3.85)***
Pantothenic acid	mg	***10***	***10***	***10***	***10***	***(1.54)***	***(1.54)***	***(1.54)***
Pyridoxine	mg	3.5	3.5	***3.0***	***3.0***	***(0.46)***	***(0.46)***	***(0.46)***
Riboflavin	mg	3.6	3.6	3.0	3.0	***(0.46)***	***(0.46)***	***(0.46)***
Thiamin	mg	***1.80***	***1.80***	***1.80***	***1.80***	***(0.28)***	***(0.28)***	***(0.28)***

[a] Values in normal font are minimum nutrient requirement recommendations based on empirical studies summarized in Table 15-2. Values that are bold and in italics are estimates based on research that did not meet the experimental or statistical criteria described in the "Nutrient Requirements" section or calculated as specified in footnote *b*. These values are considered as guidance concentrations and should be used with appropriate caution. Values in shaded areas were published previously (NRC, 1994), for which no updated estimates or confirming information are available. For references to the studies supporting shaded values, please refer to the 1994 report. The age ranges used in this table reflect a four-phase growing period and three-phase breeding period. Other age ranges are possible, and many use a three-phase growing period in which the designated pre-breeder period starts at ~19 weeks (~133 d). The use and age of implementation of a pre-breeder period phase contributes to the variability in calcium and non-phytate phosphorus suggestions. Implementation at younger ages allows for the use of lower dietary amounts.

[b] Pre-breeder values for pullets are equivalent to dietary concentrations recommended for broilers within the 2,801–3,600 g body weight range for vitamins and minerals other than calcium and phosphorus. Reports for dietary calcium concentrations during the pre-breeder stage range between 1.2 and 1.5% of diet, while non-phytate phosphorus varied from 0.35 to 0.45%. Further refinement of these ranges was not possible. During the breeder period, recommendations were given as units/hen/day unless new data provided dietary concentrations. Recommendations for daily intakes of choline, iodine, magnesium, manganese, potassium, vitamin A, and selenium were estimated by averaging recommendations for white and brown egg strains and correcting for the higher feed intake of breeders compared to layers. Daily intakes of vitamin E, thiamin, riboflavin, niacin, pantothenic acid, pyridoxine, folacin, and vitamin B_{12} were estimated using values for dietary concentrations for broilers within the 2,801–3,600 g body weight range corrected for the lower feed intake compared to broilers. The daily intake for vitamin K was identical when calculated using either broiler or layer data. Values for calcium, chloride, non-phytate phosphorous, and sodium were retained from the 1994 report as no new data was available and values did not disagree with those estimated from layer values.

[c] These are typical dietary energy concentrations used in nutrient requirement studies and expressed as kcal AMEn/kg diet. If qualitative feed restriction is used, dietary energy density will be reduced by addition of materials that do not provide usable energy in order to provide a greater volume of feed while continuing to control body weight gain.

[d] Amino acid values are on a digestible basis.

[e] Chickens do not have a requirement for crude protein per se. However, there should be sufficient crude protein to ensure an adequate amino nitrogen supply for synthesis of nonessential amino acids. The recommended crude protein values shown in this table were determined with corn-soybean meal–based diets formulated on a total amino acid basis. When synthetic amino acids, other feedstuffs, and by-products are included in the diets, or when formulating diets on a digestible amino acid basis, the crude protein values attained may differ.

[f] Amino acid values are on a total basis. Digestible Lys concentrations in Breeder I was 0.62 percent (Cerrate et al., 2019), while the concentrations were reduced by 0.04 and 0.08% in Breeder II and III periods, respectively, compared with digestible Lys concentration in Breeder I. Digestible amino acid ratios to Lys were 0.46, 0.52, 0.99, 0.76, 1.08, 0.28, 1.12, 0.91, 0.88, and 0.69 for Met, Cys, Met + Cys, Phe, Phe + Tyr, Trp, Arg, Ile, Val, and Thr, respectively (Ekmay et al., 2013).

use adult male diets lower in protein and calcium than are used in adult female diets to the benefit of male performance (Romero-Sanchez et al., 2007, 2008; Silveira et al., 2014). The previous report (NRC, 1994) provided a limited set of nutrient requirements for adult breeder males that included energy intake recommendations along with dietary protein, lysine, methionine, methionine + cyst(e)ine, calcium, and non-phytate phosphorus for grower, developer, and adult phases. However, most were estimates based on data from other ages or production types (i.e., broilers). Furthermore, dietary crude protein percentage is an index, and protein per se is not a nutrient, since birds require amino acids and amino nitrogen rather than preformed protein. Current approaches typically specify essential amino acid requirements in conjunction with amino-nitrogen requirements to support nonessential amino acid formation (Dozier et al., 2008).

Management strategies for juvenile broiler breeder females primarily aim to limit growth by limiting total nutrient intake through restricted feed allocation programs. The most efficacious feed allocation systems produce sufficient reductions in energy and nutrient intakes to limit growth rates and body weight gains to values suggested by primary breeders for maximum hatching egg production (Renema and Robinson, 2004). Two general strategies have been used to control juvenile growth: qualitative feed restriction (nutrient dilution) and quantitative feed restriction. Qualitative feed restriction typically employs increased dietary fiber with the assumption that increasing gut fill will increase satiety (Mench, 2002). However, the previous report (NRC, 1994) concluded that the capacity of modern broiler strains to consume large volumes of feed renders nutrient dilution (qualitative feed restriction) ineffective as a sole means to control body weight. Quantitative feed restriction limits the total weight of feed allocated for each bird to an amount that supports a limited growth rate consistent with improved adult reproductive performance and health. Management of pullet feed allocation delivery varies with both skip-a-day and everyday protocols described in management guides (Aviagen, 2018; Cobb-Vantress, 2022). The skip-a-day feed delivery method is most used as a means to improve flock body weight uniformity when the equipment used to distribute feed is slow or fails to distribute feed throughout all feeders simultaneously (van Emous, 2022). Breeders fed everyday experience less stress (de Beer et al., 2008; Ekmay et al., 2010), are more feed efficient (de Beer and Coon, 2007), and show equivalent body weight uniformity (Zuidhof et al., 2015) and improved reproductive performance (de Beer and Coon, 2007; Sweeney et al., 2022) compared to pullets managed with unfed days.

Diet formulations change with bird age and developmental stage, a process called phase feeding. Different producers parse the growth period of females differently, using either three or four phases in which different diets are fed. The greatest variation occurs in the pre-breeder phase, a time period in which a key goal is to increase the calcium reserves of females. When implemented early, at ~15 weeks of age, dietary calcium is increased less than when it is implemented at ~19 weeks of age (Table 15-1). In restrictive feed allocation programs, nutrient requirements may be expressed as amounts required per bird per day during each phase in addition to concentration within the formulated diet. Broiler genetics include appetitive drive (Bokkers and Koene, 2003) capable of supporting rapid growth and lean tissue gain; limitations in available feed due to feed allocation program specifications result in chronic hunger. Several groups have evaluated strategies that increase the volume of feed in quantitative restriction programs through addition of fibrous ingredients (Mohiti-Asli et al., 2012a,b; Aranibar et al., 2020; Tahamtani et al., 2020) based on the rationale that increased feed volume will reduce feed frustration (Riber and Tahamtani, 2020). Use of such hybrid approaches should consider the negative effects of elevated dietary fiber on nutrient availability to ensure that daily requirements are met in face of reduced availability (Singh and Kim, 2021). Because juvenile growth rates are limited in breeders, vitamin and mineral needs were assumed to be met by diets containing concentrations recommended for broilers of similar age that are fed to achieve their full growth potential (NRC, 1994) due to the high nutrient demands of rapid growth. However, reports of studies employing direct rigorous testing of the scalability of vitamin and mineral concentrations with growth rate and subsequent adult reproductive performance were not found.

The pullet-to-layer transition starts as pullets enter puberty at ~16 weeks of age and ends when egg production reaches 5 percent. It is a challenging period for feed allocation management task because sufficient energy and amino acids must be provided both to support growth of reproductive tissues, such as the left oviduct and ovary, and to support the hepatic synthesis of yolk precursor macromolecules (yolk-targeted very-low-density lipoprotein and vitellogenin), while responding to any stress arising from transfer from rearing to breeding facilities. Furthermore, female body weights must continue to be controlled, necessitating close monitoring of body weight, breast muscle development ("fleshing"), and adipose deposition in key indicator locations such as pelvic bones (van Emous et al., 2020) to achieve desired body compositions. While the desired physiological changes taking place in the birds are known, as noted previously, producers parse the growth period of females differently, and so flock nutritional management to achieve desired changes shows the greatest variation during the pre-breeder phase. A lack of definitive biomarkers and well-controlled studies during this period to guide recommendations increases the importance of flock manager experience and judgment to optimize adult performance following this transition period management. Pullets aged 16–20 weeks are managed to produce an accelerated body weight gain colloquially termed the "turn-up" because the recommended body weight gain of approximately 35 percent primarily comprises reproductive tissues and fat as skeletal growth is completed by 16 weeks. After placement in the breeder facility, the basis for energy recommendations shifts from body weight gains per se to body weight gain and egg production once egg production reaches 5 percent (van Emous et al., 2020). Adult breeders are fed to maintain body weight gains within breeder recommendations. Adult breeder hens that are laying eggs are no longer growing; tissue maintenance, egg production, and modest weight gain provide the basis for energy and amino acid requirements. Similarly, males need to be fed to support a limited but continuous body weight gain to maintain testicular function and flock fertility (Romero-Sanchez et al., 2008).

NUTRITIONAL RECOMMENDATIONS

Energy

Pullets

Dietary energy recommendations for broiler breeder pullets are designed to control growth to amounts needed to optimize skeletal development while limiting muscular development (fleshing) to optimize adult livability and reproductive performance. Colloquially termed "fleshing," growth of the breast muscle is the primary focus of growth limitation (Aviagen, 2018; Cobb-Vantress, 2022).

Usual management strategies provide pullets with free access to feed for the first 1 to 3 weeks of life prior to implementation of restricted feeding, while monitoring body weight gains and implementing restrictions earlier (or later) if target body weights are exceeded (or not met). Modeling studies for breeder pullets conducted during periods of quantitative feed restriction use different methods and models to estimate energy needs. A study with 5- to 20-week-old Hubbard Hi-Yield pullets managed using skip-a-day feeding at

22°C (Sakomura et al., 2003) found that metabolizable energy requirements for maintenance (ME_m) were 131.56 kcal/kg $BW^{0.75}$/day (with body weight gain requiring an additional ~2.6 kcal/g gain). Energy requirements were found to be somewhat lower in studies with slightly older, skip-a-day, and precision-fed Ross 308 pullets (Hadinia et al., 2018), albeit skip-a-day pullets were less efficient. In that study, pullets were evaluated from 10 to 23 weeks of age at 20°C. The determined ME_m was 120 kcal/kg $BW,^{0.68}$ with gains requiring an additional 1.52 kcal/g gain. Studies show that energetic needs are affected by genetics and environmental factors such as temperature and housing system. Differences in feed delivery such as skip-a-day, every day, or precision feeding affect the efficiency of energy use.

Pullets begin to enter puberty at about 16 weeks of age, and breeder companies typically recommend increased rates of body weight gain to account for development of reproductive organs and deposition of sufficient fat to support the onset of egg production. Weight management continues throughout the production cycle with exact weekly recommendations during puberty and production being strain-specific and set by breeders.

Hens

The 1994 report (NRC, 1994) noted that broiler breeder hens are fed to maintain body weight within breeder guidelines and that daily energy consumption usually ranges from 400 to 450 kcal ME per hen daily depending upon bird age, stage of production, and environmental temperature. Subsequent efforts developed models to estimate energy requirements for maintenance, growth, and production (Sakomura, 2004; Hadinia et al., 2018) and were consistent with that 400 to 450 kcal ME per hen range. Generic models have been developed from data collected at specified times and places under conditions and with bird genetics that may not be applicable to all situations. Outcomes from production settings can be expected to differ from those inherent in a model unless local conditions and genetics are used in the modeling program.

Males

The 1994 report (NRC, 1994) reported that males raised on litter required daily energy intakes of 400 to 458 kcal ME per bird. Males maintained in cages required less energy, within the range of 346 to 358 kcal ME per bird per day; reflecting the reduced physical activity of birds held in cages for ME intake determinations.

Amino Acids and Total Amino Nitrogen

Feed allocation programs are paramount to the proper development of the broiler breeder's reproductive potential; amino acid intakes specified in those programs must support proper frame and organ development while limiting excessive breast muscle development. Amino acid recommendations serve to identify amounts of requisite essential amino acids and sufficient amino nitrogen for nonessential amino acid formation. Peer-reviewed studies to define amino acid requirements for broiler breeder hens during the rearing and laying cycle of modern genetic strains are sparse. Indeed, insufficient research since the 1994 report (NRC, 1994) is available to derive definitive male breeder requirements. Table 15-1 provides estimates of digestible amino acid needs of broiler breeder pullets during rearing based on findings of nine peer-reviewed articles listed in Table 15-3 that used dietary treatments providing optimum performance. Digestible amino acid values presented in Table 15-1 were computed using averages of data from those studies. Recommendations for adult hens encompass an approximate 40-week laying cycle; defining amino acid requirements through multiple dietary phases is technically demanding and expensive due to the required labor and limited funding available to support these experiments. Dietary amino acid guidance concentrations provided in Table 15-1 show the recommended dietary digestible lysine contents as 0.62 percent, 0.58 percent, and 0.54 percent from 26 to 33 weeks of age, 37 to 46 weeks of age, and 47 to 65 weeks of age, respectively (Cerrate et al., 2019). Digestible amino acid ratios were applied to the digestible lysine values to estimate the values for the other essential amino acid values (Ekmay et al., 2013).

Amino acid requirements (mg/d/hen) of Cobb × Cobb 500 broiler breeder hens during the lay cycle calculated using a factorial model are presented in Table 15-2. Requirements in mg/d/hen are based on body weight, egg output, and their respective standard deviations, representing the intake of amino acids to meet the requirements of 85 percent of the population studied. The in-feed concentration is based on the ratio between metabolizable energy requirement (see Chapter 1) and the metabolizable energy of the feed. The requirement of breeders for metabolizable energy are calculated based on body weight, body weight gain, and egg output (Sakomura et al., 2003; Afrouziyeh et al., 2022). The Reading model (Fisher et al., 1973) as implemented by (Sakomura et al., 2015) was applied to calculate the amounts of Lys necessary to meet the requirement of 85 percent of the population. Note that this initial modeling effort is not validated and led to unusually large feed intake estimates and inappropriately low g Lys/hen for hens coming into peak egg production. Additional data from specifically designed experiments would be needed to identify the driver of model inaccuracies and validate a final model. Fisher et al. (1973) cautioned that as presented, the model assumed that egg laying was near 100 percent and would not be suitable for application to flocks containing immature or out of lay hens. Egg production is typically lower in breeder flocks than it is in layer flocks (Walzem and Chen, 2014) and so may underlie the questionable model predictions.

The estimates presented in Table 15-1 indicate guidance intake amounts for digestible lysine of 955, 864, and 783 mg/day/hen for breeders from 25 to 36, 37 to 46, and 47 to

TABLE 15-2 Amino Acid and Metabolizable Energy Requirements for Broiler Breeder Hens from 18 to 65 Weeks of Age Generated Using a Factorial Model

Age	Body Weight	Average Egg Output	Feed Intake[a]	Met. Energy[b]	Lysine Requirement[c]	
weeks	kg	g/day	g/day	kcal/day	mg/day	%
25	3.09	10.0	157	440	281	0.179[d]
26	3.22	25.7	170[d]	475	535	0.315[d]
27	3.33	38.7	175[d]	491	749	0.427[d]
28	3.42	44.8	175[d]	490	851	0.487[d]
29	3.49	47.6	171[d]	479	899	0.526[d]
30	3.54	49.2	166[d]	464	927	0.559[d]
31	3.58	50.0	163	458	941	0.576
32	3.61	50.8	161	451	955	0.594
33	3.63	50.3	157	440	948	0.604
34	3.65	50.3	157	441	949	0.602
35	3.67	50.2	158	442	948	0.601
36	3.69	50.1	158	443	947	0.599
37	3.71	50.0	159	444	946	0.597
38	3.73	49.7	159	444	942	0.593
39	3.75	49.5	159	445	939	0.591
40	3.77	49.1	159	445	933	0.587
41	3.79	48.8	159	446	929	0.584
42	3.81	48.4	159	446	924	0.580
43	3.83	47.9	159	446	916	0.575
44	3.85	47.4	159	446	909	0.570
45	3.87	46.9	159	446	901	0.566
46	3.89	46.4	159	446	894	0.561
47	3.91	45.8	159	446	885	0.556
48	3.93	45.2	159	446	876	0.550
49	3.95	44.6	159	445	867	0.545
50	3.97	44.0	159	445	858	0.540
51	3.99	43.3	159	445	847	0.534
52	4.01	42.6	159	444	837	0.527
53	4.03	41.9	158	444	826	0.521
54	4.05	41.1	158	443	814	0.514
55	4.07	40.2	158	442	800	0.507
56	4.09	39.4	158	441	788	0.500
57	4.11	38.5	157	440	774	0.492
58	4.13	37.6	157	439	761	0.485
59	4.15	36.7	157	438	747	0.477
60	4.17	35.8	156	437	733	0.469
61	4.19	34.8	156	436	718	0.461
62	4.21	33.8	155	435	702	0.452
63	4.23	32.7	155	434	685	0.443
64	4.25	31.7	154	432	670	0.434
65	4.25	30.6	146	408	653	0.448

[a] Feed intake is the amount of feed (containing 2,800 kcal of metabolizable energy/kg of feed) necessary to meet the daily ME requirement.

[b] Metabolizable energy (ME) is calculated considering body weight (BW), maintenance ($BW^{0.75}$), body weight gain (BWG), and egg output (EO). Following the equation: $ME = 113*(BW^{0.75}) + 7.62*(BWG) + 2.4*(EO)$.

[c] Reading model (Fisher et al., 1973) as implemented by (Sakomura et al., 2015) was applied to calculate Lys requirement: $Lys = 32.2*(BW) + 14.5*(EO) + \sqrt{[32.2^2 * (SD_{BW})^2 + 14.5^2 * (SD_{EOmax})^2]} * Z$. It was assumed a standard deviation (SD) for body weight as 0.300 and for maximum egg output as 0.125. Z is the area bellow a normal curve that represents the percentage of broiler breeders in a population. The amounts of Lys necessary to meet the requirement of 85% of the population ($Z = 1.04$) was applied.

[d] This initial model is not validated leading questionable prediction estimates for feed and lysine intakes as hens come into peak egg production. Additional data from specifically designed experiments would be needed to identify the driver of the overestimate to validate this model. Fisher et al. (1973) commented in their report that the model assumed a high rate of egg production; pre-peak broiler breeder hens are unlikely to achieve that assumption.

65 weeks of age, respectively. The intake of digestible lysine predicted from factorial models (Table 15-2; at 31, 42, and 56 weeks of age) are approximately 99, 107, and 101 percent of the guidance from Table 15-1. The main difference was an 8 percent higher estimate for the phase from 37 to 46 weeks of age when using the factorial model. Both tables show a linear response in digestible lysine intake in which the reduction observed in Table 15-1 is more pronounced than that predicted by the factorial model (Table 15-3).

Minerals and Vitamins

No studies meeting all the committee's specifications for nutrient requirement studies were found for minerals and vitamins specific to broiler breeders at various life stages identified in phase-feeding programs. Limitations in peer-reviewed literature prompted the committee to extrapolate most values from other literature. This approach was also followed by the 1994 NRC committee; however, the current committee also specified rationales used to develop values shown in Table 15-1. Juvenile growth rates in breeders are intentionally limited by feed allocation programs. The limited growth within the juvenile period is used by some as a rationale for the assumption that vitamin and mineral needs during this period are met by diets containing amounts recommended for broilers fed to achieve their full growth potential. The committee partially adapted this approach by using values of dietary concentrations for broilers within the 2,801–3,600 g body weight range for vitamins and minerals other than calcium and phosphorus for breeder pullets during the pre-breeder period. Reports for dietary calcium concentrations during the pre-breeder stage range between 1.2 and 1.5 percent of diet, while non-phytate phosphorus varied from 0.35–0.45 percent. Further refinement of these ranges was not possible. Adult breeder requirements differ from those of juveniles due to differences in developmental stage and reproductive status. The committee found new publications for copper, iron, zinc, and vitamin D_3 during the breeding period, albeit only tentative recommendations could be made from them. In the 1994 NRC report it was suggested that values for table-egg strain layers be used as a guide, albeit no guidance was provided as to how to use those values. The committee adopted the approach of giving recommendations as units/hen/day during the breeder period as was done in 1994 while detailing estimation methods to increase calculation transparency. In Table 15-1, the daily intake recommendations for choline, iodine, magnesium, manganese, potassium, vitamin A, and selenium were estimated by averaging recommendations for white and brown egg strains and correcting for the higher feed intake of breeders compared to layers. Breeder feed intake was estimated to average 154 g/day while that of layer-type hens averages 100 g/day. Using this calculation method, values for copper, iron, and zinc were similar to those found in new literature. This was not true for vitamin D_3 values whose literature values were 5–10-fold higher than those calculated from layer literature. Daily intakes of vitamin E, thiamin, riboflavin, niacin, pantothenic acid, pyridoxine, folacin, and vitamin B_{12} were estimated using values for dietary concentrations for broilers within the 2,801–3,600 g body weight range corrected for differences in feed intake where broiler feed intake was estimated to average 223 g/day and breeder feed intake was estimated to average 154 g/day. The 16 μg biotin/hen/day reported in the 1994 report is near to the 18 μg biotin/hen/day estimated from differences in feed intake between broilers and breeders. The units/hen/day values for vitamin K were identical when calculated using either broiler or layer data. Values for calcium, chloride, non-phytate phosphorus, and sodium were retained from the 1994 report as no new data was available and values did not disagree with those estimated from layer values.

Macromineral Requirements of Broiler Breeders for Reproduction

Calcium

As with all egg-laying hens, egg production, eggshell strength, and hen bone health are the key response criteria when evaluating the calcium requirements of broiler breeder hens. Additionally, one of the best determinants of calcium adequacy for breeder hens is egg specific gravity; McDaniel et al. (1979) found that eggs should have a specific gravity of 1.080 or greater for optimal hatchability. However, work with modern Cobb 500 birds found that eggs with specific gravities of ~1.070 supported ≥87 percent hatch of fertile hens, suggesting that acceptable specific gravity values of eggs from modern strains may need to be reevaluated (Ulmer-Franco et al., 2010). Broiler breeders are provided with a specific amount of feed each day, and so calcium requirements are generally expressed as grams of calcium provided per unit time (e.g., day). The calcium requirement of broiler breeders was estimated to be 4.0 g per day in the previous report (NRC, 1994); however, this number was not based on traditional requirement experiments as those data were lacking. Support for the 4.0 g per day estimate was provided by Wilson and coworkers (1981) as broiler breeders raised on litter did not respond to dietary calcium amounts above 3.91 g/hen/day. Experiments centered on determining the dietary calcium requirements of broiler breeder hens continue to be rare, and no reports published since the review of data for the 1994 report were found to meet the criteria currently required to establish such. However, one report was found to support the 1994 recommendation of 4.0 g of calcium daily, as broiler breeder hens fed 3.8 and 5.4 g of calcium daily had increased egg mass compared to hens fed 2.1 g of calcium daily. No reports were found to contradict the estimate provided by the 1994 committee; no change in the calcium recommendations for broiler breeder hens can be made at this time.

Both broiler breeder bone health and egg specific gravity are important indicators of hen dietary calcium status. Along with the amounts of calcium fed, the timing of calcium

TABLE 15-3 Documentation of Amino Acid, Trace Mineral, and Vitamin Requirements of Female Broiler Breeders

Age, wk	Egg Production %/d/Hen Housed[a]	Nutrient	Estimated Requirement vs. Response Criteria mg/kg Diet (mg/hen/d)	Response Criteria	Strain	Diet	Comments	Reference
Amino Acids								
0–22		Lysine, methionine, sulfur amino acids, threonine, and tryptophan		Uniformity, Carcass traits, Tibia parameters, Behavior as wing and tail feather scores	Ross 308		Values for methionine, sulfur amino acids, threonine, tryptophan, and lysine	Asensio et al. 2020
0–22		Everyday or skip-a-day feeding		Uniformity, Carcass traits, Tibia parameters, Adult reproductive performance	Ross 708	Crude protein, CP %, CP week 1–3: 18.1 CP week 3–10: 15.3 CP week 10–22; 14.6	Gives values for digestible lysine, methionine, sulfur amino acids, threonine, tryptophan, branched chain, arginine histidine, and phenylalanine	Avila et al., 2023
22–60				Reproductive performance, Back feather coverage	Ross 708	3 (Crude protein, CP, week 25–36: 14.72, 15.3, 16.37) (CP week 36–60: 13.83, 14.7, 15.33)	Provided likely energy intakes in Breeder I, II, and III for use with digestible lysine estimates from Breeder I period and ratios to other amino acids	Cerrate et al., 2019
0–19				Uniformity Mortality Carcass traits Tibia Parameters Behavior: time budget 2-minute feeding rate	Ross 308		Values for methionine, sulfur amino acids, threonine, tryptophan, and lysine. Included for 0–21, 22–42, and 43–105 d	de Los Mozos et al., 2017
30–42				Urine pH Fertility	Cobb 500	Corn–soy diet, CP 15%, plus crystalline amino acids	Digestible amino acid ratios to lysine during Breeder period Used factorial model in which egg production, egg mass, maintenance, and body weight gain were used in estimation	Ekmay et al., 2013
0–22					Pure Line A: pure broiler breeder male line of a commercial breeding company at the A position of a four-way cross that makes up a commercial broiler	Five phases CP: 21.9 (Starter 1), 19.5 (Starter 2), 15.6 (Grower), 16.2 (Pre-breeder), 16.1 (Breeder) plus a 25% CP reduction treatment starting at Starter 2	Values for sulfur amino acids, leucine, lysine, threonine, and tryptophan	Lesuisse et al., 2018

continued

TABLE 15-3 Continued

Age, wk	Egg Production %/d/Hen Housed[a]	Nutrient	Estimated Requirement vs. Response Criteria mg/kg Diet (mg/hen/d)	Response Criteria	Strain	Diet	Comments	Reference
0–22				*In F0 generation* Egg production Egg weight Body composition Body fat Albumen % in egg *In F1 generation* Chick weight Chick adiposity Egg production % Egg number Egg mass Hatch of fertile % Fertility %	Pure Line A: pure broiler breeder male line of a commercial breeding company at the A position of a four-way cross that makes up a commercial broiler	Five phases CP: 21.1 (Starter 1), 19.0 (Starter 2), 15.0 (Grower), 15.9 (Pre-breeder), 15.6 (Breeder) plus a 25% CP reduction treatment starting at Starter 2	Values for arginine, sulfur amino acids, leucine, lysine, threonine, and tryptophan	Lesuisse et al., 2018
0–3 and 3–6				Broiler performance	Ross 308	CP = 17.7% starter; 15.96% CP in grower	Values for methionine, sulfur amino acids, threonine, tryptophan, and lysine included for 0–21 and 22–42 days	Plumstead et al., 2007
2–22				Feed intake Eating time Eating rate Behavior Corticosterone Feather cover	Ross 308	Three CPs Common Starter 1 20.7% Starter 2, Grower, Pre-breeder Low = 14.5, 12.3, 12.8% Medium = 18.0, 13.0, 14.0% High = 19.2, 14.1, 14.9%	Values for methionine, sulfur amino acids, threonine, tryptophan, and lysine	van Emous et al., 2014
2–22				Body weight Body composition Egg production Egg traits Defective egg production Mortality Hatchability	Ross 308	Two CPs Common Starter 1 19.4% Starter 2, Grower, Pre-breeder Low = 13.8, 11.3, 12.6% High = 16.7, 14.2, 15.1%	Values for methionine, sulfur amino acids, threonine, tryptophan, and lysine	van Emous et al., 2015

2–40				Uniformity Mortality Body weight Body composition Egg production Egg traits Defective egg production Mortality Hatchability	Ross 308	Three CPs Common Starter 1 20.7% CP and Laying 14.5% CP Starter 2, Grower, Pre-breeder Low = 14.5, 12.3, 12.8% Medium = 18.0, 13.0, 14.0% High = 19.2, 14.1, 14.9%	Values for methionine, sulfur amino acids, threonine, tryptophan, and lysine	van Emous et al., 2013
0–3				Protein turnover using stable isotopes	Cobb 500	1 (20.9%)	Values for arginine, methionine, sulfur amino acids, threonine, and lysine included for 0 to 21 days	Vignale et al., 2017
Calcium								
36–60	67.5	Limestone	NA	Egg production Egg mass Egg weight Shell weight, g Shell weight, % Shell thickness	Ross	1.5, 2.5, and 3.5% dietary Ca	Average Ca intake (g/hen/day) of 2.14, 3.76, and 5.39 for the 1.5%, 2.5%, and 3.5% Ca diets	Moreki et al., 2011
Phosphorus								
21–45	Eggs/hen housed 96.9 eggs	Dicalcium phosphate	NA	Egg production Egg weight Mortality Egg specific gravity Prodigy weight (day 1) Hatch of fertile eggs Tibia ash	Cobb 500	0.15% and 0.40% dietary nPP	216 mg per day nPP intake from 0.15% diet 576 mg per day nPP intake from 0.40% diet	Ekmay et al., 2012
22–64	62.1	Dicalcium phosphate	NA	Egg production Feed conversion ratio Fertility chicks per hen	Ross 308 broiler breeder pullets and Ross 344 cockerels	0.19% and 0.37% dietary nPP	Pullets also exposed to dietary nPP treatments Phytase treatments were not evaluated	Plumstead et al., 2007
24–45	Eggs/hen housed 80.3 eggs	Dicalcium phosphate	NA	Egg production Egg weight Egg specific gravity Tibia ash	Cobb 500	0.20, 0.25, 0.30, 0.35 and 0.40% dietary nPP	288, 360, 432, 504, and 576 mg nPP at peak intake	Ekmay and Coon, 2011

continued

TABLE 15-3 Continued

Age, wk	Egg Production %/d/Hen Housed[a]	Nutrient	Estimated Requirement vs. Response Criteria mg/kg Diet (mg/hen/d)	Response Criteria	Strain	Diet	Comments	Reference
Copper								
25–44	75.4	CuSO4.5H2O	8.8 (1.26) 13.5 (1.92) 10.2 (1.45) 14.3 (2.03) 14.9 (2.07) 15.5 (2.22) 9.42 (1.34) 12.0 (1.70)	Egg production Total eggs/hen Settable eggs Egg weight Hen hemoglobin Hen hematocrit Chick weight Chick hematocrit	Cobb 500	6 (2.67–20.19)	Hens were fed a depletion diet (2.67 mg/kg Cu) for 4 weeks prior to the study. Hens were fed an amount of diet recommended by the breeding company Data analyzed by quadratic polynomial, exponential asymptotic, and broken-line quadratic models The requirement here is the average of the models that were significant	Berwanger et al., 2018
Iron								
47–70	53.8	FeSO4.7H2O	91.8 (13.2) 90.4 (13.0) 99.1 (14.3) 132.8 (19.1) 110.6 (15.9) 125.5 (18.1) 105.4 (15.2)	Eggs production Settable eggs Yolk Fe Hen hematocrit Hen hemoglobin Chick hematocrit Chick hemoglobin	Cobb 500	6 (24.6–148.2)	Basal diet was corn, rice, and concentrated soy protein and had 24.6 mg/kg Fe. Hens were fed an amount of diet recommended by the breeding company. Hens were Fe depleted prior to the experimental period Data analyzed by quadratic polynomial, exponential asymptotic, and broken-line quadratic models The requirement here is the average of the models that were significant	Taschetto et al., 2017
62–68	38.4	FeSO4.7H2O	65 64	Egg production Egg weight	Cobb 500	4 (37–82)	Basal diet was corn starch–soy meal and had 37 mg/kg Fe Data analyzed by broken line regression Egg production was low due to the late stage of production Regression estimates of requirement were not significant at $P < 0.05$, likely due to a low N Feed intake was not reported	Abbasi et al., 2015

Zinc								
33–36 37–40 41–44 33–44	76.4	ZnSO4.7H2O	81.0 (12.4) 63.4 (9.7) 58.2 (8.9) 53.3 (8.2) 57.8 (8.9) 64.9 (10.0) 135.1 (20.7) 82.4 (12.6) 66.1 (10.2)	Egg production Settable eggs Egg production Settable eggs Egg production Settable eggs Eggshell % Breaking strength Shell thickness	Cobb 500	7 (18.7–170.6)	Basal diet was rice, corn, and soy protein isolate and had 18.7 mg/kg Zn Hens were Zn depleted prior to the experimental period. Hens were fed an amount of diet recommended by the breeding company Data analyzed by quadratic polynomial (QP), exponential asymptotic, and broken-line quadratic models The QP model overestimated the requirement relative to AOVA and the other two models. The requirement here is the average of the other two models if they were significant. If neither were significant, the results of QP were used if significant	Mayer et al., 2019
39–47	66.7	Zn-glycinate	24–80 24–44 44–64 64–84	Eggs production Feed conversion Egg weight Fertility	Lingnan yellow feathered	5 (24–104)	Basal diet was corn–soybean meal and had 24 mg Zn/kg, 20.5 mg/kg of which was added as ZnO Diets were fed ad libitum Hens were fed a depletion diet (24 mg/kg Zn) for 4 weeks prior to the study Data was analyzed by ANOVA	Zhang et al., 2017
58–65	64.2	ZnSO4.H2O	71.1 (13.9) 92.3 (10.9) 94.4 (11.9) 98.7 (11.6)	Eggs production Yolk Zn Fertility Hatchability	Lingnan yellow feathered	5 (22.8–142.8)	Basal diet was corn–soybean meal and had 22.8 mg Zn/kg Hens were fed 118 g diet per day Data were analyzed using quadratic polynomial or exponential asymptotic regression	Li et al., 2019
60–42	Not reported	ZnO	60–90 60–90 60–90 30–60 60–90 60–90	Plasma Zn Plasma protein Plasma P Plasma Ca Plasma LDH Plasma ALT	Cobb 500		Basal diet was corn, corn starch, soybean meal and had 30 mg Zn/kg, 20.5 mg/kg of which was added as ZnO Hens were fed a depletion diet (9.5 mg/kg Zn) for 2 weeks prior to the study Data was analyzed by ANOVA	Sharideh et al., 2015

continued

TABLE 15-3 Continued

Age, wk	Egg Production %/d/Hen Housed[a]	Nutrient	Estimated Requirement vs. Response Criteria mg/kg Diet (mg/hen/d)	Response Criteria	Strain	Diet	Comments	Reference
Vitamins								
Vitamin A								
36–56 week		Retinyl acetate	<5,000 IU/ kg of supplemental vitamin A	Egg production, egg weight, egg quality, fertility, hatchability	Ross 308	5,000; 10,000; 15,000; 20,000; 25,000; 30,000; or 35,000 IU/kg	Basal diet not measured No effect on performance traits	Yuan et al., 2014
31–55 week		Retinyl acetate	No response to > 5,000 IU/ kg of supplemental vitamin A; 135,000 IU/ kg decreased breeder performance	Egg production Egg weight Egg quality Fertility Hatchability	Ross 308	5,000; 15,000; 45,000; and 135,000 IU/kg	Basal diet not measured Linear decrease in egg weight from 48 to 55 weeks with increasing vitamin A Fertility, hatchability at 43 and 55 weeks of age decreased by 135,000 IU/kg feed vs. 5,000 and 15,000 IU/kg feed	Yuan et al., 2014
Adequate Intake								
31–56 week			<5,000 IU/kg of supplemental vitamin A					
Vitamin D								
27–52 week					Ross × Ross	0, 125, 250, 500 1,000, 2,000 IU/kg diet (hens receiving 0 IU/kg switched to 4,000 IU/kg @ 36 weeks of age	Basal diet content not measured; assumptions stated regarding the lack of D3 in ingredients Conclusions regarding vitamin D requirement for maximum egg production cites SPSS abstracts Vitamin D activity in diet not measured	Atencio et al., 2005a
73–90 week	Molted breeders				Ross 508 × Ross 344	Maternal diets: 0, 3,125, 12,500, 50,000 ng/kg (0, 125, 500, 2,000 IU/kg)	Basal diet content not measured; assumptions stated regarding the lack of D3 in ingredients Not suitable for requirement determination	Atencio et al., 2005b
27–36 week	Peak		1,424 IU/kg	HD egg production	Ross 508 × Ross 344	0, 125, 250, 500, 1,000, and 2,000 IU/kg diet	Basal diet content not measured; assumptions stated regarding the lack of vitamin D activity in ingredients	Atencio et al., 2006

27–36 week	Peak		1,390 IU/kg	Maximum hatchability	Ross 508 × Ross 344	0, 125, 250, 500, 1,000, and 2,000 IU/kg diet		Atencio et al., 2006
27–36 week	Peak		1,288 IU/kg	Minimum early embryo mortality	Ross 508 × Ross 344	0, 125, 250, 500, 1,000, and 2,000 IU/kg diet		Atencio et al., 2006
27–36 week	Peak		1,130 IU/kg	Minimum mid embryo mortality	Ross 508 × Ross 344	0, 125, 250, 500, 1,000, and 2,000 IU/kg diet		Atencio et al., 2006
27–36 week	Peak		1,393 IU/kg	Minimum late embryo mortality	Ross 508 × Ross 344	0, 125, 250, 500, 1,000, and 2,000 IU/kg diet		Atencio et al., 2006
27–36 week	Peak		1,182 IU/kg	Egg weight	Ross 508 × Ross 344	0, 125, 250, 500, 1,000, and 2,000 IU/kg diet		Atencio et al., 2006
27–36 week	Peak		1,337 IU/kg	Egg SG	Ross 508 × Ross 344	0, 125, 250, 500, 1,000, and 2,000 IU/kg diet		Atencio et al., 2006
27–36 week	Peak	Overall recommendation of authors	1,400 IU/kg		Ross 508 × Ross 344	0, 125, 250, 500, 1,000, and 2,000 IU/kg diet		Atencio et al., 2006
37–66 week	Post-peak		2,804 IU/kg	HD egg production	Ross 508 × Ross 344	0, 125, 250, 500, 1,000, and 2,000 IU/kg diet		Atencio et al., 2006
37–66 week	Post-peak (52 week)		2,708 IU/kg	Maximum hatchability	Ross 508 × Ross 344	0, 125, 250, 500, 1,000, and 2,000 IU/kg diet		Atencio et al., 2006
37–66 week	Post-peak			Minimum early embryo mortality	Ross 508 × Ross 344	0, 125, 250, 500, 1,000, and 2,000 IU/kg diet	No effect observed	Atencio et al., 2006
37–66 week	Post-peak		2,568 IU/kg	Minimum mid embryo mortality	Ross 508 X Ross 344	0, 125, 250, 500, 1,000, and 2,000 IU/kg diet		Atencio et al., 2006
37–66 week	Post-peak		2,756 IU/kg	Minimum late embryo mortality	Ross 508 × Ross 344	0, 125, 250, 500, 1,000, and 2,000 IU/kg diet		Atencio et al., 2006
37–66 week	Post-peak			Egg weight	Ross 508 × Ross 344	0, 125, 250, 500, 1,000, and 2,000 IU/kg diet	No effect observed	Atencio et al., 2006
37–66 week	Post-peak		> 2,000 IU/kg	Egg SG	Ross 508 × Ross 344	0, 125, 250, 500, 1,000, and 2,000 IU/kg diet		Atencio et al., 2006
37–66 week	Post-peak	Overall recommendation of authors	2,800 IU/kg		Ross 508 × Ross 344	0, 125, 250, 500, 1,000, and 2,000 IU/kg diet		Atencio et al., 2006
24–66 week		Chick TD	> 250 IU/kg (positive response to chick weight with 2,000 IU); no effect on TD		Ross × Ross		Basal diet content not measured. Chicks fed TD-inducing diet to 16 d; test diets 250 vs. 2,000 IU diet contained 1,100 IU/kg diet	Driver et al., 2006

continued

TABLE 15-3 Continued

Age, wk	Egg Production %/d/Hen Housed[a]	Nutrient	Estimated Requirement vs. Response Criteria mg/kg Diet (mg/hen/d)	Response Criteria	Strain	Diet	Comments	Reference
Requirement								
27–36 week			1,400 IU/kg feed					
37–66 week			2,800 IU/kg feed					
Vitamin E								
25–40 week		Not reported	<25 mg supplemental vitamin E/kg diet	HHEP, Egg weight, Hatchability, Neonatal chick weight Chick performance	Ross broiler breeders	25, 50, 75, and 100 mg/kg diet	Basal diet content not measured Form of vitamin E not recorded Not useful for establishing a requirement	Hossain et al., 1998
75–87 week			200, 400, > 100 for both hatch of set and hatch of fertile	Hatch of set Hatch of fertile	Not reported	100, 200, and 400 mg/kg	Basal diet content not measured Form of vitamin E not recorded Not useful for establishing a requirement	Yang et al., 2020
24–54 week		dl-α-tocopherol acetate	300 IU/kg > 10 IU/kg	% of flock entering lay HH egg number	Not reported	10 or 300 IU/kg feed as supplement	Basal diet content not measured Diets supplemented with 10 or 300 IU/kg feed	Siegel et al., 2001
55–63 week			300 IU/kg > 10 IU/kg	HD ovulations HD normal eggs			Basal diet content not measured Feeding started at 24 week	Siegel et al., 2001
Hens			Adequate intakes; <200 IU/kg				Supplemented as dl-α-tocopherol acetate @ 10 or 300 IU/kg feed	
Vitamin K								
	No recommendation based on broiler breeder data							
Thiamin								
~36 week			32 mg/kg in breeder diet decreased broiler G:F 1–21 d vs 0 and 2 mg/kg	Chick performance to 21 d	Hubbard broiler breeders	0, 2, 8, and 32 mg thiamin/kg feed	Basal diet content not measured Hen performance not recorded No data on breeder performance reported	Olkowski and Classen, 1999
	No recommendation based on broiler breeder data							
Riboflavin								
	No recommendation based on broiler breeder data							
Niacin								
	No recommendation based on broiler breeder data							

Pantothenic acid							
	No recommendation based on broiler breeder data						
Pyridoxine							
	No recommendation based on broiler breeder data						
Folacin							
32–42 week			Egg production Feed intake FCR Fertility Hatchability Egg weight loss during incubation	Baheij (local Egyptian breed)	0, 2, 4, 8, 16, and 32 mg folic acid/kg diet	Basal diet content not measured No response to supplementation	Khalifah and Shahein, 2006 Abstract only
	No recommendation based on broiler breeder data						
Biotin							
Hens		16 µg	Egg production Fertility			Biotin is the only vitamin for which a specification is given for broiler breeders in the 1994 NRC report	NRC, 1994 (Table 2-7)
30–33 week		Egg production of 0.3 > 0 in younger birds Fertility of 0.45 > 0 (main effect)	Egg production Fertility	Arian (Iranian breed)	0, 0.3, and 0.45 mg biotin/L drinking water	Feed, water biotin content not measured	Daryabari et al., 2014
53–56 week		Fertility of 0.45 > 0 (main effect)	Egg production Fertility	Arian (Iranian breed)	0, 0.3, and 0.45 mg biotin/L drinking water	Feed, water biotin content not measured Production—no effect in older birds	Daryabari et al., 2014
30–33 week		Egg production of 0.3 and 0.45 > 0 (main effect); Fertility in sire (low fertility): 0.45 > 0; no effect in dam line	Egg production Fertility	Arian (Iranian breed)—pure sire line (low fertility), pure dam line (high fertility)	0, 0.3, and 0.45 mg biotin/L drinking water	Feed, water biotin content not measured	Daryabari et al., 2015

continued

TABLE 15-3 Continued

Age, wk	Egg Production %/d/Hen Housed[a]	Nutrient	Estimated Requirement vs. Response Criteria mg/kg Diet (mg/hen/d)	Response Criteria	Strain	Diet	Comments	Reference
Requirement								
Hens			16 μg/kg					
Vitamin B_{12}								
	No recommendation based on broiler breeder data							
Choline								
29–40 week				HDEP Fertility Hatchability	Broiler breeders	0, 760, and 1,520 mg/kg supplemental choline	Basal diet content not measured No effect observed	Rama Rao et al., 2001
	No recommendation based on broiler breeder data							
Vitamin C								
~19–60 week				Breeder production traits (multiple)	Peterson × Arbor Acres	0 vs. 75 mg of supplemental ascorbic acid	Basal diet contained 50 mg/kg (assayed). No effect observed	Creel et al., 2001
	No recommendation based on broiler breeder data							

feeding over the day affects the utilization of calcium. Calcium provided in the morning is not primarily utilized for eggshell formation and therefore can be used for bone mineralization. However, calcium provided in the afternoon may increase eggshell quality (Farmer et al., 1983; Van Wambeke and DeGroote, 1986). The particle size of calcium supplements has also been shown to have effects on the calcium status of broiler breeder hens. As with laying hens, feeding broiler breeders larger particle limestone increases limestone retention in the gizzard and increases the in vivo solubility of that limestone (Zhang and Coon, 1997). This generally translates to increased bone ash and shell quality in broiler breeders and their eggs (Reis et al., 1995; Ekmay and Coon, 2011; Manangi et al., 2018).

Phosphorus

Phosphorus comprises some 0.3 weight percent of eggshell, with the majority of egg phosphorus found in the yolk phosvitins (Li et al., 2014), which are phosphoproteins derived from hepatically synthesized vitellogenin. Dietary phosphorus is not directly utilized in eggshell formation but, as detailed in Chapter 13 (laying hens), it is required in higher concentrations in the diet of egg laying hens to replace phosphorus lost through hydroxyapatite mobilization from bone when dietary calcium is inadequate to support calcium needs. The embryo uses phosvitin phosphorus for its bone formation (Li et al., 2014). The timing of phosphorus feeding does not appear to influence eggshell quality (Van Wambeke and DeGroote, 1986). Despite the economic, environmental, and health aspects of broiler breeder phosphorus nutrition, little published research has been completed on the non-phytate phosphorus (nPP) requirements of broiler breeder hens since the 1994 report (NRC, 1994). Ekmay and Coon (2010) published the most robust evaluation of nPP requirements for broiler breeder hens feeding diets that contained 288, 360, 432, 504, or 576 mg nPP/d at peak production from 182 to 455 days of age. Their results support the 1994 recommendation of 350 mg nPP/d (NRC, 1994) as egg production, egg weights, and egg specific gravity were not significantly different when dietary nPP was greater than 360 mg nPP/d. However, hen tibia ash was reduced when hens were fed diets containing less than 360 mg nPP/d. Follow-up work compared lower dietary nPP of 188 to 228 mg nPP/d versus a control of 452 to 547 mg nPP/d from 114 to 334 days of age (Ekmay et al., 2012). The lower concentrations of nPP resulted in reduced egg production, egg weights, hen survivability, hen tibia ash, and 1 d progeny weights, although hatch of fertile eggs was not different, and egg specific gravity was increased. Plumstead and others (2007) studied whether phytase addition to the diet could be used to reduce breeder diet nPP, and subsequent phosphorus excretion, without compromising production or health of breeder hens, albeit impacts on hen tibia ash were undetermined. Because feed intake was not measured, mg of phosphorus intake per day per hen could not be calculated. Despite its limitations, the outcomes suggested that phytase can be used to reduce dietary nPP and phosphorus excretion without compromising hen performance. Additional research is needed to generate specific recommendations on the degree of nPP reduction possible with phytase addition with different diet formulations and hen genetics.

Trace Mineral Requirements of Broiler Breeders for Reproduction

The previous report (NRC, 1994) and those that preceded it did not provide separate recommendations for trace mineral requirements of growing broiler breeders. There was a presumption that due to growth limitations imposed by feed restriction, the daily requirements determined for commercial broilers are likely more than adequate for growing breeders. This assumption has not been directly tested; furthermore, trace mineral requirements of adult breeding males remain an area lacking sufficient new research to make any changes in requirement recommendations. However, new research on broiler breeder hens in the laying phase permits new recommendations for several trace minerals. Many of the studies depleted the hen's storage pools for the mineral of interest prior to the study period by feeding them a deficient diet and the exact impact of this on the requirement estimate is not known. However, depleted storage pools likely result in a slightly higher requirement during the early part of the study (Emmert and Baker, 1995). Most of the studies reviewed used mineral sulfates to determine the requirement, and if inorganic forms or organic forms with higher bioavailability are used, lower dietary concentrations may be possible. The requirement values (amounts/kg diet) described subsequently are for the mineral itself and not for mineral sulfate form.

Copper

The requirement for copper from $CuSO_4{\cdot}5H_2O$ was examined in copper-depleted Cobb 500 hens from 25 to 44 weeks of age by Berwanger and others (2018). They found requirements of 8.8, 13.5, 10.2, 14.3, and 9.42 mg copper/kg diet for egg production (hen/d), total eggs per hen, settable eggs, egg weight, and chick weight, respectively. Higher concentrations were needed to maximize hemoglobin and hematocrit concentrations in the hen, but the offspring's hematocrit concentrations were maximized at 12.0 mg Cu /kg diet. Based on dietary concentrations needed to maximize total eggs per hen and chick weight, the recommended dietary level is 13.5 mg Cu/kg diet. Because this recommendation is based on a single study, it is designated as tentative.

Iron

Taschetto and others (2017) examined the requirement for iron (as $FeSO_4{\cdot}7H_20$) of iron-depleted Cobb 500 hens from 47 to 70 weeks of age and averaging 54 percent egg production per hen housed per day. They found requirements of 91.8, 90.4, and 99.1 mg iron/kg diet/hen/d for egg production, settable eggs, and yolk iron, respectively. Higher concentrations were needed to maximize hemoglobin and hematocrit concentrations in the hen and the offspring; however, there was no indication of any iron deficiency–related pathology at lower concentrations. Abbasi and others (2015) examined the requirement for iron (as $FeSO_4{\cdot}7H_20$) of iron-depleted Cobb 500 hens during the late stage of lay (62 to 68 weeks) when the rate of lay was only 38 percent egg production per hen housed per day. They estimated the requirement to be 65 and 64 mg iron/kg diet for egg production and egg weight, respectively. Based on dietary concentrations needed to maximize egg production, the recommended dietary iron level for peak egg production is 92 mg/kg diet, declining to 65 mg/kg diet during the post-peak stage of production. Because these recommendations are based on a single study, they are designated as tentative.

Zinc

Four studies met the criteria for inclusion and examined the zinc requirement of broiler hens during various stages of lay (Sharideh et al., 2015; Zhang et al., 2017; Li et al., 2019; Mayer et al., 2019). Sharideh and others (2015) examined the requirement for zinc (as ZnO) in Cobb 500 hens in the later stage of production. Plasma Zn, Ca, P, total protein concentrations, and activities of lactase dehydrogenase (LDH), and alanine transaminase (ALT) were used to conclude a requirement between 60 to 90 mg/kg diet by ANOVA. Zhang and others (2017) estimated the requirement for zinc as zinc glycinate in zinc-depleted Lingnan yellow-feathered broiler hens from 39 to 47 weeks, although they did not fit the data using mathematical models. However, their ANOVA analysis indicated a requirement for egg production ~58 mg/kg diet, similar to that observed by Mayer and others (2019), who examined the requirement for zinc (as $ZnSO_4{\cdot}7H_2O$) of zinc-depleted Cobb 500 hens from 33 to 44 weeks of age. From 33 to 36 weeks, they estimated the requirement to be 81.0 and 83.4 ppm (mg/kg diet) for egg production and number of settable eggs, respectively. During the next 6 weeks (37 to 44 weeks) the rate of egg production declined, and the estimated requirement declined to an average of 58.0 and 59.1 mg/kg for egg production and number of settable eggs, respectively. Eggshell as percent of egg weight, shell breaking strength, and shell thickness measured at the end of each period were maximized at 135.1, 82.4, and 66.1 mg/kg, respectively.

The requirement for zinc (as $ZnSO_4{\cdot}H_2O$) was examined in Lingnan yellow-feathered hens from 58 to 65 weeks of age and was estimated to be 71.1 ,92.3, 94.4, and 98.7 for maximal egg production, yolk zinc, fertility, and hatchability, respectively (Li et al., 2019). Based on these studies, the apparent requirement for egg production in the early phase of production is 81 mg/kg diet and 71 mg/kg diet during later stages, but the recommended requirement for fertility may be higher, so these recommendations are tentative until more research on the zinc requirement for fertility and hatchability is reported.

Vitamins Requirements of Broiler Breeders for Reproduction

Few studies have been conducted that meet the inclusion criteria to determine vitamin requirements for broiler breeders, and as such, no requirements are indicated for most vitamins. In the 1994 NRC report, the only vitamin for which a specification was given was biotin (NRC, 1994), which remains the same in the present report (16 μg/day at peak production).

Recommendations for vitamin D_3 have been specified for breeders aged 27 to 36 and 37 to 66 weeks of age based on a single published study (Atencio et al., 2006). Although this study did not include analyzed dietary vitamin D_3 concentrations, diets were formulated to include ingredients devoid of vitamin D activity. For this reason, values are listed as "apparent adequate intakes" rather than requirements.

Several studies have investigated varying concentrations of vitamins A and E and biotin. However, these studies did not meet the inclusion criteria due to a limited number of dietary concentrations and a lack of analyzed dietary values. In these cases, evidence of adequate intake was provided based on a plateau in relevant response criteria. However, these values should be used with caution as the values may be higher than the actual requirements due to potentially variable vitamin contributions from other ingredients, and often wide ranges between tested concentrations.

No studies suitable to determine requirements or adequate intakes have been reported since 1994 for vitamin K, thiamin, riboflavin, niacin, pantothenic acid, pyridoxine, nor choline.

REFERENCES

Abbasi, M., M. Zaghari, M. Ganjkhanlo, and S. Khalaji. 2015. Is dietary iron requirement of broiler breeder hens at the late stage of production cycle influenced by phytase supplementation? *Journal of Applied Animal Research*. 43(2):166–176.

Afrouziyeh, M., N.M. Zukiwsky, J. You, R.P. Kwakkel, D.R. Korver, and M.J. Zuidhof. 2022. Architecture of broiler breeder energy partitioning models. *Poultry Science*. 101(1):101518.

Aranibar, C.D., C. Chen, A.J. Davis, W.I. Daley, C. Dunkley, W.K. Kim, C. Usher, A.B. Webster, and J.L. Wilson. 2020. Impact of an alternate feeding program on broiler breeder pullet behavior, performance, and plasma corticosterone. *Poultry Science*. 99(2):829–838.

Asensio, X., N. Abdelli, J. Piedrafita, M.D. Soler, and A.C. Barroeta. 2020. Effect of fibrous diet and vitamin C inclusion on uniformity, carcass traits, skeletal strength, and behavior of broiler breeder pullets. *Poultry Science*. 99:2633–2644.

Atencio, A., H.M. Edwards Jr., and G. Pesti. 2005a. Effects of vitamin D3 dietary supplementation of broiler breeder hens on the performance and bone abnormalities of the progeny. *Poultry Science*. 84(7):1058–1068.

Atencio, A., G.M. Pesti, and H.M. Edwards. 2005b. Twenty-five hydroxycholecalciferol as a cholecalciferol substitute in broiler breeder hen diets and its effect on the performance and general health of the progeny. *Poultry Science*. 84(8):1277–1285.

Atencio, A., H.M. Edwards, G.M. Pesti, and G.O. Ware. 2006. The vitamin D3 requirement of broiler breeders. *Poultry Science*. 85(4):674–692.

Aviagen. 2018. *Ross Parentstock Management Guide*. Aviagen. http://www.rosspoultrybreeders.co.za/downloads/breeder/2018RossPSHandbook.pdf.

Avila, L.P., K.M. Sweeney, C.R. Evans, D.L. White, W.K. Kim, P. Regmi, S.M. Williams, J. Nicholds, and J.L. Wilson. 2023. Body composition, gastrointestinal, and reproductive differences between broiler breeders fed using everyday or skip-a-day rearing programs. *Poultry Science*. 102(9):102853.

Berwanger, E., S.L. Vieira, C.R. Angel, L. Kindlein, A.N. Mayer, M.A. Ebbing, and M. Lopes. 2018. Copper requirements of broiler breeder hens. *Poultry Science*. 97(8):2785–2797.

Bokkers, E.A., and P. Koene. 2003. Eating behaviour, and preprandial and postprandial correlations in male broiler and layer chickens. *British Poultry Science*. 44(4):538–544.

Cash, C., K. Witherspoon, and G. Athrey. 2025. How concerned should we be about broiler breeder fertility declines? *Poultry Science*. 104(5): 104992.

Cerrate, S., J.T. Halley, A. Corzo, and B.I. Fancher. 2019. Effect of dietary amino acid density on broiler breeder reproductive performance. *Poultry Science*. 98(5):2072–2079.

Chen, S.E., J.P. McMurtry, and R.L. Walzem. 2006. Overfeeding-induced ovarian dysfunction in broiler breeder hens is associated with lipotoxicity. *Poultry Science*. 85(1):70–81.

Cobb-Vantress. 2022. *Cobb Breeder Management Guide*. https://www.cobb-vantress.com/assets/Cobb-Files/management-guides/ca1b2a76ed/Cobb-Breeder-Management-Guide.pdf.

Creel, L.H., D.V. Maurice, S.F. Lightsey, and L.W. Grimes. 2001. Stability of dietary ascorbic acid and the effect of supplementation on reproductive performance of broiler breeder chickens. *British Poultry Science*. 42(1):96–101.

Daryabari, H., A. Akhlaghi, M.J. Zamiri, G.R. Mianji, Z.A. Pirsaraei, H. Deldar, and A.N. Eghbalian. 2014. Reproductive performance and oviductal expression of avidin and avidin-related protein-2 in young and old broiler breeder hens orally exposed to supplementary biotin. *Poultry Science*. 93(9):2289–2295.

Daryabari, H., A. Akhlaghi, M.J. Zamiri, Z.A. Pirsaraei, G.R. Mianji, H. Deldar, and A.N. Eghbalian. 2015. Oral administration of supplementary biotin differentially influences the fertility rate and oviductal expression of avidin and avidin-related protein-2 in low- and high-fertility broiler line hens. *Poultry Science*. 94(2):289–295.

de Beer, M., and C.N. Coon. 2007. The effect of different feed restriction programs on reproductive performance, efficiency, frame size, and uniformity in broiler breeder hens. *Poultry Science*. 86(9):1927–1939.

de Beer, M., J.P. McMurtry, D.M. Brocht, and C.N. Coon. 2008. An examination of the role of feeding regimens in regulating metabolism during the broiler breeder grower period. 2. Plasma hormones and metabolites. *Poultry Science*. 87(2):264–275.

de Los Mozos, J., A.G. Ruiz, L.A. Hartog, and M. Villamide. 2017. Growth curve and diet density affect eating motivation, behavior, and body composition of broiler breeders during rearing. *Poultry Science*. 96:3382.

Dozier, W., M. Kidd, and A. Corzo. 2008. Dietary amino acid responses of broiler chickens. *Journal of Applied Poultry Research*. 17:157–167.

Driver, J.P., A. Atencio, G.M. Pesti, H.M. Edwards Jr., and R.I. Bakalli. 2006. The effect of maternal dietary vitamin D3 supplementation on performance and tibial dyschondroplasia of broiler chicks. *Poultry Science*. 85(1):39–47.

Ekmay, R.D., and C.N. Coon. 2010. An examination of the P requirements of broiler breeders for performance, progeny quality and P balance 1. Non-phytate phosphoros. *International Journal of Poultry Science*. 9(11):1043–1049.

Ekmay, R.D., and C.N. Coon. 2011. An examination of the P requirements of broiler breeders for performance, progeny quality and P balance 2. Ca particle size. *International Journal of Poultry Science*. 10(10):760–765.

Ekmay, R.D., M. de Beer, S.J. Mei, M. Manangi, and C.N. Coon. 2013. Amino acid requirements of broiler breeders at peak production for egg mass, body weight, and fertility. *Poultry Science*. 92(4):992–1006.

Ekmay, R.D., M. de Beer, R.W. Rosebrough, M.P. Richards, J.P. McMurtry, and C.N. Coon. 2010. The role of feeding regimens in regulating metabolism of sexually mature broiler breeders. *Poultry Science*. 89(6): 1171–1181.

Ekmay, R.D., C. Salas, J. England, S. Cerrate, and C.N. Coon. 2012. The effects of pullet body weight, dietary nonpyhtate phosphorus intake, and breeder feeding regimen on production performance, chick quality, and bone remodeling in broiler breeders. *Poultry Science*. 91(4):948–964.

Emmert, J.L., and D.H. Baker. 1995. Zinc stores in chickens delay the onset of zinc deficiency symptoms. *Poultry Science*. 74(6):1011–1021.

Farmer, M., S. Roland, D.A.J. Brake, and M.K. Eckman. 1983. Calcium metabolism in broiler breeder hens. 1. Calcium status of the digestive tract of broiler breeders throughout a 24-hour period. *Poultry Science*. 62:459–464.

Fisher, C., T.R. Morris, and R.C. Jennings. 1973. A model for the descriptions and prediction of the response of laying hens to amino acid intake. *British Poultry Science*. 14:469–484.

Hadinia, S.H., P.R.O. Carneiro, C.A. Ouellette, and M.J. Zuidhof. 2018. Energy partitioning by broiler breeder pullets in skip-a-day and precision feeding systems. *Poultry Science*. 97(12):4279–4289.

Hossain, S., S. Barreto, A. Bertechini, A. Rios, and C. Silva. 1998. Influence of dietary vitamin E level on egg production of broiler breeders, and on the growth and immune response of progeny in comparison with the progeny from eggs injected with vitamin E. *Animal Feed Science Technology*. 73(3–4):307–317.

Hudson, B. 2022. *Broiler Breeder Males: Feed and Nutrition Discussion*. Proceedings of the Arkansas Nutrition Conference. Little Rock.

Khalifah, M., and E. Shahein. 2006. Effect of dietary folic acid supplementation on production and hatching performance in Baheij chicken strain. *Egypt Poultry Science*. 26(3):843–855.

Lesuisse, J., C. Li, S. Schallier, W. L. S. Clímaco, A. Bautil, N. Everaert, and J. Buyse. 2018. Multigenerational effects of a reduced balanced protein diet during the rearing and laying period of broiler breeders. 1. Performance of the F1 breeder generation. *Poultry Science*. 97: 1651–1665.

Lewis, P.D., R.M. Gous, and T.R. Morris. 2007. Model to predict age at sexual maturity in broiler breeders given a single increment in photoperiod. *British Poultry Science*. 48(5):625–634.

Li, C., F. Geng, X. Huang, M. Ma, and X. Zhang. 2014. Phosvitin phosphorus is involved in chicken embryo bone formation through dephosphorylation. *Poultry Science*. 93(12):3065–3072.

Li, L., K.F.M. Abouelezz, Z. Gou, X. Lin, Y. Wang, Q. Fan, Z. Cheng, F. Ding, S. Jiang, and Z. Jiang. 2019. Optimization of dietary zinc requirement for broiler breeder hens of Chinese yellow-feathered chicken. *Animals*. 9(7):472.

Lin, H.Y., P.C. Chou, Y.H. Chen, L.S. Lai, T.K. Chung, R.L. Walzem, S.Y. Huang, and S.E. Chen. 2019a. Dietary supplementation of 25-hydroxycholecalciferol improves livability in broiler breeder hens: Amelioration of cardiac pathogenesis and hepatopathology. *Animals*. 9(10):770.

Lin, H.Y., T.K. Chung, Y.H. Chen, R.L. Walzem, and S.E. Chen. 2019b. Dietary supplementation of 25-hydroxycholecalciferol improves livability in broiler breeder hens. *Poultry Science*. 98(11):6108–6116.

Manangi, M.K., P. Maharjan, and C.N. Coon. 2018 Calcium particle size effects on plasma, excreta, and urinary Ca and P changes in broiler breeder hens. *Poultry Science*. 97:2798–2806.

Mayer, A.N., S.L. Vieira, E. Berwanger, C.R. Angel, L. Kindlein, I. Franca, and T.L. Noetzold. 2019. Zinc requirements of broiler breeder hens. *Poultry Science*. 98(3):1288–1301.

McDaniel, G.R., D.A. Roland, and M.A. Coleman. 1979. The effect of egg shell quality on hatchability and embryonic mortality. *Poultry Science*. 58(1):10–13.

Mench, J.A. 2002. Broiler breeders: Feed restriction and welfare. *World's Poultry Science Journal*. 58(1):23–29.

Mohiti-Asli, M., M. Shivazad, M. Zaghari, S. Aminzadeh, M. Rezaian, and G.G. Mateos. 2012a. Dietary fibers and crude protein content alleviate hepatic fat deposition and obesity in broiler breeder hens. *Poultry Science*. 91(12):3107–3114.

Mohiti-Asli, M., M. Shivazad, M. Zaghari, M. Rezaian, S. Aminzadeh, and G.G. Mateos. 2012b. Effects of feeding regimen, fiber inclusion, and crude protein content of the diet on performance and egg quality and hatchability of eggs of broiler breeder hens. *Poultry Science*. 91(12): 3097–3106.

Moreki, J.C., H.J. Van Der Merwe, and J.P. Haynes. 2011. Effect of dietary calcium level on egg production and eggshell quality in broiler breeder hens from 36 to 60 weeks of age. *Journal of Animal Feed Research*. 1(1):1–7.

Muir, W.M., G.K.-S. Wong, Y. Zhang, J. Wang, M.A.M. Groenen, R.P.M.A. Crooijmans, H.-J. Megens, H. Zhang, R. Okimoto, A. Vereijken, A. Jungerius, G.A.A. Albers, C.T. Lawley, M.E. Delany, S. MacEachern, and H.H. Cheng. 2008. Genome-wide assessment of worldwide chicken SNP genetic diversity indicates significant absence of rare alleles in commercial breeds. *Proceedings of the National Academy of Sciences*. 105(45):17312–17317.

Noetzold, T.L., E.A. Obi, B. Fancher, M. Silva, A. Thomson, and M.J. Zuidhof. 2024. Body weight optimization of broiler breeder hens. 1. Pullet growth, feed efficiency, carcass composition, and sexual maturation. *Poultry Science*. 103(12):104414.

NRC (National Research Council). 1994. *Nutrient Requirements of Poultry*, 9th rev. ed. Washington, DC: National Academy Press.

OECD (UN Organisation for Economic Co-operation and Development) and FAO (UN Food and Agriculture Organization). 2021. *OECD-FAO Agricultural Outlook 2021–2030*. Author.

Olkowski, A.A., and H.L. Classen. 1999. The effects of maternal thiamine nutrition on thiamine status of the offspring in broiler chickens. *International Journal for Vitamin and Nutrition Research*. 69(1):32–40.

Plumstead, P.W., H. Romero-Sanchez, R.O. Maguire, A.G. Gernat, and J. Brake. 2007. Effects of phosphorus level and phytase in broiler breeder rearing and laying diets on live performance and phosphorus excretion. *Poultry Science*. 86:225–231.

Rama Rao, S.V., G.S. Sunder, M.R. Reddy, N.K. Praharaj, M.V. Raju, and A.K. Panda. 2001. Effect of supplementary choline on the performance of broiler breeders fed on different energy sources. *British Poultry Science*. 42(3):362–367.

Reis, L.H., P. Feio, L.T. Gama, and M. Chaveiro Soares. 1995. Extra dietary calcium supplement and broiler breeders. *Journal Applied Poultry Science*. 4(3):276–282.

Renema, R.A., and F.E. Robinson. 2004. Defining normal: Comparison of feed restriction and full feeding of female broiler breeders. *World's Poultry Science Journal*. 60(4):508–522.

Renema, R.A., F.E. Robinson, J.A. Proudman, M. Newcombe, and R.I. McKay. 1999. Effects of body weight and feed allocation during sexual maturation in broiler breeder hens. 2. Ovarian morphology and plasma hormone profiles. *Poultry Science*. 78(5):629–639.

Renema, R.A., M.E. Rustad, and F.E. Robinson. 2007. Implications of changes to commercial broiler and broiler breeder body weight targets over the past 30 years. *World's Poultry Science Journal*. 63(3):457–472.

Riber, A.B., and F.M. Tahamtani. 2020. Motivation for feeding in broiler breeder pullets fed different types of restricted high-fibre diets. *Applied Animal Behaviour Science*. 230:105048.

Richards, M.P. 2003. Genetic regulation of feed intake and energy balance in poultry. *Poultry Science*. 82(6):907–916.

Romero-Sanchez, H., P.W. Plumstead, N. Leksrisompong, K.E. Brannan, and J. Brake. 2008. Feeding broiler breeder males. 4. Deficient feed allocation reduces fertility and broiler progeny body weight. *Poultry Science*. 87(4):805–811.

Sakomura, N.K. 2004. Modeling energy utilization in broiler breeders, laying hens and broilers. *Brazilian Journal of Poultry Science*. 6(1):1–11.

Sakomura, N.K., R. Silva, H.P. Couto, C. Coon, and C.R. Pacheco. 2003. Modeling metabolizable energy utilization in broiler breeder pullets. *Poultry Science*. 82(3):419–427.

Sakomura, N.K., E.P. Silva, J.C.P. Dorigam, R.M. Gous, and N. St-Pierre. 2015. Modeling amino acid requirements of poultry. *The Journal of Applied Poultry Research* 24:267–282.

Sharideh, H., M. Zhandi, M. Zaghari, and A. Akhlaghi. 2015. Effect of dietary zinc oxide and phytase on the plasma metabolites and enzyme activities in aged broiler breeder hens. *Iranian Journal of Veterinary Medicine*. 9(4):263–270.

Siegel, P.B., S.E. Price, B. Meldrum, M. Picard, and P.A. Geraert. 2001. Performance of pureline broiler breeders fed two levels of vitamin E. *Poultry Science*. 80(9):1258–1262.

Silva, E.P., N.K. Sakomura, C.F.S. Oliveira, F.G.P. Costa,, J.C.P. Dorigam, and E.B. Malheiros. 2015. The optimal lysine and threonine intake for Cobb broiler breeder hens using Reading model. *Livestock Science*. 174:59–65.

Silveira, M.M., A.G. Freitas, C. Moraes, F.S. Gomes, F. Litz, J.M.S. Martins, N. Fagundes, and E. Fernandes. 2014. Feeding management strategy for male broiler breeders and its effects on body weight, hatchability and fertility. *Revista Brasileira de Ciencia Avicola*. 16(4):397–402.

Singh, A.K., and W.K. Kim. 2021. Effects of dietary fiber on nutrients utilization and gut health of poultry: A review of challenges and opportunities. *Animals*. 11(1):181.

Sweeney, K.M., C.D. Aranibar, W.K. Kim, S.M. Williams, L.P. Avila, J.D. Starkey, C.W. Starkey, and J.L. Wilson. 2022. Impact of every-day versus skip-a-day feeding of broiler breeder pullets during rearing on body weight uniformity and reproductive performance. *Poultry Science*. 101(8):101959.

Tahamtani, F.M., H. Moradi, and A.B. Riber. 2020. Effect of qualitative feed restriction in broiler breeder pullets on stress and clinical welfare indicators. *Frontiers in Veterinary Science*. 7:316.

Taschetto, D., S.L. Vieira, C.R. Angel, C. Stefanello, L. Kindlein, M.A. Ebbing, and C.T. Simoes. 2017. Iron requirements of broiler breeder hens. *Poultry Science*. 96(11):3920–3927.

Ulmer-Franco, A.M., G.M. Fasenko, and E.E. O'Dea Christopher. 2010. Hatching egg characteristics, chick quality, and broiler performance at 2 breeder flock ages and from 3 egg weights. *Poultry Science*. 89(12): 2735–2742.

van Emous, R. 2022. State of the art pullet and breeder nutrition. *Proceedings of the Arkansas Nutrition Conference 2022*. 11.

van Emous, R.A., R.P. Kwakkel, M.M. van Krimpen, and W.H. Hendriks. 2015. Effects of dietary protein levels during rearing and dietary energy levels during lay on body composition and reproduction in broiler breeder females. *Poultry Science*. 94:1030–1042.

van Emous, R.A., R.P. Kwakkel, M.M. van Krimpen, H. van den Brand, and W.H. Hendriks. 2013. Effects of growth patterns and dietary crude protein levels during rearing on body composition and performance in broiler breeder females during the rearing and laying period. *Poultry Science*. 92:2091–2100.

van Emous, R., J. Holleman, and T. van Schie. 2020. *Breeder Signals. Vol. 1: Poultry Signals*. J. Lemmens, R. van Oosten, H. Rodeboog, and A. Steentjes, editors. Zutphen, The Netherlands: Roodbont Publishers B.V.

Van Wambeke, F., and G. DeGroote. 1986. L'influence du moment de l'alimentation et du replacement partiel de la craie en poudre par des coquilles d'huitres comme source de calcium sur les resultats de reproduction de deux souches de poules reproduction du type chair. *Revue de l'Agriculture*. 39:137.

Vignale, K., J.V. Caldas, J.A. England, N. Boonsinchai, P. Sodsee, M. Putsakum, E.D. Pollock, S. Dridi, and C. Coon. 2017. The effect of four different feeding regimens from rearing period to sexual maturity on breast muscle protein turnover in broiler breeder parent stock. *Poultry Science*. 96:1219–1227.

Walzem, R.L., and S.E. Chen. 2014. Obesity-induced dysfunctions in female reproduction: Lessons from birds and mammals. *Advanced Nutrition*. 5(2):199–206.

Weaver, E.A., and R. Ramachandran. 2023. Metformin improves ovarian function and increases egg production in broiler breeder hens. *Reproduction*. 165(3):289–300.

Wilson, H.R., E.R. Miller, R.H. Harms, and B.L. Damron. 1981. Hatchability of chicken eggs as affected by dietary phosphorus and calcium. *Poultry Science*. 59:1284–1289.

Yang, J., X. Ding, S. Bai, J. Wang, Q. Zeng, H. Peng, Z. Su, Y. Xuan, G.S. Fraley, and K. Zhang. 2020. Effects of maternal dietary vitamin E on the egg characteristics, hatchability and offspring quality of prolonged storage eggs of broiler breeder hens. *Journal of Animal Physiology and Animal Nutrition*. 104:1384–1391.

Yu, M.W., F.E. Robinson, R.G. Charles, and R. Weingardt. 1992. Effect of feed allowance during rearing and breeding on female broiler breeders. 2. Ovarian morphology and production. *Poultry Science*. 71(10):1750–1761.

Yuan, J., A.R. Roshdy, Y. Guo, Y. Wang, and S. Guo. 2014. Effect of dietary vitamin A on reproductive performance and immune response of broiler breeders. *PLoS One*. 9(8):e105677.

Zhang, B., and C.N. Coon. 1997. The relationship of calcium intake, source, size, solubility in vitro and in vivo, and gizzard limestone retention in laying hens. *Poultry Science*. 76(12):1702–1706.

Zhang, L., Y.-X. Wang, X. Xiao, J.-S. Wang, Q. Wang, K.-X. Li, T.-Y. Guo, and X.-A. Zhan. 2017. Effects of zinc glycinate on productive and reproductive performance, zinc concentration and antioxidant status in broiler breeders. *Biological Trace Element Research*. 178(2): 320–326.

Zuidhof, M.J., D.E. Holm, R.A. Renema, M.A. Jalal, and F.E. Robinson. 2015. Effects of broiler breeder management on pullet body weight and carcass uniformity. *Poultry Science*. 94(6):1389–1397.

Zuidhof, M.J., B.L. Schneider, V.L. Carney, D. Korver, and F. Robinson. 2014. Growth, efficiency, and yield of commercial broilers from 1957, 1978, and 2005. *Poultry Science*. 93(12):2970–2982.

16

Nutrient Requirements of Turkeys

WORLD TURKEY PRODUCTION AND THE MAJOR TURKEY BREEDS AND THEIR USES

World Turkey Meat and Egg Production

Turkey meat is a lean and healthy choice of animal protein and represented approximately 4.2 percent of total global poultry meat production in 2021 (FAOSTAT, 2022). The global production of turkeys and turkey meat increased by 4.26-fold and 6.45-fold, respectively, between 1961 and 2021 (FAOSTAT, 2022). The number of turkeys produced in the world increased by about 2.35 percent per year from 1961 to 1996 and then remained relatively consistent at about 636 million head slaughtered (Figure 16-1). In contrast, turkey meat production has increased linearly from 1961 to approximately 6 million tonnes in 2021 (Figure 16-2). Genetic selection and management practices for increased market weights have compensated for the plateau in the number of turkeys produced during the last 30 years to meet consumer demand. Moreover, meat produced per turkey (males and females

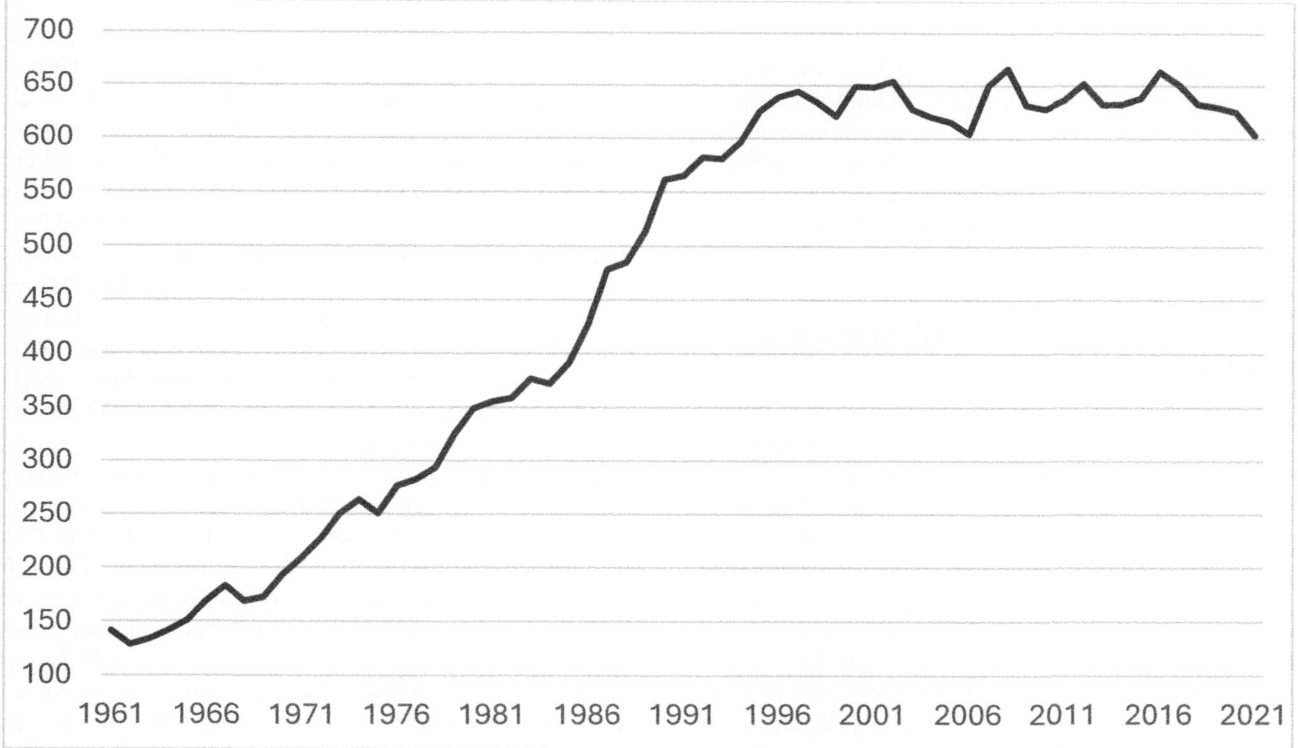

FIGURE 16-1 World turkey production (million head) from 1961 to 2021.
SOURCE: Data from FAOSTAT (2022).

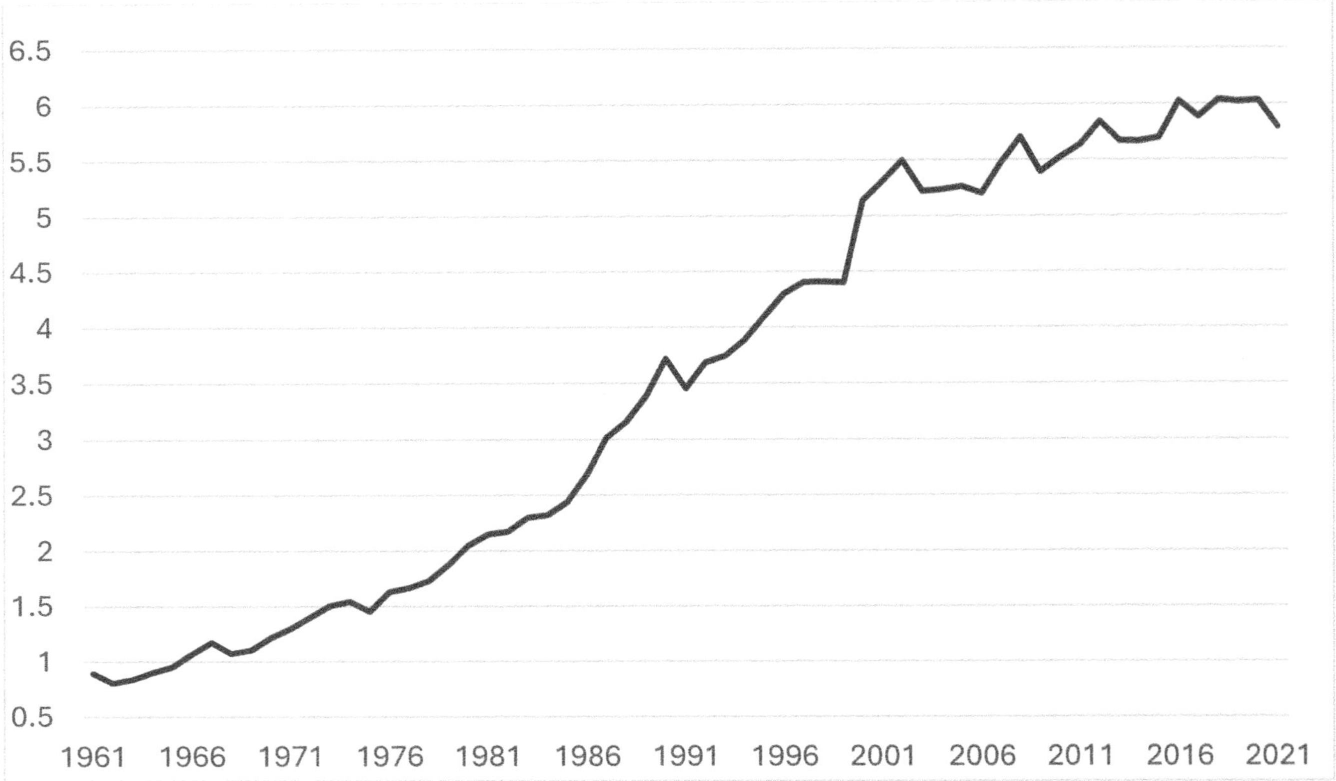

FIGURE 16-2 World turkey meat production (million tonnes) from 1961 to 2021.
SOURCE: Data from FAOSTAT (2022).

combined) increased from 6.3 kg/bird to 9.5 kg/bird over the last 60 years (FAOSTAT, 2022). Of the ~6 million tonnes of turkey meat produced in the world, 44 percent is produced in the United States, 32 percent in Europe, 16 percent in South America (~10% in Brazil), and 8 percent in the Middle East (North Africa and West Asia). Israel ranks first in per capita consumption of turkey products.

Genetic Origin of Domestic Breeds and Strain Growth Characteristics Used in Production

The wild turkey (*Meleagris gallopavo*) is indigenous to North America and had a historical range that included the United States, Mexico, and southern Canada. Domestication of wild turkeys was accomplished by Native Americans over 2,000 years ago in two locations, southern Mexico and the southwestern United States (Thornton, 2016). Turkeys are the only vertebrate of agricultural significance that was domesticated in North America. Genetic evidence shows that in Mexico the domesticated subspecies was *Meleagris gallopavo gallopavo* and that in the Four Corners region of the United States the subspecies was *Meleagris gallopavo merriam* (Manin et al., 2018). Genetic evidence also shows that all current commercial and heritage breeds of turkeys examined to date are derived from the stock domesticated in southern Mexico and not those in the southwestern United States or the eastern United States (Speller et al., 2010; Aslam et al., 2012; Manin et al., 2018). Archeological evidence suggests that domestication in Mexico occurred in agricultural regions that had permanent settlements, and the birds were raised in pens. Analysis of isotope ratios in turkey bones obtained from archeological sites indicated that their diet became corn-based over time during domestication (Manin et al., 2018). Following contact with Aztecs in southern Mexico, Hernán Cortés was credited for bringing turkeys to Spain in the early 1500s, and from there the domesticated turkey was distributed widely over Europe as there were reports of turkeys in Germany in 1530 and in England by 1541 (Brant, 1998).

The feeding preferences of the progenitors of southern Mexican domestic turkeys, *Meleagris gallopavo gallopavo*, are mostly unknown because they are now extinct in their natural habitat. However, all of the subspecies of *Meleagris gallopavo* live mostly in forested areas and depend heavily on foods produced by trees and the invertebrate inhabits of forests (Korschgen, 1967). The remaining subspecies of *Meleagris gallopavo* that live in proximity to the extinct progenitor are more insectivores than those from the United States, likely due to the higher availability of insects in their habitat.

The feeding preferences of the primary wild turkey subspecies in the United States, *Meleagris gallopavo sylvestris*, has been carefully studied because they are a popular game bird. Wild turkeys are omnivores and feed on a diverse mixture of foods, and in one large study more than 350 species of plants and 300 species of invertebrates were identified in crop contents (Mosby and Handley, 1943). Feeding methods include whole ingestion, picking, stripping, clipping, and scratching (Mosby and Handley, 1943; Korschgen, 1967). Seeds from trees are typically consumed whole without shelling, and all but the largest invertebrates are also consumed whole (Korschgen, 1967). Wild turkeys forage throughout the day, but most feeding activity occurs during the first 2 to 3 hr after leaving the roost at daybreak and 2 to 3 hr before sunset (Leopold, 1933). Depending on the subspecies, location, and season, the percentage of the diet of adult turkeys from animal sources ranges from 5 to 40 percent (Mosby and Handley, 1943; Schorger, 1966). Insects are of critical importance to poults, and survival is low in habitats that do not support large insect populations. The intake of insects by young poults has been estimated to be about 60 percent of the diet and decreases with age (Korschgen, 1967; Rumble and Anderson, 1996). As stated previously, the nutritional biology of the southern Mexican subspecies (*Meleagris gallopavo gallopavo*) that is the direct ancestor of the domestic turkey is unknown; however, the habitat of east-central Mexico, where it was indigenous, is tropical deciduous forests and mountain pine forests with highly available invertebrates (Leopold, 1933; York and Schemnitz, 2003).

The domesticated turkey was further bred for traits related to plumage coloration, growth rate, or size and eventually was dispersed throughout the world, including back to North America. Genetic selection of turkeys into the commercial strains used today occurred mostly from about 1920 until 1990 (Buss, 1989). During the early part of the 1900s, Lord Rothschild contracted Jesse Throssell to supply large turkeys at a premium price to give away at Christmas time. Throssell selectively bred for the largest turkeys and subsequently developed his own strain, the Sheffield Bronze (Brant, 1998). Following the death of Lord Rothschild, Throsell emigrated with his stock to Aldergrove, BC, Canada, and soon turkey breeders in Oregon and Washington purchased some Sheffield Bronze males and selected turkeys for increased meat production, ultimately launching the Broad-Breasted Bronze strain that gained much attention when exhibited at the 1939 VIIth World Poultry Congress in Cleveland, Ohio. But to appease many consumers who preferred smaller turkeys with better carcass appeal and finish (more subcutaneous fat and no black pin feathers and feather follicles), U.S. Department of Agriculture scientists developed the Beltsville Small White strain, which was sold to the public in 1941 at a market weight about 4.5 kg and 7.0 kg for females and males, respectively. By 1954, this strain accounted for about 30 percent of all turkeys produced in the United States. This strain remained popular for the following 20 years until the market demand shifted to large broad-breasted white strains able to supply the needs of the further processing industry. For the past 50 years, turkey live market weights have increased linearly as commercial turkey breeding companies consolidated down to two: Aviagen turkeys[1] and Hybrid turkeys,[2] each supplying various genetic lines to meet different commercial demands and production systems. Collectively, from 1966 until 2021, live weights of male turkeys marketed at 18 weeks of age increased linearly by 217 g per year, whereas female turkeys marketed at 14 weeks of age increased linearly by 87 g per year (Figure 16-3) (Ferket, 2004; Flores-Maldonado, 2021).

Turkey Meat Products

Turkey meat is a high-quality protein with several nutritional benefits in comparison to meat from other animal species. It has a low fat content and high amounts of vitamins and minerals such as iron, zinc, potassium, phosphorus, selenium, pyridoxine, and niacin (Amirkhanov et al., 2017; Bohrer, 2017). Historically, in North America, turkey has been and continues to be consumed primarily during holidays, such as Thanksgiving and Christmas, as a whole carcass product although the economics of the turkey industry are far more reliant on breast meat (products) throughout the year than whole carcass sales in the fourth quarter. However, congruent with the genetic selection for larger birds as illustrated in Figure 16-3, the turkey meat products market has evolved to be segmented based on the type, category, production system, and distribution channel. Types of turkey meat products consists of whole and carcass cuts (with or without bone) or further processed. Within each product type, categories consist of frozen, fresh, and cooked. Within each type and category, production systems include conventional, all-vegetable-fed, raised without antibiotics, organic, free range, animal welfare certified, and so forth. Finally, the distribution channel may include supermarkets, specialty stores, online retail, institutional food service, and restaurants. Turkey hens are typically marketed at 12 to 16 weeks of age to predominantly supply fresh or frozen whole bird and bone-in carcass parts market, whereas turkey toms are typically marketed at 18 to 22 weeks of age to predominantly supply the fresh or frozen boneless breast or be further processed for market. Considering this vast array of turkey meat products available to consumers, the nutritional program used for commercial turkey production will depend upon the market niches that the producer seeks to supply and the optimum return on investment. The nutrient requirements presented herein will focus on those that sustain near optimum growth and health characteristics for turkeys of different age categories.

Turkey Production and Management Systems

Turkey production and management systems employed in different parts of the world depend upon a variety of political, social, and economic constraints of the market. Depending

[1] See https://www.aviagenturkeys.us.
[2] See https://www.hybridturkeys.com.

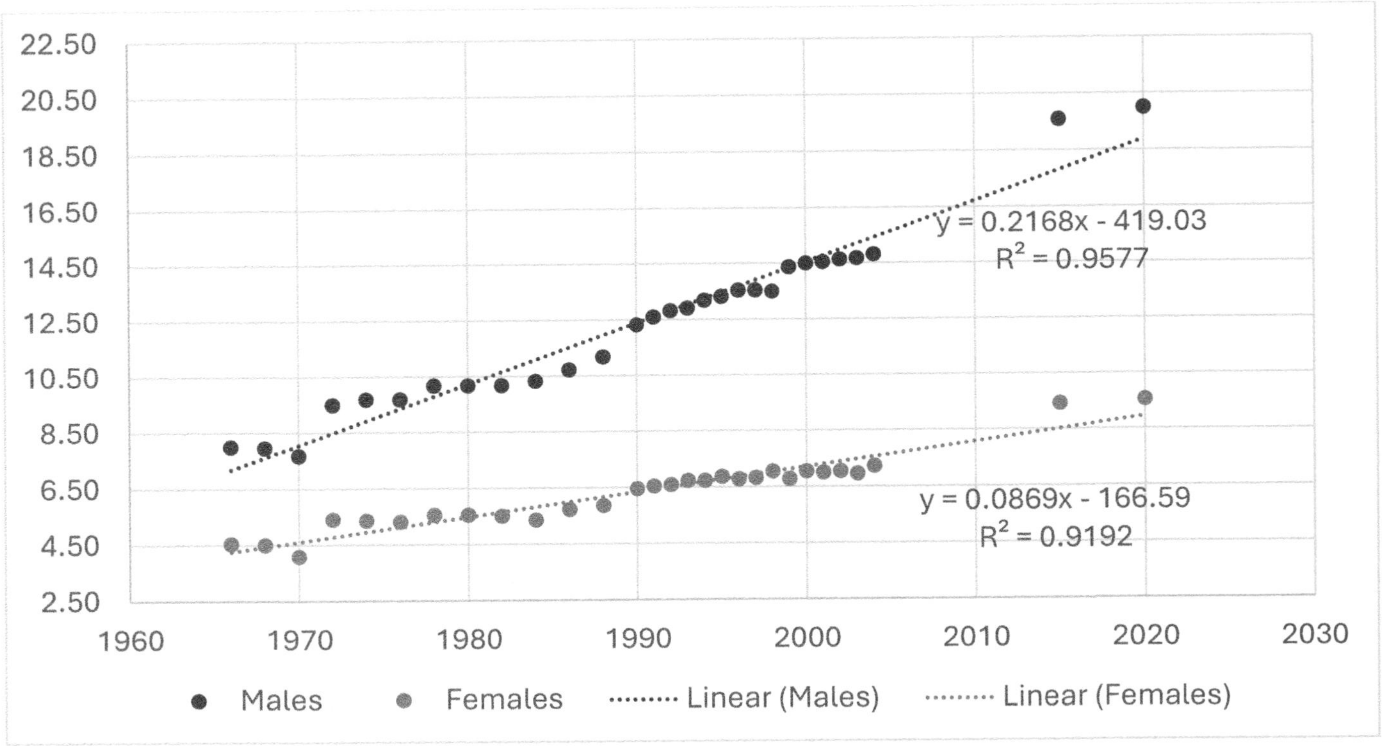

FIGURE 16-3 Live market weights (kg) of male turkeys at 18 weeks of age and female turkeys at 14 weeks of age.
SOURCES: Data from 1966 to 2003 are from Ferket (2004); subsequent data are from Flores-Maldonado (2021).

upon the country, turkeys are produced by independent or vertically integrated producers as free- or supply-managed marketing (van Staaveren et al., 2020). Because of the considerable differences in growth and behavior characteristics, turkey hens and toms are usually reared separately under different feeding and management programs to optimize production efficiency. The sex of turkey poults is typically determined by vent inspection in the hatchery by skilled technicians. However, where vent sexing is not feasible at the hatchery, hens and toms are separated on the farm when the gender attributes are more easily observed. Typically, turkey farms are multi-age production systems consisting of one brooder house to raise turkeys from placement to about 5 to 6 weeks of age paired with two sets of grow-out houses so that housing and management conditions can be optimized. However, where biosecurity and disease transmission risks are a concern in areas of concentrated integrated poultry production, turkey production farms have transitioned to single-age production hub systems. An integrated turkey production hub consists of a large single-age brooder contract farm to raise poults from all-in-all-out from 1-d-old placement until about 5 or 6 weeks of age, which is paired with several grow-finish farms that raise birds to market age.

As a means to best control biosecurity, turkey breeder parent stock are typically raised on farm hubs consisting of three production phases in separate farms: (1) a brooder farm housing poults to first selection at about 6 weeks of age; (2) a farm where breeding stock are raised sex-separately until the second selection at about market age for each, and then maintained until sexual maturity while controlling photoperiod and feed nutrient intake by either diet dilution or limiting daily feed intake; and (3) a breeder production farm with several hen houses for hatching egg production and an accompanying tom house for housing studs for semen production. All commercial turkey breeder hens are artificially inseminated every 1 to 2 weeks with fresh semen with a specified dose of sperm cells per hen, depending on the stage of egg production. Artificial insemination is necessary because of the large degree of sexual dimorphism in current highly selected turkey lines, which leads to poor reproductive success (Mohan et al., 2018).

Turkey Feeding Programs

Turkeys are fed according to production phases corresponding to physiological stages of development or reproductive status as follows: (1) brooding or starting, (2) growing or developing, (3) finishing, (4) breeder holding, and (5) breeder production. Within each feeding phase, diet nutrient contents are tailored to meet the birds' metabolic and physiological demands. As nutrient requirements to support optimum growth performance change throughout the growth period of turkeys, most flocks are provided four to eight different diets or feed phases, although it is possible

TABLE 16-1 An Example of Phase Feeding Guidelines by Age for Large White Commercial Meat Turkeys[a]

		Age (days)	
Feed	Phase	Males	Females
Starter 1	1	0 to 21	0 to 21
Starter 2	2	22 to 42	22 to 42
Grower 1	3	43 to 63	43 to 63
Grower 2	4	64 to 84	64 to 84
Grower 3	5	85 to 105	85 to 98
Finisher 1	6	106 to 126	99 to 112

[a] See Table 16-2 for diet nutrient recommendations during these phases.

to provide up to 20 feed phases when diet transitions are made in a controlled way by blending proportions of different manufactured feeds. An example of a six-phase (toms) and a six-phase (hens) feeding program by age for large white commercial turkeys (Table 16-1) represents a general guideline. In the previous version of this report (NRC, 1994), the committee recommended longer (4-week) phases and extended feeding periods (e.g., toms to 24 weeks and hens to 20 weeks) for large-type turkeys based on pre-1991 research (i.e., when their literature reviews were completed). However, in two footnotes in Table 3-1 of the 1994 report, the committee stated that genetic improvements in male and female body weight gains have led to earlier implementation of the feed phases by the industry and suggested the general sample feeding chronology shown in Table 16-1, which is more like today's commercial programs. In addition, because smaller turkey strains finish progressively earlier than larger types, adjustments in feeding programs need to be made according to the corresponding stages of development and marketing age. Moreover, in the future, feeding programs may change diets (phases) based on the turkeys' target body weights rather than specific ages, but examples of such could not be found in the literature.

Havenstein et al. (2007) stated that selection pressure applied by industry geneticists has greatly increased growth rate and feed conversion efficiency, decreased age to slaughter, and increased the yield of edible meat in commercial turkeys, and these trends continue today. However, weekly or bi-weekly body weights of modern turkeys from hatch to market age, as presented by Havenstein et al. (2007) for 2003 versus 1996 turkeys, are nonexistent in the literature since their report. The reader is thus referred to commercial management guides (e.g., Aviagen and Hybrid) for the most recent and comprehensive body weight and feed consumption information for various turkey strains, which can differ substantially in growth rate, carcass protein deposition, and rate of progression through the feeding phases.

Providing pelleted feed to meat birds increases performance and is often economically advantageous (Poholsky et al., 2023). Commercial starter feeds typically consist of 2 to 3 mm medium crumbles, grower feeds initially consist of 3 to 4 mm large crumbles later transitioning to short pellets (4 mm diameter × 6 mm long), and finisher and breeder feeds consist of 4 mm diameter pellets (13 to 16 mm long) (Hybrid, 2023). It is important to maintain good pellet durability; fines should be reduced to minimize feed wastage, especially in the finishing feed phases where feed intake (FI) increases (Flores et al., 2021). Pellet degradation can occur at the feed mill, during transport, in the feed bin, and during conveyance throughout the feeder system (Poholsky et al., 2023). Inclusion of whole grain or coarsely ground (<1,200 microns) corn before or after pelleting has been demonstrated to result in several benefits for turkeys, including reduced feed milling costs, improved gut health, better litter quality, and improved feed conversion ratios (Favero et al., 2009; Zduńczyk et al., 2013; Flores-Maldonado, 2021). Other benefits include reduced *Salmonella* cecal colonization and fecal shedding (Santos et al., 2006), and whole grains or coarsely ground corn also can be used as dietary diluents to control the growth of breeder toms and hens (Ferket and Moran, 1985).

Turkey Breeder Nutrition

Turkey breeder nutritional requirements are far less established in the literature as compared with those for starting and growing turkeys. Unlike turkeys destined for meat production, where growth rate and feed conversion must be optimized, body weight control is the objective to condition turkey breeders for optimum reproductive performance. Because of the success of body weight management of chicken broiler breeders (see Chapter 15), research on the use of feed restriction programs for juvenile female turkey breeders dates back to the early 1960s and gained interest as the industry shifted some of its selection emphasis toward growth in the female pedigree lines (Krueger et al., 1978). However, as of 1990, feed restriction of parent stock turkey breeder hens was not a routine industry practice (Hester and Stevens, 1990).

As described by Hester and Stevens (1990), there are many approaches to feed restriction of turkey breeder hens, including the feeding of high-fiber or low energy diets, feeding low amounts of dietary protein, limiting the quantity of feed provided, employing a skip-a-day feeding program, inclusion of unpalatable compounds in the feed, or any combinations of these methods. Moreover, the method of feed restriction, when it is initiated, the degree of restriction, the duration of restriction, the season of the year during restriction, the strain of the hen, the timing and length of the re-alimentation period following restricted feeding, and time of photostimulation can all affect hen body weights, efficiency of feed utilization, onset of sexual maturity, egg production, and poult yield per hen (Hester and Stevens, 1990; Crouch et al., 1999). Not surprisingly, research on the effects of feed and body weight restriction of turkey breeder hens have yielded variable results (Hester and Stevens, 1990). For example, some studies have reported a beneficial effect of controlling body weights on hen reproductive performance (e.g., egg numbers, egg weights, fertility, and hatchability), while others have found no favorable results of feed restriction, which led Crouch and others (2002b) to consider the majority of evidence to be "at best confusing and inconclusive." However, they stated that, under some conditions, "feed restriction may be justified as a means of controlling body weight and improving subsequent reproduction." Unlike ducks (Dean, 1986), turkeys do not have an exceptional capacity for compensatory growth (Leeson and Summers, 2005).

The previous version of this report (NRC, 1994) only provided one set of nutrient specifications for holding and laying turkey breeders. Yet despite a dearth of published research on this topic since then, most commercial breeding companies today have unique nutritional programs for turkey breeder candidates and breeding birds, depending upon growth targets, strain, and egg production, based on their in-house research. However, no peer-reviewed studies could be found in the literature to expand the nutrient recommendations for breeders, so specifications for only holding and laying turkey breeders are provided.

DIGESTIVE ANATOMY AND PHYSIOLOGY OF TURKEYS

Turkeys are omnivores and evolved on diets of seeds, vegetation, insects and other invertebrates, and small vertebrates (see previous discussion). Their digestive anatomy is like that of broiler chickens with the size as indicated by percent of body weight of the tract being nearly identical (Denbow, 2015). The primary difference is the relative cecal size, which is almost twice as large in turkeys. The larger ceca are likely an evolutionary adaptation to a diet that is higher in vegetation and fiber than that consumed by jungle fowl (Klasing, 2005).

As with chickens, the crop of turkeys plays an important role in food storage and hydration. The turkey's crop fills during large meals and especially in the evening, when it fills more than in the morning, apparently to store food for digestion during the night (Jackson and Duke, 1995). Emptying of the crop is at least partially under the control of the stomach, which regulates delivery of its food (Chaplin et al., 1992). When full at lights off, it can gradually empty over a period of 9 hr (Cutler et al., 2005). The proventriculus, or glandular stomach, of turkeys is histologically like that of chickens with glandular lobes in the submucosa emptying to elevated papillae facing the lumen (Madkour and Kandyel, 2022). Similarly, the muscular stomach, or gizzard, is functionally like that of the chicken (Duke, 1989, 1992). Duke and colleagues (Duke et al., 1973; Duke, 1989, 1992) have extensively studied the regulation of motility in the turkey's upper digestive tract including the proventriculus, gizzard, and proximal duodenum. They described a cycle that begins with the simultaneous contraction of gizzard's thin muscles, and as they relax, the pylorus opens to release the fluid contents of the gizzard into the duodenum, where powerful peristaltic contraction occurs to mix and move the digesta. Next, the pair of thick muscles of the gizzard contract, which grinds the luminal contents and moves some of the digesta, typically larger particles, back into the proventriculus where they can be recoated with acid and pepsin. After the thick muscles relax, the proventriculus contracts and returns its contents to the gizzard for another round of grinding. When the particles become sufficiently small, they can escape through the pylorus into the duodenum. This contraction cycle takes place up to three to four times per minute but is interrupted several times an hour by strong antiperistaltic contractions throughout the duodenum and upper ileum (Duke, 1992). It is thought that this reverse peristalsis allows additional exposure to the small intestinal digestive and absorptive processes and increases the efficiency of digestion (Basha and Duke, 1999).

Gastric motility is regulated in three stages in turkeys. The cephalic phase is initiated by the sight of food and the anticipation of eating and entails an increase in the frequency of gastric contractions (Duke et al., 1976). The gastric phase follows as food enters the proventriculus and the frequency of contractions increases (Duke et al., 1976). Lastly, the intestinal phase begins with the entrance of gastric contents into the duodenum, which slows gastric contractions to match the pace of gastric emptying to the rate of intestinal digestion (Duke and Evanson, 1972).

The ceca of turkeys are a site for fermentation of carbohydrates, such as refractory starches and hemicelloses and pectins, that escape digestion in the small intestine. The ceca are filled by colonic antiperistalsis, which primarily moves the more fluid digesta into the ceca, and the particulate material remains in the colon for defecation (Bjornhag and Sperber, 1977; Lai and Duke, 1978). Up to one-third of the excreted urine also enters the ceca (Bjornhag and Sperber, 1977) due to reflux from the cloaca into the ceca (Lai and Duke, 1978; Duke, 1989) where uric acid can provide nitrogen and energy for microbial protein synthesis. Turkeys evacuate their ceca and excrete cecal feces around dawn and again in the midafternoon (Duke et al., 1980). When cecal evacuation occurs, the two ceca evacuate simultaneously 84.6 percent of the time (Duke et al., 1983). Using indigestible markers, Hillerman et al. (1953) found that the average feed passage time through the digestive tract was about 2.5 hr in 16-week-old female turkeys.

As in chickens and quail, the particle size and fiber content of the diet influence the size of digestive organs. For example, the weight and, likely, the musculature of the gizzard increases with the particle size of the diet (Favero et al., 2009) and increasing fiber as cellulose increases the length and surface area of the small intestine (Sklan et al., 2003).

In early life, the poult's pancreatic enzymes are low at the time of hatch but increase robustly during the first few days after hatch; conversely the yolk sac, which is a rich source of fat and protein, is mostly emptied by 6 d post-hatch (Sell et al., 1991). In the case of fat digestion, the turkey appears to develop faster than chickens, especially in capacity to digest saturated fats (Mossab et al., 2000).

NUTRIENT REQUIREMENTS

The nutrient requirements of turkeys are divided into needs of birds used as a source of growth and needs of those for reproduction. These two categories differ largely in the proportion of nutrients devoted to productive use (growth or reproduction) as opposed to those used for maintenance activities. The review of literature used to compile the nutrient requirement recommendations in the 9th revised edition of this report was completed by the committee in September 1991 (NRC, 1994). Therefore, revisions in the recommended nutrient requirements of turkeys in the present report are based primarily on peer-reviewed publications that appeared after that time. Moreover, the reader is reminded that the recommended nutrient requirements may vary due to turkey strain (genetics), husbandry practices, and environmental conditions to which the birds are exposed.

***Minimum nutrient requirement* values appear in normal font in Table 16-2.** These recommendations are based on results from multiple peer-reviewed feeding studies that met the committee's inclusion criteria for a requirement study: A common basal diet was employed; sufficient dietary amounts were included in order to establish a dose–response relationship; dietary amounts of the nutrient of interest were chemically analyzed; the experimental design and statistical analyses were appropriate for estimating nutrient requirements, and sufficient information was provided to allow for replication of the experiment; management practices were employed that did not impair animal performance; and the studies were not conducted under environmental conditions that would affect production rates. The response criteria of these studies were based on the dietary amount that resulted in optimal performance and avoidance of signs of deficiency, whichever was lowest. The dietary amount of a nutrient needed to support maximal performance of starting and growing turkeys or breeders was set by averaging the estimates from all studies that met the inclusion criteria. There is a high degree of confidence that these recommendations are accurate under the conditions in which the experiments were conducted.

Studies were excluded from consideration if the experimental design or statistical analysis was not appropriate for estimating nutrient requirements or if insufficient information was provided to allow replication of the experiment or the statistical analysis of the resulting data. Other exclusion factors included failure to chemically analyze dietary amounts of the nutrient of interest; management practices employed that impacted requirements, analytical methodologies, or statistical procedures; studies conducted at environmental or management extremes that would affect production rates and nutrient requirements; or if performance rates were low for other reasons.

***Guidance* values appear in bold italic font in Table 16-2** and are provided for cases where no studies were available that met the foregoing inclusion criteria, but there were studies that met the most important criteria. Exceptions were made so that a rough estimate of a nutrient requirement could be provided. There is not a high degree of confidence that these recommendations are an optimal amount in a least-cost ration formulation, but they are likely to support adequate animal performance. Other situations from which guidance values were derived include the following:

1. For some nutrients, there were studies that did meet the foregoing inclusion criteria, but the base diet had adequate amounts of the nutrient under study as indicated by the lack of significant changes in performance by the addition of higher amounts of that nutrient. In these cases, nutrient recommendations are based on the amount of the nutrient in the base diet.
2. Many of the published experiments designed to titrate the requirement for various nutrients analyzed the results using analysis of variance (ANOVA) followed by mean comparisons and did not use analysis by broken-line regressions or other valid regression approaches to estimate a requirement more quantitatively. Thus, many of the recommendations in Table 16-2 are based on the range between treatment amounts that were significantly different.

The recommended values in Table 16-2 do not include a margin of safety to account for environmental, management, toxicological, or disease stressors or for losses during manufacture and storage of diets. Nutritionists should add a margin of safety appropriate for their specific production system to adjust for these added needs. The recommended values in Table 16-2 also assume that the dietary ingredients used have digestibilities similar to that of a typical diet based on corn and soybean meal and that supplemented amino acids, minerals, and vitamins have high bioavailability. The values for energy and crude protein are not requirements but are typical amounts used in research and commercial production. Poultry can adapt to a range of dietary energy densities by changing their feed intake and can utilize lower amounts of amino-nitrogen if an optimal balance of amino acids is maintained by supplementing diets with purified amino acids. Lastly, **nutrient requirement values carried over from the NRC (1994) report**, because there were no new studies upon which to elicit any changes, **appear in Table 16-2 in shaded normal or italic font.**

Water

Water, the largest single constituent of the body, is the most critical nutrient that is supplied to poultry, but it is not possible to state precise requirements. The amount needed depends upon environmental temperature and relative humidity, the diet composition, rate of growth or egg production, and efficiency of kidney resorption of water in individual birds (Medway and Kare, 1959). Water needs are also partially fulfilled by the feed consumed, which typically contains ~10 percent water by weight, and via oxidation of carbohydrates, protein, and fat (so-called metabolic water). It is generally assumed

TABLE 16-2 Minimum Nutrient Requirements of Turkeys as Percentages or Units per Kilogram of Diet (90% Dry Matter).[a,b]

		Growing Turkeys, Males[c] and Females[d]							
		0 to 3 Weeks[c]	3 to 6 Weeks[c]	6 to 9 Weeks[c]	9 to 12 Weeks[c]	12 to 15 Weeks[c]	15 to 18 Weeks[c]	Breeders	
		0 to 3 Weeks[d]	3 to 6 Weeks[d]	6 to 9 Weeks[d]	9 to 12 Weeks[d]	12 to 14 Weeks[d]	14 to 16 Weeks[d]	Holding	Laying Hens
Nutrient	Unit								
Metabolizable Energy (AME_n)[e]	kcal	2,800	2,860	2930	3,000	3,100	3,200	2,900	2,900
Protein and amino acids[f,g]									
Protein[e]	%	28	27	25	22	19	16.5	***14***	***16***
Arginine	%	1.6	***1.5***	***1.3***	***1.1***	***0.90***	***0.75***	***0.5***	***0.6***
Glycine + serine	%	1.0	***0.9***	***0.9***	***0.8***	***0.7***	***0.6***	***0.4***	***0.5***
Histidine	%	0.58	***0.5***	***0.5***	***0.4***	***0.3***	***0.3***	***0.2***	***0.3***
Isoleucine	%	1.1	***1.0***	***0.9***	***0.8***	***0.6***	***0.5***	***0.4***	***0.5***
Leucine	%	1.9	***1.8***	***1.7***	***1.5***	***1.3***	***1.0***	***0.5***	***0.5***
Lysine	%	1.53	1.37	1.25	0.98	***0.76***	***0.76***	***0.5***	***0.6***
Methionine	%	0.46	0.48	0.43	0.32	0.35	***0.25***	***0.2***	***0.2***
Methionine + cyst(e)ine	%	0.90	***0.85***	***0.74***	***0.62***	***0.46***	***0.42***	***0.4***	***0.4***
Phenylalanine	%	1.0	***0.9***	***0.9***	***0.8***	***0.7***	***0.6***	***0.4***	***0.55***
Phenylalanine + tyrosine	%	1.8	***1.7***	***1.5***	***1.2***	***1.0***	***0.9***	***0.8***	***1.0***
Threonine	%	0.94	0.89	0.82	***0.72***	***0.59***	***0.58***	***0.4***	***0.45***
Tryptophan	%	0.26	***0.25***	***0.23***	***0.20***	***0.18***	***0.18***	***0.1***	***0.13***
Valine	%	***0.96***	***1.1***	***1.0***	***0.9***	***0.8***	***0.7***	***0.5***	***0.58***
Fat									
Linoleic acid	%	1.0	***1.0***	***1.0***	***0.8***	***0.8***	***0.8***	***0.8***	1.1
Macrominerals									
Calcium	%	1.30	1.15	1.00	0.90	0.80	0.75	***0.65***	***2.90***
Nonphytate phosphorus	%	0.65	0.58	0.50	0.45	0.40	0.37	***0.35***	***0.35***
Potassium	%	0.7	***0.6***	***0.6***	***0.5***	***0.5***	***0.4***	***0.4***	***0.6***
Sodium	%	0.17	0.15	0.12	0.10	0.10	0.10	***0.12***	***0.12***
Chlorine	%	0.15	***0.14***	***0.14***	***0.14***	***0.12***	***0.12***	***0.12***	***0.12***
Magnesium	mg	500	***500***	***500***	***500***	***500***	***500***	***500***	***500***
Trace minerals									
Manganese	mg	***60***	***60***	***60***	***60***	***60***	***60***	***60***	***60***
Zinc	mg	***60***	***60***	***57***	***50***	***40***	***40***	***40***	***65***
Iron	mg	***80***	***70***	***60***	***60***	***60***	***50***	***50***	***60***
Copper	mg	***8***	***8***	***6***	***6***	***6***	***6***	***6***	***8***
Iodine	mg	***0.4***	***0.4***	***0.4***	***0.4***	***0.4***	***0.4***	***0.4***	***0.4***
Selenium	mg	0.3	0.3	0.3	0.3	0.3	0.3	***0.2***	***0.2***
Fat-soluble vitamins									
A	IU	5,000	5,000	***5,000***	***5,000***	***5,000***	***5,000***	***5,000***	***5,000***
D_3[h]	ICU	1,100	***1,100***	***1,100***	***1,100***	***1,100***	***1,100***	***1,100***	1,100
E	IU	***37***	***37***	***37***	***34***	***34***	***34***	***10***	25
K	mg	1.75	***1.6***	***1.3***	***1.0***	***0.75***	***0.75***	***0.50***	***1.0***
Water-soluble vitamins									
B_{12}	mg	0.003	0.003	***0.003***	***0.003***	***0.003***	***0.003***	***0.003***	***0.003***
Biotin[i]	mg	0.25	0.22	***0.18***	***0.13***	***0.13***	***0.10***	***0.10***	***0.80***
Choline	mg	1,600	1,500	***1,300***	***1,100***	***1100***	***950***	***800***	***1,000***
Folacin	mg	1.0	1.0	***0.9***	***0.8***	***0.8***	***0.8***	***0.7***	1.0
Niacin	mg	60	***86***	***86***	***86***	***86***	***86***	***40***	***40***
Pantothenic acid	mg	10.0	10.0	***9.0***	***9.0***	***9.0***	***9.0***	***9.0***	16.0
Pyridoxine	mg	***4.5***	***4.5***	***4.2***	***3.5***	***3.5***	***3.0***	***3.0***	***4.0***
Riboflavin	mg	4.0	3.7	***3.4***	***3.0***	***3.0***	***2.5***	***2.5***	***4.0***
Thiamin	mg	2.0	2.0	***2.0***	***2.0***	***2.0***	***2.0***	***2.0***	***2.0***

[a]Values in normal font are minimum nutrient requirement recommendations based on empirical studies summarized in Table 17-3. Values in bold italics are guidance amounts based on estimates due to insufficient experimental data (see front matter for details).

[b]Values in shaded areas are based on NRC (1994) recommendations adjusted to the age ranges used in this report, for which no new studies are available.

[c]The age intervals for nutrient requirements of males represent an earlier implementation than reported in NRC (1994) due to genetic improvements in body weight gain.

[d]The age intervals for nutrient requirements of females increase the rate of diet changes compared to those reported in NRC (1994) due to genetic improvements in body weight gain.

[e]These are approximate metabolizable energy values provided with corn-soybean meal-based feeds, expressed as kcal AME_n/kg diet. Such energy when accompanied by the suggested nutrient amounts is expected to provide near-maximum growth, particularly with pelleted feed.

[f]Amino acid values are on a total basis.

[g]Turkeys do not have a requirement for crude protein *per se*. However, there should be sufficient crude protein to ensure an adequate amino nitrogen supply for the synthesis of nonessential amino acids. The suggested crude protein requirements shown in this table are typical for corn-soybean meal-based diets formulated on a total amino acid basis. When synthetic amino acids, other feedstuffs, and by-products are included in the diets, or when formulating diets on a digestible amino acid basis, the crude protein values attained may differ.

[h]These concentrations of vitamin D are considered satisfactory when the associated calcium and phosphorus amounts are used.

[i]Requirement may increase in wheat-based diets.

that turkeys drink approximately twice as much water as the amount of feed consumed on a weight basis, but water intake varies greatly. The water supplied to turkeys should not contain excessive amounts of minerals or be contaminated with bacteria. Other than a report by Morehouse (1949), daily water consumption of growing or breeding turkeys from hatch to market age is nonexistent in the literature. The reader is thus referred to commercial management guides (e.g., Aviagen, Hybrid) for the most recent comprehensive information.

Dietary factors that increase water intake and feed:water ratios in broiler chickens include increased crude protein (Marks and Pesti, 1984) and increased dietary sodium chloride (Marks, 1987), and it is assumed that this is also the case for turkeys, although supporting data could not be found. In turkeys, water consumption has been reported to increase with increased ambient temperatures (Parker et al., 1972) and when the birds are fed diets containing 30 percent wheat distiller's dried grains with solubles (Opoku et al., 2015).

The poult is particularly susceptible to dehydration, and mortality can occur when poults are deprived of drinking water and conversely when drinking water is reintroduced to the poults (Marsden et al., 1965), a situation referred to as drunken syndrome or water intoxication. Chilson and Patrick (1946) reported that withholding of feed and water for 24, 48, or 72 hr post-hatch resulted in 4.8 percent, 11.5 percent, and 29.3 percent mortality, respectively, up to 12 weeks of age. Moran (1978) observed that when male poults were allowed access to feed and water either 6 hr or 24 hr post-hatch, the latter birds had reduced body weights and higher mortality, and the body weight differences between the two groups persisted through the subsequent 12 weeks of the experiment. Corless and Sell (1999) showed that, compared to poults allowed access to feed and water 6 hr post-hatch, a 54-hr delay in access to feed and water generally delayed development of the digestive system, impaired nutrient utilization, and significantly reduced 14-d or 28-d body weights.

Energy

Like other poultry, turkeys appear to eat primarily to satisfy their appetite for energy. Up until the late 1950s, the importance of maintaining desirable energy-to-crude-protein ratios was a commonly accepted principle of turkey nutrition such that varying Apparent metabolizable energy (AME)-to-crude protein ratios could be used without adversely affecting the growth performance of turkeys. However, Atkinson et al. (1957) reported that the calorie-to-protein ratio was less critical in high protein diets than when low protein diets were fed. Balloun et al. (1959) found little evidence for a significant effect of protein on energy requirements, or of energy on protein requirements, and concluded that "the two factors seemed to act relatively independently." Sell et al. (1985) concluded that the use of optimal AME-to-protein ratios as a guideline for diet formulation may be inappropriate for growing turkeys and that greater emphasis should be placed on the nutritional and economic implications of the independent effects of dietary AME and dietary protein on turkey performance traits. In a subsequent study, Sell et al. (1989) fed male Nicholas Large White turkeys 20 diets in a factorial arrangement of five protein amounts (82, 91, 100, 109, and 118% of 1984 NRC recommendations, with the low-protein diets supplemented with essential amino acids to meet the 1984 NRC recommendations) and four dietary AME_n amounts (99, 103, 107, and 111% of 1984 NRC recommendations) from 1 to 140 d of age. Neither dietary protein nor AME_n amounts significantly affected growth to 140 d, but feed per gain decreased linearly ($P<0.001$) with increasing increments of protein or AME_n. Moreover, no interaction effects between dietary protein and AME_n on performance and carcass traits were observed despite the wide ranges of each. The results of this study confirmed previous observations (Balloun et al., 1959; Sell et al., 1985) that dietary protein and AME_n exert independent effects on growth and feed efficiency of turkey toms. Gous et al. (2019) suggested that the reduced efficiency of conversion (e_p) of dietary protein to body protein in turkeys, resulting from the lack of dietary energy at high dietary protein amounts, may be overcome by supplementation of additional energy. Subsequently, Gous et al. (2023) investigated whether e_p in turkeys from 21 to 35 d of age was influenced by the dietary AME_n:digestible crude protein (DCP) ratio. They observed that turkey poults increased both body protein and lipid gains when additional dietary energy was provided to a feed with a AME_n:DCP ratio below 60 MJ/kg. However, e_p was not increased, so they concluded that the additional feed intake resulting from the higher energy content and/or lower heat increment was converted to body protein and lipid without affecting feed efficiency.

Dietary energy utilization in young poults presents some challenges. The turkey is hatched with very low available energy reserves, and although they have the capacity for considerable glycogenesis if they have access to dietary carbohydrate (Turner et al., 1999), glycogen is most likely synthesized by gluconeogenesis from protein (Leeson and Summers, 2005). Young turkeys also have a limited ability to utilize dietary saturated fats during early life (Sell et al., 1986).

The AME_n amounts presented in Table 16-2 at each age are not intended to be absolute but to establish a feed intake reference for other nutrients. The given energy and amino acid amounts would be satisfied largely when corn and soybean meal are combined with a small amount of added fat, in turn promoting near-maximum growth.

Protein and Amino Acids

Protein

The turkey's dietary protein requirement is in reality a requirement for amino acids, present in the diet in adequate quantities and in the proper balance. However, there should be sufficient crude protein to ensure an adequate amino nitrogen supply for the synthesis of nonessential amino acids. The suggested crude protein requirements shown in Table 16-2 are typical of those derived with corn-soybean meal–based diets and formulated on a total amino acid basis. When synthetic amino acids, other feedstuffs, and by-products are included in the diets, or when formulating diets on a digestible amino acid basis, the crude protein values attained may differ. Factors that have contributed to the variability in the protein requirements of turkeys include the breeds/genetic lines used, the response criterion (e.g., growth, feed efficiency, carcass yield), use of diets of widely differing composition and form (e.g., crumble and pellet sizes, use of supplemental crystalline amino acids), and husbandry conditions.

Turkeys are typically provided with between six and eight different feeds during the growing period, with diet changes every 2, 3, or 4 weeks depending upon the age of the bird and the specific breeder recommendations. The dietary crude protein and energy guidelines from the previous version of this report (NRC, 1994) for growing turkeys were 28.0 percent protein and 2,800 kcal AME/kg from 0 to 4 weeks (males and females), 26.0 percent protein and 2,900 kcal AME/kg from 4 to 8 weeks (males and females), 22.0 percent protein and 3,000 kcal AME/kg from 8 to 12 weeks (males; 8 to 11 weeks for females), 19.0 percent protein and 3,100 kcal AME/kg from 12 to 16 weeks (males; 11 to 14 weeks for females), 16.5 percent protein and 3,200 kcal AME/kg from 16 to 20 weeks (males; 14 to 17 weeks for females), and 14.0 percent protein and 3,200 kcal AME/kg from 20 to 24 weeks (males; 17 to 20 weeks for females). Since significant improvements in growth and feed conversion of commercial turkeys have been achieved over the past three decades, feeding phases have been shortened and the birds are marketed earlier. As a result, the suggested nutrient requirements for crude protein and energy presented in Table 16-2 have been carried over from the six feeding phases of the previous report (NRC, 1994) but are based on 3-week intervals for males up to 18 weeks of age and for females up to 12 weeks of age, with 2-week phases from 12 to 16 weeks of age. The recommended protein requirements for holding and laying breeder hens were increased to 14 percent and 16 percent, respectively, based on the reports of Crouch et al. (2002a) and Godwin et al. (2005).

Amino Acids

Diets for turkeys have historically been formulated based on the analyzed (total) concentration of amino acids in the feed ingredients. As in broiler nutrition, formulation based on calculated ileal digestible amino acid concentrations is increasingly being applied. This is especially important when using by-products and other ingredients that have low or variable digestibilities. However, until recently, most of the turkey research did not estimate amino acid requirements on a digestible basis, although there are some notable exceptions. For those studies that did measure digestible amino acid requirements, the digestible values were converted to a total basis so that they could be compared to the rest of the literature. Therefore, the amino acid requirements reported in Table 16-2 are provided on a total amino acid basis to permit the use of all of the peer-reviewed studies. Digestible amino acid requirements, from the subset of studies that reported them, are provided in the text when sufficient information permits. See Table 16-3 for additional details on diets, statistical analyses, and other relevant information on these studies. Many of the requirements for essential amino acids discussed subsequently indicate that requirements expressed as a percentage of the diet have decreased since the previous report (NRC, 1994), which has been a period of markedly increasing growth rates. This suggests that the efficiency of digestion and/or utilization of limiting essential amino acids has increased considerably over time.

The requirement for sulfur-containing amino acids (methionine + cyst(e)ine; Met+Cys) is first-limiting in a diet based on corn and soybean meal for growing poults. Supplementing a methionine source, such as DL-methionine or DL-2-hydroxy-(4-methylthio)-butanoic acid, allows the amount of dietary protein to be decreased. Further decreases in crude protein can be made when the second- and third-limiting amino acids (lysine and threonine, respectively) are supplemented (Waibel et al., 2000a,b). Recommended dietary requirements are provided subsequently for turkeys of various ages for lysine, methionine, Met+Cys, threonine, and valine only; no new published requirement studies were found for arginine, glycine + serine, histidine, isoleucine, leucine, phenylalanine, phenylalanine + tyrosine, or tryptophan, so those requirements were carried over from the previous report (NRC, 1994).

Lysine

0 to 3 Weeks of Age. There were four studies (Boling and Firman, 1998; Firman, 2004; Thompson et al., 2004; Linares et al., 2012) that were published since the last report (NRC, 1994) that met the committee's inclusion criteria for estimating the lysine requirement during all or part of the first 0 to 21 d of production. Boling and Firman (1998) fed six different concentrations of lysine (1.39% to 1.54% total; 1.26% to 1.41% digestible) to female Hybrid poults from 13 to 26 d and estimated the requirement to be 1.45 percent total lysine (1.32% digestible) for weight gain and 1.47 percent total lysine (1.34% digestible) for efficiency of feed conversion. Thompson et al. (2004) fed diets with eight different lysine amounts (1.20% to 1.55% total; 1.08% to 1.43% digestible) to Nicholas White female poults from 4 to 15 d and estimated the requirement to be 1.45 percent total lysine (1.29% digestible) for rate of gain. Firman (2004) fed diets with eight different lysine concentrations (1.20% to 1.55% total lysine; 1.10% to 1.45% digestible) to male Nicholas White poults from 7 to 18 d and estimated the requirement for rate of gain to be 1.41 percent total lysine (1.31% digestible). Linares et al. (2012) fed 8- to 21-d-old male Nicholas White poults seven different diets containing 1.35 to 1.71 percent total lysine (1.14% to 1.50% digestible) and estimated the requirement for rate of gain to be 1.57 percent total lysine (1.36% digestible) and the requirement for efficiency of feed conversion to be 1.60 percent total lysine (1.39% digestible).

TABLE 16-3 Documentation of Nutrient Requirements of Turkeys[a]

Age (d)	Weight, Start–End of Study, kg	Estimated Requirements per Response (%)	Response Criteria	Strain and Sex	Number of Levels, (range, %)	Comments	Reference
Lysine							
4–15	Not reported	1.41 total 1.29 digestible	Weight gain	Nicholas White female	8 (1.20–1.55 total, 1.08–1.43 digestible)	Corn–soybean meal diet containing 20.63% protein and 3,100 kcal AME/kg. Data analyzed by ANOVA and broken line regression analysis.	Thompson et al., 2004
7–18	Not reported	1.41 total 1.31 digestible	Weight gain	Nicholas White male	8 (1.20–1.55 total, 1.10–1.45 digestible)	Corn–soybean meal diet containing 21.75% protein and 3,100 kcal AME/kg. Data analyzed by ANOVA and broken-line analysis.	Firman, 2004
8–21	Not reported	1.36 digestible[c] 1.73 digestible[c]	Weight gain Feed efficiency	Nicholas White male	7 (1.14–1.50 digestible)	Corn–peanut meal–soybean meal diet containing 25.8% protein and 3,000 kcal AME/kg. Data analyzed by ANOVA, quadratic, and broken line regression analysis.	Linares et al., 2012
13–27	Not reported	≤ ***1.57*** total ≤ ***1.42*** digestible	Weigh gain Feed efficiency	Hybrid female	6 (1.57–1.87 total, 1.17–1.72 digestible)	Corn–soybean meal diet containing 21.60% protein and 3,099 kcal AME/kg. Data analyzed by ANOVA.	Boling and Firman, 1998
13–26	Not reported	≤ ***1.49*** total ≤ ***1.36*** digestible	Weight gain Feed efficiency	Hybrid female	6 (1.3–1.61 to-tal, 1.17–1.48 digestible)	Corn–soybean meal diet containing 18.36% protein and 3,192 kcal AME/kg. Data analyzed by ANOVA.	Boling and Firman, 1998
13–26	Not reported	1.45 total 1.32 digestible 1.47 total 1.34 digestible	Weight gain Feed efficiency	Hybrid female	6 (1.39–1.54 total, 1.26–1.41 digestible)	Corn–soybean meal diet containing 18.36% protein and 3,192 kcal AME/kg. Data analyzed by ANOVA and broken-line regression analysis.	Boling and Firman, 1998
23–37	Not reported	1.28 total 1.19 digestible	Weight gain	Nicholas White male	8 (1.11–146 total, 1.02–1.37 digestible)	Corn–soybean meal diet containing 20.77% protein and 3,205 kcal AME/kg. Data analyzed by ANOVA and splined-regression analysis.	Firman, 2004

continued

TABLE 16-3 Continued

Age (d)	Weight, Starts–End of Study, kg	Estimated Requirements per Response (%)	Response Criteria	Strain and Sex	Number of Levels, (range, %)	Comments	Reference
Lysine							
29–40	Not reported	1.26 total 1.16 digestible 1.24 total 1.14 digestible	Weight gain Feed efficiency	Nicholas White female	8 (0.98–1.38 total, 0.88–1.28 digestible)	Corn–soybean meal diet containing 19.32% and protein 3,200 kcal AME/kg. Data analyzed by ANOVA and broken-line regression analysis.	Thompson et al., 2004
29–42	Not reported	1.27 digestible[c] 1.29 digestible	Weight gain Feed efficiency	Nicholas White male	7 (1.06–1.42 digestible)	Corn–peanut meal–soybean meal diet containing 24.0% protein and 3,100 kcal AME/kg. Data analyzed by ANOVA, quadratic, and broken line regression analysis.	Linares et al., 2012
35–56	1.1–2.2	1.50 1.38 1.37 1.32	Weight gain, non-linear regression Weight gain, broken line Feed efficiency, nonlinear regression Feed efficiency, broken line	BUT Big 6 male	6 (1.18–1.64)	Corn–wheat–soybean meal diet containing 25.1% protein and 2,962 kcal AME/kg. Data analyzed by ANOVA, nonlinear exponential regression at 95% of asymptote, and by broken-line regression.[d]	Lemme et al., 2002
46–57	Not reported	1.12 total 1.04 digestible 1.15 total 1.07 digestible	Weight gain Feed efficiency	Nicholas White female	7 (0.93–1.29 total, 0.85–1.21 digestible)	Corn–soybean meal diet containing 17.2% protein and 3,300 kcal AME/kg. Data analyzed by ANOVA and broken-line regression analysis.	Thompson et al., 2005
49–61	Not reported	1.18 total 1.09 digestible 1.20 total 1.11 digestible	Weight gain Feed efficiency	Nicholas White male	8 (total, 0.88–1.23 digestible)	Corn–soybean meal diet containing 18.44% protein and 3,300 kcal AME/kg. Data analyzed by ANOVA and broken-line analysis.	Baker et al., 2003a
56–84	4.0–8.8	≤ ***1.17*** total ≤ ***1.07*** total	Weight gain Feed efficiency	BUT Big 6 male	6 (0.87–1.37 total)	Corn–wheat–soybean meal–corn gluten meal diet containing 22.5% protein and 3,035 kcal AME/kg. Data analyzed by ANOVA and quadratic regression analysis.	Lehmann et al., 1996
71–85	Not reported	0.93 total 0.86 digestible 0.89 total 0.82 digestible	Weight gain Feed efficiency	Nicholas White female	8 (0.70–1.05 digestible)	Corn–soybean meal diet containing 15.0% protein and 3,300 kcal AME/kg. Data analyzed by ANOVA and broken-line regression analysis.	Thompson et al., 2005

72–83	Not reported	0.96 total 0.87 digestible 0.95 total 0.86 digestible	Weight gain Feed efficiency	Nicholas White male	8 (0.68–0.96 digestible)	Corn–soybean meal diet containing 15.11% protein and 3,400 kcal AME/kg. Data analyzed by ANOVA and splined-regression analysis.	Baker et al., 2003a
72–83	Not reported	0.68 total 0.67 total	Weight gain Feed efficiency	BUT male	8 (0.54–0.75 digestible)	Corn–soybean meal diet containing 14.3% protein and 3,500 kcal AME/kg. Data analyzed by ANOVA and broken-line regression analysis.	Baker et al., 2003b
84–95	Not reported	0.53 total 0.54 total	Weight gain Feed efficiency	Nicholas male	8 (0.44–0.65 digestible)	Corn–soybean meal diet containing 13.2% protein and 3,600 kcal AME/kg. Data analyzed by ANOVA and brokenline regression analysis.	Baker et al., 2003b
91–112	8.6–3.2	***1.23*** ***0.92*** ***0.94*** ***0.91***	Weight gain, nonlinear regression Weight gain, broken line Feed efficiency, nonlinear regression Feed efficiency, broken line	BUT Big 6 male	6 (0.80–1.20)	Corn–wheat–soybean meal diet containing 19.6% protein and 2,867 kcal AME/kg. Data analyzed by ANOVA, nonlinear exponential regression at 95% of asymptote, and by broken line regression.[d] Toms had been used for a requirement study 5 weeks previously. Although they were blocked into the new treatments, the prior lysine deficient diets may have affected the requirement estimates.	Lemme et al., 2002
112–240	13.1–18.5	≤ ***0.75*** total ≤ ***0.89*** total	Weight gain Feed efficiency	BUT Big 6 male	6 (0.61–0.96 total)	Corn–wheat–full-fat soybean–corn gluten meal diet containing 16.8% protein and 3,225 kcal AME/kg. Data analyzed by ANOVA and quadratic regression analysis.	Lehmann et al., 1996
Methionine							
1–28	Not reported	≤0.47 total ≤0.47 total	Weight gain Feed efficiency	Hybrid Converter female	6 (Total Met+Cys 0.65–0.97 analyzed; supplemented as DL-met)	Wheat–corn–triticale–soybean meal diet containing 26.6% protein and 2750 kcal AME/kg. Data analyzed by ANOVA.	Kubinska et al., 2016

continued

TABLE 16-3 Continued

Age (d)	Weight, Start–End of Study, kg	Estimated Requirements per Response (%)	Response Criteria	Strain and Sex	Number of Levels, (range, %)	Comments	Reference
Methionine							
63–84	Not reported	≤0.29 total ≤0.32 total	Weight gain Feed efficiency	Hybrid Converter female	6 (Total methionine 0.41–0.75 analyzed; supplemented as DL-Met)	Wheat–corn–soybean meal diet containing 21.0% protein and 3,050 kcal AME/kg. Data analyzed by ANOVA.	Jankowski et al., 2017
Methionine + Cyst(e)ine							
1–28	Not reported	≤***1.00*** total ≤***0.88*** total	Weight gain Feed efficiency	Hybrid Converter female	6 (Met+Cys 0.81–1.16 analyzed; supplemented as DL-Met)	Wheat–corn–triticale–soybean meal diet containing 26.6% protein and 2,750 kcal AME/kg. Data analyzed by ANOVA.	Kubinska et al., 2016
7–16	Not reported	1.02 total 0.83 digestible 1.01 total 0.82 digestible	Weight gain Feed efficiency	BUT male	8 (Met+Cys 0.60–1.00 digestible; supplemented as DL-Met)	Corn–soybean meal diet containing 21.0% protein and 3,100 kcal AME/kg. Data analyzed by ANOVA and broken-line regression analysis.	Moore et al., 2001
9–23	Not reported	≤0.88 total 0.77 digestible ≤0.88 total 0.77 digestible	Weight gain Feed efficiency	BUT female	7 (Met+Cys 0.88–1.31 analyzed, 0.77–1.19 digestible); supplemented as DL-Met	Corn–soybean meal diet containing 18.4% protein and 3,172 kcal AME/kg. Data analyzed by ANOVA and broken-line regression analysis.	Boling and Firman, 1997, Experiment 1
9–22	Not reported	≤0.88 total 0.77 digestible ≤0.88 total 0.77 digestible	Weight gain Feed efficiency	Nicholas female	7 (Met+Cys: 0.61–1.45 analyzed; 0.50–1.33 digestible); supplemented as DL-Met	Corn–soybean meal diet containing 18.4% protein and 3,172 kcal AME/kg. Data analyzed by ANOVA and broken-line regression analysis.	Boling and Firman, 1997, Experiment 2
8–22	Not reported	0.87 total 0.76 digestible 0.86 total 0.75 digestible	Weight gain Feed efficiency	Nicholas female	9 (Met+Cys 0.71–0.95 analyzed; 0.59–0.83 digestible), supplemented as DL-Met	Corn–soybean meal diet containing 18.4% protein and 3,172 kcal AME/kg. Data analyzed by ANOVA and broken=line regression analysis.	Boling and Firman, 1997, Experiment 3
21–35	Not reported	0.96 total 0.78 digestible 0.85 total 0.77 digestible	Weight gain Feed efficiency	BUT male	8 (Met+Cys 0.57–0.95 digestible; supplemented as DL-Met)	Corn–soybean meal diet containing 21.0% protein and 3,100 kcal AME/kg. Data analyzed by ANOVA and broken-line regression analysis.	Moore et al., 2001

42–63	Not reported	0.80 total 0.65 digestible 0.81 total 0.66 digestible	Weight gain Feed efficiency	BUT male	8 (Met+Cys 0.50–0.82 digestible; supplemented as DL-Met)	Corn–soybean meal diet containing 14.9% protein and 3,300 kcal AME/kg. Data analyzed by ANOVA and broken-line regression analysis.	Moore et al., 2004
63–84	Not reported	0.69 total 0.56 digestible 0.69 total 0.56 digestible	Weight gain Feed efficiency	BUT male	8 (Met+Cys 0.4–0.66 digestible; supplemented, as DL-Met)	Corn–soybean meal diet containing 11.5% protein and 3,400 kcal AME/kg. Data analyzed by ANOVA and broken-line regression analysis.	Moore et al., 2004
63–84	Not reported	≤***0.62*** total ≤***0.66*** total	Weight gain Feed efficiency	Hybrid Converter female	6 (Met+Cys 0.62–0.87 analyzed; supplemented as DL-Met)	Wheat–corn–soybean meal diet containing 21.0% protein and 3,050 kcal AME/kg. Data analyzed by ANOVA.	Jankowski et al., 2017
83–93	Not reported	0.48 total, 0.39 digestible 0.49 total 0.40 digestible	Weight gain Feed efficiency	BUT male	8 (Met+Cys 0.83–0.92 digestible; supplemented as DL-Met)	Corn–soybean meal diet containing 8.92% protein and 4,025 kcal AME/kg. Data analyzed by ANOVA and broken-line regression analysis.	Moore et al., 2003
106–116	Not reported	0.46 total 0.37 digestible 0.42 total 0.34 digestible	Weight gain Feed efficiency	BUT male	8 (Met+Cys 0.30–0.50 digestible; supplemented as DL-Met)	Corn–soybean meal diet containing 8.0% protein and 3,599 kcal AME/kg. Data analyzed by ANOVA and broken-line regression analysis.	Moore et al., 2003
Threonine							
1–28	0.057–1.104	≤***0.88*** total ≤***0.88*** total	Weight gain Feed efficiency	BUT Big 6 male	6 (0.82–1.12 total)	Corn–wheat–soybean meal diet containing 26.0% protein and 2,850 kcal AME/kg. Data analyzed by ANOVA.	Lehmann et al., 1997
1–21	Not reported	0.93 total	Weight gain Feed efficiency	BUT male	7 (0.71–1.07 total)	Corn–soybean meal–peanut meal diet containing 24.0% protein and 2,900 kcal AME/kg. Data analyzed by GLM and quadratic regression using 95% of the asymptote.	Kidd et al., 1998
1–14	0.236, Final weight	≤***0.80*** total ≤***0.80*** total	Weight gain Feed efficiency	BUT Large White male	5 (0.65–1.25 total)	Corn–peanut meal–soybean meal diet containing 23.1% protein and 2,962 kcal AME/kg. Data analyzed by ANOVA.	Lilburn and Barbour, 2001

continued

TABLE 16-3 Continued

Age (d)	Weight, Start–End of Study, kg	Estimated Requirements per Response (%)	Response Criteria	Strain and Sex	Number of Levels, (range, %)	Comments	Reference
Threonine							
7–21	Not reported	≤***0.88*** total ≤***0.77*** digestible ≤***0.81*** total ≤***0.70*** digestible	Weight gain Feed efficiency	Nicholas female	8 (0.63–1.12 digestible)	Corn–soybean meal diet containing 17.9% protein and 3,188 kcal AME/kg. Data analyzed by ANOVA.	Kamyab and Firman, 2000
7–18	Not reported	0.78 total 0.67 digestible 0.85 total 0.74 digestible	Weight gain Feed efficiency	Nicholas female	8 (0.54–1.03 digestible)	Corn–soybean meal diet containing 17.9% protein and 3,188 kcal AME/kg. Data analyzed by ANOVA and broken-line regression analysis.	Kamyab and Firman, 2000
8–21	Not reported	0.74 digestible[c] 0.81, digestible[c]	Weight gain Feed efficiency	Nicholas White male	7 (0.56–1.04 digestible)	Corn–peanut meal–soybean meal diet containing 25.8% protein and 3,000 kcal AME/kg. Data analyzed by ANOVA, quadratic, and broken-line regression analysis.	Linares et al., 2012
21–35	1.384, Final weight	≤***0.80*** total ≤***0.80*** total	Weight gain Feed efficiency Pectoralis major weight	BUT Large White male	5 (0.65–1.25 total)	Corn–peanut meal–soybean meal diet containing 23.1% protein and 2,962 kcal AME/kg. Data analyzed by ANOVA.	Lilburn and Barbour, 2001
21–42	Not reported	0.88 total 0.88 total	Weight gain Feed efficiency	BUT male	7 (0.66–1.02 total)	Corn–soybean meal–peanut meal diet containing 21.8% protein and 3,128 kcal AME/kg. Data analyzed by GLM and quadratic regression using 95% of the asymptote.	Kidd et al., 1998
21–42	Not reported	0.92 total 0.87 total	Weight gain Feed efficiency	Nicholas 700 Large White male	6 (0.73–0.98 total)	Corn–soybean meal diet containing 22% protein 2,900 kcal AME/kg. Data analyzed by ANOVA and linear and quadratic regressions using 95% of the asymptote.	Waldroup et al., 1998
29–42	Not reported	0.72 digestible[c] 0.76 digestible[c]	Weight gain Feed efficiency	Nicholas White male	7 (0.51–0.99 digestible)	Corn–peanut meal–soybean meal diet containing 24.0% protein and 3,100 kcal AME/kg. Data analyzed by ANOVA, quadratic, and broken-line regression analysis.	Linares et al., 2012

42–63	Not reported	0.77 total 0.77 total	Weight gain Feed efficiency	BUT male	7 (0.51–0.87 total)	Corn–soybean meal–peanut meal diet containing 17.48% protein and 3,348 kcal AME/kg. Data analyzed by GLM and quadratic regression using 95% of the asymptote.	Kidd et al., 1998
42–63	Not reported	0.86 total 0.84 total	Weight gain Feed efficiency	Nicholas 700 Large White male	6 (0.65–0.95 total)	Corn–soybean meal diet containing 25.9% protein and 3,000 kcal AME/kg. Data analyzed by ANOVA and linear and quadratic regressions using 95% of the asymptote.	Waldroup et al., 1998
56–70	3.2–4.7	≤***0.59*** total ≤***0.72*** total	Weight gain Feed efficiency	BUT male	5 (0.59–1.11 total)	Corn–peanut meal diet containing 22.8% protein and 3,313 kcal AME/kg. Data analyzed by GLM.	Barbour et al., 2008
56–84	Not reported	16% protein diet 0.68 total 0.83 total 0.61 total 0.77 total 19.8% protein diet: 0.76 total 0.98 total 0.70 total 0.82 total	 Weight gain, broken line Weight gain, quadratic Feed efficiency, broken line Feed efficiency, quadratic Weight gain, broken line Weight gain, quadratic Feed efficiency, broken line Feed efficiency, quadratic	BUT Large White male	6 (0.50–0.85 total)	Corn–soybean meal diet containing either 16.0% protein and 3,145 kcal AME/kg or 19.8% protein and 3,149 kcal AME/kg. Data analyzed by ANOVA, quadratic, broken-line regression analysis, and quadratic regressions at that asymptote.	Waldroup et al., 2002
56–84	4.0–8.6	≤***0.69*** total ≤***0.69*** total	Weight gain Feed efficiency	BUT Big 6 male	6 (0.69–0.91 total)	Corn–wheat–soybean meal diet containing 22.8% protein and 3,070 kcal AME/kg. Data analyzed by ANOVA.	Lehmann et al., 1997
77–91	6.1–8.1	≤***0.72*** total ≤***0.72*** total	Weight gain Feed efficiency	BUT male	5 (0.59–1.11 total)	Corn–peanut meal diet containing 22.8% protein and 2,900 kcal AME/kg. Data analyzed by GLM.	Barbour et al., 2008
112–140	13.1–18.5	≤***0.49*** total ≤***0.49*** total	Weight gain Feed efficiency	BUT Big 6 male	6 (0.49–0.64 total)	Corn–wheat–soybean meal diet containing 16.4% protein and 3,225 kcal AME/kg. Data analyzed by ANOVA.	Lehmann et al., 1997

continued

TABLE 16-3 Continued

Age (d)	Weight, Start–End of Study, kg	Estimated Requirements per Response (%)	Response Criteria	Strain and Sex	Number of Levels, (range, %)	Comments	Reference
Valine							
8–21	Not reported	0.70–0.79 digestible 0.70–0.79 digestible	Weight gain Feed efficiency	Nicholas, female	7 (0.82–1.36 total; 0.70–1.24 digestible)	Corn–soybean meal containing 17.6% protein and 3,218 kcal AME/kg. Data analyzed by ANOVA and broken-line regression analysis.	Kamyab and Firman, 1999
8–21	Not reported	0.82 digestible 0.74 digestible	Weight gain Feed efficiency	Nicholas, female	7 (0.70–0.85 digestible)	Corn–soybean meal containing 17.6% protein and 3,218 kcal AME/kg. Data analyzed by ANOVA and broken-line regression analysis.	Kamyab and Firman, 1999
Calcium							
29–91	1.10–11.05	1.0	Weight gain, feed efficiency	BUT Big 6 male	4 (0.6–1.8)	Final body weights were averaged across 4 dietary P amounts; 0.3% P was considered optimal. Diets were supplemented with 150 µg/kg cholecalciferol.	Hocking et al., 2002a
57–77	8.06–12.26	0.80	Bone fracture force	Nicholas 700 male	5 (0.45–0.90)	Value is 105% of NRC (1994) recommended Ca amount adjusted to a 3-week basis. Diet contained 5,000 ICU cholecalciferol.	Roberson, 2004
78–98	12.26–16.17	0.70	Bone fracture force	Nicholas 700 male	5 (0.4–0.8)	Value is 105% of NRC (1994) recommended Ca amount adjusted to a 3-week basis. Diet contained 5,000 ICU cholecalciferol.	Roberson, 2004
64–84	5.49–9.54	1.20	Bone fracture force and bone ash percentage	BUT Big 6 male	3 (0.62–1.20)	There was no carryover effect of feeding > NRC (1994) recommended Ca amounts in the pre-starter and starter diets. Diet contained 3,500 ICU cholecalciferol.	Roberson et al., 2004

85–105	9.54–12.94	1.10		Bone fracture force and bone ash percentage	BUT Big 6 male	3 (0.56–1.10)	There was no carryover effect of feeding > NRC (1994) recommended Ca amounts in the pre-starter and starter diets. Diet contained 3,500 ICU cholecalciferol.	Roberson et al., 2004
106–119	12.94–15.68	1.00		Bone fracture force and bone ash percentage	BUT Big 6 male	3 (0.48–1.00)	There was no carryover effect of feeding > NRC (1994) recommended Ca amounts in the pre-starter and starter diets. Diet contained 3,500 ICU cholecalciferol.	Roberson et al., 2004
15–35	Not reported–1.68	1.19		Weight gain, bone health	Large White Nicho-las male	4 (0.90–1.37)	Diets contained 5,000 ICU cholecalciferol.	Roberson, 2009
36–56	1.68–4.58	1.04		Weight gain, bone health	Large White Nicho-las male	4 (0.76–1.22)	Diets contained 5,000 ICU cholecalciferol.	Roberson, 2009
57–77	4.58–8.83	0.97		Weight gain, bone health	Large White Nicho-las male	4 (0.68–1.15)	Diets contained 3,500 ICU cholecalciferol.	Roberson, 2009
78–98	8.83–12.61	0.86		Weight gain, bone health	Large White Nicho-las male	4 (0.61–1.04)	Diets contained 3,500 ICU cholecalciferol.	Roberson, 2009
99–119	12.61–17.56	0.79		Weight gain, bone health	Large White Nicho-las male	4 (0.50–0.97)	Diets contained 5,000 ICU cholecalciferol.	Roberson, 2009
120–133	17.56–20.71	0.74		Weight gain, bone health	Large White Nicho-las male	4 (0.45–0.92)	Diets contained 5,000 ICU cholecalciferol.	Roberson, 2009
Non-phytate phosphorus								
29–56	0.63–2.85	0.37		Weight gain, feed efficiency	Large White Nicho-las male	4 (0.15–0.55)	Optimal P diet contained 1.0% Ca and no phytase.	Atia et al., 2000
29–56	0.58–2.45	0.37		Weight gain, feed efficiency	Large White Nicho-las female	4 (0.15–0.55)	Optimal P diet contained 1.0% Ca and no phytase.	Atia et al., 2000
57–84	2.85–6.51	0.31		Weight gain, feed efficiency	Large White Nicho-las male	4 (0.13–0.46)	Optimal P diet contained 0.85% Ca and no phytase.	Atia et al., 2000
57–77	2.45–4.47	0.31		Weight gain, feed efficiency	Large White Nicho-las female	4 (0.13–0.46)	Optimal P diet contained 0.85% Ca and no phytase.	Atia et al., 2000
85–102	6.51–10.74	0.28		Weight gain, feed efficiency	Large White Nicho-las male	4 (0.11–0.42)	Optimal P diet contained 0.75% Ca and no phytase.	Atia et al., 2000
78–98	4.47–6.63	0.28		Weight gain, feed efficiency	Large White Nicho-las female	4 (0.11–0.42)	Optimal P diet contained 0.75% Ca and no phytase.	Atia et al., 2000

continued

TABLE 16-3 Continued

Age (d)	Weight, Start–End of Study, kg	Estimated Requirements per Response (%)	Response Criteria	Strain and Sex	Number of Levels, (range, %)	Comments	Reference
Non-phytate phosphorus							
57–77	8.06–12.26	0.40	Bone fracture force	Nicholas 700 male	5 (0.23 0.45)	Value is 105% of NRC (1994) recommended nPP amount adjusted to a 3-week basis. Diet contained 5,000 ICU cholecalciferol.	Roberson, 2004
78–98	12.26–16.17	0.35	Bone fracture force	Nicholas 700 male	5 (0.2–0.4)	Value is 105% of NRC (1994) recommended nPP amount adjusted to a 3-week basis. Diet contained 5,000 ICU cholecalciferol.	Roberson, 2004
64–84	5.49–9.54	0.60	Bone fracture force and bone ash percentage	BUT Big 6 male	3 (0.31–0.60)	There was no carryover effect of feeding > NRC (1994) recommended P amounts in the pre-starter and starter diets. Diet contained 3,500 ICU cholecalciferol.	Roberson et al., 2004
85–105	9.54–12.94	0.55	Bone fracture force and bone ash percentage	BUT Big 6 male	3 (0.28–0.55)	There was no carryover effect of feeding > NRC (1994) recommended P amounts in the pre-starter and starter diets. Diet contained 3,500 ICU cholecalciferol.	Roberson et al., 2004
106–119	12.94–15.68	0.50	Bone fracture force and bone ash percentage	BUT Big 6 male	3 (0.24–0.50)	There was no carryover effect of feeding > NRC (1994) recommended P amounts in the pre-starter and starter diets. Diet contained 3,500 ICU cholecalciferol.	Roberson et al., 2004
15–35	Not reported–1.68	0.66	Weight gain, bone health	Large White Nicholas male	4 (0.50–0.76)	Diets contained 5,000 ICU cholecalciferol.	Roberson, 2009
36–56	1.68–4.58	0.58	Weight gain, bone health	Large White Nicholas male	4 (0.42–0.68)	Diets contained 5,000 ICU cholecalciferol.	Roberson, 2009
57–77	4.58–8.83	0.54	Weight gain, bone health	Large White Nicholas male	4 (0.38–0.64)	Diets contained 3,500 ICU cholecalciferol.	Roberson, 2009
78–98	8.83–12.61	0.48	Weight gain, bone health	Large White Nicholas male	4 (0.32–0.58)	Diets contained 3,500 ICU cholecalciferol.	Roberson, 2009

99–119	12.61–17.56	0.44		Weight gain, bone health	Large White Nicholas male	4 (0.28–0.54)	Diets contained 3,500 ICU cholecalciferol.	Roberson, 2009
120–133	17.56–20.71	0.41		Weight gain, bone health	Large White Nicholas male	4 (0.25–0.51)	Diets contained 3,500 ICU cholecalciferol.	Roberson, 2009
217–434	11.0–12.0	0.17		Reproductive performance (egg production, settable eggs, egg fertility, egg hatchability, hatchability of fertile eggs, embryonic mortality)	Hybrid EURO FP female	3 (0.17–0.55)	Diets contained 16.1% protein, 2,971 kcal AME/kg, 2.90% Ca, and 8,000 ICU cholecalciferol.	Godwin et al., 2005
Sodium								
1–126	Hatch–17.98	0.13		Weight gain, feed efficiency	BUT Large White Big 6 male	3 (0.08–0.18)	Basal diets contained 0.024% sodium and 0.091% chlorine (0–28 d), 0.022% sodium and 0.084% chlorine (29–56 d), 0.017% sodium and 0.072% chlorine (57–84 d), and 0.015% sodium and 0.07% chlorine (85–126 d). Also examined three sodium sources (NaCl, $NaHCO_3$, and Na_2SO_4) which gave variable results on weight gain and feed efficiency depending upon the growth phase.	Jankowski et al., 2012a
1–42	Hatch–1.46	0.13		Weight gain, absolute and relative tibia weight, tibia length, tibia cross-sectional area, tibia bone density, tibia breaking strength	BUT-10 female	3 (0.034–0.282)	Diets contained either 0.034% sodium and 0.17% chlorine, 0.134% sodium and 0.32% chlorine, or 0.282% sodium and 0.59% chlorine. All three diets contained 1.10% potassium.	Jankowski et al., 2012b
7–126	Not reported–14.28	0.17		Weight gain, tibia density and elastic strength, cecal metabolism	BUT-9 male	4 (0.07–0.22)	Diet contained 0.07% sodium and 0.17% chlorine, 0.12% sodium and 0.25% chlorine, 0.17% and 0.33% sodium and chlorine, or 0.22% sodium and 0.40% chlorine. All four diets contained 1.10% potassium.	Jankowski et al., 2012c

continued

TABLE 16-3 Continued

Age (d)	Weight, Start–End of Study, kg	Estimated Requirements per Response (%)	Response Criteria	Strain and Sex	Number of Levels, (range, %)	Comments	Reference
Manganese							
1–98	Hatch–9.24	~61 mg/kg	Weight gain, feed efficiency	Hybrid Converter male	3 (~61–138 mg/kg)	Diets were supplemented with either 10, 50, or 100 mg/kg manganese; values in the previous column include the analyzed total amount of manganese in the diets. Feeding the lowest amount of supplemental manganese did not negatively affect growth performance.	Jankowski et al., 2019b
Selenium							
1–14	Hatch–0.535	0.21 mg/kg 0.28 mg/kg	Plasma glutathione peroxidase (GPX) Liver GPX	BUT BIG 6 male	8 (0.01–0.38 mg/kg)	Basal diet contained <0.01 mg/kg selenium and was supplemented with sodium selenate. Data analyzed by broken line regression. Pre-starter diet contained 29.7% protein, 11.2 MJ/kg AME, and 27.1 mg/kg vitamin E.	Fischer et al., 2008
15–35	Hatch–2.568	0.30 mg/kg 0.20 mg/kg	Plasma GPX Liver GPX	BUT BIG 6 male	8 (0.01–0.39 mg/kg)	Basal diet contained <0.01 mg/kg selenium and was supplemented with sodium selenate. Data analyzed by broken line regression. Pre-starter diet contained 27.1% protein, 11.8 MJ/kg AME, and 27.1 mg/kg vitamin E. Starter diet contained 27.1% protein, 11.8 MJ/kg AME, and 24.7 mg/kg vitamin E.	Fischer et al., 2008
1–35	Hatch–1.746	0.15–0.25 mg/kg 0.25–0.35 mg/kg 0.25–0.35 mg/kg	Plasma GPX Liver GPX Breast GPX	BUT BIG 6 male	4 (0.05–0.35 mg/kg)	Basal diet contained 0.05 mg/kg selenium and was supplemented with sodium selenate. Data analyzed by ANOVA and means separation. Pre-starter diet contained 29.8% protein and	Mueller et al., 2009

						11.5 MJ/kg AME. Starter diet contained 26.5% protein, 11.6 MJ/kg AME, and either 10 or 60 IU vitamin E. Vitamin E amount did not affect estimated Se requirement.	
1–28	0.063–0.769	0.05 mg/kg 0.23 mg/kg 0.33 mg/kg 0.29 mg/kg 0.24 mg/kg 0.30 mg/kg	Growth rate Liver GPX1 Liver GPX1 Plasma GPX3 Gizzard GPX4 Pancreas GPX1	Nicholas male	10 (0.005–1.0 mg/kg)	Basal diet contained 0.005 mg/kg selenium and was supplemented with sodium selenite. Data analyzed by sigmoidal or hyperbolic regression analysis. Diet contained 2899 kcal AME/kg diet and 150 mg/kg vitamin E.	Taylor and Sunde, 2016
1–28	0.063–0.769	0.25 mg/kg 0.30 mg/kg 0.35 mg/kg 0.30 mg/kg 0.32 mg/kg 0.32 mg/kg	Kidney GPX4 Kidney GPX1 Thigh GPX1 Thigh GPX4 Breast GPX1 Breast GPX4	Nicholas male	10 (0.005–1.0 mg/kg)	Basal diet contained 0.005 mg/kg selenium and was supplemented with sodium selenite. Data analyzed by sigmoidal or hyperbolic regression analysis. Diet contained 2899 kcal AME/kg diet and 150 mg/kg vitamin E.	Taylor and Sunde, 2017
Zinc							
224–392	Not reported	154 mg/kg	Fertilization ability of semen, hatchability of eggs from inseminated turkey hens	BUT BIG 6 male	3 (34–154 mg/kg)	"Standard diets" were fed to breeder toms with a base (unsupplemented) zinc content of 34 mg/kg. Diets were supplemented with either 0, 60, or 120 mg/kg zinc.	Glowgowski et al., 2004
1–98	Hatch–8.59	58 mg/kg	Weight gain and immune response	Hybrid Converter female	3 (58–148)	The unsupplemented wheat–soybean meal base diet contained ~47 mg/kg zinc. The amount of dietary zinc had no significant effect on growth performance or immune response.	Jankowski et al., 2019a
Vitamin A							
21–41	Not reported	6,667 IU/kg	T-lymphocyte proliferative responses to con A and antibody titers against Newcastle disease virus and turkey pox virus 10 d or 20 d post-inoculation	BUT male	6 (0–44,000 IU)	Poults fed the diet with no supplemental vitamin A died between 18 and 22 d.	Sklan et al., 1995

continued

TABLE 16-3 Continued

Age (d)	Weight, Start–End of Study, kg	Estimated Requirements per Response (%)	Response Criteria	Strain and Sex	Number of Levels, (range, %)	Comments	Reference
Vitamin E							
1–70	Hatch–2.01	22 IU/kg	Weight gain, feed efficiency, and antibody titers against Newcastle disease virus and turkey pox virus at various times post-inoculation or boost until d 70	BUT male	3 (22–355 IU/kg)	Unsupplemented corn–soybean meal basal diet contained 22 IU vitamin E/kg. Antibody titers were decreased in poults fed either of the diets containing supplemental vitamin E.	Friedman et al., 1998
1–64	Hatch–3.6	37 IU/kg	Weight gain	BUT BIG 6 female	3 (111–333 IU/kg of DL-α-tocopherol acetate) or 2 (37–74 IU/kg of RRR-D-α-tocopherol)	The corn–wheat–soybean meal basal diets [starter (weeks 1 to 2), grower 1 (weeks 3 to 5), and grower 2 (weeks 6 to 9)] contained 0.352, 0.371, and 0.370 mg selenium/kg, respectively; the vitamin E content of the unsupplemented basal diets was not reported.	Ognik and Wertelecki, 2012
65–112	3.6–9.13	34 IU/kg	Weight gain, carcass dressing percentage, and plasma antioxidant properties	BUT BIG 6 female	3 (100–300 IU/kg of DL-α-tocopherol acetate) or 2 (34–67 IU of RRR-D-α-tocopherol)	The corn–wheat–soybean meal basal diets [grower 3 (weeks 10 to 14) and finisher (weeks 14 to 16)] contained 0.374 and 0.319 mg selenium/kg, respectively; the vitamin E content of the unsupplemented basal diets was not reported.	Ognik and Wertelecki 2012
Biotin							
1–98	Hatch–11.74	0.2 mg/kg	Weight gain, foot pad lesions, liver fat content	BUT T8 male	4 (0–1.6 mg/kg)	There was no effect of supplementary biotin in wheat–soybean meal–full fat soy diets on poult growth foot pad lesions, or liver fat contents.	Mayne et al., 2007

217–406 (Exp. 1) and 210–399 (Exp. 2)	10.70–11.01 (Exp. 1) and 11.39–11.04 (Exp. 2)	0.80 mg/kg	Feed consumption, reproductive performance (egg production, egg mass, fertile eggs, hatchability of fertile eggs)	Large White Nicholas female	2 (0.178–0.75 mg/kg)	The corn–soybean meal-based diet contained 14.9% protein by analysis and was calculated to contain 2,938 kcal AME/kg, 2.30% calcium, 0.50% nPP, and 0.178 mg biotin/kg. Analyzed biotin contents of the basal diet were 0.088 mg/kg (Exp. 1) or 0.210 mg/kg (Exp. 2).	Chen et al., 1994
Niacin							
56–112	2.44–5.75	Between 86 and 206 mg/kg	Weight gain, feed efficiency, carcass yield, relative parts yield (breast, thigh, drumsticks)	BUT male	3 (26–206 mg/kg)	The unsupplemented corn–soybean meal basal grower and finisher diets contained 26.7 and 26.0 mg niacin/kg, respectively.	Adebowale et al., 2019a
29–56	0.94–2.45	Between 86 and 206 mg/kg	Weight gain, feed efficiency, nutrient digestibility, hematology	BUT male	3 (26–206 mg/kg)	The unsupplemented corn–soybean meal basal grower and finisher diets, contained 26.7 and 26.0 mg niacin/kg, respectively.	Adebowale et al., 2019b

[a] With few exceptions, references were published after September 1991, the date when the literature review was completed for the 9th revised edition of this report. The 1994 report should be consulted for older references that inform turkey nutrient requirements.

[b] Requirement estimates shown in bold italicized print indicate that the requirement was shown to be outside of the range of treatment amounts, there were minor deficiencies in the experimental design, or that the data was not analyzed using a broken line or quadratic regression. In some cases only digestible amino acid values were reported, and in those cases they were converted to total analyzed values using the data in Moore et al. (2001).

[c] Insufficient information was provided on the concentration of lysine in the individual feed ingredients, especially peanut meal, to permit calculation of total (analyzed) concentrations of the diet.

[d] The data were fit to broken-line analysis so that they could be compared to other studies.

Averaging the requirement estimates from the four studies gave an estimated total lysine requirement of 1.46 percent (1.32% digestible) for rate of gain. The efficiency of feed utilization was significantly affected by dietary lysine concentration in only two of the four studies (Boling and Firman, 1998; Linares et al., 2012), and the average requirement was 1.53 percent total lysine (1.36% digestible). The previous report (NRC, 1994) estimated the requirement as 1.6 percent total lysine from 0 to 4 weeks of age, based on four studies conducted in the early 1970s and one from 1957.

Based on more recent studies that used faster growing and more modern breeds, the recommended lysine requirement has been decreased to 1.53 percent total lysine (1.36% digestible) during the starting period.

3 to 6 Weeks of Age. Since the last report (NRC, 1994), there were three published studies (Firman, 2004; Thompson et al., 2004; Linares et al., 2012) that examined the lysine requirements of growing turkeys during parts of the 3-to-6-week period that met the committee's experimental design and statistical analysis inclusion criteria. Firman (2004) fed diets with eight different lysine amounts (1.11% to 1.46% total lysine; 1.02% to 1.37% digestible) to male Nicholas White poults from 23 to 37 d and estimated the requirement for rate of gain to be 1.28 percent total lysine (1.19% digestible). Thompson et al. (2004) fed eight different lysine amounts (0.98% to 1.38% total; 0.88% to 1.28% digestible) to Nicholas White female poults from 19 to 40 d and estimated the requirement to be 1.26 percent total lysine (1.16% digestible) for rate of gain and 1.24 percent total lysine (1.14% digestible) for efficiency of feed conversion. Linares et al. (2012) fed male Nicholas White poults seven different lysine amounts (1.24% to 1.60% total lysine; 1.06% to 1.42% digestible) and estimated the requirement for rate of gain to be 1.48 percent total lysine (1.27% digestible) and for efficiency of feed conversion to be 1.50 percent total lysine (1.29% digestible). Averaging the results of these three studies gives an estimated requirement of 1.34 percent total lysine (1.21% digestible) for rate of gain. The average requirement for efficiency of feed utilization from the two studies that observed a significant effect of dietary lysine (Firman, 1998; Linares et al., 2012) is 1.37 percent total lysine (1.21% digestible). The previous report (NRC, 1994) recommended 1.5 percent total lysine for growing turkeys from 4 to 8 weeks of age, but no reference was cited to support this requirement. Based on more recent research, the recommended lysine requirement has been decreased to 1.37 percent total lysine (1.21% digestible).

6 to 9 Weeks of Age. There were three studies (Lemme et al., 2002; Baker et al., 2003a; Thompson et al., 2005) published since the last report (NRC, 1994) that examined the lysine requirements of growing turkeys during parts of the 6-to-9-week period that met the committee's experimental design and statistical analysis inclusion criteria. Lemme et al. (2002) fed male BUT Big 6 turkeys six different diets containing 1.19 to 1.64 percent total lysine from 35 to 56 d of age and estimated the requirement for total lysine for rate of gain to be 1.38 percent and 1.50 percent when using broken-line and nonlinear regression analyses, respectively. The requirement for total lysine for efficiency of feed utilization was 1.32 percent and 1.37 percent for the broken line and nonlinear regression analyses, respectively. Baker et al. (2003a) fed male Nicholas White turkeys seven diets containing 0.97 percent to 1.32 percent total lysine (0.88% to 1.23% digestible) from 49 to 61 d and estimated the total lysine requirement to be 1.18 percent and 1.20 percent for rate of gain and feed efficiency, respectively. On a digestible basis, the requirements were 1.09 percent and 1.11 percent, respectively. Thompson et al. (2005) fed female Nicholas White turkeys seven diets containing 0.93 to 1.29 percent total lysine (0.85% to 1.21% digestible) from 46 to 57 d and estimated the total lysine requirement to be 1.12 percent and 1.15 percent for rate of gain and feed efficiency, respectively. On a digestible basis, the requirements were 1.04 percent and 1.07 percenr, respectively. Based on the average of the relevant studies, the recommended total lysine requirement from 6 to 9 weeks is 1.25 percent (1.09% digestible). The previous report (NRC, 1994) recommended 1.3 percent total lysine for growing male turkeys from 8 to 12 weeks of age (females, 8 to 11 weeks of age), but no referenced publications were provided to support that requirement.

9 to 12 Weeks of Age. There were four studies (Lehmann et al., 1996; Baker et al., 2003a,b; Thompson et al., 2005) published since the last report (NRC, 1994) that examined the lysine requirements of growing turkeys during parts of the 9-to-12-week period that met the committee's experimental design and statistical analysis inclusion criteria. Lehmann et al. (1996) fed male BUT Big 6 turkeys six different diets containing 0.87 percent to 1.37 percent total lysine from 56 to 84 d and estimated the total lysine requirement to be 1.05 percent for rate of gain and 1.04 percent for efficiency of feed utilization. Baker et al. (2003a) fed male Nicholas White turkeys eight different diets containing from 0.76 percent to 1.04 percent total lysine (0.68% to 0.96% digestible) from 72 to 83 d and estimated the total lysine requirement for rate of gain to be 0.96 percent and for efficiency of feed conversion to be 0.95 percent. On a digestible basis, the requirements were 0.87 percent and 0.86 percent, respectively. Baker et al. (2003b) fed male BUT turkeys eight different diets containing total lysine from 0.61 percent to 0.81 percent (0.54% to 0.75% digestible) from 72 to 83 d and estimated the total lysine requirement for rate of gain to be 0.75 percent and for efficiency of feed conversion to be 0.74 percent. On a digestible basis, the requirements were 0.68 percent and 0.67 percent, respectively. Thompson et al. (2005) fed male Nicholas White turkeys eight different diets containing from 0.77 percent to 1.12 percent total lysine (0.70% to 1.05% digestible) from 71 to 85 d and estimated the requirement for rate of gain to be 0.93 percent and for efficiency of feed conversion to be 0.89 percent. On a digestible basis, the requirements were 0.86 percent and 0.82 percent, respectively. The average total lysine requirement of the four studies is 0.93 percent for rate

of growth and 0.91 percent for efficiency of feed utilization, which is considerably lower than the recommendation of the previous report (NRC, 1994). This is partly because the report of Baker et al. (2003b) had substantially lower requirement estimates than both the other three studies and studies done prior to the last report. By not including the findings of Baker et al. (2003b) in the average, the total lysine requirement of the remaining three studies is 0.98 percent for rate of growth and 0.96 percent for efficiency of feed utilization. Based on this average, the recommended total lysine requirement from 9 to 12 weeks is 0.98 percent (0.87 percent digestible). The previous report (NRC, 1994) recommended 1.3 percent total lysine for growing male turkeys from 8 to 12 weeks of age (females, 8 to 11 weeks of age), but no referenced publications were provided to support that requirement.

12 to 15 Weeks of Age. There were two studies (Lemme et al., 2002; Baker et al., 2003b) published since the last report (NRC, 1994) that examined the lysine requirements of growing turkeys during parts of the 9-to-12-week period that met the committee's experimental design and statistical analysis inclusion criteria. Lemme et al. (2002) fed male BUT Big 6 turkeys six different diets containing between 0.80 percent to 1.20 percent total lysine from 91 to 112 d and estimated the requirement for total lysine for rate of gain to be 0.92 percent and for efficiency of feed conversion to be 0.91 percent. Baker et al. (2003b) fed male Nicholas White turkeys eight different diets containing between 0.50 percent to 0.71 percent total lysine (0.44% to 0.65% digestible) from 84 to 95 d and estimated the requirement for total lysine for rate of gain to be 0.59 percent and for efficiency of feed conversion to be 0.60 percent. On a digestible basis, the requirements were 0.53 percent and 0.54 percent, respectively. The previous report (NRC, 1994) recommended 1.0 percent total lysine for growing male turkeys from 12 to 16 weeks of age (females, 11 to 14 weeks of age), but no referenced publications were provided to support this requirement. The two studies published since the last report (NRC, 1994) had markedly different requirement estimates and were averaged, resulting in 0.76 percent total lysine for both rate of gain and efficiency of feed conversion. Because of the wide divergence in results from these two studies, the average results are provided as guidance until additional experiments are published and the recommended requirement has been decreased to 0.76 percent dietary lysine.

15 to 18 Weeks of Age. Lehmann et al. (1996) fed male BUT Big 6 turkeys six different diets containing between 0.61 percent and 0.96 percent total lysine from 112 to 140 d and estimated the requirement for total lysine for rate of gain to be 0.80 percent and for efficiency of feed conversion to be 0.79 percent. No other peer-reviewed publications that met the committee's inclusion criteria were identified. However, based on the study by Lehmann et al. (1996), the recommended requirement from 15 to 18 weeks has been decreased to 0.76 percent dietary lysine and is provided as a guidance value until additional experiments are published. The previous report (NRC, 1994) recommended 0.80 percent total lysine for growing male turkeys from 16 to 20 weeks of age (females, 14 to 17 weeks of age), but no reference was cited to support this requirement.

Methionine

Dietary methionine is used for many functions, including protein synthesis, donation of methyl groups, and conversion to cysteine. The methionine requirement accounts for the first two of these but assumes that dietary cystine and cysteine (cyst(e)ine) is sufficient in the diet and that no methionine is used for its synthesis. Behrends and Waibel (1980) found that cyst(e)ine can provide 55, 58, and 57 percent of the sulfur amino acid requirement for starting, growing, and finishing turkeys, respectively. Although there are a moderate number of studies since the last report (NRC, 1994) examining the total sulfur amino acid requirement (see next section), most of these provided dietary Met+Cys amounts but not the individual methionine and cyst(e)ine amounts. Some of these studies did report methionine requirements, but not the dietary cyst(e)ine amounts, so it was not possible to know if there was sufficient dietary cyst(e)ine to meet the cyst(e)ine requirement. For this reason, the results of those studies could not be utilized in determining the methionine requirement. In contrast, two studies were judged to be useful for examining the methionine requirement, one during the 0-to-3-week period (Kubinska et al., 2016) and one during the 9-to-12-week period (Jankowski et al., 2017). No other studies met the criteria for estimating the turkeys' methionine requirement during any other period.

0 to 3 Weeks of Age. Kubinska et al. (2016) fed six dietary amounts of methionine to female Hybrid Converter turkeys from 1 to 28 d but were not able to fit a broken-line regression analysis to the data; however, they found that the requirement for rate of gain and efficiency of feed utilization were both ≤0.47 percent on a total methionine basis. The previous report (NRC, 1994) recommended 0.55 percent total methionine for growing turkeys from 0 to 4 weeks of age and was based on three studies in the 1950s and one in 1980 (Behrends and Waibel, 1980); the latter study met the committee's experimental design and statistical analysis inclusion criteria. Behrends and Waibel (1980) estimated the methionine requirement as 0.46 percent from 0 to 4 weeks. Based on the studies by Kubinska et al. (2016) and Behrends and Waibel (1980), plus the observed decrease in the Met+Cys requirement since the last report (see subsequent discussion), the recommended methionine requirement has been lowered to 0.46 percent from 0 to 3 weeks.

9 to 12 Weeks of Age. Jankowski et al. (2017) found that the total methionine requirement for female Hybrid Converter turkeys from 63 to 84 d was ≤0.29 percent for rate of gain and ≤0.32 percent for efficiency of feed conversion, but they were unable to establish the specific requirements. The previous report (NRC, 1994) recommended 0.40 percent total methionine for growing male turkeys from 8 to 12 weeks

of age (females, 8 to 11 weeks of age) and was based on two studies, one of which (Behrends and Waibel, 1980) met the committee's experimental design and statistical analysis inclusion criteria. Behrends and Waibel (1980) estimated the methionine requirement as 0.30 percent from 8 to 12 weeks. Based on studies by Jankowski et al. (2017) and Behrends and Waibel (1980), the recommended methionine requirement has been lowered to 0.32 percent from 9 to 12 weeks.

Methionine + Cyst(e)ine

0 to 3 Weeks of Age. Three studies (Boling and Firman, 1997; Moore et al., 2001; Kubinska et al., 2016) published since the last report (NRC, 1994) that examined the Met+Cys requirements of growing turkey poults during part of the 0-to-3-week period met the committee's experimental design and analysis inclusion criteria. Boling and Firman (1997) conducted three experiments with female BUT poults (Experiment 1; 9 to 23 d) or Nicholas female poults (Experiment 2, 9 to 22 d; and Experiment 3, 8 to 22 d). In Experiment 1, seven different diets containing between 0.88 percent to 1.31 percent total Met+Cys (0.77% to 1.19% digestible) were fed. In Experiment 2, seven diets containing between 0.61 percent and 1.45 percent (0.50% to 1.33% digestible) were fed. In Experiment 3, nine diets containing between 0.71 percent and 0.95 percent total Met+Cys (0.50% to 0.83% digestible) were fed. They were unable to establish a Met+Cys requirement for either the rate of gain and efficiency of feed utilization in Experiments 1 and 2, but they estimated that both were ≤0.88 percent (≤0.77% digestible). In Experiment 3, where the methionine titration amounts were narrowed so that a requirement could be estimated more accurately, Boling and Firman (1997) reported that, on a digestible basis, the requirements for rate of gain and efficiency of feed utilization were 0.76 percent and 0.75 percent, respectively (0.87% and 0.86% total, respectively). Moore et al. (2001) fed male BTU poults eight different diets containing between 0.67 percent and 1.07 percent total Met+Cys (0.60% to 1.00% digestible) from 7 to 16 d and estimated the total Met+Cys requirement for the rate of gain to be 0.92 percent and for efficiency of feed conversion to be 0.91 percent. On a digestible basis, the requirements were 0.83 percent and 0.82 percent, respectively. Kubinska et al. (2016) fed six dietary amounts of Met+Cys to Hybrid turkeys from 1 to 28 d but were not able to fit broken-line regression to the data; however, they found that the requirements for rate of gain and efficiency of feed utilization were ≤1.00 percent and ≤0.88 percent, respectively, on a total Met+Cys basis. The previous report (NRC, 1994) recommended 1.05 percent total Met+Cys for growing turkeys from 0 to 4 weeks of age and were based on five studies conducted throughout the 1970s and early 1980s; however, most of these studies did not meet the committee's experimental design and statistical analysis inclusion criteria. Averaging the results of the previous two studies (Boling and Firman, 1997; Moore et al., 2001) that were able to establish a specific requirement, the total Met+Cys requirements for rate of gain and efficiency of feed conversion were 0.90 percent and 0.89 percent, respectively. Given that all three studies conducted since the last report found a lower requirement than was recommended in NRC (1994), the requirement for total and digestible Met+Cys has been decreased to 0.90 percent and 0.80 percent, respectively, from 0 to 3 weeks of age.

3 to 6 Weeks of Age. There was only one study (Moore et al., 2001) published since the last report (NRC, 1994) that examined the Met+Cys requirements of growing turkeys during parts of the 3-to-6-week period that met the committee's experimental design and statistical analysis inclusion criteria. They fed male BUT poults eight different diets containing between 0.64 percent and 1.01 percent total Met+Cys (0.57% to 0.95% digestible) from 21 to 35 d and estimated that the total Met+Cys requirement was 0.85 percent for the rate of gain and 0.84 percent for efficiency of feed conversion (0.76% and 0.75% digestible, respectively). The previous report (NRC, 1994) recommended 0.95 percent total Met+Cys for growing turkeys from 4 to 8 weeks of age, but no reference was cited to support this requirement. Based on the study by Moore et al. (2001), the recommended total Met+Cys requirement has been decreased to 0.85 percent from 3 to 6 weeks of age, but this is tentative until further research is published.

6 to 9 Weeks of Age. There was only one study (Moore et al., 2004) published since the last report (NRC, 1994) that examined the Met+Cys requirements of growing turkeys during all or part of the 6-to-9-week period that met the committee's experimental design and statistical analysis inclusion criteria. They fed male BTU poults eight different diets containing between 0.58 percent and 0.90 percent total Met+Cys (0.50% to 0.82% digestible) from 42 to 63 d and estimated the total Met+Cys requirement for the rate of gain to be 0.73 percent and for efficiency of feed conversion to be 0.74 percent (0.65% and 0.66% digestible, respectively). The previous report (NRC, 1994) recommended 0.95 percent total Met+Cys for growing turkeys from 4 to 8 weeks of age, but no reference was cited to support this requirement. Based on the study by Moore et al. (2001), the recommended total Met+Cys requirement has been decreased to 0.74 percent (0.66% digestible), but this is tentative until further research is published.

9 to 12 Weeks of Age. There were two studies (Moore et al., 2004; Jankowski et al., 2017) published since the last report (NRC, 1994) that examined the Met+Cys requirements of growing turkeys during all or part of the 9-to-12-week period that met the committee's experimental design and statistical analysis criteria. Moore et al. (2004) fed male BTU poults eight different diets containing between 0.46 percent and 0.72 percent total Met+Cys (0.40% to 0.66% digestible) from 63 to 84 d and estimated the total Met+Cys requirement for both the rate of gain and for efficiency of feed conversion to be 0.62 percent (0.56% digestible). Jankowski et al. (2017) reported that the Met+Cys requirement for female Hybrid Converter turkeys from 63 to 84 d was ≤0.62 percent for rate of gain and ≤0.66 percent for efficiency of feed conversion, but

they were unable to establish the specific requirements. The previous report (NRC, 1994) recommended 0.80 percent total Met+Cys for growing male turkeys from 8 to 12 weeks of age (females, 8 to 11 weeks of age), but no reference was cited to provide support for this requirement. Based on the study by Moore et al. (2004), the recommended total Met+Cys requirement has been decreased to 0.62 percent (0.56% digestible) from 9 to 12 weeks, but this is tentative until further research is published.

12 to 15 Weeks of Age. There was only one study (Moore et al., 2003) published since the last report (NRC, 1994) that examined the Met+Cys requirements of growing turkeys during all or part of the 12-to-15-week period that met the committee's experimental design and statistical analysis inclusion criteria. They fed male BTU poults eight different diets containing between 0.41 percent and 0.65 percent total Met+Cys (0.35% to 0.59% digestible) from 83 to 92 d and estimated the total Met+Cys requirement for the rate of gain and efficiency of feed conversion to be 0.45 percent and 0.46 percent, respectively (0.39% and 0.40% digestible, respectively). The previous report (NRC, 1994) recommended 0.65 percent total Met+Cys for growing male turkeys from 12 to 16 weeks of age (females, 11 to 14 weeks of age), but no reference was cited to support this requirement. Based on the study by Moore et al. (2004), the guidance value for the total Met+Cys requirement has been decreased to 0.46 percent (0.40% digestible) from 12 to 15 weeks, but this is tentative until further research is published.

15 to 18 Weeks of Age. There was one study (Moore et al., 2003) published since the last report (NRC, 1994) that examined the Met+Cys requirement of growing turkeys during all or part of the 12-to-15-week period that met the committee's experimental design and statistical analysis criteria. They fed male BTU poults eight different diets containing between 0.35 percent and 0.55 percent total Met+Cys (0.30% to 0.50% digestible) from 106 to 116 d and estimated the total Met+Cys requirement for the rate of gain to be 0.42 percent and for efficiency of feed conversion to be 0.39 percent. On a digestible basis, the requirements were 0.37 percent and 0.34 percent, respectively. The previous report (NRC, 1994) recommended 0.55 percent total Met+Cys for growing male turkeys from 16 to 20 weeks (females, 14 to 17 weeks of age), but no reference was cited to support this requirement. Based on the study by Moore et al. (2004), the guidance value for the total Met+Cys requirement has been decreased to 0.42 percent (0.37% digestible) from 15 to 18 weeks, but this is tentative until further research is published.

Threonine

0 to 3 Weeks of Age. There were four studies (Lehmann et al., 1997; Kidd et al., 1998; Kamyab and Firman, 2000; Linares et al., 2012) published since the last report (NRC, 1994) that examined the threonine requirements of growing turkey poults during all or part of the 0-to-3 week period that met the committee's experimental design and statistical analysis inclusion criteria. Lehmann et al. (1997) fed male BUT Big 6 poults six different diets containing between 0.57 percent to 1.14 percent total threonine from 1 to 21 d and estimated the total threonine requirement for the rate of gain to be 0.91 percent. Reanalysis of their data by broken-line analysis gave a requirement of 0.93 percent total threonine. Kidd et al. (1998) fed male BUT poults seven different diets containing between 0.71 percent to 1.07 percent total threonine from 1 to 21 d and estimated the total threonine requirement for the rate of gain to be 0.93 percent and for efficiency of feed conversion to be 0.97 percent. Kamyab and Firman (2000) fed female Nicholas poults eight different diets containing between 0.64 percent to 1.13 percent total threonine (0.54% to 1.03% digestible) from 7 to 18 d and estimated the total threonine requirement for the rate of gain to be 0.77 percent and for efficiency of feed conversion to be 0.84 percent (0.67% and 0.74% digestible, respectively). Linares et al. (2012) fed male Nicholas White poults seven different diets containing between 0.77 percent to 1.25 percent total threonine (0.56% to 1.04% digestible) from 8 to 21 d and estimated the total threonine requirement for the rate of gain to be 0.88 percent and for efficiency of feed conversion to be 1.02 percent (0.67% and 0.81% digestible, respectively). The study of Lilburn and Barbour (2001) provided additional information relevant to this topic. They observed that the total threonine requirement of male BUT Large White poults for both rate of gain and efficiency of feed conversion was between 0.65 percent and 0.80 percent from 0 to 14 d of age but were unable to use regression to establish specific requirements. The previous report (NRC, 1994) recommended 1.0 percent total threonine for growing turkeys from 0 to 4 weeks of age and was based on three studies conducted in the early 1970s, but none of these studies met the committee's experimental design and statistical analysis inclusion criteria. Averaging the results of the four more recent studies that met the committee's criteria for inclusion, the recommended total threonine requirements for rate of gain and efficiency of feed conversion are 0.88 percent (0.67% digestible) and 0.94 percent (0.78% digestible), respectively. As a result, the recommended dietary threonine requirement from 0 to 3 week of age has been decreased to 0.94 percent (0.78% digestible).

3 to 6 Weeks of Age. There were three studies (Kidd et al., 1998; Waldroup et al., 1998; Linares et al., 2012) published since the last report that examined the threonine requirements of growing turkeys during all or part of the 3 to 6 week period that met the committee's experimental design and statistical analysis inclusion criteria. Kidd et al. (1998) fed male BUT turkeys seven different diets containing between 0.66 percent to 1.02 percent total threonine from 21 to 32 d and estimated the total threonine requirement to be 0.88 percent for both rate of gain and efficiency of feed conversion. Waldroup et al. (1998) fed male Nicholas 700 Large White turkeys seven different diets containing between 0.73 percent to 1.03 percent total threonine from 21 to 42 d and estimated the total dietary threonine requirement to be 0.92 percent for rate of gain and 0.87 percent for efficiency of feed conversion. Linares et al. (2012) fed

male Nicholas White turkeys seven different diets containing between 0.67 percent to 1.15 percent total threonine (0.51% to 0.99% digestible) from 29 to 42 d and estimated the total dietary threonine requirement to be 0.88 percent for rate of gain and 0.92 percent for efficiency of feed conversion (0.72% and 0.76% digestible, respectively). The study of Lilburn and Barbour (2001) provided additional information relevant to this topic. The researchers determined that the total threonine requirement of male BUT Large White turkeys for both rate of gain and efficiency of feed conversion was between 0.65 percent and 0.80 percent from 21 to 35 d of age but were unable to use regression to establish specific requirements. The previous report (NRC, 1994) recommended 0.95 percent total threonine for growing turkeys from 4 to 8 week of age, but no reference was cited to support this requirement. Averaging the results of the three studies that met the committee's criteria for inclusion, the total threonine requirements for rate of gain and efficiency of feed conversion were 0.89 percent (0.72% digestible), and 0.89 percent (0.76% digestible), respectively. Thus, the recommended dietary requirement for total threonine has been decreased to 0.89 percent (0.76% digestible) from 3 to 6 weeks of age.

6 to 9 Weeks of Age. There were two studies published since the last report that examined the threonine requirements of growing turkeys during all or part of the 6-to-9-week period that met the committee's experimental design and statistical analysis inclusion criteria. Kidd et al. (1998) fed male BUT turkeys seven different diets containing between 0.51 percent to 0.87 percent total threonine from 42 to 63 d and estimated the total threonine requirement as 0.77 percent for both rate of gain and efficiency of feed conversion. Waldroup et al. (1998) fed male Nicholas 700 Large White turkeys seven different diets containing between 0.65 percent to 0.95 percent total threonine from 42 to 63 d and estimated the total threonine requirement for the rate of gain to be 0.86 percent and for the efficiency of feed conversion to be 0.84 percent. The previous report (NRC, 1994) recommended 0.80 percent total threonine for growing male turkeys from 8 to 12 weeks (females, 8 to 11 weeks of age), but no reference was cited to support this requirement. Averaging the results of the two studies that met the committee's criteria for inclusion, the total threonine requirements for rate of gain and efficiency of feed conversion are 0.82 percent and 0.81 percent, respectively. Thus, the recommended dietary requirement for total threonine has been increased to 0.82 percent from 6 to 9 weeks of age.

9 to 12 Weeks of Age. There was only one study published since the last report that examined the threonine requirement of growing turkeys during all or part of the 9-to-12-week period that met the committee's experimental design and statistical analysis inclusion criteria. Waldroup et al. (1998) fed male BUT Large White turkeys six different diets containing between 0.50 percent to 0.85 percent total threonine from 56 to 84 d using two different diets containing either 16 percent or 19.8 percent crude protein. When the turkeys were fed the low protein diet, the total threonine requirement was determined to be 0.68 percent for rate of gain and 0.61 percent for the efficiency of feed conversion. When the turkeys were fed the higher protein diet, the total threonine requirement for rate of gain was 0.76 percent and 0.71 percent for efficiency of feed conversion. The two base diets used to obtain the desired protein amounts were composed of protein sources that differed in threonine digestibility, and this could explain the apparent difference in threonine requirements due to protein amount. Two additional studies provided information relevant to estimating the threonine requirement during this period. Lehmann et al. (1997) fed six different amounts of threonine but did not obtain a significant response to supplementation for rate of gain or efficiency of feed utilization, indicating that the requirement was less than the 0.69 percent total threonine in the base diet (22.8% protein). Barbour et al. (2008) did not observe a response for weight gain when adding threonine to a base diet containing 0.59 percent total threonine and 22.8 percent crude protein. Their experiment had too few treatment amounts for a broken-line analysis, but efficiency of feed utilization was maximized at an amount between 0.59 percent and 0.72 percent total threonine when analyzed by ANOVA. The previous report (NRC, 1994) recommended 0.80 percent total threonine for growing male turkeys from 8 to 12 weeks (females, 8 to 11 weeks of age), but no reference was cited to support this requirement. The two protein amounts used in the Waldroup et al. (1998) study were in the range of those typically fed to turkeys in this age group, and the average of their two treatment groups is 0.72 percent total threonine for rate of gain and 0.66 percent for efficiency of feed conversion. Based on the studies of Barbour et al. (2008) and Lehmann et al. (1997), the recommended total dietary threonine requirement has been decreased to 0.72 percent, but this is tentative until further research is published.

12 to 15 Weeks of Age. There were no studies published since the last report (NRC, 1994) that examined the threonine requirements of growing turkeys that met the committee's experimental design and statistical analysis inclusion criteria. Due to the paucity of studies in this period, a study by Barbour et al. (2008) was considered to provide additional guidance. They fed male BUT turkeys five different diets containing between 0.59 percent to 1.11 percent total threonine from 77 to 91 d of age and estimated the requirement to be between 0.59 percent and 0.85 percent for rate of gain and between 0.59 percent and 0.72 percent for efficiency of feed utilization. The previous report (NRC, 1994) recommended 0.75 percent total threonine for growing male turkeys from 12 to 16 weeks (females, 11 to 14 weeks of age), but no reference was cited to support this requirement. Based on the study by Barbour et al. (2008), the recommended total dietary threonine requirement has been decreased to 0.59 percent, but this is tentative until further research is published.

15 to 18 Weeks of Age. There was one study published since the previous report (NRC, 1994) that examined the threonine requirements of growing turkeys that met the com-

mittee's experimental design and statistical analysis inclusion criteria. Lehmann et al. (1997) fed male BUT Big 6 poults six different diets containing between 0.49 percent to 0.64 percent total threonine from 112 to 140 d. They reported their results based on ANOVA, which does not provide a precise requirement. Reanalysis of their data by broken line analysis gave a requirement of 0.58 percent and 0.55 percent total threonine for rate of gain and efficiency of feed utilization, respectively. The previous report (NRC, 1994) recommended 0.60 percent total threonine for growing male turkeys from 16 to 20 weeks (females, 14 to 17 weeks of age), but no reference was cited to support this requirement. Based on the study by Lehmann et al. (1997), the recommended dietary total threonine requirement has been decreased to 0.58 percent, but this is tentative until further research is published.

Valine

There were no studies published since the last report (NRC, 1994) that examined the valine requirements of growing turkeys that met the committee's experimental design and statistical analysis inclusion criteria. One study by Kamyab and Firman (1999) examined the valine requirement of starting Nicholas poults, but they did not chemically analyze their diets for valine content. Their study estimated that the total valine requirement of male Nicholas poults from 7 to 21 d was 0.96 percent (0.82% digestible) for rate of gain. The previous report (NRC, 1994) recommended 1.20 percent total valine for growing poults from 0 to 4 weeks of age based on three publications from the early 1970s, but none of those publications met the committee's inclusion criteria for a requirement study. Based on the more recent study by Kamyab and Firman (1999), the guidance recommendation for poults is 0.96 percent total valine (0.82% digestible) from 0 to 3 weeks, but this is tentative until further research is published.

Amino Acid Requirement Trends

There was sufficient research to assess the amino acid requirement across all time periods for lysine, Met+Cys, and threonine, and during two periods for methionine and one period for valine. In most cases, the requirements were found to be lower than those recommended in the previous report (NRC, 1994). These decreases in amino acid requirements were not expected due to the striking increase in lean tissue accretion in today's modern turkey strains. It may be that the higher rate of feed intake in the modern breeds accounts for these differences when requirements are expressed as a percent of the diet.

Minerals

Due to the paucity of reports on the mineral requirements of turkeys that have appeared in the literature since September 1991, many of the revised recommended values in Table 16-2 are estimates. Since no new published requirement studies were found for chlorine, potassium, magnesium, copper, iron, and iodine, the previous values (NRC, 1994) were carried over in Table 16-2 and are not subsequently discussed below.

Calcium and Phosphorus

Dietary Sources

With regard to inorganic sources of dietary Ca and P, the P in anhydrous dicalcium phosphate ($CaHPO_4$) is almost completely unavailable to turkeys, whereas the dihydrate of dicalcium phosphate ($CaHPO_4 \cdot 2H_2O$) is highly available (Scott et al., 1962). The apparent cause of the poor availability of the anhydrous form was attributed to an inability to dissolve it due to the low gastric acidity of the proventriculus of the neonatal poults (Scott, 1987). Other poorly available P sources for turkeys include colloidal P and Curacao Island P, whereas defluorinated P and bone meal had greater availabilities similar to those for chicks (Scott, 1987). Although differences in P availability exist between poultry species for plant and mineral P sources, availabilities for inorganic P sources determined with broilers can be applied to turkey and duck feeds without a risk of overestimating availabilities (Rodehutscord and Dieckmann, 2005).

Dietary Calcium and Phosphorus Requirements

As was the case over 20 years ago (Hocking et al., 2002a), research on production responses to dietary Ca and P in growing turkeys, as well as Ca and P requirements for optimum turkey skeletal growth, remain somewhat scarce. Ledoux et al. (1995) did not observe growth, bone, or P digestibility responses to supplemental dietary phytase when ~70 percent of the NRC (1994) recommended amount for nonphytate P (nPP) was fed to Nicholas Large White turkey hens grown to market age. Hocking et al. (2002a) fed BUT male poults 16 diets containing four amounts of Ca (0.6, 1.0, 1.4, and 1.8%) and four amounts of nPP (0.3, 0.5, 0.7, and 0.9%) in a 4×4 factorial arrangement of treatments from 4 to 13 weeks of age. The diets also contained 150 µg/kg supplemental cholecalciferol. They concluded that the dietary amounts for optimal growth and feed efficiency were 1.0 percent Ca and 0.3 percent nPP, and that there may an opportunity to decrease the amount of P in rations for growing turkeys, provided that dietary Ca amounts are increased and that vitamin D amounts are maintained.

Others have also suggested that the NRC (1994) nPP requirement was too high. Atia et al. (2000) reported that the performance of Large White Nicholas turkey poults was not negatively affected when they were fed 73 percent of NRC (1994) recommended nPP amounts without phytase from 5 to 16 weeks of age. This equated to 0.37, 0.31, and 0.28 percent from 4 to 8, 8 to 12, and 12 to 16 weeks of age (toms) or 4 to 8, 8 to 11, and 11 to14 weeks of age (hens), respectively. When dietary Ca was sufficient, Hocking et al. (2002b) did not observe an effect on the prevalence of tibial dyschondroplasia in BUT male poults raised up to 13 weeks of age when dietary nPP amounts varied between 0.3 and 0.9 percent. Roberson (2004) evaluated the dietary Ca and P needs of Nicholas male poults from 8 to 14 wk of age by feeding diets containing 60, 75, 90, 105, or 120% of NRC (1994) recommended amounts each of Ca and nPP. In disagreement with the findings of Ledoux et al. (1995) and Atia et al. (2000), Roberson concluded that amounts of Ca and nPP that were slightly higher than those recommended by NRC (1994) were needed to increase bone strength but cautioned that because of environmental concerns related to nutrient management, more research was needed to confirm the dietary P needs growing and finishing turkeys. In other work, Roberson et al. (2004) fed BUT male poults fed either 100 percent or 125 percent of NRC (1994) recommended amounts of dietary Ca and nPP from 3 to 9 weeks of age. Each group was then fed either 75 percent (low), 100 percent (medium), or 145 percent (High) of the NRC (1994) recommended requirements for Ca and nPP from 9 to 17 weeks of age. Based primarily on bone breaking strength and bone ash values, the researchers concluded that feeding amounts of Ca and nPP that were greater than those recommended by NRC (1994) on a 3-week basis after 9 weeks of age was warranted. However, a 23 percent increase in litter P was observed when the high-P diet was fed during the grower-finisher period as compared to feeding the medium-P diet, which suggest that the physiological advantages of feeding higher nPP amounts may be somewhat offset by environmental concerns. In a subsequent study, Roberson (2009) fed Large White Nicholas male poults a low-P diet [NRC (1994) recommended nPP amounts], a medium-P diet (0.06 percentage units nPP > low-P diet), a high-P diet (0.10 percentage units nPP > medium-P diet), or a very high-P (vhigh-P) diet (0.10 percentage units nPP > high-P diet) from 2 to 19 weeks of age in a 3-week phase program. For example, starter diet nPP amounts were 0.50, 0.56, 0.66, and 0.76 percent, respectively, from 2 to 5 weeks of age and contained 5,000 ICU cholecalciferol. Dietary Ca was kept at a 1.8:1 ratio with nPP for all diets. Roberson concluded that, for the best overall growth performance and bone health, the high-P diet (0.16 percentage units > low-P diet) was the optimal dietary regimen for commercial toms grown to heavy body weights and fed pelleted feed. The respective Ca and nPP amounts associated with the high-P diets were 1.19 and 0.66 percent from 2 to 5 weeks of age; 1.04 and 0.58 percent from 5 to 8 weeks of age; 0.97 and 0.54 percent from 8 to 11 weeks of age; 0.86 and 0.48 percent from 11 to 14 weeks of age; 0.79 and 0.44 percent from 14 to 17 weeks of age; and 0.74 and 0.41 percent from 17 to 19 weeks of age. Litter P increased proportionately with increasing dietary P content, ranging from 1.28 percent (low-P diet) to 1.45 percent (medium-P diet), to 1.70 percent (high-P diet), to 2.03 percent (vhigh-P diet).

Only one report was found that centered on dietary P for turkey breeder hens. Godwin et al. (2005) investigated the effect of dietary P amounts and phytase inclusion in the breeder ration on hen reproductive performance. Corn-soybean–based mash diets containing 2.9 percent Ca and either low (0.17%), medium (0.35%), or high (0.55%) amounts of nPP, with or without 1 g/ton of added phytase from *Aspergillus niger* (containing 11.27 phytase U/g diet), were fed to Large White turkey breeder hens from 31 to 62 weeks of age. Decreasing dietary P resulted in no impairment of reproductive performance of the hens to 62 weeks of age while decreasing the total and water-soluble P in excreta. They concluded that P could be lowered in turkey breeder hen diets, as compared with current surveyed industry amounts, without impairing reproductive performance.

Taken together, the clear benefits of increasing growth performance and bone health in starting and growing Large White turkeys by increasing dietary Ca and nPP above NRC (1994) recommended amounts may outweigh environmental implications of increased litter P, providing that the latter is managed properly. Therefore, based mainly on the reports of Roberson, and in an attempt to convert from the 4-week feed phases of NRC (1994) to shorter 3-week feed phases while admittedly having limited available empirical data upon which to rely, the committee recommends the following dietary Ca and nPP requirements for starting and growing turkeys: Phase 1: 1.30 percent Ca; 0.65 percent nPP; Phase 2: 1.15 percent Ca; 0.58 percent nPP; Phase 3: 1.00 percent Ca; 0.50 percent nPP; Phase 4: 0.90 percent Ca; 0.45 percent nPP; Phase 5: 0.80 percent Ca; 0.40 percent nPP; and Phase 6: 0.75 percent Ca; 0.37 percent nPP (Table 16-2).

The committee is not aware of any Ca and P requirement studies for parent stock published in the primary literature since the previous report (NRC, 1994). Therefore, contents of the holding diets (0.65% Ca and 0.35% nPP) and laying diets (2.90% Ca and 0.35% nPP) employed by Godwin et al. (2005), which were identical to those reported by Crouch et al. (2002a) for Large White breeder hens, have been used to update the Ca and nPP requirements of breeding turkeys. Although Godwin et al. (2005) concluded that the dietary nPP amount could be lowered in turkey breeder hen diets, the committee was not comfortable reducing it below the NRC (1994)-recommended amount of 0.35 percent, particularly because the recommended dietary Ca amount has been increased from 2.25 percent to 2.90 percent. As a percentage of the diet, these figures are considerably lower than those of chickens, due largely to the greater feed intake of turkey hens compared to laying chickens, as pointed out by Scott (1987).

Use of Phytase in Turkey Diets

The majority of P in plant-derived feedstuffs occurs in the form of phytate, which is unavailable or poorly utilized by poultry. Phytate can also interact with several minerals and proteins and subsequently inhibit their absorption (Humer et al., 2015). The enzyme phytase catalyzes the stepwise hydrolysis of phytate to inorganic phosphates and inositol. Although endogenous (intestinal mucosal) phytase activity has been detected in broilers (reviewed by Humer et al., 2015) and Pekin ducks (Rush et al., 2005), but not yet investigated in turkeys, such activity is considered to be of little practical importance (Humer et al., 2015) in the context of plant phytate hydrolysis. Therefore, exogenous forms of phytase are almost universally supplemented in poultry diets as a means of improving dietary P availability and reducing P excretion in manure (Amerah et al., 2014).

Ledoux et al. (1995) reported that supplementation of a microbial phytase (1,000 U/kg), as a partial substitute for nPP in corn–soybean meal diets, improved the efficiency of feed utilization and increased body weight gains of Nicholas Large White turkey hens grown to market age but was without effect on body composition, toe ash, or bone breaking strength. In a 3-week experiment with 1-d-old BUT female poults fed three different amounts of nPP (0.27, 0.36, or 0.45%) with seven added amounts of a microbial phytase (0, 200, 400, 600, 800, 1,000, and 1,200 U/kg diet), Ravindran et al. (1995) concluded that 652 U of microbial phytase was equivalent to 1 g of P from defluorinated P in turkey starter diets using soybean meal as the only source of phytate P. Qian et al. (1996) reported that the P equivalencies of 652 and 963 U of a supplemental microbial phytase were equivalent to 1 g nPP, respectively, in diets containing 0.27 and 0.36 percent nPP, for male BUT poults from hatch to 21 d of age. Atia et al. (2000) found that exogenous phytase was effective in reducing the amounts of supplemental dietary nPP needed to support optimal growth performance and bone characteristics of Large White Nicholas turkey poults from 4 to 16 weeks of age when the dietary amounts of nPP were below the NRC (1994) recommended requirements. Godwin et al. (2005) fed Hybrid EURO FP turkey breeder hen poults three different amounts of nPP, with or without an exogenous phytase derived from *Aspergillus niger* providing 11.27 phytase U/g diet, from 31 to 62 weeks of age. By week 28 of lay, 41 percent of hens not fed phytase had gone out of production while 32 percent of the hens fed phytase had gone out of production. No quantitative equivalency of phytase U to nPP was reported. The optimal dose of phytase to employ is susceptible to change, influenced by the price of phytase and P sources, and the efficacy with which the phytase liberates P from the diet (Wealleans et al., 2016). Consideration should also be given to the value of extra-phosphoric benefits of phytase inclusion, such as improvements in crude protein, amino acid, and energy digestibilities, the mitigation of the anti-nutritional effects of phytate, and positive environmental effects due to decreased P excretion (Wealleans et al., 2016).

Sodium, Potassium, and Chlorine

In an 18-week study, Jankowski et al. (2012a) fed BUT Big 6 Large White male turkeys nine diets containing three supplemental amounts of sodium (Na) (0.08, 0.13, or 0.18 percent) and one of three Na sources (NaCl, $NaHCO_3$, or Na_2SO_4) in a 3×3 factorial arrangement. The basal diets contained ~0.02 percent Na and between 0.70 percent and 1.1 percent potassium (K), which was only supplied by the feed ingredients and decreased as lower amounts of soybean meal were used in each progressive feeding period. Based on the observed growth rate and feed conversion efficiency, Jankowski et al. (2012a) concluded that 0.08 percent dietary Na was a deficient amount and that the Na sources alternative to NaCl improved feed utilization but may adversely affect breast muscle traits. Concentration of Mg, P, Na, and Cl in serum were unaffected by the dietary sodium source but were significantly reduced in birds fed the lowest Na amount versus the highest Na amount. Jankowski et al. (2012b) subsequently reported that diets containing either a moderate (0.13%) or a high (0.28%) dietary Na amount (from supplemental NaCl) and a constant dietary amount of K (1.1%) promoted equivalent growth, tibia mineralization, and serum electrolyte concentrations in female BUT-10 turkeys from hatch to 6 weeks of age. In a third report, Jankowski et al. (2012c) fed male BUT-9 turkeys NaCl-supplemented diets containing either 0.07 percent Na and 0.17 percent Cl, 0.12 percent Na and 0.25 percent Cl, 0.17 percent Na and 0.33 percent Cl, or 0.22 percent Na and 0.40 percent Cl from hatch to 8 weeks of age. Only the lowest NaCl treatment adversely affected growth rate and feed conversion efficiency, while the two lowest NaCl treatments negatively affected the density index and elastic strength of the tibia. Elevated cecal digesta pH and ammonia concentrations were observed in birds fed the highest NaCl treatment. The authors concluded that the diet containing 0.17 percent Na and 0.33 percent Cl resulted in the most acceptable growth performance, tibia properties, and gastrointestinal function metrics. In a review of their studies investigating the effects of dietary Na on gastrointestinal function, excreta moisture content, footpad dermatitis, tibia mineralization, and growth performance of turkeys, Zduńczyk and Jankowski (2014) concluded that the NRC (1994) recommendations, as well as those of poultry breeding companies, were in excess of the young turkeys' needs and could be lowered to 0.13 percent, 0.12 percent, and 0.10 percent at 1 to 4, 5 to 8, and 9 to 12 weeks of age, respectively, and in older birds the dietary Na content can be maintained at 0.09 percent. When considering the collective work of Jankowski et al., the committee recommends the following dietary Na requirements for starting and growing turkeys: 0.17, 0.15, 0.12, 0.10, 0.10, and 0.10 for Phases 1 through 6, respectively (Table 16-2). Thus, as compared to the previous report (NRC, 1994), the recommended Na requirement values are lower for the final three feed phases. Dietary recommended Na requirements for breeder turkeys, as well as dietary recommended Cl requirements for all turkeys, remain unchanged due to a lack of peer-reviewed research.

Manganese

Jankowski et al. (2019b) fed Hybrid Converter turkeys wheat-soybean meal–based diets containing three amounts of supplemental manganese (10, 50, and 100 mg/kg) and two sources (manganese oxide and manganese nanoparticles) in a 3×2 factorial arrangement from hatch until 98 d of age. When the supplemented amount of Mn was subtracted from the reported analyzed total amount of Mn at each level and averaged across the three feeds employed during the 98-d study, it appears that the base diet contained somewhere between ~40 to 50 mg/kg Mn. Since the authors reported that turkey growth performance was unaffected by either the dietary amount or source of Mn, this suggested that the turkey required no more than ~50 to 60 mg/kg Mn (the calculated base amount + 10 mg/kg diet). The authors concluded that the amount of supplemental dietary Mn can be reduced from that recommended by the breeder (100 mg/kg). In the previous report (NRC, 1994), the recommended requirement of Mn was 60 mg/kg for all ages of starting and growing turkeys, as well as for holding and laying breeder hens. It is recommended that Mn requirement remain unchanged from the previous report.

Selenium

The previous report (NRC, 1994) suggested that poults required 0.2 mg/kg diet of inorganic selenium (Se). The response criteria used previously was the frequency of gizzard myopathies, which are a frequent pathology in selenium-deficient turkeys. However, of the two studies reviewed by the 1994 committee, one suggested a requirement of 0.20 mg Se/kg diet and the other 0.28 mg Se/kg diet. Although the growth rate of commercial turkeys has increased markedly since the previous report (Figure 16-3), only four studies investigating the turkeys' dietary Se requirement were subsequently identified that met the committee's inclusion criteria, and all examined the requirement for Se in poults during the first few weeks post-hatch.

Fischer et al. (2008) fed male BUT BIG 6 poults either a Se-deficient corn–soybean meal basal diet or the basal diet supplemented with sodium selenate to provide 0.10, 0.15, 0.20, 0.25, 0.30, 0.35, or 0.40 mg Se/kg diet from hatch to 5 weeks of age. All diets contained adequate amounts (25 to 27 mg/kg) of vitamin E. Growth performance was significantly reduced only in birds fed the Se-deficient basal diet, while supplementing Se greatly elevated Se concentrations in plasma, liver, heart, thigh, breast, and plasma (GPX3) and liver (GPX1) glutathione peroxidase (GPX) activities, although plateaus were reached at different amounts. Glutathione peroxidase requires selenium as a cofactor to provide antioxidant protection necessary to prevent gizzard myopathy and other pathologies associated with Se deficiency. Fischer et al. (2008) concluded that 0.30 mg Se/kg was necessary to ensure maximum Se accumulation in organs as well as maximum plasma (GPX3) and liver (GPX1) GPX activities. In a follow-up study, Mueller et al. (2009) investigated the influence of dietary vitamin E on the dietary Se requirement of BUT BIG-6 turkeys. Male poults were fed corn-soybean meal–based diets supplemented with 0.05, 0.10, 0.20, and 0.30 mg Se/kg diet from sodium selenate, in combination with 10 or 50 IU vitamin E/kg diet in a 4 × 2 factorial arrangement, from hatch to 5 weeks of age. There was no significant effect of diet on body weights, feed intakes, or feed efficiency values, and no signs of muscular lesions were observed in any of the groups. Liver and plasma Se concentrations paralleled dietary Se contents, while GPX activities in plasma, liver, and breast muscle were highest in poults fed either 0.20 or 0.30 mg/kg Se independent of the dietary vitamin E amount. Mueller et al. (2009) concluded that growing turkeys required at least 0.20 mg Se/kg diet.

Taylor and Sunde (2016) determined the dietary Se requirements of male Nicholas white-derived poults as indicated by selenoprotein transcript levels and enzyme activities. From hatch to 28 d of age, poults were fed a Se-deficient Torula yeast-sucrose-based diet containing 150 mg/kg all-rac-α-tocopherol acetate (12.5-times the vitamin E requirement) to prevent vitamin E deficiency related disease and supplemented with either 0.025, 0.05, 0.1, 0.2, 0.3, 0.4, 0.5, 0.75, or 1.0 mg Se/kg diet as sodium selenite. Although they found that the minimum requirement for growth to be 0.05 mg Se/kg diet, using selenoenzyme (GPX) activity, minimum Se requirements based on red blood cell GPX1, plasma GPX3, and pancreas and liver GPX1 activities were 0.29 to 0.33 mg Se/kg diet. Taylor and Sunde (2016) concluded that the turkey's dietary Se requirement should be increased from 0.2 mg Se/kg diet (NRC, 1994) to 0.3 mg Se/kg diet. In a subsequent study, Taylor and Sunde (2017) extended their analysis of the effect of dietary Se on tissue selenoprotein enzyme activity to kidney, heart, breast, and thigh. The same 10 dietary treatments described previously were fed to male Nicholas White-derived poults from hatch to 28 d of age. Increasing supplemental dietary Se resulted in hyperbolic response curves for all tissues, with breakpoint analyses indicating minimum dietary Se requirements of 0.34 to 0.36 mg Se/kg diet based on tissue Se contents of liver, kidney, and thigh. Similarly, GPX1 activity in kidney and muscle tissues also responded hyperbolically to increasing dietary Se, reaching well-defined plateaus with breakpoints at 0.30 to 0.36 mg Se/kg diet. Based on GPX4 activity of breast and thigh, minimum dietary Se requirements were between 0.30 to 0.32 mg Se/kg diet. Taylor and Sunde (2017) concluded that the young turkey's dietary requirement for Se should be raised to 0.4 mg Se/kg diet and further suggested that, based on Se response curve plateaus, selenite could even be safely supplemented to turkeys at 0.5 mg Se/kg diet.

Three of the four most recent studies described previously uniformly found that the young turkey's dietary Se requirement was higher than the 0.2 mg Se/kg recommended in the previous version of this report (NRC, 1994). Therefore, the recommended dietary amount for fast-growing turkey lines is increased to 0.3 mg (inorganic) Se/kg diet while that for breeder birds remains unchanged at 0.2 mg Se/kg diet due to a lack of any recent peer-reviewed reports.

Zinc

As zinc (Zn) is essential for spermatogenesis, Glogowski et al. (2004) investigated the effect of dietary Zn on the quantity, quality, biochemical indices, and biological value of turkey semen. BUT Big 6 breeder toms (29 weeks of age) were fed a base diet of unspecified composition containing 34 mg/kg zinc, or the base diet supplemented with an additional 60 mg/kg Zn or 120 mg/kg Zn. The Zn supplement was in the form of a chelate compound with methionine. Semen was collected twice per week from 32 to 56 weeks of age. Serum and sperm Zn concentrations, as well as sperm motility, were unaffected by dietary Zn, while ejaculate volume was highest in turkeys fed the unsupplemented (base) diet. Fertilization ability of semen, as well as hatchability of eggs from turkey hens inseminated with the semen from toms fed the highest dietary amount of Zn, were about 3 percent and 4–5 percent higher, respectively, than the other two dietary groups. This suggested that for breeder toms, the optimal dietary amount of Zn is about 150 mg/kg (base diet amount +120 mg/kg). Although a separate listing for turkey toms is not reported in Table 16-2, the committee felt that these findings were relevant and should be included in this report.

Jankowski et al. (2019a) investigated the effect of dietary Zn on growth performance and immune and antioxidant status of Hybrid Converter turkey females from hatch to 98 d of age. Three amounts of supplemental Zn (10, 50, and 100 mg/kg) and two dietary sources (Zn oxide and Zn nanoparticles) were fed in a 3×2 factorial arrangement. The Zn content of the unsupplemented wheat-soybean meal base diet was ~47 mg/kg. The amount of dietary Zn had no effect on growth performance or turkey livability, while the Zn source had no influence on growth performance or immune or antioxidant status of the turkeys. Jankowski et al. (2019a) suggested that reducing the dietary amount of Zn from that recommended by the breeder to ~60 mg/kg Zn would not compromise turkey growth performance. Because Jankowski et al. (2019a) did not include dietary amounts less than 60 mg/kg Zn in their study, the committee recommends reducing the Zn requirement for the first two feed phases to 60 mg/kg in Table 16-2, but keeping the subsequent amounts unchanged from the previous (NRC, 1994) recommendations for growing or breeding turkeys.

Vitamins

Very few peer-reviewed reports on the vitamin requirements of turkeys have appeared in the literature since the previous report (NRC, 1994), and no new published requirement studies for starting and growing turkeys were found for vitamin D, vitamin B_{12}, choline, folacin, pantothenic acid, pyridoxine, riboflavin, or thiamin. Therefore, recommended requirement values for these vitamins in Table 16-2 were carried over from the previous report (NRC, 1994). Only one report on the vitamin requirements of breeding turkey hens was found (for biotin), so all of the other recommended vitamin requirements for holding and laying hens were unchanged from the 1994 report.

Vitamin A

Sklan et al. (1995) investigated the effect of vitamin A on antibody production and T-cell response male BUT poults fed a sorghum-soybean meal–based diet supplemented with either 0, 0.25, 0.5, 2.0, 6.0, or 13.2 mg retinol equivalents from retinyl acetate/kg diet until 42 d of age. These doses equated to 0, 833, 1,667, 6,667, 20,000, and 44,000 IU vitamin A/kg diet, respectively. They were vaccinated against Newcastle disease virus (NDV; on d 5, 11, and 21) and turkey pox virus (TPV; on d 21). All of the poults fed diets without supplemental vitamin A died by d 22 post-hatch. Plasma and liver vitamin A concentrations increased with increasing dietary vitamin A, while d-31 T-lymphocyte proliferative responses to con A, a nonspecific mitogen, were maximal at 2.0 mg vitamin A/kg diet. Antibody titers 10 d or 20 d post-vaccination to both viruses increased with increasing dietary vitamin A and were maximal in poults fed 6.0 mg vitamin A/kg but were not significantly different from those fed 2.0 mg vitamin A/kg. Sklan et al. (1995) concluded that the young poult may require more than the 5,000 IU of vitamin A/kg recommended in NRC (1994). Although there is evidence to suggest that for certain nutrients increased intake above recommended requirement amounts may help to optimize immune functions (Wu et al., 2019), higher antibody production and T cell proliferation may not translate to resistance to infectious disease. The committee is therefore hesitant to recommend a change in the young turkey's dietary vitamin A requirement based on the report of Sklan et al. (1995); thus, the requirement remains at the NRC (1994) recommended amount of 5,000 IU/kg.

Vitamin E

Friedman et al. (1998) determined the effect of high dietary intakes of vitamin E on antibody production in response to NDV and TPV in male BUT poults. The birds were fed a corn-soybean meal–based diet (containing 22 IU vitamin E by analysis) that was supplemented with 0, 50, or 150 mg DL-α-tocopheryl acetate/kg diet from hatch to 84 d of age. The diets contained a total of 0, 133, or 355 IU of vitamin E/kg diet, respectively. Poults were vaccinated against NDV at 8, 21, and 49 d of age, and against TPV at 21 d of age. Supplemental dietary vitamin E was without effect on turkey weight gains or feed efficiency values, but serum α-tocopherol concentrations increased with increased vitamin E intake. Concentrations of circulating antibodies to NDV and TPV were highest in the unsupplemented group, while birds fed 50 or 150 mg vitamin

E/kg were slightly or significantly lower, respectively. Friedman et al. (1998) concluded that excessive vitamin E intake was detrimental to antibody production in turkeys, which was also observed in broiler chickens by Leshchinsky and Klasing (2001). Ognik and Wertelecki (2012) fed Big 6 BUT females poults corn-wheat-soybean meal–based diets supplemented with two sources of dietary vitamin E from hatch to 16 weeks of age as follows: DL-α-tocopherol acetate at 50, 100, or 150 mg/kg diet (weeks 1 to 9) or 45, 90, or 135 mg/kg diet (weeks 10 to 16); and RRR-D-α-tocopherol, a natural form of vitamin E, at 25 or 50 mg/kg diet (weeks 1 to 9) or 22.5 or 45 mg/kg (weeks 10 to 16). These doses equated to 111, 222, or 333 IU/kg (weeks 1 to 9) or 100, 200, or 300 IU/kg (weeks 10 to 16) of DL-α-tocopherol acetate, respectively. The RRR-D-α-tocopherol doses equated to 37 or 74 IU/kg (weeks 1 to 9) or 34 or 67 IU/kg (weeks 10 to 16), respectively. Body weight gains of birds fed the two higher amounts of DL-α-tocopherol acetate were significantly greater than the poults fed the lowest supplemental amount but similar to that of the RRR-D-α-tocopherol-fed groups. However, the latter group had greater carcass dressing percentages and lower carcass abdominal fat percentages. Plasma, liver, and breast muscle α-tocopherol contents in birds fed the two higher amounts of DL-α-tocopherol acetate and the higher amount of RRR-D-α-tocopherol were greater than those of poults fed the lowest supplemental amount, while the natural form of vitamin E was the most efficient antioxidant as gleaned from the amounts of peroxide and malondialdehyde detected in blood plasma. Based primarily on 16-week body weight gains reported by Ognik and Wertelecki (2012), the recommended guidance values for the turkeys' dietary vitamin E requirement have been increased to 37 IU/k g from hatch to 9 weeks, and to 34 IU/kg from 10 to 16 weeks of age, but these are tentative until further research is published.

Vitamin K

Jin and Sell (2001) investigated the effect of various dietary amounts and sources of dietary vitamin K on plasma prothrombin time. One-d-old Nicholas male poults were fed diets containing 0, 0.1, 0.25, 0.5, or 2.0 mg/kg of vitamin K_1 for 7 d (Experiment 1), 0, 0.10, 0.25, 0.50, or 2.0 mg/kg of vitamin K_1 for 7 d (Experiment 2), or 0 and 0.1, 0.25, or 0.50 mg of vitamin K_1/kg, 0.1, 0.25, or 0.50 mg of menadione dimethylpyrimidinol bisulfite (MPB), or 0.1, 0.25, or 0.50 mg of menadione nicotinamide disulfite (MNB) in a 3×3 factorial arrangement (Experiment 3) for 14 d. The sources of MPB and MNB contained 34 percent menadione, and the three treatment amounts were equivalent, on a molar basis, to 0.1, 0.25, and 0.50 mg of vitamin K_1/kg of diet, respectively. The researchers observed that the biopotencies of MPB and MNB were similar, but each was less than that of vitamin K_1 in terms of reducing prothrombin time. A nadir of prothrombin time and a plateau or prothrombin concentration were observed when MPB and MNB were supplemented at 0.25 mg of menadione/kg. Based on prothrombin time or plasma prothrombin concentration, Jin and Sell (2001) concluded that the young turkey requires between 0.079 to 0.13 mg/kg of dietary K_1, which is between 13 and 22 times lower than NRC (1994) recommended dietary vitamin K requirement of 1.75 mg/kg for poults from hatch to 4 weeks of age. The committee recommends that the dietary vitamin K requirement remain unchanged.

Biotin

Mayne et al. (2007) investigated the effect of dietary biotin supplementation on macroscopic and histological changes associated with foot pad dermatitis in growing turkeys. Male BUT T8 poults were fed wheat-soy–based diets containing 0, 0.2, 0.8, and 1.6 mg supplemental biotin/kg diet from hatch to 14 weeks of age. Although the base starter diet (0 to 4 weeks) contained no detectable biotin, the two grower diets (5 to 8 weeks and 9 to 14 weeks) contained 0.20 mg biotin/kg by analysis, which was equal to or exceeded the NRC (1994) dietary recommended biotin requirement for poults older than 4 weeks of age. Increasing dietary biotin was associated with higher plasma biotin concentrations, but there was no effect on external scores or histopathology of foot pad lesions or liver fat content. The researchers concluded that current recommendations for biotin in wheat–soy rations for growing turkeys were adequate for normal growth and development and that feeding higher dietary amounts of biotin were without effect on foot pad dermatitis. The recommended dietary biotin requirements for starting and growing turkeys remain unchanged from NRC (1994).

Chen et al. (1994) examined the effect of biotin supplementation on turkey breeder hen performance in two replicate 27-week feeding experiments with Large White Nicholas hens. A control corn–soybean meal diet containing (by analysis) 0.088 mg biotin/kg (Experiment 1) or 0.210 mg biotin/kg (Experiment 2), and a test diet containing 0.775 mg biotin/kg (both experiments) were fed beginning at 31 weeks of age (Experiment 1) or 30 weeks of age (Experiment 2) when the hens were light-stimulated. At that time, hens were inseminated weekly with pooled semen, and egg production and feed consumption were recorded at 4-week intervals. The researchers observed no significant effect of dietary biotin on feed consumption, egg mass, hatchability of fertile eggs, or fertility. During every 4-week period, hen daily egg production averaged almost 7 percent higher (average of 52.3% vs. 45.6%) in the high-biotin-fed hens, but the only significant difference was during weeks 50 to 54; however, the overall (34-to-58-week egg production) treatment difference approached significance ($P = 0.07$). When both experiments were analyzed together, significant improvements in egg production were observed during the first 4 weeks and after 16 weeks of production. Hatchability of eggs from breeders fed the higher amount of biotin was significantly greater than those from control birds during weeks 50 to 54 of Experiment 1. Chen et al. (1994) suggested that the increased hatchability may have been due to higher egg yolk biotin binding protein and yolk biotin content. Based on this study, the recommended dietary biotin requirement of laying turkey hens has been increased from 0.20 mg/kg to 0.80 mg/kg, but this is tentative until further research is published.

Niacin

Although most animals have the ability to synthesize niacin from tryptophan, the turkey is a relatively poor converter (Scott, 1987). Like ducks, turkeys have relatively high hepatic picolinic acid carboxylase activity, which diverts 2-amino-3-carboxymuconic-6-semialdehyde toward the formation of carbon dioxide rather than toward the formation of quinolinic acid, an intermediate in the synthesis of nicotinamide adenine dinucleotide (Chen et al., 1996). Adebowale et al. (2019a) fed male BUT poults a commercial corn-soybean meal–based diet that met their recommended nutrient requirements (NRC, 1994), including niacin (60 mg/kg) from hatch to 56 d of age. On d 56, the birds were fed diets supplemented with either 0, 60, or 180 mg niacin/kg until 112 d of age. By analysis, and averaged across the grower and finisher diets, the three experimental diets contained an average of ~26, 86, and 204 mg niacin/kg, respectively. As compared to unsupplemented controls, birds fed either of the niacin-supplemented diets had significantly improved weight gains and feed conversion values, and the turkeys fed 180 mg supplemental niacin/kg had significantly greater weight gains and feed conversion values than those fed 60 mg supplemental niacin/kg diet during both feed phases (d 56 to 84 and d 85 to 112). Carcass yields and relative parts yields of breast, thigh, and drumsticks (as a percentage of body weight) on d 112 followed a similar pattern. Although Adebowale et al. (2019a) concluded that 180 mg/kg supplemental niacin (204 mg niacin/kg total) increased production performance to the greatest extent, the turkey's dietary niacin requirement from 56 to 112 d most likely lies somewhere between 86 and 204 mg/kg. In a companion study, Adebowale et al. (2019b) fed male BUT poults a commercial corn-soybean meal–based diet that met their recommended nutrient requirements (NRC, 1994), including niacin (60 mg/kg) from hatch to 28 d of age. Beginning on d 29, the poults were fed the same three diets as described for Adebowale et al. (2019a), which contained by analysis ~26, 86, and 206 mg niacin/kg diet, from 29 to 56 d of age. Body weight gains at 56 d were significantly greater in the birds fed 180 mg supplemental niacin/kg (206 mg/kg total) versus the 0 supplemental (26 mg niacin/kg total) and the 60 mg supplemental (86 mg niacin/kg total) diets, while the feed/gain ratio was significantly poorest (highest) in the unsupplemented group versus the two niacin-supplemented groups. Turkeys fed the highest dietary niacin amount had a significantly better (lower) feed/gain ratio than birds supplemented with 60 mg niacin/kg diet. The results of the two studies by Adebowale et al. (2019a,b) suggests that growing turkeys require between 86 and 206 mg niacin/kg diet from 28 to 112 d of age. Based on this work, the recommended niacin requirement of growing turkeys from 21 to 112 d has been increased to 86 mg/kg diet.

FUTURE RESEARCH

Because of the importance of lean tissue accretion, future research should continue to refine the growing turkey's requirements for essential amino acids, not only during the early growth phases but in later life stages, as new genetic lines are introduced commercially. In addition, updating the life cycle calcium, nPP, and vitamin D requirements is necessary to continue to ensure optimal skeletal health and leg strength in order to support the turkeys' ever-increasing frame size and body weight (Figure 16-3). Lastly, since there are virtually no recent peer-reviewed studies on the nutrient requirements of parent stock (holding males and females and laying hens) and on feed restriction programs for female breeders, these are other important areas of research that should be pursued in order to improve turkey reproductive performance.

REFERENCES

Adebowale, T.O., H. Liu, A. O. Oso, O.E. Oke, T. Hussain, A.M. Bamgbose, and Y. Yulong. 2019a. Effect of dietary niacin supplementation on performance, total tract nutrient retention, carcass yield and meat lipid profile of growing turkeys. *Animal Production Science*. 59:1098–1107.

Adebowale, T.O., A. Oso, H. Liu, M. Tossou, J. Chen, H. Li, B. Kang, and K. Yao. 2019b. Effect of dietary niacin supplementation on growth performance, nutrient digestibility, hematology, and lipoprotein concentrations of young turkeys, *Meleagris gallopavo*. *Journal of Poultry Science*. 56:112–119.

Amerah, A.M., P.W. Plumstead, L.P. Barnard, and A. Kumar. 2014. Effect of calcium level and phytase addition on ileal phytate degradation and amino acid digestibility of broilers fed corn-based diets. *Poultry Science*. 93:906–915.

Amirkhanov, K., A. Igenbayev, A. Nurgazezova, E. Okuskhanova, S. Kassymov, N. Muslimova, and Z. Yessimbekov. 2017. Comparative analysis of red and white turkey meat quality. *Pakistan Journal of Nutrition*. 16:412–416.

Aslam, M.L., J.W.M. Bastiaansen, M.G. Elferink, H.-J. Megens, R.P.M.A. Crooijmans, L.A. Blomberg, R.C. Fleischer, C.P. Van Tassell, T.S. Sonstegard, S.G. Schroeder, M.A.M. Groenen, and J.A. Long. 2012. Whole genome SNP discovery and analysis of genetic diversity in turkey (*Meleagris gallopavo*). *BMC Genomics*. 13:391.

Atia, F.A., P.E. Waibel, I. Hermes, C.W. Carlson, and M.M. Walser. 2000. Effect of dietary phosphorus, calcium, and phytase on performance of growing turkeys. *Poultry Science*. 79:231–239.

Atkinson, R.L., A.A. Kurnick, T.M. Ferguson, B.L. Reid, J.H. Quisenberry, and J.R. Couch. 1957. Protein and energy levels for turkey starting diets. *Poultry Science*. 36:767–773.

Baker, K., J.D. Firman, E. Blair, J. Brown, and D. Moore. 2003a. Digestible lysine requirements of male turkeys during the 6 to 12 week period. *International Journal of Poultry Science*. 2(2):97–101.

Baker, K., J.D. Firman, E. Blair, J. Brown, and D. Moore. 2003b. Digestible lysine requirements of male turkeys during the 12 to 18 week period. *International Journal of Poultry Science*. 2(3):229–233.

Balloun, S.L., W.J. Owings, J.L. Sell, and R.E. Phillips. 1959. Energy and protein requirements for turkey starting diets. *Poultry Science*. 38(6):1328–1340.

Barbour, G.W., M.T. Farran, and M.S. Lilburn. 2008. Threonine requirement of the grower turkey tom. *Journal of Poultry Science*. 45(2):96–100.

Basha, M.E., and G.E. Duke. 1999. Effect of fasting on small intestinal antiperistalsis in the Nicholas turkey (*Meleagris gallopavo*). *Journal of Explorative Zoology*. 283(4–5):469–477.

Behrends, B.R., and P. Waibel. 1980. Methionine and cystine requirements of growing turkeys. *Poultry Science*. 59(4):849–859.

Bjornhag, G., and I. Sperber. 1977. Transport of various food components through the digestive tract of turkeys, geese and guinea fowl. *Swedish Journal of Agricultural Research*. 7:57–66.

Bohrer, B.M. 2017. Nutrient density and nutritional value of meat products and non-meat foods high in protein. *Trends in Food Science and Technology*. 65:103–112.

Boling, S.D., and J.D. Firman. 1997. Digestible sulfur amino acid requirement of starting turkeys. *Poultry Science*. 76(1):873–877.

Boling, S.D., and J.D. Firman. 1998. Digestible lysine requirement of female turkeys during the starter period. *Poultry Science*. 77(4):547–551.

Brant, A.W. 1998. A brief history of the turkey. *World's Poultry Science Journal*. 54(4):365–373.

Buss, E.G. 1989. Genetics of turkeys: Origin and development. *World's Poultry Science Journal*. 45(1):27–52.

Chaplin, S.B., J. Raven, and G.E. Duke. 1992. The influence of the stomach on crop function and feeding behavior in domestic turkeys. *Physiology and Behavior*. 52(2):261–266.

Chen, B.-J., T.-F. Shen, and R.E. Austic. 1996. Efficiency of tryptophan-niacin conversion in chickens and ducks. *Nutrition Research*. 16(1):91–104.

Chen, F., S.L. Noll, and P.E. Waibel. 1994. Dietary biotin and turkey breeder performance. *Poultry Science*. 73(5):682–686.

Chilson, W.T., and H. Patrick. 1946. Effect of withholding feed and water on early poult mortality and growth. *Poultry Science*. 25(1):86–87.

Corless, A.B., and J.L. Sell. 1999. The effects of delayed access to feed and water on the physical and functional development of the digestive system of young turkeys. *Poultry Science*. 78(8):1158–1169.

Crouch, A.N., J.L. Grimes, V.L. Christensen, and J.D. Garlich. 1999. Restriction of feed consumption and body weight in two strains of Large White turkey breeder hens. *Poultry Science*. 78(8):1102–1109.

Crouch, A.N., J.L. Grimes, V.L. Christensen, and K.K. Krueger. 2002a. Effect of physical feed restriction during rearing on Large White turkey breeder hens: 1. Growth performance. *Poultry Science*. 81(1):9–15.

Crouch, A.N., J.L. Grimes, V.L. Christensen, and K.K. Krueger. 2002b. Effect of physical feed restriction during rearing on Large White turkey breeder hens: 2. Reproductive performance. *Poultry Science*. 81(1):16–22.

Cutler, S., M. Rasmussen, M. Hensley, K. Wilhelms, R. Griffith, and C. Scanes. 2005. Effects of Lactobacilli and lactose on Salmonella typhimurium colonisation and microbial fermentation in the crop of the young turkey. *British Poultry Science*. 46(6):708–716.

Dean, W.F. 1986. Nutrient requirements of meat-type ducks. In *Duck Production Science and World Practice*. D.J. Farrell and P. Stapleton, editors. Armidale, NSW, Australia: University of New England. 31–57.

Denbow, D.M. 2015. Gastrointestinal anatomy and physiology. *Sturkie's Avian Physiology*, 6th ed. C.G.Scanes, editor. Amsterdam, The Netherlands: Elsevier. 337–366.

Duke, G., and O. Evanson. 1972. Inhibition of gastric motility by duodenal contents in turkeys. *Poultry Science*. 51(5):1625–1636.

Duke, G.E., O.A. Evanson, J.G. Ciganek, J.F. Miskowiec, and T.E. Kostuch. 1973. Inhibition of gastric motility in turkeys by intraduodenal injections of amino acid solutions. *Poultry Science*. 52(5):1749–1756.

Duke, G., O. Evanson, and P. Redig. 1976. A cephalic influence on gastric motility upon seeing food in domestic turkeys (*Melagris gallopavo*), great-horned owls (*Bubo virginianus*) and red-tailed hawks (*Buteo jamaicensis*). *Poultry Science*. 55(6):2155–2165.

Duke, G.E., O. Evanson, and B. Huberty. 1980. Electrical potential changes and contractile activity of the distal cecum of turkeys. *Poultry Science*. 59(8):1925–1934.

Duke, G., O. Evanson, and D. Epstein. 1983. Coordination of cecal motility during cecal evacuation. *Poultry Science*. 62(3):545–550.

Duke, G.E. 1989. Relationship of cecal and colonic motility to diet, habitat, and cecal anatomy in several avian species. *Journal of Experimental Zoology*. 3:38–47.

Duke, G.E. 1992. Recent studies on regulation of gastric motility in turkeys. *Poultry Science*. 71(1):1–8.

FAOSTAT. 2022. Turkeys, tonnes of meat (2021); and turkeys, head slaughtered (2021). OECD-FAO Agricultural Outlook 2021–2030.

Favero, A., A. Maiorka, F. Dahlke, R.F.P. Meurer, R.S. Oliveira, and R.F. Sens. 2009. Influence of feed form and corn particle size on the live performance and digestive tract development of turkeys. *Journal of Applied Poultry Research*. 18(4):772–779.

Ferket, P.R. 2004. Tom weights up seven percent: Turkey growth and performance survey shows that the industry's history of growing bigger birds faster is continuing. *Watt Poultry*. 32–42.

Ferket, P.R., and E.T. Moran Jr. 1985. Effect of plane of nutrition from starting through the breeder period on the reproductive performance of tom turkeys. *Poultry Science*. 64(11):2110–2118.

Firman, J. 2004. Digestible lysine requirements of male turkeys in their first six weeks. *International Journal of Poultry Science*.3(6):373–377.

Fischer, J., A. Bosse, E. Most, A. Mueller, and J. Pallauf. 2008. Selenium requirement of growing male turkeys. *British Poultry Science*. 49(5):583–591.

Flores, K.R., A. Fahrenholz, and J.L. Grimes. 2021. Effect of pellet quality and biochar litter amendment on male turkey performance. *Poultry Science*. 100(4):101002.

Flores-Maldonado, K.R. 2021. *Application of machine learning methodology on turkey gut health data.* Ph.D. dissertation, North Carolina State University.

Friedman, A., I. Bartov, and D. Sklan. 1998. Humoral immune response impairment following excess vitamin E nutrition in the chick and turkey. *Poultry Science*. 77(7):956–962.

Glogowski, J., J. Jankowski, D. Suszyńska, M. Polak, and A. Ciereszko. 2004. Quality and biological value of semen collected from turkey-toms fed diets with different zinc content. *Archiv für Geflügelkunde*. 68(5):235–238.

Godwin, J.L., J.L. Grimes, V.L. Christensen, and M.J. Wineland. 2005. Effect of dietary phosphorus and phytase levels on the reproductive performance of Large White turkey breeder hens. *Poultry Science*. 84(3):485–493.

Gous, R.M., C. Fisher, E. Tůmová, V. Machander, and D. Codová. 2023. Dietary energy:protein ratio influences the efficiency of utilisation of dietary protein by growing turkeys. *British Poultry Science*. 64(1): 116–121.

Gous, R.M., C. Fisher, E. Tůmová, V. Machander, D. Codová, J. Vlčková, L. Uhlířová, and M. Ketta. 2019. The growth of turkeys. 2. Body components and allometric relationships. *British Poultry Science*. 60(5): 548–553.

Havenstein, G.B., P.R. Ferket, J.L. Grimes, M.A. Qureshi, and K.E. Nestor. 2007. Comparison of the performance of 1966- versus 2003-type turkeys when fed representative 1966 and 2003 turkey diets: Growth rate, livability, and feed conversion. *Poultry Science*. 86(2): 232–240.

Hester, P.Y., and R.W.C. Stevens. 1990. Feed restriction of turkey breeder hens—A review. *Poultry Science*. 69(9):1439–1446.

Hillerman, J.P., F. Kratzer, and W.O. Wilson. 1953. Food passage through chickens and turkeys and some regulating factors. *Poultry Science*. 32(2):332–335.

Hocking, P.M., G.W. Robertson, and C. Nixey. 2002a. Effects of dietary calcium and phosphorus on mineral retention, growth, feed efficiency and walking ability in growing turkeys. *British Poultry Science*. 43(4):607–614.

Hocking, P.M., S. Wilson, L. Dick, L.N. Dunn, G.W. Robertson, and C. Nixey. 2002b. Role of dietary calcium and available phosphorus in the aetiology of tibial dyschondroplasia in growing turkeys. *British Poultry Science*. 43(3):432–441.

Humer, E., C. Schwarz, and K. Schedle. 2015. Phytate in pig and poultry nutrition. *Journal of Animal Physiology and Animal Nutrition*. 99(4): 605–625.

Hybrid. 2023. *Technical guide for Hybrid Turkeys commercial products.* Commercial_Management_Guide_EN.pdf.

Jackson, S., and G.E. Duke. 1995. Intestine fullness influences feeding behaviour and crop filling in the domestic turkey. *Physiology and Behavior*. 58(5):1027–1034.

Jankowski, J., J. Juśkiewicz, K. Lichtorowicz, and Z. Zduńczyk. 2012a. Effects of dietary level and source of sodium on growth performance, gastrointestinal digestion and meat characteristics in turkeys *Journal of Animal Feed Science and Technology*. 178(1–2):74–83

Jankowski, J., K. Lichtorowicz, Z. Zduńczyk, and J. Juśkiewicz. 2012b. The effect of different dietary sodium levels on blood mineral concentrations and tibia mineralization in turkeys. *Polish Journal of Veterinary Sciences*. 15(2):227–232.

Jankowski, J., Z. Zduńczyk, K. Lichtorowicz, and J. Juśkiewicz. 2012c. Effect of different levels of dietary sodium from sodium chloride on gastrointestinal tract response, tibia mineralization, and footpad dermatitis in young turkeys. *Journal of Applied Poultry Research*. 21(4):856–867.

Jankowski, J., M. Kubinska, J. Juskiewicz, A. Czech, K. Ognik, and Z. Zduńczyk. 2017. Effect of different dietary methionine levels on the growth performance and tissue redox parameters of turkeys. *Poultry Science*. 96(5):1235–1243.

Jankowski, J., K. Koslowski, K. Ognik, K. Otowski, J. Juśkiewicz, and Z. Zduńczyk. 2019a. The effect of the dietary inclusion levels and sources of zinc on the performance, metabolism, redox and immune status of turkeys. *Journal of Animal Feed Science and Technology*. 252:103–114.

Jankowski, J., K. Ognik, A. Stępniowska, Z. Zduńczyk, and K. Kozlowski. 2019b. The effect of the source and dose of manganese on the performance, digestibility and distribution of selected minerals, redox, and immune status of turkeys. *Poultry Science*. 98(3):1379–1389.

Jin, S., and J.L. Sell. 2001. Dietary vitamin K_1 requirement and comparison of biopotency of different vitamin K sources for young turkeys. *Poultry Science*. 80(5):615–620.

Kamyab, A., and J.D. Firman. 1999. Starter period digestible valine requirement of female Nicholas poults. *Journal of Applied Poultry Research*. 8(3):339–344.

Kamyab, A., and J.D. Firman. 2000. Digestible threonine requirement of female Nicholas poults during the starter period. *Journal of Applied Poultry Research*. 9(1):62–65.

Kidd, M.T., P.R. Ferket, and J.D. Garlich. 1998. Dietary threonine responses in growing turkey toms. *Poultry Science*. 77(10):1550–1555.

Klasing, K. 2005. Poultry nutrition: A comparative approach. *Journal of Applied Poultry Research*. 14(2):426–436.

Korschgen, L.J. 1967. Chapter 7: Feeding habits and foods. In *The Wild Turkey*. O.H. Hewitt, editor. Bethesda, MD: The Wildlife Society.

Krueger, K.K., J.A. Owen, C.E. Krueger, and T.M. Ferguson. 1978. Effect of feed and light regimes during the growing period on subsequent reproductive performance of Broad Breasted White turkeys fed two protein levels. *Poultry Science*. 57(1):27–37.

Kubinska, M., J. Jankowski, J. Juskiewicz, K. Ognik, A. Czech, J. Celej, and Z. Zduńczyk. 2016. Growth rate and metabolic parameters in young turkeys fed diets with different inclusion levels of methionine. *Journal of Animal Feed Science*. 25(2):152–159.

Lai, H., and G. Duke. 1978. Colonic motility in domestic turkeys. *The American Journal of Digestive Diseases*. 23:673–681.

Ledoux, D.R., K. Zyla, and T.L. Veum. 1995. Substitution of phytase for inorganic phosphorus for turkey hens. *Journal of Applied Poultry Research*. 4(2):157–163.

Leeson, S., and J.D. Summers. 2005. *Commercial Poultry Nutrition*, 3rd ed.: Nottingham, England: Nottingham University Press.

Lehmann, D., M. Pack, and H. Jeroch. 1996. Responses of growing and finishing turkey toms to dietary lysine. *Poultry Science*. 75(6):711–718.

Lehmann, D., M. Pack, and H. Jeroch. 1997. Effects of dietary threonine in starting, growing, and finishing turkey toms. *Poultry Science*. 76(5): 696–702.

Lemme, A., E. Strobel, D. Hoehler, W. Matzke, M. Pack, and H. Jeroch. 2002. Impact of graded levels of dietary lysine on performance in turkey toms 5 to 8 and 13 to 16 weeks of age. *Archiv für Geflügelkunde*. 66(3):102–107.

Leopold, A. 1933. *Game Management*. New York: Charles Scribner's Sons.

Leshchinsky, T.V., and K.C. Klasing. 2001. Relationship between the level of dietary vitamin E and the immune response of broiler chickens. *Poultry Science*. 80(11):1590–1599.

Lilburn, M., and G. Barbour. 2001. The threonine requirement of starter male turkey poults. *Lebanese Science Journal*. 2(2):25–33.

Linares, L.B., R.A. Murarolli, E.A. Guaiume, R.B. Shirley, D. Hoehler, D.R. Ledoux, and J.D. Firman. 2012. Digestible lysine and threonine requirements of male turkeys from days 8 to 21 and days 29 to 42. *Journal of Applied Poultry Research*. 21(2):384–398.

Madkour, F.A., and R.M. Kandyel. 2022. Histomorphology and ultrastructure of the proventriculus of the Broad Breasted White turkey (*Meleagris gallopavo*, Linnaeus 1758). *Microscopy Research and Technique*. 85(11):3514–3529.

Manin, A., E. Corona-M, M. Alexander, A. Craig, E.K. Thornton, D.Y. Yang, M. Richards, and C.F. Speller. 2018. Diversity of management strategies in Mesoamerican turkeys: Archaeological, isotopic and genetic evidence. *Royal Society Open Science*. 5(1):171613.

Marks, H.L. 1987. Water and feed intake, feed efficiency, and abdominal fat levels of dwarf and normal chickens selected under different water:feed ratio environments. *Poultry Science*. 66(12):1895–1900.

Marks, H.L., and G.M. Pesti. 1984. The roles of protein level and diet form in water consumption and abdominal fat pad deposition of broilers. *Poultry Science*. 63(8):1617–1625.

Marsden, S.J., G.S. McKee, and M.L. Crandall. 1965. Water deprival and replenishment in poults. *Poultry Science*. 44(3):793–797.

Mayne, R.K., R.W. Else, and P.M. Hocking. 2007. High dietary concentrations of biotin did not prevent foot pad dermatitis in growing turkeys and external scores were poor indicators of histopathological lesions. *British Poultry Science*. 48(3):291–298.

Medway, W., and M.R. Kare. 1959. Water metabolism of the growing domestic fowl with special reference to water balance. *Poultry Science*. 38(3):631–637.

Mohan, J., S.K. Sharma, G. Kolluri, and K. Dhama. 2018. History of artificial insemination in poultry, its components and significance. *World's Poultry Science Journal*. 74(3):475–488.

Moore, D.T., K. Baker, and J.D. Firman. 2001. Digestible sulfur amino acid requirements for male turkeys to five weeks of age. *Journal of Applied Poultry Research*. 10(4):363–370.

Moore, D.T., K. Baker, and J.D. Firman. 2004. Digestible sulfur amino acid requirement for male turkeys from six to twelve weeks of age. *Journal of Applied Poultry Research*. 13(2):155–162.

Moore, D.T., K. Baker, K. Thompson, E. Blair, and J.D. Firman. 2003. Digestible sulfur amino acid requirement of male turkeys during the 12 to 18 week period. *International Journal of Poultry Science*. 2(1):38–43.

Moran Jr., E.T. 1978. Performance and carcass quality of broiler tom turkeys subjected to a post-hatch fast and offered starting rations of different nutrient concentration. *Canadian Journal of Animal Science*. 58(2):233–243.

Morehouse, N.F. 1949. Water consumption in growing turkeys. *Poultry Science*. 28(1):152–153.

Mosby, H.S., and C.O. Handley. 1943. *The Wild Turkey in Virginia: Its Status, Life History and Management*. Richmond, VA: Division of Game and Inland Fisheries.

Mossab, A., J.M. Hallouis, and M. Lessire. 2000. Utilization of soybean oil and tallow in young turkeys compared with young chickens. *Poultry Science*. 79(9):1326–1331.

Mueller, A.S., J. Fischer, E. Most, and J. Pallauf. 2009. Investigation into selenium requirement of growing turkeys offered a diet supplemented with two levels of vitamin E. *Journal of Animal Physiology and Animal Nutrition*. 93(3):313–324.

NRC (National Research Council). 1994. *Nutrient Requirements of Poultry*, 9th rev. ed. Washington, DC: National Academy Press.

Ognik, K., and T. Wertelecki. 2012. Effect of different vitamin E sources and levels on selected oxidative status indices in blood and tissues as well as on rearing performance of slaughter turkey hens. *Journal of Applied Poultry Research*. 21(2):259–271.

Opoku, E.Y., H.L. Classen, and T.A. Scott. 2015. Evaluation of inclusion level of wheat distillers dried grains with solubles with and without protease or β-mannanase on performance and water intake of turkey hens. *Poultry Science*. 94(7):1600–1610.

Parker, J.T., M.A. Boone, and J.F. Knechtges. 1972. The effect of ambient temperature upon body temperature, feed consumption and water consumption using two varieties of turkeys. *Poultry Science*. 51(2):659–664.

Poholsky, C.M., L.S. Erb, A.M. Lyons, P. Rohlf, and J.W. Boney. 2023. Improving pellet quality enhances Nicholas Select turkey performance in targeted phases of production. *Journal of Applied Poultry Research*. 32(2):100340.

Qian, H., E.T. Kornegay, and D.M. Denbow. 1996. Phosphorus equivalence of microbial phytase in turkey diets as influenced by calcium to phosphorus ratios and phosphorus levels. *Poultry Science*. 75(1):69–81.

Ravindran, V., E.T. Kornegay, D.M. Denbow, Z. Yi, and R.M. Hulet. 1995. Response of turkey poults to tiered levels of Natuphos® phytase added to soybean meal-based semi-purified diets containing three levels of nonphytate phosphorus. *Poultry Science*. 74(11):1843–1854.

Roberson, K.D. 2004. Estimation of the dietary calcium and nonphytate phosphorus needs of growing commercial tom turkeys weighing four to twelve kilograms. *International Journal of Poultry Science*. 3(3):175–178.

Roberson, K.D. 2009. Growth performance and spontaneous bone fracture incidence of turkey toms fed various levels of calcium and nonphytate phosphorus to heavy market weight. *Journal of Applied Poultry Research*. 18(2):158–164.

Roberson, K.D., M.W. Klunzinger, and R.W. Charbeneau. 2004. Benefit of feeding dietary calcium and nonphytate phosphorus levels above NRC recommendations to tom turkeys in the growing-finishing phases. *Poultry Science*. 83(4):689–695.

Rodehutscord, M., and A. Dieckmann. 2005. Comparative studies with three-week-old chickens, turkeys, ducks, and quails on the response in phosphorus utilization to a supplementation of monobasic calcium phosphate. *Poultry Science*. 84(8):1252–1260.

Rumble, M.A., and S.H. Anderson. 1996. Feeding ecology of Merriam's turkeys (*Meleagris gallopavo merriami*) in the Black Hills, South Dakota. *The American Midland Naturalist*. 136(1):157–171.

Rush, J.K., C.R. Angel, K.M. Banks, K.L. Thompson, and T.J. Applegate. 2005. Effect of dietary calcium and vitamin D_3 on calcium and phosphorus retention in White Pekin ducklings. *Poultry Science*. 84(4):561–570.

Santos, F.B.O., A.A. Santos Jr., P.R. Ferket, and B.W. Sheldon. 2006. Influence of grain particle size and insoluble fiber content on salmonella colonization and shedding of turkeys fed corn-soybean meal diets. *International Journal of Poultry Science*. 5(8):731–739.

Schorger, A.W. 1966. *The Wild Turkey: Its History and Domestication*. Norman, OK: University of Oklahoma Press.

Scott, M.L. 1987. *Nutrition of the Turkey*. M.L. Scott of Ithaca, NY: Cornell University.

Scott, M.L., H.E. Butters, and G.O. Ranit. 1962. Studies on the requirement of young poults for available phosphorus. *Journal of Nutrition*. 78:223–230.

Sell, J., C. Angel, F. Piquer, E. Mallarino, and H. Al-Batshan. 1991. Developmental patterns of selected characteristics of the gastrointestinal tract of young turkeys. *Poultry Science*. 70(5):1200–1205.

Sell, J.L., P.R. Ferket, C.R. Angel, S.E. Scheideler, F. Escribano, and I. Zatari. 1989. Performance and carcass characteristics of turkey toms as influenced by dietary protein and metabolizable energy. *Nutrition Reports International*. 40:979–992.

Sell, J.L., R.J. Hasiak, and W.J. Owings. 1985. Independent effects of dietary metabolizable energy and protein concentrations on performance and carcass characteristics of tom turkeys. *Poultry Science*. 64(8):1527–1535.

Sell, J.L., A. Krogdahl, and N. Hanyu. 1986. Influence of age on utilization of supplemental fats by young turkeys. *Poultry Science*. 65(3):546–554.

Sklan, D., D. Melamed, and A. Friedman. 1995. The effect of varying dietary concentrations of vitamin A on immune response in the turkey. *British Poultry Science*. 36(3):385–392.

Sklan, D., A. Smirnov, and I. Plavnik. 2003. The effect of dietary fibre on the small intestines and apparent digestion in the turkey. *British Poultry Science*. 44(5):735–740.

Speller, C.F., B.M. Kemp, S.D. Wyatt, C. Monroe, W.D. Lipe, U.M. Arndt, and D.Y. Yang. 2010. Ancient mitochondrial DNA analysis reveals complexity of indigenous North American turkey domestication. *Proclamation of the National Academies of Sciences*. 107(7):2807–2812.

Taylor, R.M., and R.A. Sunde. 2016. Selenoprotein transcript level and enzyme activity as biomarkers for selenium status and selenium requirements in the turkey (*Meleagris gallopavo*). *PLoS One*. 11(3):e0151665.

Taylor, R.M., and R.A. Sunde. 2017. Selenium requirements based on muscle and kidney selenoprotein enzyme activity and transcript expression in the turkey poult (*Meleagris gallopavo*). *PloS One*. 12(11):e0189001.

Thompson, K.A., K.A. Baker, and J.D. Firman. 2005. Digestible lysine requirements of hen turkeys from 6 to 12 weeks of age. *International Journal of Poultry Science*. 4(9):639–644.

Thompson, K.A., E. Blair, K.A. Baker, and J.D. Firman. 2004. Digestible lysine requirement for hen turkeys from 0 to 6 weeks of age. *International Journal of Poultry Science*. 3(9):558–562.

Thornton, E.K. 2016. Introduction to the special issue—Turkey husbandry and domestication: Recent scientific advances. *Journal of Archaeological Science: Reports*. 10:514–519.

Turner, K.A., T.J. Applegate, and M.S. Lilburn. 1999. Effects of feeding high carbohydrate or high fat diets. 1. Growth and metabolic status of the posthatch poult following immediate or delayed access to feed. *Poultry Science*. 78(11):1573–1580.

van Staaveren, N., E.M. Leishman, S.M. Adams, B.J. Wood, A. Harlander-Matauschek, and C. F. Baes. 2020. Housing and management of turkey flocks in Canada. *Animals*. 10(7):1159.

Waibel, P.E., C.W. Carlson, J.A. Brannon, and S.L. Noll. 2000a. Identification of limiting amino acids in methionine- and lysine-supplemented low-protein diets for turkeys. *Poultry Science*. 79(9):1299–1305.

Waibel, P.E., C.W. Carlson, J.A. Brannon, and S.L. Noll. 2000b. Limiting amino acids after methionine and lysine with growing turkeys fed low-protein diets. *Poultry Science*. 7(9):1290–1298.

Waldroup, P.W., J.A. England, and M.T. Kidd. 1998. An evaluation of threonine requirements of young turkeys. *Poultry Science*. 77(7):1020–1023.

Waldroup, P.W., J.H. Kersey, and M.T. Kidd. 2002. Dietary interactions between threonine and crude protein in diets for growing tom turkeys 8 to 12 weeks of age. *International Journal of Poultry Science*. 1(4):74–77.

Wealleans, A.L., L.P. Barnard, L.F. Romero, and C. Kwakernaak. 2016. A value based approach to determine optimal phytase dose: A case study in turkey poults. *Journal of Animal Feed Science and Technology*. 216: 288–295.

Wu, D., E.D. Lewis, M. Pae, and S.N. Meydani. 2019. Nutritional modulation of immune function: Analysis and evidence, mechanisms, and clinical relevance. *Frontiers in Immunology*. 9:3160.

York, D.L., and S.D. Schemnitz. 2003. Home range, habitat use, and diet of Gould's turkeys, Peloncillo Mountains, New Mexico. Southwest. *National*. 48(2):231–240.

Zduńczyk, Z., and J. Jankowski. 2014. The effects of inclusion level and source of sodium in diets for growing turkeys: A review. *Journal of Animal Feed Science*. 23(1):3–12.

Zduńczyk, Z., J. Jankowski, D. Mikulski, B. Przybylska-Gornowicz, E. Sosnowska, and J. Juskiewicz. 2013. Gastrointestinal morphology and function in turkeys fed diets diluted with whole grain wheat. *Poultry Science*. 92(7):1799–1811.

17

Nutrient Requirements of Ducks

WORLD DUCK PRODUCTION AND THE MAJOR DUCK BREEDS AND THEIR USES

World Duck Meat and Egg Production

The global production of ducks and duck meat increased by over 14-fold between 1961 and 2019 (FAOSTAT, 2021) and represents ~4 percent of total poultry production (Guémené et al., 2012). According to FAOSTAT (2021), world production of duck meat exceeded 4.8 million tonnes in 2019, with over 3.3 billion head slaughtered. South and East Asia accounted for ~79 percent and ~74 percent of these figures, respectively, with China alone accounting for ~74 percent of the number of ducks slaughtered and ~67 percent of world duck meat production. The United States ranked 10th and 9th, respectively, with 27.5 million head slaughtered and 64,049 tonnes of duck meat produced in 2019 (FAOSTAT, 2021).

Asia is also the world's top producer of duck eggs (Arthur, 2017). Tao et al. (2017) stated that China, the world's leading chicken egg producing country, produced 30 billion duck eggs annually, and Hou (2016) reported that in 2015, total duck egg production in China was 4.22 million tonnes and accounted for approximately 15 to 20 percent of total poultry egg production. In agreement with Hou (2016), Arthur (2017) estimated that in China, 20 percent of table egg production is from ducks and quail. In many countries in Asia, fresh duck eggs are rarely eaten directly by consumers due to fishy off-flavors, which have been attributed to the high levels of yolk trimethylamine under even "normal dietary conditions" (Li et al., 2018). Thus, duck eggs are typically processed as "balut," "penoy," salted, and "century" eggs (Farrell, 1995; Arthur, 2017; Benjakul and Kaewmanee, 2017).

Origin of the Domestic Breeds

All ducks are classified as class *Aves*, order *Anseriformes*, and family *Anatidae* (Arthur, 2017). With the exception of the Muscovy (*Cairina moschata*), which is native to Central and South America, all of the economic breeds of ducks originated from the mallard (*Anas platyrhynchos*) (Qu et al., 2009; Li et al., 2010; Zhou et al., 2018). Additionally, several Chinese breeds have some genetic input from the closely related spot-billed duck (*Anas zonorhyncha*). In the wild, all of these closely related species are omnivores that are adapted to feeding in both aquatic and terrestrial habitats. Ducks were probably domesticated in southern China approximately 2,500 years ago (Hitosugi et al., 2007). The Pekin, which originated in China, was introduced to the United States in 1873 (Lee and Haynes, 1933; Jung and Zhou, 1980; Scott and Dean, 1991). Consequences of domestication include an increased body size and greater egg production, the loss of the ability to fly, and the almost complete cessation of brooding and nesting behavior, although domestic ducks have retained their flocking tendency on the ground (Hetzel, 1986). Today, the genetics of meat ducks are concentrated in only a few highly specialized companies (Thiele, 2016).

Ducks are raised for meat production, eggs, foie gras ("fat liver"), down, and feathers under rearing conditions that differ widely based on location, species and breeds, environment, cultural backgrounds, management systems, and production objectives (Guémené et al., 2012). In North America and Europe, ducks are primarily raised for meat, while in Asia, they are used for the production of both table eggs and meat.

Meat-Type Ducks

The absolute growth rate (g/d) of young White Pekin ducklings is greater than that of broiler chickens and turkeys of the same age (Rodehutscord, 2006). White Pekins and Muscovies, either as pure lines or crosses, are the world's preferred meat ducks, with the former accounting for ~90 percent of the global duck meat (Guémené et al., 2012). Historically, a two-phase feeding program is typically used in duck meat production, with a starter diet fed from hatch to 14 d of age and a grower diet fed until market age (Chen and Applegate, 2016). White Pekins reach their near-mature body weights at only 6 to 7 weeks of age, while Muscovy drakes are typically grown to 10 to 11 weeks, are leaner, and deposit relatively more breast muscle than Pekin drakes grown to a similar age (Scott and Dean, 1991; Guémené et al., 2012). In some parts of the world, particularly Taiwan, hybrid and infertile "Mule ducks" are produced by mating Muscovies to local domestic ducks (Ash, 1976; Scott and Dean, 1991). In France, male Mule ducks ("Mulard"), resulting from the mating of a male Muscovy duck with a female Pekin-type duck (Skippon, 2013), are also used for foie gras production. Because most ducks grown commercially in North America are White Pekins raised in intensive systems without water for swimming, and the major meat-type ducks in China are also primarily local or imported commercial strains of Pekins (Hou, 2016), this chapter will focus primarily on their nutrient requirements. There has been a paucity of studies on the nutrient requirements of other meat-type ducks (e.g., Muscovy and Mule) since the last revised edition of this report, so data from other breeds (or genetic crosses) of meat ducks are generally unavailable.

Since the first detailed report on White Pekin duck performance by Horton (1928), subsequent advances in genetics, nutrition, and health have led to major improvements in growth rate, feed conversion efficiency, breast meat deposition, carcass yield and leanness, and early feathering (Cherry and Morris, 2008). Recently, the average body weight at slaughter age (6 weeks) of intensively raised modern White Pekin strains is approximately 3.2 to 3.5 kg, with a feed conversion efficiency less than 2.0 kg feed/kg body weight, and carcass, breast muscle, and leg muscle yields of approximately 72 to 75 percent, 20 percent, and 12.5 percent, respectively (Huang et al., 2012; Chen and Applegate, 2016; Hou, 2016; Thiele, 2016). From 1928 to 1993, the amount of feed required to produce 1 kg of body weight was reduced by approximately 27 g annually (Cherry and Morris, 2008). In Pekins, the proportion of carcass breast muscle increases at a slower rate than leg muscle as the bird ages (Scott and Dean, 1991). Thus, there is a disadvantage associated with slaughtering ducks at too early an age because the breast is the most valuable part of the duck. On the other hand, it can be costly for the producer to extend the rearing period because of the large amount of feed required per kg of weight after growth rate has peaked (Scott and Dean, 1991), so economics most likely influence when the birds are marketed. Although carcass fat of Pekins has been reduced over the years through genetic selection programs, the meat of Muscovies and Mulards is leaner. Muscovy and Mule ducks grow at a slower rate than Pekins and are typically marketed at 10 to 11 weeks of age at approximately 3 kg. In addition, Muscovies are sexually dimorphic (Huang et al., 2012). In terms of body composition, Pekin and Muscovy females have greater carcass fat than males (Lin et al., 2018), mature sooner, and can be processed earlier than males in order to limit carcass fat deposition (Cherry and Morris, 2008). Unlike chickens, much of the duck's body fat is subcutaneous, and little is found in the abdominal fat pad.

Egg-Type Ducks

Among the highest egg producing breeds are Khaki Campbell and Indian Runner, both of which are prolific egg layers (>300 per year), although the Pekin also has a high level of egg production and is considered to be a multi-purpose breed (Ash, 1976; Arthur, 2017). Many Asian countries have their own indigenous breeds of egg-laying ducks, sometimes derived from crosses with Campbells and Runners (see Farrell [1995] and Arthur [2017] for more information). According to Hou (2016), typical egg production for five common Chinese breeds (Shaoxing, Jinding, Shanma, Youxian Sheldrake, and Jiangjiang) are approximately 280 to 320 eggs in 500 days, with a feed:egg mass ratio of 2.65:1 to 2.85:1. Although the Brown Tsaiya of Taiwan is a prolific egg-laying breed (Cheng et al., 2003), Tsaiya ducks are also bred with Pekins and Muscovies to produce three-way crossed Mule ducks used for meat (Farrell, 1995; Arthur, 2017).

Ducks normally begin laying eggs at approximately 16 weeks of age (Huang et al., 2012; Arthur, 2017). Similar to broiler breeders and as compared to ad libitum fed birds, restricted feeding of juvenile Pekin duck breeders from 3 to 20 weeks delays sexual maturity and, during the laying period, results in fewer small eggs, reduces multiple ovulations (double-yolked eggs) and mortality, increases rate of lay, and increases fertility and hatchability of eggs set (Olver, 1995; Leeson and Summers, 2005; Cherry and Morris, 2008).

There are some notable differences between chickens and ducks in terms of egg production and shell quality. As compared to chicken eggs, duck eggs are larger with thicker shells. Chen and Shen (1989) reported that Brown Tsaiya ducks retained more dietary calcium (Ca) and had higher plasma Ca concentrations than White Leghorn hens and suggested that these traits might facilitate the observed species differences in egg characteristics. In addition, the interval between consecutive ovipositions for Khaki Campbell ducks is about 24 hr (Shen, 1986; Farrell, 1995). Production of almost one egg per d for over 12 months has been documented in individual Khaki Campbell ducks, while flock averages in excess of 300 eggs per hen per year are common (Guémené et al., 2012). Modern meat strain breeders, such as the White Pekin, are less productive but reportedly yield ~235 ducklings in 52 weeks of production (Thiele, 2016) which, assuming an 80 percent hatchability rate, calculates to approximately 294 eggs.

Since the previous edition of this report (NRC, 1994), the primary literature on the nutrient needs of egg-type ducks have been mainly limited to various Chinese breeds. Two extensive reviews centered on the nutrient requirements of egg-laying ducks were published more than 25 years ago and focused on Asian breeds (Shen, 1986; Farrell, 1995), while more recently published nutrient requirement tables for Longyan ducks (Fouad et al., 2018b) and Pekin breeders and dual-purpose breeders (Ministry of Agriculture of China, 2012) lacked specific documentation for each nutrient requirement. Therefore, there is little basis upon which to revise the nutrient requirements of Pekin breeders from the previous NRC (1994) report. However, for interested readers, references for nutrient requirement studies of egg-type ducks of various breeds conducted since September 1991 (when the literature review for the 1994 NRC report was concluded) have been included in Table 17-3.

Management Systems for Duck Production

Ducks can be successfully raised without access to water for swimming. Duck production systems can be generally classified into intensive or extensive and are described in detail in Farrell and Stapleton (1986), Scott and Dean (1991), Cherry and Morris (2008), Guémené et al. (2012), and Onbasilar and Yalcin (2018). Intensive systems, common in the Western world, employ total confinement in insulated houses with controlled ventilation or naturally ventilated insulated houses with sidewall curtains that can be raised or lowered based on weather conditions (Pyle, 2017). Extensive duck rearing, which has endured for centuries in Asia, includes open ponds with sun shelters coupled with fish farming, as well as traditional herding systems in rice-producing wet lowlands (Scott and Dean, 1991; Men et al., 2002a,b). Whole grains of rice, snails, insects, leaf materials, crabs, and frogs were among the most common crop contents of herded ducks in Indonesia (Evans and Setioko, 1986). The quality of the meat from spent layer ducks is problematic due to "fishy taint" (Bird, 1986), similar to that noted in duck eggs and due to the presence of trimethylamine (Li et al., 2018).

DIGESTIVE ANATOMY AND PHYSIOLOGY OF DUCKS

Wild mallards are omnivorous and consume a very wide variety of foods including plankton, vegetation, seeds, fish, mollusks, and invertebrates that they find in aquatic environments and terrestrial environments. This flexible foraging strategy results in an exceptionally wide geographic distribution and population size. Adult mallards consume a greater proportion of plant-based foods compared to animal-based foods, except for breeding females, which consume considerably more of the latter (Swanson et al., 1985). With a few important exceptions, the anatomy and physiology of the digestive system of domestic ducks and chickens are similar and have been described by Koch (1973), Crompton and Nesheim (1976), Herd (1985), Elkin (1987), Adeola (2006), Gussekloo (2006), and Denbow (2015). The bill and esophagus of domestic ducks diverge in physiologically important ways from those of chickens. The bill of mallards is designed for filtering large plankton particles from water and mud when feeding in aquatic environments. It is also designed for grasping vegetation, seeds, and other foods on land. Filtration occurs via the tongue and bill acting as a suction pressure pump and sieving apparatus (Zweers, 1973). Movement of the tongue causes water to be drawn into the oral cavity through the bill tips and then forced out through the lateral edges of the bill. The upper and lower bills have interdigitating serrations (lamellae) on the sides that form a sieve that strains out potential food items. Papillae on the lateral edges of the tongue (Skieresz-Szewczyk and Jackowiak, 2016) fine tune the mesh size of the sieve to optimize food retention and also collect the food and transport it posteriorly for swallowing (Zweers et al., 1977). Taste and mechanoreceptors receptors on the lateral edges of the bill and the tongue determine if the strained particles are swallowed or if they are rejected (Berkhoudt, 1978; Dubbeldam et al., 1979). The tips of the upper and lower bills have a nail-like structure called the lingual nail, which is a keratinized projection similar to human fingernails that facilitates grasping food items on land such as seeds, vegetation, and insects and other small animals (Skieresz-Szewczyk and Jackowiak, 2016). Associated with the lingual nail are taste buds that guide food selection (Berkhoudt, 1978). Mallards are also capable of grazing grasses and forbs using the lateral rims of their bill working with their tongue to grasp the stems and break them off (Skieresz-Szewczyk and Jackowiak, 2016).

Wild Muscovy ducks are omnivorous and have a generalized feeding pattern similar to mallards, though they are larger and also consume small vertebrates. Although there have been few studies on the functional anatomy of their digestive system, the bill and tongue are anatomically similar to that described for mallards (Igwebuike and Anagor, 2013).

In contrast to chickens, domestic ducks have a spindle-shaped widening of the esophagus rather than a crop per se, and their proventriculus is cylindrical and of uniform diameter (i.e., not fusiform; Das et al., 1965). The former characteristic is most likely responsible for a faster rate of passage of ingesta in ducks compared to chickens of similar ages fed identical diets (Pisharody and Nair, 1972). Contractions of the thoracic esophagus and the glandular stomach also are more active in Pekin ducks than in chickens (Pastea et al., 1968). In addition, the glandular system of the Pekin duck's proventriculus is not as extensive as the chicken's, papillae ducting gastric juice to the lumen are less prominent, and the venous blood from the proventriculus returns directly to the vena cava rather than entering the portal system (Moran, 1986).

A feature of marine birds, including domestic ducks, is that they have a salt gland or nasal gland that is capable of eliminating sodium chloride extra-renally as a highly concentrated solution following the ingestion of seawater or feed with a high salt content (Schmidt-Nielsen and Kim, 1964; Hammel et al., 1980). Domestic ducks have salt glands of intermediate size whose development depends upon exposure to sodium chloride; in the absence of acute osmotic stress, the salt gland produces no fluid. Moreover, newly hatched Pekin ducklings are quite intolerant to salt solutions and must be gradually adapted to increasing concentrations in drinking solutions (Schmidt-Nielsen and Kim, 1964).

In a study with male mallards, White Pekins, Muscovies, and a Muscovy × White Pekin cross, Gille et al. (1999) reported that in all strains, the growth rate of all digestive organs, except for the esophagus, exhibited accelerated post-hatch growth compared to whole-body growth. Watkins et al. (2004) observed that the greatest intestinal growth in ducks occurred during the first 2 to 3 weeks post hatch, but following 5 weeks of rapid growth, the small intestines of Pekin ducks exhibited a 38 percent reduction in absolute mass.

In a comparative study of domesticated ducks (commercial strain of White Pekins) with their wild ancestor, the mallard (*Anas platyrhynchos*), Watkins et al. (2004) observed that Pekins were larger, grew faster, and had a larger gastrointestinal tract than mallards. However, while the process of domestication induced major changes in the absolute characteristics of the gastrointestinal tract, it did not have the same effect on the relative characteristics, in that the relative masses of the small intestine and gastrointestinal tract as a whole were 45 percent and 66 percent greater, respectively, in mallards than in Pekin ducks from 3 to 5 weeks of age. In contrast, as compared to mallards, Pekins had higher digestive enzyme activities and greater absolute and relative absorptive surface areas throughout development, which Watkins et al. (2004) suggested was compensatory for the failure of the relative masses of digestive organs in domesticated ducks to accompany the increased body mass and growth rate.

Wasilewski et al. (2015) evaluated the effect of genotype and sex on the digestive system of SM3 Heavy Pekin ducks (Cherry Valley) and AF51, a Polish strain. Although the SM3 Heavy ducks had greater body, gizzard, and liver weights, the AF51 strain had greater ratios of intestinal length/body length and intestinal length/trunk length. Moreover, with the exception of gizzard weight, which was greater in the SM3 Heavy males versus females, morphometric characteristics of the intestine and other digestive organs did not significantly differ between sexes. Although males had greater 49-d body weights than females, the lack of differences between sexes in relative intestinal length or digestive organ weights, with the exception of heavier gizzards in males, was confirmed in 49-d-old SM3 Heavy strain Pekin ducks by Steczny et al. (2017). A report by Qureshi et al. (2017) concluded that there was no significant effect of sex on growth and the maturation of the gastrointestinal tract, liver, pancreas, and gall bladder in White Pekin ducks from 30 d of age to >3 yr of age.

Jamroz et al. (2001) compared the development of the gastrointestinal tract in Shaver-Starbro hybrid chickens, Mulard ducks (Pekin × Muscovy cross), and Landaise geese all fed the same diets containing 400 g/kg barley (starter, 0 to 21 d; grower, 22 to 42 d). They observed that at 21 d of age, in relation to metabolic live weight ($kg^{0.67}$), the lengths of the small intestines in chickens and geese were greater than in ducks but not different among species at 42 d of age. Absolute and relative intestinal weights at both 21 and 42 d of age were smaller in ducks compared to chickens and geese. They suggested that as ducks grow, intestinal tissues represent a lesser percentage of total body tissues, which confirmed the findings of King et al., 2000a.

In a follow-up study, Jamroz et al. (2002) investigated the comparative development of digestive activity in chickens, ducks, and geese. In contrast to their previous report (Jamroz et al., 2001), separate starter and grower diets were formulated to meet the nutrient requirements of each species from 0 to 21 d and from 22 to 42 d of age, respectively. It was concluded that, as compared to ducks and geese of the same age, chickens exhibited an earlier maturing gastric and intestinal function, as well as pancreatic activity, which suggested that they may be fed more intensively at a younger age than waterfowl. However, somewhat paradoxically, the same group reported that chickens appeared to resorb yolk sac contents less rapidly than geese or ducks (Jamroz et al., 2004).

Turkey poults and White Pekin ducklings have similar incubation periods, hatch from eggs of similar weights, and have similar body weights at hatch (Applegate et al., 2005). In a comparative study of intestinal development in commercial turkey poults and White Pekin ducklings, Applegate et al. (2005) found that although the yolk-free body weight of poults was greater than ducklings at hatch, the opposite was true at 1 d of age, and by 7 d of age, ducklings were 140 g heavier. This species difference was primarily reflected in duodenal and jejunal weights, lengths, and villus heights, which apparently allowed for enhanced nutrient use and increased growth during the first week post-hatch. Applegate et al. (2005) concluded that, as compared to turkey poults, the small intestine of White Pekin ducklings develops more rapidly after hatching, based on ileal and jejunal weight, length, and villus height measurements.

NUTRIENT REQUIREMENTS

The review of literature used to compile the nutrient requirement recommendations in the 9th revised edition of this report was completed by the committee in September 1991 (NRC, 1994). Therefore, revisions in the recommended nutrient requirements of ducks in the present report are based primarily on peer-reviewed publications that appeared after that time. Moreover, the reader is reminded that the recommended nutrient requirements may vary due to duck strain (genetics), husbandry practices, and environmental conditions to which the birds are exposed (Adeola, 2006).

In determining empirical nutrient requirements, data were included if the strain and age of duck were identified, there was no dietary confounding, and at least three dietary levels of the nutrient under study were employed (the lone exception was Chen et al. [2016], who employed two dietary levels of crude protein). It was also required that analyzed dietary contents be reported for the nutrient(s) under study; the lone exceptions were some of the energy requirement studies. Lastly, the majority of the meat-type duck studies published since 1991 have centered on male White Pekins, with very few reports on female Pekins, Muscovy, or Mule ducks. Therefore, in contrast to the previous NRC (1994) report, the data presented in Tables 17-1 and 17-2 were derived primarily from male White Pekin ducks. However, some nutrient requirement studies with Muscovy, Mule, and other meat-type ducks have

TABLE 17-1 Minimum Nutrient Requirements of White Pekin Ducks as Percentages or Units per Kilogram of Diet (90% Dry Matter)[a,b]

Nutrient	Unit	0 to 2 Weeks; 2,900[c]	2 to 7 Weeks; 3,000[c]	Breeding; 2,900[c]
Protein and amino acids[d]				
Protein[e]	%	21	17	***15***
Arginine	%	***1.05***	***0.90***	?
Isoleucine	%	***0.63***	***0.46***	***0.38***
Leucine	%	***1.26***	***0.91***	***0.76***
Lysine	%	1.05	0.90	***0.60***
Methionine	%	0.52	0.43	***0.27***
Methionine + cyst(e)ine	%	0.85	0.71	***0.50***
Threonine	%	0.75	0.70	?
Tryptophan	%	0.20	***0.17***	***0.14***
Valine	%	***0.82***	***0.59***	***0.47***
Macrominerals				
Calcium	%	***0.95***	***0.85***	***2.75***
Chlorine	%	0.12	***0.12***	***0.12***
Magnesium	mg	***500***	***500***	***500***
Non-phytate phosphorus	%	***0.47***	***0.42***	?
Sodium	%	0.15	***0.15***	***0.15***
Trace minerals				
Copper	mg	***8***	***12***	?
Manganese	mg	***50***	?	?
Selenium	mg	***0.25***	***0.25***	***0.20***
Zinc	mg	***60***	***60***	?
Fat-soluble vitamins				
A	IU	***2,500***	***2,500***	***4,000***
D_3	IU	***1,500***	***1,500***	***900***
E	IU	***15***	***15***	***10***
K	mg	***0.5***	***0.5***	***0.5***
Water-soluble vitamins				
B_{12}	mg	***0.02***	***0.02***	?
Biotin	mg	***0.2***	***0.2***	?
Choline	mg	***1,000***	***1,000***	?
Niacin	mg	***55***	***55***	***55***
Pantothenic acid	mg	11	11	***11***
Pyridoxine	mg	***4***	***4***	***3***
Riboflavin	mg	4	***4***	***4***

[a] Values in normal font are minimum nutrient requirement recommendations based on empirical studies summarized in Table 17-3. Values in bold italics are guidance levels based on estimates due to insufficient experimental data (see front matter for details). Question marks indicate no values are available.

[b] Values in shaded areas are the NRC (1994) recommendations, for which no new studies are available.

[c] These are typical dietary energy concentrations used in nutrient requirement studies and expressed as kcal ME_n/kg diet.

[d] Amino acid values are on a total basis.

[e] Ducks do not have a requirement for crude protein per se. However, there should be sufficient crude protein to ensure an adequate amino nitrogen supply for synthesis of nonessential amino acids. The recommended crude protein values shown in this table were determined with corn-soybean meal–based diets formulated on a total amino acid basis. When synthetic amino acids, other feedstuffs, and by-products are included in the diets, or when formulating diets on a digestible amino acid basis, the crude protein values attained may differ.

TABLE 17-2 Approximate Body Weights and Cumulative Feed Consumption of White Pekin Ducks to 7 Weeks of Age

Age (Weeks)	Body Weight (kg)	Cumulative Feed Consumption (kg)
0	0.06	0.00
1	0.30	0.25
2	0.82	1.05
3	1.51	2.42
4	2.31	4.10
5	3.03	6.04
6	3.49	7.42
7	3.89	9.25

been included in Table 17-3 for interested readers but will be discussed sparingly, if at all, in the body of this chapter.

With regard to laying ducks, there was a paucity of Pekin breeder studies published since 1991, and virtually all of the egg-type duck studies utilized various Chinese breeds, so it was not possible to update the nutrient requirements of breeding Pekin ducks presented in the previous edition of this report. However, nutrient requirement information from studies with various breeder ducks and egg-type strains were included in Table 17-3. Interested readers are also referred to the Ministry of Agriculture of China (2012) Nutrient Requirements publication (which covers Pekin breeders), the review by Fouad et al. (2018b) for Longyan laying ducks, and various commercial breeder management guides.

With the exception of energy, the values recommended in Table 17-1 are minimum concentrations in the diet needed to support optimal performance and to prevent signs of deficiencies. These values do not include a margin of safety to account for environmental, toxicological, or disease stresses. They also assume that the dietary ingredients and supplements provide nutrients of high bioavailability. The terminology used to express recommendations for minimal dietary nutrient levels is described in detail in the front matter of this report and encompasses three categories:

1. **Minimum nutrient requirements:** Recommendations that are based on results from multiple feeding studies that meet the committee's inclusion criteria for a requirement study described above. The response criteria of these studies were based on the dietary level that resulted in optimal performance and avoidance of signs of deficiency, whichever is lowest. There is a high degree of confidence that these recommendations are accurate under conditions under which the experiments were done. Values appear in normal font in Table 17-1.
2. **Estimated nutrient requirements:** Recommendations derived from mathematical models that are based on results from empirical studies, such as those characterized previously, and data on the composition and

TABLE 17-3 Documentation of Nutrient Requirements of Ducks[a]

Nutrient and Estimated Requirement	Age Period (Days)	Response Criteria	Weight (g), at Start–End of the Study[b]	Breed	Number of Levels (Range)	Reference	Comments
AME,[c] kcal/kg							
2,700	0–14	Growth, feed efficiency	53–644	Cherry Valley Pekin	3 (2,700–3,200)	Sritiawthai et al., 2013	Dietary CP was 18%
3,018	0–21	Feed efficiency	Nr[d]-1,319	White Pekin	5 (2,450–3,050)	Xie et al., 2010	Dietary CP was 20.5%
2,953	0–21	Growth	Nr-1,147	Korean native duck	8 (2,600–3,300)	Wickramasuriya et al., 2016	Dietary CP was 18%
2,950	0–21	Feed efficiency	Nr-1,147	Korean native duck	8 (2,600–3,300)	Wickramasuriya et al., 2016	Dietary CP was 18%
2,900	1–28	Growth, feed efficiency	Nr-506	Muscovy	3 (2,700–2,900)	Ali and Sarker, 1992	Dietary CP was 20%
3,300	15–35	Growth, feed efficiency	810–3,080	White Pekin	3 (2,820–3,300)	Zeng et al., 2015a	Dietary CP was 19%
3,008	14–42	Growth	Nr-1,957	White Pekin	6 (2,600–3,100)	Fan et al., 2008	Dietary CP was 18%
3,030	14–42	Feed efficiency	Nr-1,957	White Pekin	6 (2,600–3,100)	Fan et al., 2008	Dietary CP was 18%
2,900	28–84	Growth, feed efficiency	506–1,964	Muscovy	3 (2,700–2,900)	Ali and Sarker, 1992	Dietary CP was 18%
2,700	0–28	Growth, feed efficiency	59–635	Khaki Campbell × Thai	5 (2,700–2,900)	Thongwittaya et al., 1992	Dietary CP was 17%
2,700	29–126	Growth, feed efficiency	590–1,413	Khaki Campbell × Thai	5 (2,700–2,900)	Thongwittaya et al., 1992	Dietary CP was 15.5%
2,600	112–126	Egg production, egg weight, egg mass, feed efficiency		Longyan	3 (2,400–2,600)	Xia et al., 2014	Dietary CP was 17.1% for the early laying period
2,600	127–274	Egg production, egg weight, egg mass, feed efficiency		Longyan	3 (2,400–2,600)	Xia et al., 2014	Dietary CP was 17.1% for the peak laying period
2,500	274–329	Egg production, egg weight, egg mass, feed efficiency		Longyan	3 (2,400–2,600)	Xia et al., 2014	Dietary CP was 18.3% for the late laying period
Protein, %							
18	0–14	Growth, feed efficiency	53–644	Cherry Valley Pekin	3 (17–22)	Sritiawthai et al., 2013	Dietary ME was 2,700 kcal/kg
24	0–14	Growth, feed efficiency breast muscle weight	Nr-622	White Pekin	2 (20–24)	Chen et al., 2016	Dietary ME was 3,100 kcal/kg
19.7	0–19	Feed efficiency	48–988	White Pekin	6 (16–21)	Xie et al., 2017b	Dietary ME was 2,880 kcal/kg
23.2	0–21	Growth	Nr-877	Mule	5 (15.4–25.3)	Baéza et al., 2012	Dietary ME was 2,897 kcal/kg
21.4	0–21	Feed efficiency	Nr-877	Mule	5 (15.4–25.3)	Baéza et al., 2012	Dietary ME was 2,897 kcal/kg
23.1	0–21	Breast muscle weight	Nr-877	Mule	5 (15.4–25.3)	Baéza et al., 2012	Dietary ME was 2,897 kcal/kg
23.5	0–21	Thigh muscle weight	Nr-877	Mule	5 (15.4–25.3)	Baéza et al., 2012	Dietary ME was 2,897 kcal/kg
20	1–28	Growth, feed efficiency	Nr-506	Muscovy	3 (17–22)	Ali and Sarker, 1992	Dietary ME was 2,800 kcal/kg
17.2	14–35	Growth	Nr-2,377	White Pekin	6 (13.5–17.2)	Xie et al., 2017a	Dietary ME was 2,932 kcal/kg

19	15–35	Growth, feed efficiency, carcass traits	810–3,080	White Pekin	3 (15–19)	Zeng et al., 2015a	Dietary AME_n was 3,300 kcal/kg
15.1	22–49	Growth	877–3,008	Mule	5 (10.8–18.7)	Baéza et al., 2012	Dietary ME was 2,897 kcal/kg
14.8	22–49	Feed efficiency	877–3,008	Mule	5 (10.8–18.7)	Baéza et al., 2012	Dietary ME was 2,897 kcal/kg
15.4	22–49	Breast muscle weight, thigh muscle weight	877–3,008	Mule	5 (10.8–18.7)	Baéza et al., 2012	Dietary ME was 2,897 kcal/kg
18	28–84	Growth, feed efficiency	382–1,964	Muscovy	3 (16–20)	Ali and Sarker, 1992	Dietary ME was 2,800 kcal/kg
13.6	50–70	Growth	2,702–3,830	Mule	5 (10.2–18.0)	Baéza et al., 2012	Dietary ME was 2,897 kcal/kg
13.8	50–70	Feed efficiency, breast muscle weight	2,702–3,830	Mule	5 (10.2–18.0)	Baéza et al., 2012	Dietary ME was 2,897 kcal/kg
12.5	50–70	Thigh muscle weight	2,702–3,830	Mule	5 (10.2–18.0)	Baéza et al., 2012	Dietary ME was 2,897 kcal/kg
17	0–28	Growth, feed efficiency	58–750	Khaki Campbell × Thai	4 (17.0–21.5)	Thongwittaya et al., 1992	Dietary ME was 2,700 kcal/kg
15.5	29–126	Growth, feed efficiency	738–1,485	Khaki Campbell × Thai	4 (15.5–20.0)	Thongwittaya et al., 1992	Dietary ME was 2,700 kcal/kg
16–18	0–28	Subsequent laying performance		Shaoxing	3 (12–22)	Chen and Jiang, 1998	Dietary ME was 2,751 kcal/kg
14–16	29–84	Subsequent laying performance		Shaoxing	3 (12–22)	Chen and Jiang, 1998	Dietary ME was 2,751 kcal/kg
12–13	85–126	Subsequent laying performance		Shaoxing	3 (12–22)	Chen and Jiang, 1998	Dietary ME was 2,751 kcal/kg
17.1	112–126	Egg production, egg weight, egg mass, feed efficiency		Longyan	3 (2,400–2,600)	Xia et al., 2014	Dietary ME was 2,600 kcal/kg for the early laying period
17.1	127–274	Egg production, egg weight, egg mass, feed efficiency		Longyan	3 (2,400–2,600)	Xia et al., 2014	Dietary ME was 2,500 kcal/kg for the peak laying period
18.3	274–329	Egg production, egg weight, egg mass, feed efficiency		Longyan	3 (2,400–2,600)	Xia et al., 2014	Dietary ME was 2,500 kcal/kg for the late laying period
Arginine, %							
0.95	1–21	Growth	Nr-1,116	White Pekin	10 (0.71–1.72)	Wang et al., 2013a	Basal diet contained 22.02% CP, 1.11% lysine, and 3,080 kcal ME/kg
1.16	1–21	Feed efficiency	Nr-1,116	White Pekin	10 (0.71–1.72)	Wang et al., 2013a	Basal diet contained 22.02% CP, 1.11% lysine, and 3,080 kcal ME/kg
0.99	1–21	Breast meat yield	Nr-1,116	White Pekin	10 (0.71–1.72)	Wang et al., 2013a	Basal diet contained 22.02% CP, 1.11% lysine, and 3,080 kcal ME/kg
0.85	15–35	Growth, feed efficiency	Nr-1,893	White Pekin	3 (0.60–1.70)	Wang et al., 2013b	Basal diet contained 18.51% CP, 0.85% lysine, and 3,170 kcal ME/kg
1.70	15–35	Breast and leg meat yields	Nr-1,893	White Pekin	3 (0.60–1.70)	Wang et al., 2013b	Basal diet contained 18.51% CP, 0.85% lysine, and 3,170 kcal ME/kg
0.66	119–217	Egg production		Longyan	5 (0.66–1.46)	Xia et al., 2017	Basal diet contained 17.0% CP, 0.70% lysine, and 2,486 kcal ME/kg
1.26	119–217	Egg yolk relative percentage		Longyan	5 (0.66–1.46)	Xia et al., 2017	Basal diet contained 17.0% CP, 0.70% lysine, and 2,486 kcal ME/kg

continued

TABLE 17-3 Continued

Nutrient and Estimated Requirement	Age Period (Days)	Response Criteria	Weight (g), at Start–End of the Study[b]	Breed	Number of Levels (Range)	Reference	Comments
Arginine, %							
1.46	119–217	Egg weight, shell percentage, shell thickness		Longyan	5 (0.66–1.46)	Xia et al., 2017	Basal diet contained 17.0% CP, 0.70% lysine, and 2,486 kcal ME/kg
Lysine, %							
1.10	0–14	Growth, feed efficiency	Nr-503	White Pekin	3 (0.95–1.25)	Wang et al., 2006	Basal diet contained 20.26% CP and 2,976 kcal ME/kg
1.10[c]	1–14	Growth	52–397	White Pekin	5 (0.80–1.12)	Zhou et al., 2017	Basal diet contained 20.1% CP and 2,910 kcal ME/kg
1.15[c]	1–14	Feed efficiency	51–397	White Pekin	5 (0.80–1.12)	Zhou et al., 2017	Basal diet contained 20.1% CP and 2,910 kcal ME/kg
1.12[c]	1–14	35-d Breast meat yield	51–397	White Pekin	5 (0.80–1.12)	Zhou et al., 2017	Basal diet contained 20.1% CP and 2,910 kcal ME/kg
1.17	0–21	Growth	Nr-1,313	White Pekin	6 (0.76–1.26)	Bons et al., 2002	Basal diet contained 22.4% CP, 1.10% arginine, and 3,028 kcal ME/kg
1.06	0–21	Feed efficiency	Nr-1,313	White Pekin	6 (0.76–1.26)	Bons et al., 2002	Basal diet contained 22.4% CP, 1.10% arginine, and 3,028 kcal ME/kg
0.94	0–21	Weight gain	Nr-1,197	White Pekin	6 (0.70–1.20)	Wen et al., 2017	Basal diet contained 19.6% CP and 2,750 kcal ME/kg
0.98	0–21	Weight gain	Nr-1,197	White Pekin	6 (0.70–1.20)	Wen et al., 2017	Basal diet contained 19.6% CP and 3,050 kcal ME/kg
0.90	0–21	Breast meat percentage	Nr-1,197	White Pekin	6 (0.70–1.20)	Wen et al., 2017	Basal diet contained 19.6% CP; 2,900 kcal/kg ME (averaged across basal diets)
1.10	0–21	Leg meat percentage	Nr-1,197	White Pekin	6 (0.70–1.20)	Wen et al., 2017	Basal diet contained 19.6% CP; 2,900 kcal/kg ME (averaged across basal diets)
0.84	7–21	Growth	213–1,241	White Pekin	5 (0.65–1.25)	Xie et al., 2009a	Basal diet contained 20.3% CP, 1.37% arginine, and 2,900 kcal ME/kg
0.90	7–21	Feed efficiency	213–1,241	White Pekin	5 (0.65–1.25)	Xie et al., 2009a	Basal diet contained 20.3% CP, 1.37% arginine, and 2,900 kcal ME/kg
0.97	7–21	Breast meat weight	213–1,241	White Pekin	5 (0.65–1.25)	Xie et al., 2009a	Basal diet contained 20.3% CP, 1.37% arginine, and 2,900 kcal ME/kg
0.98	7–21	Breast meat yield	213–1,241	White Pekin	5 (0.65–1.25)	Xie et al., 2009a	Basal diet contained 20.3% CP, 1.37% arginine, and 2,900 kcal ME/kg
0.99	1–35	Growth, feed efficiency	Nr-1,328	Campbell	4 (0.95–1.06)	Attia, 2003	Basal diet contained 18.6% CP and 2,844 kcal ME/kg
0.88[e]	15–35	Growth, breast meat yield	367–2,004	White Pekin	5 (0.60–0.92)	Zhou et al., 2017	Basal diet contained 16.5% CP and 2,940 kcal ME/kg
0.92[e]	15–35	Feed efficiency	367–2,004	White Pekin	5 (0.60–0.92)	Zhou et al., 2017	Basal diet contained 16.5% CP and 2,940 kcal ME/kg
0.87	22–49	Growth	733–2,419	White Pekin	6 (0.62–1.12)	Bons et al., 2002	Basal diet contained 21.5% CP, 1.02% arginine, and 3,179 kcal ME/kg
0.82	22–49	Feed efficiency	733–2,419	White Pekin	6 (0.62–1.12)	Bons et al., 2002	Basal diet contained 21.5% CP, 1.02% arginine, and 3,179 kcal ME/kg
1.00	22–49	Breast meat yield	733–2,419	White Pekin	6 (0.62–1.12)	Bons et al., 2002	Basal diet contained 21.5% CP, 1.02% arginine, and 3,179 kcal ME/kg

0.85	36–56	Growth, feed efficiency	1,325–2,189	Campbell	4 (0.73–0.94)	Attia, 2003	Basal diet contained 16.5% CP and 2,964 kcal ME/kg
0.80	154–266	Egg weight		Longyan	5 (0.75–0.95)	Fouad et al., 2018a	Basal diet contained 17% CP and 2,500 kcal ME/kg
Methionine, %							
0.46	0–14	Growth, feed efficiency	Nr-508	White Pekin	3 (0.26–0.66)	Wang et al., 2006	Basal diet contained 20.26% CP, 0.26% Met, and 2,976 kcal ME/kg; cyst(e)ine content not reported
0.53	1–9	Growth, feed efficiency	Nr-300	White Pekin	4 (0.30–0.83)	Prahkarnkaeo et al., 2017	Basal diet contained 21.9% CP, 0.30% Met, 0.35% Cys, and 2,849 kcal ME/kg
0.57	10–16	Growth, feed efficiency	204–864	White Pekin	4 (0.29–0.69)	Prahkarnkaeo et al., 2017	Basal diet contained 19.8% CP, 0.29% Met, 0.32% Cys, and 2,900 kcal ME/kg
0.48	0–21	Growth	Nr-1,004	White Pekin	5 (0.285–0.685)	Xie et al., 2004	Basal diet contained 20.1% CP, 0.29% Met, 0.33% Cys, and 2,943 kcal ME/kg
0.59	0–21	Feed efficiency	Nr-1,004	White Pekin	5 (0.285–0.685)	Xie et al., 2004	Basal diet contained 20.1% CP, 0.29% Met, 0.33% Cys, and 2,943 kcal ME/kg
0.47	15–35	Growth	792–3,234	White Pekin	5 (0.35–0.75)	Zeng et al., 2015b	Basal diet contained 18% CP, 0.30% Met, 0.25% Cys, and 3,196 kcal ME/kg
0.48	15–35	Feed efficiency, feather coverage	792–3,234	White Pekin	5 (0.35–0.75)	Zeng et al., 2015b	Basal diet contained 18% CP, 0.30% Met, 0.25% Cys, and 3,196 kcal ME/kg
0.41	15–35	Breast meat yield	792–3,234	White Pekin	5 (0.35–0.75)	Zeng et al., 2015b	Basal diet contained 18% CP, 0.30% Met, 0.25% Cys, and 3,196 kcal ME/kg
0.25	17–47	Growth, feed efficiency	681–3,329	White Pekin	4 (0.25–0.61)	Prahkarnkaeo et al., 2017	Basal diet contained 18.4% CP, 0.25% Met, 0.30% Cys, and 2,899 kcal ME/kg
0.38	21–49	Growth, breast meat yield	1,934[f]	White Pekin	6 (0.20–0.575)	Xie et al., 2006	Basal diet contained 17% CP, 0.20% Met, 0.31% Cys, and 2,909 kcal/kg
0.43	113–322	Egg weight		Longyan	6 (0.25–0.50)	Fouad et al., 2016a	Basal diet contained 17% CP, 2.6% Met, 2.7% Cys, and 2,500 kcal ME/kg
0.41	113–322	Egg mass, feed efficiency		Longyan	6 (0.25–0.50)	Fouad et al., 2016a	Basal diet contained 17% CP, 2.6% Met, 2.7% Cys, and 2,500 kcal ME/kg
0.46	113–322	Albumen weight		Longyan	6 (0.25–0.50)	Fouad et al., 2016a	Basal diet contained 17% CP, 2.7% Met, 2.6% Cys, and 2,500 kcal ME/kg
0.45	420–476	Egg production, feed efficiency		Shaoxin	5 (0.26–1.86)	He et al., 2003	Basal diet contained 17.5% CP, 2.6% Met, and 2,749 kcal ME/kg; cyst(e)ine content not reported
Methionine + cyst(e)ine, %							
0.90	1–9	Growth, feed efficiency	Nr-300	White Pekin	4 (0.67–1.08)	Prahkarnkaeo et al., 2017	Basal diet contained 21.9% CP, 0.30% Met, 0.35% Cys, and 2,849 kcal ME/kg
0.92	10–16	Growth, feed efficiency	204–864	White Pekin	4 (0.61–1.01)	Prahkarnkaeo et al., 2017	Basal diet contained 19.8% CP, 0.29% Met, 0.32% Cys, and 2,900 kcal ME/kg
0.81	0–21	Growth	Nr-1,025	White Pekin	4 (0.61–1.25)	Xie et al., 2004	Basal diet contained 20.1% CP, 0.29% Met, 0.33% Cys, and 2,943 kcal ME/kg
0.92	0–21	Feed efficiency	Nr-1,025	White Pekin	4 (0.61–1.25)	Xie et al., 2004	Basal diet contained 20.1% CP, 0.29% Met, 0.33% Cys, and 2,943 kcal ME/kg
0.73	15–35	Growth	792–3,234	White Pekin	5 (0.60–1.03)	Zeng et al., 2015b	Basal diet contained 18% CP, 0.30% Met, 0.25% Cys, and 3,196 kcal ME/kg
0.74	15–35	Feed efficiency, feather coverage	792–3,234	White Pekin	5 (0.60–1.03)	Zeng et al., 2015b	Basal diet contained 18% CP, 0.30% Met, 0.25% Cys, and 3,196 kcal ME/kg

continued

TABLE 17-3 Continued

Nutrient and Estimated Requirement	Age Period (Days)	Response Criteria	Weight (g), at Start–End of the Study[b]	Breed	Number of Levels (Range)	Reference	Comments
Methionine + cyst(e)ine, %							
0.67	15–35	Breast meat yield	792–3,234	White Pekin	5 (0.60–1.03)	Zeng et al., 2015b	Basal diet contained 18% CP, 0.30% Met, 0.25% Cys, and 3,196 kcal ME/kg
0.58	17–47	Growth, feed efficiency	681–3,329	White Pekin	4 (0.58–0.90)	Prahkarnkaeo et al., 2017	Basal diet contained 18.4% CP, 0.25% Met, 0.30% Cys, and 2,899 kcal ME/kg
0.69	21–49	Growth, breast meat yield	1,934[f]	White Pekin	6 (0.51–1.09)	Xie et al., 2006	Basal diet contained 17% CP, 0.20% Met, 0.31% Cys, and 2,909 kcal/kg
0.70	133–322	Egg weight		Longyan	6 (0.53–0.78)	Fouad, Li, et al., 2016a	Basal diet contained 17% CP, 2.7% Met, 2.6% Cys, and 2,500 kcal ME/kg
0.68	133–322	Egg mass, feed efficiency		Longyan	6 (0.53–0.78)	Fouad, Li, et al., 2016a	Basal diet contained 17% CP, 2.7% Met, 2.6% Cys, and 2,500 kcal ME/kg
0.76	133–322	Albumen weight		Longyan	6 (0.53–0.78)	Fouad, Li, et al., 2016a	Basal diet contained 17% CP, 2.7% Met, 2.6% Cys, and 2,500 kcal ME/kg
Threonine, %							
0.86	0–14	Weight gain	56–785	White Pekin	6 (0.60–1.09)	Zhang et al., 2016a	Basal diets contained 21.11% CP and either 3,050 or 3,053 kcal ME/kg; requirement averaged across experiments
0.94	0–14	Breast meat weight	56–785	White Pekin	6 (0.60–1.09)	Zhang et al., 2016a	Basal diets contained 21.11% CP and either 3,050 or 3,053 kcal ME/kg; requirement averaged across experiments
0.96	0–14	Breast meat yield	56–785	White Pekin	6 (0.60–1.09)	Zhang et al., 2016a	Basal diets contained 21.11% CP and either 3,050 or 3,053 kcal ME/kg; requirement averaged across experiments
0.93	0–14	Wet crude mucin	56–785	White Pekin	6 (0.60–1.09)	Zhang et al., 2016a	Basal diets contained 21.11% CP and either 3,050 or 3,053 kcal ME/kg; requirement averaged across experiments
0.92	0–14	Feed efficiency	56–785	White Pekin	6 (0.60–1.09)	Zhang et al., 2016a	Basal diet contained 21.11% CP and 3,053 kcal ME/kg
0.56	0–14	Growth	Nr-656	White Pekin	5 (0.36–0.61)	Jiang et al., 2018	Basal diet contained 16.13% CP and 2,904 kcal ME/kg
0.61	0–14	Growth	Nr-676	White Pekin	5 (0.41–0.69)	Jiang et al., 2018	Basal diet contained 17.65% CP and 2,902 kcal ME/kg
0.60	0–14	Growth	Nr-708	White Pekin	5 (0.46–0.74)	Jiang et al., 2018	Basal diet contained 18.18% CP and 2,904 kcal ME/kg
0.63	0–14	Growth	Nr-745	White Pekin	5 (0.52–0.80)	Jiang et al., 2018	Basal diet contained 20.81% CP and 2,901 kcal ME/kg
0.67	0–14	Growth	Nr-704	White Pekin	5 (0.59–0.867)	Jiang et al., 2018	Basal diet contained 22.29% CP and 2,900 kcal ME/kg
0.67	0–21	Growth	Nr-1,382	White Pekin	5 (0.50–0.82)	Xie et al., 2014c	Basal diet contained 19% CP and 3,021 kcal ME/kg
0.74	0–21	Feed efficiency, higher intestinal villus heights	Nr-1,382	White Pekin	5 (0.50–0.82)	Xie et al., 2014c	Basal diet contained 19% CP and 3,021 kcal ME/kg

0.56	0–21	Growth	Nr-1,243	White Pekin	6 (0.46–0.78)	Jiang et al., 2016	Basal diet contained 17.73% CP and 2,892 kcal ME/kg
0.61	0–21	Growth	Nr-1,317	White Pekin	6 (0.54–0.83)	Jiang et al., 2016	Basal diet contained 20.89% CP and 2,915 kcal ME/kg
0.79	0–21	Growth	Nr-889	White Pekin	5 (0.50–0.71)	Bi et al., 2018	Basal diet contained 18.12% CP 3,078 kcal ME/kg
0.77	0–21	Breast meat weight, leg meat weight	Nr-889	White Pekin	5 (0.50–0.71)	Bi et al., 2018	Basal diet contained 18.12% CP 3,078 kcal ME/kg
0.66	1–21	Growth	Nr-1,736	White Pekin	5 (0.46–0.74)	Jiang et al., 2017	Basal diet contained 18.18% CP and 2,900 kcal ME/kg
0.67	1–21	Breast muscle percentage	Nr-1,736	White Pekin	5 (0.46–0.74)	Jiang et al., 2017	Basal diet contained 18.18% CP and 2,900 kcal ME/kg
0.70	1–21	Growth	Nr-1,746	White Pekin	5 (0.59–0.87)	Jiang et al., 2017	Basal diet contained 22.29% CP and 2,900 kcal ME/kg
0.73	1–21	Breast muscle percentage	Nr-1,746	White Pekin	5 (0.59–0.87)	Jiang et al., 2017	Basal diet contained 22.29% CP and 2,900 kcal ME/kg
0.71	15–35	Growth	791–3,092	White Pekin	5 (0.55–0.85)	Zhang et al., 2014	Basal diets varied in CP from 16.39 to 18.46%; ME not reported
0.69	15–35	Feed efficiency, carcass weight, breast meat percentage	791–3,092	White Pekin	5 (0.55–0.85)	Zhang et al., 2014	Basal diets varied in CP from 16.39 to 18.46%; ME not reported
0.70	15–35	Breast meat weight	791–3,092	White Pekin	5 (0.55–0.85)	Zhang et al., 2014	Basal diets varied in CP from 16.39 to 18.46%; ME not reported
0.58	49–154	Growth	1,750–3,300	Cherry Valley (Pekin)	3 (0.58–0.78)	Halas et al., 2012	Basal diet contained 15% CP and 2,856 kcal ME/kg
0.57	119–315	Egg production, egg mass, feed efficiency		Longyan	6 (0.39–0.64)	Fouad et al., 2017	Basal diet contained 16% CP and 2,500 kcal ME/kg
0.58	119–315	Egg weight		Longyan	6 (0.39–0.64)	Fouad et al., 2017	Basal diet contained 16% CP and 2,500 kcal ME/kg
0.68	210–343	Egg production		Tsaiya	6 (0.50–1.00)	Shen and Li, 1997	Basal diet contained 17% CP and 2,750 kcal ME/kg
0.77	210–343	Feed efficiency		Tsaiya	6 (0.50–1.00)	Shen and Li, 1997	Basal diet contained 17% CP and 2,750 kcal ME/kg
Tryptophan, %							
<0.20	0–14	Growth, feed efficiency	Nr-493	White Pekin	3 (0.20–0.40)	Wang et al., 2006	Basal diet contained 20.26% CP, 30 mg niacin/kg, and 2,976 kcal ME/kg
0.17	0–21	Growth	Nr-1,092	White Pekin	5 (0.13–0.27)	Xie et al., 2014a	Diet contained 18.1% CP, 75 mg niacin/kg, and 3,047 kcal ME/kg
0.17	0–21	Growth	Nr-1,083	White Pekin	5 (0.13–0.27)	Xie et al., 2014a	Diet contained 18.1% CP, 115 mg niacin/kg, and 3,047 kcal ME/kg
0.17	0–21	Growth	Nr-1,092	White Pekin	5 (0.13–0.27)	Xie et al., 2014a	Diet contained 18.1% CP, 155 mg niacin/kg, and 3,047 kcal ME/kg
0.48[g]	22–42	Growth	1,220–2,630	White Pekin	4 (0.17–1.08)	Liu et al., 2015	Basal diet contained 17.01% CP, 50 mg niacin/kg, and 3,000 kcal ME/kg
0.78[g]	22–42	Feed efficiency	1,220–2,630	White Pekin	4 (0.17–1.08)	Liu et al., 2015	Basal diet contained 17.01% CP, 50 mg niacin/kg, and 3,000 kcal ME/kg

continued

TABLE 17-3 Continued

Nutrient and Estimated Requirement	Age Period (Days)	Response Criteria	Weight (g), at Start–End of the Study[b]	Breed	Number of Levels (Range)	Reference	Comments
Valine, %							
0.82	0–7	Growth	56–290	White Pekin	9 (0.68–1.27)	Timmler and Rodehutscord, 2003	Basal diet contained 18.3% CP and 2,988 kcal ME/kg
0.80	8–14	Growth	290–842	White Pekin	8 (0.71–1.27)	Timmler and Rodehutscord, 2003	Basal diet contained 18.3% CP and 2,988 kcal ME/kg
0.79	15–21	Growth	842–1,539	White Pekin	8 (0.71–1.27)	Timmler and Rodehutscord, 2003	Basal diet contained 18.3% CP and 2,988 kcal ME/kg
0.82	0–21	Feed efficiency	56–1,539	White Pekin	8 (0.71–1.27)	Timmler and Rodehutscord, 2003	Basal diet contained 18.3% CP and 2,988 kcal ME/kg
0.92	210–343	Egg production		Tsaiya	6 (0.60–1.10)	Shen et al., 1996	Basal diet contained 17.0% CP and 2,750 kcal ME/kg
0.89	210–343	Feed efficiency		Tsaiya	6 (0.60–1.10)	Shen et al., 1996	Basal diet contained 17.0% CP and 2,750 kcal ME/kg
Calcium, %							
0.81	0–14	Growth	Nr-434	White Pekin	5 (0.4–1.2) with variable nPP	Xie et al., 2009b	Basal diet contained 20.5% CP, 0.40% calcium, 0.20% non-phytate P, and 2,866 kcal ME/kg
0.80	0–14	Feed efficiency	Nr-434	White Pekin	5 (0.4–1.2) with variable nPP	Xie et al., 2009b	Basal diet contained 20.5% CP, 0.40% calcium, 0.20% non-phytate P, and 2,866 kcal ME/kg
0.95	7–17	Growth, feed efficiency	289–940	White Pekin	4 (0.6–1.2)	Rush et al., 2005	Basal diet contained 22% CP, 0.60% calcium, 0.40% non-phytate P, and 3,002 kcal ME/kg
0.85	7–17	Toe ash	289–940	White Pekin	4 (0.6–1.2)	Rush et al., 2005	Basal diet contained 22% CP, 0.60% calcium, 0.40% non-phytate P, and 3,002 kcal ME/kg
0.75	0–21	Growth, feed efficiency	Nr-1,250	Cherry Valley (Pekin)	4 (0.55–1.15) with variable P	Zhu et al., 2018	Basal diet contained 20.5% CP, 0.60% total P, 2,000 IU of vitamin D_3, and 2,895 kcal ME/kg
0.95	0–21	Serum calcium, tibia ash	Nr-1,250	Cherry Valley (Pekin)	4 (0.55–1.15) with variable P	Zhu et al., 2018	Basal diet contained 20.5% CP, 0.60% total P, 2,000 IU of vitamin D_3, and 2,895 kcal ME/kg
0.72	15–42	Growth	2,148[f]	White Pekin	4 (0.30–0.84) with variable nPP	Xie et al., 2009c	Base diet contained 18.60% CP, 0.30% Ca, 0.17% non-phytate P, and 2,940 kcal ME/kg
0.66	15–42	Tibia ash P content	2,148[f]	White Pekin	4 (0.30–0.84) with variable nPP	Xie et al., 2009c	Basal diet contained 18.60% CP, 0.30% Ca, 0.17% non-phytate P, and 2,940 kcal ME/kg
3.60	140–231	Egg production, egg quality, bone quality		Longyan	5 (2.80–3.40)	Xia et al., 2015	Basal diet contained 17.0% CP, 2.80% calcium, 0.40% non-phytate P, and 2,490 kcal ME/kg
Non-phytate phosphorus, %							
0.40	0–14	Growth	Nr-434	White Pekin	4 (0.2–0.5) with variable Ca	Xie et al., 2009b	Basal diet contained 20.5% CP, 0.40% calcium, 0.42% non-phytate P, and 2,866 kcal ME/kg
0.38	0–14	Feed efficiency	Nr-434	White Pekin	4 (0.2–0.5) with variable Ca	Xie et al., 2009b	Basal diet contained 20.5% CP, 0.40% calcium, 0.42% non-phytate P, and 2,866 kcal ME/kg

0.27	1–21	Growth	54–1,474	White Pekin	8 (0.41–0.86)	Rodehutscord et al., 2003	Basal diet contained 21.6% CP 0.56% calcium, and 0.41% non-phytate P
0.34	1–21	Phosphorus retention	54–1,474	White Pekin	8 (0.41–0.86)	Rodehutscord et al., 2003	Basal diet contained 21.6% CP 0.56% calcium, and 0.41% non-phytate P
0.37	15–42	Growth	2,148[f]	White Pekin	4 (0.17–0.41) with variable Ca	Xie et al., 2009c	Basal diet contained 18.60% CP, 0.30% Ca, 0.17% non-phytate P, and 2,940 kcal ME/kg
0.37	15–42	Tibia ash P content	2,148[f]	White Pekin	4 (0.17–0.41) with variable Ca	Xie et al., 2009c	Basal diet contained 18.60% CP, 0.30% Ca, 0.17% non-phytate P, and 2,940 kcal ME/kg
0.13	21–49	Growth	1,435–3,492	White Pekin	8 (0.38–0.95) with variable Ca	Rodehutscord et al., 2003	Basal diet contained ~19% CP 0.57–0.61% calcium, and 0.30–0.38% non-phytate P
0.23	21–49	Phosphorus retention	1,435–3,492	White Pekin	8 (0.38–0.95) with variable Ca	Rodehutscord et al., 2003	Basal diet contained ~19% CP 0.57–0.61% calcium, and 0.30–0.38% non-phytate P
Copper, mg/kg							
<7	1–28	Growth	Nr-858	White Pekin	5 (0–150)	Attia et al., 2012	Basal starter diet (1–35 d) contained 21.2% CP and 7.0 mg copper/kg and the basal grower-finisher diet (36–56 d) contained 17.36% CP and 7.0 mg copper/kg
11–19	29–56	Growth	817–1,816	White Pekin	5 (0–150)	Attia et al., 2012	Basal starter diet (1–35 d) contained 21.2% CP and 7.0 mg copper/kg and the basal grower-finisher diet (36–56 d) contained 17.36% CP and 7.0 mg copper/kg
11–15	29–56	Feed efficiency	817–1,816	White Pekin	5 (0–150)	Attia et al., 2012	Basal starter diet (1–35 d) contained 21.2% CP and 7.0 mg copper/kg and the basal grower-finisher diet (36–56 d) contained 17.36% CP and 7.0 mg copper/kg
4.6	119–315	Egg production, egg weight, and egg quality		Shanma	6 (4.6–28.6)	Fouad et al., 2016b	Basal diet contained 17.0% CP, 4.6 mg copper/kg, and 2,500 kcal ME/kg
Manganese, mg/kg							
19.1	119–252	Egg production, egg weight, and egg quality		Shanma	7 (18.1–109.1)	Fouad et al., 2016a	Basal diet contained 17.0% CP, 18.1 mg manganese/kg, and 2,500 kcal ME/kg
Selenium, mg/kg							
0.40	1–49	Growth	56–3,810	Cherry Valley (Pekin)	4 (0.04–0.64)	Baltić et al., 2015	Basal starter diet (0–14 d) contained 22.68% CP and 0.0381 mg selenium/kg, and the basal finisher diet (15–49 d) contained 17.39% CP and 0.0411 mg selenium/kg
0.15	203–238	Egg production		Linwu	5 (0.15–0.95)	Huang et al., 2017	Basal diet contained 18.0% CP, 0.15 mg selenium/kg, and 2,600 kcal ME/kg
0.55	203–238	Serum selenium and glutathione, glutathione peroxidase, and superoxide dismutase		Linwu	5 (0.15–0.95)	Huang et al., 2017	Basal diet contained 18.0% CP, 0.15 mg selenium/kg, and 2,600 kcal ME/kg

continued

TABLE 17-3 Continued

Nutrient and Estimated Requirement	Age Period (Days)	Response Criteria	Weight (g), at Start–End of the Study[b]	Breed	Number of Levels (Range)	Reference	Comments
Selenium, mg/kg							
0.18	189–245	Egg production		Shanma	6 (0.04–0.44)	Chen et al., 2015	Basal diet contained 17.0% CP, 0.042 mg selenium/kg, and 2,502 kcal ME/kg
0.21	189–245	Feed efficiency		Shanma	6 (0.04–0.44)	Chen et al., 2015	Basal diet contained 17.0% C), 0.042 mg selenium/kg, and 2,502 kcal ME/kg
0.24	246–322	Egg production		Shanma	6 (0.04–0.44)	Chen et al., 2015	Basal diet contained 17.0% C), 0.042 mg selenium/kg, and 2,502 kcal ME/kg
0.23	246–322	Feed efficiency		Shanma	6 (0.04–0.44)	Chen et al., 2015	Basal diet contained 17.0% C), 0.042 mg selenium/kg, and 2,502 kcal ME/kg
0.37	246–322	Plasma glutathione peroxidase-3		Shanma	6 (0.04–0.44)	Chen et al., 2015	Basal diet contained 17.0% C), 0.042 mg selenium/kg, and 2,502 kcal ME/kg
0.27	246–322	Erythrocyte glutathione peroxidase-1		Shanma	6 (0.04–0.44)	Chen et al., 2015	Basal diet contained 17.0% C), 0.042 mg selenium/kg, and 2,502 kcal ME/kg
0.41	246–322	Liver glutathione peroxidase-1		Shanma	6 (0.04–0.44)	Chen et al., 2015	Basal diet contained 17.0% C), 0.042 mg selenium/kg, and 2,502 kcal ME/kg
0.39	246–322	Liver glutathione peroxidase-1 mRNA expression		Shanma	6 (0.04–0.44)	Chen et al., 2015	Basal diet contained 17.0% C), 0.042 mg selenium/kg, and 2,502 kcal ME/kg
Zinc, mg/kg							
56	1–56	Growth	Nr-2,023	White Pekin	4 (26–146)	Attia et al., 2013	Basal diet contained 21.26% CP and 26.0 mg zinc/kg
84.4	161–171	Egg production, egg weight, egg quality, antioxidant capacity		Longyan	7 (37.4–127.4)	Chen et al., 2017	Basal diet contained 17.3% CP, 37.4 mg zinc/kg, and 2,510 kcal ME/kg
64.6	172–292	Egg production, egg weight, egg quality, antioxidant capacity		Longyan	7 (37.4–127.4)	Chen et al., 2017	Basal diet contained 17.3% CP, 37.4 mg zinc/kg, and 2,510 kcal ME/kg
Vitamin A, IU/kg							
<2,500	0–14	Growth	Nr-657	White Pekin (Peking)	5 (0–15,000)	Wei et al., 2008	Basal diet contained 19.8% CP and 2,808 kcal ME/kg
2,500	0–21	Weight gain, feed efficiency, intestinal villi height and crypt depth	Nr-1,261	White Pekin (Peking)	5 (0–15,000)	Wei et al., 2009	Basal diet contained 19.8% CP and 2,808 kcal ME/kg
2,606	0–21	Weight gain	Nr-1,046	White Pekin	8 (0–14,000)	Feng et al., 2019	Basal diet contained 21.3% CP and 2,850 kcal ME/kg
Vitamin D, IU/kg							
1,000	0–14	Growth	Nr-452	White Pekin	6 (0–3,000)	Wang et al., 2010	Basal diet contained 20.1% CP, 0.40% calcium, 0.20% non-phytate P, and 2,868 kcal ME/kg
2,000	0–14	Feed efficiency, plasma calcium, tibia ash calcium	Nr-452	White Pekin	6 (0–3,000)	Wang et al., 2010	Basal diet contained 20.1% CP, 0.40% calcium, 0.20% non-phytate P, and 2,868 kcal ME/kg

Vitamin E, IU/kg							
16 (12.4 mg/kg)	0–21	Growth, feed efficiency	Nr-1,205	White Pekin	6 (0–100 mg)	Xie et al., 2018	Basal diet contained 19.6% CP and 2,700 kcal ME/kg (Expt. 1)
14 (11.0 mg/kg)	0–21	Growth, feed efficiency	Nr-929	White Pekin	4 (0–100 mg)	Xie et al., 2018	Basal diet contained 19.0% CP and 2,750 kcal ME/kg (Expt. 2)
125 (100 mg/kg)	350–462	Egg production, fertility		Cherry Valley (Pekin)	3 (50–200 mg)	Lei et al., 2014	Basal diet contained 18% CP and 2,698 kcal ME/kg
250 (200 mg/kg)	350–462	Eggshell weight		Cherry Valley (Pekin)	3 (50–200 mg)	Lei et al., 2014	Basal diet contained 18% CP and 2,698 kcal ME/kg
Vitamin B_{12}, µg/kg							
20	1–49	Breast meat yield	Nr-4,008	White Pekin	4 (4–44)	Halle et al., 2011	Basal diet (starter) contained 19.9% CP, 2,957 kcal ME/kg, 48 µg/kg cobalt, and 4.0 µg/kg vitamin B_{12}; basal diet (grower) contained 17.2% CP, 3,016 kcal ME/kg, 39 µg/kg cobalt, and 4.0 µg/kg vitamin B_{12}
Biotin, mg/kg							
0.18	0–21	Growth	Nr-1,241	White Pekin	9 (0–0.21)	Zhu et al., 2012	Basal diet contained 19.84% CP and 2,882 kcal ME/kg
0.21	0–21	Minimal foot pad lesions	Nr-1,241	White Pekin	9 (0–0.21)	Zhu et al., 2012	Basal diet contained 19.84% CP and 2,882 kcal ME/kg
Choline, mg/kg							
810	0–21	Growth	Nr-993	White Pekin	8 (302–1,832)	Wen et al., 2014b	Basal diet contained 20.02% CP, 0.41% methionine, 302 mg/kg choline, and 3,148 kcal ME/kg
823	0–21	Feed intake	Nr-993	White Pekin	8 (302–1,832)	Wen et al., 2014b	Basal diet contained 20.02% CP, 0.41% methionine, 302 mg/kg choline, and 3,148 kcal ME/kg
1,182	0–21	Perosis prevention	Nr-993	White Pekin	8 (302–1,832)	Wen et al., 2014b	Basal diet contained 20.02% CP, 0.41% methionine, 302 mg/kg choline, and 3,148 kcal ME/kg
833	0–21	Liver lipid content	Nr-993	White Pekin	8 (302–1,832)	Wen et al., 2014b	Basal diet contained 20.02% CP, 0.41% methionine, 302 mg/kg choline, and 3,148 kcal ME/kg
1,311	0–21	Liver triglyceride content	Nr-993	White Pekin	8 (302–1,832)	Wen et al., 2014b	Basal diet contained 20.02% CP, 0.41% methionine, 302 mg/kg choline, and 3,148 kcal ME/kg
1,230	0–21	Liver phospholipid content	Nr-993	White Pekin	8 (302–1,832)	Wen et al., 2014b	Basal diet contained 20.02% CP, 0.41% methionine, 302 mg/kg choline, and 3,148 kcal ME/kg
980	21–42	Growth	1,063–2,769	White Pekin	6 (0–1,962)	Wen et al., 2014a	Basal diet contained 17.65% CP, 0.48% methionine, no choline, and 3,035 kcal ME/kg
950	21–42	Feed intake	1,063–2,769	White Pekin	6 (0–1,962)	Wen et al., 2014a	Basal diet contained 17.65% CP, 0.48% methionine, no choline, and 3,035 kcal ME/kg
1,130	21–42	Liver lipid content	1,063–2,769	White Pekin	6 (0–1,962)	Wen et al., 2014a	Basal diet contained 17.65% CP, 0.48% methionine, no choline, and 3,035 kcal ME/kg

continued

TABLE 17-3 Continued

Nutrient and Estimated Requirement	Age Period (Days)	Response Criteria	Weight (g), at Start–End of the Study[b]	Breed	Number of Levels (Range)	Reference	Comments
Choline, mg/kg							
946	7–28	Growth	132–617	White Pekin	5 (0–1,743)	Wen et al., 2016	Diet contained 22.12% CP, 0.48% methionine, and 3,480 kcal ME/kg
907	7–28	Feed intake	132–617	White Pekin	5 (0–1,743)	Wen et al., 2016	Diet contained 22.12% CP, 0.48% methionine, and 3,480 kcal ME/kg
Pantothenic acid, mg/kg							
10.05	0–21	Growth	53–1,120	White Pekin	7 (4.65–22.50)	Tang et al., 2020b	Basal diet contained 20.4% CP, 3,191 kcal ME/kg, and 4.65 mg/kg pantothenic acid
11.01	0–21	Feed efficiency	53–1,120	White Pekin	7 (4.65–22.50)	Tang et al., 2020b	Basal diet contained 20.4% CP, 3,191 kcal ME/kg, and 4.65 mg/kg pantothenic acid
11.24	0–21	Liver pantothenic acid content	53–1,120	White Pekin	7 (4.65–22.50)	Tang et al., 2020b	Basal diet contained 20.4% CP, 3,191 kcal ME/kg, and 4.65 mg/kg pantothenic acid
10.27	15–42	Growth	445–3,031	White Pekin	6 (4.52–14.61)	Tang et al., 2020a	Common starter diet contained 20.0% CP, 2,919 kcal ME/kg, and 18.90 mg/kg pantothenic acid; Basal grower diet contained 18.1% CP, 3,189 kcal ME/kg, and 4.52 mg/kg pantothenic acid
12.06	15–42	Plasma pantothenic acid concentration	445–3,031	White Pekin	6 (4.52–14.61)	Tang et al., 2020a	Common starter diet contained 20.0% CP, 2,919 kcal ME/kg, and 18.90 mg/kg pantothenic acid; Basal grower diet contained 18.1% CP, 3,189 kcal ME/kg, and 4.52 mg/kg pantothenic acid
10.79	15–42	Liver pantothenic acid content	445–3,031	White Pekin	6 (4.52–14.61)	Tang et al., 2020a	Common starter diet contained 20.0% CP, 2,919 kcal ME/kg, and 18.90 mg/kg pantothenic acid; Basal grower diet contained 18.1% CP, 3,189 kcal ME/kg, and 4.52 mg/kg pantothenic acid
Pyridoxine, mg/kg							
4.37	0–28	Feed efficiency	57–1,099	White Pekin	7 (1.93–5.89)	Xie et al., 2014c	Basal diet contained 20.0% CP, 3,070 kcal ME/kg, and 1.93 mg/kg pyridoxine
4.01	0–28	Plasma aspartate aminotransferase	57–1,099	White Pekin	7 (1.93–5.89)	Xie et al., 2014c	Basal diet contained 20.0% CP, 3,070 kcal ME/kg, and 1.93 mg/kg pyridoxine
Riboflavin, mg/kg							
3.31	0–21	Growth	54–1,255	White Pekin (male)	8 (1.2–15.2)	Tang et al., 2013	Basal diet contained 20.54% CP, 3,040 kcal ME/kg, and 1.2 mg/kg riboflavin
5.20	0–21	Feed efficiency	54–1,255	White Pekin (male)	8 (1.2–15.2)	Tang et al., 2013	Basal diet contained 20.54% CP, 3,040 kcal ME/kg, and 1.2 mg/kg riboflavin
3.85	0–21	Plasma riboflavin	54–1,255	White Pekin (male)	8 (1.2–15.2)	Tang et al., 2013	Basal diet contained 20.54% CP, 3,040 kcal ME/kg, and 1.2 mg/kg riboflavin
3.91	0–21	Liver riboflavin	54–1,255	White Pekin (male)	8 (1.2–15.2)	Tang et al., 2013	Basal diet contained 20.54% CP, 3,040 kcal ME/kg, and 1.2 mg/kg riboflavin

3.27	0–21	Growth	53–1,181	White Pekin (female)	8 (1.2–15.2)	Tang et al., 2013	Basal diet contained 20.54% CP, 3,040 kcal ME/kg, and 1.2 mg/kg riboflavin
3.33	0–21	Feed efficiency	53–1,181	White Pekin (female)	8 (1.2–15.2)	Tang et al., 2013	Basal diet contained 20.54% CP, 3,040 kcal ME/kg, and 1.2 mg/kg riboflavin
3.84	0–21	Plasma riboflavin	53–1,181	White Pekin (female)	8 (1.2–15.2)	Tang et al., 2013	Basal diet contained 20.54% CP, 3,040 kcal ME/kg, and 1.2 mg/kg riboflavin
3.28	0–21	Liver riboflavin	53–1,181	White Pekin (female)	8 (1.2–15.2)	Tang et al., 2013	Basal diet contained 20.54% CP, 3,040 kcal ME/kg, and 1.2 mg/kg riboflavin
3.01	0–21	Growth	51–1,090	White Pekin	7 (1.69–9.26)	Tang et al., 2014	Basal diet contained 20.02% CP, 2,920 kcal ME/kg, and 1.69 mg/kg riboflavin
2.79	0–21	Feed efficiency	51–1,090	White Pekin	7 (1.69–9.26)	Tang et al., 2014	Basal diet contained 20.02% CP, 2,920 kcal ME/kg, and 1.69 mg/kg riboflavin
4.18	0–21	Plasma riboflavin	51–1,090	White Pekin	7 (1.69–9.26)	Tang et al., 2014	Basal diet contained 20.02% CP, 2,920 kcal ME/kg, and 1.69 mg/kg riboflavin
3.63	0–21	Liver riboflavin	51–1,090	White Pekin	7 (1.69–9.26)	Tang et al., 2014	Basal diet contained 20.02% CP, 2,920 kcal ME/kg, and 1.69 mg/kg riboflavin
2.43	15–35	Growth	1,724[f]	White Pekin	6 (1.38–6.38)	Tang et al., 2015	Common starter diet contained 20.0% CP, 2,920 kcal ME/kg, and 5.69 mg/kg riboflavin; Base grower diet contained 17.6% CP, 3,021 kcal ME/kg, and 1.38 mg/kg riboflavin
2.31	15–35	Feed efficiency	1,724[f]	White Pekin	6 (1.38–6.38)	Tang et al., 2015	Common starter diet contained 20.0% CP, 2,920 kcal ME/kg, and 5.69 mg/kg riboflavin; Base grower diet contained 17.6% CP, 3,021 kcal ME/kg, and 1.38 mg/kg riboflavin
3.57	15–35	Plasma riboflavin	1,724[f]	White Pekin	6 (1.38–6.38)	Tang et al., 2015	Common starter diet contained 20.0% CP, 2,920 kcal ME/kg, and 5.69 mg/kg riboflavin; Base grower diet contained 17.6% CP, 3,021 kcal ME/kg, and 1.38 mg/kg riboflavin
2.78	15–35	Liver riboflavin	1,724	White Pekin	6 (1.38–6.38)	Tang et al., 2015	Common starter diet contained 20.0% CP, 2,920 kcal ME/kg, and 5.69 mg/kg riboflavin; Base brower diet contained 17.6% CP, 3,201 kcal ME/kg, and 1.38 mg/kg riboflavin

[a] With few exceptions, references were published after September 1991, the date which the literature review was completed for the 9th revised edition of this report. The 1994 report should be consulted for older references that inform duck nutrient requirements.

[b] Data provided for meat-type ducks only.

[c] Apparent metabolizable energy. Although performance was maximized at the dietary AME levels listed for the cited studies, because ducks can reach maximal performance over a range of dietary AME contents, there is no specific requirement, as mentioned in Table 17-1.

[d] Nr = not reported. In those studies, the authors reported average daily gains but did not disclose the starting (hatch) weight. Thus, add ~50 g to the end of study values to estimate the final BW.

[e] Data converted from a digestible amino acid basis to a total amino acid basis based on Hong et al. (2002).

[f] No starting or final weight was provided; value is maximum g of BW gained from 21–49 d of age (Xie et al., 2006), 15–35 d of age (Tang et al., 2015), or 15–42 d of age (Xie et al., 2009c).

[g] Ducks were subjected to higher than normal stocking density.

efficiency of tissues accreted during production (e.g., meat or eggs). Unlike those for chickens and turkeys, none of the recommended duck nutrient requirements were derived from models.

3. **Guidance levels:** These are based on the results of empirical studies that did not meet the inclusion criteria of the previous recommendations or studies, or when there was insufficient experimental data. There is not a high degree of confidence that this recommendation is the optimal level in a least cost ration formulation, but it is likely to give expected animal performance. Values appear in bold italics font in Table 17-1.

Nutrient requirement values carried over from the NRC (1994) report, because there were no new studies upon which to elicit any changes, appear in Table 17-1 in normal or italic font but shaded.

Water

Ducks typically consume water at a rate of approximately 4 to 5 times the weight of the daily feed intake, which is almost twice that of chickens (Scott and Dean, 1991). For example, Siregar and Farrell (1980) reported that, in White Pekin ducks and commercial broiler chickens fed high- or low-energy corn-wheat-soybean meal–based diets, the average ratio of water:food was 4.2:1 for 5- to 22-d-old ducklings and 2.3:1 for 11- to 32-d-old chickens. In addition, duck excreta has a very high water content as compared to chicken excreta, averaging approximately 88 percent (Davis and Dean, 1968; Lin et al., 2011) versus 74 to 80 percent (Henuk and Dingle, 2003; Hoeven-Hangoor et al., 2013), respectively.

According to Farrell (1990), ducks also appear to overconsume water beyond that needed for maximum growth rate by about 20 percent, and that the high water requirement is partly a reflection of the inherent characteristic of ducks to "shovel" or "filter" water for feed residues. Citing unpublished work by Siregar, Siregar and Farrell (1980) stated that water consumption of ducklings could be reduced to 85 percent of ad libitum without a depression in food intake or growth rate. Veltmann and Sharlin (1981) observed under laboratory conditions that White Pekin ducks provided water (from bell waterers) for either 6 or 8 hr consumed approximately 14 percent less water than ducks provided water for either 16 or 24 hr but exhibited normal growth and carcass development. Taken together, these observations support the belief that ducks consume water in excess of the amount needed for normal performance (Scott and Dean, 1991). Although restricting access to water would be advantageous in terms of water usage and litter conditions, it is not known if this practice can be safely applied to commercial production practices, especially under situations of high ambient temperatures and/or dietary salt levels that would require consumption of additional water to maintain optimal plasma electrolyte balance (Scott and Dean, 1991).

While ducks should be provided with a constant supply of clean drinking water, drinker design can affect water consumption (Cherry and Morris, 2008). Rauch et al. (2016a) reported that Pekin ducks favored round bell drinkers over nipple drinkers. The former enable the birds to immerse their heads, to drink and "filter" in a species-specific manner, to care for their plumage, and to clean their bills and eyes. Heyn et al. (2006) reported that Pekin ducks with access only to nipple drinkers exhibited a significantly higher percentage of plugged-up nostrils compared to ducks from pens with open water drinkers (troughs and bells). Rauch et al. (2016b) confirmed this finding in that Pekin ducks with access to round drinkers had fewer eye infections or plugged nostrils than ducks with access only to nipple drinkers.

Energy

Ducks appear to eat primarily to satisfy their appetite for energy (Dean, 1986a). Provided the proper ratio of energy to protein and all other nutrients is maintained, the White Pekin duck can obtain sufficient energy for near normal weight gain with pelleted diets ranging from 2,200 to 3,300 kcal of metabolizable energy (ME) per kg of diet (Scott and Dean, 1991), while the comparative energy range for the Muscovy duck is 2,500 to 3,000 kcal ME/kg diet (Baéza, 2016). Thus, although duck growth will be unaffected within the preceding dietary energy ranges, feed consumption and the feed conversion ratio will be less efficient with low dietary energy levels than with high dietary energy levels (Baéza, 2016). Conversely, feed efficiency and carcass fat are usually increased when dietary nitrogen-corrected ME (ME_n) is increased (Wilson, 1975; Scott and Dean, 1991). Thus, a very different carcass composition within a strain can be obtained simply by altering the dietary ME/crude protein (CP) ratio or amino acid profile (Chen and Applegate, 2016).

Although growing ducks can adjust feed intake across a wide range of energy levels without sacrificing performance, they are typically fed two feeds during the growing period. The reference values from the previous version of this report (NRC, 1994) for White Pekin ducks were 2,900 kcal ME/kg (22% CP) for starter diets (0 to 2 weeks) and 3,000 kcal ME/kg (16% CP) for grower-finisher diets (2 to 7 weeks). Adeola (2006) recommended that starter diets (0 to 2 weeks) for White Pekin ducks contain 2,825 kcal ME/kg (23% CP) or 2,875 kcal ME/kg (20.5% CP), and grower-finisher diets (2 to 6 weeks) contain 3,050 kcal ME/kg (17.5% CP) or 3,075 kcal ME/kg (15% CP). A caveat is that these recommendations were not based on specific empirical studies per se, but rather represent a synthesis of literature information and a plethora of experiments conducted in the author's laboratory (Adeola, 2006).

Other reviews by Baéza (2016) and Fouad et al. (2018b), the report of the Ministry of Agriculture of China (2012), and commercial information sources, such as those of Grimaud Frères Company, cited in Baéza (2016), recommended dietary ME contents of 2,900 kcal/kg and 3,000 kcal/kg for starter ducks (0 to 2 weeks) and grower-finisher ducks (2 to 7 weeks), respectively. However, because supplying energy is a major determinant of the cost of a diet, decisions on the optimal energy level to employ are often driven by local feed costs in addition to the level that maximizes growth rates, efficiency, or product quality. In this regard, the estimated AME requirements of White Pekin ducks listed in Table 17-3, derived from the empirical studies of Fan et al. (2008), Xie et al. (2010), Sritiawthai et al. (2013), and Zeng et al. (2015a), are somewhat variable. Therefore, the energy densities provided in Table 17-1 are not meant to be optimal for either diet cost or productive performance, and when other energy densities are used, the level of the various other nutrients should be adjusted to account for different rates of feed intake that will occur.

In well-balanced diets, the metabolizability of the energy of feedstuffs by ducks has been found to be approximately equal to that of chickens (Shen and Dean, 1982; Ostrowski-Meissner, 1984; Scott and Dean, 1991), although species differences for some feedstuffs have been reported (Siregar and Farrell, 1980; Mohamed et al., 1984; Elkin, 1987; Adeola et al., 1997; Baéza, 2016). Nevertheless, ME values of feedstuffs for chickens are often used in duck diet formulations (Scott and Dean, 1991), not only because they are somewhat interchangeable but also because of the comparative paucity of determined feedstuff energy values for ducks.

Adeola et al. (1997) and Adeola (2006) described feeding and excreta collection techniques that enhanced the accuracy of determining AME, true metabolizable energy (TME), and nitrogen-corrected AME (AME_n) and TME (TME_n) values versus conventional feeding methodologies using collection pans. By surgically attaching an apparatus around the animal's vent area, the excreta collected was free of contamination from spilled feed, dander, and feathers. In addition, as duck excreta is quite watery and ducks forcefully eject excreta far and wide (Ostrowski-Meissner, 1984), sample losses due to splatter were eliminated (Adeola, 2006). Using this technique, Adeola (2006) reported AME, AME_n, TME, and TME_n values of 11 grains and seven meals and by-products based on experiments conducted in his laboratory with White Pekin ducks (Adeola et al., 1997; King et al., 1997; Ragland et al., 1997; Hong et al., 2001, 2002; Adeola, 2003). Additionally, AME, AME_n, TME, and TME_n values for spray-dried eggs, plasma protein, and soybean meal were reported for male White Pekin ducks (Norberg et al., 2004) using the excreta collection device of Adeola et al. (1997), while those for corn, broken rice, rice bran, wheat bran, cassava meal, soybean meal, and two fishmeals were reported by Hoai et al. (2011) employing ad libitum–fed Cherry Valley Pekin ducks and a total collection technique. Taken together, for all feedstuffs, TME_n values were higher than AME_n values while, with a few exceptions, AME > AME_n and TME > TME_n. Moreover, it was shown that AME, AME_n, TME, and TME_n values of individual feed ingredients (e.g., barley and canola meal [Hong et al., 2001] and corn, wheat red dog, and soybean meal (Hong et al., 2002) were additive when included in complete diets for ducks.

With regard to predicting the ME content of a feedstuff for ducks from its chemical composition, Zhao et al. (2008) determined the AME, AME_n, TME, and TME_n contents of 30 corn calibration samples with White Pekin drakes and found that they were negatively correlated with crude fiber, acid detergent fiber (ADF), and neutral detergent fiber (NDF), moderately correlated with gross energy, and not significantly correlated with CP, ether extract (EE), or ash contents. The researchers reported that the maximum absolute differences between the predicted and in vivo determined values of AME, AME_n, TME, and TME_n were 58, 35, 61, and 33 kcal/kg, respectively, which represented 1.5 percent or less of the determined ME values. As a result, Zhao et al. (2008) concluded that the accuracy of the prediction model for duck feedstuffs was acceptable. In a similar study, Wan et al. (2009) developed a model to predict the TME content of wheat by-products from their chemical composition utilizing male Cherry Valley Pekin ducks. Seven representative wheat by-product samples averaged 2,872 kcal TME/kg, and TME was negatively correlated to crude fiber, NDF, and ADF, with the greatest correlation coefficient ($r = -0.969$) observed between NDF and TME. Similar to the report of Zhao et al. (2008), no significant correlations were found between TME and CP, EE, or ash, as well as gross energy. Wan et al. (2009) concluded, based on their prediction equation ($R^2 = 0.99$, $P < 0.01$), that NDF was an effective predictor of the TME value of wheat by-products for Cherry Valley ducks. Lastly, Garcia et al. (2012) utilized data from a previous publication on the AME of 12 meat and bone meal (MBM) samples and evaluated whether particle size and protease resistance, which are two noncompositional attributes of MBM, would have utility in improving AME predictions based on in vitro measurements. Utilizing a stepwise construction of multiple linear regression models, the researchers found that the former measure was useful in building AME models, but the latter was not. Garcia et al. (2012) concluded that proximate composition, along with particle size data, can be used to construct accurate AME and AME_n prediction models for MBM, thereby allowing greater confidence regarding its contribution to the energy content of diets for White Pekin ducks.

Protein and Amino Acids

Protein

The duck's dietary protein requirement is essentially a requirement for amino acids, present in the diet in adequate quantities and in the proper balance, and typically expressed in terms of the ratio of ME to protein. Much of the disagreement in the literature regarding the protein requirement of ducks is the result of their exceptional capacity for compensatory growth (Dean, 1986a). In this regard, it has been shown that ducklings can overcome an early growth retardation due to underfeeding of protein (or another nutrient) by the time they reach market age, although if the protein deficiency is too severe, they do not always recover (Scott and Dean, 1991). Moreover, it is important to feed a starter diet that is adequate in protein (essential amino acids and amino nitrogen for synthesis of nonessential amino acids) in order to ensure proper feather cover, and underfeeding protein during this time could cause ducks to engage in feather pulling (Scott and Dean, 1991). Other factors that have contributed to the variability in the protein requirements of ducks include the breeds or genetic lines used, the response criterion (e.g., growth, feed efficiency, carcass yield), use of diets of widely differing composition and form (e.g., pelleted vs. mash, use of supplemental crystalline amino acids), and husbandry conditions (Scott and Dean, 1991). Elkin (1987) reviewed the protein requirements of several major duck breeds/genetic crosses, beginning with the first investigation with White Pekin ducks by Horton (1928). In general, the dietary protein requirements of White Pekins are slightly higher than those of Muscovies and similar to Mule ducks.

Ducks are typically fed two or three feeds during the growing period, and the nutrient requirements from the previous version of this report for White Pekin ducks were 22 percent protein (2,900 kcal ME/kg) for starter diets (0 to 2 weeks) and 16 percent protein (3,000 kcal ME/kg) for grower-finisher diets (2 to 7 weeks). Although several reviews and research articles on the nutrient requirements of meat-type ducks (e.g., Adeola, 2006; Ministry of Agriculture of China, 2012; Baéza, 2016; Fouad et al., 2018b) have been published since the literature review of the previous report (NRC, 1994) was completed in September 1991, they contain little, if any, documentation for their recommendations. Therefore, based upon recent published empirical data from studies where White Pekin ducks were fed corn-soybean meal–based diets (Zeng et al., 2015a; Chen et al., 2016; Xie et al., 2017a,b), the recommended CP requirements of starter (0 to 2 week) and grower (2 to 7 week) ducks have been revised to 21 percent and 17 percent, respectively (Table 17-1). As compared to the 1994 NRC recommendations, the small reduction in dietary CP for starter ducks is most likely a reflection of the current use of supplemental synthetic amino acids (methionine, lysine, threonine, and tryptophan), while the slightly higher dietary CP for grower-finisher ducks most likely relates to their relatively greater growth rate and increased breast meat and carcass yields as compared to strains utilized at the time that the previous report (NRC, 1994) was published.

Since the previous edition of this report (NRC, 1994), there also have been very few reports on the dietary CP requirements of egg-type ducks, and virtually none on White Pekins. Farrell (1995) reported recommended nutrient requirements based on the work of T.F. Shen (e.g., see Shen, 1986) from the National Taiwan University as follows: For starter (0 to 4 week), grower (4 to 9 week), developer (9 to 14 week), and layer/breeder (>14 week) ducks that are appropriate for high-producing breeds such as Khaki Campbell or Brown Tsaiya, he recommended 20 percent protein (2,796 to 2,892 kcal ME/kg), 17 percent protein (2,749 to 2,844 kcal ME/kg), 15 percent protein (2,653 to 2,749 kcal ME/kg), and 17–19 percent protein (2,653 to 2,749 kcal ME/kg), respectively. In contrast, Thongwittaya et al. (1992) found that a native Thai duck × Khaki Campbell strain required 17 percent CP during the starting phase (0 to 4 week) and 15.5 percent CP during the growing phase (4 to 18 week) with 2,700 kcal ME/kg promoting optimal growth at both phases. Xia et al. (2014) investigated the CP and ME requirements of Longyan ducks during the early (16 to 18 weeks of age), peak (19 to 39 weeks of age), and late (40 to 47 weeks of age) laying periods. Within each period, the ducks were fed nine corn-soybean meal–based diets containing different combinations of CP and ME content in a 3×3 factorial arrangement of treatments (16.42, 17.07, and 18.26% CP and 2,600, 2,500, and 2,400 kcal ME/kg). Based on egg production, egg weights, egg mass, and feed efficiency values, Xia et al. (2014) concluded that the respective optimal CP and ME requirements for the early, peak, and late laying periods were 17.07 percent CP and 2,600 kcal ME/kg, 17.07 percent CP and 2,500 kcal ME/kg, and 18.26 percent CP and 2,500 kcal ME/kg, respectively. Because the previous edition of this report (NRC, 1994) included nutrient recommendations for White Pekin breeder ducks, these data cannot be updated due to a lack of empirical studies published since 1991. However, since several reports with various Asian laying/breeding duck strains have appeared in the literature since 1991, these empirical studies are referenced in Table 17-3.

Amino Acids

Research on the amino acid requirements of ducks has primarily centered on those that are most likely to be limiting in practical diets (Scott and Dean, 1991). For poultry fed corn-soybean meal–based diets, the order of limitation of amino acids is generally considered to be methionine, lysine, and threonine (Kidd, 2000; Bi et al., 2018). However, when corn-soybean meal–based diets were formulated on an ideal protein basis with suboptimal protein levels and supplemented with crystalline amino acids in order to limit amino acid excesses and imbalances, Fernandez et al. (1994) reported the order of limitation for broiler chickens as methionine, threonine, and lysine. Therefore, there is some disagreement in the literature as to the second- and third-limiting amino acids for poultry. Other amino acids of interest in duck nutrition are arginine, which is a dietary essential amino acid for domestic fowl (Scott et al., 1982), and tryptophan, which also is poorly converted to niacin by ducks in comparison to chickens (Chen et al., 1996). As discussed subsequently, the recommended requirements for lysine, methionine, methionine + cyst(e)ine (i.e., total sulfur amino acids; TSAA), and valine have been increased over NRC (1994) values, arginine and tryptophan have been lowered, and a requirement for threonine has been added. No revised requirements for isoleucine or leucine, or any requirements for histidine, phenylalanine, or phenylalanine + tyrosine, are provided due to the continued lack of experimental data. Compared to the previous edition of this report, the increased requirements for lysine, TSAA, and valine most likely reflect the enhanced rate of growth and breast meat deposition of today's modern White Pekin strains.

Diets for ducks are typically formulated on either a total amino acid basis or employ amino acid digestibility coefficients determined for broiler chickens; however, the latter may be lower for ducks compared to broilers (Baéza, 2016). This was shown in a study comparing amino acid digestibilities of rapeseed and soybean meal in ducks, chickens, and turkeys, in which average digestibility across all amino acids were lower in ducks versus chickens and turkeys (Kluth and Rudehutscord, 2006). Moreover, Kong and Adeola (2013a) reported that ducks fed a nitrogen-free diet had approximately 3-fold higher basal endogenous amino acid losses compared with broiler chickens fed a nitrogen-free diet. Both Kluth and Rodehutscord (2006) and Kong and Adeola (2013a) concluded that amino acid digestibilities determined with broilers should not be used in formulating diets for ducks. Since the previous edition of this report, numerous studies have been conducted to determine amino acid digestibility values for feedstuffs for ducks (Hong et al., 2001, 2002; Norberg et al., 2004; Adeola, 2005, 2006; Adeola et al., 2008; Kong and Adeola, 2010, 2013b). However, with one exception (Zhou et al., 2017), every duck amino acid requirement study cited in this report was obtained employing diets formulated on a total amino acid basis. Therefore, although an ideal protein approach to diet formulation for ducks is possible, all of the dietary amino acid requirements in Table 17-1 are reported on a total basis, in concert with the literature at the time that this report was published.

Revisions in the recommended dietary amino acid requirements of White Pekin ducks were based primarily on the empirical data cited subsequently and benchmarked against the recommendations of Adeola (2006), Baéza (2016), and Fouad et al. (2018b), as well as the report of the Ministry of Agriculture of China (2012), and commercial recommendations, such as those of Grimaud Frères Company, cited in Baéza (2016).

Arginine

Ducks, like other *Aves*, are uricotelic. Although arginase activities have been detected in avian liver and kidney, the hepatic activity in the domestic fowl is very low compared to ureotelic species (Stevens, 1996) and at least two separate enzymes required for arginine synthesis are lacking in avian kidney (Leeson and Summers, 2001). Therefore, birds are considered to be unable to synthesize arginine de novo due to the lack of a complete urea cycle (Wu et al., 2011). Dietary supplemental arginine also has been reported to influence both the feeding behavior of White Pekin ducks (Wang et al., 2014a,b) and body fat deposition (Wu et al., 2011).

Wang et al. (2013a) investigated the dietary arginine requirement of male White Pekin ducks from 1 to 21 d of age and concluded that the optimal requirements (as a percent of the diet, total arginine) from 1 to 21 d were 0.95 percent for weight gain, 1.16 percent for feed efficiency, and 0.99 percent for breast meat yield. Leg meat percentage was not significantly affected by dietary arginine. In a follow-up study, Wang et al. (2013b) showed that 0.85 percent total dietary arginine was sufficient for maximum growth and optimal feed efficiency from 15 to 35 d of age. However, for maximal breast and leg muscle yields, 1.70 percent arginine was required.

The recommended total dietary arginine requirements of starter and grower-finisher White Pekin ducks have been reduced slightly from the NRC (1994) levels to 1.05 percent and 0.90 percent, respectively. For arginine and all subsequent nutrients discussed in this chapter, the recommended requirements were generally obtained by averaging the estimated requirements for various response criteria from the cited studies within a growth period (e.g., averaging the estimated requirements from all of the studies reporting data from 0 to 21 d of age) in Table 17-3, with the exception of values that appeared to be outliers (e.g., 1.70 percent arginine requirement for breast and leg muscle yields reported by Wang et al., 2013b, were excluded).

Lysine

Over the past two decades, the growth performance and breast meat yield of White Pekin strains has markedly increased while carcass fatness has decreased (Huang et al., 2012). Considering the high lysine content of breast muscle protein, Bons et al. (2002) suggested that the Pekin's lysine requirement was higher than that reported in NRC (1994) and, based on 95 percent of the asymptote in a nonlinear regression model, maximal weight gain and optimal feed efficiency during the starter phase were obtained with dietary lysine levels of 1.17 and 1.06 percent, respectively, while 0.87 and 0.82 percent lysine were required to achieve 95 percent of the asymptote for weight gain and feed efficiency during the grower-finisher phase. A dietary lysine content of 1.00 percent was needed to achieve 95 percent of the asymptote for optimal breast meat weight. Based on the starter phase growth rate and breast meat yield for the grower phase, Bons et al. (2002) recommended a dietary lysine content of 1.17 percent until d 21, and 1.00 percent from d 22 onward. Both values were higher than the NRC (1994) recommendations, and Bons et al. (2002) suggested that the requirements for other amino acids should likewise be assessed for modern strains of Pekins.

Wang et al. (2006) concluded that from hatch to 14 d of age, White Pekin ducks require 1.10 percent lysine for optimal body weights and feed efficiency. Based on a broken-line regression analysis, Xie et al. (2009a) determined that the lysine requirements from day 7 to 21 for weight gain, feed efficiency, breast meat weight, and breast meat yield were 0.84, 0.90, 0.97, and 0.98 percent, respectively. Since White Pekins are raised primarily for meat production, Xie et al.

(2009a) recommended a lysine requirement of 0.98 percent for ducklings during the starter phase.

Based on a broken-line regression model, Wen et al. (2017) concluded that the lysine requirements for 21-d weight gain of White Pekin ducklings were 0.94 and 0.98 percent in diets containing 2,750 and 3,050 kcal AME/kg, respectively.

Zhou et al. (2017) determined the digestible lysine requirements of male Cherry Valley (Pekin) ducks from hatch to 35 d of age. Based on a quadratic regression model, the researchers determined that the digestible lysine requirements for average daily gain, feed efficiency, and breast meat yield were 0.948, 0.986, and 0.961 percent, respectively, for 1 to 14 d and 0.758, 0.792, and 0.761 percent, respectively, for 15 to 35 d. Subcutaneous fat decreased linearly with increasing dietary digestible lysine content. Assuming that lysine was 86 percent digestible in a corn–soybean meal diet for ducks (Hong et al., 2002), the digestible lysine requirements determined by Zhou et al. (2017) would equate to total dietary lysine requirements for average daily gain, feed efficiency, and breast meat yield to be 1.10, 1.15, and 1.12 percent, respectively, for 1 to 14 d and 0.88, 0.92, and 0.88 percent, respectively, for 15 to 35 d.

The recommended total dietary lysine requirements of starter and grower-finisher White Pekin ducks have been increased from the NRC (1994) levels to 1.05 percent and 0.90 percent, respectively.

Methionine

The previous revised edition of this report (NRC, 1994) listed the methionine requirement of starting White Pekin ducks as 0.40 percent (0.70 percent for methionine + cyst(e)ine), which was based on data reported by Elkin et al. (1986). Subsequent studies have suggested that this requirement was too low. Similar conclusions have been made about the dietary requirements of starting and growing White Pekins for other amino acids, which most likely are a reflection of today's duck genotypes and the associated advances in bird performance that have occurred since the early 1990s (Xie et al., 2006).

Employing a corn-peanut meal–based diet similar to that of Elkin et al. (1986), Xie et al. (2004) concluded that the methionine requirement for optimal growth of White Peking ducklings from hatch to 21 d of age was 0.481 percent (95% of the maximum growth response). They also reported that the methionine requirement for optimal feed efficiency was 0.585 percent and that the dietary cyst(e)ine requirement for both variables was not more than 0.325 percent. Thus, the total sulfur amino acid (TSAA) requirement for growth from hatch to 21 d of age was 0.806 percent. In a subsequent study from the same laboratory, Xie et al. (2006) conducted a dose–response experiment with male White Pekin ducks and observed that daily weight gain and breast meat yield both increased and then decreased as the level of dietary methionine increased, resulting in a significant quadratic response and requirements of 0.377 percent methionine (95% of the maximum growth response) and 0.379 percent (95% of the maximum breast meat yield response), respectively. Xie et al. (2006) concluded that the TSAA requirements of White Pekin ducks from 21 to 49 d of age for maximum growth and breast meat yield were 0.687 percent and 0.689 percent, respectively. Adeola (2007) reported that male White Pekin ducks exhibited a constant utilization of methionine from 22 to 42 d of age, over a range of 0.24 to 0.39 percent digestible methionine, with an efficiency of deposition into the body of 31 percent, and retention of 15 mg of methionine in each g of body protein accreted.

Wang et al. (2006) concluded that White Pekin ducks require 0.46 percent methionine for optimal body weights and feed efficiency from hatch to 14 d of age. Since contents of the basal diet were not reported, the duck's dietary TSAA requirement was not determined.

Zeng et al. (2015b) investigated the dietary methionine requirement of White Pekin ducks from 15 to 35 d of age. They concluded that the optimal methionine requirement for 35-d body weight, 15- to 35-d feed efficiency, 35-d breast meat yield, and 35-d feather coverage were 0.468, 0.476, 0.408, and 0.484 percent, respectively, as determined by quadratic broken-line model, and 0.605, 0.591, 0.564, and 0.612 percent, respectively, by quadratic polynomial regression. Since each diet contained approximately 0.26 percent cyst(e)ine, the TSAA requirements for 35-d body weight, 15- to 35-d feed efficiency, 35-d breast meat yield, and 35-d feather coverage were approximately 0.728, 0.736, 0.668, and 0.744 percent, respectively, as determined by quadratic broken-line model, and 0.865, 0.851, 0.824, and 0.872 percent, respectively, by quadratic polynomial regression. The maximum responses obtained with the quadratic broken-line model, as compared to the quadratic polynomial regression model, seemed to be more in line with other recent reports, as well as with older recommendations for a commercial strain (Cherry Valley, 2004).

Prahkarnkaeo et al. (2017) determined the methionine requirements of male Cherry Valley (Pekin) ducks from 1 to 47 d of age and reported that there was no beneficial effect on feed efficiency or carcass characteristics of feeding TSAA levels greater than 100 percent of the commercial recommendation (starter 1, 0.90% TSAA [0.53% methionine + 0.37%cyst(e)ine]; starter 2, 0.84% TSAA [0.49% methionine + 0.35% cyst(e)ine], and grower (0.75% TSAA [0.42% methionine + 0.33%cyst(e)ine], respectively).

The recommended total dietary methionine (and TSAA) requirements of starter and grower-finisher White Pekin ducks have been increased from NRC (1994) levels to 0.52 percent (0.85 percent) and 0.43 percent (0.71 percent), respectively. In many of the studies, no distinction was made between cystine (dimer) and cysteine (monomer), so in Tables 17-1 and 17-3, the TSAA requirement is denoted as methionine + cyst(e)ine. Moreover, ducks do not appear to utilize the D- and L-isomers of methionine with equal efficacy (see next section).

Methionine Sources

With regard to sources of methionine activity, the two most common forms of supplemental methionine activity in poultry diets are DL-methionine and DL-2-hydroxy-4-(methylthio)butanoic acid (HMB). D-methionine and both the D- and L-isomers of HMB require *in vivo* conversion to L-methionine in order to be used for body protein synthesis.

In contrast to studies with chickens, reports assessing the relative methionine activities of these sources for ducks are very limited. Elkin et al. (1988a) fed male White Pekin ducklings a corn-peanut meal–based mash diet supplemented with 0, 0.04, 0.08, or 0.16 percent dietary D- or L-methionine from hatch to 12 d of age and reported that weight gains of ducks fed the D-isomer were approximately 78 percent lower than that of ducks fed an equivalent amount of L-methionine. This contrasted with the conclusion that bioavailability of D-methionine was usually equal to that of the L-isomer when relatively small amounts of the D-isomer were fed in practical diets (Lewis and Baker, 1995) and suggested that the level of DL-methionine may need to be adjusted accordingly when supplemented in starter duck diets (NRC, 1994). Although the substrate specificity of duckling hepatic and kidney D-amino acid oxidase (DAAO) is relatively high for methionine compared to most of the other amino acids (Elkin and Lyons, 1988), it is possible that the oxidative deamination step catalyzed by DAAO, the reamination of the keto-acid to the L-isomer, or both reactions, are not robust enough to completely convert D- to L-methionine in the young duck. More recent studies by Zhang et al. (2019) and Wu et al. (2021) confirmed that the bioavailability of L-methionine was greater than that of D-methionine when fed to White Pekin ducks.

Xue et al. (2018) conducted a dose–response experiment with male White Pekin ducks fed a corn-soybean meal–based control diet supplemented with five different levels each of DL-methionine or L-methionine from 7 to 21 d of age. Since the control diet exceeded the Pekin's methionine requirement (NRC, 1994), this was essentially a toxicity study, and all supplemental levels beyond 0.25 percent depressed weight gain and feed intake, but not feed efficiency, which suggested that the growth depression resulted from a reduction in feed intake. Moreover, DL-methionine and L-methionine were found to be equally growth depressing.

Elkin et al. (1988b) determined the relative response of White Pekin ducklings to various supplemental dietary levels of HMB-Ca salt (HMBC) and DL-methionine from hatch to 12 d of age. When growth data were pooled across all levels of four experiments, mean 12-d weight gains were similar for HMBC and DL-methionine. Kluge et al. (2016) investigated the biological efficacy of HMB compared to an equivalent molar amount of DL-methionine as sources of methionine activity for male White Pekin ducks from hatch to 21 d of age and, based on 21-d body weights, weight gains, feed intake, and feed conversion efficiency, also concluded that DL-methionine and HMB have similar efficacies as sources of dietary methionine for starter ducks. Zhao et al. (2018) reported a lack of significant differences between methionine sources on duck growth performance, carcass traits, and feather growth, thereby confirming the reports of Elkin et al. (1988b) and Kluge et al. (2016).

In a study assessing the relative toxicities of DL-methionine versus HMB in male White Pekin ducks from 21 to 42 d of age, Xie et al. (2007) supplemented a corn-soybean meal–based diet with either 1 or 2 percent DL-methionine or the molar equivalent of HMB. The researchers reported that HMB was much less growth-depressing than DL-methionine, particularly at the 2 percent level of supplementation. Taken together, these studies suggest that DL-methionine and HMB have similar efficacies in ducks, and at toxic levels, HMB appears to have a less of a negative impact on productivty.

Threonine

Threonine is generally considered to be the third-limiting amino acid after methionine and lysine in corn-soybean meal diets for poultry (Kidd, 2000). It plays an important role in the structure of mucus proteins and digestive enzymes (Horn et al., 2009, 2010) and is also an important component of feathers and immunoglobulins (Halas et al., 2012; Zhang et al., 2014). Moreover, because mucin proteins are resistant to digestion and their amino acids are not efficiently reutilized, intestinal mucin secretion represents a significant net loss of body threonine (Zhang et al., 2014). Thus, a dietary deficiency of threonine would be expected to not only impair duck growth, feed efficiency, feathering, and carcass characteristics but might also adversely affect gut protective functions and nutrient retention due to alterations in the mucus layer. Despite its important functions, recommended threonine requirements of ducks were not included in the previous NRC (1994) report.

Using a quadratic broken-line regression model, Xie et al. (2014c) estimated that White Pekin ducklings required 0.672 percent threonine for optimal weight gain from hatch to 21 d of age. Feed efficiency appeared to plateau at 0.74 percent dietary threonine and, at that level, ducks also had higher villus heights in the duodenum, jejunum, and ileum.

Two experiments to determine the dietary threonine requirement of male White Pekin ducks from hatch to 14 d of age were carried out by Zhang et al. (2016a). Based on a quadratic broken-line regression model, the threonine requirements for optimal 14-d weight gain, breast meat weight, breast meat yield, and wet crude mucin and dry crude mucin (dry matter intake [DMI] basis) were 0.86, 0.94, 0.97, 0.93, and 0.94 percent, respectively (Experiment 1), and 0.86, 0.93, 0.95, 0.92, and 0.93 percent, respectively (Experiment 2). The threonine requirement for optimal feed efficiency was only reported for Experiment 2 and was determined to be 0.92 percent.

Jiang et al. (2016) reported that, based on quadratic broken-line regression, the dietary requirement for threonine for optimal 21-d weight gain of White Pekin ducklings was 0.610 percent at 20.89 percent CP and 0.556 percent at

17.73 percent CP. The threonine requirement was influenced by CP when expressed as a percentage of CP but was similar when expressed as a percentage of dietary lysine. In a follow-up study, Jiang et al. (2017) investigated the effects of dietary CP content and threonine supplementation on 21-d White Pekin duck growth, carcass traits, plasma chemistries, and liver lipid contents. For optimal average daily weight gain and breast muscle percentage, 0.66 and 0.67 percent threonine, respectively, were optimal for a 19 percent CP diet and 0.70 and 0.73 percent threonine, respectively, were optimal for a 22 percent CP diet. These values were higher than those of their previous study (Jiang et al., 2016) and the divergence was attributed to possible differences in the Pekin strains employed. The researchers concluded that, with threonine supplementation, the level of dietary CP could be reduced to 19 percent in White Pekin ducks from 1–21 d of age. Dietary threonine was also observed to lower serum and liver lipid contents when supplemented in 16 and 19 percent CP diets but was without effect in 22 percent CP diets.

Zhang et al. (2014) investigated the dietary threonine requirement of male White Pekin ducks from 15 to 35 d of age and reported that, as determined by a quadratic broken-line regression model, the dietary threonine requirements for 15- to 35-d weight gain, feed efficiency, mortality, carcass weight, breast meat weight, and breast meat percentage, were 0.707, 0.690, 0.620, 0.686, 0.695, and 0.686, respectively. The level of dietary threonine had no significant effect on jejunal and ileal villus height, crypt depth, goblet cells, and MUC2 gene expression.

Jiang et al. (2018) conducted an experiment to determine the effects of dietary CP on the threonine requirement of White Pekin ducks from hatch to 14 d of age. As observed in their previous study (Jiang et al., 2016), a significant CP × threonine interaction was observed for weight gain and feed efficiency. When determined by broken-line regression analysis and expressed as percent of diet, percent of CP, and percent of lysine, the threonine requirement increased, decreased, and increased, respectively, as the dietary CP content increased. As a percentage of the diet, the threonine requirement for optimal daily gain was 0.56, 0.61, 0.60, 0.63, and 0.67 percent for the 16.0, 17.5, 19.0, 20.5, and 22.0 percent CP diets, respectively.

Bi et al. (2018) investigated the dietary threonine requirements of male White Pekin ducks from hatch to 21 d of age. Five graded levels of dietary threonine were fed to ducks that were injected intramuscularly in the left leg with either 1 mL of a 5 mg/mL solution of bovine serum albumin (BSA) in normal saline or 1 mL of normal saline only at 11 d of age. No significant interaction of dietary threonine and BSA injection was observed for weight gain or feed efficiency, while BSA injection significantly decreased growth but increased feed efficiency. Both BSA and saline-only injected ducks had optimal weight gain and feed efficiency with 0.76 percent dietary threonine. Based on linear broken-line analysis, the dietary threonine requirements for optimal weight gain, breast weight, and leg weight in the saline-injected group were 0.705, 0.600, and 0.694 percent, respectively, while the dietary threonine requirements based on quadratic broken-line analyses were 0.793. 0.765, and 0.772 percent, respectively. Because the quadratic model provided a better fit to the data, Bi et al. (2018) considered it better than the linear model to estimate dietary threonine requirements.

Based on these studies, the recommended total dietary threonine requirements of starter and grower-finisher White Pekin ducks, which were not reported in NRC (1994), are 0.75 percent and 0.70 percent, respectively.

Tryptophan

In addition to serving as component of protein, the indispensable amino acid tryptophan is a precursor of both nicotinic acid (niacin) and serotonin, a neurotransmitter and potent vasoconstrictor and stimulator of smooth muscle contraction (Rodwell, 2012). Using male White Pekin duckling feed efficiency values and weight gains from 4 to 7 weeks of age, Adeola (1998) estimated the bioavailability of tryptophan in soybean meal relative to L-tryptophan to be 92 to 94 percent, respectively.

Chen et al. (1996) fed White Pekin ducks, Mule ducks, and broiler chickens a common diet that was marginal in tryptophan content and deficient in niacin, and supplemented that diet with three levels of dietary tryptophan and four levels of dietary niacin in a 3 × 4 factorial arrangement of treatments. They calculated that the efficiency of conversion of tryptophan to niacin in Pekin ducklings and Mule ducklings averaged 177 mg of tryptophan equivalent to 1 mg of niacin, while broiler chicks had a conversion efficiency of 47:1. On a molar basis, this equated to approximately 1 percent and 3.5 percent of dietary tryptophan converted to niacin in ducks and chicks, respectively. This species difference can be explained by ducks having three to four times greater hepatic picolinic acid carboxylase activities than chickens (Scott et al., 1982; Chen et al., 1996), which essentially diverts an intermediate in the conversion pathway (2-amino-3-acroleyl fumaric acid) toward the glutaryl-CoA pathway instead of allowing the compound to condense to quinolinic acid, the immediate precursor of nicotinic acid (McDowell, 2008).

Wang et al. (2006) concluded that the White Pekin duck requires no more than 0.20 percent tryptophan for optimal body weights and feed efficiency values from hatch to 14 d of age when the diet also contained 30 mg supplemental niacin/kg. Total dietary niacin contents were not reported.

Xie et al. (2014a) examined the effects of niacin status on the dietary tryptophan requirements of male White Pekin ducks from hatch to 21 d of age. They estimated, via broken-line regression, that the dietary tryptophan requirements for weight gain at each supplemental niacin level were 0.24 percent, 0.17 percent, 0.17 percent, and 0.17 percent for ducks fed the basal diet supplemented with 0, 40, 80, and 120 mg niacin/kg, respectively. Since producers and researchers would typically not feed a diet devoid of supplemental tryptophan, the

0.24 percent estimate was not considered when establishing the recommended dietary tryptophan requirement.

Liu et al. (2015) investigated whether dietary tryptophan could alleviate the detrimental effects of high stocking density (11 birds/m^2) on male White Pekin ducks raised in pens from 22 to 42 d of age. Since weight gain and feed efficiency were significantly increased, they suggested that supplemental dietary tryptophan could potentially alleviate stress, enhance antioxidative activity, and improve meat quality in ducks raised under high stocking density conditions.

The recommended total dietary tryptophan requirements of starter and grower-finisher White Pekin ducks are 0.20 percent and 0.17 percent, respectively, which are lower and unchanged, respectively, from the previous report (NRC, 1994).

Valine

According to Fernandez et al. (1994), after methionine, threonine, and lysine, valine was the fourth-limiting amino acid in a corn–soybean meal diet formulated on an ideal protein basis using digestible amino acid values and fed to broiler chicks from 8 to 22 d of age. However, only one study on the valine requirements of White Pekin ducks has been published since the last revised edition of this report. Timmler and Rodehutscord (2003) determined the response of White Pekin ducks to supplemental dietary valine from hatch to 3 weeks of age and reported that 95 percent of y_{max} for body weight in their exponential model was achieved with 0.82, 0.80, and 0.79 percent valine at the end of weeks 1, 2, and 3, respectively. Over the entire 21-d study, feed efficiency appeared to be optimal at 0.82 percent dietary valine, while the estimated valine requirements needed to achieve 95 percent of y_{max} for fat and protein accretion were 0.79 and 0.77 percent, respectively. The amounts of dietary valine needed to achieve 95 percent of y_{max} for gain and efficiency were slightly higher than the NRC (1994) recommended level of 0.78 percent, which was determined in Mule ducks (Yu and Shen, 1984).

The recommended total dietary valine requirements of starter and grower-finisher White Pekin ducks are 0.82 percent and 0.59 percent, which are both 5 percent greater than the NRC (1994) recommendations. Despite the lack of empirical studies, the rationale for increasing the latter requirement by the same amount as the starter value is that current duck genotypes perform at a higher level than 1994 genotypes and most likely have a higher dietary requirement for valine from 2 to 7 weeks of age.

Minerals

Due to the paucity of reports on the mineral requirements of White Pekin ducks that have appeared in the literature since September 1991, most of the recommended values in Table 17-1 are considered to be estimates. In addition to empirical data cited subsequently, all of the revisions in the recommended dietary mineral requirements of White Pekin ducks were benchmarked against information from the reviews of Adeola (2006), Baéza (2016), and Fouad et al. (2018b), as well as the report of the Ministry of Agriculture of China (2012), and commercial recommendations, such as those of Grimaud Frères Company, cited in Baéza (2016). Since no new published studies were found for chlorine, magnesium, sodium, or manganese, the previous values (NRC, 1994) were carried over in Table 17-1, and they are not subsequently discussed.

Calcium and Phosphorus

Ca and phosphorus (P) are typically discussed together because they are closely associated in metabolism, particularly with regard to their roles in bone formation (Scott et al., 1982). Prior to 2004, there were no reports on the availability of inorganic P sources for ducks. In a study with male White Pekin ducks from 10 to 23 d of age, Wendt and Rodehutscord (2004) determined the percent utilization of dietary P from the following seven sources: monobasic sodium phosphate (NaH_2PO_4), 100 percent available; Ca-sodium magnesium-phosphate [(Ca, Na, Mg)$PO_4 \cdot nH_2O$], 92 percent; anhydrous dibasic Ca phosphate ($CaHPO_4$), 95 percent; dibasic Ca phosphate ($CaHPO_4 \cdot 2H_2O$), 86 percent; monodibasic Ca phosphate [$CaHPO_4 \cdot Ca(H_2PO_4) \cdot 2H_2O$], 91 percent; Ca-sodium-phosphate ($CaNaPO_4$), 77 percent; and dibasic Ca phosphate ($CaHPO_4 \cdot 2H_2O$), 86 percent. They also reported that the P availability of corn-wheat gluten-soybean meal–based diets was 40 to 44 percent but varied when the Ca:P ratio was altered (49 percent at Ca:P of 1.2:1 versus 40 percent at Ca:P of 2:1). Although Wendt and Rodehutscord (2004) suggested that P availability values for inorganic P sources may not differ much between ducks and chickens, this was not confirmed in subsequent work from the same laboratory (Rodehutscord and Dieckmann, 2005). In this study, P utilization of a corn-potato protein-dried egg white–based diet (0.70% Ca and 0.29% total P) was determined in 3-week-old male White Pekin ducks, male broiler chickens (Ross), male turkeys (British United Turkeys Big 6), and unsexed Japanese quail. When the diet was unsupplemented, the utilization of P was 58, 46, 39, and 55 percent for broilers, ducks, turkeys, and quail, respectively. When the basal diet was supplemented with monobasic Ca phosphate [$Ca(H_2PO_4)_2$]), P accretion was significantly increased in all species to between 61 and 66 percent. In terms of marginal efficiency of P utilization in the monobasic Ca phosphate-supplemented diets, ducks were most efficient (96 percent), followed by turkeys (81 percent), Japanese quail (77 percent), and broilers (74 percent). Rodehutscord and Dieckmann (2005) concluded that differences in P availability of plant and mineral P sources exist among poultry species.

Although both phytase (Rush et al., 2005) and alkaline phosphatase (King et al., 2000a) activities have been detected in the mucosa of the small intestine of Pekin ducks, many studies have shown that, as in chickens, supplemental exogenous microbial phytase typically increases P retention, tibia ash, feed intake, and growth rate in both starter and grower ducks (Farrell et al., 1993; Farrell and Martin, 1998; Martin et al., 1998; Orban et al., 1999; Rodehutscord et al., 2006; Adeola, 2018). Moreover, increased dietary P availability is associated with lower feed costs (i.e., decreased inorganic P addition) and environmental benefits via decreased P excretion.

Using regression analyses based on plasma P concentration and bone properties, Orban et al. (1999) estimated that the additional available P due to the dietary addition of 750 units of microbial phytase/kg diet ranged from 0.05 to 0.07 percent of the diet and from 0.06 to 0.08 percent of the diet, respectively, in finisher ducks (3 to 6 weeks of age). In a short-term P-balance study, Wendt et al. (2003) utilized a non-linear regression analysis to estimate P utilization in male White Pekin ducks from 15 to 20 d of age. When fed a non-P-supplemented corn-soybean meal-based diet, P utilization was estimated to be 28 percent, while phytase supplementation (from 100 up to 1,500 units/kg) increased P utilization to an asymptotic value of 63 percent. Wendt et al. (2003) also estimated microbial phytase available P equivalency, ranging from 0.012 percent of the diet with 100 units phytase/kg diet to 0.09 percent of the diet for 1,500 units phytase/kg diet. Adeola (2010) used linear regression analyses to determine mean P equivalency values of an *E. coli* phytase in starter ducks (7 to 17 d of age) and reported that for 500, 1,000, and 1,500 phytase units/kg diet, the dietary P equivalency values were 0.045, 0.085, and 0.124 percent of the diet, respectively.

Rodehutscord et al. (2003) were the first to establish dietary P dose–response curves in White Pekin ducks with data from both growth and balance trials as response criteria. Corn-soybean meal–based diets, fed from hatch to 21 d of age (Experiments 1 and 2) and from 21–49 d of age (Experiment 3), were supplemented with graded levels of monobasic Ca phosphate at the expense of sand. Based on P retention data, the researchers recommended dietary levels of 0.34 percent available P and 0.23 percent available P, from d 1 to 21 and d 21 to 49, respectively. For optimal growth rate, Rodehutscord et al. (2003) recommended dietary levels of 0.27 percent available P and 0.13 percent available P, from d 1 to 21 and d 21 to 49, respectively. However, since Rodehutscord et al. (2003) employed suboptimal levels of dietary Ca (0.56 to 0.61 percent), Xie et al. (2009b) suggested that their figures, which are quite low, may be underestimated.

Rush et al. (2005) investigated the effect of dietary Ca on Ca and P retention in male White Pekin ducklings from 7 to 17 d of age. Body weight gain, feed intake, and toe ash percentage exhibited a quadratic response to increasing dietary Ca contents and were found to be maximal at 0.95, 0.95, and 0.85 percent, respectively. The Ca requirements for maximal growth and feed intake exceeded the 1994 NRC recommendations of 0.65 percent for 0 to 14 d of age.

Xie et al. (2009b) studied the interaction of dietary Ca and nonphytate P in male White Pekin ducklings from hatch to 14 d of age. Various amounts of limestone, dicalcium phosphate, and monobasic sodium phosphate were supplemented to a corn-soybean meal–based diet. The researchers observed a significant Ca × nonphytate P interaction for weight gain, feed intake, feed efficiency, and tibia ash but nonsignificant main effects for each mineral, which suggested that Ca and P be considered simultaneously when determining their requirements. Using a quadratic response surface regression analysis, Xie et al. (2009b) predicted that starter Pekin ducks required 0.806 percent dietary Ca and 0.403 percent dietary nPP for maximal weight gain and 0.796 percent dietary Ca and 0.379 percent nPP for optimal feed efficiency when the Ca:nPP ratio was approximately 2:1. Tibia ash was also maximized at the 2:1 ratio, which was similar to the 2.1:1 ratio for maximal P utilization in 30-35 d old Pekin ducks from a balance trial reported by Rodehutscord et al. (2003). The dietary Ca requirement was greater than the 1994 NRC requirement (0.65 percent).

Xie et al. (2009c) determined the Ca and P requirements of male White Pekin ducks from 15 to 42 d of age. Ducks were fed a corn-soybean meal–based diet supplemented with nonphytate P and Ca in a 4×4 factorial arrangement of treatments. There were significant Ca × P interactions for weight gain, feed intake, tibia ash, and tibia P. At the lowest dietary P level, weight gain, feed intake, tibia ash, and tibia P significantly decreased with increasing Ca. However, at the higher dietary nonphytate P levels, weight gain and feed intake exhibited a quadratic trend, tibia ash increased, and tibia P was not significantly affected at the two highest dietary Ca levels. Feed efficiency was not significantly affected by either dietary Ca or P. Xie et al. (2009c) concluded that, based on quadratic regression analysis, the dietary Ca requirements of Pekin ducks for optimal weight gain and tibia P percent were 0.72 and 0.66 percent, while the respective dietary available P requirements were 0.37 and 0.37 percent.

Zhu et al. (2018) investigated the influence of dietary Ca levels on growth and bone mineralization in male Cherry Valley (Pekin) ducklings from hatch to 21 d of age. Two corn-soybean meal–based diets (both containing 2,000 IU of vitamin D_3/kg but with different levels of total P) were supplemented with four levels of limestone to create eight diets in a 4×2 factorial arrangement of treatments. Dietary Ca, dietary P, and their interaction were all significant for final body weight, average daily gain, average daily feed intake, tibia ash, and tibia mineral density, while feed efficiency, mortality, percent tibia ash Ca, and percent tibia ash P were only significantly affected by dietary P. Increasing dietary Ca from 0.55 to 0.95 percent had no significant effects on duck growth or feed intake; however, the diet highest in Ca depressed growth, feed intake, tibia ash Ca percent, and tibia mineral density compared to the other three Ca diets within each P level. Tibia ash Ca percent was greatest in ducks fed the 1.15 percent Ca diet. Ducks fed the low-P diets exhibited poorer growth and feed efficiency, lower tibia ash percentage, tibia ash Ca percentage, tibia ash P percentage, serum Ca, serum P, and tibia mineral density, and higher mortality, compared to birds fed the P-adequate diets. In the context of a dietary Ca requirement, when considering only the P-adequate diets, ducks fed 0.75 percent Ca had the numerically greatest ($P > 0.05$) final body weights, average daily gain, and feed efficiency, and the lowest mortality, while birds fed 0.95 percent had significantly the highest serum Ca concentrations and percentage tibia ash.

The recommended dietary Ca and non-phytate P requirements of White Pekin ducks are estimated to be 0.95 and 0.47 percent, respectively, for the starter period and 0.85 and 0.42 percent, respectively, for the grower-finisher period, each of which are greater than the NRC (1994) recommended levels. The recommended dietary Ca requirement for starting ducks was based primarily on the amount of Ca required for maximal growth and feed efficiency reported by Rush et al. (2005) and maximal serum Ca and tibia ash reported by Zhu et al. (2018) and were benchmarked against both review articles and current industry standards. The recommended level of 0.47 percent non-phytate P for starting ducks was made in order to maintain the Ca:P ratio at 2:1. Moreover, since studies on the Ca and non-phytate P requirements of growing ducks were extremely limited (one or two experiments total), the Ca and non-phytate P recommended dietary requirements were arbitrarily set at 90 percent of the starting duck levels, which were more in line with both review article recommendations and current industry standards.

Since virtually no Pekin layer/breeder studies were reported since September 1991, no changes were made in the recommended nutrient levels for laying/breeding ducks in Table 17-1. Moreover, there is no specific information in the 1994 NRC report on Ca sources and particle sizes for laying ducks. Considering that ducks and chickens differ with regard to the presence of a crop, Wang et al. (2014c) suggested that simple extrapolation of Ca source and particle size findings obtained with laying hens may not be valid for ducks. In a study with 24-week-old Longyan laying ducks, the researchers fed diets containing two dietary Ca sources (limestone and oyster shell) and two particle sizes (<0.1 mm and 0.85 to 2.00 mm) for 12 weeks, utilizing a 2×2 factorial arrangement of treatments. The limestone and oyster shells were ground coarsely or finely to obtain similar particle size distributions. There were no significant two-way interactions of particle size and Ca source on either productive performance or egg quality measures. In addition, Ca source had no significant effects on productive performance of the ducks. However, ducks fed the large particle sources had greater egg production ($P = 0.11$), and significantly greater egg mass, feed efficiency, egg breaking strength, albumen height, Haugh units, shell P percent, tibia breaking strength, and tibia ash Ca percent and decreased shell magnesium (Mg) percent. As compared to oyster shell, limestone increased albumen height, Haugh units, eggshell Ca percent and P percent, and tibia breaking strength. Wang et al. (2014c) concluded that limestone of larger particle size was the more appropriate Ca source for laying ducks. If oyster shell was substituted for limestone, the authors cautioned that a small decrease in egg quality and tibia characteristics could result.

Copper

Attia et al. (2012) investigated the effect of dietary inorganic and organic copper sources on the productive performance of male White Pekin ducklings from 1 to 56 d of age. Corn-soybean meal-based mash starter and grower-finisher diets (both containing 7.0 mg copper/kg) were fed from 1 to 35 d and 36 to 56 d, respectively, and were supplemented with either copper sulfate pentahydrate ($CuSO_4 \cdot 5H_2O$) or a commercial organic copper–lysine source to provide an additional 4, 8, 12, and 150 mg copper/kg diet. No copper source × level interaction was observed for weight gain or feed efficiency. Moreover, as compared to ducks fed the organic copper source, ducks fed the inorganic copper source exhibited significantly lower weight gains from 1 to 28 d of age but significantly greater weight gains from 29 to 56 d of age; thus, overall (1 to 56 d) there was no effect of copper source on duck growth. From 29 to 56 d of age there was a significant effect of copper level with ducks fed 12 mg/kg added copper growing faster than those fed 4 mg/kg. Also, during this period there was a significant effect of copper level on feed conversion values, with ducks fed 8 mg/kg added copper having better conversion than those fed 4 mg/kg. Ducks fed the organic copper source consumed more feed and had poorer feed efficiency values compared to birds fed inorganic copper from 29 to 56 d of age and from 1 to 56 d of age. Copper excretion and liver copper contents were higher in birds fed the organic source versus the inorganic source. Attia et al. (2012) concluded that the dietary copper requirement for adequate growth of male White Pekin ducks from 1 to 29 d of age was met by the 7 mg copper in their basal diet. However, their results show that the requirements for maximal growth and feed efficiency from 29 to 56 d of age were between 11 and 19 mg/kg diet and between 11 and 15 mg/kg diet, respectively.

Based on this single study, the recommended requirement for starting ducks is 8 mg copper/kg diet, and the level for growing ducks is 12 mg/kg diet. No recommendations were provided in the previous report (NRC, 1994).

Selenium

Since the last NRC report (1994), there has been only one study using inorganic selenium to determine the requirement for ducks. Feng (2013) added five levels of selenium as sodium selenite to a corn-soybean meal–based diet (0.04 mg selenium/kg and 15 IU added vitamin E) and fed to Jinding laying ducks from 5 to 12 weeks of age. Ducks fed 0.12 mg selenium/kg diet had significantly increased rates of gain and feed efficiency and greater relative bursal weights compared to those fed 0.04 mg/kg.

Baltić et al. (2015) and Marković et al. (2017) assessed the influence of dietary selenium from an organically bound selenium yeast product, which contained selenomethionine as the predominant selenocompound. Four levels of dietary selenium were each added to corn-soybean meal–based starter and finisher diets (containing 0.0381 mg and 0.0411 mg selenium/kg, respectively), which also contained 20 IU/kg of added vitamin E, and were fed to Cherry Valley (Pekin) ducks from hatch to 14 d of age and from 15 to 49 d of age, respectively. Dietary selenium level did not affect gain or efficiency of feed conversion from 1 to 14 d or from 1 to 49 d or the yield of breast or drumsticks and thighs (Marković et al., 2017). However, 15- to 49-d body weight gain and feed efficiency values were greater in ducks fed 0.44 mg selenium/kg diet compared to those fed either 0.04 or 0.64 mg/kg, suggesting that the dietary selenium requirement was between 0.04 and 0.44 mg/kg diet. Ducks fed 0.44 or 0.64 mg selenium/kg had higher 14-d and 49-d plasma glutathione peroxidase (GSH-Px) activities than those fed 0.04 or 0.24, but not those fed 0.64 mg/kg, suggesting that higher selenium levels are needed to maximize GSH-Px than growth (Baltić et al., 2015).

The recommended selenium level from 0 to 14 d of age in the 1994 NRC report was 0.20 mg/kg diet and was based on a study that used sodium selenite (Dean and Combs, 1981). That report did not make a recommendation for growing or breeding diets. The dietary level of selenium that promoted the highest GSH-Px in the study by Baltić et al. (2015) exceeded both the 0.3 ppm level currently permitted by the Food and Drug Administration (2017) and the NRC (1994) requirement (0.2 mg/kg). Therefore, it is recommended that the selenium requirement for growing ducks be slightly increased from the previous edition of this report to 0.25 mg/kg diet. The previous report (NRC, 1994) did not include a recommendation for breeding ducks. Therefore, since maximal egg production in two Chinese laying breeds (Shanma and Linwu) was obtained with either 0.15 mg selenium/kg diet (Huang et al., 2017) or between 0.18 and 0.24 mg selenium/kg diet (Chen et al., 2015), a breeder requirement of 0.20 mg/kg diet was recommended.

Zinc

Attia et al. (2013) investigated the influence of dietary inorganic and organic zinc on the performance of male White Pekin ducks from 1 to 56 d of age. Corn-soybean meal–based mash starter and grower-finisher diets (both containing 26.0 mg zinc/kg), fed from 1 to 35 d and 36 to 56 d, respectively, were supplemented with three levels of either zinc oxide or a commercial organic zinc source. Significant two-way interactions of zinc source × level were observed for both 1- to 56-d and 29- to 56-d weight gain and feed efficiency. Increasing dietary zinc depressed weight gains and increased feed conversion ratios from 1 to 56 d and from 29 to 56 d, while birds fed inorganic zinc had greater weight gains and feed efficiency values than ducks fed organic zinc. The addition of 30 mg zinc/kg increased weight gains, but further additions were either without effect or depressed growth. Attia et al. (2013) concluded that a dietary level of 56 mg zinc/kg promoted optimal growth of male White Pekin ducks from 1 to 56 d of age.

Since the value for starting ducks reported by Attia et al. (2013) was close to the recommended requirement of 60 mg/kg in the NRC (1994) report, the recommendation therefore remains unchanged (but is not shaded in Table 17-1), while a recommended requirement of 60 mg/kg was included for growing ducks. The previous report (NRC, 1994) did not provide an estimate for growing ducks.

Vitamins

Very few reports on the vitamin requirements of White Pekin ducks have appeared in the literature since September 1991, and no new published studies were found for vitamin K. Therefore, every vitamin requirement value in Table 17-1 was either carried over from the previous report (NRC, 1994) or is considered to be an estimate, with the exceptions of riboflavin for starter ducks and niacin and pantothenic acid for starter and grower ducks.

Vitamin A

Wei et al. (2008) investigated the vitamin A requirement of Pekin ducks from hatch to 2 weeks of age and reported that the dietary requirement of vitamin A for optimal growth and feed conversion appeared to be 2,500 IU/kg or less. In a follow-up study from the same laboratory, Wei et al. (2009) concluded that, based on weight gain and feed efficiency, a dietary vitamin A level of 2,500 IU/kg could meet the requirements for (Beijing) Pekin ducks from hatch to 3 weeks of age. Feng et al. (2019) conducted a vitamin A dose–response study with male White Pekin ducks from hatch to 21 d of age and observed xerophthalmia and depressed growth in the unsupplemented ducks. Based on a broken-line regression estimate, they suggested that the vitamin A requirement for optimal White Pekin duck weight gain from hatch to 21 d of age was 2,606 IU/kg.

The recommended dietary vitamin A requirement remains unchanged from the previous report (NRC, 1994) at 2,500 IU/kg for both starter and grower-finisher White Pekin ducks, but only the grower value is shaded in Table 17-1.

Vitamin D

Wang et al. (2010) determined the effect of different dietary contents of vitamin D on growth, feed efficiency, blood biochemistry, and tibia ash Ca in male White Pekin ducks from hatch to 14 d of age and concluded that the vitamin D requirement was between 1,000 and 2,000 IU/kg diet.

The estimated vitamin D_3 requirement has been increased to 1,500 IU/kg for both starter and grower-finisher White Pekin ducks. The previous report (NRC, 1994) recommended 400 IU/kg for each phase.

Vitamin E

Xie et al. (2018) conducted two experiments to determine the vitamin E requirement of male starter White Pekin ducklings from hatch to 21 d of age. In both studies, the unsupplemented ducks exhibited the poorest weight gains, lowest feed intakes, lowest plasma and liver α-tocopherol contents, and highest liver malondialdehyde levels, each of which were improved by the lowest level of vitamin E supplementation (5 mg/kg in Experiment 1; 10 mg/kg in Experiment 2). No further increases in weight gain or feed efficiency were observed beyond the first supplemental level. Xie et al. (2018) concluded that for starter Pekin ducks, 12.41 mg vitamin E/kg (Experiment 1) or 11.04 mg vitamin E/kg (Experiment 2) were adequate for supporting optimal growth. They also suggested that plasma and liver α-tocopherol concentrations could both serve as sensitive indicators of vitamin E status of ducks. According to the Dietary Supplement Ingredient Database (2021), 1 IU of vitamin E activity = 0.67 mg for d-α-tocopherol (natural) or 0.90 mg for d-α-tocopherol (natural) or 0.90 mg (synthetic). Since the requirement studies of Xie et al. (2018) involved both forms of the vitamin, a hybrid conversion factor of 0.80 was used.

Therefore, the estimated vitamin E requirement has been increased to 15 IU (12 mg/kg) for both starter and grower-finisher White Pekin ducks. The previous report (NRC, 1994) recommended 10 IU/kg for each phase.

Vitamin B_{12}

Although there is no evidence that poultry require cobalt other than through vitamin B_{12}, Halle et al. (2011) investigated the effects of vitamin B_{12} and cobalt supplementation on growth performance and carcass composition of male broiler chickens and White Pekin ducks. They concluded that broiler chickens and ducks both require 20 μg/kg of vitamin B_{12} for optimal growth, and that there was no overall advantage of supplementing diets with cobalt.

The recommended dietary vitamin B_{12} requirement for both starter and grower-finisher White Pekins ducks is 0.02 mg/kg. No recommendation was provided in the previous report (NRC, 1994).

Biotin

Zhu et al. (2012) investigated the biotin requirements of male White Pekin ducks from hatch to 21 d of age. Based on a broken-line regression model, 95 percent of the maximum body weight was obtained at a total dietary biotin content of between 0.26 and 0.27 mg/kg (supplemental dietary biotin level of between 0.175 and 0.185 mg/kg, respectively). Ducks fed the unsupplemented diet (0.085 mg biotin/kg) exhibited foot pad dermatitis (FPD) affecting the plantar region of the feet, and there appeared to be a major reduction in the incidence of FPD between 0.235 and 0.265 mg total biotin/kg, and then a further smaller reduction at 0.295 mg total biotin/kg. Since Zhu et al. (2012) considered both the biotin content of liver meal and the bioavailability of biotin in wheat to be low, they suggested that the actual amount of available biotin in the basal diet might be negligible. Therefore, the researchers concluded that the supplemental dietary biotin requirements of starter ducklings for optimal body weight gain and minimize foot pad lesions to be 0.180 and 0.210 mg/kg, respectively.

The recommended biotin requirement for both starter and grower-finisher White Pekin ducks is 0.2 mg/kg. No recommendation was provided in the previous report (NRC, 1994).

Choline

Like chicks and poults, ducklings are unable to synthesize choline but apparently develop this ability in later life (Scott and Dean, 1991; Leeson and Summers, 2001); however, the extent is limited and supplementation is necessary when demand exceeds biosynthesis capacity. Wen et al. (2014b) determined the choline requirements of male Z4 White Pekin ducks from hatch to 21 d of age. Weight gain and feed intake were maximized at 990 mg choline/kg diet, while 1,182 mg choline/kg diet was required to completely prevent the incidence of perosis. When data were analyzed by a broken-line regression, the choline requirements for weight gain and feed intake were 810 and 823 mg/kg, respectively. Ducks fed the unsupplemented diet (302 mg choline/kg) had the largest relative liver weights and the highest liver lipid and triglyceride contents, all of which were inversely related to the dietary choline content. In contrast, hepatic phospholipid content increased with increasing dietary choline. When subjected to broken-line regression analyses, it was determined that the starter Pekin duck required 833, 1,311, and 1,230 mg/kg choline to minimize liver lipid and triglyceride contents and to maximize hepatic total phospholipid levels, respectively. Moreover, citing a prior report in egg-type Tsaiya ducks (Lien and Jan, 1999), in which supplemental dietary choline increased serum very-low-density lipoprotein (VLDL) and apolipoprotein B during both the growing period (8 to 12 weeks of age) and the laying period (26 to 30 weeks of age), they suggested that the excessive accumulation of liver fat in choline-deficient ducks may result from impaired VLDL synthesis and secretion. Wen et al. (2014b) concluded that since the amount of dietary choline needed to prevent perosis and excess liver lipid accumulation exceeded that for growth and feed intake, the higher levels should be considered when establishing the duck's dietary choline requirement.

In a follow-up study, Wen et al. (2014a) investigated the dietary choline requirements of male White Pekin ducks from 21 to 42 d of age. Using a broken-line regression model, the researchers reported that the choline requirements of the White Pekin duck for weight gain and feed intake from 21 to

42 d of age were 980 and 950 mg/kg, respectively. As in their previous study with starter ducks, the choline requirement for minimizing total liver lipid content was higher than that for growth or feed intake and was determined by broken-line regression to be 1,130 mg/kg. Hepatic triglyceride and phospholipid contents tended to decrease and increase, respectively, with increased dietary choline, but requirements were not reported.

Wen et al. (2016) conducted an experiment to determine the effect of dietary methionine levels on the choline requirement of male White Pekin ducklings from 7 to 28 d of age. Increasing dietary levels of both methionine and choline significantly increased growth, but only the former significantly increased feed efficiency. In addition, excess choline was observed to efficiently compensate for a dietary methionine deficiency, which had been previously reported with chicks. Based on broken-line regression analyses, White Pekin ducks required 1,472 and 1,424 mg/kg choline with 0.28 percent methionine for optimal weight gain and feed intake, respectively, while with 0.48 percent methionine, the respective requirements for weight gain and feed intake were 946 and 907 mg/kg, respectively. In addition, with 0.28 percent methionine, 1,743 mg choline/kg was required to completely prevent perosis whereas 1,239 mg/kg was required with 0.48 percent methionine. Since the 0.28 percent dietary level of methionine was deficient, the requirement for choline was only considered with the methionine-adequate (0.48 percent) diets.

The recommended choline requirement for both starter and grower-finisher White Pekin ducks is 1,000 mg/kg. No recommendation was provided in the previous report (NRC, 1994).

Niacin

As previously mentioned, although tryptophan serves as a precursor for niacin synthesis, ducks are very poor converters as compared to chickens. Xie et al. (2014a) examined the effects of niacin status on the dietary tryptophan requirements of male White Pekin ducks from hatch to 21 d of age. Without supplemental niacin, ducks exhibited bowed leg deformities regardless of dietary supplemental tryptophan level, although the incidence was less with each incremental increase in dietary tryptophan. Tryptophan was more growth promotive than niacin, and niacin deficiency markedly increased the ducks' dietary tryptophan requirements. Weight gain appeared to be maximal with 40 mg supplemental nicotinamide/kg, which also completely prevented the bowed leg deformity. Feed efficiency did not appear to be affected by supplemental nicotinamide when averaged across dietary tryptophan levels. However, because of the uncertainty regarding the availability of niacin in the basal diet ingredients, it was not possible to estimate a dietary niacin requirement from this study.

Therefore, the recommended level of dietary niacin for both starter and grower-finisher White Pekin ducks was unchanged from the previous report (NRC, 1994) at 55 mg/kg.

Pantothenic Acid

Two recent studies have confirmed the NRC (1994) dietary pantothenic acid requirements of 11 mg/kg diet for starter (0 to 21 d of age; Tang et al., 2020b) and grower (15 to 42 d of age; Tang et al., 2020a) male White Pekin ducks. In both reports, ducks were fed a basal diet supplemented with crystalline d-Ca pantothenate. Starter ducks fed the lowest dietary level of pantothenic acid (4.65 mg/kg) exhibited poor growth, eyelid exudate, dermatosis, poor feathering, and 100 percent mortality, all of which were alleviated with supplementation of higher dietary levels of the vitamin. Based on average daily gain, feed conversion efficiency, and liver pantothenic acid content, the pantothenic acid requirements for starter ducks, determined by quadratic broken-line regression analysis, were 10.05, 11.01, and 11.24 mg/kg, respectively (Tang et al., 2020b). Based on a broken-line regression model, Tang et al. (2020a) concluded that the pantothenic acid requirements of growing White Pekin ducks from 15 to 42 d of age were 10.27, 12.06, and 10.79 for average daily gain, plasma pantothenic acid concentration, and liver pantothenic acid content, respectively (Tang et al., 2020a).

Although the NRC (1994) recommended pantothenic acid requirements for starter and grower ducks were not changed, they were substantiated by these new empirical studies and hence the shading and bold italics have been removed from the values in Table 17-1.

Pyridoxine

Xie et al. (2014b) conducted a pyridoxine dose–response study in male White Pekin ducks from hatch to 28 d of age. Ducks fed the unsupplemented basal diet (1.93 mg/kg) exhibited the poorest growth and feed efficiency values, which was alleviated by pyridoxine-HCl supplementation. The latter variable showed a significant quadratic response to increasing supplemental pyridoxine. Convulsions, paralysis, and poor feather development, which were three pyridoxine deficiency symptoms reported by Hegsted and Rao (1945), were not observed while severe anemia, which is another pyridoxine deficiency symptom, was not assessed by Xie et al. (2014b). Plasma alanine aminotransferase activity was not significantly affected by dietary pyridoxine, but aspartate aminotransferase activity showed a quadratic response to increasing amounts of the vitamin. Plasma levels of homocysteine, a potential biomarker for pyridoxine status, were highest in ducks fed the unsupplemented diet and were reduced by pyridoxine supplementation, most likely due to pyridoxine's role in the conversion of homocysteine to cystathionine as a cofactor of cystathionine β-synthase (D'Mello, 2003). Based on quadratic regressions, Xie et al. (2014b) predicted that the dietary pyridoxine requirement for White Pekin ducks from hatch to 28 d of age was 4.37 mg/kg for optimal feed efficiency and 4.01 mg/kg for maximal plasma aspartate aminotransferase activity.

The estimated pyridoxine requirement has been increased to 4 mg/kg for both starter and grower-finisher White Pekin ducks. The previous report (NRC, 1994) recommended 2.5 mg/kg for each phase. Based on its important role in amino acid metabolism, an increased requirement for pyridoxine would be in concert with the concomitant increased requirements for lysine, methionine, and TSAA.

Riboflavin

Tang et al. (2013) determined the dietary riboflavin requirements of male and female White Pekin ducks from hatch to 21 d of age. Ducks fed 1.2 mg riboflavin/kg in a corn-corn gluten meal–based diet grew very poorly, exhibited a leg paralysis, and had high mortality, all of which were overcome with riboflavin supplementation. Based on broken-line regression analyses, Tang et al. (2013) reported that for optimal weight gain and feed efficiency, male White Pekin ducklings required 3.31 and 5.20 mg riboflavin/kg diet, respectively, while females required 3.27 and 3.33 mg riboflavin/kg diet, respectively. As a result, a significant sex × riboflavin interaction was observed for feed efficiency, which was attributed by the authors to the higher riboflavin requirement of males versus females. According to broken-line regression analysis, riboflavin requirements for saturation of plasma and liver riboflavin concentrations/contents were 3.85 and 3.91 mg/kg for males, respectively, and 3.84 and 3.28 mg/kg for females, respectively.

In a follow-up study, Tang et al. (2014) determined the dietary riboflavin requirements of male White Pekin ducks fed more practical corn–soybean meal diets from hatch to 21 d of age. Ducks fed 1.69 mg riboflavin/kg diet grew poorly and exhibited higher mortality than birds fed supplemental riboflavin, but the leg paralysis observed previously by the researchers in ducks fed 1.2 mg riboflavin/kg diet was absent. Based on broken-line regression analyses, Tang et al. (2013) concluded that for optimal weight gain, feed efficiency, plasma riboflavin concentration, and liver riboflavin contents, male White Pekin ducks from hatch to 21 d of age required 3.01, 2.79, 4.18, and 3.63 mg/kg, respectively.

Tang et al. (2015) conducted a study to determine the riboflavin requirements of growing male White Pekin ducks from 15 to 35 d of age. Ducks fed 1.38 mg riboflavin/kg diet exhibited the poorest growth and feed efficiency and the lowest breast meat yield, but leg meat yield was not affected. Unlike starter ducks fed riboflavin-deficient diets (Tang et al., 2013), high mortality and leg paralysis were not observed. It is likely that the riboflavin-adequate starter diets allowed the ducklings to temporarily build up body reserves so that when they were presented with the unsupplemented grower diet, the deficiency was not as severe as anticipated. Based on broken-line regression analyses, Tang et al. (2015) stated that the riboflavin requirements for growing White Pekin ducks (15 to 35 d of age) were 2.43, 2.31, 3.57, and 2.78 for weight gain, feed efficiency, plasma free riboflavin, and liver free riboflavin, respectively. The researchers suggested that growing ducks required less riboflavin than starter ducks.

The recommended riboflavin requirement remains unchanged from the NRC (1994) values for both starter and grower-finisher White Pekin ducks at 4 mg/kg, although the value for starter ducks is no longer an estimate.

Feedstuffs and Form of Feed

Ducks are less affected by dietary fiber than chickens (Scott and Dean, 1991) and thus can better utilize lower energy/lower cost feedstuffs such as wheat mill by-products, distiller's dried grains and solubles, etc. The ability to tolerate high dietary levels of salt, which can be present in bakery wastes, is due to the presence of a salt gland, which was previously described. Compared to chickens and turkeys, ducks appear to be more sensitive to dietary aflatoxin (in peanut meal, corn, and other feedstuffs) and erucic acid (in rapeseed oil) (Scott and Dean, 1991).

Feedstuffs and Exogenous Enzymes

Similar to broiler chickens, feed accounts for approximately 65 to 70 percent of the cost of duck production (Chen and Applegate, 2016). Although many duck diets closely resemble diets for chickens, being largely comprised of corn and soybean meal, numerous nontraditional nutrient sources are used in duck diets worldwide and were recently reviewed by Baéza (2016). These include broken rice, full fat rice bran, extracted rice bran, palm kernel cake, distiller's dried grains with solubles (at levels higher than are fed to chickens), wheat milling by-products, coconut meal (copra meal), cottonseed and peanut meals (tested for aflatoxin content), alfalfa meal, rapeseed meal, sunflower meal, seaweed (*Polysiphonis* spp.), sweet lupins, and bitter lupins (soaked and micronized before use). Scott and Dean (1991) described other nontraditional feedstuffs, including bakery waste, fish meal, cassava, molasses, water hyacinth (*Eichhornia crassipes*) and water spinach (*Ipomoea aquatica*), duckweeds and aquatic weeds, crustacean meals, and insects. As in broiler chickens, condensed tannins in sorghum grain depressed growth of White Pekin ducks (Elkin et al., 1990, 1991; King et al., 2000b); however, as compared to chickens, ducks were less affected by dietary sorghum tannins (Elkin et al., 1990, 1991). Adeola et al. (1994) reported that pearl millet had a higher energy value than corn and appeared to be equal to corn in promoting duck weight gains. Moreover, since ducks are affected less by dietary bulk than chickens, some lower cost ingredients can be included in duck feeds at levels that might not be economically advantageous in feeds for chickens (Scott and Dean, 1991). Exogenous enzymes (phytase, xylanase, β-glucanase, amylase, cellulose, and protease) are also used in duck diets to increase nutrient utilization and digestibility of feedstuffs by reducing digesta viscosity (Adeola and Bedford, 2004; Adeola et al., 2007, 2008; Baéza, 2016; Zhang et al., 2016b).

Form of Feed

The structure of the duck's bill is not well designed for the consumption of mixed feeds in the form of a dry mash because, when mixed with saliva, a sticky paste forms and adheres to the papillae and other structures bordering the outer margin of the tongue and the upper and lower bill (Scott and Dean, 1991). Feeding wet mash diets overcame the problem of feed caking on various mouthparts, but the practice was very labor intensive since the wet mash diets were fed multiple times daily and the feed became moldy if mixed too far in advance or left in the troughs for too long a period (Scott and Dean, 1991).

Pelleting obviated the problems associated with feeding ducks moist mash diets, and although pelleted diets were fed to ducks as early as 1932 (Tallent, 1932), several decades elapsed before the practice was uniformly adopted by the North American industry. McMurray (1935) reported greater weight gains in ducks fed pelleted diets versus wet mash, which was subsequently confirmed by Heuser and Scott (1951). It has since been demonstrated that pelleting diets reduces feed spillage (Wilson, 1973; Ash, 1976; Baéza, 2016) and greatly increases duck growth and feed efficiency as compared to feeding mash diets (Wilson, 1973; Patrick and Schaible, 1980; Dean, 1986b; Elkin et al., 1991). The two most common pellet sizes for starter diets (0 to 2 weeks) are 3.18 mm (1/8 in) and 3.97 mm (5/32 in) in diameter and no more than 7.94 mm (5/16 in) in length. After 2 weeks of age, Pekin ducks can consume pellets that are 4.76 mm (3/16 in) diameter and 12.7 mm (1/2 in) in length without difficulty (Scott and Dean, 1991; Baéza, 2016). The pellet durability index (ASAE, 1997) should be ≥ 98 (Chen, 2018, personal communication).

INFLUENCE OF DIET ON CARCASS COMPOSITION AND MEAT AND EGG QUALITY

Factors Affecting Meat Quality

Numerous studies have shown that duck carcass composition and meat yield can be altered by breed, age, sex, weight, grade, temperature, and diet (reviewed by Elkin, 1987). Both wild-type mallard and Pekin ducks achieve mature body weight by about 12 weeks, when both primary and secondary wing feathering is complete (Cherry and Morris, 2008). Moreover, duck meat is all dark meat (Farrell, 1990) and leg bone and leg muscle mature earlier than breast muscle (Cherry, 1993; Gille and Salomon, 1998). As Pekin ducks age, the content of edible components as a percent of body weight increases, mostly due to a significant increase in the amount of muscle tissue; in contrast, the content of nonedible components decreases with age, due mostly to a reduction in slaughter offal with minimal changes in relative bone content (Murawska, 2012). Skin and subcutaneous fat contents also increase during the entire rearing period and stabilize (relative to body weight) after 4 weeks of age (Murawska, 2012). Muscovies are slower growing than Pekins, contain less body fat, and exhibit sexual dimorphism in body weight (males > females).

Typical table (meat-type) ducklings have higher body fat contents and lower body protein contents as compared to chickens and turkeys (Scott and Dean, 1991; Lin et al., 2018), although today there are select commercial strains of ducks with carcass fat contents similar to that of chickens. According to Farrell (1990), fat serves a different function in ducks compared to chickens. In the former, it is mainly subcutaneous and serves as insulation against cold, while in the latter, fat is a reserve of energy closely associated with lean tissue. In addition, very little body fat is found in the abdominal fat pad of ducks versus chickens (Farrell, 1990). Since most of the duck's body fat is found in the skin and subcutaneous fat, changes in this body component are highly correlated with changes in total fat content of the eviscerated carcass (Scott and Dean, 1991). Although palmitic (16:0), oleic (18:1 n-9), linoleic (18:2 n-6), and palmitoleic (16:1 n-7) acids predominate in the body fat of ducks fed common (i.e., corn-soy based) diets (Scott and Dean, 1991), carcass, egg, and hepatic fatty acid composition can be altered by changes in dietary fatty acid contents (El-Deek et al., 1997; Chen and Hsu, 2003; Baéza, 2016; Chen et al., 2017b; Du et al., 2017). Varying dietary n-6:n-3 levels were recently reported to be without effect on the growth of Cherry Valley (Pekin) ducks (Li et al., 2017). However, serum and tissue lipid levels, as well as serum, hepatic, and muscle fatty acid profiles and hepatic expression of elongase and desaturase genes, were significantly influenced by the dietary n-6:n-3 ratio (Li et al., 2017). The influence of dietary energy, protein, and individual amino acid contents on carcass traits were previously discussed in the respective sections of this chapter.

In general, ducks and chickens both exhibit a reduction in body fat and an increase in body moisture and protein as the energy/protein ratio is lowered (Dean, 1986a). Due to its tremendous ability to consume feed, the duck, unlike the chicken, can grow normally when fed pelleted very-low-energy feeds provided that the ratio of energy to protein (ME:P) and all other nutrients is maintained (Scott and Dean, 1991). Based on the work of Scott and Dean (1991), Cherry and Morris (2008) provided estimates (equations) for the effect of ME:P ratios on breast meat yield, breast and leg meat yield with or without bone, and skin and subcutaneous fat. They concluded that reducing the ME:P ratio resulted in increased breast meat and reduced skin and subcutaneous fat percentages. Interestingly, although Mule and Muscovy ducks are leaner than White Pekins, they do not apparently require higher levels of dietary protein for maximal weight gain, possibly due to their slower growth rate (Scott and Dean, 1991).

In addition, many duck strains will readily develop fatty livers when fed an excess of carbohydrates, due to an inability to adequately transport lipids out of the liver, a trait that was originally exploited by the French, who developed the Mulard duck for the purpose of producing foie gras (Scott and Dean, 1991). Typically, 12-week-old Mulard drakes are force-fed 2 to 3 corn-based meals per d for 12 to 15 d. By the end of the force-feeding period, the bird's liver is 7 to 10 times the size of a normal liver with an average weight of 550 to 982 g and a lipid content of 55.8 percent versus 76 g and 6.6 percent, respectively, in the liver of a non-force-fed drake (Skippon, 2013). A recent study by Tavernier et al. (2017) determined the genes involved in establishing hepatic steatosis in overfed Muscovy, Pekin, and Mule ducks. In Mule and Muscovy ducks, increased hepatic expression of fatty acid synthase and di-acylglycerol acyltransferase 2 genes, combined with attenuated apolipoprotein B and microsomal triglyceride transfer protein (MTTP) expression, resulted in decreased VLDL formation and lipid accumulation in the liver. MTTP expression was upregulated in Pekins, suggesting that they were able to better export lipids from the liver. France is the world's largest producer of foie gras and most likely will remain so in the foreseeable future, since production in other countries has decreased or ceased due to animal health and welfare concerns raised by animal advocates and the veterinary profession (Skippon, 2013).

With the exception of thiamine, riboflavin, and niacin, which are respectively higher, higher, and lower, duck carcass vitamin, mineral, and indispensable amino acid contents (as percent of total carcass protein) are similar to broiler chickens and turkeys (Scott and Dean, 1991). Body weight and cumulative feed consumption data for White Pekin ducks from the time of hatching to 7 weeks of age are shown in Table 17-2 and, when compared to data from NRC (1994), reflect the advances in duck growth performance that have occurred during the past quarter century (Huang et al., 2012).

FUTURE RESEARCH

When the topics or focus of the empirical research studies conducted with ducks since 1991 and cited in this report are considered, they break out approximately as follows: metabolizable energy (ME) requirements, ME contents of common feedstuffs, and fatty acid nutrition and metabolism, 14.2 percent; protein and amino acid requirements and amino acid metabolism, 38.4 percent; vitamin requirements, 11.1 percent; mineral requirements and supplemental dietary phytase, 13.1 percent; and other (anatomy and morphology, husbandry, breeding, antinutrients, egg and meat quality, etc.), 23.2 percent. Since the greatest amount of duck nutrient requirement studies conducted since the conclusion of the literature review of the previous report (NRC, 1994) has centered on protein and amino acids, this is the area where the most changes in minimum nutrient requirement recommendations were made, which likely was a reflection of the concomitant increased leanness and breast meat yield of "modern" meat-type ducks versus. those of 30 years ago. Nevertheless, studies on the minimum nutrient requirements of meat-type ducks for arginine and the branched-chain amino acids (isoleucine, leucine, and valine) remain scarce as compared to the other essential amino acids (17-1) and are targets for future duck nutrition research. It is also anticipated that research on the ducks' requirements for methionine (plus cyst[e]ine), lysine, and threonine) will continue to dominate amino acid requirement studies as new genetic lines of meat-type ducks are developed and released commercially. Also greatly lacking in the present report are empirical studies centered on vitamins and minerals, so almost all of the associated recommended nutrient requirements for meat-type ducks are listed as guidance values (Table 17-1). Of these, calcium, phosphorus, vitamin D, niacin, and choline might be considered to be the most important needs for future requirement studies. With regard to laying (egg-type) and breeder ducks, virtually no utilizable peer-reviewed reports existed. Laying ducks are not of much importance in the United States since duck egg consumption is virtually nil, but certainly future research on the nutrient needs of meat-type duck breeder hens is needed and warranted.

REFERENCES

Adeola, O. 1998. Bioavailability of tryptophan in soybean meal and tryptophan retention in the carcasses of four-week-old ducks. *Poultry Science*. 77(9):1312–1319.

Adeola, O. 2003. Energy values of feed ingredients for White Pekin ducks. *International Journal of Poultry Science*. 2(5):318–323.

Adeola, O. 2005. Metabolisable energy and amino acid digestibility of high-oil maize, low-phytate maize and low-phytate soybean meal for White Pekin ducks. *British Poultry Science*. 46(5):607–614.

Adeola, O. 2006. Review of research in duck nutrient utilization. *International Journal of Poultry Science*. 5(3):201–218.

Adeola, O. 2007. Efficiency of methionine retention in ducks. *British Journal of Nutrition*. 97(3):478–483.

Adeola, O. 2010. Phosphorus equivalency value of an *Escherichia coli* phytase in the diets of White Pekin ducks. *Poultry Science*. 89(6): 1199–1206.

Adeola, O. 2018. Phytase in starter and grower diets of White Pekin ducks. *Poultry Science*. 97:592–598.

Adeola, O., and M.R. Bedford. 2004. Exogenous dietary xylanase ameliorates viscosity-induced anti-nutritional effects in wheat-based diets for White Pekin ducks (*Anas platyrinchos domesticus*). *British Journal of Nutrition*. 92(1):87–94.

Adeola, O., C.M. Nyachoti, and D. Ragland. 2007. Energy and nutrient utilization responses of ducks to enzyme supplementation of soybean meal and wheat. *Canadian Journal of Animal Science*. 87(2):199–205.

Adeola, O., D. Ragland, and D. King. 1997. Feeding and excreta collection techniques in metabolizable energy assays for ducks. *Poultry Science*. 76(5):728–732.

Adeola, O., J.C. Rogler, and T.W. Sullivan. 1994. Pearl millet in diets of White Pekin ducks. *Poultry Science*. 73:425–435.

Adeola, O., D.J. Shafer, and C.M. Nyachoti. 2008. Nutrient and energy utilization in enzyme-supplemented starter and grower diets for White Pekin ducks. *Poultry Science*. 87(2):255–263.

Ali, M.A., and G.A. Sarker. 1992. A study on the protein and energy requirements of Muscovy ducklings. *Asian-Australasian Journal of Animal Science*. 5(1):69–73.

Applegate, T.J., D.M. Karcher, and M.S. Lilburn. 2005. Comparative development of the small intestine in the turkey poult and Pekin duckling. *Poultry Science*. 84(3):426–431.

Arthur, J. 2017. Duck eggs. *In Egg Innovations and Strategies for Improvements*. P.Y. Hester, editor. Amsterdam, The Netherlands: Elsevier 23–32.

ASAE (American Society of Agricultural Engineering). 1997. ASAE S269.4. Cubes, pellets, and crumbles—Definitions and methods for determining density, durability, and moisture. Standards 1997. American Society of Agricultural Engineering.

Ash, W.J. 1976. *Raising Ducks (Farmers' Bulletin No. 2215)*. U.S. Department of Agriculture.

Attia, Y.A. 2003. Performance, carcass characteristics, meat quality and plasma constituents of meat type drakes fed diets containing different levels of lysine with or without a microbial phytase. *Archives of Animal Nutrition*. 57(1):39–48.

Attia, Y.A., A.E. Abd Al-Hamid, H. S. Zeweil, E. M. Qota, F. Bovera, G. Monasatra, and M. D. Sahledom. 2013. Effect of dietary amounts of inorganic and organic zinc on productive and physiological traits of White Pekin ducks. *Animal*. 7(6):895–900.

Attia, Y.A., E.M. Qota, H.S. Zeweil, F. Bovera, A.E. Abd Al-Hamid, and M.D. Sahledom. 2012. Effect of different dietary concentrations of inorganic and organic copper on growth performance and lipid metabolism of White Pekin male ducks. *British Poultry Science*. 53(1):77–88.

Baéza, E. 2016. Nutritional requirements and feed management of meat type ducks. *World's Poultry Science Journal*. 72(1):5–19.

Baéza, E., M.D. Bernadet, and M. Lessire. 2012. Protein requirements for growth, feed efficiency, and meat production in growing mule ducks. *Journal of Applied Poultry Research*. 21(1):21–32.

Baltić, M.Ž., M.D. Starčević, M. Bašić, A. Zenunović, J. Ivanović, R. Marković, J. Janjić, and H. Mahmutović. 2015. Effects of selenium yeast level in diet on carcass and meat quality, tissue selenium distribution and glutathione peroxidase activity in ducks. *The Journal of Animal Feed Science and Technology*. 210:225–233.

Benjakul, S., and T. Kaewmanee. 2017. Sodium chloride preservation in duck eggs. In *Egg Innovations and Strategies for Improvements*. P.Y. Hester, editor. Amsterdam, The Netherlands: Elsevier 415–426.

Berkhoudt, H. 1978. Tastebuds in the bill of the mallard duck (*Anas platyrhynchos*). *Acta Morphologica Neerlando-Scandinavica*. 133–134.

Bi, Y., X.M. Nan, S.S. Zheng, L.S. Jiang, and B.H. Xiong. 2018. Effects of dietary threonine and immune stress on growth performance, carcass trait, serum immune parameters, and intestinal muc2 and NF-κb gene expression in Pekin ducks from hatch to 21 days. *Poultry Science*. 97(1):177–187.

Bird, R.S. 1986. The future of modern duck production, breeds, and husbandry in South-East Asia. In *Duck Production Science and World Practice*. D.J. Farrell and P. Stapleton, editors. Armidale, NSW, Australia: University of New England. 229–237.

Bons, A., R. Timmler, and H. Jeroch. 2002. Lysine requirement of growing male Pekin ducks. *British Poultry Science*. 43(5):677–686.

Chen, A.G., and Z.J. Jiang. 1998. Effect of dietary protein levels in the rearing period on egg laying performance of Shaoxing ducks. *Journal of Zhejiang University (Agriculture and Life Sciences)*. 24(5):647–652.

Chen, B.J., T.F. Shen, and R.E. Austic. 1996. Efficiency of tryptophan-niacin conversion in chickens and ducks. *Nutrition Research*. 16(1):91–104.

Chen, T.F., and J.C. Hsu. 2003. Incorporation of n-3 long-chain polyunsaturated fatty acids into duck egg yolks. *Asian-Australasian Journal of Animal Science*. 16(4):565–569.

Chen, W., S. Wang, H.X. Zhang, D. Ruan, W.G. Xia, Y.Y. Cui, C.T. Zheng, and Y.C. Lin. 2017a. Optimization of dietary zinc for egg production and antioxidant capacity in Chinese egg-laying ducks fed a diet based on corn-wheat bran and soybean meal. *Poultry Science*. 96(7):2336–2343.

Chen, W., H.X. Zhang, S. Wang, D. Ruan, X.Z. Xie, D.Q. Yu, and Y.C. Lin. 2015. Estimation of dietary selenium requirement for Chinese egg-laying ducks. *Animal Production Science*. 55(8):1056–1063.

Chen, W.L., and T.F. Shen. 1989. Comparative studies on the utilization of calcium between laying Tsaiya duck and Leghorn hen. *Asian-Australasian Journal of Animal Sciences*. 2(2):67–75.

Chen, X., and T.J. Applegate. 2016. Meat duck nutrition—formulation considerations across genetic and feedstuff resources. *Proceedings of the XXV World's Poultry Congress. Invited Lecture Papers* (N. Yang, L. Lian, J. Zheng, X. Liu, and C. Wu, editors). September 5–9, 2016, Beijing, China. 213–216.

Chen, X., X. Du, J. Shen, L. Lu, and W. Wang. 2017b. Effect of various dietary fats on fatty acid profile in duck liver: Efficient conversion of short-chain to long-chain omega-3 fatty acids. *Experimental Biology and Medicine*. 242(1):80–87.

Chen, X., R. Murdoch, Q. Zhang, D.J. Shafer, and T.J. Applegate. 2016. Effects of dietary protein concentration on performance and nutrient digestibility in Pekin ducks during aflatoxicosis. *Poultry Science*. 95(4): 834–841.

Cheng, Y.S., R. Rouvier, Y.H. Hu, J.J.L. Tai, and C. Tai. 2003. Breeding and genetics of waterfowl. *World's Poultry Science Journal*. 59(4):509–519.

Cherry, P. 1993. *Sexual Maturity in the Domestic Duck*. Ph.D. thesis, University of Reading.

Cherry, P., and T.R. Morris. 2008. *Domestic Duck Production: Science and Practice*. Wallingford, Oxfordshire, UK: CAB International.

Cherry Valley. 2004. *SM3 Commercial Management Manual*. Laceby, UK: Cherry Valley Farms.

Crompton, D.W.T., and M.C. Nesheim. 1976. Host-parasite relationships in the alimentary tract of domestic birds. *Advanced Parasitology*. 14: 95–194.

Das, L.N., D.B. Mishra, and G. Biswal. 1965. Comparative anatomy of the domestic duck (*Anas boscas*). *Indian Veterinary Journal*. 42:320–326.

Davis, H.R., and W.F. Dean. 1968. Environmental control of ducklings. *Transactions of the American Society of Agricultural and Biological Engineers*. 11(5):736–738.

Dean, W.F. 1986a. Nutrient requirements of meat-type ducks. In *Duck Production Science and World Practice*. D.J. Farrell and P. Stapleton, editors. Armidale, NSW, Australia: University of New England. 31–57.

Dean, W.F. 1986b. Nutrition of the Pekin duck in North America: An update. In *Proceedings of the Cornell Nutrition Conference for Feed Manufacturers*, Cornell University, Ithaca, New York. 44–51.

Dean, W. F. and G.F. Combs. 1981. Influence of dietary selenium on performance, tissue selenium content, and plasma concentrations of selenium-dependent glutathione peroxidase, vitamin E, and ascorbic acid in ducklings. 60(12):2655–2663.

Denbow, D.M. 2015. Gastrointestinal anatomy and physiology. In *Sturkie's Avian Physiology*, 6th ed. C.G. Scanes, editor. Amsterdam, The Netherlands: Elsevier 331–336.

Dietary Supplement Ingredient Database. 2021. https: //www.dietary supplementdatabase.usda.nih.gov.

D'Mello, J.P.F. 2003. Responses of growing poultry to amino acids. In *Amino Acids in Animal Nutrition*. J.P.F. D'Mello. CABI Digital Library. 237–263.

Du, X., Y. Liu, L. Lu, W. Wang, T. Zeng, Y. Tian, X. Xu, J. Shen, D. Niu, and Y. Lu. 2017. Effects of dietary fats on egg quality and lipid parameters in serum and yolks of Shan partridge duck. *Poultry Science*. 96(5): 1184–1190.

Dubbeldam, J.L., E.R. Brus, S.B. Menken, and S. Zeilstra. 1979. The central projections of the glossopharyngeal and vagus ganglia in the mallard, *Anas platyrhynchos* L. *Journal of Comparative Neurology*. 183(1):149–168.

El-Deek, A.A., M.O. Barakat, Y.A. Attia, and A.S. El-Sebeay. 1997. Effect of feeding Muscovy ducklings different protein sources: Performance, omega-3 fatty acid contents, and acceptability of their tissues. *Journal of the American Oil Chemists' Society*. 74(8): 999–1009.

Elkin, R.G. 1987. A review of duck nutrition research. *World's Poultry Science Journal*. 43(2):84–106.

Elkin, R.G., and M.L. Lyons. 1988. Substrate specificity of duckling hepatic and renal D-amino acid oxidase. *Poultry Science*. 67(5):851–854.

Elkin, R.G., M.L. Lyons, and J.C. Rogler. 1988a. Comparative utilization of D- and L-methionine by the White Pekin duckling (*Anas platyrhynchos*). *Comparative Biochemistry and Physiology*. 91(2):325–329.

Elkin, R.G., M.L. Lyons, and J.C. Rogler. 1988b. Utilization of DL-2-hydroxy-4-(methylthio) butanoic acid as a dietary source of methionine activity for White Pekin ducks. *Nutrition Reports International*. 37(6): 1189–1198.

Elkin, R.G., J.C. Rogler, and T.W. Sullivan. 1990. Comparative effects of dietary tannins in ducks, chicks, and rats. *Poultry Science*. 69(10): 1685–1693.

Elkin, R.G., J.C. Rogler, and T.W. Sullivan. 1991. Differential response of ducks and chicks to dietary sorghum tannins. *Journal of the Science of Food and Agriculture*. 57(4):543–553.

Elkin, R.G., T.S. Stewart, and J.C. Rogler. 1986. Methionine requirement of male White Pekin ducklings. *Poultry Science*. 65(9):1771–1776.

Evans, A.J., and A.R. Setioko. 1986. Traditional systems of layer flock management in Indonesia. In *Duck Production Science and World Practice*. D.J. Farrell and P. Stapleton, editors. Armidale, NSW, Australia: University of New England. 306–322.

Fan, H.P., M. Xie, W.W. Wang, S.S. Hou, and W. Huang. 2008. Effects of dietary energy on growth performance and carcass quality of White growing Pekin ducks from two to six weeks of age. *Poultry Science*. 87(6):1162–1164.

FAOSTAT. 2021. Ducks, tonnes of meat (2019); and ducks, head slaughtered (2019). Ducks | Gateway to poultry production and products | FAO.

Farrell, D.J. 1990. Nutrition and carcass quality in ducks. In *Recent Advances in Animal Nutrition*. W. Haresign and D.J.A. Cole, editors. London, UK: Butterworths. 177–197.

Farrell, D.J. 1995. Table egg laying ducks: Nutritional requirements and current husbandry systems in Asia. *Avian and Poultry Biology Reviews*. 6(1):55–69.

Farrell, D.J., and E.A. Martin. 1998. Strategies to improve the nutritive value of rice bran in poultry diets. III. The addition of inorganic phosphorus and a phytase to duck diets. *British Poultry Science*. 39(5):601–611.

Farrell, D.J., E. Martin, J.J. Du Preez, M. Bongarts, M. Betts, A. Sudaman, and E. Thomson. 1993. The beneficial effects of a microbial feed phytase in diets of broiler chickens and ducklings. *Journal of Animal Physiology and Animal Nutrition*. 69(1–5):278–283.

Farrell, D.J., and P. Stapleton, editors. 1986. *Duck Production: Science and World Practice*. Armidale, NSW, Australia: University of New England.

Feng, J. 2013. *Effects of Trace Element Selenium on Growth Performance and Biochemical Indexes for Cages-Rearing Growing Layer Ducks*. Thesis. Harbin, Heilongjiang, China: Northeast Agricultural University.

Feng, Y.L., M. Xie, J. Tang, W. Huang, Q. Zhang, and S.S. Hou. 2019. Effects of vitamin A on growth performance and tissue retinol of starter White Pekin ducks. *Poultry Science*. 98(5):2189–2192.

Fernandez, S.R., S. Aoyagi, Y. Han, C.M. Parsons, and D.H. Baker. 1994. Limiting order of amino acids in corn and soybean meal for growth of the chick. *Poultry Science*. 73(12):1887–1896.

Food and Drug Administration. 2017. Food additives permitted in feed and drinking water of animals. Code of Federal Regulations, Title 21, Vol. 6, 21CFR573.920. Revised April 1, 2017.

Fouad, A.M., W. Chen, D. Ruan, S. Wang, W. Xia, and C. Zheng. 2018a. Effects of dietary lysine supplementation on performance, egg quality, and development of reproductive system in egg-laying ducks. *Journal of Applied Animal Research*. 46(1):386–391.

Fouad, A.M., Y. Li, W. Chen, D. Ruan, S. Wang, W.G. Xia, Y.C. Lin, and C.T. Zheng. 2016b. Effects of dietary copper supplementation on laying performance, egg quality and plasma cholesterol fractions in laying ducks. *Pakistan Journal of Nutrition*. 15(9):878–882.

Fouad, A.M., D. Ruan, Y.C. Lin, C.T. Zheng, H.X. Zhang, W. Chen, S. Wang, W.G. Xia, and Y. Li. 2016a. Effects of dietary methionine on performance, egg quality and glutathione redox system in egg-laying ducks. *British Poultry Science*. 57(6):818–823.

Fouad, A.M., D. Ruan, S. Wang, W. Chen, W. Xia, and C. Zheng. 2018b. Nutritional requirements of meat-type and egg-type ducks: What do we know? *Journal of Animal Science and Biotechnology*. 9(1):1.

Fouad, A.M., H.X. Zhang, W. Chen, W.G. Xia, D. Ruan, S. Wang, and C.T. Zheng. 2017. Estimation of L-threonine requirements for Longyan laying ducks. *Asian-Australasian Journal of Animal Science*. 30(2):206–210.

Garcia, R.A., J.G. Phillips, and O. Adeola. 2012. Improved prediction of meat and bone meal metabolizable energy content for ducks through in vitro methods. *Poultry Science*. 91(8):1854–1859.

Gille, U., and F.-V. Salomon. 1998. Muscle growth in wild and domestic ducks. *British Poultry Science*. 39(4):500–505.

Gille, U., F.-V. Salomon, and J. Ronnert. 1999. Growth of the digestive organs in ducks with considerations on their growth in birds in general. *British Poultry Science*. 40(2):194–202.

Guémené, D., Z.D. Shi, and G. Guy. 2012. Production systems for waterfowl. In *Alternative Systems for Poultry—Health, Welfare and Productivity*. V. Sandilands and P.M. Hocking, editors. Wallingford, Oxfordshire, UK: CAB International. 128–154.

Gussekloo, S.W.S. 2006. Feeding structures in birds. In *Feeding in Domestic Vertebrates: From Structure to Behaviour*. Wallingford, Oxfordshire, UK: CAB International. 14–32.

Halas, V., Z. Szegedi, and J. Nagy. 2012. Preliminary study to determine the threonine requirement of specific humoral immune response in rearing ducks. *Acta Agriculturae Slovenica*. 100(3):269–273.

Halle, I., M. Henning, and P. Köhler. 2011. Influence of vitamin B_{12} and cobalt on growth of broiler chickens and Pekin ducks. *Landbauforschung*. 61(4):299–306.

Hammel, H.T., C. Simon-Oppermann, and E. Simon. 1980. Properties of body fluids influencing salt secretion in Pekin ducks. *The American Journal of Physiology*. 239(5):R489–R496.

He, J.H., J.B. Li, F.X. Gao, Q.H. Liu, J.C. Shu, and D.J. Liu. 2003. Dietary methionine requirement of the Chinese egg-laying duck. *British Poultry Science*. 44:741–745.

Hegsted, D.M., and M.N. Rao. 1945. Nutritional studies with ducks. II. Pyridoxine deficiency. *Journal of Nutrition*. 30(5):367–374.

Henuk, Y.L., and J.G. Dingle. 2003. Poultry manure: Source of fertilizer, fuel, and feed. *World's Poultry Science Journal*. 59(3):350–360.

Herd, R.M. 1985. Anatomy and histology of the gut of the emu *Dromaius novaehollandiae*. *Emu—Australian Ornithology*. 85(1):43–46.

Hetzel, D.J.S. 1986. Domestic ducks: An historical perspective. In *Duck Production Science and World Practice*. D.J. Farrell and P. Stapleton, editors. Armidale, NSW, Australia: University of New England.

Heuser, G.F., and M.L. Scott. 1951. Studies in duck nutrition. 1. Methods of feeding. *Poultry Science*. 30(2):161–163.

Heyn, E., K. Damme, M. Manz, F. Remy, and M.H. Erhard. 2006. Adequate water supply for Peking ducks—alternatives for bathing. *Deutsche Tierärztliche Wochenschrift*. 113(1):90–93.

Hitosugi, S., K. Tsuda, H. Okabayashi, and Y. Tanabe. 2007. Phylogenetic relationships of mitochondrial DNA cytochrome *b* gene in East Asian ducks. *Journal of Poultry Science*. 44(2):141–145.

Hoai, H.T., L.V. Kinh, T.Q. Viet, P.Y. Sy, N.V. Hop, D.K. Oanh, and N.T. Yen. 2011. Determination of the metabolizable energy content of common feedstuffs in meat-type growing ducks. *Journal of Animal Feed Science and Technology*. 170(1–2):126–129.

Hoeven-Hangoor, E. van der, N.D. Paton, I.B. van de Linde, M.W.A. Verstegen, and W.H. Hendriks. 2013. Moisture content in broiler excreta is influenced by excreta nutrient contents. *Journal of Animal Science*. 91(12):5705–5713.

Hong, D., D. Ragland, and O. Adeola. 2001. Additivity and associative effects of metabolizable energy and amino acid digestibility in barley and canola meal for White Pekin ducks. *Poultry Science*. 80(11):1600–1606.

Hong, D., D. Ragland, and O. Adeola. 2002. Additivity and associative effects of metabolizable energy and amino acid digestibility of corn, soybean meal, and wheat red dog for White Pekin ducks. *Journal of Animal Science*. 80(12):3222–3229.

Horn, N.L., S.S. Donkin, T.J. Applegate, and O. Adeola. 2009. Intestinal mucin dynamics: Response of broiler chicks and White Pekin ducklings to dietary threonine. *Poultry Science*. 88(9):1906–1914.

Horn, N.L., J.S. Radcliffe, T.J. Applegate, and O. Adeola. 2010. Gut morphology and nutrient retention responses of broiler chicks and White Pekin ducklings to dietary threonine deficiency. *Canadian Journal of Animal Science*. 90(4):513–520.

Horton, D.H. 1928. The growth of White Pekin ducklings. *Poultry Science*. 7(4):163–167.

Hou, S.S. 2016. Duck production in China. *Proceedings of the XXV World's Poultry Congress. Invited Lecture Papers* (N. Yang, L. Lian, J. Zheng, X. Liu, and C. Wu, editors). September 5–9, 2016, Beijing, China. 209–212.

Huang, J.F., H. Pingel, G. Guy, E. Lukaszewicz, E. Baéza, and S.D. Wang. 2012. A century of progress in waterfowl production, and a history of the WPSA Waterfowl Working Group. *World's Poultry Science Journal*. 68(3):551–563.

Huang, X., C. Li, G. Jiang, X. Zhang, X. Wang, Y. Hu, B. Liu, and Q. Dai. 2017. Effects of selenium supplemental level on laying performance, egg quality, serum antioxidant indices and dynamic change of egg selenium content of Linwu ducks during peak laying period. *Chinese Journal of Animal Nutrition*. 29(1):4334–4341.

Igwebuike, U.M., and T.A. Anagor. 2013. The morphology of the oropharynx and tongue of the Muscovy duck (*Cairina moschata*). *Veterinarski arhiv*. 83(6):685–693.

Jamroz, D., K. Jakobsen, J. Orda, J. Skorupinska, and A. Wiliczkiewicz. 2001. Development of the gastrointestinal tract and digestibility of dietary fibre and amino acids in young chickens, ducks, and geese fed diets with high amounts of barley. *Comparative Biochemistry and Physiology Part A: Molecular Integrative Physiology*. 130(4):643–652.

Jamroz, D., T. Wertelecki, A. Wiliczkiewicz, J. Orda, and J. Skorupińska. 2004. Dynamics of yolk sac resorption and post-hatching development of the gastrointestinal tract in chickens, ducks and geese. *Journal of Animal Physiology and Animal Nutrition*. 88(5–6):239–250.

Jamroz, D., A. Wiliczkiewicz, J. Orda, T. Wertelecki, and J. Skorupinska. 2002. Aspects of development of digestive activity of intestine in young chickens, ducks and geese. *Journal of Animal Physiology and Animal Nutrition*. 86(11–12):353–366.

Jiang, Y., J. Tang, M. Xie, Z.G. Wen, S.Y. Qiao, and S.S. Hou. 2017. Threonine supplementation reduces dietary protein and improves lipid metabolism in Pekin ducks. *British Poultry Science*. 58(6):687–693.

Jiang, Y., M. Uzma, J. Tang, Z.G. Wen, S.S. Hou, W. Huang, and M. Xie. 2016. Effects of dietary protein on threonine requirements of Pekin ducks from hatch to 21 days of age. *Journal of Animal Science and Technology*. 217:95–99.

Jiang, Y., Y.W. Zhu, M. Xie, J. Tang, Z.G. Wen, S.Y. Qiao, and S.S. Hou. 2018. Interactions of dietary protein and threonine on growth performance in Pekin ducklings from 1 to 14 days of age. *Poultry Science*. 97(1):262–266.

Jung, Y., and Y.P Zhou. 1980. The Pekin duck in China. *World Animal Review*. 34:11–14

Kidd, M.T. 2000. Nutritional considerations concerning threonine in broilers. *World's Poultry Science Journal*. 56(2):139–151.

King, D.E., E.K. Asem, and O. Adeola. 2000a. Ontogenetic development of intestinal digestive functions in White Pekin ducks. *Journal of Nutrition*. 130(1):57–62.

King, D., M.Z. Fan, G. Ejeta, E.K. Asem, and O. Adeola. 2000 b. The effects of tannins on nutrient utilization in the White Pekin duck. *British Poultry Science*. 41(5):630–639.

King, D., D. Ragland, and O. Adeola. 1997. Apparent and true metabolizable energy values of feedstuffs for ducks. *Poultry Science*. 76(10): 1418–1423.

Kluge, H., D.K. Gessner, E. Herzog, and K. Eder. 2016. Efficacy of DL-methionine hydroxy analogue-free acid in comparison to DL-methionine in growing male white Pekin ducks. *Poultry Science*. 95(3):590–594.

Kluth, H., and M. Rodehutscord. 2006. Comparison of amino acid digestibility in broiler chickens, turkeys, and Pekin ducks. *Poultry Science*. 85(11):1953–1960.

Koch, T. 1973. *Anatomy of the Chicken and Domestic Birds*. Ames, IA: Iowa State University Press.

Kong, C., and O. Adeola. 2010. Apparent ileal digestibility of amino acids in feedstuffs for White Pekin ducks. *Poultry Science*. 89(3):545–550.

Kong, C., and O. Adeola. 2013a. Comparative amino acid digestibility for broiler chickens and White Pekin ducks. *Poultry Science*. 92(9): 2367–2374.

Kong, C., and O. Adeola. 2013b. Additivity of amino acid digestibility in corn and soybean meal for broiler chickens and White Pekin ducks. *Poultry Science*. 92(9):2381–2388.

Lee, A.R., and S. Haynes. 1933. *Duck Raising (Farmers' Bulletin No. 697)*. U.S. Department of Agriculture.

Leeson, S., and J.D. Summers. 2001. *Nutrition of the Chicken*, 4th ed. University Books.

Leeson, S., and J.D. Summers. 2005. *Commercial Poultry Nutrition*, 3rd ed. Nottingham, England: Nottingham University Press.

Lewis, A.J., and D.H. Baker. 1995. Bioavailability of D-amino acids and DL-hydroxy-methionine. In *Bioavailability of Nutrients for Animals: Amino Acids, Minerals, and Vitamins*. C.B. Ammerman, D.H. Baker, and A.J. Lewis, editors. San Diego, CA: Academic Press. 67–81.

Li, H.F., W.Q. Zhu, W. T. Song, J.T. Shu, W. Han, and K.-W. Chen. 2010. Origin and genetic diversity of Chinese domestic ducks. *Molecular Phylogenetics and Evolution*. 57(2):634–640.

Li, M., S. Zhai, Q. Xie, L. Tian, X. Li, J. Zhang, H. Ye, Y. Zhu, L. Yang, and W. Wang. 2017. Effects of dietary n-6: n-3 PUFA ratios on blood lipid levels and fatty acid profile of Cherry Valley ducks at 15–42 days of age. *Journal of Agricultural and Food Chemistry*. 65(46):9995–10002.

Li, X., G. Yuan, X. Chen, Y. Guo, N. Yang, J. Pi, H. Zhang, and J. Zheng. 2018. Fishy odor and TMA content levels in duck egg yolks. *Journal of Food Science*. 83(1):39–45.

Lien, T.F., and D.F. Jan. 1999. The effect on the lipid metabolism of Tsaiya ducks when high levels of choline or methionine are added to the ducks' diet. *Asian-Australasian Journal of Animal Science*. 12(7):1090–1095.

Lin, F.B., F. Zhu, J.P. Hao, F.X. Yang, and Z.C. Hou. 2018. In vivo prediction of the carcass fatness using live body measurements in Pekin ducks. *Poultry Science*. 97(7):2361–2365.

Lin, J.H., C.H. Su, S.R. Lee, T.M. Su, S.C. Chang, and M.D. Koh. 2011. A survey of duck excreta: Quantity and composition. *Journal of Taiwan Livestock Research*. 44(1):175–188.

Liu, Y., J.M. Yuan, L.S. Zhang, Y.R. Zhang, S.M. Cai, J.H. Yu, and Z.F. Xia. 2015. Effects of tryptophan supplementation on growth performance, antioxidative activity, and meat quality of ducks under high stocking density. *Poultry Science*. 94(8):1894–1901.

Marković, R., M. Glišić, M. Bošković, and M.Ž. Baltić. 2017. New scientific challenges–the possibilities of using selenium in poultry nutrition and impact on meat quality. *IOP Conference Series: Earth and Environmental Science*. 85(1):012032.

Martin, E.A., J.V. Nolan, Z. Nitsan, and D.J. Farrell. 1998. Strategies to improve the nutritive value of rice bran in poultry diets. IV. Effects of addition of fish meal and a microbial phytase to duckling diets on bird performance and amino acid digestibility. *British Poultry Science*. 39(5):612–621.

McDowell, L.R. 2008. *Vitamins in Animal and Human Nutrition*. Hoboken, NJ: Wiley-Blackwell.

McMurray, M. 1935. Table poultry experiments. Table ducklings. Growers ration versus fattening ration. *Harper Adams Utility. Poultry Journal*. 20:498–499.

Men, B.X., R.B. Ogle, and J.E. Lindberg. 2002a. Studies on integrated duck–rice systems in the Mekong Delta of Vietnam. *Journal of Sustainable Agricultural Sciences*. 20(1):27–40.

Men, B.X., B. Ogle, and J.E. Lindberg. 2002b. Effect of diet and management system on growing duck performance in the Mekong Delta of Vietnam. *Journal of Sustainable Agricultural Sciences*. 20(3):21–32.

Ministry of Agriculture of China. 2012. *Nutrient Requirements of Meat-Type Ducks*. Beijing, China: Agriculture Press.

Mohamed, K., B. Leclercq, A. Anwar, H. El-Alaily, and H. Soliman. 1984. A comparative study of metabolizable energy in ducklings and domestic chicks. *Journal of Animal Science and Technology*. 11(3):199–209.

Moran, E.T. 1986. Digestive physiology of the duck. In *Duck Production Science and World Practice*. D.J. Farrell and P. Stapleton, editors. Armidale, NSW, Australia: University of New England. 6–15.

Murawska, D. 2012. The effect of age on the growth rate of tissues and organs and the percentage content of edible and nonedible carcass components in Pekin ducks. *Poultry Science*. 91(8):2030–2038.

Norberg, S.E., R.N. Dilger, H. Dong, B.G. Harmon, O. Adeola, and M.A. Latour. 2004. Utilization of energy and amino acids of spray-dried egg, plasma protein, and soybean meal by ducks. *Poultry Science*. 83(6): 939–945.

NRC (National Research Council). 1994. *Nutrient Requirements of Poultry*, 9th rev. ed. Washington, DC: National Academy Press.

Olver, M.D. 1995. Effect of restricted feeding during the rearing period and a "forced moult" at 40 weeks of production on the productivity of Pekin breeder ducks. *British Poultry Science*. 36(5):737–746.

Onbasilar, E.E., and S. Yalcin. 2018. Fattening performance and meat quality of Pekin ducks under different rearing systems. *Worlds' Poultry Science Journal*. 74(1):61–68.

Orban, J.I., O. Adeola, and R. Stroshine. 1999. Microbial phytase in finisher diets of White Pekin ducks: Effect on growth performance, plasma phosphorus concentration, and leg bone characteristics. *Poultry Science*. 78(3):366–377.

Ostrowski-Meissner, H.T. 1984. A method for simultaneous measurement of apparent (AME) and true (TME) metabolizable energy with ducks and comparisons of data obtained with drakes and cockerels. *Nutrition Reports International*. 29(1):1239–1248.

Paştea, E., A. Nicolau, V. Popa, and I. Roşca. 1968. Dynamics of the digestive tract in hens and ducks. *Acta Physiologica Academiae Scientiarum Hungaricae*. 33(1):305–310.

Patrick, H., and P.J. Schaible. 1980. Duck and geese feeds-feeding. In *Poultry: Feeds and Nutrition*, 2nd ed. Roslyn, NY: Avi Publishing.

Pisharody, C.R.R., and S.G. Nair. 1972. A comparative study on certain aspects of digestive physiology of chicks and ducklings. *Indian Veterinary Journal*. 49(1):890–897.

Prahkarnkaeo, K., K. Choowongkomon, B. Chomtee, C. Rakangthong, and C. Bunchasak. 2017. Re-estimation of supplemented methionine as total sulfur amino acid requirement for commercial male meat-type ducks. *Thai Journal of Veterinary Medicine*. 47(4):457–464.

Pyle, D.A. 2017. Biosecurity considerations for commercial waterfowl. In *A Practical Guide for Managing Risk in Poultry Production*, 2nd ed. R.L. Owen, editor. Jacksonville, FL: American Association of Avian Pathologists. 301–309.

Qu, L., W. Liu, F. Yang, Z. Hou, J. Zheng, G. Xu, and N. Yang. 2009. Origin and domestication history of Peking ducks determined through microsatellite and mitochondrial marker analysis. *Science China. Series C: Life Sciences*. 52(11):1030–1035.

Qureshi, A.S., T. Faisal, M.K. Saleemi, and M.Z. Ali. 2017. Histological and histometric alterations in the digestive tract and accessory glands of duck (*Anas platyrhynchos*) with sex and progressive age. *Journal of Animal Plant Science*. 27(1):1528–1533.

Ragland, D., D. King, and O. Adeola. 1997. Determination of metabolizable energy contents of feed ingredients for ducks. *Poultry Science*. 76(9): 1287–1291.

Rauch, E., N. Firnkäs, N. Hirsch, K. Damme, P. Schmidt, M.H. Erhard, and S. Bergmann. 2016a. Use of the AquaDuc T® round drinker with Pekin ducks under field conditions—behaviour as one indicator of welfare. *Berliner und Munchener Tierarztliche Wochenschrift*. 129(1–2):28–39.

Rauch, E., N. Hirsch, N. Firnkäs, K. Damme, M.H. Erhard, and S. Bergmann. 2016b. Animal hygiene, water quality and animal health using round drinkers as an animal-friendly water supply for Pekin ducks under practical conditions. *Berliner und Munchener Tierarztliche Wochenschrift*. 129(1–2):15–27.

Rodehutscord, M. 2006. Optimising the use of phosphorus sources in growing meat ducks. *World's Poultry Science Journal*. 62(3):513–523.

Rodehutscord, M., and A. Dieckmann. 2005. Comparative studies with three-week-old chickens, turkeys, ducks, and quails on the response in phosphorus utilization to a supplementation of monobasic calcium phosphate. *Poultry Science*. 84(8):1252–1260.

Rodehutscord, M., R. Hempel, and P. Wendt. 2006. Phytase effects on the efficiency of utilisation and blood concentrations of phosphorus and calcium in Pekin ducks. *British Poultry Science*. 47(3):311–321.

Rodehutscord, M., R. Timmler, and P. Wendt. 2003. Response of growing Pekin ducks to supplementation of monobasic calcium phosphate to low-phosphorus diets. *Poultry Science*. 82(2):309–319.

Rodwell, V.W. 2012. Conversion of amino acids to specialized products. In *Harper's Illustrated Biochemistry*, 29th ed. R.K. Murray, D.A. Bender, K.M. Botham, P.J. Kennelly, V.W. Rodwell, and P.A. Weil, editors. Columbus, OH: McGraw Hill Medical. 297–306.

Rush, J.K., C.R. Angel, K.M. Banks, K.L. Thompson, and T.J. Applegate. 2005. Effect of dietary calcium and vitamin D_3 on calcium and phosphorus retention in White Pekin ducklings. *Poultry Science*. 84(4): 561–570.

Schmidt-Nielsen, K., and Y.T. Kim. 1964. The effect of salt intake on the size and function of the salt gland of ducks. *The Auk*. 81(2):160–172.

Scott, M.L., and W.F. Dean. 1991. *Nutrition and Management of Ducks*. M.L. Scott & Associates.

Scott, M.L., M.C. Nesheim, and R.J. Young. 1982. *Nutrition of the Chicken*, 3rd ed. M.L. Scott & Associates.

Shen, T.F. 1986. Nutrient requirements of egg-laying ducks. In *Duck Production Science and World Practice*. D.J. Farrell and P. Stapleton, editors. Armidale, NSW, Australia: University of New England. 16–30.

Shen, T.F., and W.F. Dean. 1982. True metabolizable energy value of corn and soybean meal for ducks. *Poultry Science*. 61(7):1543.

Shen, T.F., and Y.T. Li. 1997. Threonine requirement of laying Tsaiya ducks. *Journal of Taiwan Livestock Research*. 30(1):387–394.

Shen, T.F., Y.T. Li, and C.M. Pan. 1996. Valine requirements of laying Tsaiya ducks. *Journal of the Chinese Society of Animal Science*. 25(1): 373–380.

Siregar, A.P., and D.J. Farrell. 1980. A comparison of the energy and nitrogen metabolism of fed ducklings and chickens. *British Poultry Science*. 21(3):213–227.

Skieresz-Szewczyk, K., and H. Jackowiak. 2016. Morphofunctional study of the tongue in the domestic duck (*Anas platyrhynchos f. domestica*, Anatidae): LM and SEM study. *Zoomorphology*. 135(2):255–268.

Skippon, W. 2013. The animal health and welfare consequences of foie gras production. *Canadian Veterinary Journal*. 54(4):403–404.

Sritiawthai, E., S. Sakulthai, J. Sakdee, C. Bunchasak, C. Kaewtapee, and T. Poeikhampha. 2013. Effect of protein level and dietary energy on production, intestinal morphology and carcass yield of meat duck during starter phase of 14 days. *Journal of Applied Science*. 13(1):315–320.

Steczny, K., D. Kokoszynski, Z. Bernacki, R. Wasilewski, and M. Saleh. 2017. Growth performance, body measurements, carcass composition and some internal organ characteristics in young Pekin ducks. *South African Journal of Animal Science*. 47(3):399–406.

Stevens, L. 1996. *Avian Biochemistry and Molecular Biology*. Cambridge, UK: Cambridge University Press.

Swanson, G.A., M.I. Meyer, and V.A. Adomaitis. 1985. Foods consumed by breeding mallards on wetlands of south-central North Dakota. *The Journal of Wildlife Management*. 49(1):197–203.

Tallent, V.K. 1932. Pellets versus mash for table duck production. *Harper Adams Utility Poultry Journal*. 17(1):539–543.

Tang, J., J. Hu, Z. G. Wen, Y. Jiang, H. Al-Kateb, W. Huang, Y.M. Guo, M. Xie, and S.S. Hou. 2015. Effects of riboflavin supplementation on growth performance, carcass traits, and riboflavin status of growing male White Pekin ducks. *Journal of Animal Feed Science and Technology*. 209:274–279.

Tang, J., M. Xie, J. Yang, Z.G. Wen, Y.W. Zhu, W. Huang, and S.S. Hou. 2013. Riboflavin requirements of White Pekin ducks from hatch to 21 d of age. *British Poultry Science*. 54(3):407–411.

Tang, J., Z.G. Wen, Z.B. Guo, W. Huang, Y.M. Guo, M. Xie, and S.S. Hou. 2014. Dietary riboflavin supplementation improves the growth performance and antioxidant status of starter White Pekin ducks fed a corn-soybean meal diets. *Livestock Science*. 170:131–136.

Tang, J., B. Zhang, S. Liang, Y. Wu, Y. Feng, Z. Guo, G. Xing, J. Jiao, Z. Zhou, M. Xie, and S. Hou. 2020a. Effects of pantothenic acid on growth performance and antioxidant status of growing male White Pekin ducks. *Poultry Science*. 99(9):4436–4441.

Tang, J., B. Zhang, M. Xue, W.B. Shi, Y.B. Wu, Y.L. Feng, W. Huang, Z.K. Zhou, M. Xie, and S.S. Hou. 2020b. Pantothenic acid requirement of male White Pekin ducks from hatch to 21 days of age. *Journal of Animal Science and Technology*. 269:114637.

Tao, Z., W. Song, C. Zhu, W. Xu, H. Liu, S. Zhang, and H. Li. 2017. Comparative transcriptomic analysis of high and low egg-producing duck ovaries. *Poultry Science*. 96(12):4378–4388.

Tavernier, A., S. Davail, K. Ricaud, B. M.-D., and K. Gontier. 2017. Genes involved in the establishment of hepatic steatosis in Muscovy, Pekin and mule ducks. *Molecular and Cellular Biochemistry*. 424(1): 147–161.

Thiele, H.H., 2016. Breeding Pekin ducks for meat production. VOL49-THIELE-Breeding-Pekin-Ducks.pdf.

Thongwittaya, N., P. Pleusamran, N. Choktaworn, and I. Tasaki. 1992. Energy and protein requirements of Khaki Campbell × Thai native growing ducks. *Asian-Australasian Journal of Animal Science*. 5(2):357–363.

Timmler, R., and M. Rodehutscord. 2003. Dose–response relationships for valine in the growing White Pekin duck. *Poultry Science*. 82(11): 1755–1762.

Veltmann, J.R., and J.S. Sharlin. 1981. Influence of water deprivation on water consumption, growth, and carcass characteristics of ducks. *Poultry Science*. 60(3):637–642.

Wan, H.F., W. Chen, Z.L. Qi, P. Peng, and J. Peng. 2009. Prediction of true metabolizable energy from chemical composition of wheat milling by-products for ducks. *Poultry Science*. 88(1):92–97.

Wang, C., S.S. Hou, W. Huang, T.S. Xu, G.H. Rong, and M. Xie. 2014a. Arginine affects appetite via nitric oxide in ducks. *Poultry Science*. 93(8):2048–2053.

Wang, C., M. Xie, W. Huang, J.J. Xie, J. Tang, and S.S. Hou. 2013a. Arginine requirements of White Pekin ducks from 1 to 21 days of age. *Poultry Science*. 92(4):1007–1010.

Wang, C., M. Xie, W. Huang, J.Y. Yu, Y.L. Feng, and S.S. Hou. 2013b. Dietary arginine from 15 to 35 days of age affects feed intake and carcass yield in White Pekin ducks. *Animal Production Science*. 53(10): 1041–1045.

Wang, C., A.J. Zheng, M. Xie, W. Huang, J.J. Xie, and S.S. Hou. 2014b. Hypothalamic protein profiles associated with inhibited feed intake of ducks fed with insufficient dietary arginine. *Animal*. 8(7):1113–1118.

Wang, S., W. Chen, H.X. Zhang, D. Ruan, and Y.C. Lin. 2014c. Influence of particle size and calcium source on production performance, egg quality, and bone parameters in laying ducks. *Poultry Science*. 93(10): 2560–2566.

Wang, S., S.S. Hou, M. Xie, W. Huang, and J. Yu. 2010. Effect of different concentrations of vitamin D on growth performance, blood biochemistry and tibia parameter for White Pekin ducklings. *Acta Veterinaria et Zootechnica Sinica*. 41:1414–1420.

Wang, Y.S., S.S. Hou, W. Huang, L. Zhao, H.P. Fan, M. Xie, and L.H. Wang. 2006. Lysine, methionine and tryptophan requirements of Beijing ducklings of 0–2 weeks of age. *Agricultural Sciences in China*. 5(3):228–233.

Wasilewski, R., D. Kokoszyński, A. Mieczkowska, Z. Bernacki, and A. Górska. 2015. Structure of the digestive system of ducks depending on sex and genetic background. *Acta Veterinaria Brno*. 84(2):153–158.

Watkins, E.J., P.J. Butler, and B.P. Kenyon. 2004. Posthatch growth of the digestive system in wild and domesticated ducks. *British Poultry Science*. 45(3):331–341.

Wei, L.M., S.S. Hou, W. Huang, M. Xie, T. Zhang, and F.Z. Liu. 2009. Effects of vitamin A on performance and intestinal development of Beijing ducks from 0 to 3 weeks of age. *Chinese Journal of Animal Nutrition*. 21(1):462–467.

Wei, L.M., F.Z. Liu, S.S. Hou, W. Huang, and M. Xie. 2008. Vitamin A requirement of Peking ducks from hatch to 2 weeks of age. *Acta Agriculturae Boreali-Occidentalis Siniaca*. 98(5):26–28.

Wen, Z.G., S.S. Hou, J. Tang, Y.L. Feng, W. Huang, Y.M. Guo, and M. Xie. 2014a. Choline requirements of male White Pekin ducks from 21 to 42 d of age. *British Poultry Science*. 55(4):548–552.

Wen, Z.G., T.J. Rasolofomanana, J. Tang, Y. Jiang, M. Xie, P.L. Yang, and S.S. Hou. 2017. Effects of dietary energy and lysine levels on growth performance and carcass yields of Pekin ducks from hatch to 21 days of age. *Poultry Science*. 96(9):3361–3366.

Wen, Z.G., J. Tang, S.S. Hou, Y.M. Guo, W. Huang, and M. Xie. 2014b. Choline requirements of White Pekin ducks from hatch to 21 days of age. *Poultry Science*. 93(12):3091–3096.

Wen, Z.G., J. Tang, M. Xie, P.L. Yang, and S.S. Hou. 2016. Effects of dietary methionine levels on choline requirements of starter white Pekin ducks. *Asian-Australasian Journal of Animal Science*. 29(12):1742–1747.

Wendt, P., and M. Rodehutscord. 2004. Investigations on the availability of inorganic phosphate from different sources with growing White Pekin ducks. *Poultry Science*. 83(9):1572–1579.

Wendt, P., R. Timmler, and M. Rodehutscord. 2003. Dose–response relationship between a supplemented microbial phytase and utilisation of phosphorus from a maize/soybean meal-based diet in pekin ducks. *Archiv für Geflügelkunde*. 67(5):193–197.

Wickramasuriya, S.S., J. Yoo, J.C. Kim, and J.M. Heo. 2016. The apparent metabolizable energy requirement of male Korean native ducklings from hatch to 21 days of age. *Poultry Science*. 95(1):77–83.

Wilson, B.J. 1973. Effects of diet form on the performance of table ducklings. *British Poultry Science*. 14(6):589–593.

Wilson, B.J. 1975. The performance of male ducklings given starter diets with different concentrations of energy and protein. *British Poultry Science*. 16(6):617–625.

Wu, L. Y., Y.J. Fang, and X.Y. Guo. 2011. Dietary L-arginine supplementation beneficially regulates body fat deposition of meat-type ducks. *British Poultry Science*. 52(2):221–226.

Wu, Y., J. Tang, J. Cao, B. Zhang, Y. Chen, M. Xie, Z. Zhou, and S. Hou. 2021. Effect of dietary L-methionine supplementation on growth performance, carcass traits, and plasm parameters of starter Pekin ducks at different dietary energy levels. *Animals*. 11(1):144.

Xia, W., A.M. Fouad, W. Chen, D. Ruan, S. Wang, Q. Fan, Y. Wang, Y. Cui, and C. Zheng. 2017. Estimation of dietary arginine requirements for Longyan laying ducks. *Poultry Science*. 96(1):144–150.

Xia, W.G., H.X. Zhang, Y.C. Lin, and C.T. Zheng. 2015. Evaluation of dietary calcium requirements for laying Longyan shelducks. *Poultry Science*. 94(12):2932–2937.

Xia, W.G., H.X. Zhang, Y.C. Lin, C.T. Zheng, L. Wang, W. Chen, D. Ruan, and S.A. Wang. 2014. Effects of dietary metabolizable energy and crude protein levels on laying performance of laying ducks. *Chinese Journal of Animal Nutrition*. 26(1):3599–3607.

Xie, M., Y. Guo, T. Zhang, S. Hou, and W. Huang. 2009a. Lysine requirement of male White Pekin ducklings from seven to twenty-one days of age. *Asian-Australasian Journal of Animal Science*. 22(10):1386–1390.

Xie, M., X.F. Han, W. Huang, and S.S. Hou. 2014a. Effects of niacin status on tryptophan requirements of starter white Pekin ducks. *Livestock Science*. 159:75–78.

Xie, M., S.S. Hou, and W. Huang. 2006. Methionine requirements of male White Peking ducks from twenty-one to forty-nine days of age. *Poultry Science*. 85(4):743–746.

Xie, M., S.S. Hou, W. Huang, and H.P. Fan. 2007. Effect of excess methionine and methionine hydroxy analogue on growth performance and plasma homocysteine of growing Pekin ducks. *Poultry Science*. 86(9): 1995–1999.

Xie, M., S.S. Hou, W. Huang, L. Zhao, J.Y. Yu, W.Y. Li, and Y.Y. Wu. 2004. Interrelationship between methionine and cystine of early Peking ducklings. *Poultry Science*. 83(10):1703–1708.

Xie, M., Y. Jiang, J. Tang, Z.G. Wen, Q. Zhang, W. Huang, and S.S. Hou. 2017a. Effects of low-protein diets on growth performance and carcass yield of growing White Pekin ducks. *Poultry Science*. 96(5):1370–1375.

Xie, M., Y. Jiang, J. Tang, Q. Zhang, W. Huang, and S.S. Hou. 2017b. Starter and subsequent grower response of Pekin ducks to low-protein diets in starter phase. *Livestock Science*. 203:92–96.

Xie, M., J. Tang, Z. Wen, W. Huang, and S. Hou. 2014b. Effects of pyridoxine on growth performance and plasma aminotransferases and homocysteine of White Pekin ducks. *Asian-Australasian Journal of Animal Science*. 27(12):1744–1748.

Xie, M., S.X. Wang, S.S. Hou, and W. Huang. 2009b. Interaction between dietary calcium and non-phytate phosphorus on growth performance and bone ash in early White Pekin ducklings. *Journal of Animal Science and Technology*. 151(1–2):161–166.

Xie, M., S.X. Wang, S.S. Hou, W. Huang, L. Zhao, and J.Y. Yu. 2009c. Calcium and phosphorus requirements of Pekin ducks from 3 to 6 week of age. *Chinese Journal of Animal Nutrition*. 21:25–30.

Xie, M., S. Wang, W. Huang, and S.S. Hou. 2018. Effects of vitamin E on growth performance, tissue α-tocopherol, and lipid peroxidation of starter White Pekin ducks. *Poultry Science*. 97(6):2139–2143.

Xie, M., L. Zhang, Z. G. Wen, J. Tang, W. Huang, and S.S. Hou. 2014c. Threonine requirement of White Pekin ducks from hatch to 21 d of age. *British Poultry Science*. 55(4):553–557.

Xie, M., J.N. Zhao, S.S. Hou, and W. Huang. 2010. The apparent metabolizable energy requirement of White Pekin ducklings from hatch to 3 weeks of age. *Journal of Animal Science and Technology*. 157(1–2):95–98.

Xue, J.J., M. Xie, J. Tang, W. Huang, Q. Zhang, and S.S. Hou. 2018. Effects of excess DL- and L-methionine on growth performance of starter Pekin ducks. *Poultry Science*. 97(3):946–950.

Yu, R, and T. Shen. 1984. Leucine, isoleucine and valine requirements of Mule ducklings. *Journal of the Chinese Society of Animal Science*. 13:1–15.

Zeng, Q.F., P. Cherry, A. Doster, R. Murdoch, O. Adeola, and T.J. Applegate. 2015a. Effect of dietary energy and protein content on growth and carcass traits of Pekin ducks. *Poultry Science*. 94(3):384–394.

Zeng, Q.F., Q. Zhang, X. Chen, A. Doster, R. Murdoch, M. Makagon, A. Gardner, and T.J. Applegate. 2015b. Effect of dietary methionine content on growth performance, carcass traits, and feather growth of Pekin duck from 15 to 35 days of age. *Poultry Science*. 94(7): 1592–1599.

Zhang, Q., L. Xu, A. Doster, R. Murdoch, P. Cotter, A. Gardner, and T.J. Applegate. 2014. Dietary threonine requirement of Pekin ducks from 15 to 35 days of age based on performance, yield, serum natural antibodies, and intestinal mucin secretion. *Poultry Science*. 93(8):1972–1980.

Zhang, Q., Q.F. Zeng, P. Cotter, and T.J. Applegate. 2016a. Dietary threonine response of Pekin ducks from hatch to 14 d of age based on performance, serology, and intestinal mucin secretion. *Poultry Science*. 95(6):1348–1355.

Zhang, X., H. Li, X. Huang, G. Jiang, X. Wang, C. Li, D. Wu, and Q. Dai. 2016b. Effects of compound enzymes on nutrient utilization rate and metabolizable energy of several meal type feed ingredients of Linwu ducks. *Chinese Journal of Animal Nutrition*. 28:135–141.

Zhang, Y.N., R.S. Xu, L. Min, D. Ruan, H. Y. Kim, Y.G. Hong, W. Chen, S. Wang, W.G. Xia, X. Luo, C.Y. Xie, X.G. Shang, and C.T. Zheng. 2019. Effects of L-methionine on growth performance, carcass quality, feather traits, and small intestinal morphology of Pekin ducks compared with conventional DL-methionine. *Poultry Science*. 98(12): 6866–6872.

Zhao, F., H.F. Zhang, S.S. Hou, and Z.Y. Zhang. 2008. Predicting metabolizable energy of normal corn from its chemical composition in adult Pekin drakes. *Poultry Science*. 87(8):1603–1608.

Zhao, L., N.Y. Zhang, Y.X. Pan, L.Y. Zhu, D.I. Batonon-Alavo, L.B. Ma, M.M. Khalil, D.S. Qi, and L.V. Sun. 2018. Efficacy of 2-hydroxy-4-methylthiobutanoic acid compared to DL-methionine on growth performance, carcass traits, feather growth, and redox status of Cherry Valley ducks. *Poultry Science*. 97(9):3166–3175.

Zhou, Y. F., Y.Q. Liu, H.K. Wei, and J. Peng. 2017. Estimation of the optimum digestible lysine level for Cherry Valley ducks. *Poultry Science*. 96(4):838–843.

Zhou, Z., M. Li, H. Cheng, W. Fan, Z. Yuan, Q. Gao, Y. Xu, Z. Guo, Y. Zhang, J. Hu, H. Liu, D. Liu, W. Chen, Z. Zheng, Y. Jiang, Z. Wen, Y. Liu, H. Chen, M. Xie, Q. Zhang, W. Huang, W. Wang, S. Hou, and Y. Jiang. 2018. An intercross population study reveals genes associated with body size and plumage color in ducks. *Nature Communications*. 9(1):2648.

Zhu, Y.W., J. Wen, X.X. Jiang, W.C. Wang, and L. Yang. 2018. High calcium to phosphorus ratio impairs growth and bone mineralization in Pekin ducklings. *Poultry Science*. 97(4):1163–1169.

Zhu, Y.W., M. Xie, W. Huang, L. Yang, and S.S. Hou. 2012. Effects of biotin on growth performance and foot pad dermatitis of starter White Pekin ducklings. *British Poultry Science*. 53(5):646–650.

Zweers, G., G. Kunz, and J. Mos. 1977. Functional anatomy of the feeding apparatus of the mallard (*Anas platyrhynchos* L.): Structure, movement, electro-myography and electro-neurography (author's transl.). *Anatomischer Anzeiger*. 142(1–2):10–20.

Zweers, G.A. 1973. Structure, movement, and myography of the feeding apparatus of the mallard (*Anas platyrhynchos* L.): A study in functional anatomy. *Netherlands Journal of Xoology*. 24(4):323–467.

18

Nutrient Requirements of Geese

INTRODUCTION

Geese are raised for meat, eggs, foie gras, down, and feathers. Asia, especially China, is by far the largest producer of geese, followed by Europe (FAOSTAT, 2016). The dozens of breeds found throughout the world originated from two species, the greylag goose (*Anser anser*) and the swan goose (*Anser cygnoïdes*). These two species are very closely related and in the wild have overlapping ranges, similar morphologies and diets, and sometimes hybridize. The greylag goose is the ancestor of domestic breeds of European origin (Heikkinen et al., 2015), and the swan goose is the ancestor of breeds of Asian origin (except Yili breeds of Western China; Lee et al., 2017). Breeds originating from these two species are distinguishable by the presence of a bump or, in some breeds, a knob at the base of the upper bill in those breeds derived from swan geese but not those derived from graylag geese (FAO, 2012). The body mass of these two species is similar at slightly over 3 kg, with males being slightly larger than females (Fox and Kahlert, 2005). The exact times and places of domestication are not known. Greylag geese were used for food and sacrifice in Mesopotamia more than 4,500 years ago (Zeuner, 1963), and both graylag and swan geese were likely fully domesticated about 2,500 years ago (Larson and Fuller, 2014). Many domestic European breeds are fast-growing and used mostly for meat or foie gras production, whereas many Asian breeds are more prolific egg layers and used for both table egg and meat production. European breeds dominate production throughout Europe, the Middle East, and the Americas. Asian breeds dominate worldwide production of both meat and eggs, and the vast majority of this occurs in China and other East Asian countries. Regardless of breed, a combination of large size and selective breeding restricts the ability of most domestic goose breeds to fly.

Most domestic breeds derived from these two species are considerably larger than their wild relatives. The heaviest breeds weigh 8 kg or more and include the European breeds Emden and White Roman and the Asian breed Shitou, which is also called Lions Head in some markets. These breeds are used for meat production because they typically begin to lay eggs at relatively older ages and produce fewer eggs per year than smaller breeds. Among Asian breeds, small breeds (e.g., Wulong, Wuzhong, Taihu) and medium-sized breeds (e.g., Yangzhou, Wanxi White, Xingguo Grey, Fencgheng Grey) are predominantly produced because of their robust ability to tolerate extensive production systems and their high egg laying capacity. For example, Wulong geese, which also go by the name Huoyan, lay 100–200 eggs per year that weigh about 130–200 g/egg with smaller eggs produced during the first laying season. Many of these breeds are also considered dual-use because they are raised both for meat and egg markets (FAO, 2012). Depending on the breed, egg production typically begins at between 20 and 35 weeks of age under natural lighting. Geese have a long lifespan relative to other poultry, and for egg production they are commonly maintained for two to five years.

MANAGEMENT SYSTEMS

Management systems for geese range widely and include intensive systems that are completely indoors, outdoor production in pens, extensive systems where all or the majority of nutrients come from foraging, and various combinations of these systems (Buckland and Guy, 2002). Intensive systems include the use of curtain-sided houses containing either deep litter bedding, slatted floors, or cages. Extensive systems vary considerably with location, but typically goslings are brooded in houses and provided a complete diet for several weeks before permitting goslings to forage for grasses and other vegetation. Extensively produced geese are often provided grain or a complete feed for a few weeks prior to slaughter in order to increase their growth rate and fat content. Geese are produced extensively in a greater proportion than any other categories of poultry because of their abilities to utilize high fiber foods and to fend off predators. However, the growth rate of geese decreases the more extensive the production system.

Complete feeds for geese contain much higher levels of feedstuffs containing high levels of fiber, including forages, silages, and plant processing by-products, compared with ingredients used for the other types of poultry. Even young geese grow considerably faster when dietary fiber levels in their diet exceed that provided by a simple corn and soybean meal diet (Li et al., 2017). This is likely a result of their digestive and metabolic strategy resulting from the feeding habits of their nondomestic ancestors, which eat an almost 100 percent plant-based diet from hatching through breeding and migration. The natural diets of the other poultry species in this report are highly enriched in invertebrates during the early growth phase (Klasing, 2005).

Historically, research on geese, including nutrition research, has utilized European breeds because most of this research was done in North America and Europe where these breeds are widely used. For this reason, recommendations for nutrient requirements in the previous report (NRC, 1994) were based predominantly on European breeds due to the paucity of nutritional information in the peer reviewed literature on Asian breeds. Nutrition research on Asian breeds has increased markedly since the last report and now dominates the primary literature. For this reason, most of the new recommendations in Table 18-1 are based on results using Asian breeds derived from swan geese.

TABLE 18-1 Minimum Nutrient Requirements of Geese as Percentages or Units per Kilogram of Diet (90% Dry Matter)

Nutrient	Units	0–4 Weeks, 2,800[a]	After 4 Weeks, 2,700[a]	Laying Female, 2,900[a]
Protein[b]	%	***20***	***15***	***15***
Lysine[c]	%	1.0	***0.75***	***0.6***
Methionine + cyst(e)ine[d]	%	***0.85***	***0.64***	***0.50***
Threonine	%	***0.73***	***0.54***	
Tryptophan	%		***0.20***	
Linoleic acid	%	***1.1***	***1.1***	
Calcium	%	0.65	***0.60***	***2.25***
Phosphorus (non-phytate)	%	***0.40***	***0.3***	***0.3***
Copper	mg		10	
Iron	mg		***100***	
Manganese	mg	***115***	***115***	
Selenium	mg	***0.2***	***0.2***	
Vitamin A	IU	***1,500***	***1,500***	***4,000***
Vitamin D_3	ICU	***1,000***	***1,000***	***200***
Choline	mg	***1,500***	**1,500**	***1,500***
Niacin	mg	80	80	***20***
Pantothenic acid	mg	26	21	***10***
Riboflavin	mg	***3.8***	***2.5***	***4.0***

NOTES: Values that are bold and in italics are estimates based on research that did not meet the experimental or statistical criteria desired described in the Nutrient Requirements section. These values are considered as guidance levels and should be used with appropriate caution. Values in shaded areas are the NRC (1994) recommendations, for which no updated estimates or confirming information are available. For references to the studies supporting shaded values, please refer to the 1994 report.

[a] These are typical dietary energy concentrations expressed as kcal ME_n/kg of diet and are not intended to be a specific recommendation.

[b] The values for protein are not requirements but are typical levels used in research and commercial production. Lower levels can be used when the appropriate essential amino acids are supplemented to meet the requirements.

[c] Amino acid requirements are expressed on a total amino acid basis.

[d] Methionine plus cyst(e)ine recommendations are listed as provisional because the research on which the recommendations rely did not demonstrate that cyst(e)ine was limiting in the experimental diets.

NUTRITIONAL AND DIGESTIVE STRATEGY

Natural Diets

Geese (Anserinae) are primarily herbivorous and differ from most duck species in that they graze on terrestrial more than aquatic vegetation. Wild greylag geese feed primarily on grasses, tubers, seeds, berries, and aquatic vegetation (Olsen, 2015). In agricultural settings they also consume grains, legumes, and potatoes. Swan geese feed primarily on a similar selection of plants; however, their bill is relatively longer and heavier, which is especially effective at grubbing roots and tubers (Fox et al., 2013; Olsen, 2015). Geese can survive on a much greater range of food quality than other categories of poultry. The exact mix of the foods consumed depends on the location and season. Geese typically select higher protein foods (e.g., young grasses and sedges) when growing or reproducing and higher energy foods (e.g., tubers, rhizomes, and farmed grains) prior to migration and when over-wintering (Sedinger, 1997). Both species have bills that permit them to filter aquatic plants and invertebrates from water, but they are relatively unspecialized at this method of feeding and utilize it much less than many other species of geese and ducks (van der Leeuw et al., 2003). Goslings have a well-developed capacity to graze grasses at the time of hatching, but they lack a mature mechanism for filtering plants and invertebrates from the water until 2 weeks after hatching (Skieresz-Szewczyk and Jackowiak, 2017). The almost exclusive herbivorous dietary niche of greylag and swan geese differentiates their digestive strategy from all other poultry, including domestic ducks.

Digestion

The digestive anatomy and physiology of domestic geese reflect their herbivorous diet and diverge from chickens, turkeys, and quail in several important respects. First, the bill and tongue are designed for procuring grasses and sedges, and they work together to accomplish this task (Jackowiak et al., 2011). The bill is also relatively short, which results in high clamping force for procuring grasses. The apex of both the maxilla (upper) and mandibular (lower) bill have a grasping and cutting surface referred to as a nail due to functional and morphological similarities with human fingernails. The lateral edges of the upper and lower bill have deep serrations that resemble blunt teeth. The ventral surface of the apex of the tongue has a prominent triangular scale composed of hardened keratin called the lingual nail. It ensheathes the apex of the tongue and extends past the end, providing an edge for grasping seeds and other foods in cooperation with the nails on the upper and lower beaks. It also pinches blades of grass against the edges of the bill so that grass blades can be broken off. The lingual nail is thought to be sufficiently flexible to stretch into a spoon-like shape to facilitate the scooping of grains (Jackowiak et al., 2011). The tongue also has keratinized conical papillae on the lateral edges. These hardened ridged structures interdigitate with the serrations on the lateral edges of the upper and lower bill to provide additional grasping and shearing forces. This arrangement permits geese to break off vegetation close to the ground by gripping the stems of grasses and tearing off the entire blade. This form of grazing keeps grasses short and actively growing, which maximizes their nutritional value. The lateral edges of the tongue also have hair-like filiform papillae that work with the conical papillae to permit filtration of small plants and invertebrates suspended in water, though the nutritional importance of this has not been quantified (van der Leeuw et al., 2003). The tongue has a large number of widely distributed lingual glands that secrete mucous, which covers the tongue and decreases friction during food collection (Jackowiak et al., 2011). Taken together, the anatomy of the beak and tongue of the domestic goose is indicative of feeding on terrestrial plants much more than aquatic plants (van der Leeuw et al., 2003).

The esophagus of domestic geese broadens out prior to entering the proventriculus but lacks a diverticula (crop) for storage and softening of food (Mattocks, 1971; Kierończyk et al., 2016). The wide bore of the esophagus and proventriculus provides an area for temporary storage of food. As described subsequently, food intake of wild geese is often limited by the time required to process their food through the gizzard and intestines and not by food procurement rate. This difference in procurement versus processing rates diminishes the need for a crop.

The morphology of the esophagus and gizzard is designed to process grasses and sedges to a critical size before leaving the gizzard (Clemens et al., 1975; Moore, 1998, 1999). This is because the fiber bundles in grass blades are aligned by grooves in the esophagus and transported into the gizzard along their long axis, enabling the opposing gizzard muscles to effectively strip the nutrient-filled cells of the mesophyll and palisade layers from the fibrous layers. Grazing domestic geese (*Anser anser*) cut grasses at lengths that accommodate the functional anatomy of the esophagus and gizzard (Moore, 1999). Interestingly, artificially cutting grass shorter (2 mm) than that normally cropped by grazing geese results in larger pieces leaving the gizzard, relative to consumption of typically consumed grass fragments (10 mm). This is likely due to poor alignment of the grass fragments within the gizzard and ineffective stripping of nutrient filled layers from structural components. When 10 mm pieces of grass were fed, the size of processed pieces leaving the gizzard was consistent and averaged about 0.1 mm × 0.07 mm (Moore, 1999). The folds and the small opening of the pylorus function as a sieve that prevents larger pieces from leaving the gizzard during contractions but permits smaller particles and cell contents to quickly escape into the duodenum. Larger particles that cannot escape may be refluxed back to the proventriculus where additional pepsin and acid are applied and then reenter the gizzard for another grinding cycle (Clemens et al., 1975). Very long pieces evidently escape the gizzard with little grinding, are quickly passed through the rest of the digestive system, and can be observed relatively intact in the excreta within a few hours of consumption (Mattocks, 1971).

Compared to ducks and swans, the gizzard of domestic geese has a large muscle mass relative to body mass or the volume of the lumen, which results in greater grinding pressure (Moore, 1998). The asymmetry of the muscle pairs generates lateral shearing forces during compression, which is necessary for processing terrestrial grasses and other vegetation. An examination of several breeds of domestic geese revealed that the gizzard contained an average of 15 g of small stones and sand (Mattocks, 1971), which presumably aids the grinding process. Domestic geese preferentially retain grit in the range of 0.45 to 2 mm in diameter (Ling et al., 2014). As in other poultry, the size and musculature of the gizzard increases with the fiber content of the feed (Li et al., 2017).

Following enzymatic digestion in the small intestine, the remaining indigestible components enter the colon where fluid and very small particles (<0.2 mm) are swept into the ceca by retrograde flow of urine from the cloaca (Bjornhag, 1977). The mucosa at the proximal end of the ceca possesses villi that likely act as a sieve to restrict the entrance of larger particles (Chen et al., 2002). The middle and distal end of the ceca lack villi and possess a more protective epithelium organized into parallel ridges that are thought to trap viscous fluids and small particles so that their exit is delayed, and they remain subjected to fermentation. Nevertheless, the average retention time of fluids and small particles in the ceca is short relative to turkeys or chickens (Clemens et al., 1975). The ceca of domestic geese are also proportionally shorter and have less volume than those of chickens and turkeys despite their considerably more herbivorous diet (Hallsworth and Coates, 1962; Jamroz et al., 2002b; Frei et al., 2017;). The relatively small ceca and overall rapid rate of passage of digesta through them precludes a substantial contribution of fermentation to their energy requirements (see subsequent discussion). Despite this, the ceca are the predominant fermentation area in geese as indicated by the high populations of fiber fermenting microbes, high concentrations of short chain fatty acids, and diminished fiber digestion when they are removed (Chen et al., 2003; Yang et al., 2016). Acetate, propionic acid, and butyric acid are the end-products of anaerobic fermentation of fiber in geese, and their concentrations are an order of magnitude higher in the ceca than in the proventriculus, gizzard, duodenum, jejunum, or ileum (Yang et al. 2018a). Relative to the rest of the digestive tract, the ceca maintain abundant microbes in the phyla Firmicutes and Bacteroidetes, both of which are comprised of genera known to be important for fermentation of fiber (Yang et al. 2018a; Zhou et al., 2018).

Compared to a low fiber diet based on corn and soybean meal, increasing fiber from either the addition of cellulose (Li et al., 2018) or from feeding grass (Xu et al., 2017) increases numbers of microbial species, diversity of microbiota, and abundance of fermenting microbiota in the cecum. The role of the ceca in fiber digestion has been demonstrated by experiments that have either ligated the entrance to the ceca or have removed the ceca altogether. For example, there is a 90 percent decrease in methane production following cecectomy, indicating that most fermentation occurs in this organ (Chen et al., 2003). Cecectomy also decreases the digestibility of NDF by 17 percent and ADF by over 200 percent (Yang et al., 2009). Although the preceding changes are large, they do not translate into correspondingly large changes in dietary energy digestibility because the proportion of fiber in the diet that is fermented is relatively small. For example, in geese fed moderate levels of dietary fiber provided by rice hulls (12.6% NDF and 7.3% ADF), cecectomy decreased dietary apparent and true metabolizable energy of the diet by 7.4 and 4.3 percent, respectively (Yang et al., 2016). At high levels of dietary fiber provided by rice hulls (15.2% NDF and 9.0% ADF), cecectomy decreased AME by only 3.6 and 1.3 percent, respectively (Yang et al., 2016), which is consistent with observations that geese adjust to high levels of dietary fiber by increasing digesta passage rate and decreasing fermentation (see subsequent discussion). The ceca have a much greater nutritional impact on the digestibility of phosphorus, increasing digestibility by 35 percent (Yang et al., 2009), presumably by hydrolysis of phytate due to microflora phytase activity. Ceca have no general effect on the availability of dietary protein and amino acids (Gruhn et al., 1975; Wang et al., 2008a; Yang et al., 2009, 2016), and there is little evidence that essential amino acids synthesized by cecal microflora from non-protein nitrogen provides a nutritional contribution (Gruhn et al., 1975; Jiang et al., 2013). However, research on the role of ceca on the synthesis and uptake of amino acids in geese has not been the primary focus of the aforementioned research, and this question requires additional attention before strong conclusions can be drawn.

The rate of passage of digesta through the entire GI tract is very fast in geese. For example, domestic geese (*Anser anser*) fed a meal of grass excrete around half of the resulting excreta (dry matter) within 4 hr (Moore, 1999). This rate of passage is faster than that of other poultry and results in lower fiber digestibility. For example, domestic geese consuming alfalfa pass solutes and small particles through their GI tract more than three times faster than turkeys, and this results in 41 percent lower digestibility of NDF (Frei et al., 2017). The combination of relatively small ceca and short retention time results in a low reliance on fermentation of fiber as an important energy source for these exceptionally herbivorous birds. It has been speculated that a limitation on GI tract size and weight in wild geese is due to their need to migrate long distances (Sedinger, 1997; Frei et al., 2017). Domestic geese have apparently retained the poor fermentation efficiency of their ancestors.

When eating vegetation in the summer when day length is long, vegetation has lower fiber content, and feeding occurs over a long period of time, geese have a strategy of increasing retention time and consequently the efficiency of digestion. However, when food quality and hours of daylight and food intake decline in the winter, geese are capable of increasing their consumption by increasing their rate of passage. This results in lower efficiency of digestion (kcal ME/kg consumed), but because of the large increase in food intake, the net gain in energy acquired per day (kcal ME/d) remains similar (Prop and Vulink, 1992). In other words, the strategy of eating more and skimming readily digestible nutrients (i.e., cell contents) from low-quality vegetation and excreting the fiber (i.e., cell walls) can, in some circumstances, be more nutritionally rewarding than attempting to increase nutrient extraction by slowed digestion and prolonged fermentation. This strategy results in very inefficient feed utilization but allows an herbivorous diet throughout the seasons, and in wild geese it facilitates flight by decreasing the weight of the GI tract and its contents.

The foregoing adaptations that evolved in wild-type geese also occur in domestic geese fed high fiber diets due to a high proportion of by-products. For example, Lu and Hsu (1990) added rice hulls at 10–60 percent of the diet fed to both European and Chinese breeds. They observed a corresponding increase in feed intake and decrease in efficiency of feed utilization with increasing fiber content, but body weight gains were not significantly affected in either breed. As in wild geese, the order of fiber digestibility in domestic geese is hemicellulose >> cellulose >> lignin. For example, the digestibility of hemicellulose, cellulose, and lignin by adult European geese fed a corn–soybean meal diet with 27.5 percent alfalfa meal are approximately 46.3, 17.9, and 5.5 percent, respectively (Hsu et al., 1996a). Although there is considerable variation in the digestibility of these fiber fractions across feedstuffs depending on variables, including the physical form of the feedstuff and the total amount of fiber, the relative digestibilities of the fiber fractions remain relatively constant (Jamroz et al., 1992; Timmler and Jeroch, 1994; Zhang et al., 2011). The high digestibility of hemicellulose relative to cellulose and lignin is likely due to the very short residence time of digesta in the ceca (Clemens et al., 1975). Taken together, geese are better able to tolerate diets with higher levels of structural fiber than other poultry due to their large and effective gizzard and flexible digestive strategy. Interestingly, the addition of some purified soluble fiber sources to the diet of growing geese results in a marked decrease in growth and efficiency of feed utilization (Hsu et al., 1996b; Zhao et al., 2018), likely due to their high viscosity.

Goslings grow very fast following hatching. This is accommodated by the use of residual egg yolk during the first 5 d post-hatch (Jamroz et al., 2004) and the rapid increase in enzymatic activity in the small intestine during the first week post-hatch (Shih and Hsu, 2006). Intestinal activity of the pancreatic enzymes amylase, trypsin, and chymotrypsin increase from the time of hatching until plateauing around day 7, while lipase plateaus around day 11. Microbial cellulase activity in the ceca plateaus around day 14.

RATE AND COMPOSITION OF GROWTH

Geese of moderately sized European and Asian breeds attain 50 percent of their adult weight at around 6 wk of age (Knížetová et al., 1994, 1995; Ibtisham et al., 2017). Despite a fast growth rate, the efficiency of feed utilization of domestic geese is less than that of chickens or turkeys due to their differing digestive anatomy and physiology, which results in lower digestibility of energy and amino acids. The composition of growth of geese also differs from chickens and turkeys because they deposit more fat, especially subcutaneous fat, and have a higher proportion of feathers (Nitsan et al., 1981). The amino acid composition of feathers and skin of domestic geese is similar to chickens and turkeys, but their carcass composition contains slightly lower proportions of lysine, methionine, and cysteine.

Meat birds are typically marketed between 8 to 20 weeks of age, depending on the intended markets and whether the production system is intensive or extensive. Because of the value of goose feathers and down, the timing of marketing is often matched to the timing of feather maturation, which differs somewhat across the various breeds. In general the initial feather growth, first molt, and second molt are complete around 9, 16, and 20 weeks of age, respectively, and these are common times for slaughter, but the exact timing is breed dependent.

NUTRIENT REQUIREMENTS

The *Nutrient Requirements of Poultry* was first published in 1944 and has been periodically updated since then. Recommendations on nutrient requirements of geese were first were made in 1977 and last updated in 1994 (NRC, 1994). Updated recommendations in Table 18-1 are based on the peer-reviewed scientific literature published since those considered for the last report, the details of which are summarized in Table 18-2. The previous report can be consulted for references to the older literature that was used for previous recommendations. At this time, there is insufficient evidence to determine if the nutrient requirements of breeds derived from greylag geese (European breeds) diverge from those derived from swan geese (Asian breeds). It is likely that breed characteristics such as growth rates, adult body weights, and egg production are more important factors than ancestry. For this reason, the recommended requirements in Table 18-1 are based on breeds derived from both species, and further information on the specific breeds and production rates is provided in Table 18-2. The recommended requirements in Table 18-1 are primarily applicable to geese raised in confinement and fed nutritionally balanced diets. Nutritionists feeding geese raised extensively should modify these recommendations based on the information in the preceding "Digestive and Nutritional Strategy" section.

Peer-reviewed research publications were used for estimating nutrient requirements. Studies were excluded from consideration if the experimental design or statistical analysis was not appropriate for estimating nutrient requirements or if insufficient information was provided to allow replication of the experiment or the statistical analysis of the resulting data. Exclusion factors included insufficient dietary concentrations for establishing a dose–response relationship, failure to chemically analyze dietary levels of the nutrient of interest, and failure to provide key information on the experimental design, diet nutrient levels, and management practices that impact requirements, analytical methodologies, or statistical procedures. Studies were also excluded if they were conducted at environmental or management extremes that would affect production rates and nutrient requirements or if performance rates were low for other reasons. Studies on amino acid, vitamin, and trace mineral requirements that did not use a common basal diet were also excluded. Recommendations (Table 18-1) on the dietary level needed to support maximal performance and absence of signs of deficiencies were set by averaging the estimates from all studies that met the inclusion criteria.

In cases where no studies that met the inclusion criteria were available, but there were studies that met the most important criteria, exceptions were made so that a rough estimate could be provided, and the resulting recommendation guidance is listed in Table 18-1 in bolded and italicized font and should be considered provisional. For some nutrients, there were studies that did meet the foregoing inclusion criteria, but the basal diet had adequate levels of the nutrient under study as indicated by the lack of significant changes in performance by the addition of higher levels of that nutrient. In these cases, nutrient recommendations are based on the level of the nutrient in the basal diet and are designated as guidance on an adequate dietary level in Table 18-1 (bolded and italicized font).

Many of the published experiments designed to titrate the requirement for various nutrients analyzed the results using ANOVA followed by mean comparisons and did not use analysis by broken-line regressions or other valid regression approaches to more quantitatively estimate a requirement. Thus, many of the recommendations in Table 18-1 are based on the range between treatment levels that were significantly different. In these cases, the new guidance provided is also provisional as indicated by bolded and italicized text.

The recommended values in Table 18-1 do not include a margin of safety to account for environmental, management,

TABLE 18-2 Documentation of Nutrient Requirements of Geese for Growth

Age (Days)	Weight, Start–End of Study, g	Estimated Requirement per Response Criteria	Response Criteria	Breed[a]	Number of Levels (Range)	Comments	Reference
Lysine, %							
1–21	132–1,393	***0.90–1.05***	Gain, efficiency	Lion head (AC, large)	6 (0.75–1.50)	Data analyzed by ANOVA and means separation. Basal diet had 2,894 kcal/kg and 18.6% protein.	Wang et al. 2014a
1–42	89–1,559	***<1.03***	Gain, efficiency	Wulong (AC, small)	3 (1.03–1.23)	Data analyzed by ANOVA and means separation. Basal diet had 2,808 kcal/kg and 20.2% protein.	Li et al., 2004b
35–70	1,670–3,150	***<0.65***	Gain, breast yield, leg yield	Yangzhou, AC (medium)	4 (0.65–1.1)	Data analyzed by ANOVA and means separation. Basal diet had 2,700 kcal/kg and 15.0% protein.	Wang et al., 2012
42–90	1,602–3,218	***<0.7***	Gain, efficiency	Wulong (AC, small)	3 (0.7–0.9)	Data analyzed by ANOVA and means separation. Basal diet had 2,761 kcal/kg and 15.7% protein.	Li et al., 2004a
Methionine + cyst(e)ine, %							
1–42	90–1,681	***0.8–1.0***	Gain, efficiency	Wulong (AC, small)	3 (0.8–1.0)	Data analyzed by ANOVA and means separation. Basal diet had 2,808 kcal/kg and 20.2% protein.	Li et al., 2004b
28–56		0.66	Gain	Yangzhou (AC, medium)	5 (0.48–0.88)	Data analyzed by GLM and by linear and quadratic regressions. Basal diet had 15.8% protein and 2,597 kcal/kg.	Wang et al., 2010b
28–70	1,069–3,565	***0.48–0.88***	Gain, efficiency, serum total protein, serum globulin	Jilin (AC)	4 (0.48–1.08)	Data analyzed by ANOVA and means separation. Basal diet had 0.23% met, 0.48% met+cys, 2,486 kcal/kg and 18.9% protein.	Du et al., 2017
28–70	Unknown	***≤0.43 0.43–0.73 0.83–0.93***	Gain, efficiency, glutathione, glutathione peroxidase	Lyon head (AC, large)	6 (0.43–0.93)	Geese had access to a pond—amount of supplemental food is unknown. Data analyzed by ANOVA and means separation. Basal diet had 0.26% met, 0.438% met+cys, 2,796 kcal/kg and 14.2% protein.	Lin et al., 2012
35–70	1,190–3,066	***0.33–0.43***	FE	Yangzhou (AC, medium)	5 (0.39–0.79)	Analyzed by ANOVA. Basal diet had 15.8% protein and 2,599 kcal/kg.	Zhou et al., 2008
42–90	1,600–3,200	***≤0.59***	Gain, efficiency	Wulong (AC, small)	3 (0.59–0.79)	Data analyzed by ANOVA and means separation. Basal diet had 2,761 kcal/kg and 15.7% protein.	Li et al., 2004a
Threonine, %							
1–28 1–56	Unknown Unknown	0.54–0.74 0.73	Efficiency Eviscerated carcass %	Yangzhou (AC, medium)	5 (0.54–0.94)	Data analyzed by GLM and by linear and quadratic regressions. Basal diet had 19.8% protein and 2,806 kcal/kg from 0 to 28d and 17.7% and 2,617 from 28 to 56 d.	Shi et al., 2010
Tryptophan, %							
21–63	Unknown	***0.14–0.22***	Gain, breast and thigh protein deposition	Yangzhou (AC, medium)	3 (0.14–0.30)	Data analyzed by ANOVA and means comparisons. Basal diet had 15.2% protein and 2,667 kcal/kg.	Wei et al., 2012; Yu et al., 2014

continued

TABLE 18-2 Continued

Age (Days)	Weight, Start–End of Study, g	Estimated Requirement per Response Criteria	Response Criteria	Breed[a]	Number of Levels (Range)	Comments	Reference
Linoleic acid, %							
38–119	1,477–3,960	***0.72–1.12*** ***0.72–0.92*** ***0.92–1.12*** Quadratic max 1.02	Gain Efficiency Breast yield Dressed, %	Wulong (AC, small)	6 (0.52–1.52)	Data analyzed by ANOVA and means comparisons and by linear and quadratic regressions. Basal diet had 16.2% protein and 2,725 kcal/kg.	Zhang, 2016
Calcium, %							
1–28	85–1,171	***≤0.65***	Gain, efficiency, tibia ash	Wulong (AC, small)	4 (0.65–0.95)	Ca added as Ca CO_3. aP was 0.30 or 0.40. Data analyzed by ANOVA. Basal diet had 19.9% protein and 2,830 kcal/kg.	Li et al., 2005
1–28	Unknown	***0.25–0.45*** ***0.45–0.65***	Gain, efficiency Tibia ash	White Roman (AA)	7 (0.25–1.45)	Ca added as limestone. aP was 0.62. Data analyzed by ANOVA and means comparisons. Basal diet had 19.5% protein and 2,901 kcal/kg.	Lu and Hsu, 1994
Non-phytate phosphorus, %							
1–28	Unknown	***0.22–0.32*** ***0.0.32–0.42*** Regression 0.46	Gain Tibia ash Gain	White Roman (AA)	7 (0.12–0.72)	P added as disodium hydrogen phosphate. Ca was 0.85%. Data analyzed by ANOVA and broken line regression. Basal diet had 19.5% protein and 2,903 kcal/kg.	Lu and Hsu, 1994
1–21	128–1,291	***≤0.40***	Gain, efficiency, tibia ash	Lion-head	3 (0.40–0.60)	P added as dicalcium phosphate. Data analyzed by GLM. Ca was 0.95%. Basal diet had 20% protein and 2,900 kcal/kg	Zhu et al., 2018
29–72	1,530–5,800	***≤0.30***	Gain, efficiency, tibia ash	Lion-head	3 (0.30–0.70)	P added as dicalcium phosphate. Data analyzed by GLM. Ca was 0.85%. Basal diet had 15% protein and 2,800 kcal/kg.	Zhu et al., 2018
Zinc, mg/kg							
1–28		***31–81***	Gain	Qingnonghui	6 (31–281)	Zn added as $ZnSO_4{\cdot}7H_2O$. Data analyzed by ANOVA and means comparison. Basal diet had 18.0% protein and 2,700 kcal/kg.	Chen et al., 2013
29–105		***27–77***	Gain, efficiency	Qingnonghui	6 (27–277)	Zn added as $ZnSO_4{\cdot}7H_2O$. These are the same geese used from 1 to 28 so the data is confounded by the previous Zn level. Data analyzed by ANOVA and means comparison. Basal diet had 16.0% protein and 2,749 kcal/kg.	Chen et al., 2013
Manganese, mg/kg							
1–35	Unknown –1,410	ANOVA: ***80–140*** ***80–110*** ***110–140*** Quadratic max: 117 105	Gain, efficiency Bone breaking strength, bone ash Breast muscle %, leg muscle % Gain Efficiency	Wulong (AC, light breed)	6 (20–170)	Mn added as $MnSO_4.H_2O$. Data analyzed by ANOVA and by linear and quadratic regressions. Basal diet had 20 mg/kg Mn by calculation. Basal diet had 18.9% protein and 2,809 kcal/kg.	Wang et al. 2014b

35–126	1,652–4,000	ANOVA: ***20–80*** ***80–140*** ***20–140*** ***110–140*** Quadratic max: 105 118	Gain Efficiency Breast muscle % Leg muscle % Gain Efficiency	Wulong (AC, light breed)	6 (20–170)	Mn added as $MnSO_4.H_2O$. Data analyzed by ANOVA and by linear and quadratic regressions. Temperature was 33°C. Basal diet had 20 mg/kg Mn by calculation. Basal diet had 16% protein and 2,697 kcal/kg.	Zhang et al., 2014
Iron, mg/kg							
35–119	1,361–3,800	***92–112***	Gain, efficiency, dressed %	Wulong (AC, light breed)	6 (72–172)	Iron added as $FeSO_4 \cdot 7H_2O$. Data analyzed by ANOVA and means comparisons (Xu et al., 2013). Basal diet had 72 mg/kg Fe. Basal diet had 16.3% protein and 2,697 kcal/kg.	Ma et al., 2015
Copper, mg/kg							
35–112	1,369–4,170	***5.9–20.9*** ***20.9–35.9***	Gain, efficiency, leg muscle %	Wulong (AC, light breed)	6 (5.9–245.9)	Cu supplemented as $CuSO_4.5H_2O$. Data analyzed by GLM and quadratic and linear regression. Basal diet had 16.0% protein and 2,698 kcal/kg.	Xu et al., 2013
28–70	1,381–3,727	8.8 9.2 11.6	Gain Efficiency Carcass yield	Yangzhou (dual purpose)	4 (4.2–16.2)	Cu supplemented as $CuSO_4.5H_2O$. Data analyzed by ANOVA and by curve fitting (procedure unknown; P values not provided). Basal diet had 15.9% protein and 2,643 kcal/kg.	Yang et al. 2018b
Selenium, mg/kg							
1–63	96–4,190	***≤0.08*** ***0.08–0.18***	Gain, efficiency Carcass water loss	Mirande x Atiquere gray (AA)	4 (0.08–0.58)	Se added as selenized yeast. Data analyzed by ANOVA. Starter basal had 19.0% protein and 2,810 kcal/kg and grower had 17.1% protein and 2,717 kcal/kg.	Baowei et al., 2011
1–28 1–70 28–70 1–28 28–70 1–70	Unknown –1,452 Unknown –3,737 1,452–3,737	ANOVA: ***0.20–0.35*** ANOVA ***0.05–0.20*** ***0.05–0.20*** ***0.05–0.20*** ***0.05–0.35*** Quadratic max: 0.29 0.35 0.33 0.36 0.31	Gain Gain Gain Glutathione peroxidase Malonaldehyde Total antioxidant capacity Gain Gain Gain Glutathione peroxidase Malonaldehyde	Yangzhou (dual purpose)	5 (0.05–0.65)	Se added as sodium selenite. Data analyzed by ANOVA and by linear and quadratic regressions (procedure unknown; ***P values not provided***). Basal starter diet had 18.9% protein and 2,773 kcal/kg. Grower had 16.6% and 2,627 kcal/kg.	Yang et al., 2015
Choline, mg/kg							
1–28 28–105	Unknown –1,735 1,735–4,166	***0–1,800*** ***600–1,800*** ***0–1,800*** ***600–1,800***	Gain Efficiency Gain Efficiency	Qingnong grey	6 (0–4,000 added)	Basal level of choline was not provided. Choline was supplemented as choline chloride. Data analyzed by ANOVA and quadratic regression. Starter basal diet had 18.0% protein and 2,700 kcal/kg; grower diet had 16.0% protein and 2,750 kcal/kg.	Zhang et al., 2013

continued

TABLE 18-2 Continued

Age (Days)	Weight, Start–End of Study, g	Estimated Requirement per Response Criteria	Response Criteria	Breed[a]	Number of Levels (Range)	Comments	Reference
Vitamin D, IU/kg							
35–77	775–2,848	Added D ***0–500*** ***1,000–2,000*** ***0–1,000***	Gain Efficiency Eviscerated yield, plasma Ca	Yangzhou	5 (0–4,000 added)	Basal level of vitamin D was not provided. Vitamin D was supplemented as Cholecalciferol. Data analyzed by ANOVA and means comparisons. Basal diet had 16.3% protein and 2,595 kcal/kg.	Tian et al., 2013
1–28 28–56 56–112 56–112 1–112	Unknown –1,245 1,245–2,968 2,968–4,100	ANOVA ***45.5–85.5*** ***43.5–65.5*** ***43.5–103.5*** ***65.5–85.5*** ***43.5–85.5*** ***63.5–83.5*** Quadratic max: 81.4 82.7	Efficiency, dressing % Dressing % Gain, efficiency Breast muscle %, leg muscle % Gain, efficiency Dressing %, breast muscle %, leg muscle % Efficiency Drip loss	Wulong (AC, light breed)	Starter: 6 (45.5–125.5) Grower: 5 (43.5–125.5)	Analyzed levels of niacin in basal diets were 25.5 and 23.5 for starter and grower, respectively. Niacin was supplemented as nicotinic acid. Data analyzed by ANOVA and by linear and quadratic regressions. Starter basal diet had 18.9% protein and 2,811 kcal/kg; grower had 16.3% and 2,698 kcal/kg.	Duan et al., 2014
Pantothenic acid, mg/kg							
1–28	120–1,280	ANOVA ***9.1–24.1*** Curve fitting 26.2 15.5	Gain, carcass, efficiency, leg muscle %, low abdominal fat Gain Efficiency	Wulong (AC, light breed)	6 (9.1–120)	Analyzed levels of pantothenic acid in basal diet was 9.1 mg/kg. Basal diet was supplemented with Ca pantothenate. Data analyzed by ANOVA and by curve fitting. Basal diet had 18.9% protein and 2,811 kcal/kg; grower had 16.3% and 2,698 kcal/kg.	Wang et al., 2016
35–112	1,096–3,830	ANOVA ***19.5–29.5*** ***14.5–29.5*** ***9.5–29.5*** Curve fitting 21.2 18.5	Gain, carcass drip loss Efficiency Breast muscle yield, leg muscle yield Gain Efficiency	Wulong (AC, light breed)	6 (9.5–89.5)	Basal was supplemented with Ca pantothenate. Data analyzed by ANOVA and by curve fitting. Basal diet had 16.0% protein and 2,698 kcal/kg.	Zhang et al., 2015

NOTE: Requirement estimates shown in bold italicized print indicate that the requirement was shown to be outside of the range of treatment levels or that the data was not analyzed using a broken line or quadratic regression.

[a] Lion head is also called Sihtou breed. Wulong is also called Huanyan breed.

toxicological, or disease stressors or due to losses during manufacture and storage of diets. Nutritionists should add a margin of safety appropriate for their specific production system to adjust for these added needs (see Chapters 9 and 11). The recommended values in Table 18-1 also assume that the dietary ingredients used have digestibilities similar to a typical diet based on corn and soybean meal and that supplemented amino acids, minerals, and vitamins have high bioavailability (see Chapter 8). The values for energy and protein are not requirements but are typical levels used in research and commercial production. Poultry can adapt to a range of energy levels by changing their feed intake, and they can utilize lower protein levels if optimal levels and balance of amino acids are maintained when supplementing purified amino acids.

Nutrient Requirements for Growth

Geese grow at a very fast rate after hatch, and their growth rate decelerates quickly after around 4 weeks of age. For this reason, starter (1 to 4 weeks) and grower (after 4 weeks) diets are typically fed to intensively raised geese. Geese marketed at a large body size are commonly fed a third diet with lower nutrient levels after 8–10 weeks. Due to the lack of research on geese older than 10 weeks of age, recommendations on nutrient requirements are provided only for starter and grower diets (Table 18-1).

Energy

The ME requirement (kcal/kg body weight/d) for maintenance, growth, activity, and thermoregulation of domestic geese has received little attention, and this is an area that needs focused research. However, many experiments have titrated the dietary energy density (kcal/kg diet) threshold that maximizes growth rates in a variety of production systems (Min et al., 2006, 2007; Wei et al., 2010; Yu et al., 2010; Wang et al., 2012; Li et al., 2016). This research indicates that when dietary ME concentrations are low, goslings respond with increased growth rate compared with gosling fed diets with higher levels of ME. This response reaches a plateau where gain does not change as long as other nutrients are kept at a constant proportion of ME. For example, goslings fed diets with energy levels of either 2,598, 2,717, 2,837, 2,957, or 3,076 kcal ME/kg had growth rates from 0 to 4 weeks that were maximized at the 2,837 kcal/kg level and then plateaued (Min et al., 2007). Abdominal fat increased at the two higher energy levels but was similar at the three lower levels. This study, along with others (Roberson and Francis, 1963; Stevenson, 1985), indicate that feed consumption by growing geese decreases as dietary ME level increases, but not in direct proportion. Consequently, geese fed high energy diets consume greater amounts of energy and deposit more body fat than those fed lower energy diets. Given that growing geese can adapt to a range of dietary energy densities partly by adjusting their feed intake, in practice the optimal level selected is typically the one that minimizes feed costs (Li et al., 2016). For this reason, the energy values shown in Table 18-1 are not specific recommendations but are provided because they are common energy densities used in research on nutrient requirements and are known to be in the range that geese can adapt to and still obtain maximal growth.

Determination of the ME content of forages, silages, and by-products for use in formulation of diets for geese is in its nascent stage. The available studies indicate that the ME values of feedstuffs are typically lower when fed to geese than those obtained using chickens or turkeys (Sheng and Wang, 2006; Hui et al., 2012). Unfortunately, the ME of a feedstuff is difficult to predict based on its percentages of NDF, ADF, and crude fiber (Wang et al., 2008b; Zhang et al., 2011). This research also shows that there are large differences between measured AME and TME values, especially for feedstuffs with high levels of cellulose and lignin. The development, standardization, and validation of best techniques for determining ME values of common feedstuffs in geese is very much needed. In the interim, the ME content of feedstuffs measured in chickens are often used for estimating the ME content of those feedstuffs for geese.

Protein

Like other poultry, geese do not require protein per se, but they require the essential amino acids and nitrogen from nonessential amino acids that are provided by protein. A recommendation for protein levels of 20 percent from 0 to 4 weeks and 15 percent after 4 weeks was provided in the previous report (NRC, 1994) and was based on five studies that utilized diets consisting primarily of corn, soybean meal, a fiber source, and supplemental methionine or animal protein. More recent research (Min et al., 2007; Wang et al., 2012) supports the earlier results. In addition to supplemental methionine, many commercial diets are also supplemented with lysine, and sometimes threonine, which allows use of lower protein levels for starting and growing geese. Additionally, after 10 weeks of age, geese fed a diet with only 10 percent dietary protein have rates of gain and efficiencies of feed utilization that are similar to those fed 12.5 or 15 percent protein, although they have lower body fat (Sahin et al., 2008). This study indicates that protein levels can be lowered considerably in the finishing phase of production when the rate of growth is relatively slow and feed intake is high due to its high fiber content.

Amino Acids

Methionine and lysine are the first and second limiting amino acids, respectively, in a diet based on corn and soybean meal for growing geese. Presumably threonine is the third limiting amino acid, but this has not been clearly established for geese.

Diets for chickens and turkeys are usually formulated using the ileal digestible amino acid concentrations in feedstuffs rather than total amino acid concentrations. This is especially important when using by-products and other ingredients that have low or variable digestibilities. Such an approach should be especially valuable when formulating diets for geese because they are fed higher levels of by-products, forages, and silages than other poultry species. Currently, however, there is insufficient information on the ileal digestibility of ingredients to apply this approach beyond a few common ingredients. Reliance on ileal digestibility values derived from chickens is unsuitable for growing geese as indicated by studies that compared these two species fed a common diet (Jamroz et al., 2001, 2002a). When measured at three ages (14, 28, and 42 d), the apparent ileal digestibility of amino acids in a diet based on corn, barley, and soybean meal averaged 12.2 percent lower in geese than in chickens. For methionine, cysteine, and lysine, apparent ileal digesibilities were less in geese by 18.2, 31.2, and 24.9 percent, respectively. Interestingly, the digestibility of cyst(e)ine was almost identical in these two species, indicating that the application of a single adjustment factor to data generated in chickens does not result in useful information for geese.

Several investigators have estimated the protein and amino acid needs for maintenance or for tissue accretion by conducting balance and comparative slaughter experiments in geese fed diets with normal or zero protein levels (Nitsan et al., 1983; Shi et al., 2007; Sun et al., 2010; Wang et al., 2010b). For example, Wang et al. (2010a) estimated the daily loss of amino acids for maintenance, including losses via intestinal excretion and from feathers, and the daily accretion of amino acid in the carcass and feathers of Yangzou geese from 5 to10 weeks of age. They then used the factorial method (Hauschild et al., 2010) to estimate amino acid requirements based on accretion rates, maintenance losses, and amino acid digestibility. Amino acid requirements (% of the diet at 2,600 kcal/kg) were estimated as follows: lysine, 0.78, methionine + cyst(e)ine, 0.50, and threonine, 0.76. Their estimates for lysine and methionine + cyst(e)ine are very similar to requirement estimates based on empirical studies (Table 18-1), but their estimate for threonine is considerably lower. A similar study by Sun et al. (2010) resulted in an overestimation of the requirement for many amino acids at variable amounts. It is likely that the precision of this approach for estimating requirements could be greatly improved if ileal digestible amino acids coefficients were used when correcting for digestive losses instead of an average value derived from chickens.

Lysine

Wang, Xu, et al. (2014b) examined the lysine requirement during the starting phase of Lion Head geese, which is a very large and fast-growing meat breed, and found that levels between 0.90 and 1.05 percent maximized gain and efficiency of feed utilization. Li et al. (2004a) examined the lysine requirement from 1–42 d of Wulong geese, which are a small breed and did not get a response from lysine addition to their basal diet that contained 1.03 percent lysine. These studies support the recommendation of the previous report (NRC, 1994) that was based on five studies. For this reason, the recommended level for the starter diet of 1.0 percent lysine in that report is not changed.

Two studies were conducted on the lysine requirement during the growing phase. In Yangzhou geese, which is a medium-sized breed, there was not a growth or efficiency response to additional lysine added to basal diets that contained 0.65 percent lysine (Wang et al., 2012). Likewise, the growth and efficiency of Wulong geese, which are a small breed, did not respond to lysine added to a basal diet containing 0.7 percent (Li et al., 2004b). The previous report recommended a lysine level of 0.85 percent for geese after 28 d of age based on a single report that examined the requirement of a large breed from 0 to 56 d of age using different levels in starter and grower diets (Mateova et al., 1980). Because the performance during the growing phase was confounded with lysine levels fed during the starter phase, this earlier study may have overestimated the requirement for grower diets. Consequently, the recommended lysine level has been decreased to 0.75 percent, although the possibility that large breeds may have a higher requirement requires further investigation.

Methionine + Cyst(e)ine

Li et al. (2004b) examined the methionine + cyst(e)ine requirement of Wulong geese from 1 to 42 d of age and found that a dietary level between 0.8 and 1.0 percent maximized gain and efficiency of feed utilization. This range is higher than the recommendation of 0.6 percent in the previous report, which was based on two conflicting studies; consequently, the recommendation has been increased to 0.85 percent.

Since the previous report (NRC, 1994), there have been six studies examining the methionine + cyst(e)ine requirement of geese during the growing phase. A study examining the requirement using regression analysis of gain between 28–56 d estimated the requirement to be 0.66 percent for medium-bodied Yangzhou geese (Wang et al., 2010b). Two studies examined the methionine + cyst(e)ine requirement from 28 to 70 d. Based on maximizing gain and efficiency, Jilan geese required between 0.48 and 0.88 percent (Du et al., 2017), and based on maximizing efficiency, Lion Head geese required between 0.43 and 0.73 percent (Lin et al., 2012). Given that these studies find the requirement in the same range, the quantitatively estimated value of 0.66 percent (Wang et al., 2010b) was adjusted to the typical energy value of 2,700 kcal/kg used in Table 18-1 to raise the previous recommendation to 0.64 percent for methionine + cyst(e)ine.

The dietary concentration of cyst(e)ine that meets the full requirement for this semi-essential nutrient is not known in geese, and if the dietary cyst(e)ine levels in the preceding studies were actually greater than needed, the methionine + cyst(e)ine requirement would have been underestimated. For this reason, the methionine + cyst(e)ine recommendations provided in Table 18-1 are provisional. New research that determines the level of dietary cysteine that meets the metabolic need for this semi-essential amino acid is much needed. Until this information is known, the requirements for methionine and for methionine + cyst(e)ine cannot be estimated with confidence.

Threonine

Shi et al. (2010) examined the threonine requirement of Yangzhou geese from 1 to 28 d and observed that maximal efficiency of feed conversion occurred between 0.54 and 0.74 percent. The geese were followed for an additional 28 d on the same diets. Increasing dietary threonine levels did not affect gain or efficiency from day 28 to 56, indicating a requirement of 0.54 percent or below during this period. Using regression analysis, they estimated the requirement to be 0.73 percent based on maximal carcass yield at 58 d. The above observations indicate that the effect of threonine on carcass yield occurred mostly during the starting phase. The previous report (NRC, 1994) did not provide a recommendation for the threonine requirement, so the provisional values in Table 18-1 are based on this single study.

Tryptophan

A study on the tryptophan requirement of Yangzhou geese from 21 to 63 d of age found the requirement for growth rate (Wei et al., 2012) and protein deposition in breast and thigh (Yu et al., 2014) to be between 0.14 and 0.22 percent. The previous report (NRC, 1994) did not provide a recommendation for the tryptophan requirement, so based on this single study, a provisional recommended tryptophan requirement is 0.20 percent for the grower phase.

Fat

As in other poultry, the amount and composition of tissue and adipose lipid reserves are affected by the amount and composition of dietary lipids. This relationship has been studied in relation to lipid deposition in edible tissues, including muscle, liver, and adipose (Tang Junwang, 2018). Like other poultry, the fatty acid composition of storage fat depots in geese correlate with the fatty acid composition of their diet, and this is particularly true for fatty acids that geese are unable to synthesize, such as long-chain polyunsaturated fatty acids.

In the wild, many species of geese develop a general fattening prior to migration, and the fatty liver serves as an energy storage organ (Hermier, 1997). Taking advantage of this physiological trait, the ancient Egyptians were the first to force-feed geese, and this practice was later adopted by Greeks and Romans for both meat and foie gras ("fat liver") production (Guémené and Guy, 2004). In domestic geese, fatty liver occurs when the increase in lipogenesis exceeds the capacity of synthesis and secretion of lipoproteins (Hermier, 1997), and the liver weight may increase from 100 g to 1 kg in 2 weeks (Hermier et al., 1994). However, in the context of developing fatty liver, not all breeds of geese are responsive to overfeeding. For example, the French Landes grey goose is highly susceptible to fatty liver production, while the Poland white goose is only moderately responsive (Fournier et al., 1997). The latter breed has been shown to channel hepatic triglyceride toward secretion rather than in situ storage (Fournier et al., 1997; Davail et al., 2000). France is the world's largest producer of foie gras and most likely will remain so in the foreseeable future, since production in other countries has decreased or ceased due to animal health and welfare concerns raised by animal advocates and the veterinary profession (Skippon, 2013).

Linoleic Acid

Zhang et al. (2016) examined the linoleic acid requirement of Wulong geese during the growing phase by changing the amounts of corn oil and palm oil in their diets. They found that maximal gain, efficiency of feed utilization, and breast yield required between 0.72 and 1.12, 0.72 and 0.92, and 0.92 and 1.12 percent linoleic acid, respectively. The composition of the starter diet was not provided, but presumably the birds used in this study had replete linoleic acid levels in their adipose at the time the experimental diets were introduced. Thus, triglyceride turnover in this storage pool would provide a nondietary source of linoleic acid to supply metabolic needs during this experiment, and the requirement was likely underestimated. The previous report did not make a recommendation for the linoleic requirement and based on the study by Zhang et al. (2016), a provisional recommended requirement is 1.1 percent, which is at the upper end of the range that they observed.

Minerals

Since the last report, new research on calcium and phosphorus greatly increased our confidence in past recommendations. Novel research on several trace minerals allows new recommendations on these important nutrients that are commonly deficient in diets based on grains and soybean meal.

Calcium

Two studies examined the calcium requirements of geese during the starting phase (1–28 d). Wulong geese (low adult body weight) did not respond to calcium supplementation of a basal diet that contained 0.65 percent calcium (Li et al., 2005). White Roman geese (high adult body weight) required between 0.25 and 0.45 percent calcium to maximize weight gain, and maximal tibia ash required 0.45–0.65 percent calcium (Lu and Hsu, 1994). These results confirm the provisional recommendation of 0.65 percent from the previous report (NRC, 1994), and the recommendations in Table 18-1 have not been changed.

Phosphorus

The non-phytate phosphorus requirement for high adult body weight breeds were examined in three studies. From 1 to 28 d of age, White Roman geese had maximal rate of gain with 0.46 percent non-phytate phosphorus as determined by broken-line regression and 0.32–0.42 percent for tibia ash (Lu and Hsu, 1994). The level of calcium in this experiment was 0.85 percent, which is considerably above the requirement and may have resulted in a compensatory high need for phosphorus. From 1 to 21 and from 29 to 72 d of age, Lion Head geese did not respond to additions of phosphorus to basal diets containing 0.40 and 0.30 percent non-phytate phosphorus, respectively (Zhu et al., 2018). The recommended requirement in the previous report (NRC, 1994) was based on a single study that indicated 0.46 percent total phosphorus supported optimal growth, but the level of non-phytate phosphorus was not reported. Taken together, the research suggests that a dietary level of 0.40 percent available phosphorus is adequate for growth and tibia ash during the starter phase and that 0.3 percent is adequate during the grower phase. The recommendation in Table 18-1 for 1–28 d old goslings has been increased from that in the previous report.

Bone development in goslings is sensitive to both the level and the ratio of calcium and phosphorus in the diet (Salyi, 2003). The study by Lu and Hsu (1994) shows that body weight gain and tibia ash are maximal at a calcium to phosphorus ratios between 1.2:1 and 3.4:1 and that both are impaired at 6.25:1. Although there is insufficient information to recommend an acceptable range of calcium to phosphorus ratios, the range between 1.5:1 and 3.0:1 appears to be an appropriate target for goslings.

Copper

When Wulong geese in the growing phase of production (28–70 d) were fed diets with increasing levels of copper, gain, efficiency, and carcass yield were estimated to be maximized at 8.8, 9.2, and 11.6 mg copper/kg diet using quadratic regression (Yang et al., 2018a). When Yangzhou geese were fed diets with increasing copper levels during the growing phase, gain and efficiency were maximized at between 5.9 and 20.9 mg/kg diet (Xu et al., 2013). Use of quadratic regression gave estimates of 21.9 and 16.3 mg/kg. However, the increments of copper addition were very large (15 mg/kg) relative to the likely requirement, so this study lacked the precision of the more recent study by Yang et al. (2018b), which used much smaller increments of copper addition (4 mg/kg). For this reason, the recommended copper requirement for the growth phase is 10 mg/kg. Due to the lack of experimental studies conducted during the starter phase, the requirement for this phase is unknown.

Iron

Ma et al. (2015) examined the effect of iron supplementation on Wulong geese from 35 to 119 d of age. They found maximal growth, efficiency of feed utilization, and dressing percentage when geese were fed 92–112 mg/kg diet. Using the midpoint of this range, the provisional recommendation for the growing phase is 100 mg/kg.

Manganese

Wang et al. (2014a) examined the manganese requirement of Wulong geese fed a diet based on corn and soybean meal during the starter phase. Using quadratic regression, they found that gain and efficiency of feed conversion were maximized with 117 and 105 mg/kg manganese, respectively. These same investigators (Zhang et al., 2014) also examined the manganese requirement of Wulong geese fed a diet based on corn and soybean meal during the growing phase and found that gain and efficiency of feed conversion were maximized with 105 and 118 mg/kg manganese, respectively. The previous report did not provide an estimate of the manganese requirement due to a lack of research. Based on the two studies cited previously, a provisional requirement for manganese has been set at 115 mg/kg for both the starter and grower phase.

Zinc

Chen et al. (2013) conducted an experiment to titrate the zinc response of 1–28 d goslings and found that the rate of gain on the corn–soymeal basal diet of 31 mg Zn/kg was increased by increasing the zinc level to 81 mg/kg, and further additions did not have an effect. When these geese were switched to a grower diet with stepped down levels of all nutrients, including zinc, those fed 77 mg Zn/kg had improved growth and efficiency than those fed the basal diet with 27 mg Zn/kg. This study clearly shows that adding zinc to a diet based on corn and soybean meal increases performance, but the large increments of zinc addition (50 mg/kg) preclude making a recommendation until further research using smaller levels of zinc addition becomes available.

Selenium

Baowei et al. (2011) did not see a growth or efficiency response of 1–63 d old geese to the addition of organic selenium to a basal diet containing 0.08 mg selenium/kg; however, a diet containing 0.18 mg/kg decreased water loss from the carcass. Yang et al. (2015) observed increased gain of geese from 1–28 d of age when sodium selenite was added to provide between 0.20 and 0.35 mg selenium/kg diet. These geese continued on the experiment until slaughter at 70 d and 0.05–0.20 mg of selenium/kg maximized gain. The different estimates of selenium requirements by these two studies cannot be explained by growth rate because the Yangzhou geese used by Yang et al. (2015) were slower growing. For this reason the provisional recommended requirement for is 0.20 mg/kg for both starter and grower diets.

Vitamins

Vitamin D

Tian et al. (2013) investigated the vitamin D requirement of Yangzhou geese from 35 to 77 d of age by supplementing cholecalciferol to a basal diet based on corn, soybean meal, and alfalfa. By calculation, the basal diet had approximately 4 IU D_3/kg. Gain, efficiency of feed utilization, eviscerated yield, breast yield, leg yield, and plasma calcium concentrations were highest with the addition of 0–500, 1,000–2,000, 0–1,000, 0, and 0–1,000 IU of supplemental D_3/kg, respectively. Based on this study the recommended requirement for vitamin D_3 in the growing period is increased from the provisional level of 200 IU/kg in the previous report (NRC, 1994) to 1,000 IU/kg. It is not known if geese can utilize the vitamin D_2 (ergosterol) found in plants to meet their metabolic needs for cholecalciferol. Because chickens are very inefficient at using the plant form of this vitamin, it should be presumed that geese are as well until research indicating otherwise becomes available.

Choline

Zhang et al. (2013) examined the choline requirement of Qingnong grey geese during the starting and growing phases. During both the starting and growing phases, gain and efficiency of feed utilization were maximized by the addition of 0–1,800 and 600–1,800 mg choline/kg, respectively. The choline level in the basal diet was not reported, but back calculation using typical choline concentrations in their ingredients indicates about 900 mg choline/kg diet. Consequently, the total dietary levels of choline that maximized growth and feed efficiency were 900–2,700 and 1,500–2,700 mg/kg. The previous report (NRC, 1994) recommended 1,500 and 1,000 mg/kg for the starter and grower phases, respectively, based on a single study with geese from 1 to 21 d of age. Based on these two studies the provisional recommended choline requirement is 1,500 mg/kg diet during both the starting and growing periods.

Niacin

Duan et al. (2014) examined the niacin requirement in starting (1–18 d), growing (28–56 d), and finishing (56–112 d) Wulong geese. During all three phases, gain and efficiency of feed utilization were maximal at between 43.5 and 103.5 mg niacin/kg. Using quadratic regression, the efficiency of feed utilization and carcass drip loss were optimized at 81.4 and 82.7 mg niacin/kg. The previous version of this report recommended 65 mg/kg based on two publications measuring growth and incidence of perosis during the starting phase. Considering all three of these studies, the recommended dietary level of niacin for starting and growing geese is increased to 80 mg/kg.

Pantothenic Acid

Two recent studies have examined the pantothenic acid requirements of Wolong geese. During the starting phase, Wang et al. (2016) used quadratic regression to estimate a requirement of 26.2 and 15.5 mg/kg diet for gain and efficiency of feed utilization, respectively. During the growing phase, pantothenic acid requirement for gain and efficiency were 21.1 and 18.5 for gain and efficiency, respectively, when predicted by quadratic regression (Zhang et al., 2015). The previous report (NRC, 1994) recommended 15.0 and 10.0 mg pantothenic acid/kg diet based on a single study (Serafin, 1981). However, that study used only two replicates of six goslings per treatment and lacked statistical power to observe small differences and was considered provisional. For this reason, the recommended pantothenic acid requirements for starting and growing geese are raised to 26 and 21 mg/kg diet, respectively.

Nutrient Requirements for Egg Production

There were no published research reports on geese producing eggs that meet the criteria described in the "Nutrient Requirements" section for use in revising nutrient requirements. The recommendations in Table 18-1 are based on recommendations from the previous report (NRC, 1994), and that report should be consulted for additional information.

REFERENCES

Baowei, W., H. Guoqing, W. Qiaoli, and Y. Bin. 2011. Effects of yeast selenium supplementation on the growth performance, meat quality, immunity, and antioxidant capacity of goose. *Journal of Animal Physiology and Animal Nutrition*. 95(4):440–448.

Bjornhag, C. 1977. Transport of various feed components through the digestive tract of turkey, geese and guinea fowl. *Swedish Journal of Agricultural Research*. 7:57–66.

Buckland, R., and G. Guy. 2002. *Goose Production*. Food and Agriculture Organization of the United Nations.

Chen, M., B. Wang, M. Zhang, B. Yue, Ge, Wenhua, D. Wang, W. Wang, and Y. Meng. 2013. Effects of dietary zinc levels on growth performance, serum biochemical parameters and hormone content of geese. *Journal of Animal Nutrition* 25:1105–1112.

Chen, Y., S. Wang, and J. Hsu. 2003. Effect of caecectomy on body weight gain, intestinal characteristics and enteric gas production in goslings. *Asian Australasian Journal of Animal Sciences*. 16(7):1030–1034.

Chen, Y.H., H.K. Hsu, and J.C. Hsu. 2002. Studies on the fine structure of caeca in domestic geese. *Asian Australasian Journal of Animal Sciences*. 15(7):1017–1021.

Clemens, E.T., C.E. Stevens, and M. Southworth. 1975. Sites of organic acid production and pattern of digesta movement in the gastrointestinal tract of geese. *Journal of Nutrition*. 105(10):1341–1350.

Davail, S., G. Guy, J. André, D. Hermier, and R. Hoo-Paris. 2000. Metabolism in two breeds of geese with moderate or large overfeeding induced liver-steatosis. *Comparative Biochemistry and Physiology Part A: Molecular & Integrative Physiology*. 126(1):91–99.

Du, X., Y. Lou, X. Liu, and Z. Teng. 2017. Effect of dietary methionine on the growth performance, biochemical indexes and serum antioxidant capacity of Jilin geese at the age 28–70d. *Chinese Journal of Animal Science*. 53:84–87.

Duan, C., B. Wang, W. Ge, M. Zhang, and W. Li. 2014. Effects of dietary nicotinic acid supplemental level on growth performance, slaughter performance and nutrient apparent availability of Wulong geese. *Chinese Journal of Animal Nutrition*. 26:2136–2144.

FAO (Food and Agriculture Organization of the United Nations). 2012. *Goose Production*. FAO. 144.

FAOSTAT. 2016. Agriculture data. In *Agricultural Production*. FAO.

Fournier, E., R. Peresson, G. Guy, and D. Hermier. 1997. Relationships between storage and secretion of hepatic lipids in two breeds of geese with different susceptibility to liver steatosis. *Poultry Science*. 76(4): 599–607.

Fox, A.D., R.D. Hearn, L. Cao, P.H. Cong, X. Wang, Y. Zhang, S.T. Dou, M. Barter, and E.C. Rees. 2013. Preliminary observations of diurnal feeding patterns of Swan geese *Anser cygnoides* using two different habitats at Shengjin Lake, Anhui Province, China. *Wildfowl*. 58(58):20–30.

Fox, A.D., and J. Kahlert. 2005. Changes in body mass and organ size during wing moult in non-breeding greylag geese *Anser anser*. *Journal of Avian Biology*. 36(6):538–548.

Frei, S., S. Ortmann, M. Kreuzer, J.M. Hatt, and M. Clauss. 2017. Digesta retention patterns in geese (*Anser anser*) and turkeys (*Meleagris gallopavo*) and deduced function of avian caeca. *Comparative Biochemistry and Physiology. A, Molecular & Integrative Physiology*. 204(1): 219–227.

Gruhn, K., A. Hennig, D. Jamroz, and M. Zieger. 1975. [Influence of the caecum in geese fed diets containing urea on nitrogen balance and digestibility of nutrients, also on the urea content of faeces and blood]. *Archiv fur Experimentelle Veterinarmedizin*. 29(1):199–205.

Hallsworth, E.G., and J.I. Coates. 1962. The growth of the alimentary tract of the fowl and the goose. *The Journal of Agricultural Science*. 58(2): 153–163.

Hauschild, L., C. Pomar, and P. Lovatto. 2010. Systematic comparison of the empirical and factorial methods used to estimate the nutrient requirements of growing pigs. *Animal*. 4(5):714–723.

Heikkinen, M., M. Ruokonen, M. Alexander, J. Aspi, T. Pyhäjärvi, and J. Searle. 2015. Relationship between wild greylag and European domestic geese based on mitochondrial DNA. *Animal Genetics*. 46(5):485–497.

Hermier, D. 1997. Lipoprotein metabolism and fattening in poultry. *The Journal of Nutrition*. 127(5):805S–808S.

Hsu, J.-C., T.-W. Lu, P.W. Chiou, and B. Yu. 1996a. Effects of different sources of dietary fibre on growth performance and apparent digestibility in geese. *Animal Feed Science and Technology*. 60(1–2):93–102.

Hsu, J., T. Lu, P.W.S. Chiou, and B. Yu. 1996b. Effects of different sources of dietary fibre on growth performance and apparent digestibility in geese. *Animal Feed Science and Technology*. 60(1-2):93–102.

Hui, Y., Z. Yongwen, D. Yuanfan, and Y. Lin. 2012. Comparative study on the utilization of dry matter and crude protein in different diets of ducks and geese. *Guangdong Agricultural Sciences*. 39(1):161–163.

Ibtisham, F., L. An, T. Li, Y. Niu, M. Xiao, L. Zhang, and R. Jia. 2017. Growth patterns of two Chinese native goose breeds. *Revista Brasileira de Ciência Avícola*. 19(2):203–210.

Jackowiak, H., K. Skieresz-Szewczyk, S. Godynicki, S.I. Iwasaki, and W. Meyer. 2011. Functional morphology of the tongue in the domestic goose (*Anser anser* f. domestica). *The Anatomical Record: Advances in Integrative Anatomy and Evolutionary Biology*. 294(9):1574–1584.

Jamroz, D., K. Jakobsen, J. Orda, J. Skorupinska, and A. Wiliczkiewicz. 2001. Development of the gastrointestinal tract and digestibility of dietary fibre and amino acids in young chickens, ducks and geese fed diets with high amounts of barley. *Comparative Biochemistry and Physiology Part A: Molecular & Integrative Physiology*. 130(4):643–652.

Jamroz, D., T. Wertelecki, A. Wiliczkiewicz, J. Orda, and J. Skorupińska. 2004. Dynamics of yolk sac resorption and post-hatching development of the gastrointestinal tract in chickens, ducks and geese. *Journal of Animal Physiology and Animal Nutrition*. 88(5–6):239–250.

Jamroz, D., A. Wiliczkiewicz, J. Orda, T. Wertelecki, and J. Skorupinska. 2002a. Aspects of development of digestive activity of intestine in young chickens, ducks and geese. *Journal of Animal Physiology and Animal Nutrition*. 86(11–12):353–366.

Jamroz, D., A. Wiliczkiewicz, J. Orda, T. Wertelecki, and J. Skorupińska. 2002b. Aspects of development of digestive activity of intestine in young chickens, ducks and geese. *Journal of Animal Physiology and Animal Nutrition*. 86(11–12):353–366.

Jamroz, D., A. Wiliczkiewicz, and J. Skorupinska. 1992. The effect of diets containing different levels of structural substances on morphological changes in the intestinal walls and the digestibility of the crude fibre fractions in geese (Part 3). *Journal of Animal and Feed Sciences*. 1(3):37–50.

Jiang, Y., Z. Wang, C. Zhou, and H. Yang. 2013. Effects of nitrogen-free diet method and fasting method on endogenous amino acid losses of Yangzhou geese. *Chinese Journal of Animal Nutrition*. 25(1):427–432.

Kierończyk, B., M. Rawski, J. Długosz, S. Swiatkiewicz, and D. Józefiak. 2016. Avian crop function—A review. *Annals of Animal Science*. 16: 653-678.

Klasing, K. 2005. Poultry nutrition: a comparative approach. *Journal of Applied Poultry Research*. 14(2):426–436.

Knížetová, H., J. Hyánek, L. Hyankova, and P. Bělíček. 1995. Comparative study of growth curves in poultry. *Genetics Selection Evolution*. 27(4):365.

Knížetová, H., J. Hyanek, and A. Veselsky. 1994. Analysis of growth curves of fowl. III. Geese. *British Poultry Science*. 35(3):335–344.

Larson, G., and D.Q. Fuller. 2014. The evolution of animal domestication. *Annual Review of Ecology, Evolution, and Systematics*. 45(1):115–136.

Lee, M.-Y., H.S. Jeon, Y.-S. Choi, S. Joo, and J. An. 2017. Sequencing and analyzing complete mitochondrial genome of *Anser cygnoides* (Anserini: Anserinae). *Mitochondrial DNA Part B*. 2(1):228–229.

Li, Q., X. Zhao, W. Liu, W. Liu, D. He, and X. Peng. 2016. Effects of dietary metabolizable energy level on growth performance and serum biochemical indices of Sichuan white gees aged from 1 to 3 weeks. *Journal of Animal Nutrition*. 28(1):3076–3083.

Li, W., B. Wang, Y. Lin, and G. Liu. 2004a. Effects of different dietary levels of methionine and lysine on the production and contents of plasma uric acid and free amino acids in Wulong geese. *Acta Zoonutrimenta Sinica*. 16(1):33–38.

Li, W., B. Wang, Y. Lin, and G. Liu. 2004b. Study on the appropriate levels of methionine and lysine in pre-growth diet of Wulong goose. *Journal of Nanjing Agricultural University*. 27(1):68–71.

Li, W., B. Wang, G. Liu, and Y. Lin. 2005. Effects of different levels of calcium and phosphorus on the production and biochemistry indexes of plasma and tibia in young Wulong geese. *Journal of Anhui Agricultural University*. 32(3):283–288.

Li, Y., H. Yang, L. Xu, Z. Wang, Y. Zhao, and X. Chen. 2018. Effects of dietary fiber levels on cecal microbiota composition in geese. *Asian-Australasian Journal of Animal Sciences*. 31(8):1285.

Li, Y.P., Z.Y. Wang, H.M. Yang, L. Xu, Y.J. Xie, S.L. Jin, and D.F. Sheng. 2017. Effects of dietary fiber on growth performance, slaughter performance, serum biochemical parameters, and nutrient utilization in geese. *Poultry Science*. 96(5):1250–1256.

Lin, Z., K. Feng, H. Ye, Y. Deng, L. Xu, and Y. Lu 2012. Effects of dietary methionine levels on serum biochemical parameters and antioxidant function in Lions Heads from 28 to 70 days old. *Journal of Animal Nutrition*. 24(1):2126–2132.

Ling, J., Y.-Y. Gao, Y. Hui, W.-C. Wang, Z.-P. Lin, H.-Y. Yang, S.-B. Huang, and Y. Lin. 2014. Effects of dietary fiber and grit on performance, gastrointestinal tract development, lipometabolism, and grit retention of goslings. *Journal of Integrative Agriculture*. 13(12):2731–2740.

Lu, J.J., and A. Hsu. 1990. Studies on rice hull as feedstuffs for geese, Proceedings 5th AAAP Animal Science Congress. 3:220.

Lu, J.J., and A. Hsu. 1994. Calcium and available phosphorus requirements of goslings. *Journal of the Chinese Society of Animal Science*. 23(1): 139–150.

Ma, C., B. Wang, W. Ge, M. Zhang, M. Kong, X. Zhang, and L. Sui. 2015. Effects of iron on growth performance, slaughter performance, meat quality and nutrient utilization of Wulong geese aged from 5 to 16 weeks. *Chinese Journal of Animal Nutrition*. 27(2):3420–3428.

Mateova, Z., S. Koci, V. Gergelyiova, and R. Svec. 1980. Lysine requirements of broiler goslings. *Biologizace a Chemizace Zivocisne Vyroby, Veterinaria*. 16(1):555–562.

Mattocks, J.G. 1971. Goose feeding and cellulose digestion. *Wildfowl*. 22(22):107–113.

Min, Y., S. Hou, Y. Gao, W. Huang, F. Liu, L. Zhao, and J. Yu. 2006. Study on energy and protein nutrient requirements of 5–8 week old goose. *Journal of Northwest A&F University (Natural Science Edition)*. 34(1):34–40.

Min, Y.N., S.S. Hou, Y.P. Gao, W. Huang, and F.Z. Liu. 2007. Effect of dietary crude protein and energy on gosling growth performance and carcass trait. *Poultry Science*. 86(4):661–664.

Moore, S.J. 1998. The comparative functional gizzard morphology of several species of birds. *Australian Journal of Zoology*. 46(4):359–368.

Moore, S.J. 1999. Food breakdown in an avian herbivore: Who needs teeth? *Australian Journal of Zoology*. 47(6):625–632.

Nitsan, Z., A. Dvorin, and I. Nir. 1981. Composition and amino acid content of carcass, skin and feathers of the growing gosling. *British Poultry Science*. 22(1):79–84.

Nitsan, Z., A. Dvorin, and I. Nir. 1983. Protein, essential amino acids and glycine requirements of the growing gosling (*Anser cireneus*). *British Journal of Nutrition*. 50(2):455–461.

NRC (National Research Council). 1994. *Nutrient Requirements of Poultry*, 9th rev. ed. Washington, DC: National Academy Press.

Olsen, A.M. 2015. Exceptional avian herbivores: Multiple transitions toward herbivory in the bird order Anseriformes and its correlation with body mass. *Ecology and Evolution*. 5(21):5016–5032, suppl. 5011.

Prop, J., and T. Vulink. 1992. Digestion by barnacle geese in the annual cycle: The interplay between retention time and food quality. *Functional Ecology*. 6(1):180–189.

Roberson, R.H., and D.W. Francis. 1963. The effect of energy and protein levels of the ration on the performance of White Chinese geese. *Poultry Science*. 42:867–871.

Sahin, T., M. Tilki, İ. Kaya, Y. Unal, and D.A. Elmali. 2008. Effect of different protein levels for finishing period on fattening performance and carcass traits in native Turkish geese. *Journal of Animal and Veterinary Advances*. 7(11):1364–1369.

Salyi, G. 2003. Data to the optimalisation of calcium and phosphorus needs of young geese and the development of mineral substance metabolism disturbance. *Magyar Allatorvosok Lapja*. 125(1):88–93.

Sedinger, J.S. 1997. Adaptations to and consequences of an herbivorous diet in grouse and waterfowl. *Condor*. 99(2):314–326.

Serafin, J. 1981. Studies on the riboflavin, pantothenic acid, nicotinic acid, and choline requirements of young Embden geese. *Poultry Science*. 60(8):1910–1915.

Sheng, D., and Z. Wang. 2006. Metabolic energy values and utilization of some nutrients in six common raw materials of goose. *Chinese Poultry*, 28(1):112–115.

Shi, S., Z. Wang, H. Yang, and Y. Zhang. 2007. Nitrogen requirement for maintenance in Yangzhou goslings. *British Poultry Science*. 48(2):205–209.

Shi, S.R., Z.Y. Wang, J.M. Zou, H.M. Yang, and N. Jiang. 2010. Effects of dietary threonine on growth performance and carcass traits of Yangzhou geese. *Czech Journal of Animal Science*. 55(1):382–387.

Shih, B., and J. Hsu. 2006. Development of the activities of pancreatic and caecal enzymes in White Roman goslings. *British Poultry Science*. 47(1):95–102.

Skieresz-Szewczyk, K., and H. Jackowiak. 2017. Development of mechanical papillae of the tongue in the domestic goose (*Anser anser* f. domestica) during the embryonic period. *Protoplasma*. 254(1):147–160.

Skippon, W. 2013. The animal health and welfare consequences of foie gras production. *Canadian Veterinary Journal*. 54(4):403–404.

Stevenson, M.H. 1985. Effects of diets of varying energy concentrations on the growth and carcase composition of geese. *British Poultry Science*. 26(4):493–504.

Sun, G., S. Gong, Shen Hongmin, D. He, H. Yu, L. Xia, S. Zhang, H. Xue, and Y. Hang. 2010. Development of mathematical model to estimate the amino acid requirement of meat goose. *Journal of Animal Nutrition*. 22(1):1717–1723.

Tang Junwang, F.Q., S. Rongyi, L. Jun, X. Huaxin, S. Junda, N. Dong, and L. Lizhi. 2018. Effects of different oils on the performance, serum index and fatty acid composition of Langde goose during the filling period. *Journal of Animal Nutrition*. 30(1):2550–2560.

Tian, X., Z. Wang, H. Yang, D. Jiang, H. Gu, and Y. Zhao. 2013. Effects of dietary vitamin D_3 supplemental level on growth performance, slaughter performance and serum biochemical indexes of Yangzhou geese. *Chinese Journal of Animal Nutrition*. 25(1):978–984.

Timmler, R., and H. Jeroch. 1994. Structure and physiology of the digestive tract of the goose and their influence on the ability to digest fibre. (A review). *Archiv fur Geflugelkunde*. 58(1):97–102.

van der Leeuw, A.H., K. Kurk, P.C. Snelderwaard, R.G. Bout, and H. Berkhoudt. 2003. Conflicting demands on the trophic system of Anseriformes and their evolutionary implications. *Animal Biology*. 53(3):259–301.

Wang, B.-W., J. Li-Zhen, Q. Zhang, Q.-L. Wang, Y.-C. Fan, B. Yue, P. Sun, X.-X. Jiang, and N. Wang. 2008a. Digestibility of goose fed with different levels of corn straw silage [J]. *Chinese Journal of Animal Nutrition*. 2(1):176–182

Wang, B., X. Zhang, W. Ge, M.a. Zhang, B. Yue, X. Shi, X. Wang, and C. Xu. 2014a. Effects of manganese on growth performance, slaughter performance, tibia development and enzyme activities of Wulong geese aged from 1 to 4 weeks. *Chinese Journal of Animal Nutrition*. 26(1): 877–884.

Wang, B., X. Zhang, B. Yue, W. Ge, M. Zhang, C. Ma, and M. Kong. 2016. Effects of pantothenic acid on growth performance, slaughter performance, lipid metabolism, and antioxidant function of Wulong geese aged one to four weeks. *Animal Nutrition*. 2(4):312–317.

Wang, W., L. Xu, Z. Lin, B. Xiong, and L. Yang. 2014b. Effects of different dietary lysine levels on growth performance, serum biochemical indices and protein deposition of Lion Head geese aged from 1 to 21 days. *Journal of South China Agricultural University*. 35(6):1–7.

Wang, X., Z. Wang, H. Yang, Y. Cao, and X. Zhu. 2012. Effects of dietary energy and protein level and lysine level on body weight and slaughter performance of Yangzhou geese at the age of 5 to 10 weeks. *Chinese Journal of Animal Nutrition*. 6(1):1044–1051.

Wang, Z., S. Shi, Q. Zhou, L. Fan, Y. Shi, and M. Xu. 2008b. The influence of caecectomy on amino acid availability of three feedstuffs for ganders. *British Poultry Science*. 49(2):181–185.

Wang, Z., S. Shi, Q. Zhou, Y. Haiming, Z. Kangning, and H. Han. 2010a. Ideal amino acid pattern and amino acid requirements of 5–10 weeks Yangzhou geese. *Chinese Journal of Animal Nutrition*. 1(1):31–38.

Wang, Z.Y., S.R. Shi, Q.Y. Zhou, H.M. Yang, J.M. Zou, K.N. Zhang, and H.M. Han. 2010b. Response of growing goslings to dietary methionine from 28 to 70 days of age. *British Poultry Science*. 51(1):117–121.

Wei, Z., J. Lu, Y. Yu, H. Zhang, S. Zhu, and H. Wang. 2010. Effects of dietary energy levels on slaughter performance and biochemical index of Yangzhou geese. *China Poultry*. 8(1):32–35.

Wei, Z., L. Wang, H. Pan, X. Yu, and Y. Ji. 2012. Effects of feeding regimen and dietary tryptophan level on immune function and antioxidant indices of Yangzhou geese. *Chinese Journal of Animal Nutrition*. 12(1):2356–2365

Xu, C., B. Wang, W. Ge, M. Zhang, B. Yue, and X. Shi. 2013. Effects of copper on growth performance, slaughter performance, nutrient availability and serum hormone content of Wulong geese aged 5 to 16 weeks. *Chinese Journal of Animal Nutrition*. 25(9):1989–1997.

Xu, Q., X. Yuan, T. Gu, Y. Li, W. Dai, X. Shen, Y. Song, Y. Zhang, W. Zhao, G. Chang, and G. Chen. 2017. Comparative characterization of bacterial communities in geese fed all-grass or high-grain diets. *PLoS One*. 12(10):e0185590

Yang, H., Z. Wang, H. Sun, X. Ju, S. Jin, and L. Yu. 2015. Effects of selenium on growth performance, serum biochemical indices, antioxidant capacity, slaughter performance and meat quality of goslings. *Chinese Journal of Animal Nutrition*. 27(1):3699–3707.

Yang, H., Z. Wang, J. Wang, S. Shi, and X. Zhu. 2009. Effects of caecectomy on digestibility of crude protein, calcium, phosphorus, neutral detergent fibre and acid detergent fibre in geese. *Archiv Geflugelk*. 73(3):189–192.

Yang, H., Y. Xiao, G. Gui, J. Li, J. Wang, and D. Li. 2018a. Microbial community and short-chain fatty acid profile in gastrointestinal tract of goose. *Poultry Science*. 97(4):1420–1428.

Yang, J., S.S. Zhai, Y.C. Wang, S.S. Wang, Z.P. Yang, L. Yang, and W.C. Wang. 2016. Effects of graded fiber level and caecectomy on metabolizable energy value and amino acid digestibility in geese. *Journal of Integrative Agriculture*. 15(1):629–635.

Yang, Z., X. Qi, H. Yang, H. Dai, C. Xu, and Z. Wang. 2018b. Effects of dietary copper on growth performance, slaughter performance and nutrient content of fecal in growing goslings from 28 to 70 days of age. *Revista Brasileira de Ciência Avícola*. 20(1):45–52.

Yu, L., M. Wang, H. Wang, and Z. Wei. 2010. Effects of metabolic energy levels on weight gain efficiency and body fat deposition of Yangzhou goose aged 6–10 weeks. *Chinese Journal of Animal Science*. 21(1).

Yu, L., W.Z. Zhang Jun, H. Wang, X. Pan, X. Liang, and X. Yang. (喻礼怀, 张军, 魏宗友, 王洪荣, 潘晓花, 梁祥焕, and 杨雪). 2014. Effects of dietary tryptophan levels on protein metabolism in Yangzhou goose. *Journal of Animal Nutrition*. 26(1):81–89.

Zeuner, F.E. 1963. *A History of Domesticated Animals*. London, UK: Hutchinson.

Zhang, S., J. Guo, Q. Tang, J. Zou, H. Li, C. Zhu, H. Lu, J. Su, W. Tan, and G. Xu. 2011. Metabolizable energy and fiber digestibility values of 11 unconventional feedstuffs for geese [J]. *Chinese Journal of Animal Nutrition*. 23(11):1925–1931.

Zhang, W., B. Wang, W. Ge, M. Zhang, and W. Li. 2013. Effects of choline on growth performance, slaughter performance and nutrient apparent availability of geese. *Chinese Journal of Animal Nutrition*. 25(11):778–784.

Zhang, X., B. Wang, W. Ge, M. Zhang, B. Yue, X. Shi, C. Xu, and X. Wang. 2014. Effects of manganese on growth performance, slaughter performance, nutrient availability and enzyme activity of Wulong geese aged from 5 to 16 weeks. *Chinese Journal of Animal Nutrition*. 26(1): 106–114.

Zhang, X., B. Wang, B. Yue, W. Ge, M. Zhang, C. Ma, and M. Kong. 2015. Effects of pantothenic acid on growth performance, slaughter performance, meat quality, nutrient availability and serum biochemical parameters of Wulong geese aged from 5 to 16 weeks. *Chinese Journal of Animal Nutrition*. 27(1):3411–3419.

Zhang, Y., W. Baowei, G. Wenhua, Z. Mingai, Y. Bin, Z. Huiwen, R. Min, and Z. Zenan. 2016. Effects of linoleic acid on growth performance, slaughter performance, muscle quality and nutrient utilization rate of 5–16 weeks old goose. *Journal of Animal Nutrition*. 28(1):3473–3482.

Zhao, F., H. Yang, L. Pei, X. Wan, and W. Zhiyue. 2018. Effects of pectin in diet on growth performance, body size traits, slaughter performance and organ index of Yangzhou goose from 28 to 70 days old. *Journal of Animal Nutrition*. 30(4):2716–2724.

Zhou, H., Y. Gao, B. Xu, D. Tao, Z. Teng, and Y. Lou. 2018. Effects of different dietary fiber levels on bacteria diversity in cecum of goose. *Acta Veterinaria et Zootechnica Sinica*. 49(2):565–571.

Zhou, Q., Z. Wang, H. Yang, S. Shi, K. Zhang, and H. Han. 2008. Study on optimum dietary methionine levels of 5–10 week old Yangzhou gosling. *Chinese Journal of Animal Nutrition*. 20(1):34–39.

Zhu, Y.W., C.Y. Wang, J. Wen, W.C. Wang, and L. Yang. 2018. Effect of dietary high non-phytate phosphorus level on growth performance and metabolism of calcium and phosphorus in Lion-head geese. *Animal Feed Science and Technology*. 236(1):115–121.

19

Nutrient Requirements of Japanese Quail, Bobwhite Quail, and Ring-Necked Pheasants

INTRODUCTION

This chapter provides nutrient recommendations for species that are less economically important than those in the prior chapters. However, their propagation is locally important for meat, eggs, and, in the case of ring-necked pheasants and bobwhite quail, for restocking depleted wild populations and hunting preserves.

The previous edition of this report (NRC, 1994) provides documentation for peer-reviewed studies published prior to September 1991 that were used to estimate nutrient requirements and can be consulted for references to the older literature. There has been relatively little new information on the nutrient requirements of Bobwhite quail and ring-necked pheasants since the previous report, and the changes to the respective requirement tables are relatively minor. Considerable research on nutrient requirements of Japanese quail has been conducted since the previous report, and this is reflected in changes to, or confirmation of, previous recommended values. As in the other chapters, research studies were excluded from consideration for estimating nutrient requirements if the experimental design or statistical analysis was not appropriate for estimating nutrient requirements or if insufficient information was provided to allow replication of the experiment. Exclusion factors included insufficient dietary levels for establishing a dose–response relationship, failure to chemically analyze dietary levels of the nutrient of interest, and failure to provide key information on the experimental design, housing, diet nutrient levels, management practices that impact requirements, analytical methodologies, or statistical procedures. Studies were also excluded if they were conducted at environmental temperature extremes that would affect production rates and nutrient requirements. Studies on amino acid, vitamin, and trace mineral requirements that did not use a common basal diet were also excluded. In some cases where no qualifying studies were available for a specific nutrient, exceptions were made in order to provide a rough estimate based on a dietary concentration demonstrated to be adequate for maximal productivity. In those cases, any resulting nutrient recommendations are listed as provisional in the tables (bold and italicized font).

With the exception of energy and crude protein, recommended requirements are minimum concentrations in the diet needed to support optimal performance or absence of deficiency signs. The values for energy and crude protein are not requirements but are typical levels used in research and commercial production. Poultry can adapt to a range of energy levels by changing their feed intake, and they can utilize lower crude protein levels if an optimal balance of amino acids is provided.

Because commercial production of these species uses a wider range of genetic stock than the production of chickens or turkeys, genetic differences in growth or egg production rates should be factored into the recommendations provided. In addition, the recommended values do not include a margin of safety to account for environmental, management, toxicological, or disease stressors. They also assume that the dietary ingredients used have digestibilities or bioavailabilities similar to a typical diet based on corn and soybean meal and supplemented with additional nutrients of high bioavailability (see Chapter 8).

JAPANESE QUAIL

Japanese quail (*Coturnix coturnix japonica*) are native to East Asia and were domesticated around the 12th century CE. Captive Japanese quail were kept in China for fighting and other entertainment purposes before 1300 CE and for meat production within a few hundred years thereafter (Chang et al., 2005). Strains were developed in Japan for egg production in the early 20th century (Minvielle, 2004). Strains from Japan were then introduced into North America, Europe, and the Near and Middle East in the first half of the 20th century. Japanese quail are the most common species of quail propagated in captivity worldwide. They are used commercially for the purposes of meat and egg production. They are also used as a laboratory model in a wide variety of biological and medical areas of investigation due to their rapid reproductive maturation, small size, ease of breeding, and fully sequenced genome (Baer et al., 2015).

There are many strains of Japanese quail worldwide. Domestication resulted in an increased body weight, markedly increased egg production, loss of nesting and brooding behaviors, and decreased ability to fly. Most domestic strains kept for egg production produce at rates similar to chickens (~300 in a year). Compared to strains bred for egg production, those bred specifically for meat production (meat strains) grow faster and have a greater adult body weight, which often exceeds 300 g. Japanese quail are sexually dimorphic in plumage and also in size, with females being larger due mostly to the weight of the reproductive tract, ovaries, and their relatively larger liver. Because of this, the carcass yields of females are similar to or slightly greater than that of males. Japanese quail reach adult body weight at about 7–10 weeks of age (Portillo-Loera et al., 2014; Gürcan and Kaplan, 2017) and sexual maturity at 5–9 weeks of age, depending on the lighting schedule (Baer et al., 2015).

Information on the number of Japanese quail produced by each country is imprecise because most countries do not keep records on minor species, but some generalizations are possible (Minvielle, 2004). Worldwide, Japanese quail are utilized in considerably greater numbers for egg production than meat production. Egg production dominates their use in Asia, where most Japanese quail production occurs, as well as in Brazil. Meat production is more important in the United States and Europe. The use of the slightly larger common quail (*Coturnix coturnix*; sometimes called European quail) for meat production has been increasing in some countries. Production of all strains of quail has notably increased in parts of the world where high temperatures, prevalence of infectious diseases, or lack of capital for infrastructure limit the feasibility of chickens or other large galliformes.

Nutritional and Digestive Strategy

Adult wild-type Japanese quail weigh about 90 g for males and 100 g for females and are migratory (Kawahara, 1973). Wild Japanese quail hatchlings eat considerable amounts of insects, but as they mature they consume much greater amounts of food from plants (Kawahara, 1967). The digestive tract of Japanese quail has a crop, glandular proventriculus, muscular gizzard, and paired ceca typical of Galliformes. Their crop facilitates the consumption of larger meals, especially immediately prior to nighttime, but removal of the crop does not decrease daily feed consumption when nutrient-dense diets are provided for ad libitum consumption (Savory, 1985). In female quail, intestinal villi reach their maximal length at about 42 d of age (De Souza Andrade et al., 2019). The density of villi (n/area) and expression of glucose transporters are highest soon after hatch and decline with age until egg production begins, suggesting that their expression matches metabolic demand. The ceca are short relative to many of the larger members of the Phasianids family. However, the ceca have an anatomy (Fenna and Boag, 1974) and contain a microbiota (Wilkinson et al., 2016) consistent with a primary role of fermentation. Energy balance studies using intact and cecectomized Japanese quail indicate that the ceca contribute 5.7 percent of the daily energy requirement (Thompson and Boag, 1975). Japanese quail respond to increases in dietary fiber with a considerable increase in the size of the gizzard along with moderate increases in the length of their small intestines (Starck and Rahmaan, 2003) and ceca (Savory and Gentle, 1976).

Nutrient Requirements of Japanese Quail

The requirement recommendations in the previous report (NRC, 1994) were based predominantly on random bred strains housed at research institutions under highly controlled environmental conditions. Recent research on nutrient requirements has been done primarily on strains selected for much faster rates of gain (i.e., meat strains) or egg production (i.e., egg strains). Much of the research has been conducted in Brazil, India, Iran, Turkey, and Pakistan. The growth rates, adult body weights, rates of egg production, and egg sizes differ considerably based on the selection criteria. The environmental conditions among these locations also vary, which makes it difficult to generalize nutrient requirements. Therefore, nutritional practitioners should examine Table 19-1 and make modifications that fit their production rates and environmental conditions. The vast majority of research conducted on the nutritional requirements of Japanese quail since the previous report (NRC, 1994) has centered on macronutrients, with very few studies focused on micronutrients. This disparity is reflected in the changes in recommended nutrient requirements provided in Table 19-1. Compared to egg strains, the requirements of meat strains for several nutrients are slightly greater for growth and feed efficiency, and new recommendations for several nutrients are provided for these strains.

Some recent studies with Japanese quail have expressed levels of amino acids in feedstuffs as well as dietary amino acid requirements on a digestible basis. This was done by using conversion factors compiled from research on chickens. Japanese quail diverge in their digestive physiology, especially with regard to rate of passage of feed through the GI tract, and the use of chicken conversion ratios is not clearly supported by empirical experiments across a robust range of feed ingredients. For this reason, the recommended amino acid requirements of Japanese quail will continue to be expressed on a total amino acid basis. Establishing the digestible amino acid concentrations in common feedstuffs used for quail is an important area of future research for refining amino acid requirements. Similarly, the metabolizable energy content of feedstuffs used for Japanese quail is typically based on values compiled using chickens. The empirical evidence to

TABLE 19-1 Nutrient Requirements of Japanese Quail (90% Dry Matter)

Nutrient	Units	Egg Type Starter, Grower 2,900[a]	Meat Type Starter, Grower 2,900[a]	Laying Female 2,900[a]
Protein[b]	%	24	24	20
Arginine[c]	%	1.25		***1.21***
Glycine + serine	%	1.15		***1.17***
Histidine	%	0.36		***0.42***
Isoleucine	%	0.98		***0.70***
Leucine	%	1.69		***1.42***
Lysine	%	1.35	1.40	1.16
Methionine	%	0.50	0.54	0.45
Methionine + cyst(e)ine[d]	%	***0.85***	***0.92***	***0.86***
Phenylalanine	%	0.96		***0.78***
Phenylalanine + tyrosine	%	1.80		***1.40***
Threonine	%	1.02	***1.12***	0.74
Tryptophan	%	0.22		0.19
Valine	%	0.95		0.66
Linoleic acid	%	1.1		0.70
Calcium	%	0.80	***0.80***	3.0
Chlorine	%	***0.13***	***0.13***	0.15
Magnesium	mg/kg	300		***500***
Phosphorus (non-phytate)	%	0.30		0.35
Potassium	%	***0.40***		***0.44***
Sodium	%	0.18		0.15
Copper	mg/kg	***5***		***5***
Iodine	mg/kg	***0.3***		***0.3***
Iron	mg/kg	***120***		***60***
Manganese	mg/kg	***60***	***60***	***60***
Selenium	mg/kg	***0.2***		***0.2***
Zinc	mg/kg	25		***50***
A	IU	1,650		***3,300***
D_3	ICU	750		***900***
E	IU	***12***		***25***
K	mg/kg	***1***		***1***
B_{12}	mg/kg	***0.003***		***0.003***
Biotin	mg/kg	***0.3***		***0.15***
Choline	mg/kg	***2,000***		***1,500***
Folacin	mg/kg	***1***		***1***
Niacin	mg/kg	40		***20***
Pantothenic acid	mg/kg	***10***		***15***
Pyridoxine	mg/kg	***3***		***3***
Riboflavin	mg/kg	***4***		***4***
Thiamin	mg/kg	***2***		***2***

NOTES: Values that are bold and in italics are estimates based on adequate dietary levels shown to support optimal performance in Japanese quail or studies that did not meet the inclusion criteria described in the introduction to this report. Shaded values were unchanged or unconfirmed from the 1994 recommendations (NRC, 1994). For references to the studies supporting shaded values, please refer to the 1994 report.

[a] These are typical dietary energy concentrations expressed as kcal ME_n/kg of diet.

[b] The values for protein are not requirements but are typical levels used in research and commercial production. Lower levels can be used when the appropriate essential amino acids are supplemented to meet the requirements.

[c] Amino acid requirements are expressed on a total amino acid basis.

[d] Methionine plus cyst(e)ine recommendations are listed as provisional because the research on which the recommendations rely did not demonstrate that cysteine was limiting in the experimental diets.

TABLE 19-2 Nutrient Requirements of Bobwhite Quail (90% Dry Matter)

Nutrient	Units	0 to 6 Weeks 2,800[a]	After 6 Weeks 2,800[a]	Breeding, Female 2,800[a]
Protein[b]	%	***26.0***	***20.0***	***24.0***
Methionine	%	***0.60***	***0.50***	***0.45***
Methionine + cyst(e)ine	%	***1.0***	***0.77***	***0.90***
Lysine	%	***1.35***	***0.95***	***1.0***
Linoleic acid	%	***1.0***	***1.0***	***1.0***
Calcium	%	0.65	***0.65***	***2.4***
Phosphorus (non-phytate)	%	***0.45***	***0.30***	***0.7***
Sodium	%	***0.15***	***0.15***	***0.15***
Chlorine	%	***0.11***	***0.11***	***0.11***
Iodine	mg/kg	0.30	***0.30***	***0.30***
Choline	mg/kg	**1,500**	***1,500***	**1,000**
Niacin	mg/kg	30.0	***30.0***	**20.0**
Pantothenic acid	mg/kg	12.0	***9.0***	**15.0**
Riboflavin	mg/kg	3.8	***3.0***	**4.0**

NOTES: Values that are bold and in italics are estimates based on research for other ages or on other species or are based on adequate dietary levels shown to support maximal performance. Shaded values were unchanged or unconfirmed from the 1994 recommendations (NRC, 1994). For references to the studies supporting shaded values, please refer to the 1994 report.

[a] These are typical dietary energy concentrations expressed as kcal ME_n per kg of diet.

[b] The values for protein are not requirements but are typical levels used in research and commercial production. Lower levels can be used when the appropriate essential amino acids are supplemented to meet the requirements.

support this is lacking for most feedstuffs, and further research is needed to verify this approach.

Nutrient requirement research on common quail (*Coturnix coturnix*) has accumulated since the last edition of this publication. This species has different growth characteristics and nutrient requirements, but at this time there is insufficient information to make comprehensive nutritional recommendations based on the peer-reviewed literature.

The previous addition of this report (NRC, 1994) based many of the recommended nutrient requirements of Japanese quail on research done with random bred quail. Therefore, as a reflection of the higher-producing birds used in research studies since 1991, the recommended nutrient requirements in the current edition of this report are generally somewhat greater than those in NRC (1994).

Nutrient Requirements for Growth

This section considers nutrient requirements for growth of strains selected for egg or meat production. Egg strains of quail are typically grown in colony cages, although sometimes they are started in floor pens and then moved to laying cages after a few weeks. Meat strains of quail are reared in either floor pens or in colony cages. It is not known if the type of housing influences nutrient requirements. In commercial production, quail are commonly fed two to four different diets during their growing period to adjust for their declining fractional growth rates. For most nutrients, there is insufficient information to provide requirement recommendations for these separate periods (Tables 19-1 and 19-4).

Energy

Japanese quail adjust their feed intake to obtain a level of metabolizable energy intake that approximates their energy expenditure. Growing Japanese quail can adapt to a range of metabolizable energy concentrations if the ratio of energy to all other nutrients is maintained. Because supplying energy is a major determinant of the cost of a diet, decisions on the optimal energy level are often driven by the relative prices of feedstuffs. The metabolizable energy density values (2,900 kcal/kg diet) provided in Table 19-1 are not meant to be optimal for either diet cost or productive performance but are provided as a reference value due to common use in research experiments. In practice, the concentration of the required nutrients should be adjusted to account for different rates of feed intake that occur at other energy levels.

The metabolizable energy needs of Japanese quail are high compared to chickens. Jordão Filho et al. (2011b) utilized the comparative slaughter method to determine the dietary metabolizable energy requirements for maintenance and for tissue deposition from 16 to 36 d of age. The quail were housed in cages that were held at three different temperatures (18, 24, and 28°C). At 24°C, the maintenance energy requirement was 93.10 kcal/kg$^{0.75}$/day and the energy required to deposit body tissue was 8.28 kcal/g. The efficiency of use of dietary metabolizable energy for body tissue deposition was 23 percent. From their results, the daily energy requirement needed for maintenance plus weight gain at temperatures between 18 and 28°C can be calculated using the following formula: Metabolizable energy requirement (kcal/quail/day) = body

TABLE 19-3 Nutrient Requirements of Ring-Necked Pheasants (90% Dry Matter)

Nutrient	Units	0 to 4 Weeks 2,800[a]	4 to 8 Weeks 2,800[a]	9 to 17 Weeks 2,700[a]	Breeding, Female 2,800[a]
Protein[b]	%	28	24	18	15
Lysine	%	***1.5***	***1.3***	***0.8***	***0.68***
Methionine	%	***0.50***	***0.47***	***0.3***	***0.3***
Methionine + cyst(e)ine	%	***1.0***	***0.93***	***0.6***	***0.6***
Linoleic acid	%	***1.0***	***1.0***	***1.0***	***1.0***
Calcium	%	1.0	***0.85***	***0.53***	***2.5***
Phosphorus (non-phytate)	%	0.55	***0.5***	***0.45***	***0.40***
Sodium	%	***>0.15***	***0.15***	***0.15***	***0.15***
Chlorine	%	***0.11***	***0.11***	***0.11***	***0.11***
Manganese	mg/kg	***70***	***70***	***60***	***60***
Zinc	mg/kg	60	***60***	***60***	***60***
Choline	mg/kg	**1,430**	**1,300**	**1,000**	**1,000**
Niacin	mg/kg	70	**70**	**40**	**30**
Pantothenic acid	mg/kg	10	**10**	**10**	**16**
Riboflavin	mg/kg	3.4	**3.4**	**3.0**	**4.0**

NOTES: Values that are bold and in italics are estimates based on research for other ages or on other species or are based on adequate dietary levels shown to support maximal performance. Shaded values were unchanged or unconfirmed from the 1994 recommendations (NRC, 1994). For references to the studies supporting shaded values, please refer to the 1994 report.

$\text{weight}^{0.75} \times (98.37 - 0.205 \times \text{temperature}) + 8.28 \times \text{weight gain}$, where body weight is in kg, temperature is °C, and weight gain is in g/d. These results are similar to those of Silva et al. (2004b), who measured the same variables from 15 to 32 d of age. However, Silva et al. (2004a) found that 1- to 12-d old chicks are more efficient at using energy for body weight gain, needing only 4.64 kcal/g. This may be due to the composition of gain, which is leaner in the hatchlings, in addition to the use of retained egg yolk by the hatchling.

Protein

Poultry do not require protein per se, but they require the essential amino acids and amino nitrogen for synthesis of other nutrients that are provided by protein. A recommendation for protein was provided in the previous report (NRC, 1994), but the basis for this recommendation was not clearly defined. The values in Table 19-1 are not recommendations but are provided as reference values that are typically used in research when a diet is based on corn and soybean meal and supplemented with methionine. Many commercial diets are supplemented with methionine, lysine, and, occasionally, threonine, which allows the use of lower protein levels.

Jordão Filho et al. (2012) used a diet with 23.5 percent protein and based on corn, soybean meal, methionine, and threonine to estimate the protein requirement of growing quail between 16 and 36 d of age. They found that the requirement for maintenance was 4.78 $g/kg^{75}/d$ and the efficiency of deposition of dietary protein for body protein accretion was 25 percent. These values are similar to those found in a previous study by several of the same authors (Silva et al., 2004b). Thus, the efficiency of protein utilization for growth is considerably lower than that of chickens, and this is reflected in their higher dietary protein needs.

Amino Acids

As mentioned previously, most of the amino acid requirement studies conducted since the previous report (NRC, 1994) have examined strains selected specifically for meat or egg production and not random bred or dual-purpose strains of Japanese quail. Laying strains have been selected for large egg sizes in addition to maximum rates of egg production. This has resulted in a larger mature body size and a faster growth rate. Thus, the amino acid requirements for growing laying strains are likely to be greater than those recommended in 1994. As described above, meat strains grow considerably faster than laying strains, so they are likely to have even greater requirements.

Methionine and lysine are the first and second limiting amino acids for growing Japanese quail, respectively (Cavalcante, 2013). Many recent publications report amino acid levels in diets as digestible amino acids and not as total amino acids. As mentioned previously, this approach remains to be validated across a range of feed ingredients before it can be implemented with confidence. For this reason, recommendations for amino acid levels continue to be provided on a total basis and not on a digestible basis. For reports that provided only digestible amino acid levels, those levels were converted back to total amino acid levels prior to consideration for requirement recommendations. Both total and digestible levels are reported in Table 19-4.

TABLE 19-4 Documentation of Nutrient Requirements of Japanese Quail for Growth

Age (Days)	Weight, Start–End of Study, g	Estimated Requirement	Response Criteria	Strain, Sex	Number of Levels (Range)	Comments	Reference
Lysine, %							
1–14	6–30	0.87–0.99 0.99–1.11	Gain Feather pigmentation, pectoralis and gastrocnemius weights	Random bred, mixed	4 (0.87–0.23%)	Data analyzed by GLM and means comparisons. Protein and energy levels not reported.	Barbour et al., 1993
1–14	9–49	0.99–1.11 1.11–1.23	Gain Feather pigmentation, pectoralis and gastrocnemius weights	Body weight selected, mixed	4 (0.87–0.23%)	Data analyzed by GLM and means comparisons. Protein and energy levels not reported.	Barbour et al., 1993
3–10	17–111	1.30–1.45	Gain, Feed conversion	Meat, mixed	6 (1.00–1.75%)	Data analyzed by GLM and means comparisons. Basal diet had 24% CP and 2,900 kcal ME/kg.	Hajkhodadadi et al., 2013
3–24		1.30–1.45 1.15–1.60 1.45–1.60	Gain, Breast yield, Thigh yield Gain, Breast yield, Thigh yield				
7–21	23–113	1.24 1.36 1.38	Gain, feed conversion Breast meat yield Carcass yield	Meat, mixed	5 (0.94–1.54%)	Data analyzed six types of linear, quadratic, and broken-line regressions. Basal diet had 24% CP and 2,950 kcal ME/kg.	Hasanvand et al., 2018
7–21	37–122	1.27 1.21 1.34 1.32	Gain Feed conversion Breast meat yield Thigh meat yield	Meat, mixed	6 (0.91–1.51%)	Data analyzed by broken-line, linear, and quadratic regression. Basal diet had 23% CP and 2,940 kcal ME/kg.	Mehri et al., 2013
21–35	79–160	1.29 1.31	Gain Feed conversion	Meat, mixed	6 (0.84–1.59%)	Data analyzed as average of linear, quadratic, and quadratic broken-line regressions. Basal diet had 25% CP and 2,950 kcal ME/kg.	Mehri et al., 2015
21–42	178–314	1.49 1.45 1.51 1.54	Gain Feed conversion Breast weight Thigh weight	Meat, females	6 (1.0–1.75%)	Data analyzed as broken-line and quadratic regressions. Basal diet had 24% CP and 2,900 kcal ME/kg.	Hajkhodadadi et al., 2014
1–40	8–126	1.36 (1.21) 1.35 (1.20) 1.29 (1.14)	Gain Subsequent egg production Subsequent feed conversion/dozen eggs	Laying, mixed	5 (0.88–1.28%)	Data analyzed by broken-line, linear, and quadratic regressions. Results reported as digestible lysine and converted to total lysine. Basal diet had 20.1% CP and 2,900 kcal ME/kg.	Lima et al., 2016

Methionine + Cyst(e)ine, %							
1–7	8–25	0.85 0.86	Gain Feed conversion	Laying, mixed	5 (0.54–0.78%)	Data analyzed by broken-line and quadratic regressions. Basal diet had 1.05% lysine, 20.5% CP, and 2,900 kcal ME/kg.	Lima et al., 2011
1–21	8–93	0.85–0.90	Gain, feed conversion	Meat, mixed	5 (0.35–0.60%)	Data analyzed by ANOVA and means comparisons. Basal diet had 1.29% lysine, 0.40% cysteine, 23.5% CP, and 2,900 kcal ME/kg.	Parvin et al., 2010
7–21	31–139	0.92 0.91 0.94 0.99	Gain Feed conversion Breast meat yield Leg meat yield	Meat, mixed	(0.38–0.68%)	Data was analyzed as the average of broken-line and quadratic polynomials. Basal diet 1.38% lysine, 0.39% cysteine, 25% CP, and 2,900 kcal ME/kg.	Khosravi et al., 2016
7–42	22–129	0.85 (0.76)	Feed conversion	Laying, mixed	5 (0.48–0.75%)	Data analyzed by quadratic polynomials. Lys was 1.3, Cys was 0.32. Results reported as digestible Met + Cys and converted to total Met + Cys. Basal diet had 20.2% CP and 2,900 kcal ME/kg	Pinto et al., 2003a
Threonine, %							
0–21		0.96–1.02 1.02–1.12	Gain Feed conversion	Meat, mixed	3 (0.96–11.2%)	Data analyzed by ANOVA and means comparisons. Basal diet had 23.3% CP and 2,900 kcal ME/kg.	Mandal et al., 2006
1–35		<0.81	Gain, feed conversion	Meat, mixed	6 (0.81–1.06%)	Data analyzed by ANOVA and means comparisons. No response to added levels of threonine. Basal diet had 23% CP and 3,000 kcal ME/kg.	Baylan et al., 2006
21–35		0.96–1.02	Gain	Meat, mixed	3 (0.96–11.2%)	Data analyzed by ANOVA and means comparisons. Basal diet had 23.3% CP and 2,900 kcal ME/kg.	Mandal et al., 2006
Tryptophan, %							
30–42		<0.27	Gain, feed conversion	Laying, mixed	4 (0.27–1.02%)	Data analyzed by ANOVA and means comparisons. No response to added levels. Basal diet had 20.0% CP and 2,800 kcal ME/kg.	Rizzo et al., 2008

continued

TABLE 19-4 Continued

Age (Days)	Weight, Start–End of Study, g	Estimated Requirement	Response Criteria	Strain, Sex	Number of Levels (Range)	Comments	Reference
Linoleic acid, %							
7–28	20–85	1.1	Gain, liver weight	Laying, mixed	4 (0.0–1.5%)	Data analyzed with a nonlinear regression model. Basal diet had 26.1% CP and 3,321 kcal ME/kg.	Murai et al., 1994
Calcium, %							
1–21	8–61	1.2%	Bone ash, Ca in tibia plus femur	Laying, mixed	7 (0.85–1.75%)	Data was analyzed with linear and quadratic regression. Ca was supplemented as limestone. Available P was 0.41% and held constant; basal diet had 21.3% CP and 2,900 kcal ME/kg.	Costa et al., 2009a
7–35	N/A	0.5–0.8	Gain, feed conversion	Growth selected line, mixed	3 (0.50–1.1%)	Data was analyzed by ANOVA and means comparisons. Ca was supplemented as $CaCO_3$. Basal diet had 27.5%. CP and ME were not reported.	Lalmingthanga and Shrivastav, 1998
Sodium, %							
1–21 22–40	8–61 101–166	0.18 0.17 0.14 0.15	Gain Feed conversion Gain Efficiency	Unknown, mixed	5 (0.04–0.36%)	Data was analyzed with linear and quadratic regression. Data reanalyzed by broken-line regression. Na was supplemented as $NaHCO_4$ and NaCl. Cl was 0.15% and held constant. Base excess ranged from 218 to 357 mEq/kg. Water was deionized. Basal diet had 23.3% CP and 2,900 kcal ME/kg.	Costa et al., 2012a
Chlorine, %							
1–42	7–135	<0.07	Gain, feed and water intake, efficiency, digestibility, excreta moisture	Laying, mixed	6 (0.07–0.32%)	Data was analyzed with quadratic regression. Cl^- was supplemented as NH_4Cl. Na was at 0.14%, and base excess ranged from 234 to 286 mEq/kg; water quality not reported. Basal diet had 23.8% CP and 2,900 kcal ME/kg.	Bezerra et al., 2013

Manganese, mg/kg							
7–35	8–123	60	Gain, efficiency, tibia bone ash	Growth selected line, mixed	3 (60–90 mg/kg)	Data was analyzed by ANOVA and means comparisons. Mn was supplemented as $MnSO_4$ $5H_2O$. CP and ME were not reported.	Lalmingthanga and Shrivastav, 1998
Choline, mg/kg							
1–14		1,500–2,000	Gain, feed conversion	Meat, mixed	3 (1,500–2,500 mg/kg)	Choline added as choline chloride. Diet was formulated to meet the methionine requirement with minimal excess. Basal diet had 23.9% CP and 2,818 kcal ME/kg.	Shrivastav et al., 2004
Water							
1–42	7–134	25.2 ml/d 2.4 ml water/ g feed	Water consumption Ratio of water to feed consumption	Unknown, mixed		Temperature fluctuated between 28 and 33°C; 75% relative humidity. No effect of dietary Cl^- or electrolyte balance. Na was at 0.15%; basal diet had 23.8% CP and 2,900 kcal ME/kg.	Bezerra et al., 2013
1–21 22–40	8–61 61–166	22.2 ml/d 3.7 ml water/g feed 31.1 ml/d 5.3 ml water/g feed	Water consumption Ratio of water to feed consumption Water consumption Ratio of water to feed consumption	Unknown, mixed		Consumption reported is at the Na requirement. Consumption increased linearly with diet Na level. Water was deionized. Basal diet had 23.3% CP and 2,900 kcal ME/kg.	Costa et al., 2012b

NOTE: All amino acid levels are expressed as total amino acids and, when calculated by the authors, digestible amino acid levels are provided in parentheses.

Methionine + cyst(e)ine

Lima et al. (2011) examined the methionine plus cyst(e)ine requirements of a laying strain of Japanese quail from 1 to 7 d and estimated the requirement for gain and feed conversion as 0.85 and 0.86 percent, respectively. Also in a laying strain, Pinto et al. (2003a) estimated the methionine plus cyst(e)ine requirement to be 0.85 percent based on feed conversion. These results indicate a requirement that is greater than recommended for growing quail of laying strains in the previous report, and the recommended requirement for methionine plus cyst(e)ine has been increased to 0.85 percent. Neither study established that the cysteine level in the basal diet was adequate to meet metabolic cysteine needs so that added methionine would be used to fulfil both the methionine and cysteine requirements. The dietary concentration of cysteine that meets the full requirement for this semi-essential nutrient is not known in quail, and if the dietary cysteine level was actually below that needed, the methionine plus cysteine requirement would have been overestimated. For this reason, the methionine plus cyst(e)ine recommendations provided in Table 19-1 are provisional.

Khosravi et al. (2016) examined the methionine plus cyst(e)ine requirements in a meat strain of Japanese quail from 7 to 21 d. Using the average of broken line analysis and quadratic polynomials they estimated a methionine plus cyst(e)ine requirement of 0.92 percent for body weight gain and 0.94 percent for breast meat yield. Parvin et al. (2010) also examined the methionine plus cyst(e)ine requirement for a growing meat strain; however, they did not apply a regression approach to their data analysis, and their estimate based on ANOVA was between 0.85 and 0.90 percent. As in the studies on laying strains of quail, it is not known if cyst(e)ine levels used were limiting, so the recommendations provided in Table 19-1 are provisional (Lima et al., 2015a).

Lysine

Using a laying strain of Japanese quail, Lima et al. (2016) estimated that 1.36 percent lysine maximized the rate of gain from 1 to 40 d of age. Furthermore, 1.35 percent lysine maximized subsequent egg production. Thus, the recommended lysine requirement has been increased slightly to 1.35 percent of the diet.

The previous report (NRC, 1994) did not provide recommendations for strains selected for meat production. Barbour et al. (1993) demonstrated that the dietary lysine concentration that supported maximal gain and feather pigmentation through 14 d of age in a line of Japanese quail selected for 4-week body weight was slightly greater than that needed by the random-bred population from which it was initially developed. Two studies by Hajkodadadi et al. (2013, 2014), found that the lysine requirement of their meat strain is greater than the 1.30 percent recommended in NRC (1994). Using ANOVA for data analysis, they found that the requirement of meat strain from day 3 to 24 was between 1.30 and 1.45 percent (Hajkhodadadi et al., 2013). In a subsequent study centered on females during the grower phase (21–42 d), Hajkhodadadi et al. (2014) used a broken-line regression analysis to estimate the requirement for lysine to be 1.49 or 1.45 percent, respectively, for body weight gain and feed efficiency. Based on maximal breast and thigh weight, the requirement was 1.51 and 1.54 percent, respectively. Also using a meat strain, Hasanvand et al. (2018) estimated the requirement from 7 to 21 d to be 1.24 percent for gain and feed conversion, 1.36 percent for breast meat yield, and 1.38 percent for carcass yield. In their data analysis, they used six types of linear, quadratic, and broken-line regressions and chose the one that gave the best fit. Using a meat strain, Mehri et al. (2013) found lysine requirements from 7–21 d to be 1.27 percent for gain and 1.34 percent for breast meat yield. In a subsequent study conducted during the grower phase from d 21–35, Mehri et al. (2015) reported that the requirement was 1.29 percent and 1.30 percent for body weight gain and feed conversion, respectively, using an average of linear and quadratic broken-line analysis. The lower requirement estimated in this study might be attributable to the slower rate of growth of their strain of quail compared to those used in the above studies. Because most studies indicate that broiler strains of Japanese quail require higher levels of lysine, especially to maximize meat yield, the recommended requirement of 1.40 percent for these strains is higher than for strains grown for egg production.

Threonine

Mandal et al. (2006) reported that the threonine requirement of a fast-growing strain of Japanese quail was 0.90–1.02 percent for body weight (BW) gain and 1.02–1.12 percent for feed conversion. The dietary levels needed to maximize feed conversion are slightly higher than that needed in laying strains of quail. However, their study did not have enough treatment levels to do a regression analysis to estimate an exact requirement; consequently, the threonine recommendation provided in Table 19-1 is provisional.

Linoleic acid

Murai et al. (1994) fed quail chicks a semi-purified grower diet lacking linoleic acid from hatch to 1 week of age, and then switched them to diets containing between 0.0 and 1.5 percent linoleic acid. Based on maximum body weight gain and minimum values for liver weight, the linoleic acid requirement was estimated to be 1.1 percent of the diet. Because the recommended dietary linoleic acid requirement for Japanese quail in the previous edition of this report (NRC, 1994) was based on results from growing chickens, the recommended requirement is increased slightly to 1.1 percent of the diet.

Minerals and Vitamins

There was insufficient research on vitamin requirements of Japanese quail conducted since the previous edition of this report that met the inclusion criteria described previously to modify the recommendations in the previous report (NRC, 1994), so those values remain unchanged.

Calcium

Using quadratic regression, Costa et al. (2009a) examined the response of bone ash to increasing calcium levels and estimated the requirement to be 1.2 percent of the diet for a laying strain from 1 to 21 d. This is considerably higher than all of the previous studies, some of which have shown decreased feed intake and gain (Bisoi et al., 1980) or decreased tibial ash at that level (Lalmingthanga and Shrivastav, 1998). The high requirement obtained by Costa et al. may have been due to a high level of available phosphorus in the basal diet (0.41 percent) that was not increased with increasing calcium supplementation. A study using a growth selected strain of quail confirmed that 0.8 percent dietary calcium was adequate for gain, efficiency and tibial ash (Lalmingthanga and Shrivastav, 1998). For these reasons, the recommended calcium level was not changed.

Sodium and Chlorine

Several studies have examined the requirements for sodium and chlorine, but the estimates varied considerably, possibly due to differences in the contribution of water to the intake of this nutrient. Only one study published since the last report reported the source and sodium content of the water; consequently, only that report was considered. Costa et al. (2012a) used deionized water to examine the sodium requirement from 1 to 21 d in a laying strain and found that the requirement was 0.21 percent to maximize gain and 0.22 percent to maximize feed conversion. However, they analyzed their data using quadratic regression, which sometimes overestimates requirements (Pesti et al., 2009; Hasanvand et al., 2018) and, at 0.24 percent sodium, bird performance was clearly depressed in their study. Reanalysis of their data using a broken-line regression indicates a requirement of 0.18 for gain and 0.17 for efficiency. Based on this study, the recommended sodium level was increased from 0.15 to 0.18 percent. However, this assumes very low sodium intake from the water and dietary levels can be decreased as sodium from water contributions increases, as for example by providing softened water.

Bezerra et al. (2013) examined the chlorine requirement of an egg laying strain of Japanese quail from 1 to 42 d of age using six diets ranging from 0.07 to 0.32 percent, and base excess ranged from 215 to 286 mEq/kg. Dietary chlorine level did not affect gain, feed or water intake, efficiency, nutrient digestibility, or excreta moisture, indicating that the requirement is 0.07 percent or less. However, the source of water and its chlorine levels used for this experiment was not provided. The reported ratio of water to feed consumption in this study was 2.4. Fairfield and Ritz (2009) recommended a maximal acceptable chlorine concentration of 250 mg/L in water for poultry. Assuming that the water in this study had the maximum acceptable level, it would have contributed an additional 0.06 percent chlorine for a total dietary intake of less than 0.13 percent. Although the requirement is likely considerably less than this, the recommended level was decreased only slightly from NRC (1994) to 0.13.

Manganese

A study using a growth selected strain of quail found that manganese at 60 mg/kg was adequate for gain, efficiency, and tibia ash (Lalmingthanga and Shrivastav, 1998). This confirms the recommendation provided in the previous report.

Requirements for Egg Production

Most of the research on requirements for egg production have utilized laying strains of Japanese quail and considered rates and efficiency of production, egg size, and shell quality as the performance metrics. Very little recent research has examined requirements for fertility, hatchability, or chick quality in either laying or meat strains. For this reason, the requirement estimates in Table 19-1 are for table egg production. Requirements for some micronutrients may be higher for reproduction. Unfortunately, there has been very little research on micronutrient requirements since the previous report, even for table egg production.

Energy

Belo et al. (2000) studied five levels of energy (2,600 to 3,000 kcal ME/kg, with increasing increments of 100 kcal/kg) in isonitrogenous diets (19%) for laying Japanese quail. These authors observed that an increase in the energy concentration of the feed affected neither the energy consumption nor the rate of egg production, although they detected a tendency of reduced feed conversion ratio as the energy density increased. Pinto et al. (2002) tested energy concentrations of 2,850, 2,950, and 3,050 kcal ME/kg and obtained the highest production responses with energy levels of 2,850 kcal/kg, which resulted in the consumption of 77 kcal/d/quail with an average weight was 149 g.

Jordão Filho et al. (2011a) used the comparative slaughter technique to examine the maintenance energy requirement of laying quail. They found that the requirement for maintenance was 92.34 kcal/kg body weight$^{0.75}$/d. They also reported that laying quail required 6.23 kcal/g body weight gain. A subsequent experiment examining energy balance during egg production showed that the metabolizable energy required to produce one g of egg was 4.19 kcal. Thus the metabolizable energy requirement of a laying quail (kcal/bird/d) can be estimated as (body weight$^{0.75} \times 92.34$) + ($6.23 \times$ weight gain) + ($4.19 \times$ egg mass), where body weight is in kg and weight gain and egg mass are in g/bird/d.

Protein

The crude protein level provided in Table 19-1 (20%) is not a recommended level but is provided because it was the most common level used in published research on amino acid requirements and is adequate in a diet based on corn and soybean meal and supplemented with methionine at a level that meets the methionine plus cyst(e)ine requirement. Lower levels of crude protein can be used when additional essential amino acids are supplemented. For example, Abdel-Mageed et al. (2009) observed that an 18 percent, but not a 16 percent, crude protein level resulted in egg production and feed conversion similar to 20 percent protein as long as methionine, lysine, and threonine were supplemented to meet the hen's requirement. Santos et al. (2016) observed that a corn-soybean meal–based diet with 16 percent protein and supplemented with lysine, methionine, threonine, tryptophan, isoleucine, arginine, valine, and glutamic acid gave equivalent performance and egg quality as a 20 percent protein diet supplemented with amino acids so that both diets had the same essential amino acid contents. These studies indicate that laying Japanese quail can tolerate lower protein diets when their essential amino acid requirements are met.

Amino Acids

Most of the amino acid recommendations in the 1994 report (NRC, 1994) were tentative (bold italics), and many were based on studies designed to give a rough estimate of adequate dietary levels for several essential amino acids in a single experiment. In addition, regression approaches to estimate the requirements generally were not undertaken. Therefore, in the present report, studies that employed a basal diet supplemented with the amino acid of interest at multiple levels and data analysis by broken-line or quadratic regressions are used as the basis for recommendations to the extent possible. When requirements were reported as digestible amino acids using recommendations for chickens, they were converted back to total amino acids. This is because of the lack of evidence currently supporting such an approach; however, both are provided in Tables 19-1 and 19-5.

Methionine + cyst(e)ine. Methionine is the first limiting amino acid for egg production in a diet based on corn and soybean meal. Unlike the other essential amino acids, egg weight and egg mass typically respond positively to increasing methionine + cyst(e)ine, up to a plateau. Furthermore, the level that maximizes egg weight is often higher than that needed to maximize egg production and feed conversion. Methionine is a precursor for choline synthesis, and dietary choline increases egg size when methionine is low (Castro et al., 2011), most likely by sparing methionine.

Four studies on the methionine plus cyst(e)ine requirement that met the criteria described previously were published since the last report (Pinto et al., 2003c; Costa et al., 2009b; Reis et al., 2011; Scotta et al., 2011). The average egg production ranged from 89 percent to 96 percent, and the estimated requirement to support egg production ranged from 0.75 percent to 0.86 percent, with the highest estimate being in the highest-producing flock (Reis et al., 2011). Estimated requirements to support egg mass ranged from 0.77 percent to 0.91 percent, again with the highest requirement observed in the highest producing flock. Taken together, these experiments support raising the recommended requirement from those of NRC (1994). After adjusting dietary energy levels to the reference level of 2,900 kcal/kg, the average estimated requirement to support optimal egg mass and egg production in these four studies is 0.843 percent and 0.858 percent, respectively. These studies support raising the recommended methionine plus cyst(e)ine requirements to 0.86 percent, and higher levels are likely needed during the peak of production in high producing flocks. None of these studies established that the cyst(e)ine level in the basal diet was adequate, so that added methionine most likely was used to fulfil both the methionine and the cyst(e)ine requirements. For this reason, the methionine plus cyst(e)ine recommendation provided in Table 19-1 is provisional.

Lysine. At the time of the last report (NRC, 1994), there were no published reports that examined the requirement for lysine that met the criteria described previously. Since then, there have been five peer-reviewed studies that meet those criteria. One of these studies (Moura et al., 2008) was excluded because it was conducted in conditions that were sufficiently hot and humid that the rate of egg production and the estimated requirement were much lower than all of the other studies. Three of the studies were conducted in the first half of lay and they predicted similar lysine requirements of 1.15 percent (Pinto et al., 2003b), 1.17 percent (Hurtado Nery et al., 2015), and 1.15 percent (Costa et al., 2008) for the rate of egg production. A study conducted during the second half of lay estimated a higher requirement of 1.26 percent for feed conversion, egg weight, and egg mass. This study also used a higher dietary protein level than the others, so it is not clear if the difference was due to the protein level or to the older age of the birds. The previous report recommended 1.0 percent lysine, but this is clearly too low based on the foregoing studies. The recommended requirement is increased to the average of the above three studies conducted during the first part of lay (1.16%).

Threonine. Four studies on the threonine requirement for egg-producing strains of Japanese quail were conducted since the last report (NRC, 1994). Three of these were unable to establish a requirement because the additions of threonine to the basal diet did not result in significant increases in egg

TABLE 19-5 Documentation of Nutrient Requirements of Japanese Quail for Egg Production

Age (Weeks)	Average Egg Production % ave	Estimated Requirement[a] (%)	Response Criteria	Strain	Number of Levels (Range)	Comments	Reference
Amino Acids, Total and (Digestible)							
Arginine, %							
10–18	82%	<1.24	Egg production, egg weight, feed conversion, shell thickness	Laying	3 (1.24–1.44%)	Data analyzed by ANOVA with means separation. No response above lowest level. Basal diet had 0.99% lysine, 20.6% crude protein (CP), and 2,914 kcal ME/kg.	Bulbul et al., 2015
20–32	94%	<1.22 (<1.15)	Egg production, egg weight, egg mass, feed conversion, shell weight	Laying	5 (1.15–1.47% digestible)	Data analyzed by ANOVA with means separation. No response above lowest level. Basal diet had 1.08% lysine, 20% CP and 2,800 kcal ME/kg.	Mauricio et al., 2016
23–32	95%	<1.16	Egg production, egg weight, feed conversion, shell weight and thickness	Laying	5 (1.16–1.36%)	Data analyzed by ANOVA with means separation. No response above lowest level. Basal diet had 1.00% lysine, 20% CP, and 2,800 kcal ME/kg.	Reis et al., 2012
Lysine, %							
7–23	85%	1.15 (1.05) 1.23 (1.12) 1.16 (1.05)	Egg production Egg mass Feed conversion	Laying	6 (0.80–1.30%)	Met+cys varied with lys to keep same ratio; data analyzed by quadratic regression. Hot and humid. Arginine levels were 1.17–1.64%. Basal diet had 19.56% CP and 2,900 kcal ME/kg.	Pinto et al., 2003b
9–22	79%	0.75 1.06	Egg production Egg weight	Laying	5 (0.60–1.10%)	Met+cys was 0.75%; data analyzed by broken-line regression. Hot and humid. Arginine levels not reported. Basal diet had 18.02% CP and 2,900 kcal ME/kg.	Moura et al., 2008
10–36	84%	1.17 (1.08)	Egg production	Laying	4 (0.92–1.28% digestible)	Met was 0.274%; Data analyzed by quadratic regression. Arginine levels not reported. Basal diet had 19.9% CP and 2,830 kcal ME/kg.	Hurtado Nery et al., 2015
13.4–25.4	94%	1.15 (1.03%)	Egg production	Laying	5 (0.55–1.20% digestible)	Met+cys was 0.72%; data analyzed by quadratic regression. Arginine levels not reported. Basal diet had 18.2% CP and 2,791 kcal ME/kg.	Costa et al., 2008
29.6–41.6	86%	1.26 (1.12)	Egg weight, mass, feed conversion	Laying	6 (0.95–1.20%)	Met+cys varied w lys to keep same ratio; data analyzed by quadratic regression. Results reported as digestible Lys and converted to total Lys. Hot and humid. Arginine levels were 1.25–1.39%. Basal diet had 20.37% CP and 2,800 kcal ME/kg.	Ribeiro et al., 2013
Methionine + cyst(e)ine							
5–9	91%	0.84 (0.77)	Mean egg weight, egg weight, egg mass	Laying	5 (0.60–0.88% digestible)	Data analyzed by quadratic regression. Hot and humid. Egg production and FCR did not respond to methionine. Basal diet had 19.5 and 21.5% CP and 2,900 kcal ME/kg.	Scotta et al., 2011
7–23	90%	0.77 (0.71) 0.81 (0.74) 0.80 (0.73)	Egg production Egg weight Egg mass	Laying	6 (0.60–0.85% digestible)	Data analyzed by quadratic regression. Warm and humid. Basal diet had 19.3% CP and 2,900 kcal ME/kg.	Pinto et al., 2003c

continued

TABLE 19-5 Continued

Age (Weeks)	Average Egg Production % ave	Estimated Requirement[a] (%)	Response Criteria	Strain	Number of Levels (Range)	Comments	Reference
Methionine + cyst(e)ine							
13.4–18.2	96%	0.86 (0.79) 0.91 (0.84) 0.89 (0.82)	Egg production Egg weight, egg mass Feed conversion	Laying	6 (0.12–0.90% digestible)	Data analyzed by quadratic regression. Warm and humid. Basal diet had 19.66% CP and 2,900 kcal ME/kg.	Reis et al., 2011
13.4–25.4	90%	0.75 (0.67) 0.77 (0.70) 0.75 (0.68)	Egg production Egg mass Feed conversion	Laying	5 (0.55–0.79% digestible)	Data analyzed by quadratic regression. Basal diet had 20.0% CP and 2,800 kcal ME/kg.	Costa et al., 2009b
Threonine, %							
7.9–16.9	93%	<0.63 (<0.55)	Egg production, weight, feed conversion	Laying	5 (0.55–0.75% digestible)	Data analyzed by linear and quadratic regression. No response to added levels. Basal diet had 17.8% CP and 2,950 kcal ME/kg.	Umigi et al., 2012
8.1–17.1	93%	<0.73 (<0.65)	Egg production, weight, feed conversion	Laying	5 (0.65–0.85% digestible)	Data analyzed by linear and quadratic regression. No response to added levels. Warm and humid. Basal diet had 20.0% CP and 2,910 kcal ME/kg.	Umigi et al., 2007
8–17	95%	<0.74	Egg production, weight, feed conversion, BW	Meat, 316 g	4 (0.74–1.04%)	Data analyzed by GLM. The authors found nonsignificant trends with higher levels, but none were significant at <0.05. Statistical power was very low because only 10 birds per treatment. Basal diet had 19.5 and 20% CP and 2,800 kcal ME/kg.	Canogullari et al., 2009
17.9–32.9	82%	0.77 0.75 0.78 0.80	Egg production Egg weight Egg mass Feed conversion	Laying	6 (0.66–0.86%)	Data analyzed by quadratic regression. Warm and humid. Basal diet had 18.6% CP and 2,950 kcal ME/kg.	Lima et al., 2013
Tryptophan, %							
6.4–18.4	85%	***<0.23***	Egg production, feed conversion	Laying	4 (0.27–1.02%)	No response to added levels. Basal diet had 20.0% CP and 2,800 kcal ME/kg.	Rizzo et al., 2008
21–30	87%	0.23 (0.21)	Egg production	Laying	5 (0.12–0.28% digestible)	Linear discontinuous response plateau model. Basal diet had 20.0% CP and 2,900 kcal ME/kg.	Pinheiro et al., 2008
Isoleucine							
15.2–35	90%	*<0.67*	Egg production, feed conversion, egg weight, egg mass	Laying	4 (0.67–1.25)	Basal diet had 16.0% CP and 2,800 kcal ME/kg.	Santos et al., 2016
Valine, %							
31–35	90%	0.66	Egg production, egg weight, feed intake, body weight	Laying	7 (0.20–0.94)	Basal diet had 3,011 kcal ME/kg.	Martinez et al., 2021
40.6–49.6	85%	<1.03 (<0.75)	Egg production, egg weight, feed conversion	Laying	6 (1.03–1.28%)	No response to added levels. Basal diet had 18.6% CP and 2,900 kcal ME/kg.	Paula et al., 2017
Linoleic acid, %							
8–10	88%	0.70	Egg weight	Laying	5 (0.0–1.5%	Data analyzed with a nonlinear regression model. The five levels of linoleic acid were fed to growing chicks. Basal diet had 26.1% CP and 3,321 kcal ME/kg.	Murai et al., 1994

Calcium, %							
6.1–26.1	92%	<2.0%	Egg production, egg weight, % shell, shell thickness, specific gravity	Laying	4 (2.0–3.5)	No information on temperatures or housing type. Calcium supplemented as limestone. Basal had 0.55% available P. Body weight not given. Basal diet had 18% CP and 2,700 kcal ME/kg	Masukawa, 2001
6.6–18.6	92%	3.45	Egg production	Laying	7 (2.95–3.85)	Data analyzed by linear and quadratic regression. Warm and humid. Calcium supplemented as limestone. Basal had 0.40% non-phytate phosphorus. Did not report change in BW. Basal diet had 18.2% CP and 2,800 kcal ME/kg.	Brandão et al., 2007
		3.34	Feed conversion/dozen				
		>3.85	Egg weight				
		3.27	Egg mass				
		3.51	Specific gravity				
8–24.2	90%	3.16	Weight gain	Laying	6 (1.6–3.6)	Data analyzed by linear and quadratic regression. Warm and humid. Calcium supplemented as limestone. Basal had 0.35% non-phytate phosphorus. Basal diet had 20.0% CP and 2,860 kcal ME/kg.	Barreto et al., 2007b
		3.20	Egg production				
		3.24	Feed conversion				
		<1.6	Bone ash				
		3.24	Egg mass				
		3.34	Specific gravity				
		3.42	Shell weight				
		>3.6	Shell thickness				
		<1.6	Cracked or defective eggs				
26–28	92%	<2.0	Egg production, egg weight, egg mass/dozen feed conversion, cracks and defects	Laying	3 (2.0–3.0)	Data analyzed by linear regression. Hot and humid. Calcium supplemented as supplemented as dicalcium phosphate and limestone. Basal had either 0.10, 0.17, 0.24, or 0.31% non-phytate phosphorus. Did not report change in BW. Basal diet had 20.0% CP and 2,800 kcal ME/kg.	Vieira et al., 2012
		>3.0	Shell weight, egg specific gravity				
37.4–46.4	85%	<2.0	Egg production, egg weight, % shell, shell thickness, specific gravity	Laying	4 (2.0–3.5)	No information on temperatures or housing type. Calcium supplemented as limestone. Basal had 0.55% non-phytate phosphorus. Body weight not given. Basal diet had 18% CP and 2,700 kcal ME/kg.	Masukawa, 2001
45–57	78%	<2.2	Egg production, egg weight, egg mass/d	Laying	5 (2.2–3.8)	Data analyzed by linear and quadratic regression. Warm and humid. Calcium supplemented as limestone. Basal had 0.31% non-phytate phosphorus. Basal diet had 20.0% CP and 2,800 kcal ME/kg.	Costa et al., 2010a
		>3.8	Feed conversion				
		2.9	Shell weight				
		>3.8	Shell thickness				
45–57	82%	<2.0	Egg production, feed conversion, egg weight, egg mass, cracks and defects	Unknown	4 (2.0–3.5)	Data analyzed by linear and quadratic regression. No information on temperatures or housing type. Calcium supplemented as limestone. Basal had 0.15 or 0.35% non-phytate phosphorus. Only linear effects found. No quadratic effects. Body weight not given. Basal diet had 19.3% CP and 2,800 kcal ME/kg.	Costa et al., 2010b
		>3.5	Feed conversion/dozen, egg specific gravity, shell weight				
Non-phytate phosphorus, %							
8–16	91%	<0.15	BW change, egg production, egg weight and mass, feed conversion/egg mass, shell weight	Laying	3 (0.15–0.35)	Data analyzed by ANOVA. Phosphorus supplemented as dicalcium phosphate. Basal had 2.5% Ca, constant across treatments. Basal diet had 19.31% CP and 2,800 kcal ME/kg.	Sharifi et al., 2012
		0.15–0.25	Shell % and thickness				

continued

TABLE 19-5 Continued

Age (Weeks)	Average Egg Production % ave	Estimated Requirement[a] (%)	Response Criteria	Strain	Number of Levels (Range)	Comments	Reference
Non-phytate phosphorus, %							
8.7–17.7	91%	<0.15	BW change, egg production, egg weight and mass, feed conversion/egg mass, shell % and weight	Laying	5 (0.15 -0.55)	Data analyzed by linear and quadratic regression. Warm and humid. Phosphorus supplemented as dicalcium phosphate. Basal had 2.5 or 3.0% Ca. Did not report change in BW. Basal diet had 20.0% CP and 2,900 kcal ME/kg.	Costa et al., 2007
		0.45	Cracks and defects				
		>0.55	Feed conversion/dz				
		0.46	Bone Ca at 2.5% Ca				
		0.51	Bone P at 2.5% Ca				
26–28	94%	<0.10	Egg production, egg weight and mass, feed conversion, shell weight	Laying	4 (0.10–0.31)	Data analyzed by linear and quad regression. Hot and humid. Phosphorus supplemented as dicalcium phosphate. Basal had 2.0, 2.5, or 3.0% Ca. Did not report change in BW. Basal diet had 20.0% CP and 2,800 kcal ME/kg.	Vieira et al., 2012
		>0.31	Cracks and defects, bone Ca and ash				
45–57	82%	< 0.15	Egg production, Egg weight and mass, feed conversion, cracks and defects	Laying	5 (0.15–0.45)	Data analyzed by linear and quadratic regression. Warm and humid. Phosphorus supplemented as dicalcium phosphate. Basal had 3.0% Ca, constant across treatments. Did not report change in BW. Basal diet had 20.0% CP and 2,900 kcal ME/kg.	Costa et al., 2011a
		0.43	Specific gravity				
		0.41	Shell weight				
		0.42	% Shell				
		0.27	Shell Ca				
		0.42	Bone P				
Sodium, %							
9.9–18.9	94%	0.20	Shell thickness	Laying	4 (0.017–0.028)	Data analyzed by linear and quadratic regression. Warm and humid. Sodium supplemented as NaCl. Birds on the basal diet (0.07%) became dehydrated and stopped laying. The dose response of the remaining four levels was bi-phasic for several end points, and all of the response was between 0.08 and 0.15. The use of quadratic regression gave unrealistically high requirement estimates for most variables, and the range of dose responsiveness is reported here. Basal diet had 20.0% CP and 2,900 kcal ME/kg.	Barreto et al., 2007a
		0.08–0.15	Egg production, feed conversion, egg weight, egg mass, egg specific gravity				
13–25	83%	<0.10	Egg production, feed consumption, feed conversion, egg weight, egg mass, final weight, shell %, egg specific gravity	Laying	7 (0.15–0.45)	Data analyzed by regression and ANOVA. Hot and humid. Sodium supplemented as NaCl. Chloride level (0.03–0.30). Electrolyte balance 178–180. Heat stress may have resulted in respiratory alkalosis. Basal diet had 21.5% CP and 2,850 kcal ME/kg.	Murakami et al., 2006
26.7–35.7	90%	<0.10	Egg production, feed conversion, egg weight, egg mass, shell weight, egg specific gravity	Laying	5 (0.10–0.30)	Data analyzed by linear and quadratic regression. Hot and humid. Sodium supplemented as NaCl. Chloride level not provided. Electrolyte balance 224–227. Basal diet had 19.3% CP and 2,800 kcal ME/kg.	Lima et al., 2015b

54–66	88%	<0.10	Egg production, feed consumption, feed conversion, egg weight, egg mass, final weight, shell thickness, shell %, egg specific gravity	Laying	4 (0.10–0.22)	Data analyzed by ANOVA. Sodium supplemented as NaCl. Chloride level not reported. Basal diet had 21.5% CP and 2,850 kcal ME/kg.	Pizzolante et al., 2006
Chlorine, %							
14–26	94%	<0.08 0.15 0.14 0.14 0.28	*Egg weight, specific gravity* Egg production Feed conversion Egg mass Shell thickness	Laying	(0.08–0.40)	Data analyzed by linear and quadratic regression. Hot and humid. Chloride added as NH_3Cl. Sodium was 0.015% and electrolyte balance was 171–261 mEq/kg. Basal diet had 20.0% CP and 2,900 kcal mEq/kg.	Costa et al., 2012b
Potassium							
43–55	51%	<0.25 0.43 0.44 >0.65	Egg weight, specific gravity Feed intake Egg production, egg mass, feed conversion Shell weight	Laying	5 (0.25–0.65)	Data analyzed by linear and quadratic regression Temperature not reported. Potassium supplemented as K_2CO_3. Electrolyte balance was 68–170 mEq/kg. Water source and K level not reported. Basal diet had 17.1% CP and 2,800 kcal ME/kg.	Costa, Rodrigues, et al., 2011
Water							
17–30	82%	70.6 ml/d 2.24 ml water/g feed	Water consumption Ratio of water to feed consumption	Laying		Temperature ave low of 22°C and high of 25°C; Humidty ave low of 78% and high of 88%. Basal diet had an electrolyte balance of 140 mEq/kg 18.0% CP.	Rodrigues et al., 2016

[a] All amino acid levels are expressed as total amino acids and, when calculated by the authors, digestible amino acid levels are provided in parentheses.

production or feed conversion. The basal diets contained 0.74 percent (Canogullari et al., 2009), 0.73 percent (Umigi et al., 2007), or 0.63 percent (Umigi et al., 2012) total threonine and supported high rates of egg production (90–93%), indicating that those levels are at or above the requirement. A fourth study using a basal diet with 0.66 percent threonine found responses to added threonine and estimated a requirement of 0.77 percent, 0.80, 0.75 percent, and 0.78 percent for rate of egg production, feed conversion, egg weight, and egg mass, respectively (Lima et al., 2013). The three studies that found low requirements used quail in the early phase of production (8–17 weeks of age) whereas the study that found a higher requirement occurred late in production (17–33 weeks of age) with a relatively lower rate of production (82%). It is possible that the requirement increases during the late phase of production, but this goes against trends in other poultry species and requires confirmation. The estimate of the threonine requirement in the last report (0.74%) has not been changed.

Tryptophan. Pinheiro et al. (2008) estimated the total tryptophan requirement for maximum rate of egg production using a linear discontinuous regression model to be 0.23 percent. In this study, five levels of tryptophan, ranging from 0.14 to 0.30 percent, were fed. The rate of egg production did not increase significantly with levels above 0.18 percent, so their estimated requirements may be an artifact of the statistical approach employed. Refitting their data using a broken line regression indicates a requirement of 0.19 percent. Rizzo et al. (2008) found that tryptophan additions to a basal diet containing 0.23 percent total tryptophan did not result in increased egg production or feed conversion, indicating that the requirement is below that level. The recommended requirement of 0.19 percent remains unchanged from the previous report (NRC, 1994).

Arginine. Three experiments examined the arginine requirement for hens using diets based mostly on corn and soybean meal. In each experiment, there was no response to adding arginine to the basal diets containing 1.24 percent (Bulbul et al., 2015), 1.22 percent (Mauricio et al., 2016), or 1.16 percent (Reis et al., 2012) total arginine and 20.6 percent, 20 percent, and 20 percent protein, respectively. Averaging these three studies indicates that 1.21 percent is adequate to support maximal reproduction, and the requirement is likely lower. Because 1.21 percent is slightly lower than the estimate in the last report that was based on the study of Allen and Young (1980), which did not meet the inclusion criteria described previously, the provisional arginine requirement has been lowered accordingly.

Valine. Laying quail (31 to 35 weeks of age) fed a corn–soybean meal diet increased egg production, egg weight, feed intake, and body weight with increasing valine content up to 0.66 percent valine (154 mg/bird/day) with no further increase at higher levels (Martinez et al., 2021). In another study, laying quail did not respond to additional valine added to a basal diet based on corn and soybean meal (1.03% total valine; 18.6% protein) and supplemented with lysine, methionine, arginine, and tryptophan (Paula et al., 2017). The previous recommendation (NRC, 1994) of 0.92 percent valine was based on studies that were not specifically designed to determine the valine requirement and, consequently, were tentative. For this reason, the recommended valine requirement for egg production is lowered to 0.66 percent of the diet based on the study by Martinez et al. (2021).

Isoleucine. Laying quail did not respond with increased egg production, feed conversion, egg weight, or mass to isoleucine added to a basal corn–soybean meal diet containing 0.67 percent isoleucine (Santos et al., 2016). The basal diet had only 16 percent protein and was supplemented with lysine, methionine, threonine, tryptophan, arginine, valine, and glutamic acid. Even though the protein level was very low in these diets the performance was high (90% egg production) and was not different from a control group fed 20 percent protein diet supplemented with amino acids so that all diets had the same amino acid concentrations except isoleucine. The dietary energy level in this study (2,800 kcal/kg) was lower than the reference level of 2,900 kcal/kg used in Table 19-1. Thus, this study indicates that the requirement is less than 0.70 percent when adjusted to the reference energy level. The recommended isoleucine requirement in the previous report of 0.90 percent was based on studies that do not meet the inclusion criteria described previously and were designed to demonstrate adequate levels but not requirements. For this reason, the recommendation has been decreased to 0.70 percent.

Linoleic acid. Murai et al. (1994) fed quail chicks semi-purified grower diets containing linoleic acid at levels ranging from 0.0 percent to 1.5 percent from one to six week of age. They were then switched to semi-purified layer diets containing the same levels of linoleic acid. Based on maximum egg size from 8–10 weeks of age, the linoleic acid requirement was estimated to be 0.70 percent of the diet. Because the previous report's recommendation (1.0%) was based on results from laying chickens, the recommended requirement has been lowered to 0.70 percent of the diet.

Minerals

Since the last report, an adequate amount of published research on macrominerals has allowed the refinement of the recommended requirements of these nutrients, which are provided in Table 19-1. No studies on microminerals were found that meet the criteria described previously for updating nutrient requirements of Japanese quail. Thus, the estimated requirements for micronutrients remain unchanged from the previous report (NRC, 1994).

Calcium. Since the last report, there were four studies on the dietary calcium requirements of laying quail conducted during the first half of lay (>90% egg production). One study (Masukawa, 2001) found that the inclusion of calcium to a basal diet containing 2.0 percent calcium did not affect egg production, feed conversion, egg size, or shell quality. A second study (Vieira et al., 2012) also found that egg production, feed conversion, and egg size were not improved at levels above 2.0 percent calcium, but for maximal shell weight and egg specific gravity ≥3.0 percent calcium was required. The other two studies found that levels of calcium above 3.0 percent were required to maximize laying performance and shell quality. Using linear and quadratic regressions, Brandão et al. (2007) found that levels of 3.45, 3.34, ≥3.85, 3.27, and 3.51 percent calcium were needed to maximize egg production, feed conversion per dozen eggs, egg weight, egg mass, and egg specific gravity, respectively. Similarly, Barreto et al. (2007b) found that levels of 3.20, 3.24, 3.24, 3.42, 3.60, and 3.34 percent calcium were needed to maximize egg production, feed conversion (feed/egg mass), egg mass, shell weight, shell thickness, and egg specific gravity, respectively.

Three studies evaluated calcium requirements of quail during the latter part of the laying cycle when egg production is lower. One study (Masukawa, 2001) reported that the inclusion of calcium to a basal diet containing 2.0 percent calcium did not affect egg production, feed conversion, egg size, or shell quality, while two other studies found that a basal diet containing 2.0 percent (Costa et al., 2010b) or 2.2 percent (Costa et al., 2010a) calcium supported maximum egg production, egg weight, and egg mass. However, levels higher than 3.5 percent were needed to maximize several indices of shell quality in both studies.

The previous report recommended a calcium requirement of 2.5 percent based on a single study (Nelson, 1964). Although the more recent research indicates a range of requirement findings, the majority of the studies indicate that the requirement for calcium to maximize egg shell quality is about 3.0 percent, and several of the studies examining birds near the peak of lay also indicate the requirements for maximal production and feed conversion are higher than 3 percent. For this reason, the recommendation for dietary calcium has been increased to 3.0 percent, and higher levels may be needed during peak lay or if maximal shell quality is desired.

Phosphorus. Sharifi et al. (2012) examined the non-phytate phosphorus requirement during the early phase of production (8–16 weeks) and found that a basal diet containing of 0.15 percent non-phytate phosphorus maximized performance variables with the exception of shell thickness and percent shell, which required between 0.15 percent and 0.25 percent. Costa et al. (2007) also examined the requirement in the early phase of production and found that their lowest experimental level of 0.15 percent non-phytate phosphorus was sufficient to maximize egg production, feed conversion (feed/egg mass), egg mass and weight, shell weight, and percent shell. However, greater levels were needed to minimize cracks and defects in eggs (0.45 percent) and to maximize feed conversion per dozen eggs (> 0.55 percent), bone calcium (0.46 percent), and bone phosphorus (0.51 percent). Vieira et al. (2012) obtained similar results in hens from 26–28 weeks of age. In their study, their lowest experimental level of 0.10 percent P was sufficient to maximize egg production, feed conversion, egg mass, weight, and specific gravity, and shell weight. Greater levels (>0.31%) were needed to maximize bone calcium and minimize cracks and defects in the eggs.

Costa et al. (2011a) examined the non-phytate phosphorus requirement of hens during post-peak production (45–57 weeks) and found that their lowest experimental level of 0.15 percent was sufficient to maximize egg production, feed conversion, egg mass and weight, and incidence of cracks and defects. The requirements for shell weight and percent shell were 0.41 and 0.42 percent, respectively, but this did not affect the proportion of commercially viable eggs. Thus, such high levels were without economic consequence.

These studies are in agreement that the requirement of phosphorus during the entire laying period is ≤0.15 percent to maximize the rate and efficiency of production, but higher levels are needed to increase shell quality and bone mineralization. Based on maintaining bone calcium, the non-phytate phosphorus requirement (0.35%) is not changed from the previous report (NRC, 1994), but higher levels are recommended if additional improvements in shell quality are desired. Lower levels are likely appropriate following peak lay.

Electrolytes. Three studies examined the sodium requirement for laying strains of quail around peak lay. Barreto et al. (2007a) showed that most indices of production were maximized at between 0.08 and 0.15 percent sodium, with the exception of shell thickness, which required 0.20 percent. Murakami et al. (2006) and Lima et al. (2015b) did not observe an increase in egg production, efficiency, or shell quality by addition of NaCl to their basal diet that contained 0.10 percent sodium. Pizzolante et al. (2006) examined the requirement during the last phase of lay (54–66 week) and also did not see a benefit of adding NaCl to a basal diet containing 0.10 percent sodium. Because of the disparity between the various studies during peak production, the tentative requirement estimate of 0.15 percent from (NRC, 1994) was not changed. However, a level of 0.10 percent dietary sodium appears to be adequate during the later phase of egg production.

Costa et al. (2012b) examined the chloride requirement of laying quail during the early phase of production and found that 0.15 percent chloride was optimal for most measures of production, but a much higher level of 0.28 percent was needed to maximize shell thickness. This high level for shell quality may be related to the hot and humid conditions during this experiment. The chloride requirement has been increased slightly from the last report to 0.15 percent.

The requirement for potassium during the late phase of production was examined in a study by Costa et al. (2011b) who found 0.44 percent was required to maximize egg production, egg mass, and feed conversion. Much greater levels (≥0.65%) were needed for indices of shell quality; however, this may be an artifact of the low electrolyte balance of the experimental diets. Electrolyte balance increased with increasing inclusion of potassium carbonate from a low of 68 mEq/kg in the basal diet with 0.25 percent potassium to 170 mEq/kg at the highest level of 0.65 percent potassium. The recommended potassium requirement for egg production has been increased slightly to 0.44 percent because the previous recommendation had little experimental support.

Vitamins

Since the conclusion of the review of literature for the previous edition of this report in September 1991, there have not been any publications on the vitamin requirements of laying Japanese quail that meet the inclusion criteria described in the early part of this chapter. Thus, an update of the requirement estimates for vitamins in (NRC, 1994) was precluded.

BOBWHITE QUAIL

Bobwhite quail (*Colinus virginianus*) are native to North and Central America. Commercial propagation is primarily to restock native populations for conservation and hunting and to a lesser extent for meat and egg production. Japanese quail (*Coturnix coturnix japonica*) and common quail (*Coturnix coturnix*) are more frequently used for food production. Because bobwhite quail are raised predominantly for release to wild habitats, there has been limited intensive breeding for meat or egg production. Among the strains that have been bred for a high growth or reproductive rate, little nutrition information is available, and the information that follows is based on strains that are phenotypically close to the wild type. Among the strains raised for release to the wild, their geographic origin has resulted in a range of body styles and flight characteristics. The nutrition literature rarely documents these characteristics, so the following recommendations may not be valid for all strains.

Nutritional and Digestive Strategy

Because of their value to hunters, numerous studies have been conducted on the food habits of bobwhite quail in natural habitats (Martin et al., 1951). Bobwhite quail chicks feed almost exclusively on invertebrates. Adults feed predominantly on foods of plant origin, especially seeds, young leaves, and small fruits. When seasonably available, the diet of adults may include 30 percent or more of invertebrates. In agricultural areas corn, wheat, sunflowers, and sorghum are highly consumed. Bobwhite quail possess a muscular gizzard and ceca that are relatively effective at fermenting cellulose for their relatively small size (Inman, 1973). On a relative basis, the size of their ceca is slightly larger than that of chickens and slightly smaller than that of Ring-necked pheasants (DeGolier et al., 1999).

Nutrient Requirements

Since the previous NRC report (1994), very little nutrient requirement work has been done in bobwhite quail. From 2–4 weeks of age, Blake and Hess (2013) found that a corn–soy diet with 24 percent protein and supplemented with DL-methionine and L-lysine-HCl to give 0.57 percent methionine, 0.92 percent methionine + cyst(e)ine, and 1.29 percent lysine supported growth and feed efficiency that was not different from greater levels of protein and amino acid supplementation. From 4–10 weeks of age they found that a corn–soybean meal diet with 18 percent protein, 0.41 percent methionine, 0.68 percent methionine + cyst(e)ine, and 0.95 percent lysine was adequate to support growth and efficiency but not feather quality (Blake et al., 2013). The addition of DL-methionine to provide 0.50 percent total methionine and 0.77 percent methionine + cyst(e)ine, and L-lysine-HCl to provide 0.95 percent lysine resulted in feather quality similar to chicks fed a diet with 22 percent protein. This study indicated that the protein level during the later stage of growth may be decreased from 20 percent to 18 percent with supplemental methionine and lysine.

In laying females, Dabbert et al. (1996) found that a corn–soybean diet that provides 18 percent protein was sufficient to maintain egg production and egg weight but not albumin weight. This study indicated that protein levels for layers are likely lower than recommended in this report, but research defining the requirement is needed to be able to make new recommendations.

The foregoing studies have been used to adjust recommendations for starting and growing and laying bobwhite quail (Table 19-2). The previous version of this report (NRC, 1994) can be consulted for reference to studies used to make the previous recommendations.

RINGED-NECKED PHEASANTS

Ringed-necked pheasants (*Phasianus colchicus*), also called common pheasants, are native to Asia but have been widely introduced into suitable habitats for hunting. They are the most common pheasant propagated in captivity. Ring-necked pheasants are produced commercially for the purposes of meat production and for release into the wild to restock native populations for conservation and hunting. Because of this dual use, there has not been intensive breeding for meat production, and the majority of commercially produced pheasants approximate a wild-type phenotype.

Nutritional and Digestive Strategy

In northeastern China, the natural habitat of ring-necked pheasants (hereafter referred to as pheasants) is found primarily along the forest edge, meadows, grassland, or in shrubby terrain (Robertson, 1997). Pheasants are very adaptable and can colonize farmland and pastures. In natural habitats, pheasants are omnivorous and feed opportunistically, changing their food items based on their availability, size, and nutritional quality (Doxon and Carroll, 2010). In agricultural habitats cereal grains and other seeds dominate their diet with small invertebrates accounting for around 10 percent of the diet of adults, except in the spring when invertebrate intake increases considerably. Hatchlings consume diets that can exceed 90 percent small invertebrates (Savory, 1989). Females, during egg laying, also consume a large proportion of invertebrates.

The digestive tract of pheasants is typical of medium-sized Galliformes. Adults have a large and muscular gizzard and moderate-sized ceca. On a relative basis, their ceca are larger than those of chickens but smaller than those of turkeys (DeGolier et al., 1999). The ceca retain the portion of dietary fiber that is soluble or has a very small particle size for a long period for fermentation (Duke et al., 1968; Kokoszyński et al., 2011). Very little work has been conducted on the ability of pheasants to digest energy or protein. Sultan et al. (2014) found that the apparent metabolizable energy for corn, wheat, and sorghum in adult pheasants is slightly lower than that reported for chickens (3,179, 3,011, and 2,891 kcal/kg, respectively). This study also indicates that fat and protein digestibilities are also lower than those of chickens.

Nutrient Requirements

The previous edition of this report (NRC, 1994) provided documentation for studies conducted prior to September 1991 that were used to estimate nutrient requirements and can be consulted for historical references. That report points out that pheasants are especially prone to leg disorders and abnormal feather growth when certain nutrients such as niacin, riboflavin, choline, manganese, and zinc are inadequate. High levels of calcium, as in a breeder ration, can cause leg problems and high mortality if fed to chicks.

A very small number of studies on nutrient requirements of ring-necked pheasants has been published since the 1994 NRC report. Studies by Li et al. (2001a,b) indicated that the lysine and methionine requirement from 0 to 4 weeks of age were similar to those in the last report (NRC, 1994), which were estimates based on research from other species. In their study, the methionine requirement for feathering was higher than that for protein deposition. They also found that the lysine requirement from 4–8 weeks of age is slightly lower than recommended in the last report, which also was based on inference from other species.

Salyi and Sztojkov (1994) observed that chicks fed 0.15 percent sodium grew slower than those fed 0.25 percent or 0.35 percent, indicating that the requirement is greater than 0.15 percent of the diet.

A study on the calcium requirement for egg production in yearling ring-necked pheasants showed that egg production is very low if the hens are fed a low calcium diet (0.02%) prior to the beginning of lay (Jones et al., 2010). Even high levels of dietary calcium (1.7–4.5) introduced at the beginning of lay did not increase egg production when very low levels were fed prior to production. This is likely due to the importance of the accretion of medullary bone prior to lay for subsequent egg production in this species.

Due to limited research on pheasant nutrition, most of the nutrient requirements listed in Table 19-3 for growing birds and laying pheasants are tentative. Where experimental data are lacking, recommended levels for turkeys appear to provide adequate intakes for pheasants.

REFERENCES

Abdel-Mageed, M.A.A., S.A.M. Shabaan, and M.A.E. Nadia. 2009. Effect of threonine supplementation on Japanese quail fed various levels of protein and sulfur amino acids. 2. Laying period. *Egyptian Poultry Science Journal*. 29(1):805–819.

Allen, N.K., and R.J. Young. 1980. Studies on the amino acid and protein requirements of laying Japanese quail (*Coturnix coturnix japonica*). *Poultry Science*. 59(9):2029–2037.

Baer, J., R. Lansford, and K. Cheng. 2015. Japanese quail as a laboratory animal model. *In Laboratory Animal Medicine*, 3rd ed. Amsterdam, The Netherlands: Elsevier. 1087–1108.

Barbour, G.W., K.E. Nestor, and M.S. Lilburn. 1993. The effect of different levels of dietary lysine on growth and development of a random-bred population of Japanese quail (R1) and a subline of R1 selected for increased body weight (HW). *Comparative Biochemistry and Physiology A: Comparative Physiology*. 105(1):183–186.

Barreto, S.L.T., M.S.d. Araujo, R.T. Umigi, W.C.O. Moura, C.H.R. Costa, and M.F. Sousa. 2007a. Sodium levels in laying Japanese quail diet on the peak production. *Revista Brasileira de Zootecnia*. 36(1):1559–1565.

Barreto, S.L.T., C.A. Pereira, R.T. Umigi, T.C.d. Rocha, M.S.d. Araujo, C.S. Silva, and R.d.A. Torres Filho. 2007b. Determination of nutritional calcium requirement of Japanese quails in the initial production phase. *Revista Brasileira de Zootecnia*. 36(1):68–78.

Baylan, M., S. Canogullari, T. Ayasan, and A. Sahin. 2006. Dietary threonine supplementation for improving growth performance and edible carcass parts in Japanese quails, *Coturnix coturnix japonica*. *International Journal of Poultry Science*. 5(7):635–638.

Belo, M., J.d.B. Cotta, and A. de Oliveira. 2000. Levels of methionine in rations of the Japanese quails (*Coturnix coturnix japonica*) in the laying initial period. *Ciencia e Agrotecnologia*. 24:1068–1078.

Bezerra, R.M., E.R. Freitas, G.A.J.d. Nascimento, N.d.M. Braz, N.N.P. Farias, and R.P.d.S. Xavier. 2013. Chlorine levels for Japanese quails in growing and its effects on the production phase. *Semina: Ciencias Agrarias*. 34(5):853–861.

Bisoi, P., B. Panda, V. Reddy, and R. Singh. 1980. Calcium and phosphorus requirements of starter Japanese quail. *Indian Journal of Animal Sciences*. 50(1):357–361.

Blake, J.P., and J.B. Hess. 2013. Changes in protein level for bobwhite quail. *Journal of Applied Poultry Research*. 22(3):511–515.

Blake, J.P., J.B. Hess, and W.D. Berry. 2013. Influence of hatch pull time, protein, and methionine on bobwhite quail performance. *Journal of Applied Poultry Research*. 22(3):523–528.

Brandão, P.A., F.G.P. Costa, J.H.V.d. Silva, J.S. Brandão, J.G.S. Nobre, and C.d.C. Goulart. 2007. Calcium requirement for Japanese quails (*Coturnix coturnix japonica*) in the laying phase. *Acta Scientiarum—Animal Sciences*. 29:16–21.

Bulbul, T., E. Ulutas, and M. Evcimen. 2015. Effects of dietary supplementation of arginine and lysine on performance and egg quality characteristics of laying quails. *Ankara Universitesi Veteriner Fakultesi Dergisi*. 62(4):307–312.

Canogullari, S., M. Baylan, and T. Ayasan. 2009. Threonine requirement of laying Japanese quails. *Journal of Animal and Veterinary Advances*. 8(2):1539–1541.

Castro, S.d.F., B.D.A. Fortes, J.C.C.d. Carvalho, A.G. Bertechini, L.S.B.d. Queiroz, and A.A.P. Garcia Juniorr. 2011. Dietary methionine and choline relation on the performance of Japanese quails (*Coturnix coturnix japonica*). *Ciencia Animal Brasileira*. 12(1): 635–641.

Cavalcante, D.T. 2013. Determinação do segundo aminoácido limitante para codornas japonesas. In *Zootecnia*. Vol. MS. Universidade Federal da Paraiba, Centro de Ciencias Agrarias.

Chang, G., H. Chang, X. Liu, W. Xu, H. Wang, W. Zhao, and O. Olowofeso. 2005. Developmental research on the origin and phylogeny of quails. *World's Poultry Science Journal*. 61(1):105–112.

Costa, C.H.R., S.L.d.T. Barreto, P.C. Gomes, L.H. Hosoda, C.A. Lipari, and H.J.D. Lima. 2011a. Available phosphorus levels for Japanese quail diets from 45 to 57 weeks of age. *Revista Brasileira de Zootecnia*. 40(2):2152–2160.

Costa, C.H.R., S.L.d.T. Barreto, W.C.O. Moura, R.d.S. Reis, C.D.S. Leite, and G.V.C. Maia. 2007. Níveis de fósforo e cálcio em dietas para codornas japonesas em postura. *Revista Brasileira de Zootecnia*. 36:2037–2046.

Costa, C.H.R., S.L.T. Barreto, P.C. Gomes, G.V.C. Maia, C.A. Lipari, and L.H. Hosoda. 2010a. Levels of calcium in diets for Japanese quails in the third end of posture (45–57 weeks of age). *Arquivo Brasileiro de Medicina Veterinaria e Zootecnia*. 62:1225–1231.

Costa, C.H.R., S.L.T. Barreto, R.T. Umigi, H.J.D. Lima, M.S. de Arujo, and M. Paulo. 2010b. Balance of calcium and phosphorus and study of these minerals in diets for Japanese quails (from 45 to 57 weeks of age). *Revista Brasileira de Zootecnia*. 39:1748–1755.

Costa, F.G.P., P.A. Brandão, J.H.V.d. Silva, R.d.C. Lima Neto, C.d.C. Goulart, and W.E. Pereira. 2009a. Calcium levels for 1 to 35-days-old female Japanese quails. *Acta Scientiarum—Animal Sciences*. 31:7–12.

Costa, F.G.P., D.F. Figueiredo-Lima, M.R. Lima, S.G. Pinheiro, C.C. Goulart, J.H.V. Silva, S.M. Baraldi-Artoni, F.R. Barreiro, and P.E.N. Givisiez. 2012a. Sodium levels for Japanese quail at initial phases. *Poultry Science*. 91:1128–1134.

Costa, F.G.P., F. Júnior, J. Pessoa, D.F.F. Lima, C.d.C. Goulart, J.H.V.d. Silva, M.R.d. Lima, S.G. Pinheiro, and V.P. Rodrigues. 2012b. Chlorine requirement for Japanese laying quails. *Revista Brasileira de Zootecnia*. 41:2289–2293.

Costa, F.G.P., L.R. Rodrigues, C.d.C. Goulart, C.F.S.d. Oliveira, V.P. Rodrigues, and J.H.V.d. Silva. 2011b. Nutritional potassium requirement for laying Japanese quails. *Revista Brasileira de Zootecnia*. 40: 2754–2759.

Costa, F.G.P., V.P. Rodrigues, C.d.C. Goulart, R.d.C. Lima Neto, J.G.d. Souza, and J.H.V.d. Silva. 2008. Digestible lysine requirements for laying Japanese quails. *Revista Brasileira de Zootecnia*. 37:2136–2140.

Costa, F.G.P., V.P. Rodrigues, C.d.C. Goulart, J.G. Vargas Junior, J.H.V.d. Silva, and J.G.d. Souza. 2009b. Nutritional requirements of digestible methionine+cystine for Japanese quails in production phase. *Revista Brasileira de Zootecnia*. 38:2389–2393.

Dabbert, C.B., R.L. Lochmiller, P.W. Waldroup, and R.G. Teeter. 1996. Examination of the dietary methionine requirements of breeding Northern bobwhite, *Colinus virginianus*. *Poultry Science*. 75:991–997.

De Souza Andrade, M.d.F., A.L. de Barros Moreira Filho, E.F.A. da Silva, H.B. de Oliveira, F.G.P. Costa, R.R. Guerra, and P.E.N. Givisiez. 2019. Expression of glucose transporters and morphometry in the intestine of Japanese quails after hatch. *Journal of Comparative Physiology B*. 189:61–68.

DeGolier, T.F., S.A. Mahoney, and G.E. Duke. 1999. Relationships of avian cecal lengths to food habits, taxonomic position, and intestinal lengths. *Condor*. 101(3):622–634.

Doxon, E.D., and J.P. Carroll. 2010. Feeding ecology of ring-necked pheasant and northern bobwhite chicks in conservation reserve program fields. *Journal of Wildlife Management*. 74:249–256.

Duke, G., G. Petrides, and R. Ringer. 1968. Chromium-51 in food metabolizability and passage rate studies with the ring-necked pheasant. *Poultry Science*. 47:1356–1364.

Fairchild, B.D., and C.W. Ritz. 2009. Poultry drinking water primer. Poultry Drinking Water Primer | The Poultry Site.

Fenna, L., and D. Boag. 1974. Adaptive significance of the caeca in Japanese quail and spruce grouse (Galliformes). *Canadian Journal of Xoology*. 52:1577–1584.

Gürcan, E., and S. Kaplan. 2017. Flexible alternatives to models widely used for describing growth in Japanese quail. *JAPS: Journal of Animal & Plant Sciences*. 27(1):48–56.

Hajkhodadadi, M. Shivazad, H. Moravvej, and A. Zare-Shahneh. 2013. Effect of dietary lysine on performance and immunity parameters of male and female Japanese quails. *African Journal of Agricultural Research*. 8(1):113–118.

Hajkhodadadi, I., H. Moravej, M. Shivazad, H. Ghasemi, and A. Zareh-Shahneh. 2014. Lysine requirements of female Japanese quails based on performance and carcass variables from twenty-one to forty-two days of age. *Iranian Journal of Applied Animal Science*. 4(3):629–635.

Hasanvand, S., M. Mehri, F. Bagherzadeh-Kasmani, and M. Asghari-Moghadam. 2018. Estimation of lysine requirements for growing Japanese quails. *Journal of Animal Physiology and Animal Nutrition*. 102(2):557–563.

Hurtado Nery, V.L., L. Gutiérrez Castro, and D.M. Torres Novoa. 2015. Digestible lysine levels for Japanese quail in laying phase. *Revista de la Facultad de Medicina Veterinaria y de Zootecnia*. 62(3):49–57.

Inman, D.L. 1973. Cellulose digestion in ruffed grouse, chukar partridge, and bobwhite quail. *Journal of Wildlife Management*. 37:114–121.

Jones, L.R., H.L. Black, C.M. White, N.P. Johnston, M.E. McGee, S.W. Donahue, and D.L. Eggett. 2010. Effects of calcium-loading on egg production in ring-necked pheasants. *Journal of Wildlife Management*. 74(6):1295–1300.

Jordão Filho, J., J.H.V.d. Silva, F.G.P. Costa, L.F.T. Albino, T.d.S. Melo, P.B.d. Lacerda, G.M. Dantas, and R.P. Soares. 2012. Requirement for maintenance and gain of crude protein for two genotypes of growing quails. *Revista Brasileira de Zootecnia*. 13(3):2048–2054.

Jordão Filho, J., J.H.V.d. Silva, F.G.P. Costa, N.K. Sakomura, C.T. Silva, and N.A. Chagas. 2011a. Prediction equations to estimate the demand of energy and crude protein for maintenance, gain and egg production for laying Japanese quails. *Revista Brasileira de Zootecnia*. 40: 2423–2430.

Jordão Filho, J., J.H.V.d. Silva, C.T. Silva, F.G.P. Costa, J.M.B.d. Sousa, and P.E.N. Givisiez. 2011b. Energy requirement for maintenance and gain for two genotypes of quails housed in different breeding rearing systems. *Revista Brasileira de Zootecnia*. 13(3):2415–2422.

Kawahara, T. 1967. Wild coturnix quail in Japan. *Quail Quarterly*. 4(11): 62–63.

Kawahara, T. 1973. Comparative study of quantitative traits between wild and domestic Japanese quail (*Coturnix coturnix japonica*). *Jikken dobutsu. Experimental Animals*. 22:139–150.

Khosravi, H., M. Mehri, F. Bagherzadeh-Kasmani, and M. Asghari-Moghadam. 2016. Methionine requirement of growing Japanese quails. *Animal Feed Science and Technology*. 212:122–128.

Kokoszyński, D., Z. Bernacki, and A. Cisowska. 2011. Growth and development of young game pheasants (*Phasianus colchicus*). *Archives Animal Breeding*. 54(1):83–92.

Lalmingthanga, and A.K. Shrivastav. 1998. Effect of supplemental manganese in relation to dietary calcium levels on the performance of growing Japanese quail. *Indian Journal of Poultry Science*. 33(1):166–172.

Li, Z., K. Zhang, and S. Zhou. 2001a. Influence of lysine and methionine levels in the diet on performance of young pheasants. *Journal of Sichuan Agricultural University*. 19(1):168–171.

Li, Z., K. Zhang, and S. Zuo. 2001b. Influence of dietary lysine and methionine levels on protein deposition and feather growth of ring-necked pheasants. *Journal of Sichuan Agricultural University*. 19(2):86–90.

Lima, H.J.D., S.L.T. Toledo Barreto, J. Donzele, G. Souza, R. Almeida, I. Tinoco, and L. Albino. 2016a. Digestible lysine requirement for growing Japanese quails. *The Journal of Applied Poultry Research*. 25(4): 483–491.

Lima, H.J.D., S.L.T. Barreto, J.L. Donzele, I.F.F. Tinoco, and N.S. Ribas. 2015a. Ideal ratio of digestible methionine plus cystine to digestible lysine for growing Japanese quails. *Revista Colombiana de Ciencias Pecuarias*. 28(4):313–322.

Lima, H.J.D., S.L.T. Barreto, E.d. Paula, D.R. Dutra, S.L.d. Costa, and W.d.S. Abjaude. 2015b. Sodium levels in diet of laying Japanese quails. *Revista Brasileira de Saude e Producao Animal*. 16(1):73–81.

Lima, H.J.D., S.L.T. Barreto, D. Lacerda, R. Henriques, and R. Ferraz. 2011. Relação entre metionina+cistina digestíveis e lisina digestível em dieta para codornas japonesas na fase pré-inicial. *Enciclopédia Biosfera*. 7(2):926–931.

Lima, M.R.d., F.G.P. Costa, R.R. Guerra, J.H.d. Silva, C.B.V. Rabello, M.A. Miglino, G.B.V. Lobato, S.B.S. Netto, and L.d.S. Dantas. 2013. Threonine: lysine ratio for Japanese quail hen diets. *Journal of Applied Poultry Research*. 22(2):260–268.

Mandal, A.B., S. Kaur, A.K. Johri, A.V. Elangovan, C. Deo, and H.P. Shrivastava. 2006. Response of growing Japanese quails to dietary concentration of L-threonine. *Journal of the Science of Food and Agriculture*. 86(5):793–798.

Martin, A.C., H.S. Zim, and A.L. Nelson. 1951. *American Wildlife and Plants: A Guide to Wildlife Food Habits*. New York: Dover.

Martinez, K., G.d.S. Viana, E.d. Silva, M. Reis, R. Gous, and N. Sakomura. 2021. The response of reproducing Japanese quail to dietary valine. *British Poultry Science*. 62(5):726–730.

Masukawa, Y., E. Fernandes, V. Moraes, J. Ariki, and L. Bruno. 2001. Níveis de cálcio da dieta sobre o desempenho e a qualidade da casca de ovos de codornas japonesas. *Ars Veterinária*. 17(1):144–148.

Mauricio, T.V., J.G.d.V. Junior, M.F.d. Souza, W.A. Barboza, L.d.C. Nunes, R.T.R.N. Soares, and H.d.S. Nascimento. 2016. Digestible arginine concentrations in the diet of Japanese quails. *Semina: Ciencias Agrarias*. 37(4):2453–2461.

Mehri, M., G. Jalilvand, M. Ghazaghi, A.H. Mahdavi, and F.B. Kasmani. 2013. Estimation of optimal lysine in quail chicks during the second and third weeks of age. *Italian Journal of Animal Science*. 12(4):e84

Mehri, M., F.B. Kasmani, and M. Asghari-Moghadam. 2015. Estimation of lysine requirements of growing Japanese quail during the fourth and fifth weeks of age. *Poultry Science*. 94(8):1923–1927.

Minvielle, F. 2004. The future of Japanese quail for research and production. *World's Poultry Science Journal*. 60(4):500–507.

Moura, A.M.A., R.T.R.N. Soares, V.L.H. Nery, H.P. Couto, J.B. Fonseca, and R.A.M. Vieira. 2008. Lysine requirement for Japanese quail (*Coturnix japonica*) in egg laying phase. *Archivos de Zootecnia*. 57(1): 439–448.

Murai, A., M. Furuse, and J. Okumura. 1994. Linoleic acid requirement for growth and reproduction in Japanese quail. *Japanese Poultry Science*. 31:109–118.

Murakami, A.E., M.I. Sakamoto, L. de Souza, J.R.G. Franco, and M.A.O. Mituo. 2006. Determinação do melhor nível de sal comum para codornas japonesas em postura. *Revista Brasileira de Zootecnia*. 35: 2333–2337.

Nelson, F. 1964. Calcium and phosphorus rearing for the breeding couturnix quail. *Poultry Science*. 43(1):1346–1349.

NRC (National Research Council). 1994. *Nutrient Requirements of Poultry*, 9th rev. ed. Washington, DC: National Academy Press.

Parvin, R., A.B. Mandal, S.M. Singh, and T. Rakesh. 2010. Effect of dietary level of methionine on growth performance and immune response in Japanese quails (*Coturnix coturnix japonica*). *Journal of the Science of Food and Agriculture*. 90(3):471–481.

Paula, E.d., S.L.T. Barreto, R.d.S. Reis, R.T. Umigi, H.J.D. Lima, and D.C.O. Pereira. 2017. Ratio of valine to lysine in diets for laying Japanese quail. *Pubvet*. 11(1):306–312.

Pesti, G., D. Vedenov, J. Cason, and L. Billard. 2009. A comparison of methods to estimate nutritional requirements from experimental data. *British Poultry Science*. 50(1):16–32.

Pinheiro, S.R.F., S.L.T. Barreto, L.F.T. Albino, H.S. Rostagno, R.T. Umigi, and C.O. Brito. 2008. Efeito dos níveis de triptofano digestível em dietas para codornas japonesas em postura. *Revista Brasileira de Zootecnia*. 1012–1016.

Pinto, R., A.S. Ferreira, L.F.T. Albino, P.C. Gomes, and J.G.d. Vargas Júnior. 2002. Protein and energy levels for laying Japanese quails. *Revista Brasileira de Zootecnia*. 31:1761–1770.

Pinto, R., A.S. Ferreira, J.L. Donzele, L.F.T. Albino, M.d. Almeida e Silva, R.d.T.R.N. Soares, and C.A. Pereira. 2003a. Methionine plus cystine requirement for growing Japanese quails. *Revista Brasileira de Zootecnia*. 32:1174–1181.

Pinto, R., A.S. Ferreira, J.L. Donzele, M.d. Almeida e Silva, R.d.T.R.N. Soares, G.S. Custódio, and K.d.S. Pena. 2003b. Lysine requirement for laying Japanese quails. *Revista Brasileira de Zootecnia*. 32:1182–1189.

Pinto, R., A.S. Ferreira, R.d.T.R.N. Soares, L.F.T. Albino, M.d.A. Silva, J.L. Donzele, and T.A. Pereira. 2003c. Exigência de metionina mais cistina para codornas japonesas em postura. *Revista Brasileira de Zootecnia*. 32:1166–1173.

Pizzolante, C.C., É.S.P.B. Saldanha, E.A. Garcia, A.d.P. Deodato, H.B.A.d. Souza, A.M. Scatolini, M.M. Boiago, M.d. Castro, T. Saccardo, and F.E. Dias. 2006. Common salt levels in Japanese quail rations (*Coturnix coturnix japonica*) in the end of laying period. *Ciência Animal Brasileira*. 7:123–130.

Portillo-Loera, J.J., F.G. Ríos-Rincón, C.B. Castro-Tamayo, C. Angulo-Montoya, and G. Contreras-Pérez. 2014. Características de la canal en grupos mixtos de codorniz japonesa (*Coturnix coturnix japonica*) en engorde beneficiados a diferentes edades. *Revista Científica*. 24:E.

Reis, R.d.S., S.L.T. Barreto, W.d.S. Abjaude, D.R. Dutra, M. Santos, and E.d. Paula. 2012. Relationship of arginine with lysine in diets for laying Japanese quails. *Revista Brasileira de Zootecnia*. 41:106–110.

Reis, R.d.S., S.L.T. Barreto, P.C. Gomes, H. Jose, D. Lima, P.M. Medina, and F. Ferreira. 2011. Relationship of methionine plus cystine with lysine in diets for laying Japanese quails. *Revista Brasileira de Zootecnia*. 40:1031–1037.

Ribeiro, C.L.N., S.L.T. Barreto, R.d.S. Reis, J.C.L. Muniz, J.L. Donzele, P.C. Gomes, J.G.d. Vargas Júnior, and L.F.T. Albino. 2013. Digestible lysine levels in diets for laying Japanese quails. *Revista Brasileira de Zootecnia*. 42:489–495.

Rizzo, P.V., G.C. Guandolini, L. Amoroso, R.D. Malheiros, and V.M.B.d. Moraes. 2008. Triptofano na alimentação de codornas japonesas nas fases de recria e postura. *Revista Brasileira de Zootecnia*. 1017–1022.

Robertson, P. 1997. *A Natural History of the Pheasant*. Shrewsbury, Shropshire, UK: Swan Hill Press.

Rodrigues, L.R., D.A. Furtado, F.G.P. Costa, J.W.B.d. Nascimento, and E.d.A. Cardoso. 2016. Thermal comfort index, physiological variables and performance of quails fed with protein reduction. *Revista Brasileira de Engenharia Agrícola e Ambiental*. 20:378–384.

Salyi, G., and V. Sztojkov. 1994. Sodium chloride deficiency of growing pheasants. Case report and experimental investigation. *Magyar Allatorvosok Lapja*. 49(1):497–500.

Santos, G.C.d., E.A. Garcia, J.A. Vieira Filho, A.d.B. Molino, K. Pelicia, and D.A. Berto. 2016. Performance of Japanese quails fed diets with low-protein and isoleucine. *Acta Scientiarum—Animal Sciences*. 38(1): 219–225.

Savory, C. 1985. An investigation into the role of the crop in control of feeding in Japanese quail and domestic fowls. *Physiology & Behavior*. 35(6): 917–928.

Savory, C. 1989. The importance of invertebrate food to chicks of gallinaceous species. *Proceedings of the Nutrition Society*. 48(1):113–133.

Savory, C., and M. Gentle. 1976. Effects of dietary dilution with fibre on the food intake and gut dimensions of Japanese quail. *British Poultry Science*. 17(6):561–570.

Scotta, B.A., J.G. Vargas Junior, F.B. Petrucci, L.F. Demuner, F.G.P. Costa, W.A. Barbosa, and J.F.V. Marin. 2011. Digestible methionine plus cystine and relation digestible methionine plus cystine: Lysine for Japanese quails. *Revista Brasileira de Saude e Producao Animal*. 12:729–738.

Sharifi, M., M.S. Shargh, S. Hassani, H. Senobar, and S. Jenabi. 2012. The effects of dietary nonphytate phosphorus levels and phytase on laying performance and egg quality parameters of Japanese quails (*Coturnix coturnix Japonica*). *Archiv Geflügelkunde*. 76(1):13–19.

Shrivastav, A.K., B.B. Dash, and T.S. Johri. 2004. Effect of dietary levels of choline in relation to sulphur amino acid for growth and immune response in quail broilers. *Indian Journal of Poultry Science*. 39(1):9–14.

Silva, J., M. Silva, J. Jordão Filho, E. Silva, I.S. Andrade, D. Melo, M.L.G. Ribeiro, M. Rocha, F.G.P. Costa, and W. Dutra Junior. 2004a. Exigências de mantença e de ganho de proteína e de energia em codornas japonesas (*Coturnix coturnix japonica*) na fase de 1 a 12 dias de idade. *Revista Brasileira de Zootecnia*. 33:1209–1219.

Silva, J.H.V.d., M.B.d. Silva, J. Jordão Filho, E.L.d. Silva, I.S. Andrade, D.A.d. Melo, M.L.G. Ribeiro, M.d.R.d.F. Rocha, F.G.P. Costa, and W.M. Dutra Júnior. 2004b. Maintenance and weight gain in crude protein and metabolizable energy requirements of Japanese quails (*Coturnix coturnix japonica*) from 15 to 32 days of age. *Revista Brasileira de Zootecnia*. 33: 1220–1230.

Starck, J.M., and G.H.A. Rahmaan. 2003. Phenotypic flexibility of structure and function of the digestive system of Japanese quail. *Journal of Experimental Biology*. 206(11):1887–1897.

Sultan, A., R. Ullah, S. Khan, M.T. Khan, H. Khan, and I.U. Zaman. 2014. Nutrient digestibility values and apparent metabolizable energy of corn, wheat and sorghum by pheasants (*Phasianus colchicus*). *Pakistan Veterinary Journal*. 34(4):479–483.

Thompson, D., and D. Boag. 1975. Role of the caeca in Japanese quail energetics. *Canadian Journal of Xoology*. 53(2):166–170.

Umigi, R.T., S.L.T. Barreto, J.L. Donzele, R.d.S. Reis, M.F. Sousa, and C.D.S. Leite. 2007. Digestible threonine levels in diets for laying Japanese quail. *Revista Brasileira de Zootecnia*. 36:1868–1874.

Umigi, R.T., S.L.T. Barreto, R.S. Reis, R.M. Mesquita Filho, and M.S. Araújo. 2012. Digestible threonine levels for laying Japanese quail. *Arquivo Brasileiro de Medicina Veterinária e Zootecnia*. 64: 658–664.

Vieira, D.V.G., S.L.T. Barreto, M.H. Valeriano, L.F.D.d. Jesus, L.F.F.d. Silva, R. Mencalha, K.S. Barbosa, R.K.V. Mendes, M.R. Cassuce, and T.d.S. Melo. 2012. Requirements for calcium and phosphorus available for Japanese quail of the 26 to 38 weeks of age. *Revista Brasileira de Saude e Producao Animal*. 13:204–213.

Wilkinson, N., R.J. Hughes, W.J. Aspden, J. Chapman, R.J. Moore, and D. Stanley. 2016. The gastrointestinal tract microbiota of the Japanese quail, *Coturnix japonica*. *Applied Microbiology and Biotechnology*. 100(9):4201–4209.

20

Ingredients

INTRODUCTION

Feed formulation is the process of selecting the kinds and amounts of available ingredients to supply nutrients in the correct concentrations and proportions that meet or exceed requirements for each poultry species. In a digital age, least-cost formulation software can combine real-time pricing and nutrient composition data to generate an optimized blend of ingredients for the class of poultry being fed. However, this process should not be trivialized because physicochemical characteristics of feeds such as physical form, sensory properties, naturally occurring refractory or anti-nutritive compounds, chemical contaminants, as well as conditions of feed ingredient storage all affect feed quality and acceptance. As such, the feed formulation process is of primary importance to minimize detrimental effects of poor feed palatability; subpar nutrient digestibility, absorption, and utilization; and excessive nutrient excretion, all of which will directly influence the ability to match dietary nutrient supply with physiological requirements of poultry.

When formulating poultry diets, it is important to consider both analytical values and what may be colloquially referred to as nutritive values. Analytical values are those derived from analytical chemistry (i.e., wet chemistry) or near-infrared reflectance spectroscopy (NIRS) methods, whereas nutritive values may be loosely defined as species-specific estimates generated through empirical approaches and requiring the ingredient to be consumed by an animal. Nutritive values generally include digestibility and bioavailability estimates, as described in other chapters, with the metabolizable energy, available phosphorus, and standardized ileal digestible amino acid systems being of primary importance. Details on how metabolizable energy values were determined for select ingredients are provided in the energy chapter (Chapter 1). Summary tables of literature-derived metabolizable energy values, along with associated ingredient composition data, are available in Appendix C. Furthermore, analytical values provide information on the total or gross nutrient content of an ingredient, while nutritive values effectively serve as correction factors for inefficiencies associated with digestion, absorption, metabolism, and maintenance functions in the bird (i.e., net concentrations). Considering both analytical and nutritive values simultaneously in the feed formulation process enables more personalized nutrition to be delivered to the bird, with implications toward benefiting economic, environmental, and sustainability goals of poultry production.

VARIABILITY IN NUTRIENT COMPOSITION OF INGREDIENTS

Ingredients used in poultry feed formulations vary widely in composition due to myriad factors, not limited to ingredient processing methods and geographical differences in environmental conditions for some crops. Wet chemistry analytical methods remain the hallmark source of nutrient composition data, but even use of standardized and validated methods results in compositional variability due to feedstuffs being heterogeneous substrates (Llames and Fontaine, 1994). It is not practical to analyze every ingredient to be used in diet formulation and manufacture, even when proper sampling techniques are used, so ultimately, all analytical results should be considered as estimates of true nutrient composition. Additionally, nutritionists must consider the contribution of nutrients arising from carrier and anti-caking agents associated with feed ingredients, mineral supplements, and feed additives, all of which will add variation in the final nutrient content of a diet. This point is especially pertinent for minerals (e.g., calcium) added as anti-caking agents or as carriers with specialty additives as these sources will provide usable nutrients that should be considered in an ingredient matrix for feed formulation purposes.

In general, variation in nutrient composition between and within ingredients typically used to feed poultry species must be considered by nutritionists when creating balanced feed formulations. Variation in the nutrient composition of ingredients is natural and unavoidable, so emphasis must be placed on understanding, controlling, and reporting sources of variability. Moreover, differences in analytical methods employed by independent laboratories arguably represents the largest and most important source of extrinsic variation (Gizzi and Givens, 2004), which is why there exist strict standards for validating all aspects of a method (Llames and Fontaine, 1994). Because variation is inherently related to the sensitivity of a measurement, complex methods (e.g., wet chemistry techniques) will always be more variable than estimates generated by predictive or modeling approaches. As such, application of standardized analytical techniques will accurately identify natural variance of individual nutrients within an ingredient, whereas predictive technologies, like NIRS, will produce a single value that does not capture intrinsic compositional variation. However, many benefits exist for NIRS methods, which include being low cost, high throughput, and nondestructive (Cen and He, 2007) and providing the ability to generate real-time nutrient composition data permitting adaptive formulation and manufacture processes for poultry feeds.

Another point to consider is that the accuracy of nutrient supply to the animal is not only a function of intrinsic dietary concentrations (i.e., analytical values) but also of nutrient utilization by the animal (i.e., nutritive values). While nutrient composition estimates represent the total or gross amount of an individual nutrient presented to the animal, any deviations in availability due to digestive, absorptive, or metabolic processes will be realized as a functional reduction in dietary supply. In particular, anti-nutritional factors in certain ingredients can significantly alter the efficiency of nutrient utilization by poultry species, as can ingredient processing methods designed to mitigate those anti-nutritional factors (i.e., reduced lysine availability due to generation of Amadori products through overheating of soybean meal). Therefore, ingredient composition data only represent the dietary potential (or gross concentration) for any individual nutrient, and this information must be paired with nutritional values derived from empirical animal studies conducted to generate nutrient digestibility and availability estimates. These nutritional values are presented elsewhere in this report and are specific to each category, age, weight, and physiological status of poultry species being fed. Specific emphasis is placed on nutritional values for amino acids and phosphorus as these nutrients are crucial for both economic and environmental reasons.

ANTI-NUTRITIONAL FACTORS

Anti-nutritional factors are broadly defined as naturally occurring compounds that impair the efficiency of nutrient utilization and reduce feed intake and/or growth of species that consume them (Hathcock and Rader, 1994). A wide variety of anti-nutritional factors exist in virtually every feedstuff used in poultry nutrition (Cheeke, 1998), and the reader is referred to a number of extensive reviews that have been published for poultry species (Ferket and Middleton, 1998; Elkin, 2017). In the case of poultry, anti-nutritional factors may be found in plant-based feedstuffs including cultivars of cereal grains, legumes, oilseed meals, roots and tubers, and seeds. Additionally, animal-derived by-product ingredients may include toxic metabolites formed via bacterial degradation during decomposition of offal or carcasses, including microbial decarboxylation or deamination of amino acids to yield biogenic amines or ammonia, respectively.

The main classes of plant anti-nutritional factors affecting poultry species include the following: (1) phytate (discussed further in the minerals chapter, Chapter 6); (2) mycotoxins; (3) alkaloids (pyrrolizidine, tropane, and steroid); (4) glycosides (goitrogenic, cyanogenic, vicine and convicine, and saponins); (5) proteins (proteinase inhibitors and lectins) and amino acid derivatives (amino nitriles and lathyrogens); (6) lipids (erucic acid and cyclopropenoid fatty acids); and (7) phenolics (condensed tannins, gossypol, and sinapine) (Elkin, 2017). Of these classes, phytate and mycotoxins are arguably the most important anti-nutritional factors from both economic and environmental perspectives, and industry-wide efforts to mitigate effects of these anti-nutritional factors are pervasive (Woyengo and Nyachoti, 2013; Humer et al., 2015; Murugesan et al., 2015; Cowieson et al., 2016; Guerre, 2016). Generally speaking, feedstuff processing methods and dietary mitigation strategies exist for many anti-nutritional factors commonly found in feedstuffs used in the poultry industry, so a basic understanding of anti-nutritional factors associated with each feedstuff is sufficient to allow feed formulators to avoid or minimize adverse effects (Ferket and Middleton, 1998). By controlling dietary inclusion levels and processing and handling methods associated with feedstuffs, and through use of exogenous enzymes or other dietary additives, risk of detrimental effects of anti-nutritional factors on the growth and health of poultry species can be managed. However, poultry with access to outdoor environments are at greater risk, albeit minor, of exposure to toxicants when consuming plants, insects, or other articles beyond the tightly controlled and highly regulated poultry feed provided by the producer. Additional information about specific anti-nutritional factors can be found in Chapter 12.

INGREDIENT NUTRIENT COMPOSITION DATABASE

With this context in mind, identifying sources of accurate nutrient composition data is critical. An optimal strategy for overcoming such problems is to use a "big data" approach, whereby all available data are consolidated using a standardized framework and then established statistical approaches are applied to clean the data set in an objective and transparent manner. As such, the committee decided to partner with the National Animal Nutrition Program (NANP; https://animalnutrition.org), a National Research Support Project (NRSP-9) administered by the U.S. Department of Agriculture National Institute of Food and Agriculture, to leverage a consolidated nutrient composition data set for ingredients pertinent to poultry species.

TABLE 20-1 Peer-Reviewed Journals from Which Nutrient Composition Data Were Derived

Journal Name	Journal Abbreviation	ISSN[a]
Animal Feed Science and Technology	*Anim. Feed Sci. Technol.*	0377-8401
Animal Science Journal	*Anim. Sci. J.*	1740-0929
Archives of Animal Nutrition	*Arch. Anim. Nutr.*	1477-2817
Asian-Australasian Journal of Animal Sciences	*Asian-Australas. J. Anim. Sci.*	1976-5517
British Journal of Nutrition	*Br. J. Nutr.*	1475-2662
British Poultry Science	*Br. Poult. Sci.*	1466-1799
Canadian Journal of Animal Science	*Can. J. Anim. Sci.*	1917-1825
Current Developments in Nutrition	*Curr. Dev. Nutr.*	2475-2991
Journal of Agricultural and Food Chemistry	*J. Agric. Food Chem.*	1520-5118
Journal of Animal Physiology and Animal Nutrition	*J. Anim. Phys. Anim. Nutr.*	1439-0396
Journal of Animal Science	*J. Anim. Sci.*	1525-3163
Journal of Applied Animal Nutrition	*J. Appl. Anim. Nutr.*	2049-257X
Journal of Applied Poultry Research	*J. Appl. Poult. Res.*	1537-0437
Journal of Food Science	*J. Food Sci.*	1750-3841
Journal of Nutrition	*J. Nutr.*	1541-6100
Livestock Production Science	*Livest. Prod. Sci.*	0301-6226
Livestock Science	*Livest. Sci.*	1871-1413
Poultry Science	*Poult. Sci.*	1525-3171
Translational Animal Science	*Transl. Anim. Sci.*	2573-2102
World's Poultry Science Journal	*Worlds Poult. Sci. J.*	1743-4777

[a] International standard serial number for the print version of each journal.

The NANP was founded in 2010 and serves, in part, to implement integrated and systemic approaches for collecting, assembling, synthesizing, and disseminating science-based information and educational tools, and enabling technologies on animal nutrition that facilitate high-priority research among multiple animal species. The NANP Feed Composition subcommittee has created a cross-species database of nutrient composition in feedstuffs to standardize the approach in evaluating data derived from academic laboratories, commercial laboratories, and the peer-reviewed literature. The committee's decision to employ the NANP feed composition resources was based on a number of benefits, including:

1. Perennial addition of up-to-date nutrient composition data by the NANP for all ingredients
2. Robust nature of the consolidated data set, including millions of unique records spanning ingredients commonly used by agriculturally relevant species across many years
3. Ability to convert nutrient composition data between dry matter and as-is bases in real time
4. Flexibility to summarize subsets of ingredient records based on input criteria including
   a. Reported year of ingredient analysis
   b. Source of data (peer-reviewed literature, commercial laboratory, academic laboratory)
   c. Constraints on any combination of individual nutrient inputs (i.e., only display records where corn contains lysine and methionine ranging between 0.25–0.30% and 0.16–0.20%, respectively)
5. Web-based and mobile-friendly display of summarized nutrient composition data including mean value, sample size, standard deviation, and range (minimum and maximum) in a format optimized for multiple devices
6. Convenient display of nutrient composition data alongside species-specific, animal-derived nutrient and energy digestibility values (e.g., standardized ileal amino acid digestibility values for poultry)

In addition to drawing upon the existing NANP feed composition data set, the committee contributed to NANP efforts by performing its own exhaustive literature review for ingredients commonly used in poultry feed formulations. As such, the committee accepted the nutrient composition data used for all ingredients as listed in the swine NRC (2012) report and went on to extract records for the same set of ingredients published in a specified set of journals (Table 20-1) since December 31, 2010 (i.e., the last date on which the 2012 Swine NRC committee queried records). Only quantitative nutrient composition data were captured where adequate methodological detail was provided, and data were recorded exactly as published as catalogued on either an as-is or dry matter basis.

The swine NRC (2012) committee searched the literature for a specified list of ingredients from 1995 to 2010. Whereas year of analysis was captured in their data set for each ingredient record, no information about geographic source of ingredient was available. In the vast majority of peer-reviewed publications, no geographic origin for ingredient samples was provided by authors, which means it was not possible to associate nutrient composition with a specific geographic location. Therefore, while the committee recognizes that ingredient composition varies based on geographic origin of feedstuff, there does not presently exist a robust and quantitative approach for identifying how this source of variation influences the health, well-being, growth, and productivity of poultry species. The committee recommends that specific emphasis be

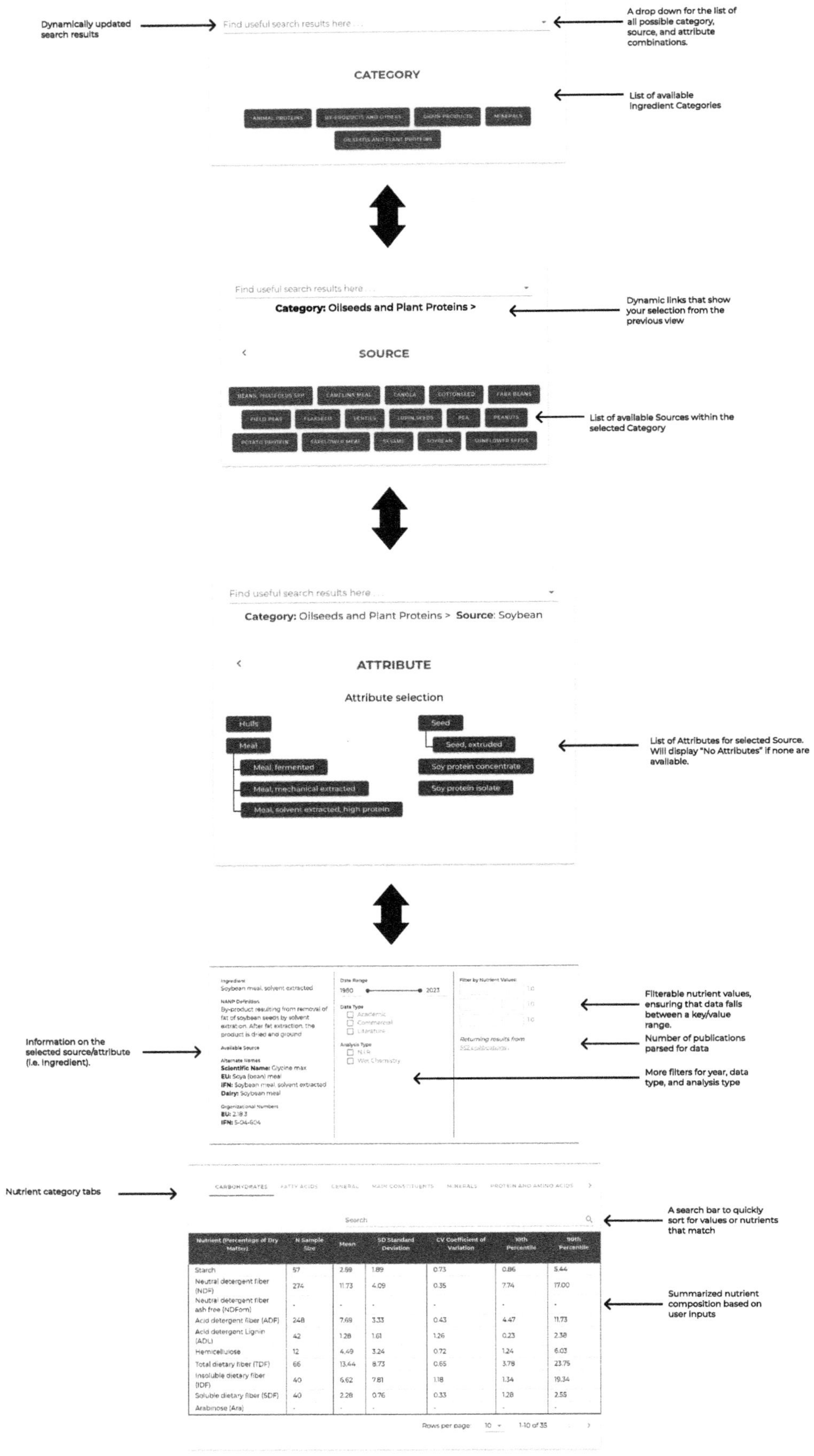

FIGURE 20-1 Workflow for using the NANP feed composition database (https://animalnutrition.org/feed-composition-database).

given by the scientific community to curate a data set targeted at answering this important question.

Considering the NANP feed composition database is designed for multiple animal species, many challenges existed for ingredient name alignment. As such, the most basic ingredient name is used and no distinction among specialized forms of that ingredient was made based on nutrient composition. Subcategories of ingredients were only used when a well-defined processing method was implemented (e.g., solvent-extracted soybean meal). This approach is based on procedures set forth by the American Association of Feed Control Officials (AAFCO) in their official publication. That annually refreshed database is used by U.S. regulatory agencies as a recognized authoritative source. Therefore, specialty and novel ingredients, including those arising from the biofuels industry, may not be found in the ingredient composition database if they have not received an official AAFCO definition and number. The NANP database will be continually updated with new ingredients and their associated composition data as solicited from a variety of sources, so there exists an impetus for manufacturers to seek AAFCO approval for ingredient definitions and share available composition data with the NANP committee to incorporate into the database.

Regarding use of the NANP feed composition database, a user starts by visiting the NANP website (https://animalnutrition.org) and clicking on "Feed Composition" either in the left navigation or in the upper-right corner of the page. Alternatively, the database can be accessed directly at https://animalnutrition.org/feed-composition-database and a bookmark can be set for this resource by the user. At the top of the database landing page, there are background details about this resource and links to additional tutorials for how to use the database. Below that information, the user can either search for an ingredient directly or select an ingredient category to view individual feedstuffs (Figure 20-1). When a category is selected, a list of ingredients is displayed, and clicking on the feedstuff name will display all available data for that particular ingredient.

As an example, selecting "soybean" as the source and the combination of "meal, solvent extracted" as attributes, the user would be taken to the compositional summary display for that ingredient, which is divided into multiple areas. In this summarized composition display, the header area serves both to display metadata about the chosen ingredient (e.g., definition, scientific name, organizational numbers) and allow the user to further manipulate the data records returned and summarized lower on the page. Navigation options present within the header box include the ability to return to the ingredient categories view or to the source plus attributes selection process.

The header box displays additional input fields that allow customization of the returned data set. Composition data can be displayed in the absence or presence of moisture; selecting the "As Fed" option requires the user to enter a custom dry matter value that will be used to recalculate nutrient concentrations displayed lower on the page. Other features allow entry of a range of years associated with feed composition records, choosing the source of data and type of analysis, and the ability to filter nutrient values based on submission of up to three inputs applied with Boolean logic. A user can easily apply these filters based on ranges set for any combination of individual nutrients, thereby allowing ultimate flexibility in defining ingredient identities and nutrient profiles most pertinent for a particular formulation effort. Each time the user changes an input in the header box and clicks the "Filter Composition" button, an amended set of feed composition records is summarized and displayed lower on the page and the number of "Returned Results" is updated. Clicking "Clear All Filters" below the "Filter Composition" button will return the user to the default data display, which includes ingredient composition summarized on a dry matter basis as analyzed by wet chemistry methods, regardless of data type. Once the desired inputs have been applied, the resulting tables are displayed on the same page, and they can be easily printed using the button located just below the header box.

Once user inputs are applied (i.e., once the "Filter Composition" button is clicked in the header box), the summarized nutrient composition details can be found on the lower part of the page. Across the top of this display are the major categories of nutrients, with the "Main Constituent" tab being selected by default. Also available are citations for any peer-reviewed publications from which data had been extracted (see "References" tab). For each nutrient, the sample size, mean, standard deviation, coefficient of deviation, and 10th and 90th percentiles are displayed based on inputs submitted by the user. Nutrient names, along with their common abbreviation and units, are displayed in the first column, and clicking on a nutrient name displays additional information including definitions and references for that analytical technique.

A listing of ingredients deemed pertinent to poultry production can be found in Table 20-2, and the nutrient composition of these ingredients can be found in the NANP database (https://animalnutrition.org) along with descriptors including the NANP ingredient definition, alternate ingredient names, and standardized nomenclature published by AAFCO, the European Union catalogue of feed materials (EU 575/2011), and the International Feed Databank System (i.e., International Feed Numbers).

TABLE 20-2 Specifications for Ingredients Pertinent to Poultry Feed Formulations Included in the Nutrient Composition Database[a]

Ingredient Name	AAFCO #
Alfalfa hay	—
Alfalfa meal	3.2
Bakery by-product meal	40.15
Barley distiller's dried grains with solubles	27.6
Barley grain	—
Beans, *Phaseolus* spp	—
Blood cells, spray dried	9.24
Blood meal	9.61
Blood plasma, spray dried	9.72
Brewer's grains, dry	15.1
Brewer's yeast, dry	96.4
Camelina meal, mechanical extracted	71.3
Canola meal (rapeseed), mechanical extracted	71.77

continued

TABLE 20-2 Continued

Ingredient Name	AAFCO #
Canola meal (rapeseed), solvent extracted	71.77
Canola seed (rapeseed)	—
Casein	54.16
Cassava (tapioca) meal	60.74
Citrus pulp, dry	21.1
Copra meal, mechanical extracted	71.6
Copra meal, solvent extracted	71.61
Corn bran	48.2
Corn distiller's dried grains	27.5
Corn distiller's dried grains with solubles	27.6
Corn distiller's solubles	27.4
Corn germ	—
Corn germ meal	48.23
Corn gluten feed, dry	48.13
Corn gluten meal	48.14
Corn grain, dry	48.4
Corn hominy feed	48.16
Cottonseed	—
Cottonseed meal, mechanical extracted	24.1
Cottonseed meal, solvent extracted	—
Dried yeast	96.1
Egg, whole spray dried	—
Faba beans	—
Feather meal	—
Field peas	—
Fish meal, mechanical extracted	51.15
Fish protein hydrolyzed	51.11
Flaxseed	—
Flaxseed meal, mechanical extracted	71.1
Gelatin	40.29
Lentils	—
Lupin seed	—
Meat and bone meal	9.41
Meat meal	9.4
Millet grain	—
Oat grain	—
Oat groats	69.1
Oat hulls	69.2
Palm kernel meal, mechanical extracted	—
Pea hulls	—
Pea protein concentrate	—
Peanut meal, mechanical extracted	71.9
Peanut meal, solvent extracted	71.9
Peas	—
Pet food by-products	60.108
Potato protein	60.94
Poultry by-product meal	9.1
Rice bran	75.7
Rice bran defatted	75.3
Rice grain	—
Rice protein concentrate	—
Rye grain	—
Safflower meal, solvent extracted	71.131
Sesame meal, mechanical extracted	—
Skimmed milk, powder	54.3
Sorghum distiller's dried grains with solubles	27.6
Sorghum grain, dry	—
Soy protein concentrate	84.12
Soy protein isolate	84.62
Soybean hulls	84.3
Soybean meal, fermented	—
Soybean meal, mechanical extracted	84.6
Soybean meal, solvent extracted	84.61
Soybean meal, solvent extracted, high protein	84.61
Soybean seed	—
Soybean, extruded	—
Sugarbeet molasses	63.1
Sugarbeet pulp, dry	63.36
Sugarcane molasses	63.7
Sunflower meal, solvent extracted	71.221
Sunflower seed	—
Triticale distiller's dried grains with solubles	27.6
Triticale grain	—
Wheat bran	93.1
Wheat distiller's dried grains with solubles	27.6
Wheat gluten	T93.9
Wheat grain	—
Wheat middlings	93.5
Whey permeate, 80% lactose	—
Whey powder	54.7
Whey powder, protein concentrate	54.25

[a] Summarized nutrient composition data for these ingredients can be found in the National Animal Nutrition Program (NANP) database located at https://animalnutrition.org. Reference is made to the American Association of Feed Control Officials (AAFCO) Official Publication, which can be accessed at https://www.aafco.org.

REFERENCES

Cen, H., and Y. He. 2007. Theory and application of near infrared reflectance spectroscopy in determination of food quality. *Trends in Food Science & Technology*. 18(2):72–83.

Cheeke, P.R. 1998. *Natural Toxicants in Feeds, Forages, and Poisonous Plants*, 2nd ed. Prairie Village, KS: Interstate Publishers.

Cowieson, A.J., J.P. Ruckebusch, I. Knap, P. Guggenbuhl, and F. Fru-Nji. 2016. Phytate-free nutrition: A new paradigm in monogastric animal production. *Animal Feed Science and Technology*. 222:180–189.

Elkin, R.G. 2017. Influence of plant toxins on laying hen performance and egg quality. In *Egg Innovation and Strategies for Improvement*. P.Y. Hester, editor. Cambridge, MA: Academic Press.

Ferket, P.R., and T. Middleton. 1998. Antinutrients in poultry feeds. *Proceedings of the WPSA 10th European Poultry Conference*, Jerusalem, Israel.

Gizzi, G., and D.I. Givens. 2004. Variability in feed composition and its impact on animal production. In *Assessing Quality and Safety of Animal Feeds*. S. Jutzi, editor. Food and Agriculture Organization of the United Nations. 39–54.

Guerre, P. 2016. Worldwide mycotoxins exposure in pig and poultry feed formulations. *Toxins*. 8(12):350.

Hathcock, J.N., and J.I. Rader. 1994. Food additives, contaminants, and natural toxins.In *Modern Nutrition in Health and Disease*, 8th rev. ed. M.E. Shils, J.A. Olson, and M. Shike, editors. Philadelphia, PA: Lea and Febiger. 1593–1611.

Humer, E., C. Schwarz, and K. Schedle. 2015. Phytate in pig and poultry nutrition. *Journal of Animal Physiology and Animal Nutrition*. 99(4): 605–625.

Llames, C.R., and J. Fontaine. 1994. Determination of amino acids in feeds: Collaborative study. *Journal of the Association of Official Analytical Chemists*. 77:1362–1402.

Murugesan, G.R., D.R. Ledoux, K. Naehrer, F. Berthiller, T.J. Applegate, B. Grenier, T.D. Phillips, and G. Schatzmayr. 2015. Prevalence and effects of mycotoxins on poultry health and performance, and recent development in mycotoxin counteracting strategies. *Poultry Science*. 94(6):1298–1315.

NRC (National Research Council). 2012. *Nutrient Requirements of Swine*. 11th rev. ed. Washington, DC: National Academy Press.

Woyengo, T.A., and C.M. Nyachoti. 2013. Review: Anti-nutritional effects of phytic acid in diets for pigs and poultry—current knowledge and directions for future research. *Canadian Journal of Animal Science*. 93(1):9–21.

21

Recommendations for Future Research

INTRODUCTION

Although the present report updates many of the minimum nutrient requirements for the major and minor poultry species, many knowledge gaps remain. Moreover, as bird performance metrics continue to increase, due mainly to advances in the areas of genetics, genomics, nutrition, management, microbiomics, metabolomics, ethology, and health, concomitant changes in the minimum requirements for many nutrients are expected. At this time, consideration of the effect of poultry nutrition on the environment has become a dominant goal. Therefore, presented here is a brief overview of suggested future research needs based on critical gaps in nutritional knowledge for the poultry species of highest economic importance to environmental health and the sustainability of poultry production. Many chapters provide specific recommendation for future research that will advance poultry nutrition, and the reader should consult each chapter for specific details.

Modeling. It has been 30 years since the National Academies of Sciences, Engineering, and Medicine last updated the *Nutrient Requirements of Poultry* series, and it is clear that empirical approaches for recommending requirements will never keep up with the rate of change in poultry genetics and management objectives ranging from least-cost through free and organic management systems. Dynamic computer models need to be further developed so that nutrient requirements can be predicted based on specifics of genetics, desired growth rates, body compositions, and management and marketing considerations. These models should allow the nutritionist to enter the preceding information and obtain optimal requirements for net energy (NE), apparent metabolizable energy (AME), and digestible amino acids, calcium, phosphorus, and other nutrients of concern. The model should be based on research done specifically for parametrizing such models, which is mostly different than research for determining nutrient requirements empirically. There are many large gaps in such research in the poultry literature.

Research to support modeling. Modeling is based on the integration of empirical research, but the initial information for supporting model development requires new information on the growth curves; nutritional content of accreted material including body tissues, eggs, and feathers; and their interrelationships. This information is needed for breeds that span those in use for different production goals. In the longer term, model development will also need to include research done to parametrize the effects of different environments, production systems, and marketing goals.

Energy value of feedstuffs. At the time of the previous report, most metabolizable energy research published was based on the true metabolizable energy (TME) system, which is relativity quick and inexpensive to determine (Chapter 1). Currently, TME is falling out of favor due to concerns regarding its applicability to fast-growing broilers and about the welfare of the birds used in the assay. Presently, assays on young growing birds are used to measure apparent metabolizable energy without nitrogen correction (i.e., AME, or with nitrogen correction [AME_n] to account for nitrogen retention). The committee found that the research on AME and AME_n in poultry is plagued with problems related to experimental design, ingredient analysis, and procedural errors that together result in an inaccurate estimates and high variability within and between studies. The committee finds that published estimates of AME in important of ingredients, such as soybean meal, to be sufficiently variable and inaccurate as to be unacceptable for accurate diet formulation. Please see Chapter 1 for an in-depth description of these problems. These sources of variability and inaccuracy could be minimized or eliminated by adoption of consensus analytical methodologies, study design, and reporting requirements. Methods that are validated through round-robin testing should be encouraged. Given the importance of usable dietary energy estimates for poultry, establishment of valid consensus methodology and reporting requirements should be a near-term priority. Ultimately, the poultry industry may be best served by moving away from an AME system to an NE "or novel systems," and research that supports or refutes such a change is highly important.

Environmental impacts and sustainability of poultry nutrition. Minimizing the environmental impacts due to the production of ingredients, selection of ingredients that minimize environmental impacts, and maximizing the efficiency of use of those ingredients for poultry production is required for this goal. Many commercially available feed formulation programs have options to enable or add constraints to the formulation matrix to minimize dietary nitrogen and phosphorus excretion and these efforts should enhanced. Research is also needed to inform feed formulation software developers to provide constraints on the lifetime impacts of greenhouse gasses for each feed ingredient and ultimately, the final least cost diet that meets the requirement of the target bird. Nutrition research that helps parameterize these constraints continues to be needed.

Maximizing the efficiency of nutrient utilization of ingredients with low environmental impacts during their production is the area where poultry nutritionists can have their highest impact. Continued focus on the efficiency of the bird's utilization of an ingredient hinges on increasing its digestibility and bioavailability of energy, amino acids, phosphorus, and salts, and trace minerals of concern will ultimately minimize their environmental impact. The weak link in this approach is knowing the digestible concentration of important nutrients and energy in feed ingredients utilized in poultry nutrition. Research to determine the concentrations of AME or NE and digestible amino acids, calcium, and phosphorus in feed ingredients is greatly needed.

Most research in poultry nutrition has focused on the "starting period" when birds are 0–4 weeks old because it is an easy and inexpensive testing period relative to later periods. The sustainability of poultry production requires attention to the entire production cycle including the finisher, maintenance, and breeding periods, which are much more difficult to study using empirical experiments because of the need for expensive facilities and labor. Research efforts designed to inform models that predict nutrient requirements during these later periods of production should be rewarding because of the greater inputs and outputs occurring during later phases. To date, almost all research to develop predictive parameters for models has been done in broilers, and that focus needs to continue, but research specific to laying hens, turkeys, and breeders of each type has been modest and should be increased. Continued genetic progress in growth-related traits will create new challenges for breeder management. With a few exceptions, there is a paucity of research using modern poultry strains as test animals to determine vitamin and mineral requirements, or the bioavailability of differing vitamin and mineral forms throughout the productive lifespan.

The reliance on feed restriction to increase reproductive performance provides a rationale to pursue research in the fundamental biology controlling appetitive drive and satiety with a view toward creating new nutritional management strategies to improve reproductive performance and bird welfare. We should also consider the carry-over effects of hen nutrition on the progeny, including their health, nutrient requirements, and performance.

Supplementing exogenous enzymes has been an immensely successful area of research for increasing the efficiency of feed utilization and minimizing nutrient excretion. Almost all of this research has been funded or conducted by industry, and this will likely continue. An important area of research that could synergize with current commercial research is the modification of the bird's digestive enzyme repertoire using advanced genomic editing techniques followed by scrupulous assessment of their impact on the health of poultry and human populations. Unlocking the energy in cellulose, almost all of which is excreted, should be a long-range goal. Similarly, genomic editing techniques could be applied to permit the synthesis of essential amino acids, which would lead to reduction in the amounts of ingredients that need to be grown or synthesized for poultry production.

APPENDIXES

Appendix A

Statement of Task

The committee will prepare a report that reviews the scientific literature on the nutrition of poultry and make recommendations on nutrient requirements. All productive phases and types of production will be addressed. New quantitative information on recommendations will be made with consideration for the increased performance of different types of poultry, including chickens, turkeys, and ducks with updates to additional species, as appropriate.

The report will include a comprehensive analysis of recent research on feeding and nutrition of poultry with a focus on energy, amino acids, lipids, minerals, vitamins, and water needed by poultry; a summary of the composition of feed ingredients, mineral supplements, and feed additives routinely fed to poultry; information about variability in feed ingredients sourced from different regions; and information about feed ingredients from the biofuels industry and other new ingredients (e.g., novel soybean products).

The report will examine requirements for digestible phosphorus and amino acids and update concentrations of digestible phosphorus and amino acids in feed ingredients. The report will examine new information on the bioavailability, metabolism, and utilization of various nutrients, and provide a review of nutritional and feeding strategies to minimize nutrient excretion.

A discussion of the effect of dietary components on the nutritional quality and functional properties of poultry meat and eggs will be included. Effects of environmental stressors, feed processing, and other production aspects including antibiotics on nutrient responses will be addressed.

The report also will include a discussion of modelling approaches for estimating nutrient responses and depending on the extent of new information available, a model suitable for estimating nutrient responses may be developed. Future areas of needed research also will be identified.

Appendix B

Acronyms and Abbreviations

AAFCO	Association of American Feed Control Officials
ADF	acid detergent fiber
ALA	α-linolenic acid, 18:3 n-3
AME	apparent metabolizable energy
AME_n	apparent metabolizable energy corrected to zero N retention
AMS	Agricultural Marketing Service
ANOVA	analysis of variance
ARA	arachidonic acid, 20:4 n-6
BSA	bovine serum albumin
Ca	calcium
Cal	calories
CF	crude fiber
CP	crude protein
CSFII	Continuing Survey of Food Intakes by Individuals 1994–1996
DAAO	D-amino acid oxidase
DC	digestibility coefficient
DE	digestible energy
DHA	docosahexaenoic acid, 22:6 n-3
DIAAS	digestible indispensable amino acid score
DM	dry matter
DPA	docosapentaenoic acid, 22:5 n-3
EE	ether extract
EPA	eicosapentaenoic acid, 20:5 n-3
ER	energy retained
FI	feed intake
FPD	foot pad dermatitis
GE	gross energy
GIT	gastrointestinal tract
GSH-Px	glutathione peroxidase
GWP	global warming potential

HI	heat increment
HMB	DL-2-hydroxy-4-(methylthio)butanoic acid
HMBC	HMB-Ca
J	joules
LC-PUFA	long-chain polyunsaturated fatty acids
LNA	linoleic acid, 18:2 n-6
MBM	meat and bone meal
ME	metabolizable energy
ME_n	nitrogen-corrected ME
MTTP	microsomal triglyceride transport protein
MUFA	monounsaturated fatty acids
NDF	neutral detergent fiber
NE	net energy
NFE	nitrogen free extract
NFNAP	National Food and Nutrient Analysis Program
NHANES	National Health and Nutrition Examination Survey
NIRS	near infrared reflectance spectroscopy
NRC	National Research Council
P	phosphorus
PDCAAS	protein digestibility-corrected amino acid score
PUFA	polyunsaturated fatty acid
SAT	saturated fatty acid
TAG	triacylglycerol
TME	true metabolizable energy
TME_n	nitrogen-corrected TME
TSAA	total sulfur amino acids
VLDL	very-low density lipoprotein

Appendix C

Literature Estimates of Ingredient Energy Content

In Chapter 1 on energy, it was related that the committee determined that the industry would be well served by moving to an NE system; however, there are scant publications that characterize NE for ingredients. Nevertheless, the committee felt that it was important to provide energy values for principal ingredients used to formulate broiler diets. For this reason, the committee sought to develop values for AME and AME_n using rigorous criteria for included literature. The committee notes that nitrogen correction may be inappropriate in fast-growing broilers, turkeys, and high-producing laying hens. Furthermore, the committee was mindful that ingredients could vary in their energy content depending upon point of origin and processing, with additional variation arising from the analytical methods used to characterize ingredient composition and experimental methodology to evaluate energy value in live birds. Further comments on the factors that influence the values obtained from bioassay methods can be found in Chapters 1, 8, and 20.

Published AME/AME_n studies varied considerably in design and rigor, and the committee established specific criteria for study inclusion. Frequently encountered defects included inadequate compositional information including compositions that did not sum to 100 percent, inadequate numbers of replicates or intervention levels, inconsideration of digestive physiology factors such as fat and ANF effects on digestibility of other dietary components. Criteria for the literature searched and studies included are detailed in Table C-1. The committee noted that some flexibility in the application of criteria was needed for some of the ingredients as completely perfect studies were not always available. The limitations in studies are noted in Table C-4 that summarizes all the studies used for tabular values.

TABLE C-1 Criteria for Publication Inclusion in Literature Estimates of Ingredient Energy Content[a]

Variable	Specification
Publication dates	2010 to present
Bird	Broiler, strain specified, 14 d of age or older
Diets	Basal diet of appropriate nutrient composition and ME concentration. At least three dietary treatments are used to determine a regression equation, all of which are on the linear part of the response.
Ingredient	International with source specified. The nutrient composition of the ingredient should be well characterized by best methodology; approved analytical methods should be stated.
Experimental design	• Basal diet should meet or exceed nutrient recommendations. • At least three dietary treatments must be used to determine a regression equation, all of which must be within the linear part of the response. • Birds should be adapted to experimental diets for a minimum of 5 d (Zhong and Adeola, 2019a,b), after which birds should be 14 d old. • If a marker is used, marker recovery should be included.[a] Note this specification was not always adhered to as actual recovery values were infrequently reported. • Body weight and feed intake throughout the test period should be provided. • Should be stated that excreta are free from foreign matter such as feathers, feed, or litter.[a] • Regression studies should contain at least 6 pens per dietary treatment; Substitution designs should include at least 8 pens per treatment.

[a] In some instances, a well-designed study that has minor insufficiencies may be included, and the specific issues with the study are described in Table C-4 so that the user can exclude their use if desired.

TABLE C-2 Search Strategy for Publications Used in Literature Estimates of Ingredient Energy Content[a]

Search Strategy[a]	Database: CAB Abstracts <1910 to 2023 Week 14>
1	metaboli*sable energy.mp. [mp=abstract, title, original title, broad terms, heading words, identifiers, cabicodes] (3,724)
2	ame.mp. [mp=abstract, title, original title, broad terms, heading words, identifiers, cabicodes] (1,862)
3	broiler*.mp. [mp=abstract, title, original title, broad terms, heading words, identifiers, cabicodes] (77,521)
4	1 or 2 (5,379)
5	corn.mp. [mp=abstract, title, original title, broad terms, heading words, identifiers, cabicodes] (404,166)
6	maize.mp. [mp=abstract, title, original title, broad terms, heading words, identifiers, cabicodes] (368,483)
7	5 or 6 (449,424)
8	3 and 4 and 7 (432)
9	limit 8 to yr="2010–2023" (237)
10	limit 9 to english language (202)

[a] The table shows an example search for studies determining the energy content to broilers of the ingredient corn. Search terms are shown in text while the number of references retrieved from each search term is shown in parentheses. The numbered search terms were combined in later search steps to refine the results; see, for example, search step 8 that combines search steps 3, 4, and 7.

TABLE C-3 Literature Estimates of Ingredient Energy Content for Broilers[a]

Table C-3a	AME by Regression Study Design[a,b]				
Ingredient	Number of Studies	Ave Dry Matter (%)	AME, Mean (kcal/kg)	Std. Dev.	Range
Corn grain, as-fed	—	—	—	—	—
Wheat grain, higher protein	—	—	—	—	—
Sorghum	—	—	—	—	—
Barley grain	3	—	2,740	140	2,622–2,894
Rye grain	1		3,053		
Soybean meal, solvent extracted	—	—	—	—	—
Canola meal (rapeseed meal), solvent extracted	—	—	—	—	—
Sunflower meal	—	—	—	—	—
Field peas	—	—	—	—	—
Meat and bone meal	6	96	2,903	245	2,512–3,125
Meat meal	—	—	—	—	—
Poultry by-product meal	—	—	—	—	—
Fish meal, mechanical extracted	—	—	—	—	—
Corn distiller's dried grains with solubles	1	NA	2,800	NA	NA
Fats and oils:					
Soy oil	—	—	—	—	—
Palm oil	—	—	—	—	—
Poultry fat	—	—	—	—	—

[a] Values in normal font are from studies meeting committee rigor specifications listed in Table C-1.

[b] Values in bold italics are guidance levels based on estimates due to insufficient experimental rigor. Limitations of individual studies in source literature are detailed in Table C-4.

The committee used specific search methodology listed in Table C-1 to identify suitable publications. The searches were carried out in the CAB Abstracts database using the following specifications in Table C-2. This limited scoping review was then hand-searched for studies that most completely met the committee's preset criteria.

The 15 ingredient classes initially selected by the committee for AME compilation were corn, wheat, sorghum, barley, rye, solvent-extracted soybean meal, solvent extracted canola (rapeseed) meal, sunflower meal, field peas, meat and bone meal, meat meal, poultry by-product meal, mechanically extracted fish meal, corn distiller's dried grains with solubles, fats, and oils. Searches were completed for corn, wheat, sorghum, barley, rye, solvent-extracted soybean meal, solvent extracted canola (rapeseed) meal, field peas, meat and bone meal, corn distiller's dried grains with solubles, soybean oil, poultry fat, and palm oil (Table C-3a–d).

Table C-3b	AME by Substitution Study Design[a,b]				
Ingredient	Number of Studies	Ave Dry Matter (%)	AME, Mean (kcal/kg)	Std. Dev.	Range
Corn grain, as-fed	7	89.26	3,341	75	3,213–3,408
Wheat grain, higher protein	***9***	***89.5***	***3,301***	***142***	***2,958–3,470***
Sorghum	3	89.9	3,099	330	2,815–3,750
Barley grain	17	—	3,051	264	2,447–3,461
Rye grain	—	—	—	—	
Soybean meal, solvent extracted	—	—	—	—	—
Canola meal (rapeseed meal), solvent extracted	8		3,279	129	3,061–3,411
Sunflower meal	—	—	—	—	—
Field peas	***7***	***87.7***	***2,675***	***327.8***	***2,245–3,296***
Meat and bone meal	3	96	2,332	151	1,353–4,070
Meat meal	—	—	—	—	—
Poultry by-product meal	—	—	—	—	—
Fish meal, mechanical extracted	—	—	—	—	—
Corn distiller's dried grains with solubles	0	NA	NA	NA	NA
Fats and oils:					
Soy oil	—	—	—	—	—
Palm oil	—	—	—	—	—
Poultry fat	—	—	—	—	—

[a] Values in normal font are from studies meeting committee rigor specifications listed in Table C-1.
[b] Values in bold italics are guidance levels based on estimates due to insufficient experimental rigor. Limitations of individual studies in source literature are detailed in Table C-4.

Table C-3c	Nitrogen corrected AME by Regression Study Design[a,b]				
Ingredient	Number of Studies	Ave Dry Matter (%)	AME_n, Mean (kcal/kg)	Std. Dev.	Range
Corn grain, as-fed	—	—	—	—	—
Wheat grain, higher protein	—	—	—	—	—
Sorghum	—	—	—	—	—
Barley grain	3		2,629	188	2,481–2,841
Rye grain	—	—	—	—	—
Soybean meal, solvent extracted	2		2,617	84	2,557–2,677
Canola meal (rapeseed meal), solvent extracted	3	90	2,067	115	2,000–2,200
Sunflower meal	—	—	—	—	—
Field peas	—	—	—	—	—
Meat and bone meal	8	96	2,263	718	1,088–2,887
Meat meal	—	—	—	—	—
Poultry by-product meal	—	—	—	—	—
Fish meal, mechanical extracted	—	—	—	—	—
Corn distiller's dried grains with solubles	1	NA	2,688	NA	NA
Fats and oils:					
Soy oil	—	—	—	—	—
Palm oil	—	—	—	—	—
Poultry fat	—	—	—	—	—

[a] Values in normal font are from studies meeting committee rigor specifications listed in Table C-1.
[b] Values in bold italics are guidance levels based on estimates due to insufficient experimental rigor. Limitations of individual studies in source literature are detailed in Table C-4.

REFERENCES

Adeola, O., and C. Kong. 2020. Energy values of triticale or sorghum distillers' dried grains with solubles and rye fed to broiler chickens. *Journal of Animal Science.* 98(2):skaa018

Adeola, O., and H. Zhai. 2012 Metabolizable energy value of dried corn distillers grains and corn distillers grains with solubles for 6-week-old broiler chickens. *Poultry Science.* 91(3):712–718.

Adeola, O., M.N. Anwar, M.R. Abdollahi, and V. Ravindran. 2018. Age-related energy values of meat and bone meal for broiler chickens. *Poultry Science.* 97(7):2516–2524.

Agyekum, A.K., and T.A. Woyengo. 2022. Nutritive value of expeller/cold-pressed canola meal and pre-pressed solvent-extracted carinata meal for broiler chicken. *Poultry Science.* 101(1):101528.

Azhar, M.R., S.P. Rose, A.M. Mackenzie, S.C. Mansbridge, M.R. Bedford, A. Lovegrove, and V.R. Pirgozliev. 2019. Wheat sample affects growth

Table C-3d AME$_n$ by Substitution Study Design[a,b]

Ingredient	Number of Studies	Ave Dry Matter (%)	AME$_n$, Mean (kcal/kg)	Std. Dev.	Range
Corn grain, as-fed	12	88.57	3,165	175	2,911–3,370
Wheat grain, higher protein	***15***	***89.6***	***3,241***	***134.4***	***2,882–3,411***
Sorghum	3	89.9	2,975	352	2,653–3,681
Barley grain	17		2,965	270	2,371–3,377
Rye grain	—	—	—	—	—
Soybean meal, solvent extracted	21	90	2,477	203	1,928–2,680
Canola meal (rapeseed meal), solvent extracted	10	90	2,821	550	1,608–3,151
Sunflower meal	—	—	—	—	—
Field peas	***12***	***87.5***	***2337***	***264.9***	***2,109–2,938***
Meat and bone meal	1	96	4,070	—	4,070
Meat meal	—	—	—	—	—
Poultry by-product meal	—	—	—	—	—
Fish meal, mechanical extracted	—	—	—	—	—
Corn distiller's dried grains with solubles—46 samples	5	NA	2,554	351	1,869–3,634
Fats and oils:					
Soy oil	1	NA	8,809	302	8,470–9,049
Palm oil	2	NA	6,651	1430	5,320–8,164
Poultry fat	1	NA	8,179	304	7,837–8,418

[a] Values in normal font are from studies meeting committee rigor specifications listed in Table C–1.

[b] Values in bold italics are guidance levels based on estimates due to insufficient experimental rigor. Limitations of individual studies in source literature are detailed in Table C–4.

performance and the apparent metabolisable energy value for broiler chickens. *British Poultry Science.* 60(4):457–466.

Bolarinwa, O.A., and O. Adeola. 2012. Energy value of wheat, barley, and wheat dried distillers grains with solubles for broiler chickens determined using the regression method. *Poultry Science.* 91(8): 1928–1935.

Broch, J., N.T.E. de Oliveira, R.V. Nunes, J.R. Henz, I.M. da Silva, R. Frank, and R.A. Schone. 2015. Chemical composition and energetic values of wheat and its sub-products for broiler chicken. *Semina: Ciências Agrárias.* 36(5):3481–3488.

Caldas-Cueva, J.P., and C.M. Owens. 2020. A review on the woody breast condition, detection methods, and product utilization in the contemporary poultry industry. *Journal of Animal Science.* 98(8):skaa207

Córdova-Noboa, H.A., E.O. Oviedo-Rondón, A. Ortiz, Y. Matta, S. Hoyos, G.D. Buitrago, J.D. Martinez, J. Yanquen, L. Peñuela, J.O.B. Sorbara, and A.J. Cowieson. 2020. Corn drying temperature, particle size, and amylase supplementation influence growth performance, digestive tract development, and nutrient utilization of broilers. *Poultry Science.* 99(11): 5681–5696.

Gehring, C.K., M.R. Bedford, A J. Cowieson, and W.A. Dozier. 2012. Effects of corn source on the relationship between in vitro assays and ileal nutrient digestibility. *Poultry Science.* 91(8):1908–1914.

Gorski, M., C. Foran, P. Utterback, and C.M. Parsons. 2017. Nutritional evaluation of conventional and increased-protein, reduced-fiber canola meal fed to broiler chickens. *Poultry Science.* 96(7):2159–2167.

Hejdysz, M., S.A. Kaczmarek, and A. Rutkowski. 2015. Factors affecting the nutritional value of pea (*Pisum sativum*) for broilers. Journal of Animal Feed Science and Technology. 24:252–259.

Hejdysz, M., S.A. Kaczmarek, and A. Rutkowski. 2016. Effect of extrusion on the nutritional value of peas for broiler chickens. *Archives of Animal Nutrition.* 70(5):364–377.

Kanji Kato, R., A.G. Bertechini, E.J. Fassani, J.A.G. de Brito, S. de Faria Castro. 2011. Metabolizable energy of corn hybrids for broiler chickens at different ages. *Zootecnia E Medicina Veterinária. Ciência e agrotecnologia.* 35(6).

Karunaratne, N.D., D.A. Abbott, P.J. Hucl, R.N. Chibbar, C.J. Pozniak, and H.L. Classen. 2018. Starch digestibility and apparent metabolizable energy of western Canadian wheat market classes in broiler chickens. *Poultry Science.* 97(8):2818–2828.

Khalil, M.M., M.R. Abdollahi, F. Zaefarian, and V. Ravindran. 2020. Measurement of ileal endogenous energy losses and true ileal digestible energy of cereal grains for broiler chickens. *Poultry Science.* 99(12): 6809–6817.

Khalil, M.M., M.R. Abdollahi, F. Zaefarian, P.V. Chrystal, and V. Ravindran. 2021. Apparent metabolizable energy of cereal grains for broiler chickens is influenced by age. Poultry Science. 100(9):101288.

Khalil, M.M., M.R. Abdollahi, F. Zaefarian, P.V. Chrystal, and V. Ravindran. 2022. Influence of broiler age on the apparent metabolizable energy of cereal grains determined using the substitution method. *Animals.* 12(2):183.

Koivunen, E., E. Talvio, E. Valkonen, T. Tupasela, P. Tuunainen, and J. Valaja. 2016. Use of semi-leafless peas (Pisum sativum L) with enzyme addition in diets for broilers. *Agricultural and Food Science.* 25(2):90–98.

Leung, H., and E.G. Kiarie. 2020. Standardized ileal digestibility of amino acids and apparent metabolizable energy in corn and soybean meal for organic broiler chicken production in Ontario. *Canadian Journal of Animal Science.* 100(3):447–454.

Meloche, K.J., B.J. Kerr, G.C. Shurson, and W.A. Dozier III. 2013. Apparent metabolizable energy and prediction equations for reduced-oil corn distillers dried grains with solubles in broiler chicks from 10 to 18 days of age. *Poultry Science.* 92(12):3176–3183.

Meloche, K.J., B.J. Kerr, N. Billor, G.C. Shurson, and W.A. Dozier III. 2014. Validation of prediction equations for apparent metabolizable energy of corn distillers dried grains with solubles in broiler chicks. *Poultry Science.* 93(6):1428–1439.

Nalle, C.L., V. Ravindran, and G. Ravindran. 2010. Nutritional value of faba beans (Vicia faba L.) for broilers: Apparent metabolisable energy, ileal amino acid digestibility and production performance. *Animal Feed Science and Technology.* 156(3):104–111.

TABLE C-4 Documentation of Literature Used for Table C-3

Ingredient	Dry Matter	Location	Age Period	Strain	AME by Regression	AME by Substitution	AME_n by Regression	AMEn by Substitution	Reference
	(%)		(days)		(kcal/kg)	(kcal/kg)	(kcal/kg)	(kcal/kg)	
Corn grain, dry	85.55	Local	16	Cobb 500				2911	Cordova-Noboa et al., 2020
	88.8	Local	16	Cobb 500				3,003	Cordova-Noboa et al., 2020
	86.5	Brazil	21	Ross 308		3,320		3,370	Kanji Kato et al., 2011
	86.5	Brazil	28	Ross 308		3,370		3,282	Kanji Kato et al., 2011
	90.9	New Zealand	22	Ross 308		3,213		3,102	Khalil et al., 2022
	90.9	New Zealand	29	Ross 308		3,374		3,226	Khalil et al., 2022
	86.72	United States	30	Ross 708				3,298	Gehring et al., 2012
	90.9	New Zealand	21	Ross 308				3,126	Khalil et al., 2020
	90.9	New Zealand	21	Ross 308		3,360		3,307	Khalil et al., 2021
	89.5	Canada	21	Cobb 500		3,408		3,066	Leung and Kiarie, 2020
Wheat Grain, Higher Protein									
Wheat	87.29	Brazil	22–32 d (5 d adaptation)	Cobb 500		2,958		2,882	Broch et al., 2015
Canadian Prairie Spring	90	Canada	0–21 d (20 d adaptation)	Ross 308				3,306	Karunaratne et al., 2018
Canadian Western Amber Durum	89.3	Canada	0–21 d (20 d adaptation)	Ross 308				3,400	Karunaratne et al., 2018
Canadian Western General Purpose	89.4	Canada	0–21 d (20 d adaptation)	Ross 308				3,332	Karunaratne et al., 2018
Canadian Western Hard White Spring	89.8	Canada	0–21 d (20 d adaptation)	Ross 308				3,411	Karunaratne et al., 2018
Canadian Western Red Spring	89.6	Canada	0–21 d (20 d adaptation)	Ross 308				3,203	Karunaratne et al., 2018
Canadian Western Soft White Spring	89.6	Canada	0–21 d (20 d adaptation)	Ross 308				3,241	Karunaratne et al., 2018
Feed wheat (soft)	88.7	United Kingdom	0–21 d (mash 0–7, pellet 8–21 d; 3 d collection)	Ross 308		3,372		3,286	Azhar et al., 2019
Feed wheat (hard)	89.2	United Kingdom	0–21 d (mash 0–7, pellet 8–21 d; 3 d collection)	Ross 308		3,363		3,267	Azhar et al., 2019
Milling wheat (hard)	90	United Kingdom	0–21 d (mash 0–7, pellet 8–21 d; 3 d collection)	Ross 308		3,408		3,315	Azhar et al., 2019
Milling wheat (soft)	89.3	United Kingdom	0–21 d (mash 0–7, pellet 8–21 d; 3 d collection)	Ross 308		3,470		3,370	Azhar et al., 2019
Wheat	90.28 (calculated)	Australia	14–21 (3 d adaptation)	Ross 308		3,191		3,047	Khalil et al., 2022
Wheat	90.28 (calculated)	Australia	22–28 d (3 d adaptation)	Ross 308		3,308		3,181	Khalil et al., 2022
Wheat	90.28 (calculated)	Australia	29–35 d (3 d adaptation)	Ross 308		3,353		3,222	Khalil et al., 2022
Wheat	90.28 (calculated)	Australia	36–42 d (3 d adaptation)	Ross 308		3,289		3,155	Khalil et al., 2022

Sorghum	N/A								
Barley grain	—	Unknown	15–22	Ross 708	2,894	—	2,841	—	Bolarinwa and Adeola, 2012
	—	New Zealand	14–21	Ross 308	—	3,195	—	3,117	Khalil et al., 2021
	—	New Zealand	21–28	Ross 308	—	3,164	—	3,090	Khalil et al., 2021
	—	New Zealand	28–35	Ross 308	—	2,890	—	2,739	Khalil et al., 2021
	—	New Zealand	35–42	Ross 308	—	2,882	—	2,763	Khalil et al., 2021
	—	New Zealand	14–21	Ross 308	—	2,904	—	2,770	Khalil et al., 2022
	—	New Zealand	21–28	Ross 308	—	2,925	—	2,789	Khalil et al., 2022
	—	New Zealand	28–35	Ross 308	—	3,415	—	3,315	Khalil et al., 2022
	—	New Zealand	35–42	Ross 308	—	3,461	—	3,377	Khalil et al., 2022
	—	New Zealand	14–21	Ross 308	—	2,447	—	2,371	Khalil et al., 2020
	—	New Zealand	21–28	Ross 308	—	3,176	—	3,098	Khalil et al., 2021
	—	New Zealand	21–28	Ross 308	—	3,243	—	3,164	Khalil et al., 2021
	—	Unknown	14–21	Cobb 500	2,703	—	2,565	—	Olukosi, 2021
	—	Unknown	18–21	Cobb 500	2,622	—	2,481	—	Olukosi, 2021
	—	Unknown	14–24	Cobb 500	—	3,059	—	3,035	Olukosi et al., 2017
	—	Unknown	17–24	Cobb 500	—	2,964	—	2,916	Olukosi et al., 2017
	—	Unknown	21–24	Cobb 500	—	2,772	—	2,725	Olukosi et al., 2017
	—	New Zealand	14–21	Ross 308	—	3,267	—	3,200	Perera et al., 2022
Rye grain	—	Unknown	18–23	Ross 308	—	3,053	—	—	Adeola and Kong, 2020
Field Peas									
Field peas—hulled	86.9	Australia	32	Ross 308				2,938	Nalle et al., 2010
Field peas—dehulled	87.4	Australia	32	Ross 308				2,842	Nalle et al., 2010
Field peas—non-extruded	Not reported	Unknown	32	Ross 308		2,795			Nalle et al., 2011a
Field peas—Santana (Yellow)	88.7	New Zealand	36	Ross 308		2,574		2,427	Nalle et al., 2011b
Field peas—Miami (Yellow)	88	New Zealand	36	Ross 308		2,424		2,343	Nalle et al., 2011b
Field peas—Courier (Brown)	87.1	New Zealand	36	Ross 308		2,482		2,319	Nalle et al., 2011b
Field peas—Rex (Yellow)	86.9	New Zealand	36	Ross 308		2,245		2,176	Nalle et al., 2011b
Turnia (Colored-flowered)	—	Poland	20	Ross 308				2,221	Hejdysz et al., 2015
Sokolik (Colored-flowered)	—	Poland	20	Ross 308				2,109	Hejdysz et al., 2015
Milwa (Colored-flowered)	—	Poland	20	Ross 308				2,114	Hejdysz et al., 2015
Muza (White-flowered)	—	Poland	20	Ross 308				2,138	Hejdysz et al., 2015
Cysterski (White-flowered)	—	Poland	20	Ross 308				2,259	Hejdysz et al., 2015
Trachalska (5 locations)—raw peas	—	Poland	20	Ross 308				2,162	Hejdysz et al., 2016
Sohvi	87.1	Unknown	32	Ross		2,914			Koivunen et al., 2016
Karita	87.5	Unknown	32	Ross		3,296		2,938	Koivunen et al., 2016

continued

TABLE C-4 Continued

Ingredient	Dry Matter	Location	Age Period	Strain	AME by Regression	AME by Substitution	AME_n by Regression	AMEn by Substitution	Reference
	(%)		(days)		(kcal/kg)	(kcal/kg)	(kcal/kg)	(kcal/kg)	
Canola Meal (Rapeseed Meal), Solvent Extracted									
Canola meal (*Brassica napus*)	88.2	United States	21	Ross 308		.	.	1,608	Agyekum and Woyengo, 2022
(*Brassica napus*)	93.1	United States	21	Ross 309		.	.	1,994	Agyekum and Woyengo, 2022
Canola meal (*Brassica napus*)	92.12	Canada	26	Cobb 500		3,214	.	3,044	Zhong and Adeola, 2019
Canola meal (*Brassica napus*)	91.83	Canada	26	Cobb 500		3,061	.	2,885	Zhong and Adeola, 2019
(*Brassica napus*)	92.26	Canada	26	Cobb 499		3,277	.	3,114	Zhong and Adeola, 2019
(*Brassica napus*)	92.41	Canada	26	Cobb 500		3,324	.	3,151	Zhong and Adeola, 2019
Canola meal (*Brassica napus*)	87.7	United Kingdom	21	Ross 308		.	.	3,071	Pirgozliev et al., 2022
(*Brassica napus*)	88.4	United Kingdom	21	Ross 308		.	.	3,119	Pirgozliev et al., 2022
Canola meal (*Brassica napus*)	88.3	United States	28	Ross 708		3,411	.	3,142	Zhang and Adeola, 2017
Canola meal (*Brassica napus*)	89.3	United States	28	Ross 708		3,386		3,079	Zhang and Adeola, 2017
Canola meal (*Brassica napus*)	90.2	United States	37	Ross 308		.	2,000	.	Gorski et al., 2017
Canola meal (*Brassica napus*)	89.4	United States	37	Ross 308		.	2,200	.	Gorski et al., 2017
Canola meal (*Brassica napus*)	88.9	United States	37	Ross 308		.	2,000	.	Gorski et al., 2017
Meat and Bone Meal									
	92.5	New Zealand	0–7	Ross 308	2,909		2,687		Adeola et al., 2018
			6–11		3,125		2,887		
			10–16		3,083		2,839		
			15–21		3,075		2,845		
	96.9	USA	14–21	Ross 308	2,512		2,345		Bolanrinwa et al., 2012
	96.7				2,711		2,127		
	94.5	Brasil	14–24			1,574	1,283		Troni et al., 2016
	95.8					1,353	1,088		
	96.8	New Zealand	21–28	Ross 308	2,909	4,070	2,687	4070	Khalil et al., 2021
Corn Distiller's Dried Grains with Solubles		USA	15–22	Ross 708			2,685		Rochell et al., 2011
		USA	15–22	Ross 708			2,146		Rochell et al., 2011
		USA	15–22	Ross 708			2,628		Rochell et al., 2011

	USA	15–22	Ross 708		3,098	Rochell et al., 2011
	USA	15–22	Ross 708		2,583	Rochell et al., 2011
	USA	15–22	Ross 708		2,903	Rochell et al., 2011
	USA	15–22	Ross 708		2,708	Rochell et al., 2011
	USA	15–22	Ross 708		2,932	Rochell et al., 2011
	USA	10–16	Ross 708		3,634	Meloche et al., 2014
	USA	10–16	Ross 708		2,553	Meloche et al., 2014
	USA	10–16	Ross 708		2,869	Meloche et al., 2014
	USA	10–16	Ross 708		2,781	Meloche et al., 2014
	USA	10–16	Ross 708		2,523	Meloche et al., 2014
	USA	10–16	Ross 708		2,535	Meloche et al., 2014
	USA	13–19	Ross 708		2,903	Meloche et al., 2014
	USA	13–19	Ross 708		2,640	Meloche et al., 2014
	USA	13–19	Ross 708		2,461	Meloche et al., 2014
	USA	13–19	Ross 708		3,120	Meloche et al., 2014
	USA	13–19	Ross 708		3,111	Meloche et al., 2014
	USA	13–19	Ross 708		2,581	Meloche et al., 2014
	USA	22–28	Ross 708		1,975	Meloche et al., 2014
	USA	22–28	Ross 708		2,644	Meloche et al., 2014
	USA	22–28	Ross 708		3,137	Meloche et al., 2014
	USA	10–16	Ross 708		1,869	Meloche et al., 2013
	USA	10–16	Ross 708		2,551	Meloche et al., 2013
	USA	10–16	Ross 708		2,487	Meloche et al., 2013
	USA	10–16	Ross 708		2,103	Meloche et al., 2013
	USA	10–16	Ross 708		2,401	Meloche et al., 2013
	USA	10–16	Ross 708		2,526	Meloche et al., 2013
	USA	10–16	Ross 708		2,309	Meloche et al., 2013
	USA	10–16	Ross 708		2,068	Meloche et al., 2013
	USA	10–16	Ross 708		2,273	Meloche et al., 2013
	USA	10–16	Ross 708		2,012	Meloche et al., 2013
	USA	10–16	Ross 708		2,418	Meloche et al., 2013
	USA	10–16	Ross 708		2,074	Meloche et al., 2013
	USA	10–16	Ross 708		2,032	Meloche et al., 2013
	USA	10–16	Ross 708		2,824	Meloche et al., 2013
	USA	10–16	Ross 708		2,687	Meloche et al., 2013
	USA	12–21	Ross 708		2,493	Caldas et al., 2020
	USA	12–21	Ross 708		2,549	Caldas et al., 2020
	USA	12–21	Ross 708		2,588	Caldas et al., 2020
	USA	12–21	Ross 708		2,404	Caldas et al., 2020
	USA	12–21	Ross 708		2,292	Caldas et al., 2020
	USA	12–21	Ross 708		2,515	Caldas et al., 2020
	USA	12–21	Ross 708		2,341	Caldas et al., 2020
	USA	12–21	Ross 708		2,540	Caldas et al., 2020
	USA	35–38	Ross 708	2,800	2,688	Adeola and Zhai, 2012

Nalle, C.L., V. Ravindran, and G. Ravindran. 2011a. Nutritional value of peas (Pisum sativum L.) for broilers: Apparent metabolisable energy, apparent ileal amino acid digestibility and production performance. *Animal Production Science*. 51(2):150–155.

Nalle, C. L., G. Ravindran, and V. Ravindran. 2011b. Extrusion of Peas (Pisum sativum L.): Effects on the Apparent Metabolisable Energy and Ileal Nutrient Digestibility of Broilers. *American Journal of Animal and Veterinary Sciences*. 6(1):25–30.

Olukosi, O.A. 2021. Investigation of the effects of substitution levels, assay methods and length of adaptation to experimental diets on determined metabolisable energy value of maize, barley and soya bean meal. *British Poultry Science*. 62(2):278–284.

Olukosi, O.A., S.A. Adedokun, and J.O. Agboola. 2017. Species-dependent response to the influence of adaptation length during assay for metabolisable energy of cereal grains employing the difference method. *Animal Feed Science and Technology*. 231:111–118.

Perera, W., U. Nipuna, M. Reza Abdollahi, F. Zaefarian, T.J. Wester, and V. Ravindran. 2022. Barley, an Undervalued Cereal for Poultry Diets: Limitations and Opportunities. *Animals*. 12(19):2525.

Pirgozliev, V., S.P. Rose, M.W. Mirza, I.M. Whiting, H. Malins, L. Bauer, and A. Lemme. 2022. Feeding guanidinoacetic acid to broiler chickens can compensate for low dietary metabolisable energy formulation. *British Poultry Science*. 63(3):368–374.

Rochell, S.J., B.J. Kerr, and W.A. Dozier III. 2011. Energy determination of corn co-products fed to broiler chicks from 15 to 24 days of age, and use of composition analysis to predict nitrogen-corrected apparent metabolizable energy. *Poultry Science*. 90(9):1999–2007.

Troni, A.R., D.M.B. Campos, R.M. Suzuki, H.S. Doreto, N.K. Sakomura, and J.A. Bendassolli. 2016. Use of stable isotopes as a tracer for broiler chickens. In *Energy and Protein Metabolism and Nutrition*. Wageningen Academic.

Zhang, F., and O. Adeola. 2017. Energy values of canola meal, cottonseed meal, bakery meal, and peanut flour meal for broiler chickens determined using the regression method. *Poultry Science*. 96(2):397–404.

Zhong, R., and O. Adeola. 2019. Energy values of solvent-extracted canola meal and expeller-derived canola meal for broiler chickens and growing pigs determined using the regression method. *Journal of Animal Science*. 97(8):3415–3425.

Appendix D

Committee Biographies

Kirk C. Klasing (*Chair*) is a distinguished professor of avian nutrition in the Department of Animal Science at the University of California, Davis. His research interests include the interactions between nutrition and the immune system of animals, for which he has received the Poultry Science Research Award from the Poultry Science Association, the BioServ and Sunde Award from the American Institute of Nutrition, and the Lilly Animal Scientist Award. He holds a B.S. from Purdue University and a Ph.D. from Cornell University. Dr. Klasing serves on the editorial boards of *Poultry Science*, *Animal Biotechnology*, *Journal of Comparative Immunology*, and *Journal of Nutrition*. Other areas of interest include immunophysiology and disease resistance, the immunologic basis of stress, and nutrition and metabolic adaptation required by carnivorous, granivorous, and nectivorous animals. He was a member of the National Research Council (NRC) Subcommittee on Metabolism Modifiers and Nutrient Requirements in Food Producing Animals and the National Academies' Division on Earth and Life Studies' Board on Agriculture and Natural Resources. He also served as the chair of the NRC Committee on Minerals and Toxic Substances in Diets and Water for Animals, which completed its report in 2005.

Ryan N. Dilger is a professor in the Department of Animal Sciences at the University of Illinois Urbana-Champaign. His research includes integrating the interdisciplinary fields of animal nutrition, immunophysiology, and neuroscience, including the ability of nutrients to impact metabolic, immunologic, and developmental patterns. Dr. Dilger has received many honors, including the American Society for Nutrition Bio-Serv Award in Experimental Animal Nutrition in 2016, International Life Sciences Institute North America Future Leader Award in 2015, and Poultry Science Association Hy-Line International Research Award in 2014. He received his B.S. and M.S. in animal science from Purdue University and a Ph.D. in animal science from the University of Illinois Urbana-Champaign.

William Dozier is a professor in the Department of Poultry Science in the College of Agriculture at Auburn University. His research efforts have focused on defining nutritional needs of broilers and the use of alternative ingredients and exogenous enzymes, which can lower feed costs. Dr. Dozier's emphasis on broiler research has been practical and application oriented, so as to benefit the broiler industry both in the United States and abroad. Through his numerous papers on the metabolizable energy and amino acid requirements of broilers, the industry has been able to formulate diets closer to the birds' requirement, improving performance and saving considerably in the cost of production. Dr. Dozier has received many honors including the National Chicken Council Broiler Research Award in 2013 and the American Feed Industry Association Poultry Nutrition Research Award in 2012. He received his B.S. in animal and dairy science from Auburn University, an M.S. in animal science from the University of Kentucky–Lexington, and a Ph.D. in poultry science from Auburn University.

Robert G. Elkin is a professor emeritus of avian nutritional biochemistry in the Department of Animal Science at Pennsylvania State University. His research program spanned a number of areas, including avian amino acid nutrition and metabolism, nutritional utilization of sorghum grain, amino acid analysis of feedstuffs and physiological fluids by high-performance liquid chromatography, nutritional/pharmacological reduction of egg cholesterol content, and most recently, novel dietary approaches to enrich eggs and poultry meat with "heart healthy" very-long-chain omega-3 fatty acids. Dr. Elkin's work has been recognized nationally and internationally, as evidenced by his receipt of the American Feed Industry Association Nutrition Research Award (1997) and the American Egg Board Research Award (1991) from the Poultry Science Association, a Fulbright Scholar Research Award (Austria, 1992–1993), an OECD Research Fellowship (Austria, 1995), and a Pioneer of the Mid-Atlantic Nutrition Conference Award (2008) for support and dedication to the establishment of the conference. Dr. Elkin served as head of the Department of Poultry Science at Pennsylvania State University from 1999 until 2012, when it was merged with the former Dairy & Animal Science Department to form the Department of Animal Science. He received his B.S. in animal science from Pennsylvania State University and an M.S. in nonruminant nutrition and a Ph.D. in nutritional biochemistry from Purdue University.

Peter Ferket is the William Neal Reynolds Professor– Extension nutritionist at North Carolina State University. His extension program deals with issues related to nutrition and feeding practices of commercial poultry, with an emphasis on meat-type turkeys and broilers. Dr. Ferket's research interests complement his extension activities in applied nutrition research of commercial meat-type and egg-laying poultry and the utilization of feed additives and agricultural co-products as feed ingredients for poultry. He is internationally recognized for his research on the nutritional requirements of turkeys, perinatal nutrition, and nutritional factors that affect gut health. His teaching expertise is in vitamin and mineral metabolism, and feed formulation and simulations modeling. He is a member of the American Registry of Professional Animal Scientists, Poultry Science Association, Southern Poultry Science Society, and World's Poultry Science Association. Dr. Ferket received his B.S. in agriculture and M.S. in poultry nutrition from the University of Guelph, and a Ph.D. in animal nutrition from Iowa State University.

Doug Korver is a professor of poultry nutrition at the University of Alberta. He studied agriculture at the University of Saskatchewan and earned a Ph.D. from the University of California, Davis. His research in feed ingredients has included a substantial effort in evaluation of feed enzymes (phytase, nonstarch polysaccharide degrading enzymes) to increase the utilization of nutrients contained within the feed. Additionally, his program has been increasingly involved in studies investigating approaches to reduce the use of prophylactic antibiotics in poultry feed. He has been a leader in the use of quantitative computed tomography to assess in vivo changes over time in broiler, laying hen, and turkey bone mineralization, as well as assessing the effects of feed ingredients on bone mineralization and metabolism. Finally, his research program also investigates the effects of feed ingredients, supplements, and novel ingredients on the effect of the inflammatory response on poultry productivity and metabolism.

Jason Lee is an associate professor of poultry science at Texas A&M University and holder of the Emanuel and Elaine Glockzin Professorship of Poultry Science. He earned a B.S. in poultry science, an M.S. in nutrition, and a Ph.D. in poultry science, all from Texas A&M University. His research focuses on increasing nutrient utilization, reducing nutrient excretion and environmental impact, and maximizing poultry performance while balancing dietary costs. Specific areas include the use of exogenous enzymes to increase nutrient digestibility (mineral, amino acid, and energy), inclusion of probiotics and prebiotics to improve poultry health as the industry shifts toward antibiotic free production, evaluation of amino acid concentration on product yield, and investigation of mineral source and concentration to improve skeletal development in rapidly growing broilers and laying hens. Dr. Lee has served on the Foundation for Food and Agriculture Research Expert Review Panel and chaired the U.S. Department of Agriculture National Program 108 Panel 10: Committee on Poultry Production and Processing Review.

Gonzalo G. Mateos is a professor of animal nutrition at the Universidad Politécnica de Madrid. His research areas include evaluation of quality of soybean meal and other protein sources; feed technology; processing of feeds and ingredients and biotechnology diets; design of feeds for piglet, pullets, and broiler chicks; and influence of nutrition and management on quality of eggs and meat products. Dr. Mateos was the co-editor in chief of *Animal Feed Science and Technology* until 2014 and has been a member of the editorial board of *Poultry Science* on several occasions. He is also the vice-president and founder of FEDNA, the Spanish Association of Animal Nutritionists. Dr. Mateos received his B.S. in animal production from Universidad Politécnica de Madrid, a D.V.M. from Universidad Complutense of Madrid, a Ph.D. in veterinary science from Barcelona University, and a Ph.D. in animal nutrition from Iowa State University.

Michael Persia is the John W. Hancock Professor in the School of Animal Sciences at Virginia Tech. His research is to increase the efficiency of poultry production through the characterization, understanding, and manipulation of bird digestive processes. The three major aims of Dr. Persia's research are to understand the effects of stress, both disease and heat, on nutrient absorption and utilization; to evaluate the efficient utilization of nutrients in poultry production to minimize environmental impacts; and to research areas within nutrition and poultry management critical to the direct needs of the commercial industry in Virginia. Dr. Persia is a member of Gamma Sigma Delta and has previously served as the president of the Southern Poultry Science Society and a 3-year term as the director for the Poultry Science Association. Dr. Persia received his B.S. in animal and biological sciences from Pennsylvania State University, an M.S. in poultry nutrition from Ohio State University, and a Ph.D. in poultry nutrition from the University of Illinois.

Sheila Purdum is a professor of poultry nutrition and the chair of the Animal Science Graduate Committee at the University of Nebraska–Lincoln. Her research interests include omega egg research and patent; nutritional factors affecting calcium utilization and shell quality in laying hens; use of exogenous feed enzymes in layer and pullet feed; and laying hen behavior. She has received many honors, including the

Nebraska Poultry Industries Hall of Fame Award in 2008 and the Helen Cecil Leadership Award from the Poultry Science Association in 2007. Dr. Purdum received her B.S. in animal science and M.S. in animal nutrition from the University of Nebraska, and a Ph.D. in animal nutrition/human nutrition from Iowa State University.

Rosemary L. Walzem is a professor in the Department of Poultry Science and a member of the Graduate Faculty of Nutrition at Texas A&M University. Her core research focus in poultry is directed toward understanding lipoprotein structure and biology and bird health. Her past research identified the key structural features that enable yolk-targeting of very-low-density lipoproteins in laying hens as well as identification of the transient high-density lipoproteins (HDLs) involved in ovarian regression. She identified the similarity in metabolic features in obese breeder hens to those of women with polycystic ovary syndrome; ultimately identifying changes in lipid metabolism that result in ovarian atresia and cardiac dysfunction in obese hens. Her new work focuses on dietary features that influence gut health, microbiota acquisition, and the role of apolipoprotein A1 in innate immune function and its relationship to HDL functionality. As a comparative nutritionist, she works across several species and has answered questions in vitamin deficiency, energy excess, protein and lipid nutrition as well as oxidative stress and phytochemical metabolism and bioactivity. Dr. Walzem received her B.S. in clinical dietetics, M.S. in nutrition, and Ph.D. in nutritional biochemistry from the University of California, Davis. She is a registered dietitian, interning at the Sutter Hospitals in Sacramento. She serves as subject expert in nutrition for the Institute of Food Technologists and has been their delegate to the Codex Committee on Nutrition and Foods for Special Dietary Purposes since 2004. She has served on the National Institute on Aging's Intervention Testing Program access panel since 2001 and several National Institutes of Health study sections where expertise on the interface between agricultural practices and food quality is required.